INTRODUCTION TO FOOD- AND AIRBORNE FUNGI

Reprint of the sixth revised edition (with some corrections)

Cover design: *Penicillium roqueforti* (magnification 5000 x).

Published and distributed by Centraalbureau voor Schimmelcultures, P.O.Box 85167, 3508 AD UTRECHT, The Netherlands. Internet: www.cbs.knaw.nl. Email info@cbs.knaw.nl.

ISBN 90-70351-42-0

Printed by Ponsen & Looyen, Wageningen, The Netherlands

INTRODUCTION TO FOOD- AND AIRBORNE FUNGI

edited by

Robert A. SAMSON and Ellen S. HOEKSTRA
Centraalbureau voor Schimmelcultures,
P.O.Box 85167, 3508 AD Utrecht
The Netherlands

Jens C. FRISVAD and Ole FILTENBORG,
BioCentrum DTU - Mycology Group,
Søltofts Plads, Building 221, Technical University of Denmark,
DK-2800 Kgs. Lyngby, Denmark

With contributions by
Enne de Boer (Zutphen)
Jan Dijksterhuis (Utrecht)
Hans P. van Egmond (Bilthoven)
Flemming Lund (Lyngby)
Per V. Nielsen (Lyngby)
M.J. Robert Nout (Wageningen)
Vincent Robert (CBS, Utrecht)
Rob P.M. Scholte (Veenendaal))
Keith A. Seifert (Ottawa)
Maudy Th. Smith (CBS, Utrecht)
Richard Summberbell (CBS, Utrecht)
Ulf Thrane (Lyngby)
David Yarrow (CBS, Utrecht)

SIXTH EDITION

CENTRAALBUREAU VOOR SCHIMMELCULTURES – Utrecht

An institute of the Royal Netherlands Academy of Arts and Sciences

CONTENTS

PREFACE

The interest of microbiologists in fungal contaminants of food is still increasing and there is a need to study these micro-organisms in more detail. Although fungi, producing toxins or which cause health hazards, are ubiquitous and belong to the common contamination flora, their recognition is hampered by incomplete and often confusing literature. This book originally was intended as a guide to the CBS course on food-borne fungi and contains mainly keys and morphological descriptions of the most common species. However in the last decade much interest has grown for the fungi present in indoor environments. Since this mycoflora (or mycobiota) is very similar to that of food, the course book combines now both food- and airborne fungi.

The sixth edition is revised extensively in close collaboration with Ole Filtenborg, Jens Frisvad, Fleming Lund, Ulf Thrane and Per Nielsen from the Technical University of Denmark at Lyngby. In Chapter 1 the taxonomy is updated and some important species were added, while the keys to the taxa were improved. The taxonomy and the nomenclature of *Fusarium* were revised. Keith Seifert (Agriculture Canada, Ottawa) together with Ulf Thrane kindly helped us with modifying the key to the *Fusarium* species. The number of *Penicillium* species has been increased because we have frequently encounter these taxa in our laboratories. The identification of *Penicillium* based on morphological characters remains difficult and therefore synoptic keys are added. Our CBS colleagues Vincent Robert, Maudy Smith and David Yarrow rewrote the text for the identification of the yeasts.

The authors, including a new chapter on indoor moulds together with Richard Summerbell, updated all chapters on applied aspects. The chapter of Spoilage Fungi in industrial processing of foods has been expanded with more details of heat resistant moulds with the help of Jan Dijksterhuis. The appendix includes a glossary and a list of media, currently recommended for isolation and identification of food-and airborne fungi.

This publication would not have been possible without the helpful cooperation of the authors of the various chapters on applied aspects. We are grateful for their contributions. Mrs Ans Spaapen, Marjan Erich-Vermaas and particularly Karin van den Tweel-Vermeulen helped us with typing the text, making and improving line drawings, preparation of the camera-ready layout and the photographs.

Rob Samson, Ellen Hoekstra, Jens Frisvad and Ole Filtenborg, Baarn and Lyngby, November 2000.

CHAPTER 1

IDENTIFICATION OF THE COMMON FOOD- and AIRBORNE FUNGI

In the following chapter we have compiled a selection of the most common food- and airborne fungi, based on our identifications of the numerous strains regularly sent to the Centraalbureau voor Schimmelcultures. Members of the Zygomycetes, Ascomycetes and Deuteromycetes are represented. Each species is illustrated in detail, but only the major characteristics are described for recognition. For a more detailed account the reader is referred to the given literature. Access to additional mycological handbooks will promote precise identification. The drawings were made with a Camera lucida, while the micrographs were photographed with Phase contrast or Nomarski interference contrast.

Note that only the common food- and airborne species are treated and that they may be abundant on certain food products or in certain indoor environments. Always consult other keys, for example von Arx (1981), Domsch *et al.* (1993), Carmichael *et al.* (1980) and Pitt and Hocking (1997), Samson *et al.* (1994) or send important isolates to the identification services of the Mycological Centres.

When examining and identifying fungi line drawings may help in structure elucidation and may be employed as later reference. Examination of many food- and airborne fungi should be done with a good microscope provided with oil-immersion (1000 x). The use of the optical equipment is discussed in Gams *et al.* (1998).

Note that the identification of the yeasts requires special training and that physiological criteria are used. In this book we compiled merely 25 common species which can be determined with a simple key. For a more detailed information see also Barnett *et al.* (2000).

CULTIVATION

It is important to culture each species on an appropriate medium to achieve typical growth and sporulation. Malt extract agar (MEA) and/or Oatmeal agar (OA) are suitable media for most species. MEA with additional (20 or 40%) sucrose used at CBS is prepared according to the CBS formula mentioned in the Appendix. A number of species require special media such as Potato-Carrot agar (PCA), Czapek Yeast autolysate agar (CYA), Carnation leaf agar (CLA). Creatine agar (CREA) is helpful for distinguishing some species of *Penicillium.* Recommendations and instructions for cultivation are given with every genus description. Formulae of most media recommended for isolation and identification of food-and airborne fungi are also provided in the Appendix, p. 378. For other mycological media see Gams *et al.*. (1998), Stevens (1974) and Samson *et al.* (1990).

Most fungi can be incubated in light or in the dark at 25°C and identified after 5-10 days. *Fusarium, Trichoderma* and *Epicoccum* show typical sporulation in diffuse daylight. Fungi such as *Phoma* should be cultivated in darkness followed by a period of alternating darkness/diffuse (day)light. To stimulate sporulation irradiation with near UV ('black' light: greatest effect at 310 nm with a maximum emission about 360 nm) can be useful.

Basidiomycetous fungi e.g. *Serpula lacrymans* (dry rot fungus) often can be identified directly from the material (wood) on basis of the fruitbody, spores and hyphal structures. Also the occurrence of *Stachybotrys* or *Memnoniella* in indoor environments growing on surfaces easily can be identified by sellotape preparations. In those cases cultivation on agar media can be omitted.

EXAMINATION

Direct macroscopical examination

The presence of fungal growth on or in food products and on surfaces or in a Petri-dish can often be confirmed directly with the naked eye and subsequently with a dissecting microscope. Fungal tissue (mycelium, fruit-bodies, sporulating structures) can be taken with a preparation needle or glass needle. Glass needles are very suitable and they can be prepared from a glass rod, which after heating can be pulled into the shape of a needle.

Preparation of a microscopical slide

A piece of fungus is transferred to a microscope slide with a sterilized needle and a drop of water or mounting fluid. The preparation is covered with a cover glass. Always use clean slides, cover slips and flame-sterilized needles.

Fruit-bodies have to be squashed first, often mycelium has to be teased out with 2 needles. It is often useful to take sporulating structures, including some agar, from young areas of the colonies. The agar can be gently melted above a spare flame (never boil!!).

For the preparation of *Aspergillus* and *Penicillium* species and other fungi producing many dry conidia it is recommended to add a drop of alcohol to wash away the mass of hydrophobic conidia prior to cover the preparation by a cover slip.

Fungal cultures grown on a transparent agar in a Petridish can be cut out in a square and put onto the microscope slide. Then a drop of water is added and covered by a cover-slip. The slide can be examined under the microscope. This technique is very helpful for Deuteromycetes to observe the development of dry chains or wet conidial heads, but is also useful for other fungi with fragile structures (e.g. *Mortierella*).

Transparent tape preparation

For some species with very fragile and complex sporulating structures (e.g. *Cladosporium, Botrytis*) preparation of a microscope slide using transparent adhesive tape may be helpful. The colonies are lightly touched with the adhesive tape, after which the tape is transferred to a microscope slide with a drop of mounting medium. This type of microscope slide is suitable for examining at low power (100 and 400 times). After placing a cover glass with a drop of mounting fluid on top of the adhesive tape, the preparation can also be examined using oil-immersion.

The 'adhesive tape' method can also be very useful for examining fungal contamination of surfaces. It does not only allow the detection of mould growth but also can provide additional ecological information e.g. the presence of other micro-organisms and/or mites.

Mounting fluids

Always make the first preparation in water. The fungus is then in its natural shape and size (no shrinkage), also the colours are best to be observed. For the Zygomycetes, Coelomycetes, and yeasts it is recommended to examine the preparation in a water mount or Shear's. If the fungus produces many dry conidia (e.g. *Penicillium*), a detergent or ethanol can be added (see above). The disadvantage of a water preparation is the rapid desiccation.

Deuteromycetes and Ascomycetes can be mounted in **lactic acid** with or without some dye (cotton blue or aniline blue). This mounting medium is prepared by adding 1 g cotton blue (aniline blue) to 1 litre DL-lactic acid (ca. 85%). Addition of a little drop of alcohol or very gently heating of the slide above a spare flame may help to remove the excess adherence of spores or air bubbles.

Also commonly used is **Shear's mounting fluid**: 3 g potassium acetate, 150 ml water, 60 ml glycerin, 90 ml ethanol (95%). This mounting medium is very suitable for microphotography.

A slide mounted with lactic acid can be made (semi)-permanent by surrounding the cover slip with glycerin-gelatine. Before applying the glycerine-gelatine the excess of mounting fluid should be removed and the edges of the cover slip cleaned with alcohol. After drying for one to several days a layer of nail varnish can be added covering the glycerine-gelatine completely. Other techniques for preserving microscopical slides are described by Gams *et al.* (1998).

HANDLING CULTURES

Many food-and airborne fungi can be toxin producers or may cause allergic responses. Be careful with colonies in Petri dishes and contaminated products and avoid inhalation of spores and other fungal products such as volatiles (e.g. *Aspergillus fumigatus, Paecilomyces variotii, Stachybotrys chartarum, Penicillium* and *Fusarium* spp.).

KEY TO THE GROUPS TREATED (See Plates 1 and 2).

1 a. Colonies consisting of loose budding cells, pseudomycelium sometimes present but true vegetative mycelium usually absent .... **YEASTS**
1 b. Colonies with abundant vegetative mycelium; conidia or spores borne in or on special cells.. 2

2 a. Mycelium without or with a few septa, often broad; asexual spores mostly formed endogenously in sporangia .... **ZYGOMYCETES**
2 b. Mycelium with regular septation; asexual or sexual spores not formed in sporangia...3

3a. Asexual spores (conidia) produced from special cells (conidiogenous cells)....... **DEUTEROMYCETES**
3b. Sexual spores formed in asci or on basidia..4

4 a. Spores formed in asci .... **ASCOMYCETES**
4 b. Spores formed on basidia .... **BASIDIOMYCETES**

Pl. 1 a. Vegetative mycelium, sporangiophore with sporangium, *Rhizopus oligosporus*, x 175; b. vegetative mycelium, fertile hyphae breaking up into conidia, *Wallemia sebi*, x 700; c. asci with ascospores, *Eurotium herbariorum*, x 1700; d. budding cells of *Debaryomyces hansenii*, x 2030; e. conidiophores and conidia, *Aspergillus flavus*, x 70; f. sporangium and sporangiospores, *Mucor racemosus*, x 800.

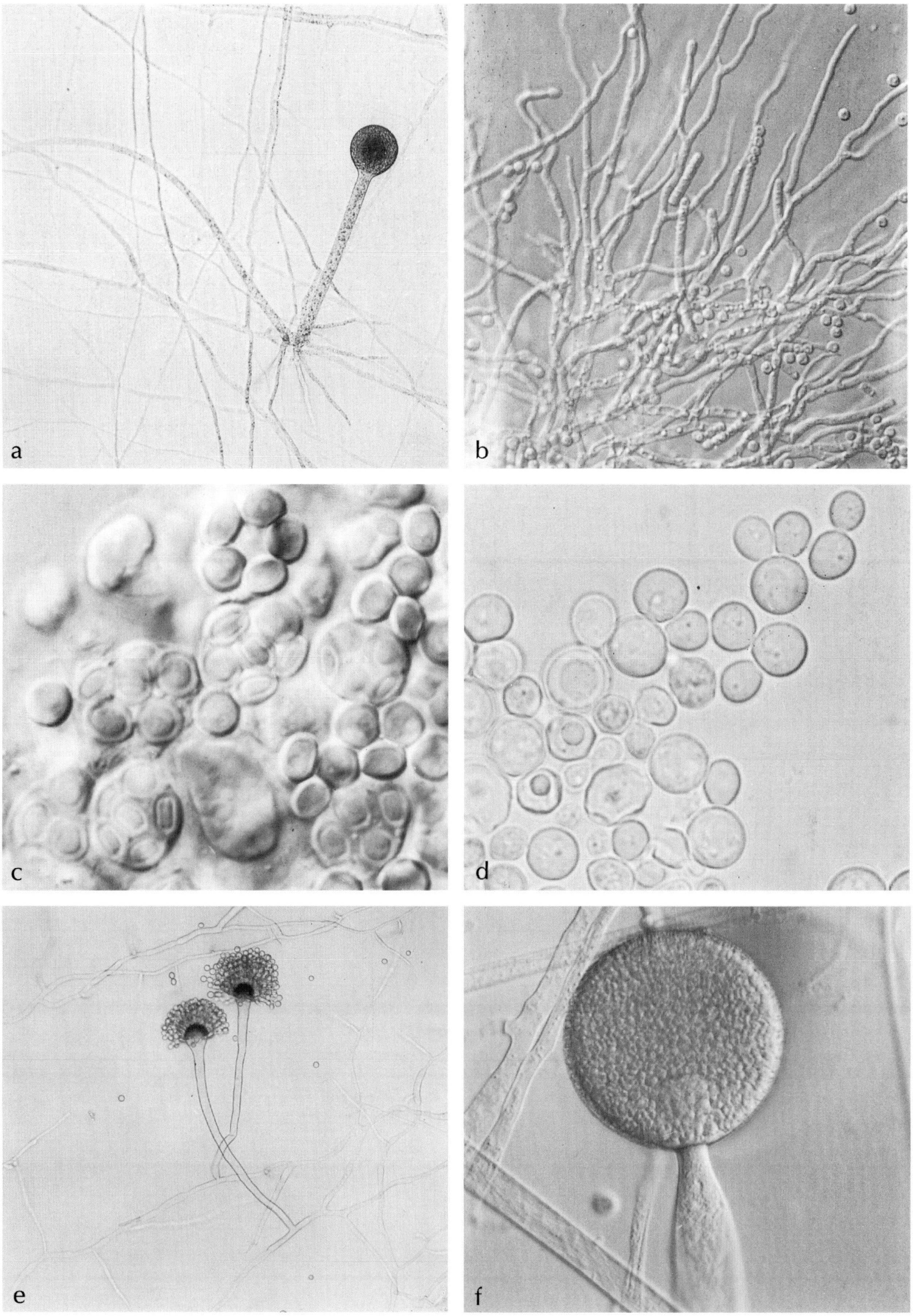
a
b
c
d
e
f

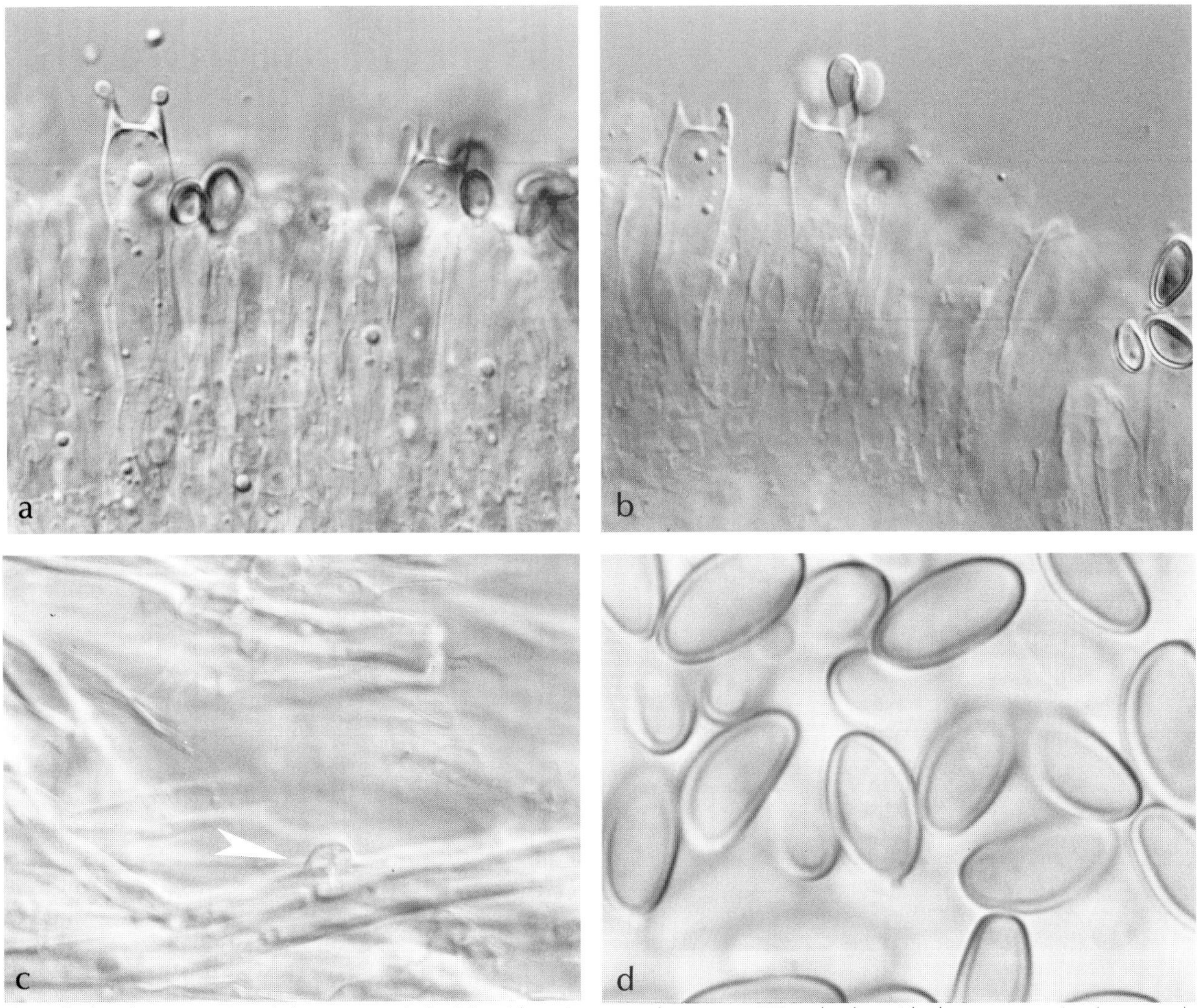

Pl. 2. The basidiomycete *Serpula lacrymans*; a-b. basidia with sterigmata and spores, x 850; c. hypha with clamp connection (see arrow), x 850; d. basidiospores, x 2000.

List of species treated in this book (● = food; ⌂ = indoor environment).

Name species	food	indoor	page
Absidia corymbifera	●	⌂	8
Acremonium butyri	●		58
Acremonium charticola	●		60
Acremonium strictum	●	⌂	60
Alternaria alternaria	●	⌂	62
Aspergillus candidus	●		68
Aspergillus clavatus	●		70
Aspergillus flavus	●		72
Aspergillus fumigatus	●	⌂	74
Aspergillus niger	●	⌂	76
Aspergillus ochraceus	●	⌂	78
Aspergillus oryzae	●		80
Aspergillus parasiticus	●		82
Aspergillus penicillioides	●	⌂	84
Aspergillus sydowii	●	⌂	86
Aspergillus tamarii	●	⌂	88
Aspergillus terreus	●	⌂	90

Name species	food	indoor	pag.
Aspergillus ustus	●	⌂	92
Aspergillus versicolor	●	⌂	94
Aspergillus wentii	●	⌂	96
Aureobasidium pullulans	●	⌂	98
Botrytis aclada	●		100
Botrytis cinerea	●	⌂	102
Byssochlamys fulva	●		28
Byssochlamys nivea	●		30
Chaetomium globosum		⌂	32
Chrysonilia crassa	●	⌂	104
Chrysonillia sitophila	●	⌂	104
Cladosporium cladosporioides	●	⌂	108
Cladosporium herbarum	●	⌂	110
Cladosporium macrocarpum	●		112
Cladosporium sphaerospermum	●	⌂	114
Curvularia geniculata		⌂	116
Emericella nidulans	●	⌂	34

Name species	food	indoor	pag.
Epicoccum nigrum	x	x	118
Eurotium amstelodami	x	x	36
Eurotium chevalieri	x	x	38
Eurotium herbariorum	x	x	40
Fusarium acuminatum	x		126
Fusarium avenaceum	x		128
Fusarium crookwellense	x		130
Fusarium culmorum	x	x	132
Fusarium equiseti	x		134
Fusarium graminearum	x		136
Fusarium oxysporum	x	x	138
Fusarium poae	x		140
Fusarium proliferatum	x		142
Fusarium sambucinum	x		144
Fusarium semitectum	x		146
Fusarium solani	x	x	148
Fusarium sporotrichioides	x		150
Fusarium subglutinans	x		152
Fusarium tricinctum	x		154
Fusarium verticillioides	x		156
Geomyces pannorum	x	x	158
Geotrichum candidum	x		160
Lecythophora hoffmannii	x		244
Memnoniella echinata		x	162
Monascus ruber	x		42
Moniliella acetoabutens	x		164
Moniliella suaveolens	x		166
Mucor circinelloides	x		10
Mucor hiemalis	x	x	12
Mucor plumbeus	x	x	14
Mucor racemosus	x	x	16
Neosartorya fischeri	x		44
Neosartorya glaber	x		46
Neosartorya pseudofischeri	x		46
Neosartorya spinosa	x		46
Oidiodendron griseum		x	168
Paecilomyces variotii	x	x	170
Penicillium allii	x		218
Penicillium aethiopicum	x		184
Penicillium atramentosum	x		186
Penicillium albocoremium	x		218
Penicillium aurantiogriseum	x	x	188
Penicillium aurantiocandidum	x		204
Penicillium brevicompactum	x	x	190
Penicillium camemberti	x		192
Penicillium carneum	x		232
Penicillium chrysogenum	x	x	194
Penicillium caseifulfum	x		192
Penicillium citrinum	x	x	196
Penicillium commune	x	x	198
Penicillium crustosum	x		202
Penicillium corylophilum	x	x	200
Penicillium cyclopium	x		204
Penicillium digitatum	x		206
Penicillium discolor	x		208
Penicillium echinulatum	x		208
Penicillium expansum	x		210
Penicillium funiculosum	x	x	212
Penicillium freii	x		204
Penicillium glabrum	x	x	214
Penicillium griseofulvum	x		216
Penicillium hirsutum	x		218
Penicillium hordei	x		218
Penicillium italicum	x		220
Penicillium melanoconidium	x		228
Penicillium nalgiovense	x		222
Penicillium nordicum	x		240
Penicillium olsonii	x	x	224
Penicillium palitans	x		198
Penicillium paneum	x		232
Penicillium polonicum	x		228
Penicillium roqueforti	x		230
Penicillium rugulosum	x		234
Penicillium solitum	x		236
Penicillium tricolor	x		242
Penicillium ulaiense	x		220
Penicillium variabile	x	x	238
Penicillium verrucosum	x		240
Penicillium viridicatum	x		242
Phialophora fastigiata	x	x	247
Phoma glomerata	x	x	248
Phoma macrostoma	x	x	250
Rhizopus oligosporus	x		18
Rhizopus oryzae	x		20
Rhizopus stolonifer	x	x	22
Scopulariopsis brevicaulis	x	x	252
Scopulariopsis candida	x		254
Scopulariopsis fusca	x	x	256
Stachybotrys chartarum		x	258
Syncephalastrum racemosum	x	x	24
Talaromyces macrosporus	x		48
Trichoderma harzianum	x	x	260
Trichoderma viride	x	x	263
Trichothecium roseum	x		264
Ulocladium chartarum	x	x	266
Wallemia sebi	x	x	268
Xeromyces bisporus	x		50
YEASTS	x	x	270

ZYGOMYCETES

GENERAL CHARACTERISTICS (Fig. 1)

This group has coenocytic mycelium. The only septa formed, separate special organs like sporangia and zygospores from the sterile mycelium. Septa sometimes occur in mature parts of the mycelium.

Asexual reproduction occurs by means of sporangiospores (single-celled aplanospores) produced endogenously in globose or pyriform sporangia with or without a columella (vesicle or central part inside the sporangium and continuous with the sporangiophore), or in merosporangia (cylindrical sporangium which breaks up into a row of merospores). A merosporangium does not have a columella.

Sporangioles (small, usually globose sporangia with one or a few spores and reduced columella) are found in the Thamnidiaceae, Cunninghamellaceae and Choanephoraceae.

Some genera are characterized by an apophysis (a swelling of the sporangiophore just below the sporangium). Stolons may occur, that are specialized hyphae which run over the surface and bear sporangiophores. These stolons may adhere to the substrate by means of rhizoids (root-like structures). Some species produce chlamydospores (thick-walled cells) or oidia (thin-walled cells, generally globose in shape), which may be terminal or intercalary, single or in chains.

Sporangiospores, chlamydospores and oidia germinate to form new mycelium.

Sexual reproduction occurs by fusion of two multinucleate gametangia and results in a thick-walled, yellow to brown or black zygospore which is often covered by spines or other projections. The two hyphal parts supporting the zygospore at each end are called suspensors. The suspensors can be equal or unequal in shape and size. Sometimes the suspensor forms appendages (e.g. some species of *Absidia*).

Most members of this group are saprobic, though some attack other fungi as well as animals and plants. Some species of *Rhizopus* and *Absidia* are widely distributed on stored grain, fruit and vegetables, in the air or in compost. Some species of *Mucor* and *Rhizopus* are important for their use in fermented foods (tempe, sufu and lao-chao) and the production of organic acids, but also cause rot in ripe and harvested fruit and vegetables. In this guide we have omitted *Thamnidium* and *Cunninghamella*, though species of these genera do occasionally contaminate food.

In indoor environments *Mucor plumbeus*, *Rhizopus stolonifer* and *Absidia corymbifera* are commonly occurring. In hospitals also *Mucor* species may contaminate infusion tubes. The genus *Rhizomucor* is distinct from *Mucor* by the formation of rhizoids. The species distinguished are thermophilic and therefore occur in warmer environments (e.g. compost).

For more detailed descriptions and keys to the Zygomycetes the reader is referred to Zycha *et al.* (1969), Hanlin (1973), Hesseltine and Ellis (1973), O'Donnell (1979) and von Arx (1981). Keys to the species of *Mucor* are presented by Schipper (1978), while the genus *Rhizopus* has been treated by Scholer and Müller (1971), Schipper (1984) and Liou *et al.* (1990), *Rhizomucor*, Schipper (1978).

CULTIVATION FOR IDENTIFICATION

Isolates are grown on MEA (4%) in plastic Petri-dishes. In *Mucor* crystallizing dishes allow undisturbed upward development (Schipper, 1973). Temperature varies from 20°C for *Mucor*, to 20-30°C for *Rhizopus* and *Syncephalastrum* to approx. 36°C for *Absidia*. All species are cultivated in darkness for one week. For zygospore development an alternate medium and temperature are often required. Two strains (+ and -, when heterothallic) are inoculated 0.5-1.0 cm apart in the centre of the Petri-dish. The zygospores will develop at the zone of contact of the mated cultures.

All forms should be studied under low magnification to determine the general form of growth. Microscopic mounts can be made by removing a piece of mycelium with sporulating structures with a (glass) needle (possibly wetting in 70% alcohol) and adding distilled water. For permanent slides lactic acid can be used but this may cause slight shrinkage of the fungal structures.

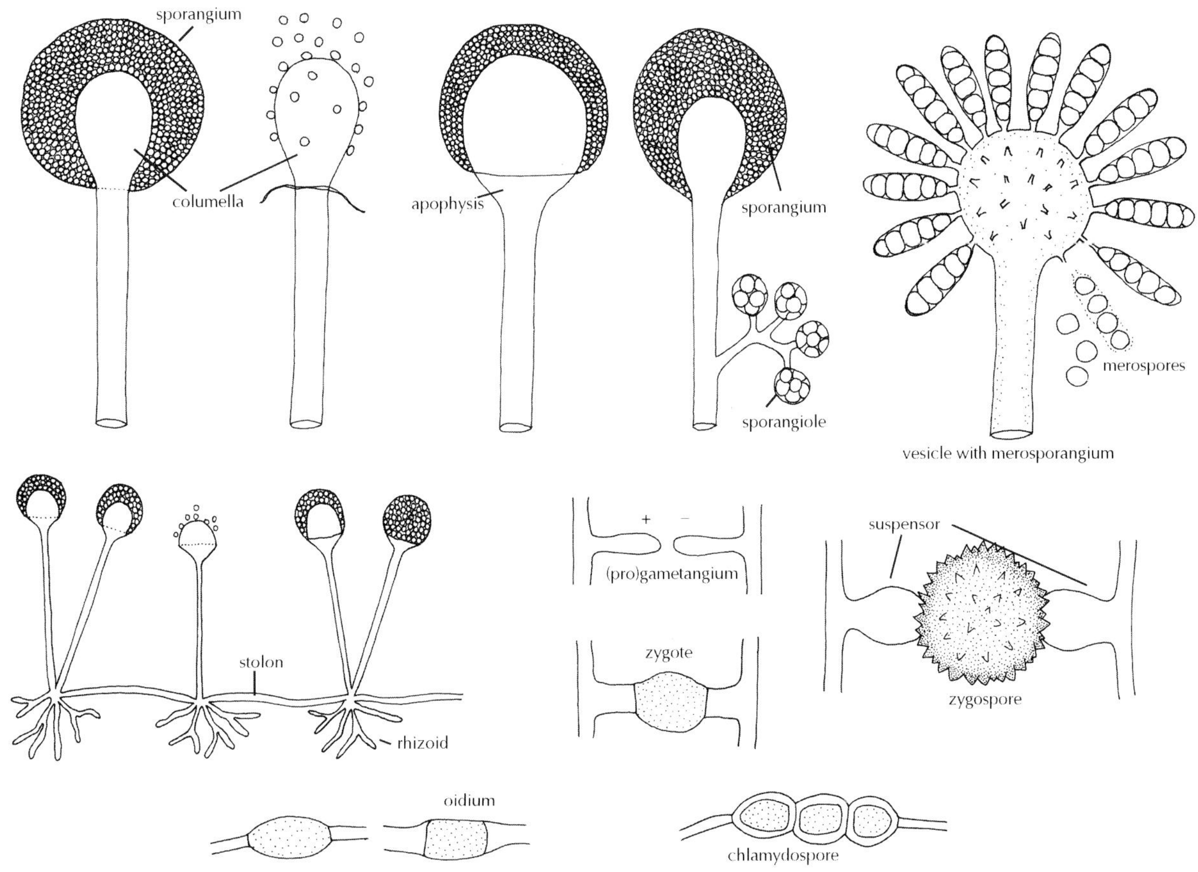

Fig. 1. Morphological structures in Zygomycetes.

KEY TO THE GENERA TREATED

1a. Sporangiospores formed in merosporangia which cover the swollen end of the sporangiophore .... *Syncephalastrum*

1b. Sporangiospores formed in globose or pyriform sporangia, with a columella 2

2a. Sporangia and sporangiophores usually dark pigmented, sporangiophores mostly unbranched, often occurring in groups. Sporangia vary from 50-360 µm in diam. Spores often striate .... *Rhizopus*

2b. Sporangia and sporangiophores not pigmented or only faintly coloured, often frequently branched. Sporangia never exceeding 100 µm in diam. Spores not striate 3

3a. Sporangia pyriform with a distinct apophysis, 10-40 µm in diam (terminal sporangia up to 80 µm in diam).. *Absidia*

3b. Sporangia globose without an apophysis, mostly larger than 40 µm in diam .....*Mucor*

Absidia corymbifera (Cohn) Sacc. & Trotter

= *Absidia ramosa* (Lindt) Lendner

Colonies floccose, light greyish, 5-7 mm high (up to 15 mm), growing rapidly, covering the whole Petri-dish within one week. Stolons 5-20 µm in diam, often with adhering droplets, hyaline to brownish, smooth-walled, with an occasional septum, often terminating in a large sporangium. Rhizoids borne on a swollen area of the stolon, up to 370 µm long, infrequently branched. Sporangiophores (40)80-450 (500) x (3)4-8(13) µm, hyaline to faintly pigmented, smooth to slightly roughened, simple or sometimes branched, arising solitarily from the stolons, in groups of three or up to 7 in a whorl below the large terminal sporangium at some distance from the rhizoids (smaller sporangiophores also bent or circinnate). Sporangia more or less pyriform, 10-40 µm in diam, up to 80 µm when terminating a stolon, many-spored, hyaline at first, becoming grey to greyish-brown when mature; wall transparent, smooth to slightly roughened. Columella hyaline to greyish, with or without a collar, (7)10-30 µm in diam, up to 60 µm when terminating a stolon, globose to short ovoid (especially larger ones). The apex of smaller columellae often with one to several projections. Sporangiospores varying from subglobose to oblong-ellipsoidal, (2.5)3-6(7) µm x 2.5-4.5 µm, hyaline to light grey, smooth-walled. Heterothallic. Zygospores (after 10 days on MYA at 33°C) borne in the aerial mycelium between the mated cultures, globose to (slightly) flattened, 40-85 (95) µm in diam, brown, slightly roughened, thick-walled with one or several equatorial ridges. Suspensors more or less equal, one frequently slightly larger and roughened near the zygospore, hyaline (roughened suspensor sometimes light brown in (roughened area), tubular and tapering. No appendages.

Temperature: optimum 35-37°C; maximum 45°C.

HABITAT: food, indoor
World-wide distribution in soil, stored grain, decaying vegetables and fruit, air, compost, animals and man.

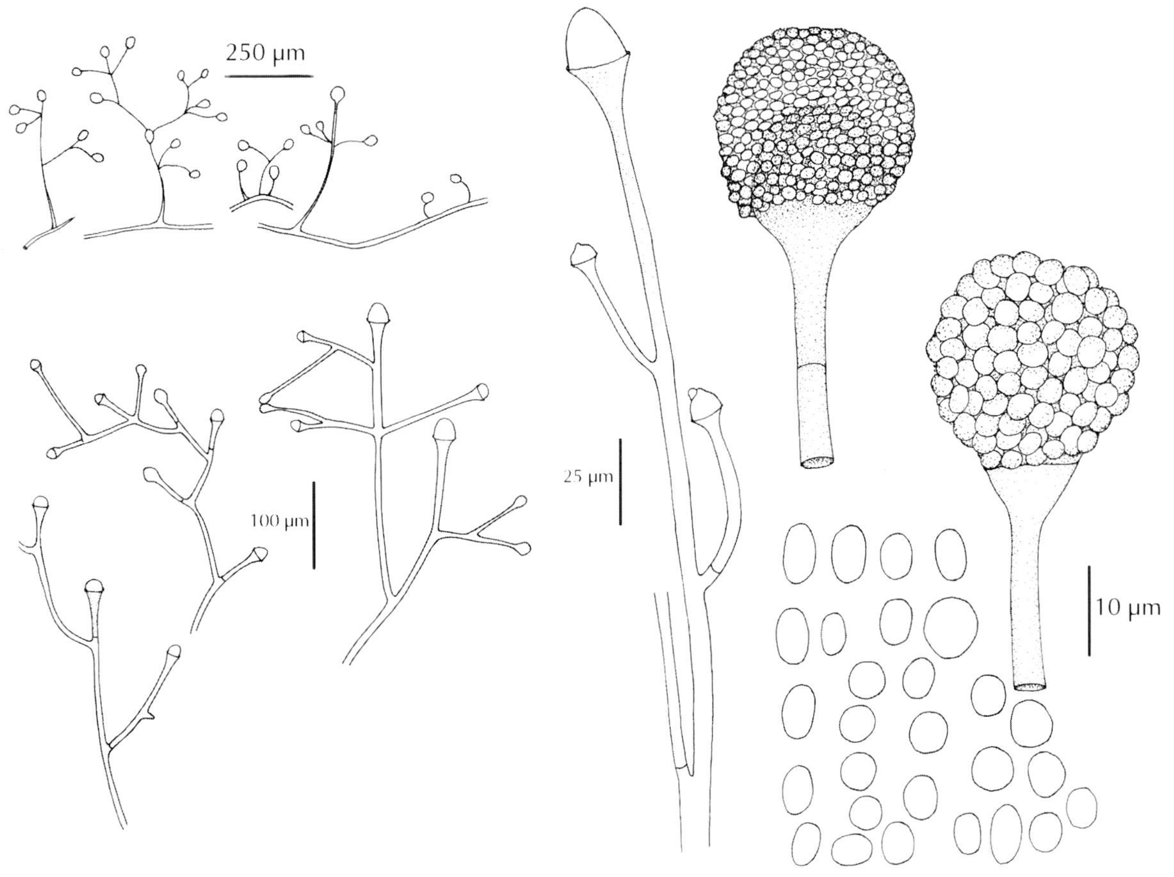

Fig. 2. *Absidia corymbifera*. Sporangiophores, sporangia, columellae and sporangiospores.
Pl. 3. *Absidia corymbifera*. a. Colony on MEA (4%) after one week; b. sporangiophores bearing sporangia, in Petri-dish, x 50; c. sporangiophore with sporangium, x 1700; d. sporangiospores, x 1500; e. zygospore after 10 days on MYA, x 200; f. zygospore, x 800.

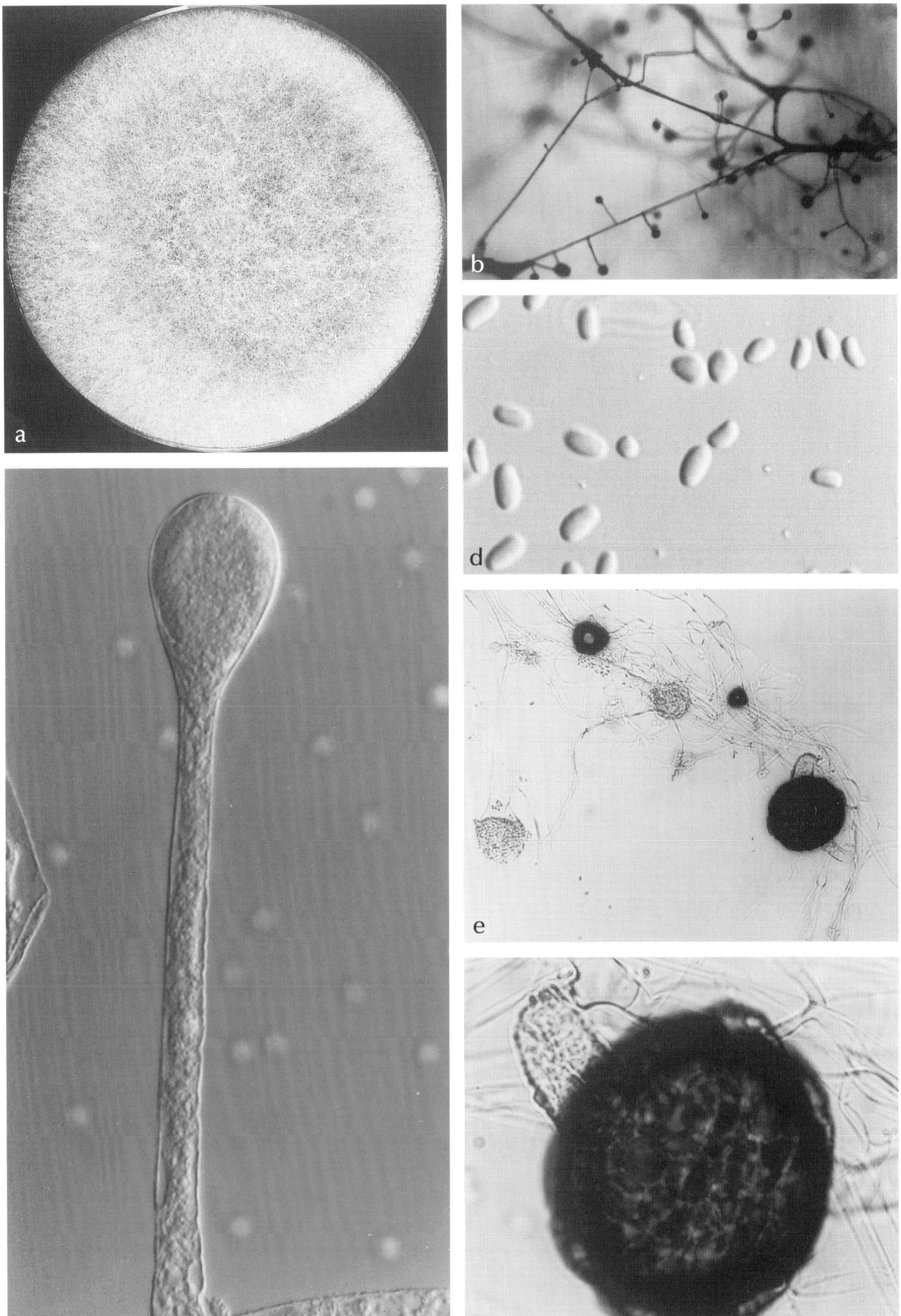
a
b
c
d
e
f

MUCOR Mich.

Hyphae white or coloured, varying from a few millimetres to some centimetres in height. Sporangiophores often branched, always ending in a many-spored sporangium without an apophysis. Sporangia varying in size; columellae well-developed; sporangium wall breaking or dissolving. Spores variable in shape, smooth-walled or slightly ornamented. Zygospores without appendages on the suspensors. Chlamydospores in some species present.

KEY TO THE SPECIES TREATED

1a. Columellae usually with one or several projections. Sporangiospores slightly roughened .... *M. plumbeus*
1b. Columellae without projections. Sporangiospores smooth-walled 2

2a. Colony composed of sporangiophores which are unbranched at first, then slightly branched sympodially. Chlamydospores absent, oidia sometimes present in substrate hyphae .... *M. hiemalis*
2b. Colony composed of both tall and short sporangiophores which are branched in either a mixed sympodial and monopodial fashion or merely in a sympodial fashion. Chlamydospores absent or present 3

3a. Chlamydospores numerous in sporangiophores and sometimes even in the columella .... *M. racemosus*
3b. Chlamydospores usually absent, if present then only a few in and on the substrate .... *M. circinelloides*

Mucor circinelloides v. Tieghem

Colonies greyish-green, up to 6(7) mm in height, composed of tall and short sporangiophores. Sporangiophores either tall, sympodially branched with long and short branches, the latter seldom circinnate, or short and often circinnate branches, with slightly encrusted walls, up to 17 µm in diam. Sporangiophores filled with droplets in younger parts. Sporangia whitish to yellowish at first becoming brownish-grey when mature, up to 80 µm in diam (rarely 100 µm), walls slightly roughened (spinulose), walls of larger sporangia leaving small collars after breaking up, walls of smaller ones persistent and leaving large basal membranes. Columella obovoid to ellipsoidal in larger sporangia and globose in smaller sporangia, up to 50 µm greyish-brown. Sporangiospores ellipsoidal, 4.4-6.8 x 3.7-4.7 µm, mostly 5.4 x 4 µm, smooth-walled. Chlamydospores few, in and on the substrate. Heterothallic. Zygospores (after 10 days on MEA at 25°C), orange-brown to dark-brown, globose to somewhat compressed, up to 100 µm in diam, with up to 7 µm long spines, suspensors usually short and slightly unequal.

Temperature: good growth and sporulation at 5-30°C (length and number of tall sporangiophores decreasing at lower temperatures); maximum 37°C.

HABITAT: food, indoor
World-wide distribution, mostly in soil, on dung, decaying potatoes, in animals and human.

NOTE:
Growth, sporulation, and presence of few or many tall and short sporangiophores can be influenced by temperature. However shape, size and uniformity (or variability) of sporangiophores are strain characteristics not influenced by temperature.

Pl. 4. *Mucor circinelloides*. a. Colony on MEA (4%) after one week; b. sporangiophores with sporangia, in Petri-dish, x 50; c. sporangiospores, x 1500; d. sporangiophore with sporangia, columella and part of sporangium wall, x 450; e. sporangiophore with columella and remnant of sporangium wall after dehiscence, x 450; f. zygospore on MEA (4%) after 10 days, x 500.

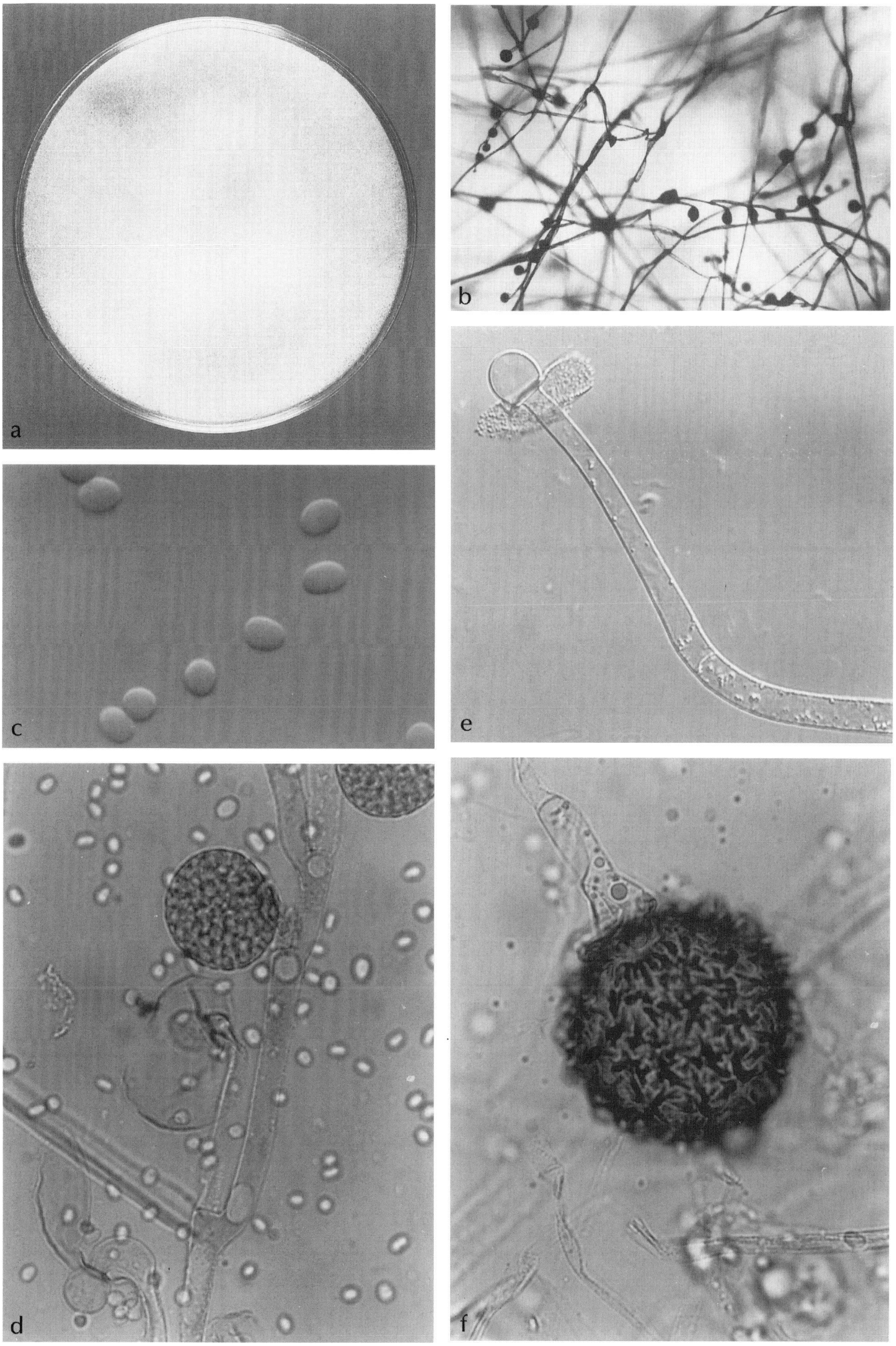
a
b
c
d
e
f

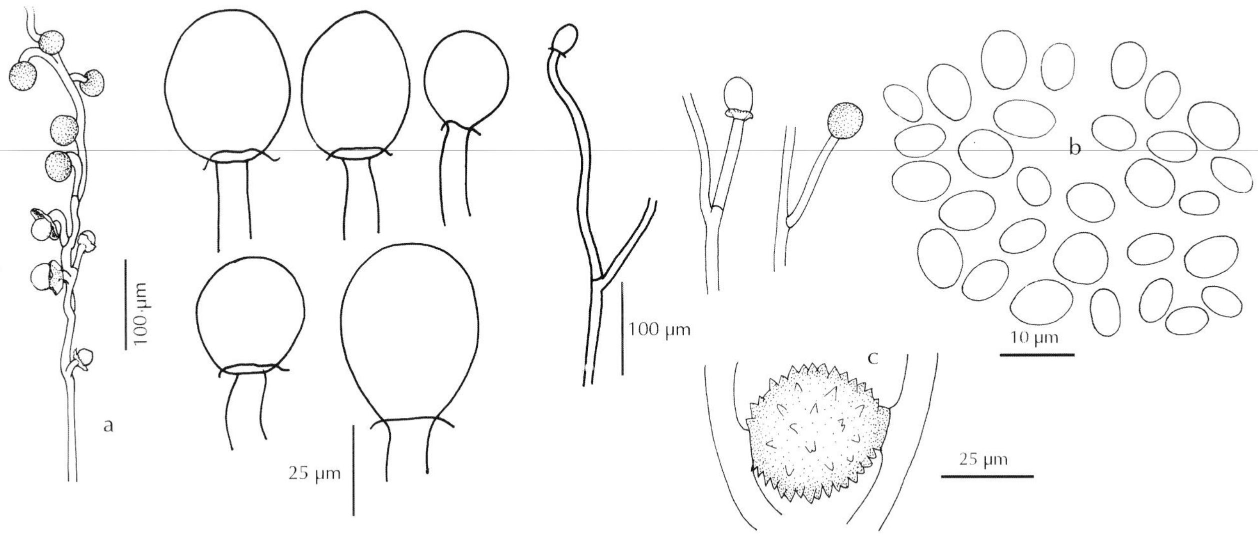

Fig. 3. a-c. *Mucor circinelloides*. a. Sporangiophores and columellae; b. sporangiospores; c. zygospore.

Mucor hiemalis Wehmer

Colony creamish-yellow in daylight, more greyish in darkness, up to 15 (20) mm in height. Sporangiophores simple at first, later slightly sympodially branched, often with yellowish contents, up to 10-12(14) µm in diam. Sporangia creamish-yellow, becoming dark brown, up to 70(85) µm in diam with deliquescent walls. Columella ellipsoidal, truncate at base, globose when young, up to 30-38 µm, hyaline, sometimes with yellowish contents. Sporangiospores ellipsoidal or somewhat kidney-shaped, varying in size, 5.7-8.7 x 2.7-5.4 µm, smooth. Oidia present in substrate hyphae. Chlamydospores absent. Heterothallic. Zygospores (after 10 days on MEA at 20°C) in a broad line between the + and - strain, yellow at first, becoming blackish-brown with age, globose or subglobose, up to 65-70 µm (90-100 µm), roughened by 3.5-4 µm long spines, suspensors equal or subequal, often with yellowish contents.

Temperature: optimum growth and sporulation at 5-25°C; maximum 30°C.

HABITAT: food, indoor
World-wide distribution, one of the commonest soil fungi. Isolated from soil, air, dung, beans, bananas, sugar cane, rice, corn, some vegetables, stored grain, cotton seed, potatoes.

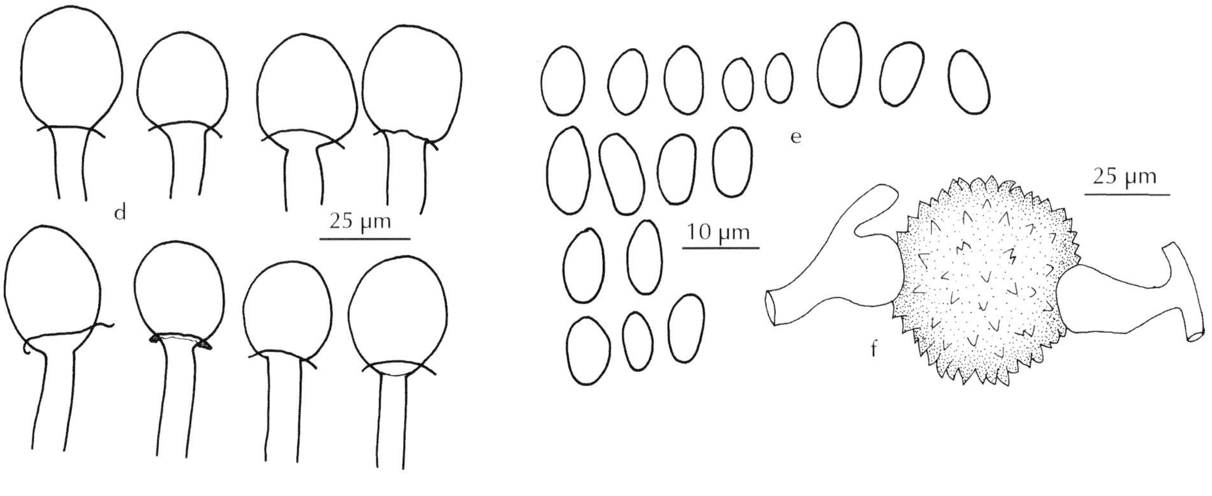

Fig 3. d-f. *Mucor hiemalis*. d. sporangiophores and columellae; e. sporangiospores; f. zygospore.
Pl. 5. *Mucor hiemalis*. a. Colony on MEA (4%) after one week; b. sporangium containing sporangiospores, x 750; c. sporangiophore with columella, x 1100; d. zygospores with suspensors in various stages of development after 10 days on MEA (4%), x 150; e. zygospore, x 400.

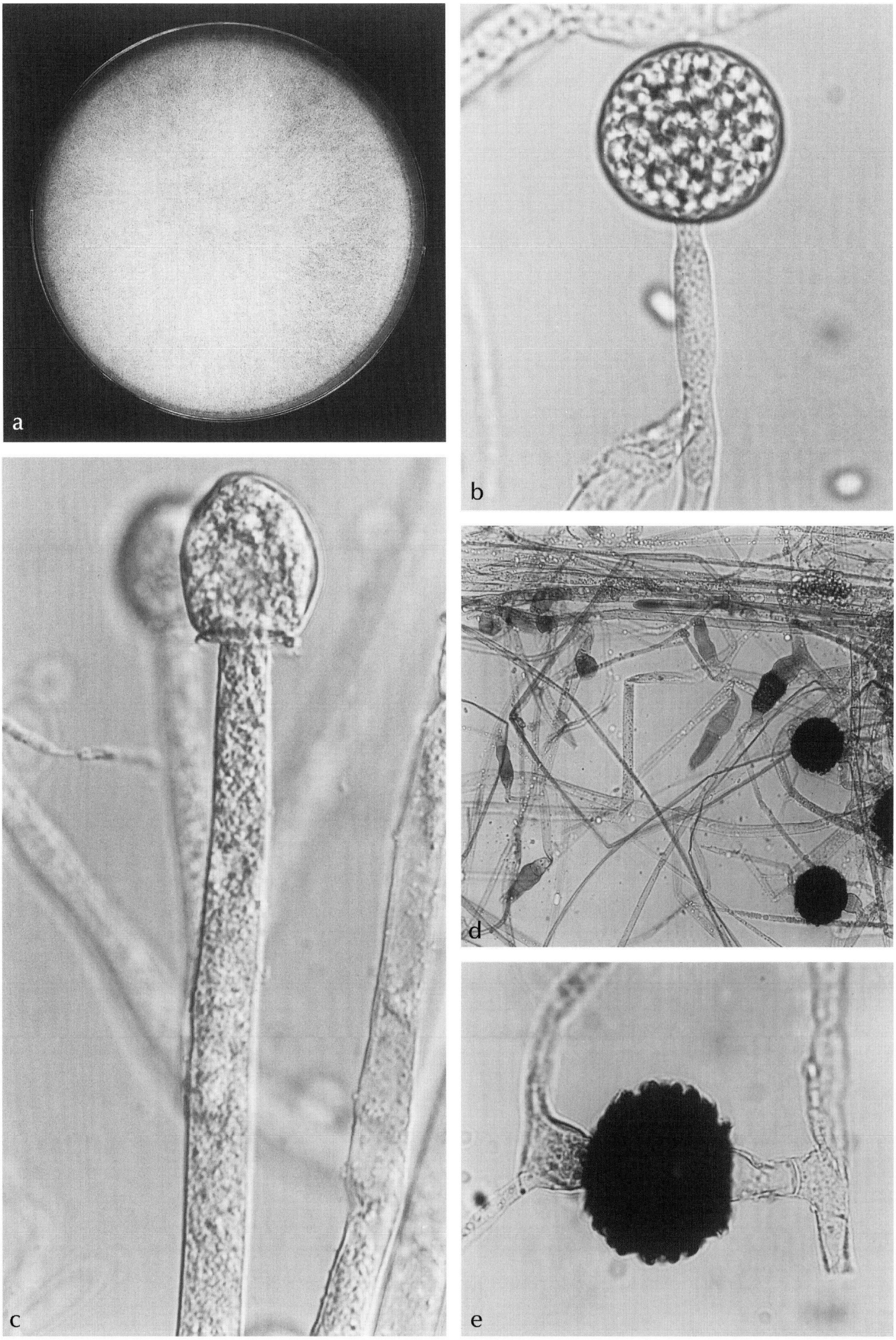
a
b
c
d
e

Mucor plumbeus Bon.

Colony grey or light olive-grey, 2-20 mm in height, Sporangiophores with slightly encrusted walls, branching (sympodial and monopodial) up to 20(21) µm in diam, constricted and somewhat recurved below the sporangium. Sporangia hyaline becoming dark-brown to brownish-grey with age, up to 80 (100) µm in diam with spinulose walls. Columellae pyriform, obovoid with a truncate base or ellipsoidal to cylindrical-ellipsoidal; smaller columellae often pointed-conical, up to 25-50 µm (seldom up to 50-70 µm), hyaline becoming brown to grey, smooth or roughened often with several projections, collar present. Sporangiospores globose, sometimes more or less ellipsoidal or irregularly shaped, (5)7-8(10) µm .
in diam, yellowish-brown, slightly rough-walled. Chlamydospores present, especially in older cultures. Heterothallic. Zygospores rare.

Temperature: growth and sporulation at 5-20°C (at 5°C slow); growth but poor sporulation at 20-30°C; no growth at 37°C.

HABITAT: food, indoor, often as air contaminant. World-wide distribution, e.g. soils, hay, dung, stored seeds of wheat, oats.

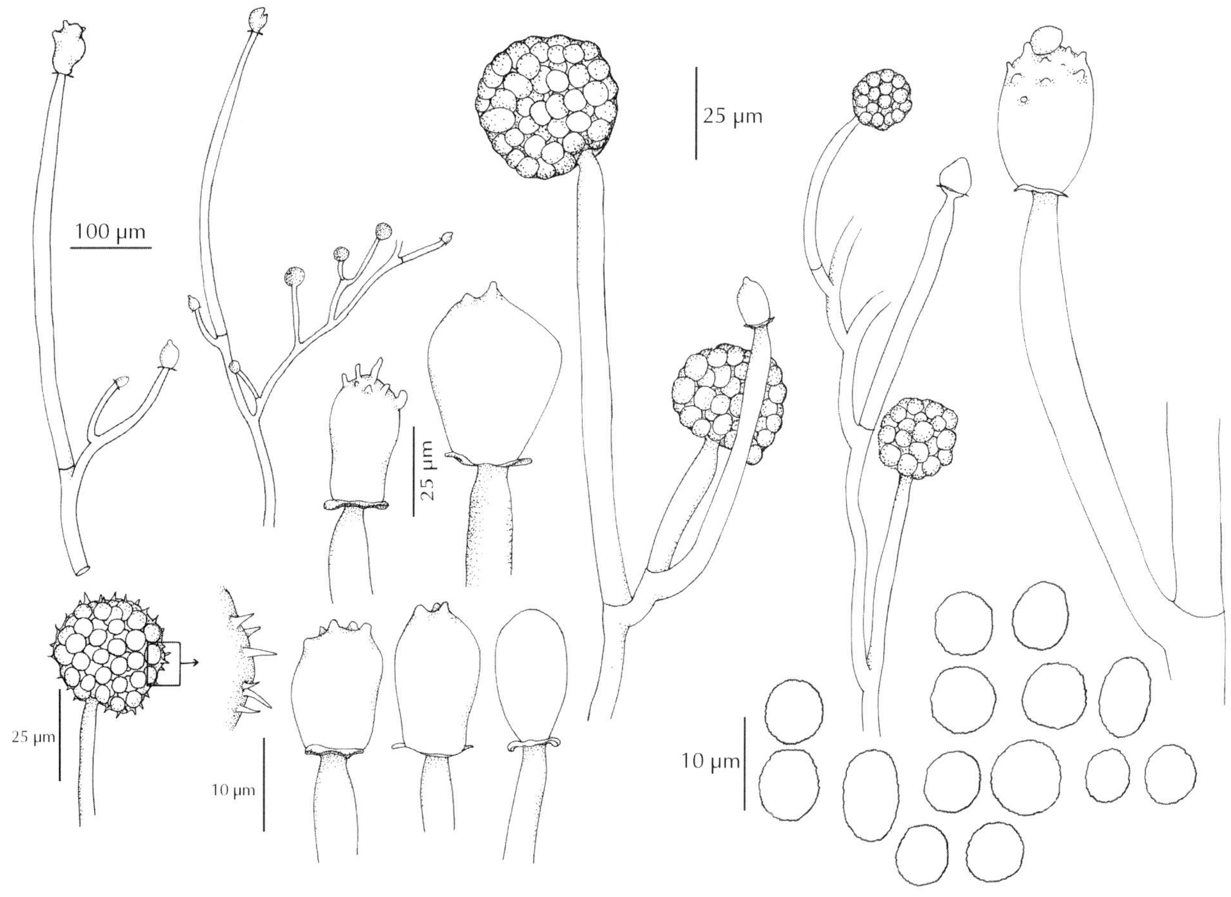

Fig. 4. *Mucor plumbeus.* Sporangiophores, sporangia, columellae and finely rough-walled sporangiospores.
Pl.6. *Mucor plumbeus.* a. Colony on MEA (4%) after one week; b. sporangiophore with sporangium, x 750; c-e, g. sporangiophores, young sporangia and columellae, x 500; f. sporangiospores, x 1000.

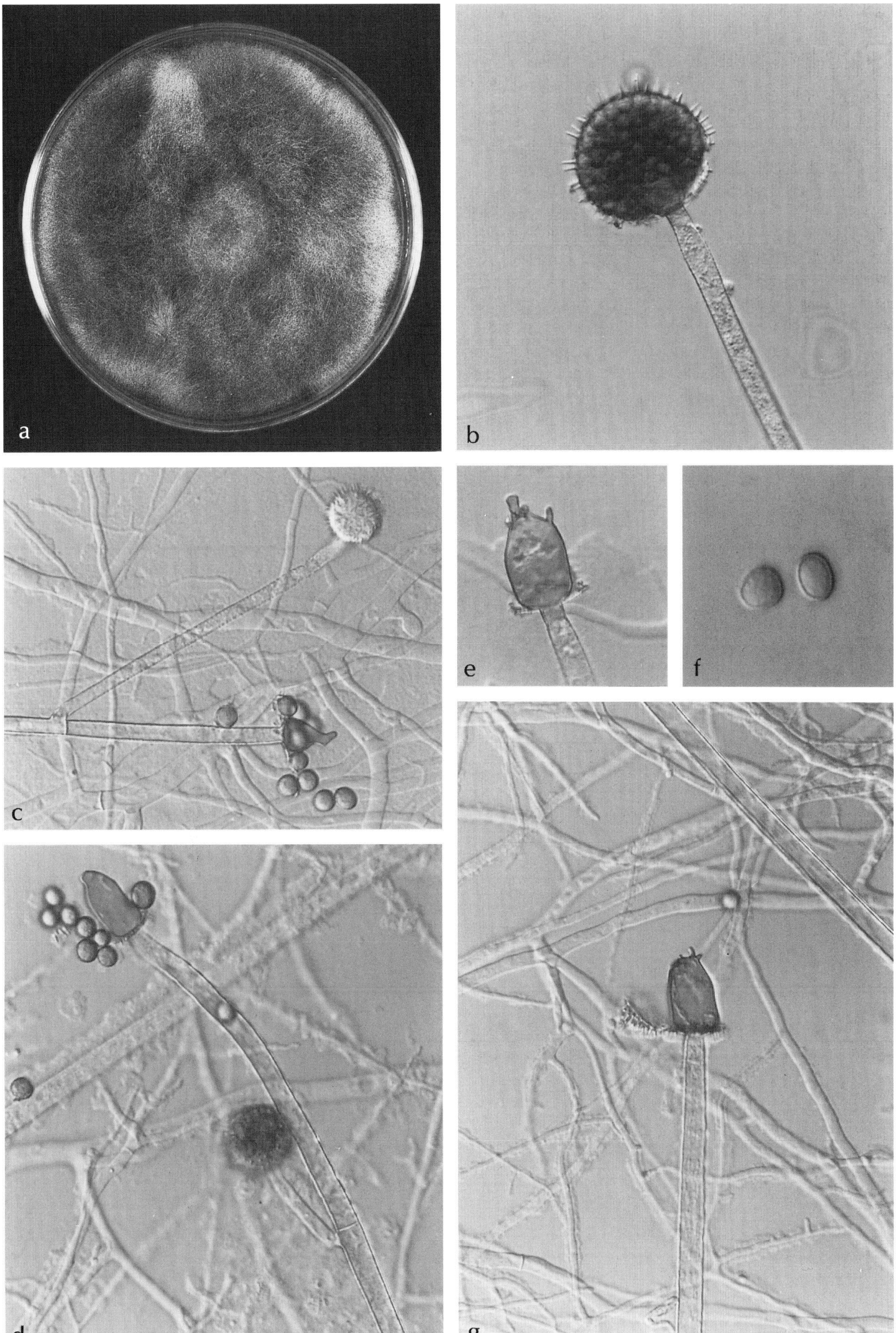
a
b
c
d
e
f
g

Mucor racemosus Fres.

Colonies white, becoming brownish-grey with age, 2-20(30) mm in height, consisting of tall and short sporangiophores. Sporangiophores branched (in a mixed sympodial and monopodial way), the short branches sometimes recurved, with encrusted walls. Sporangia hyaline becoming brownish to grey with age, up to 80(90) µm in diam but mostly 70 µm in diam, wall spinulose (small sporangia up to 20 µm in diam with persistent walls). Columella obovoid, ellipsoidal, cylindrical-ellipsoidal, slightly pyriform, usually with truncate base, up to 37-55 µm, light brown with collar. Sporangiospores broadly ellipsoidal to subglobose, 5.5-8.5(10) x 4-7 µm or 5.5-7 µm in diam, smooth walled, greyish. Chlamydospores numerous, in sporangiophores and sometimes in columellae, barrel-shaped when young, subglobose in older cultures, yellowish. Heterothallic. Zygospores (after 10 days on cherry agar at 10°C) in aerial mycelium, reddish-brown to bright-brown, globose or subglobose, up to 110 µm in diam, roughened by short spines (5 µm in length); suspensors equal or subequal, up to 33 µm long and 35 µm wide; zygospores slightly roughened. Chlamydospores can be present in zygophores and suspensors.

Temperature: good growth and sporulation at 5-30°C; optimum 20-25°C; maximum 38°C.

HABITAT: food, indoor
World-wide distribution in soil, dung, food (often in milk), seeds of wheat, barley, rice, tomatoes, tea groundnuts, animals.

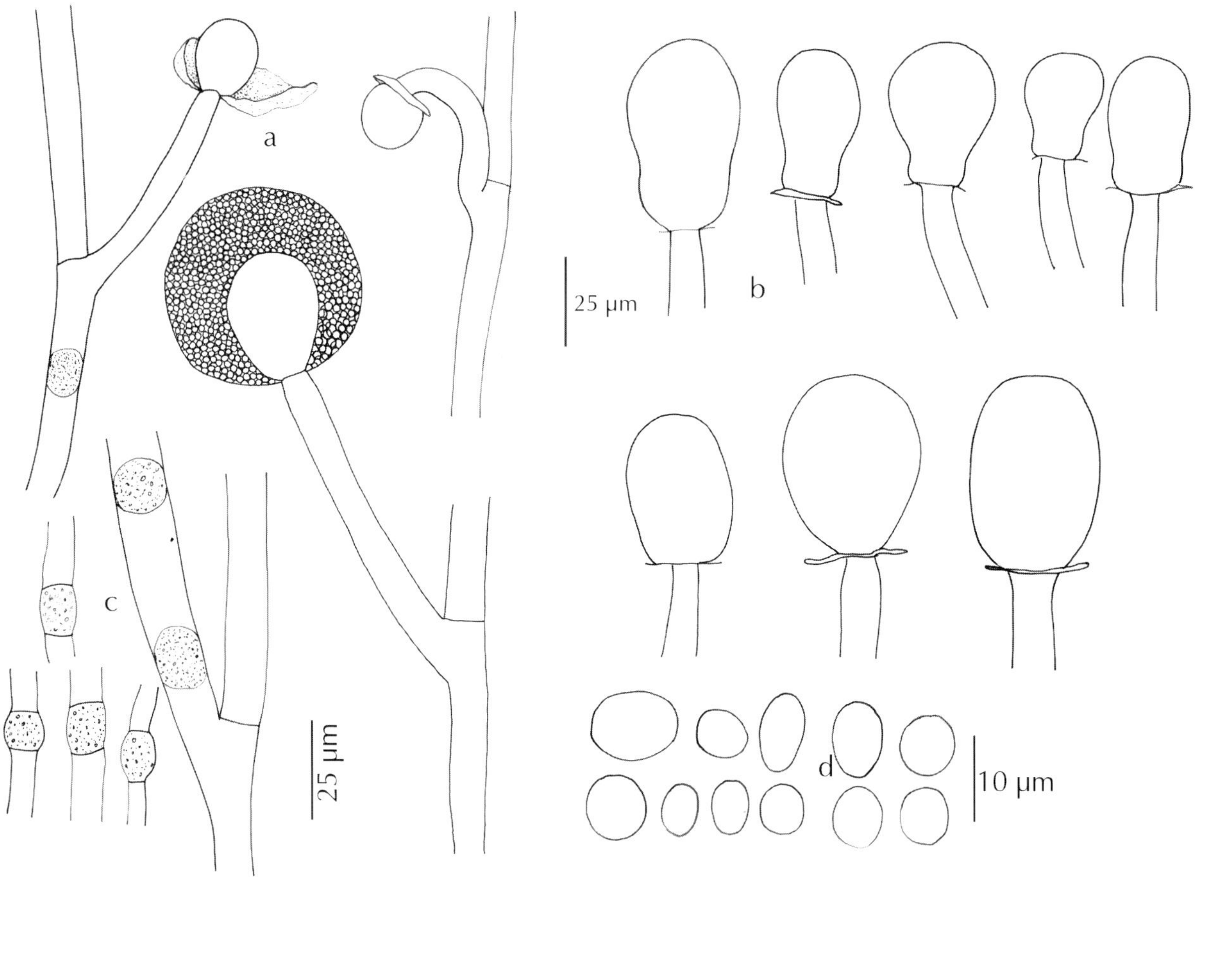

Fig. 5. *Mucor racemosus*. a. Sporangiophores with sporangium and columellae; b. columellae; c. chlamydospores; d. sporangiospores.
Pl. 7. *Mucor racemosus*. a. Colony on MEA (4%) after one week; b. sporangiophore with sporangium, x 200; c. sporangiophore with columella and hyphae, both containing chlamydospores, x 450; d. sporangiophore tip with sporangium, x 800; e. zygospores after 10 days, x 150.

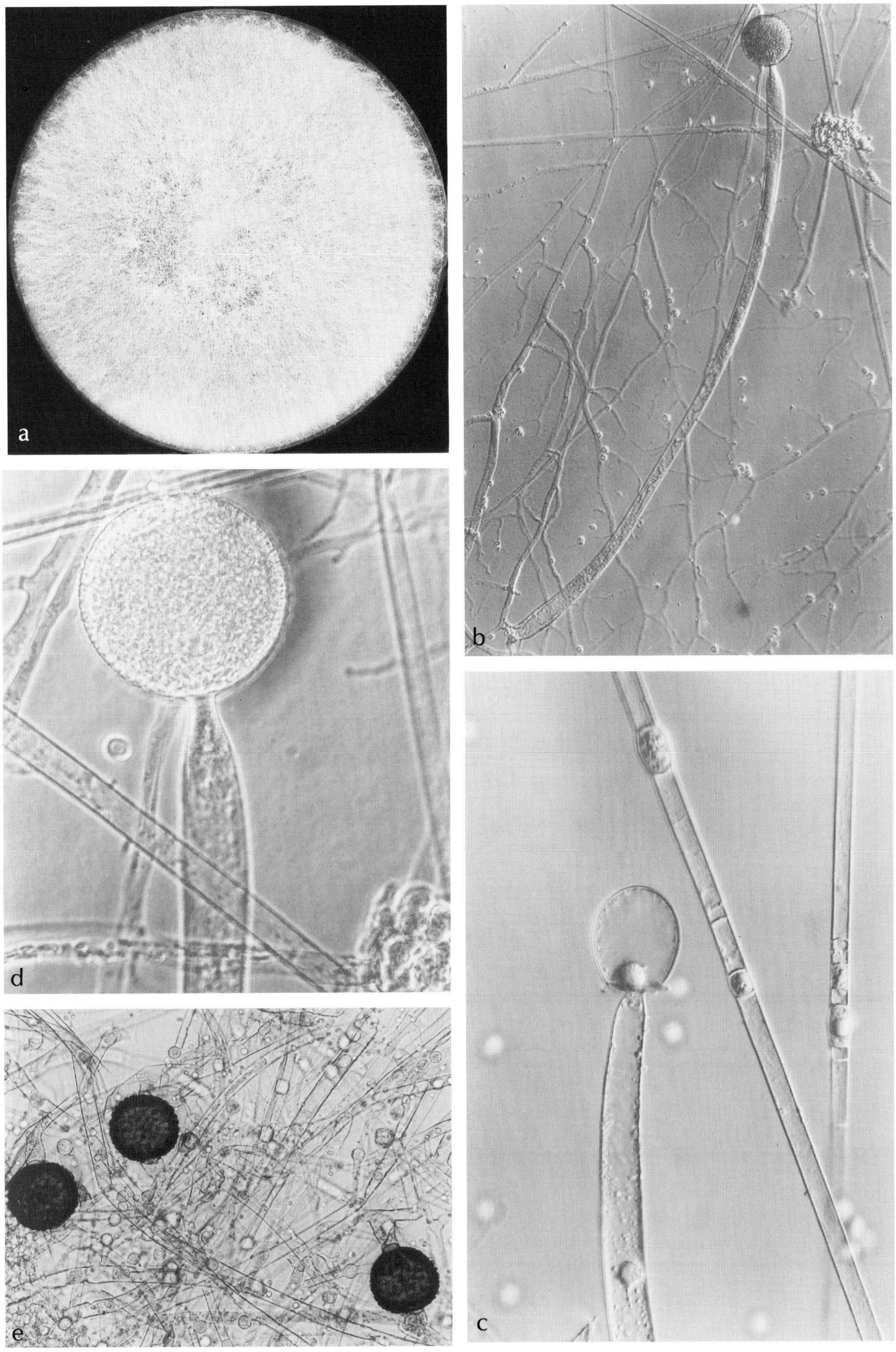
a
b
d
e
c

RHIZOPUS Ehrenb.

Colonies growing fast, with stolons, pigmented rhizoids and sporangiophores. Sporangiophores solitary or in groups, usually unbranched. Sporangia many-spored, mostly big, whitish when young, becoming blackish-brown with age. Columella brown globose or half-globose, with an apophysis. Spores short-ellipsoidal, usually irregularly angled, often striate (striations sometimes disappearing in water). Chlamydospores present in some species. Zygospores *Mucor*-like. Most species are heterothallic.

KEY TO THE SPECIES TREATED

1a. Sporangiospores non-striate, subglobose or irregular in shape. Sporangiophores not exceeding 1 mm in length. Chlamydospores abundant ....*R. oligosporus*
1b. Sporangiospores striate. Sporangiophores variable in length, up to (3)4 mm long. Chlamydospores absent or present...2

2a. No growth at 37°C, stolons without chlamydospores. Sporangiophores mostly 1.5-3(4) mm long ....*R. stolonifer*
2b. Growth at 37°C, stolons with chlamydospores. Sporangiophores mostly 1-1.5(2.5) mm long ....*R. oryzae*

Rhizopus oligosporus Saito

Colony pale brownish-grey, up to 1 mm high. Sporangiophores solitary or in groups of up to 4(6), arising as subhyaline to brownish hyphae opposite very short rhizoids, smooth or slightly rough-walled, up to 1000 µm long and 10-18 µm in diam. Sporangia globose, brownish-black at maturity, (50)100-180 µm in diam. Columella globose to subglobose with funnel-shaped apophysis. Sporangiospores irregularly shaped, globose, ellipsoidal, 7-10(24) µm in length, brownish in mass, single, subhyaline, smooth-walled. Chlamydospores abundant, single or in short chains, colourless, with granular contents, occurring in hyphae, sporangiophores and sporangia, globose, ellipsoidal, cylindrical, 7-30 µm or 12-45 x 7-35 µm.

Temperature: optimum 30-35°C; minimum 12°C; maximum 42°C.

HABITAT: food
Known from Japan, China, Indonesia. Isolated from tempe.

NOTE:
Schipper (1984) considers this taxon as a variety of *R. microsporus*.

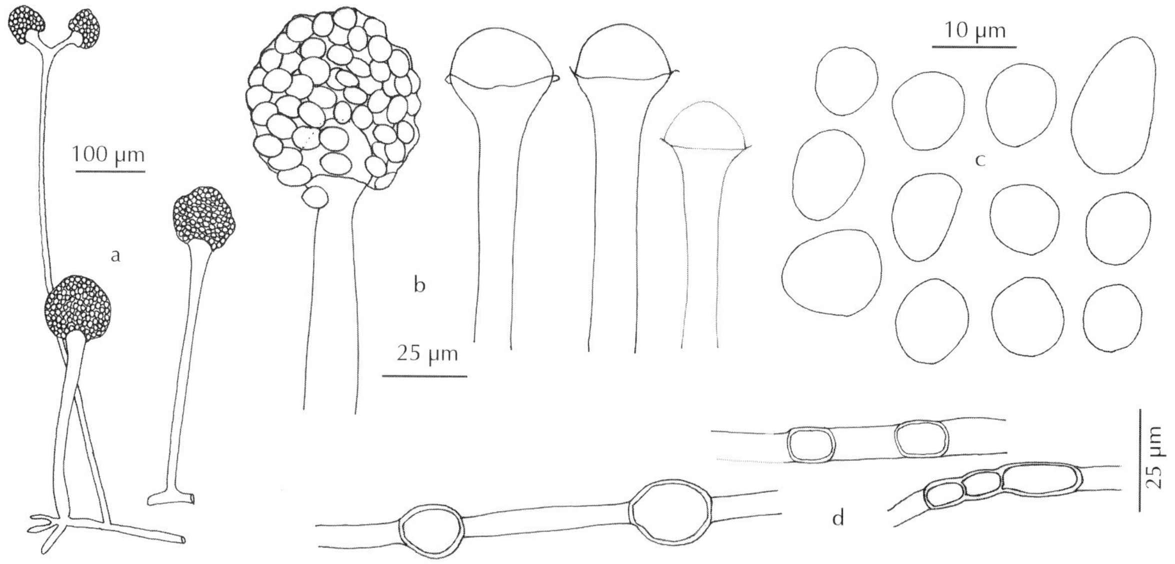

Fig. 6. a-d. *Rhizopus oligosporus*. a. Sporangiophores; b. sporangium and columellae; c. sporangiospores; d. chlamydospores.
Pl. 8. *Rhizopus oligosporus*. a. Colony on MEA (4%) after one week; b, e. sporangiophore with sporangium arising opposite the rhizoids, x 175; e. x 260; c. sporangium, x 1500; d. sporangiospores, x 800.

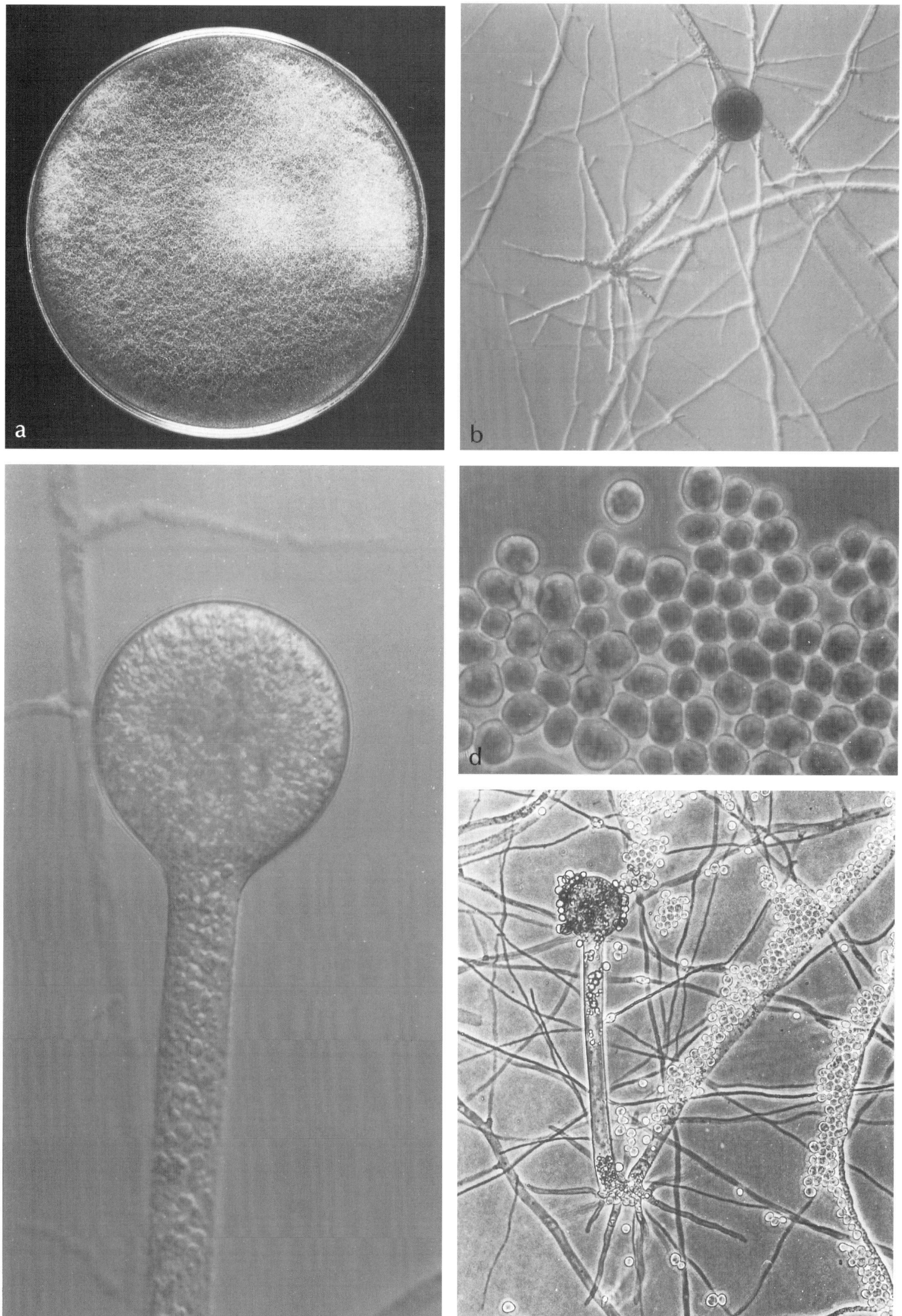
a
b
c
d
e

Rhizopus oryzae Went & Prinsen Geerlings

= *Rhizopus arrhizus* Fischer

Colony whitish becoming brownish-grey with age, about 10 mm high. Stolons smooth or slightly rough, almost colourless to yellowish-brown. Rhizoids brownish, opposite the sporangiophores or sporangiophores arising directly from stolons without rhizoids. Sporangiophores solitary or in groups of up to 5, sometimes forked, smooth-walled, 150-2000 µm long and 6-14 µm in diam. Sporangia globose or subglobose with spinulose wall, becoming dark-brown to black-brown, 50-200 µm in diam. Columella ovoid or globose, 30-120 µm in diam, smooth- or slightly rough-walled. Sporangiospores globose, ovoid or irregularly shaped, often polygonal, striate, 4-10 µm long. Chlamydospores globose, 10-35 µm in diam, ellipsoidal or cylindrical, 8-13 x 16-24 µm in diam.

Temperature: optimum at 30-35°C; minimum 5-7°C; maximum 44 (-49)°C.

HABITAT: food

World-wide distribution, though mainly in tropical and subtropical zones. Isolated from soils, grains, groundnuts, polluted water, vegetables, decaying fruit.

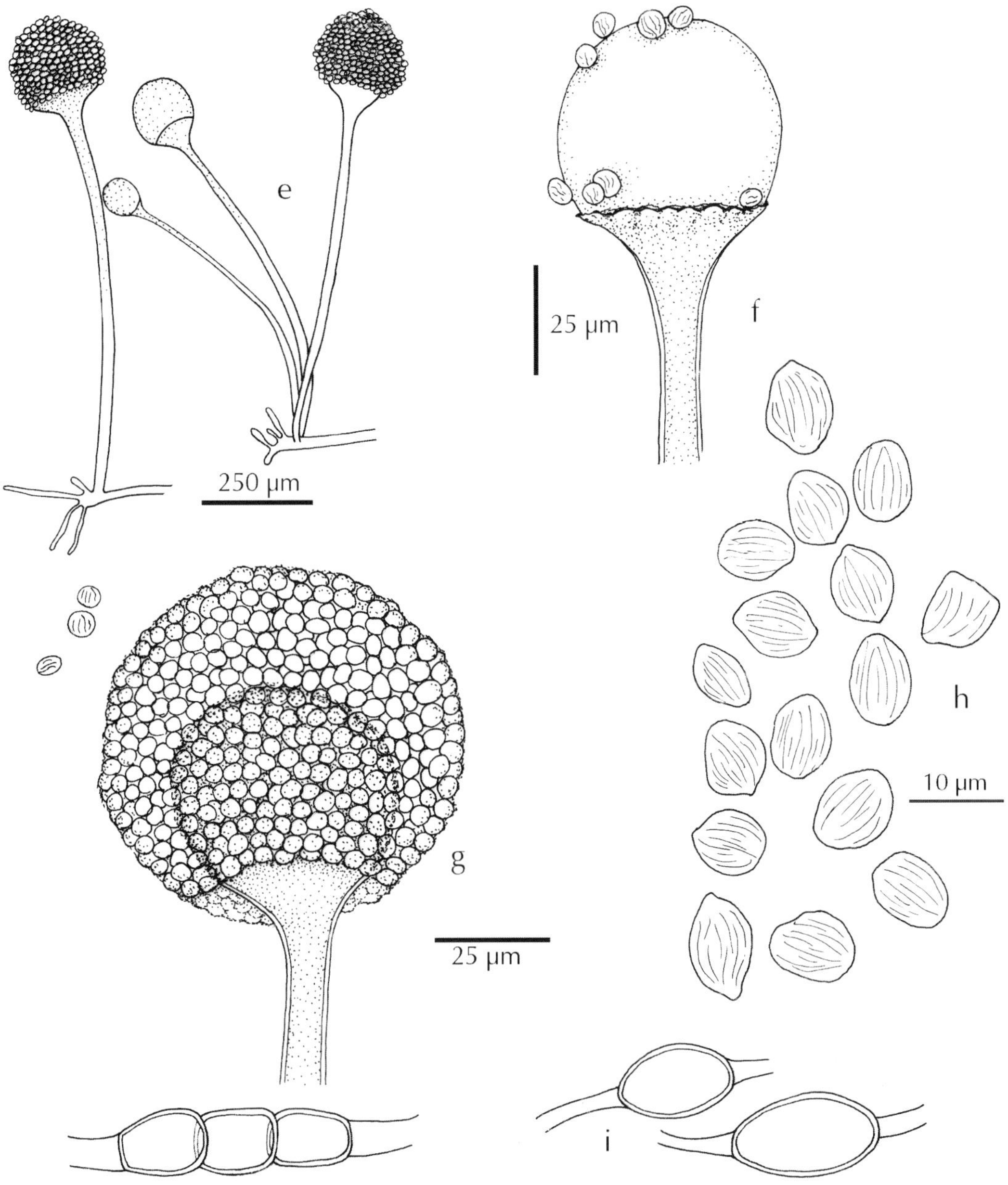

Fig.6. e-i. *Rhizopus oryzae*. e. Sporangiophores; f. columella and apophysis; g. sporangium; h. sporangiospores; i. chlamydospores.
Pl. 9. *Rhizopus oryzae*. a. Colony on MEA (4%) after one week; b. sporangium with spinulose wall, x 800; c. sporangiospores, x 1600; d. chlamydospore, x 500; e, f. sporangiophores with columellae, e. x 300, f. x 500.

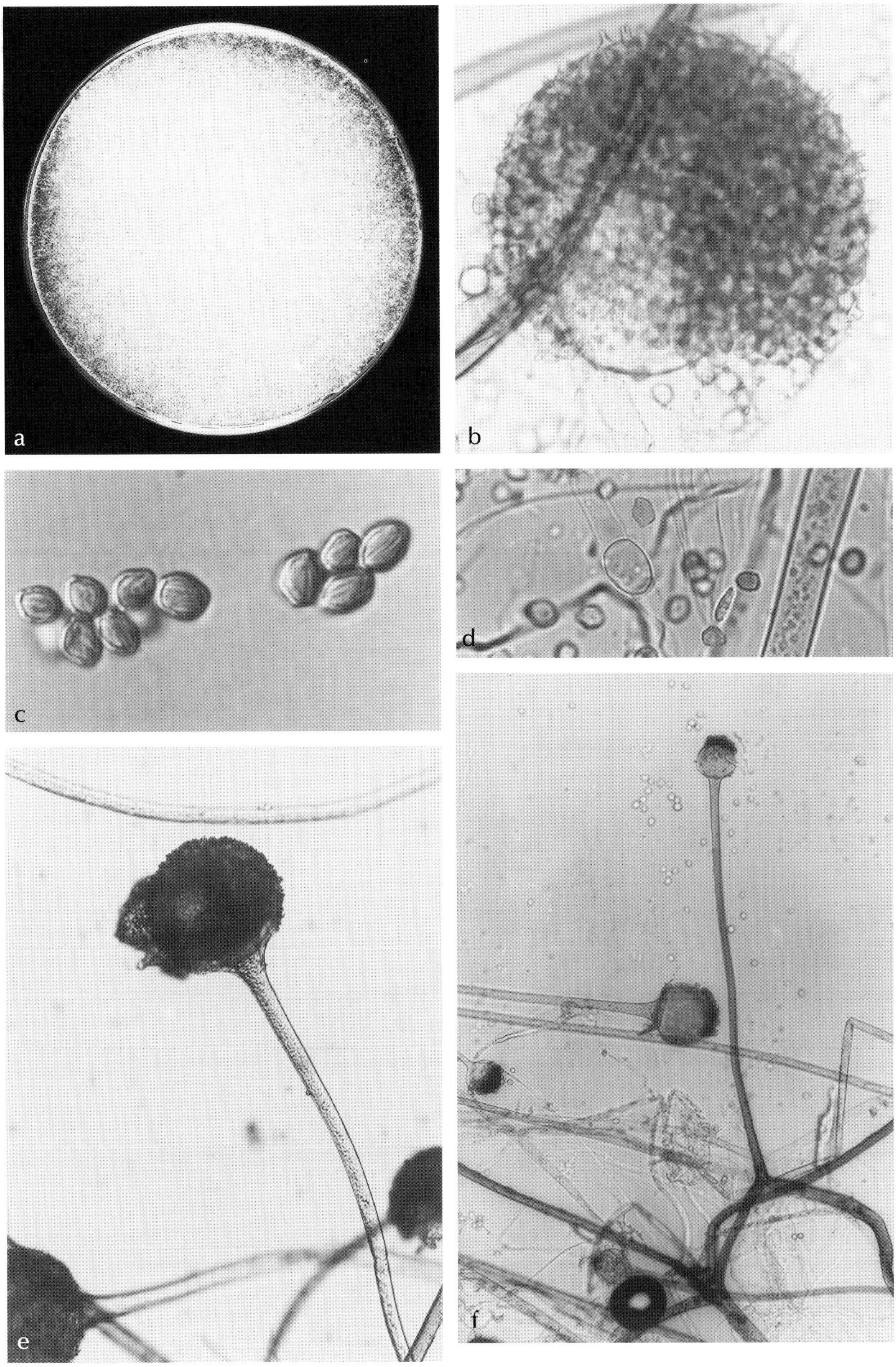
a
b
c
d
e
f

Rhizopus stolonifer (Ehrenb.) Lind.

= *Rhizopus nigricans* Ehrenb.

Colony whitish becoming greyish-brown due to brownish sporangiophores and brown-black sporangia, often over 20 mm high. Sporangiophores 1.5-3(4) mm tall, solitary or in groups of 2-7 (usually 3-4) from the almost colourless to dark-brown, smooth or slightly rough-walled stolons opposite the branched rhizoids. Sporangia globose to subglobose, (50)150-360 µm in diam, blackish-brown at maturity. Columella globose, subglobose, ovoid (40)70-160(250) µm in diam. Sporangiospores irregular in shape often polygonal or ovoid, globose, elliptical, striate, (4)7-15 x 6-8(129 µm. Chlamydospores absent in stolons, sometimes present in submerged hyphae. Heterothallic. Zygospores brownish-black, warted, with unequal suspensors, (75)150-200 um in diam.

Temperature: optimum 25-26°C; minimum about 5°C; maximum 32-33°C.

HABITAT: food, indoor (commonly air-borne) World-wide distribution but more often occurring in warmer zones. Isolated from soil, grain, vegetables, fruit, nuts etc.

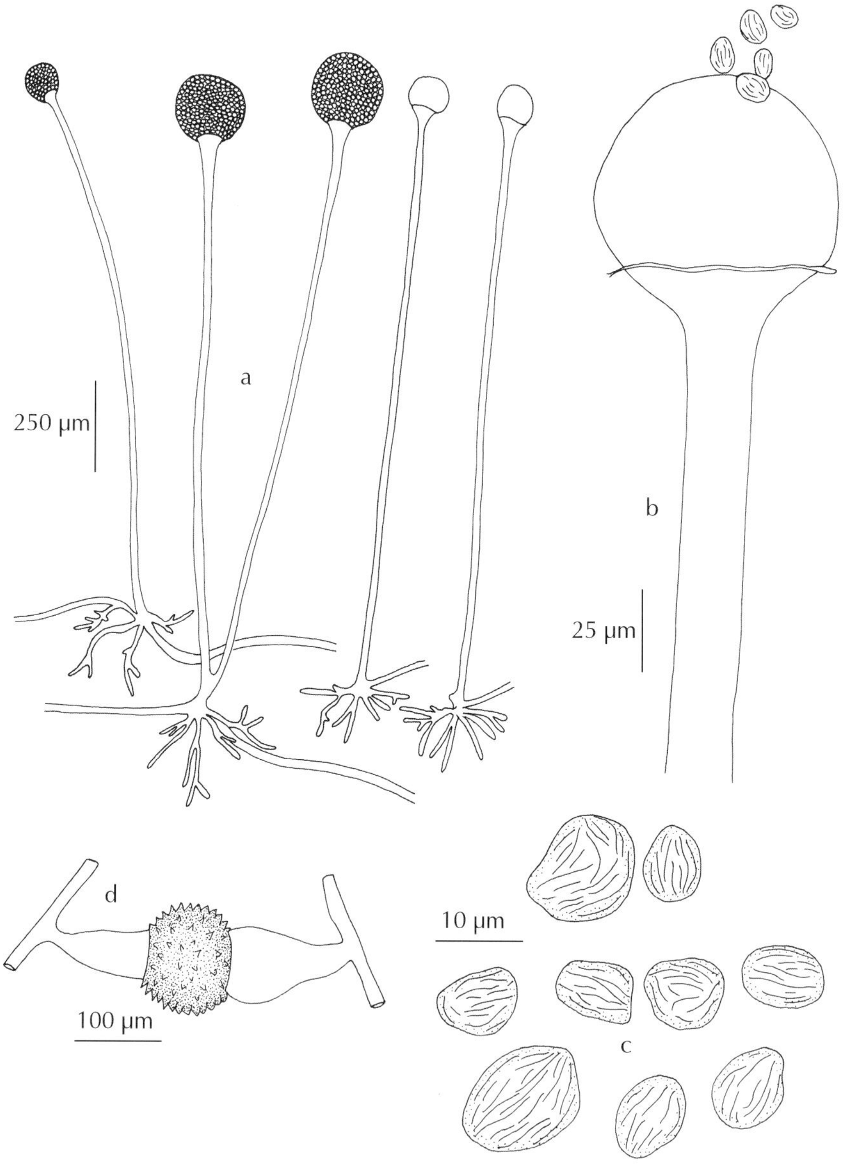

Fig. 7. *Rhizopus stolonifer.* a. Sporangiophores with sporangia, opposite the rhizoids; b. sporangiophore with columella and sporangiospores; c. sporangiospores; d. zygospore.
Pl.10. *Rhizopus stolonifer.* a. Colony on MEA (4%) after one week; b. sporangiophores with sporangia and rhizoids, x 15; c. sporangiospores, x 800; d. sporangiophore with columella and sporangiospores, x 200; e, f. zygospore with suspensors on MEA (4%) after 10 days, e. x 250, f. x 75.

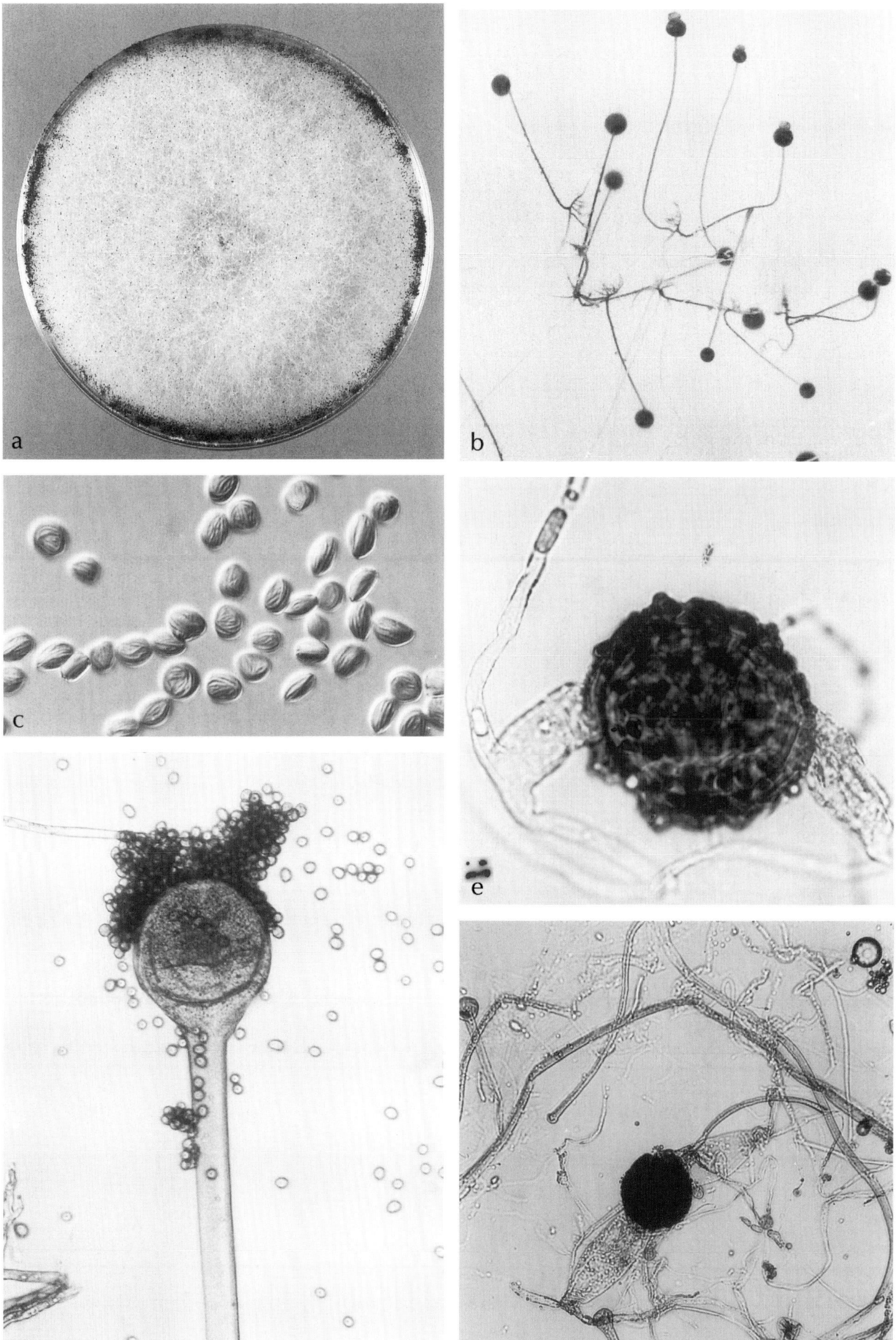
a
b
c
e
d
f

Syncephalastrum racemosum Cohn

Colony white becoming grey with age, growing rapidly, covering the Petri-dish within a week, about 4-5(15) mm high. Vegetative hyphae hyaline, branched, septa produced in old cultures to delimit reproductive structures. Main sporangiophores (with rhizoids) and their often strongly curved lateral branches each bearing a terminal vesicle which forms merosporangia all over its surface. Vesicles brownish, globose or subglobose to ovoid; terminal vesicles 30-80 µm, lateral 10-40 µm in diam, with or without a septum. Merosporangia grey, cylindrical, 15-33 µm long and 4 µm wide, breaking up into 5-10(18) merospores. Merosporangium wall thin and disappearing. Merospores globose or ovoid, 2.5-5(6.5) µm in diam, usually faintly roughened, pale brown. Chlamydospore-like swellings more or less spherical, 12-13 µm in diam. Heterothallic. Zygospores (after 10 days on MEA at 20-24°C) mainly in substrate mycelium, orange-brown, with conical projections, 50-90 µm in diam. Suspensors hyaline, smooth, subequal.

Temperature: optimum 17-40°C.

HABITAT: food (storage), indoor
Main distribution in tropical and subtropical regions; isolated from soil, dung, grain.

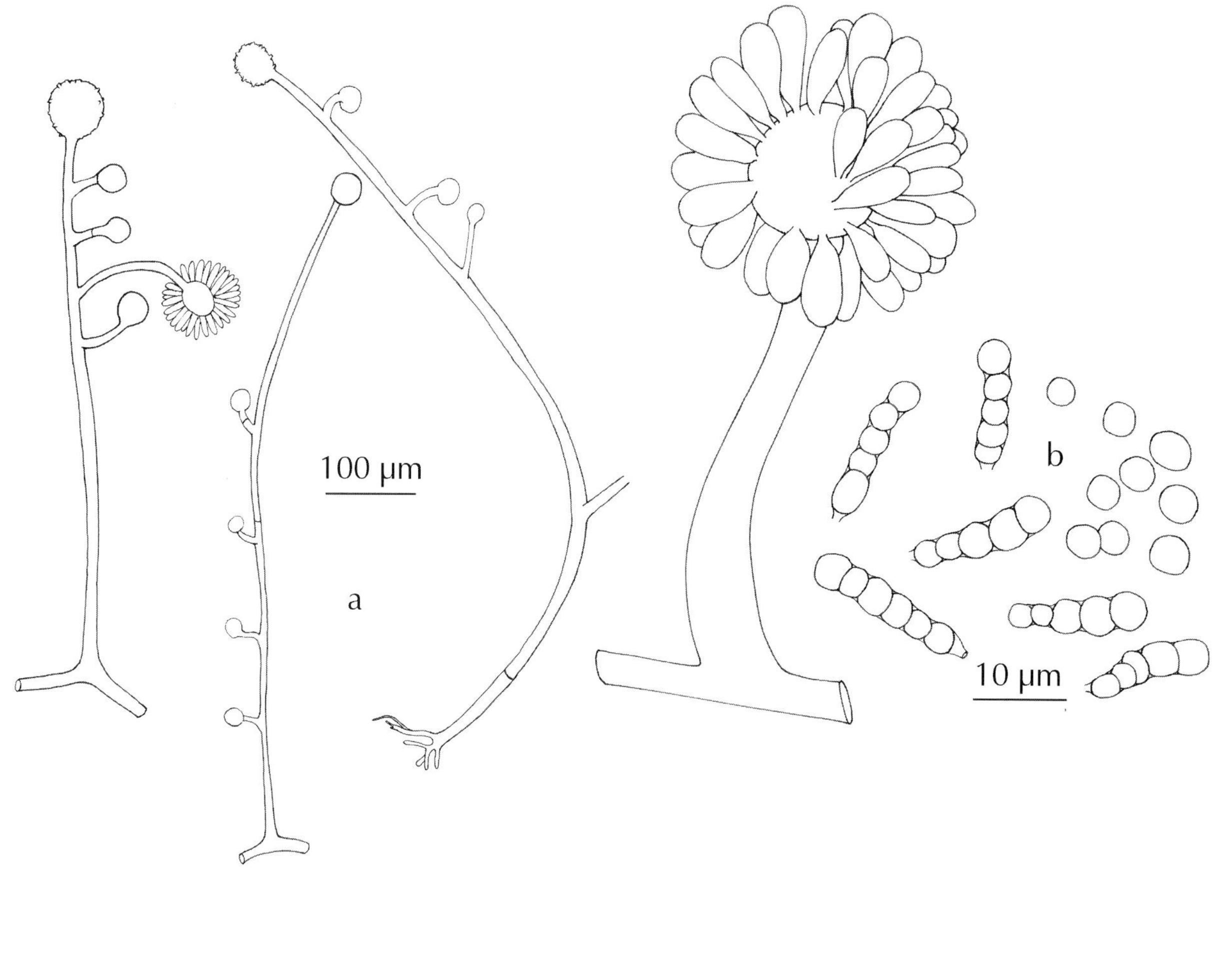

Fig. 8. *Syncephalastrum racemosum*. a. Sporangiophores bearing vesicles which produce merosporangia all over their surface; b. merosporangia and merospores.
Pl. 11. *Syncephalastrum racemosum*. a. Colony on MEA (4%) after one week; b. sporangiophores with merosporangia, in Petri-dish, x 15; c,d. sporangiophore with vesicles bearing merosporangia, x 800; e. merosporangia and merospores, x 800; f. zygospore on MEA (4%) after 10 days, x 800.

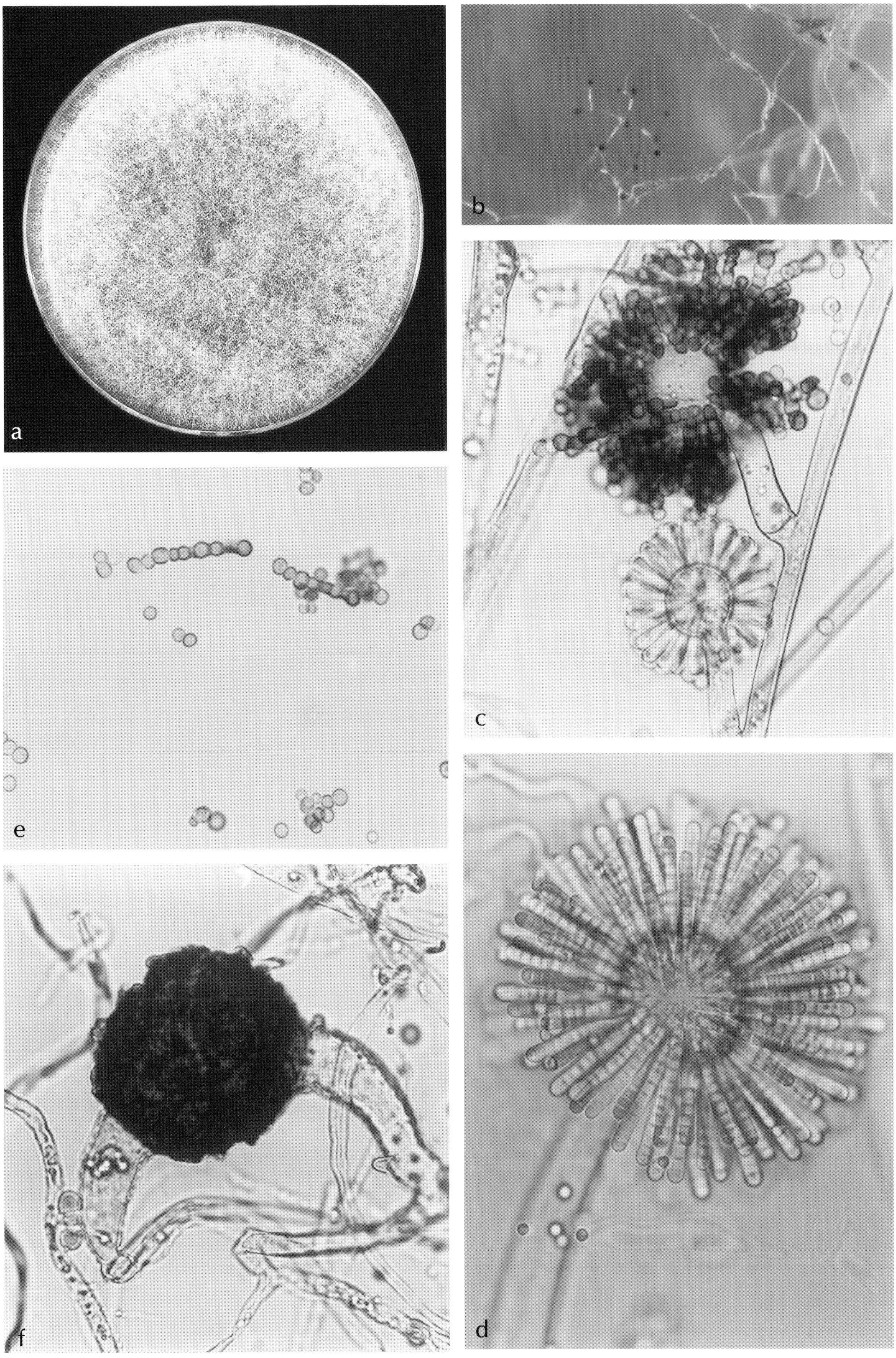
a
b
c
e
f
d

ASCOMYCETES

GENERAL CHARACTERISTICS (Fig. 10)

The vegetative mycelium of the Ascomycetes is septate and haploid. Sexual propagation takes place by means of asci which are often formed by karyogamy of two nuclei from different gametangia (male: antheridium, female: ascogonium).

Mostly 8 (sometimes 2 or 4) **ascospores** are formed in the asci after meiosis. **Asci** are usually enclosed in **ascomata** (ascocarps), that is fruit-bodies which occur singly or aggregated in or on a **stroma** (mass of vegetative hyphae). The morphology of the fruit-bodies is important for a systematic division. We can distinguish cupulate (**apothecia**), globose or subglobose non-ostiolate (**cleistothecia**), or more or less flask-shaped, mostly ostiolate (**perithecia**) fruit-bodies. Asci can also develop in cavities of stromata or arise from cushion-like structures (**pseudothecia**).

The ascigerous state or **teleomorph** is often accompanied by one or more asexual reproductive states, the **anamorph** (Fig. 9). The majority of the Deuteromycetes (*Fungi imperfecti*) are anamorphs in life-cycles of Ascomycetes. The Ascomycetes treated in this guide belong to the Eurotiales and Sordariales. The Eurotiales are characterized by ascomata with many small, globose to subglobose asci, which have single walls and one-celled often ornamented ascospores. The anamorphs of the Eurotiales treated in this guide belong to *Aspergillus, Basipetospora, Fraseriella, Paecilomyces* and *Penicillium.* Within the Sordariales the species *Chaetomium globosum* is treated. Occasionally species of *Peziza* and *Pyronema* (both belonging to the Pezizales) are encountered in indoor environments or on building materials.

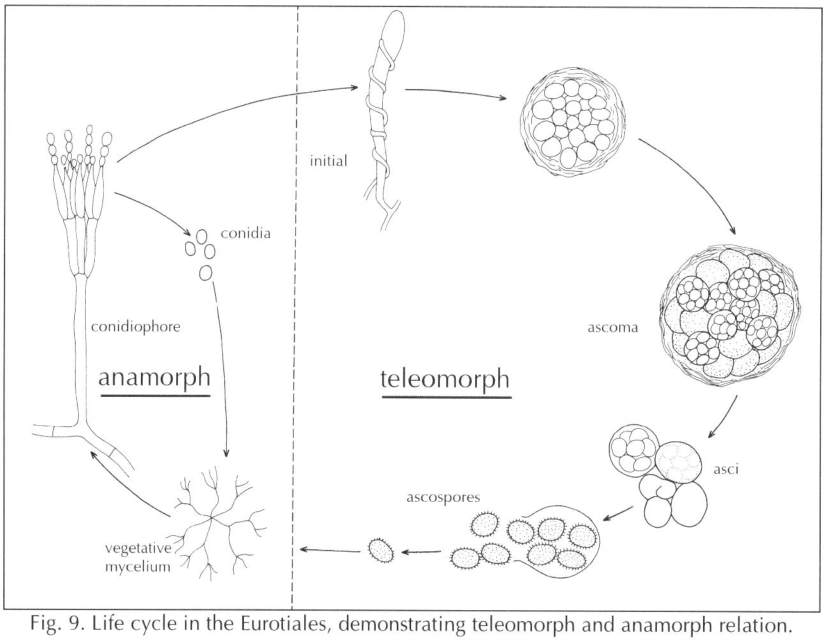

Fig. 9. Life cycle in the Eurotiales, demonstrating teleomorph and anamorph relation.

For general references to the Ascomycetes see von Arx (1981). For special publications on *Byssochlamys* see Stolk and Samson (1971) and Samson (1974), on *Eurotium* see Blaser (1975) and Pitt (1985), on *Talaromyces* see Stolk and Samson (1972). For keys to the Ascomycetes with *Aspergillus* anamorphs see also Raper and Fennell (1965). For *Chaetomium* is referred to Von Arx *et al.* (1986). The ascomycetous species are often found as storage fungi or develop after heat-treatment of a product. *Chaetomium* occurs in indoor environments on walls and cellulosic materials, often after water damages.

CULTIVATION FOR IDENTIFICATION

Isolates should be cultivated on appropriate media. *Byssochlamys, Monascus, Emericella, Talaromyces* and *Neosartorya* sporulate best on MEA and OA at 24-30°C. *Eurotium* isolates develop slowly on these media and atypical conidiophores and abortive fruit-bodies are developed.

For optimal sporulation *Eurotium* should be grown on MEA or Cz with additional sucrose (20-40%) or NaCl (10-30%). For detection Dichloran 18% Glycerol agar (DG18) can be used. Most of the ascomycetous species treated here mature after 10-14 days. Note that for some Ascomycetes (*Talaromyces* and *Neosartorya*) the anamorph is sometimes difficult to find and it occurs mostly in young cultures or when grown on poor media (potato-carrot or hay infusion agar). *Chaetomium* grows well on Oatmeal agar; a filterpaper on top of the medium often stimulates the development of fruit-bodies.

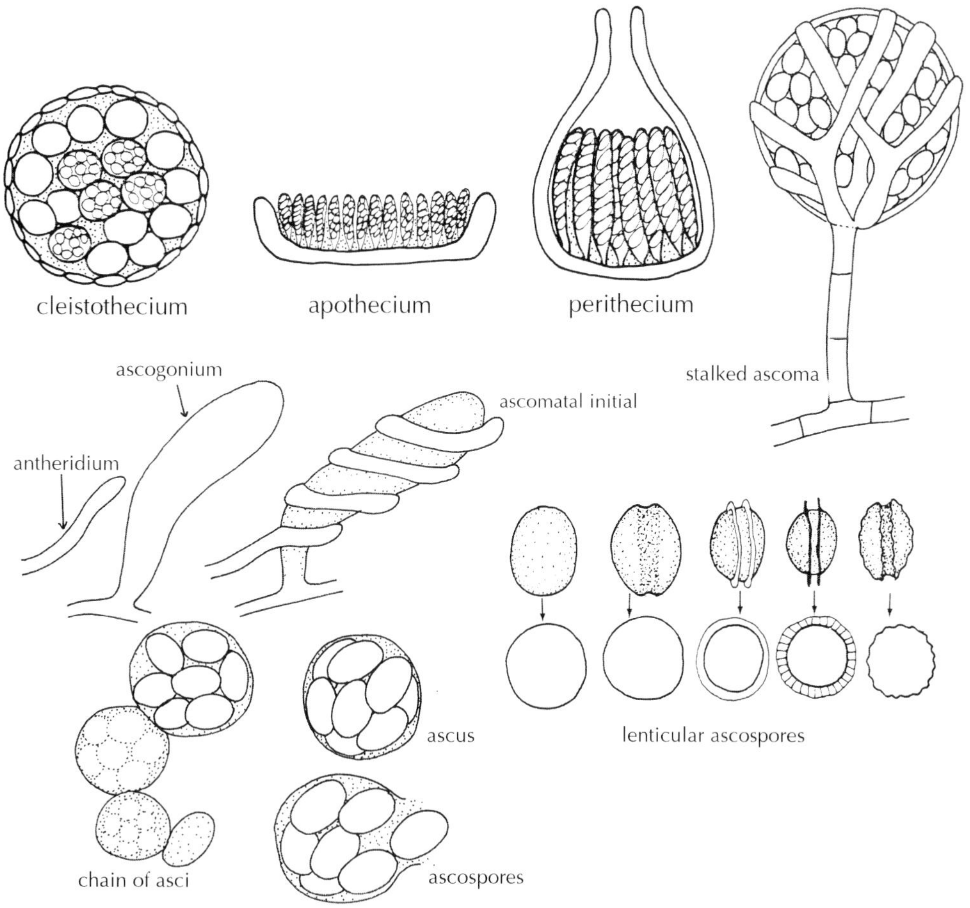

Fig. 10. Morphological structures in Ascomycetes.

KEY TO THE GENERA TREATED

1a. Ascomata distinctly stalked; stalk composed of a single hypha.
Anamorph *Basipetospora* .... *Monascus*
1b. Ascomata not stalked, sessile or almost so 2

2a. Colonies extremely slow growing, only on MY50G (extremely xerophilic)
Asci containing 2 lunate, plano-convex ascospores. Anamorph *Fraseriella* .... *Xeromyces*
2b. Colonies growing faster. Asci generally containing 8 ascospores. 3

3a. Ascomata (perithecia) covered with hairs; asci clavate .... *Chaetomium*
3b. Ascomata (cleistothecia or naked asci), without hairs, asci globose 4

4a. Ascomata without a distinct wall. Anamorph *Paecilomyces* .... *Byssochlamys*
4b. Ascomata with a wall. Anamorph *Aspergillus* or *Penicillium* 5

5a. Anamorph *Penicillium*, ascomata with a distinct covering, yellow,
sometimes becoming reddish .... *Talaromyces*
5b. Anamorph *Aspergillus*, ascomata with one-to several layered wall.
Hülle cells present or absent........ 6

6a. Wall of the ascomata surrounded by thick-walled "Hülle cells"
(see also p.64). Anamorph *Aspergillus nidulans* group........ *Emericella*
6b. Wall of the ascomata not surrounded by "Hülle cells" 7

7a. Ascomata yellow, wall consisting of one layer of flattened cells.
Anamorph: *Aspergillus glaucus* group .... *Eurotium*
7b. Ascomata white to creamish, wall consisting of several layers of
flattened cells. Anamorph: *Aspergillus fumigatus* group .... *Neosartorya*

BYSSOCHLAMYS Westling

Anamorph: *Paecilomyces* Bain.

Ascomata discrete and confluent. Wall lacking or very scanty, composed of loose wefts of hyaline, thin, twisted hyphae. Ascomatal initials consisting of ascogonia coiled around swollen antheridia. Asci globose to subglobose, 8-spored, stalked. Ascospores ellipsoidal, smooth, pale-yellowish.

KEY TO THE SPECIES TREATED

1a. Ascospores 5.2-6.5 x 3.2-4 µm; conidia cylindrical with both ends flattened, 4-8.7 x 1.5-5 µm. Chlamydospores absent.. *B. fulva*

1b. Ascospores 4-5.5 x 2.5-3.5 µm; conidia globose to broadly ellipsoidal, 3-5.7 x 2.2-4 µm. Chlamydospores abundantly produced .... *B. nivea*

Byssochlamys fulva Olliver & G. Smith

Anamorph: *Paecilomyces fulvus* Stolk & Samson

Colonies on MEA at 30°C attaining a diameter of 9 cm in 7 to 14 days, composed of a basal felt with white ascomata, occasionally in localized sectors, covered by a velvety or floccose overgrowth of the conidial state, which gives the colony a yellowish-brown (fulvus) appearance. Vegetative hyphae 0.5-5 µm in diam; submerged hyphae usually thick-walled, up to 10 µm in diam. Ascomata white, up to 150 µm in diam, ripening in 7 to 10 days at 30°C. Wall lacking or very scanty, composed of loose wefts of hyaline, thin hyphae of about 1 µm in diam. Asci globose to subglobose, 9-12.5 µm in diam. Ascospores ellipsoid, 5.2-6.5 x 3.4-4 µm in diam, smooth, thick-walled, pale-yellowish. Anamorph with septate conidiophores up to 150 µm tall, smooth with phialides borne in groups of two to five on short metulae. Single phialides sometimes borne directly on aerial hyphae. Phialides with cylindrical basal portion, 12.5-17 x 2.5-3.5 µm, tapering abruptly to a long thin neck, 3-8.5 µm long and 1-1.2 µm wide. Conidia usually cylindrical with flattened ends, 4-8.7 x 1.5-5 µm, in dry chains, yellowish. No chlamydospores.

Temperature: optimum 30-35°C; minimum about 10°C; maximum 45°C.

Important (toxic)metabolites: **patulin**, byssochlamic acid, byssotoxin.

HABITAT: food

World-wide distribution: bottled fruit, harvested grapes, soil (especially in orchards).

NOTE:

The production of asci and ascospores can be permanently lost when cultures are grown at or below 24°C and tend to become entirely conidial. By growing cultures at temperatures between 30-37°C for 10-14 days, ascospore production can be ensured. After this period cultures can be stored at a lower convenient temperature.

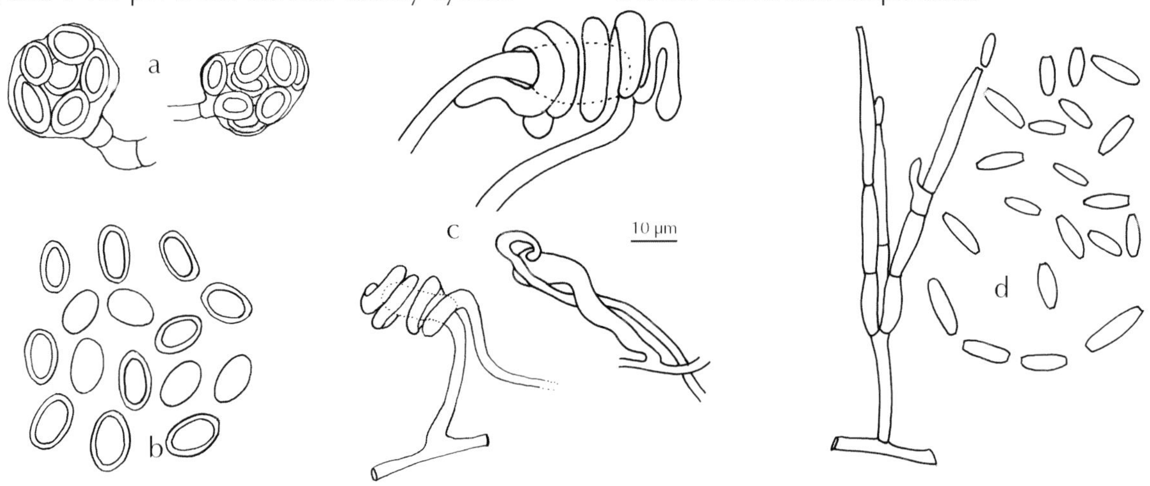

Fig. 11. a-d. *Byssochlamys fulva*. a. Asci with ascospores; b. ascospores; c. ascomatal initials; d. Conidiophore and conidia of *Paecilomyces fulvus*.

Pl. 12. *Byssochlamys fulva*. a. Colony on MEA (4%) after one week; b, c. asci with ascospores, b. x 150, c x 1500; d. ascospores, x 1700; e,f,g. conidia and conidiophores of *Paecilomyces fulvus*, x 1000.

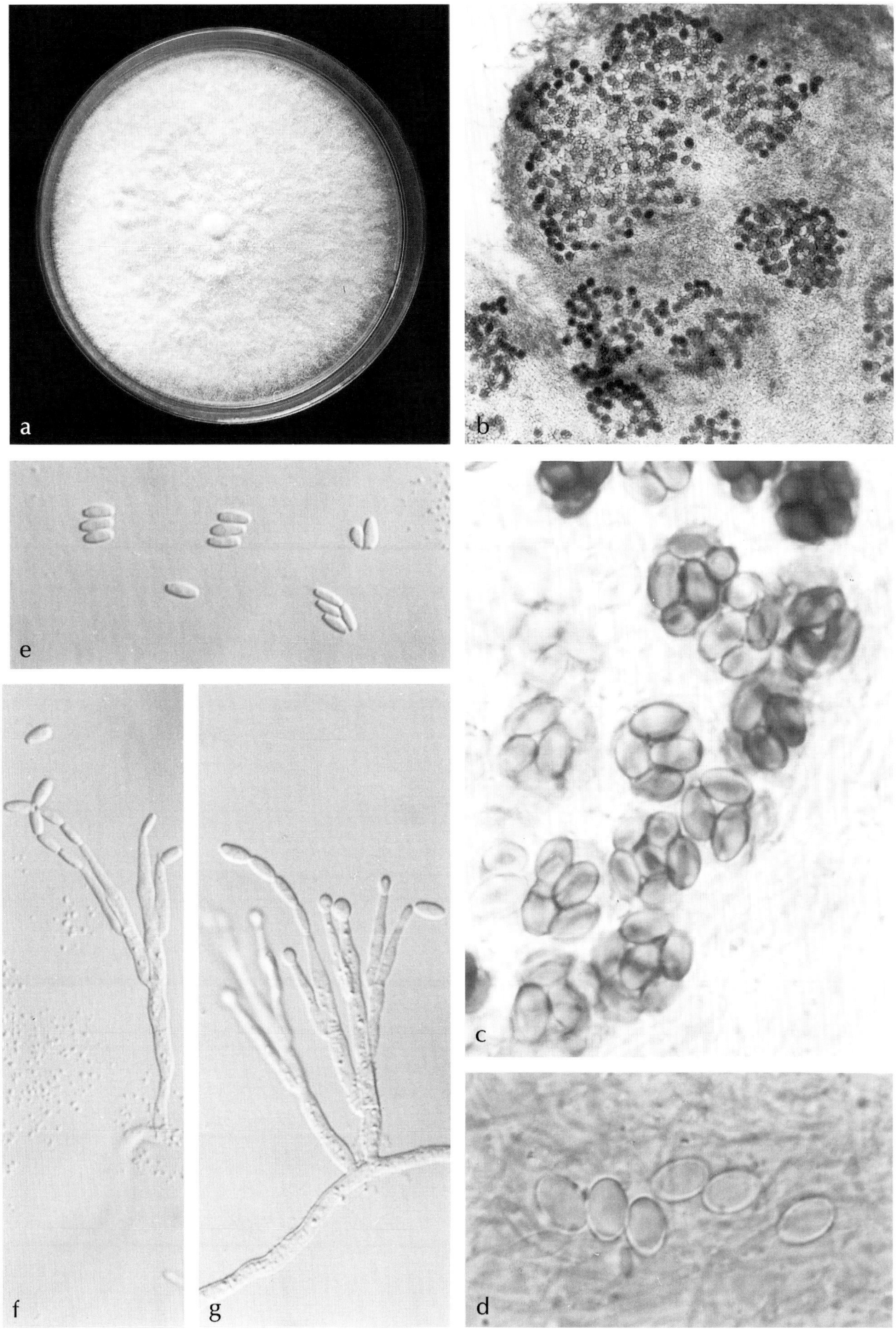
a
b
e
c
f
g
d

Byssochlamys nivea Westling

Anamorph: *Paecilomyces niveus* Stolk & Samson

Colonies on MEA at 30°C spreading broadly, attaining a diameter of 7 to 9 cm in 7 to 14 days, white ascomata in the basal felt obscured by floccose to funiculose, overgrowth, creamish. With age the colony appears almost granular by the dying down of the loosely funiculose tufts. Reverse pale brown to yellowish-cream. No exudate or few colourless drops. Vegetative hyphae hyaline, 0.5-5.5 µm in diam, submerged hyphae mostly thick-walled, up to 8 (-10.5) µm in diam. Ascomata white, up to 350 µm in diam, ripening in 7 to 10 days at 30°C. Wall lacking or very scanty, composed of hyaline, thin hyphae with a diam of 0.5-1 µm. Asci globose, to subglobose, 8-11 µm in diam. Ascospores ellipsoid, 4-5.5 x (2)2.5-3.5 µm, smooth, thick-walled, pale-yellowish. Conidiophores rare, smooth, up to 300 µm tall and 2-3 µm wide, bearing phialides in groups of two or three. Phialides usually single, borne directly on hyphae, 12.5-20 x 2-3.5 µm, smooth, with a cylindrical basal part, abruptly tapering into a long 2.5-7.5 x 0.7-1.5 µm neck. Conidia globose to broadly ellipsoid, usually with flattened base, 3-5.7 6.8 x 2.2-4(4.5) µm, smooth, hyaline to pale yellowish in dry divergent chains. Chlamydospores usually abundantly produced, single or in short chains, thick-walled, yellow-brown to brown, globose, ovoid to pyriform, up to 10 µm in diam, smooth to slightly roughened.

Temperature: optimum 30-35°C; minimum about 10°C; maximum 40°C.

Important (toxic) metabolites: **patulin**, byssochlamic acid, malformins.

HABITAT: food
World-wide distribution: soil, grain, fruit, especially damaged fruit and fruit-juices. Ascus development high in commercial samples of prune-, grape- and pineapple-juices, whereas less in apple-, orange- and tomato-juices (Beuchat and Rice, 1979).

NOTE:
B. nivea and *B. fulva* are important contaminants of canned fruit and fruit-juices, the ascospores being resistant to heat (Bayne and Michener, 1979; Beuchat and Rice, 1979; Eckhardt and Ahrens, 1977). Both species are also able to produce mycotoxins (e.g. byssochlamic acid, patulin).

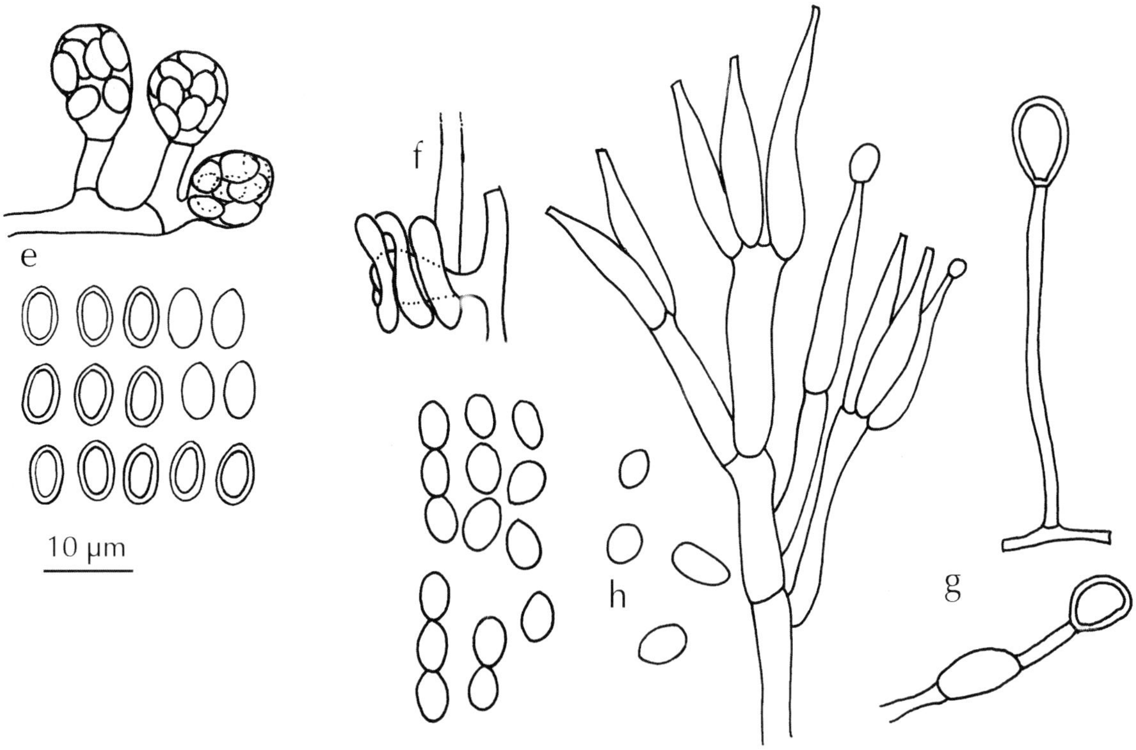

Fig. 11. e-h. *Byssochlamys nivea*. e. asci and ascospores; f. ascomatal initial; g. chlamydospores; h. conidiophore and conidia of *Paecilomyces niveus*.
Pl. 13. *Byssochlamys nivea*. a. Colony on MEA (4%) after one week; b,d. asci with ascospores, b. x 600, d. x 1500; c. ascospores, x 1500; e. ascomatal initials x 1500; f. conidial structures and chlamydospore of *Paecilomyces niveus*, x 1000.

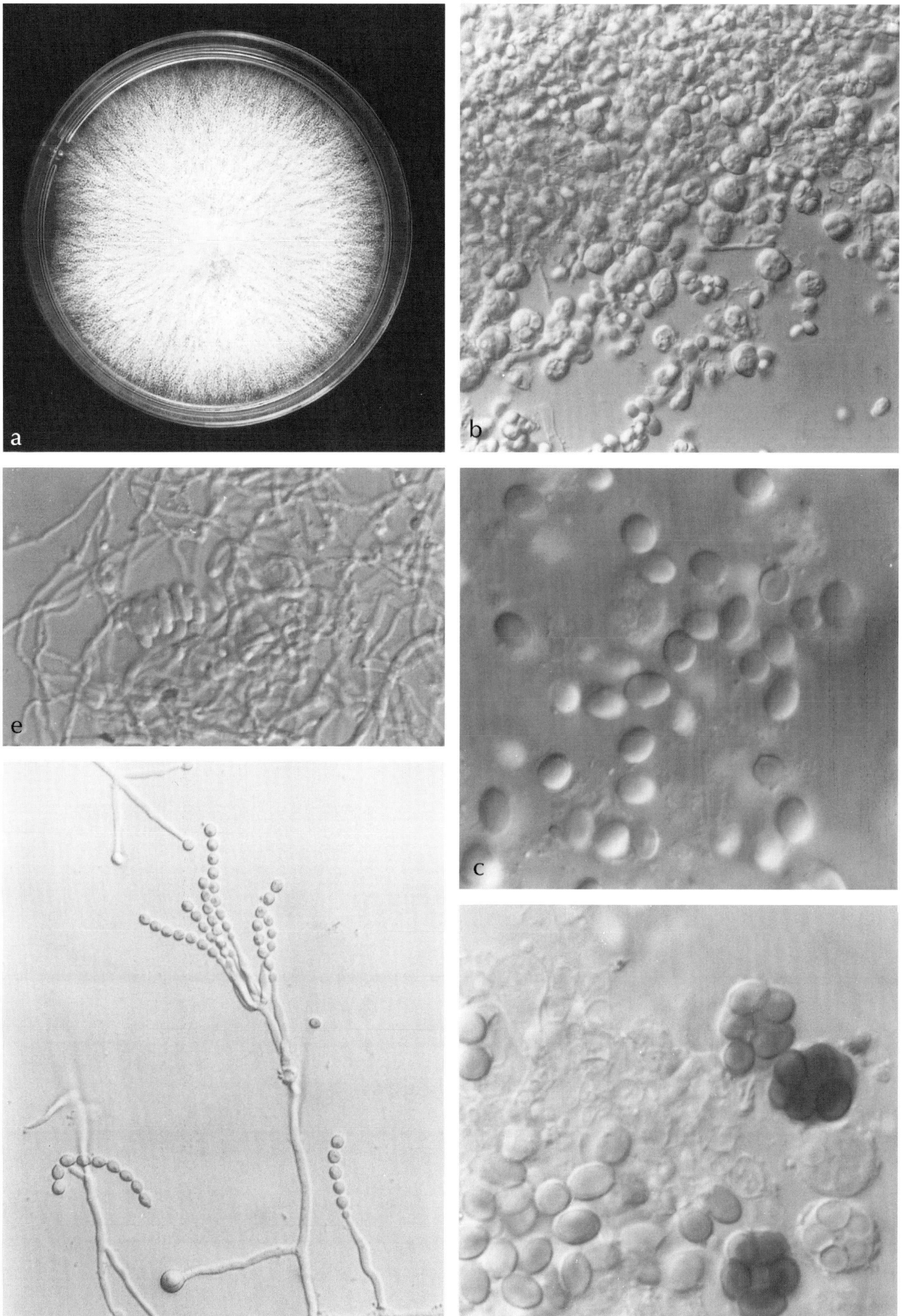
a
b
e
c
f
d

Chaetomium globosum Kunze

Colonies on OA at 24°C, attaining a diameter of 5.0-7.5 cm in 10 days, consisting of a dense layer of fruit-bodies (ascomata) giving the colony an olive-greyish appearance. Ascomata (perithecia) dark brown to blackish, globose to ovoid, mostly 150-220(-350) µm in diam with dark hyphal appendages (hairs). Perithecial hairs, numerous, septate, unbranched, flexuous or even coiled, pigmented, tapering and less pigmented towards the tip, rough-walled. Asci clavate, 8-spored. Ascospores brown, lemon- shaped, 9-11 x 7-8.5 µm, with an apical germ- pore. Homothallic.

For more detailed descriptions is referred to Von Arx (1986).
Temperature: optimum for the production of ascomata in the range of 18-20 (-24)°C.

HABITAT: indoor, often after water damages.
World-wide distribution: in soil, on plant remains and plant materials, known as a soft-rot fungus from softwood and hardwood timber, frequently encountered in archives, on wall paper, textiles.

NOTE: the ascospores are very resistant to dry circumstances, and UV irradiation.

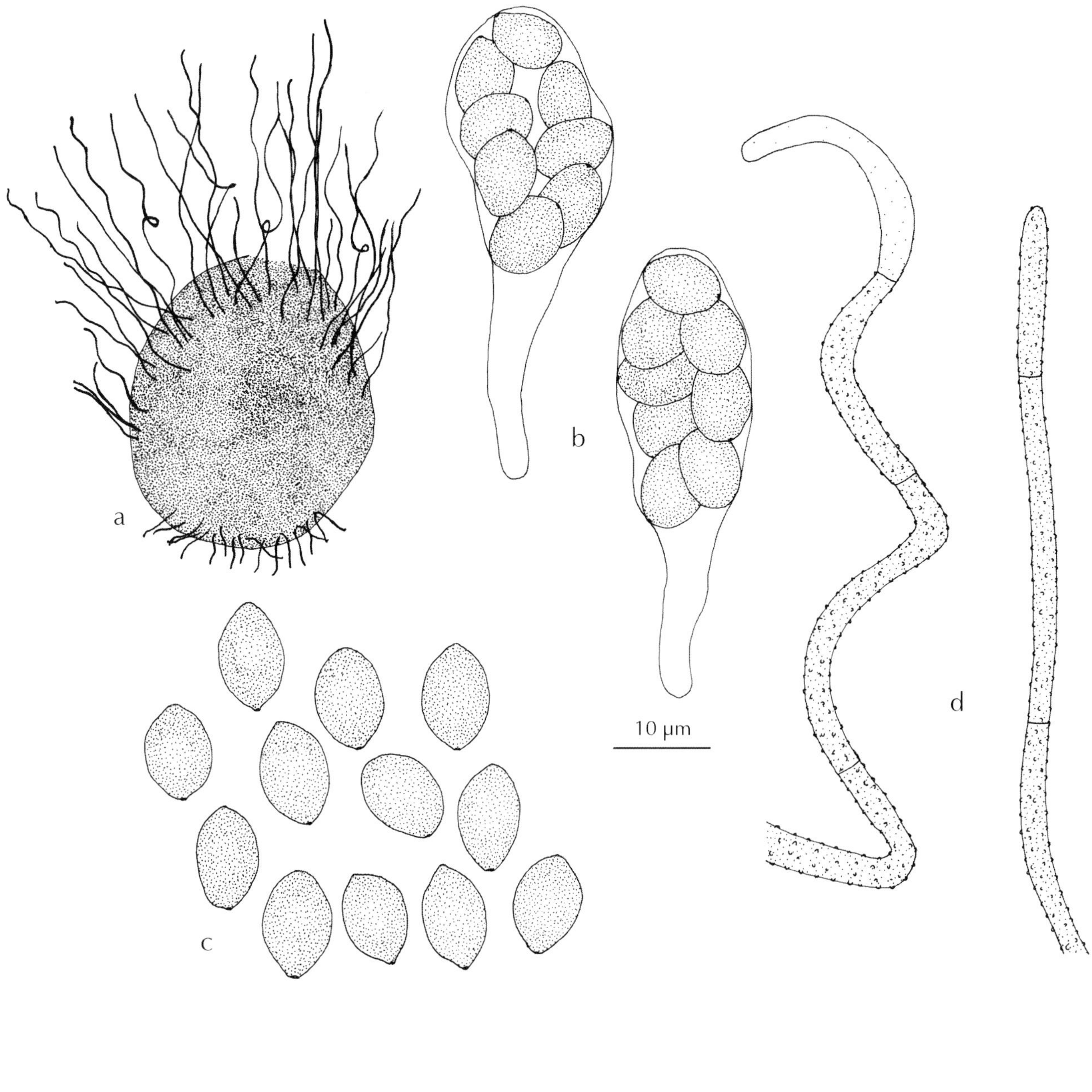

Fig. 12. *Chaetomium globosum*. a. Perithecium; b. asci with ascospores; c. ascospores; d. perithecial hairs.
Pl. 14. *Chaetomium globosum* . a. Colony on OA after 10 days; b. perithecia, x 190; c,e. clusters of clavate asci with ascospores, x 725; d. ascospores, x 725.

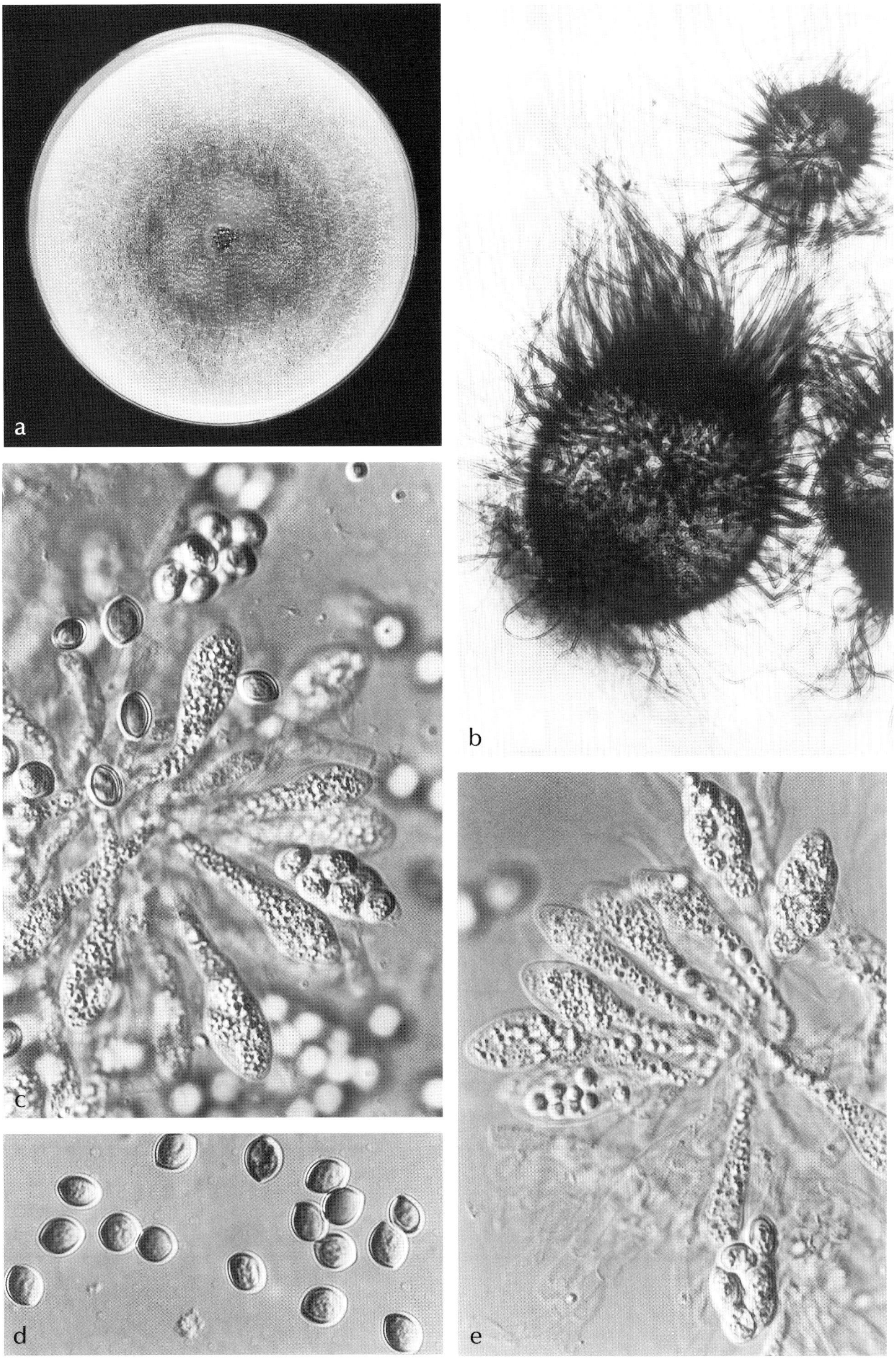
a
b
c
d
e

Emericella nidulans (Eidam) Vuill.

Anamorph: *Aspergillus nidulans* (Eidam) Wint.

Colonies on MEA at 24°C growing rapidly, attaining a diameter of 6-7 cm in 7 days, velvety, dark yellow-green shades with a brighter almost whitish margin, conidial state predominant, scattered ascomata developing within and upon the conidial layer. Reverse olive to drab-grey or purple-brown. Ascomata sometimes abundant, globose to subglobose, solitary, ranging from 100-300 µm in diam, commonly 125-150 µm, surrounded by a yellowish to cinnamon-coloured layer of scattered hyphae bearing "Hülle cells" up to 25 µm in diam, ascomawall composed of one cell layer. The yellowish appearance of the fruit-bodies turns to reddish brown from the maturing red coloured ascospores. Ascus 8-spored, globose to subglobose, (8)10-12 µm in diam. Ascospores lenticular (lens-shaped), 3-4.5 x 3.2-3.5 µm, purple-red, smooth-walled with two equatorial crests. Conidial heads in dark yellow-green shades, columnar, biseriate (see also *Aspergillus* p. 64). Conidiophores brownish, 60-130 µm tall, smooth-walled. Vesicles more or less globose, 8-10 µm in diam. Conidia rough-walled, globose, 3-3.5 µm in diam.

Important (toxic) metabolites: **sterigmatocystin**, nidulotoxin, penicillin.

HABITAT: food, indoor (compost)
World-wide distribution: especially found in soil, also in potatoes, grain, citrus and stored seeds of oats, wheat, corn, sorghum, rice, cotton. This species can be pathogenic to man and animals.

NOTE:
In some isolates fruit-bodies are not produced or only after prolonged optimal conditions e.g. at 30°C. Raper and Fennell (1965) distinguished three additional rare varieties based on different ornamentation of the ascospores. These varieties are now regarded as species (Samson, 1992, 1994 a and b).

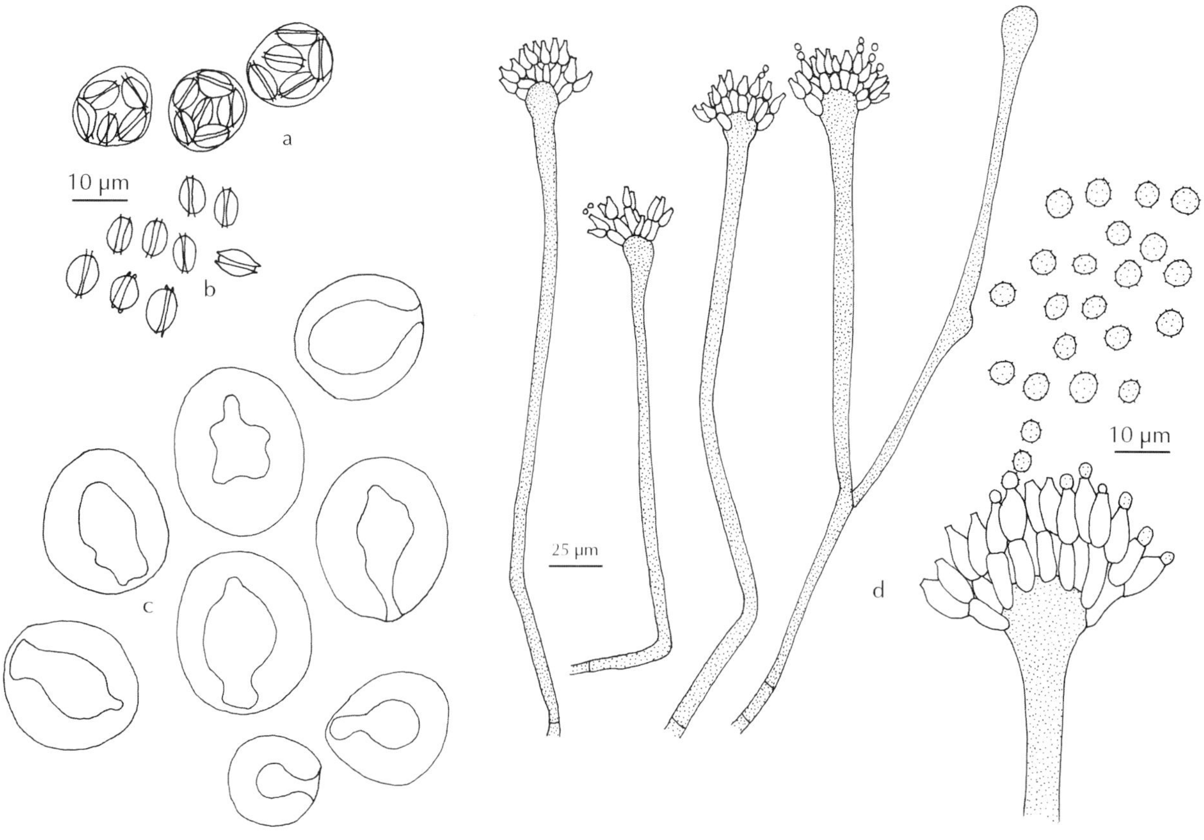

Fig. 13. *Emericella nidulans*. a. Asci with ascospores; b. ascospores; c. Hülle cells; d. conidiophores and conidia of *Aspergillus nidulans*.
Pl. 15. *Emericella nidulans*. a. Colonies on OA after one week; b. part of colony showing dark coloured conidial heads and (sub)globose bright coloured ascomata, x 40; c. asci and ascospores, x 600; d. conidiophores of *Aspergillus nidulans* x 500; e. ascospores, x 1500; f. Hülle cells, x 1500.

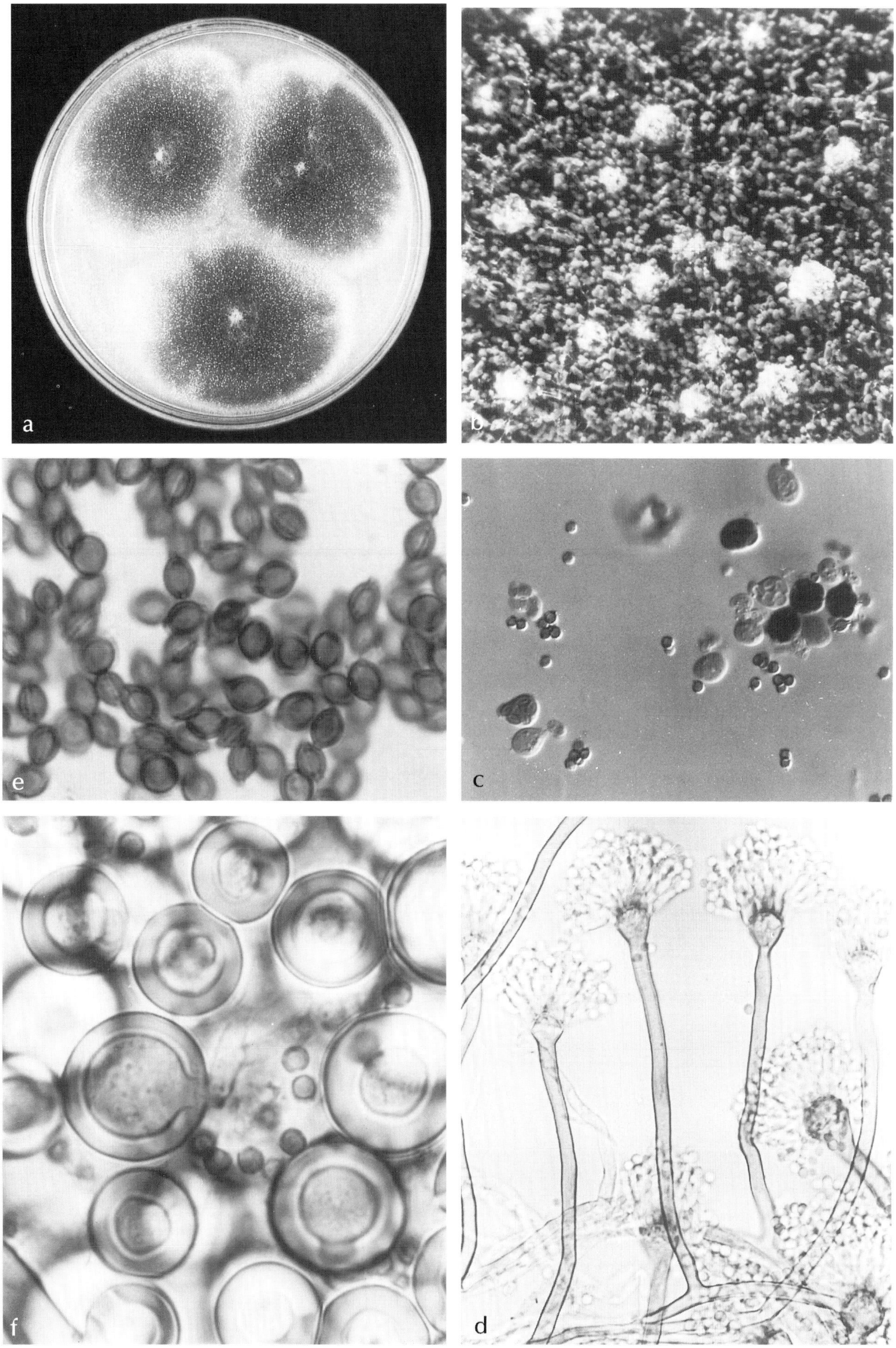
a
b
e
c
f
d

EUROTIUM Link: Fr.

Anamorph: *Aspergillus glaucus* group

Ascomata globose to subglobose, non-ostiolate (cleistothecia), often surrounded by a network of hyphae, scattered or more or less clustered, commonly accompanied by the conidial state. Asci globose to subglobose, thin-walled often dissolving at an early stage. Ascospores one-celled, lenticular, small, surface smooth or roughened. Conidiophores smooth-walled, distinctly roughened conidia (see also *Aspergillus* p. 52). Xerophilic, optimally growing on low wateractivity media.

Important (toxic) metabolites: physcion, echinulin.

KEY TO THE SPECIES TREATED

1a. Ascospores smooth, equatorial ridges lacking, furrow absent or indistinct .... *E. herbariorum*
1b. Ascospores with prominent equatorial ridges or furrows 2

2a. Ascospores with distinct equatorial furrows (V-shaped) and a roughened surface .... *E. amstelodami*
2b. Ascospores with thin distinct flexuous equatorial crests and a smooth to slightly roughened surface .... *E. chevalieri*

Eurotium amstelodami L. Mangin

Anamorph: *Aspergillus vitis* Novobr.

Colonies on MEA + 20% sucrose at 24°C attaining a diameter of 7-8 cm in 14 days and (5)8-10 cm on Cz + 20% sucrose, plane or wrinkled, yellow to dull yellow-grey from ascomata admixed with sterile hyphae and developing conidia heads. Reverse yellowish becoming tawny with age. Ascomata very abundant, clustered in masses forming a dense layer on the agar surface, bright yellow, more abundant on Cz + 20% sucrose than on MEA + 20% sucrose, globose to subglobose. Mostly 115-140 μm in diam, occasionally up to 160 μm, not covered by or embedded within a felt of sterile hyphae. Asci globose to subglobose, 10-12 μm in diam. Ascospores lenticulate, 4-4.7(5) x 3.6-3.8 μm, roughened, with a V-shaped equatorial furrow. Conidial heads olive-green to colourless, 275-350 μm tall. Vesicles globose, 18-25 μm in diam. Conidia echinulate, subglobose to ellipsoidal, often with both ends more or less flattened, varying from 3.5-5.2 μm in length.

HABITAT: food, indoor
Xerophilic. World-wide distribution, particularly in tropical and subtropical areas. Soils, stored and/or decaying food products, fruit-juices, grain, nuts, milled rice, various dried foodstuffs. Also occurring in indoor environments.

Pl. 16. *Eurotium amstelodami.* Colonies after one week, a. on MEA + 20% sucrose, b. on Cz + 20% sucrose; c. ascomata, x 60; d. asci with ascospores, x 1750; e. part of ascoma with asci, x 2000; f. ascospores, x 1750.

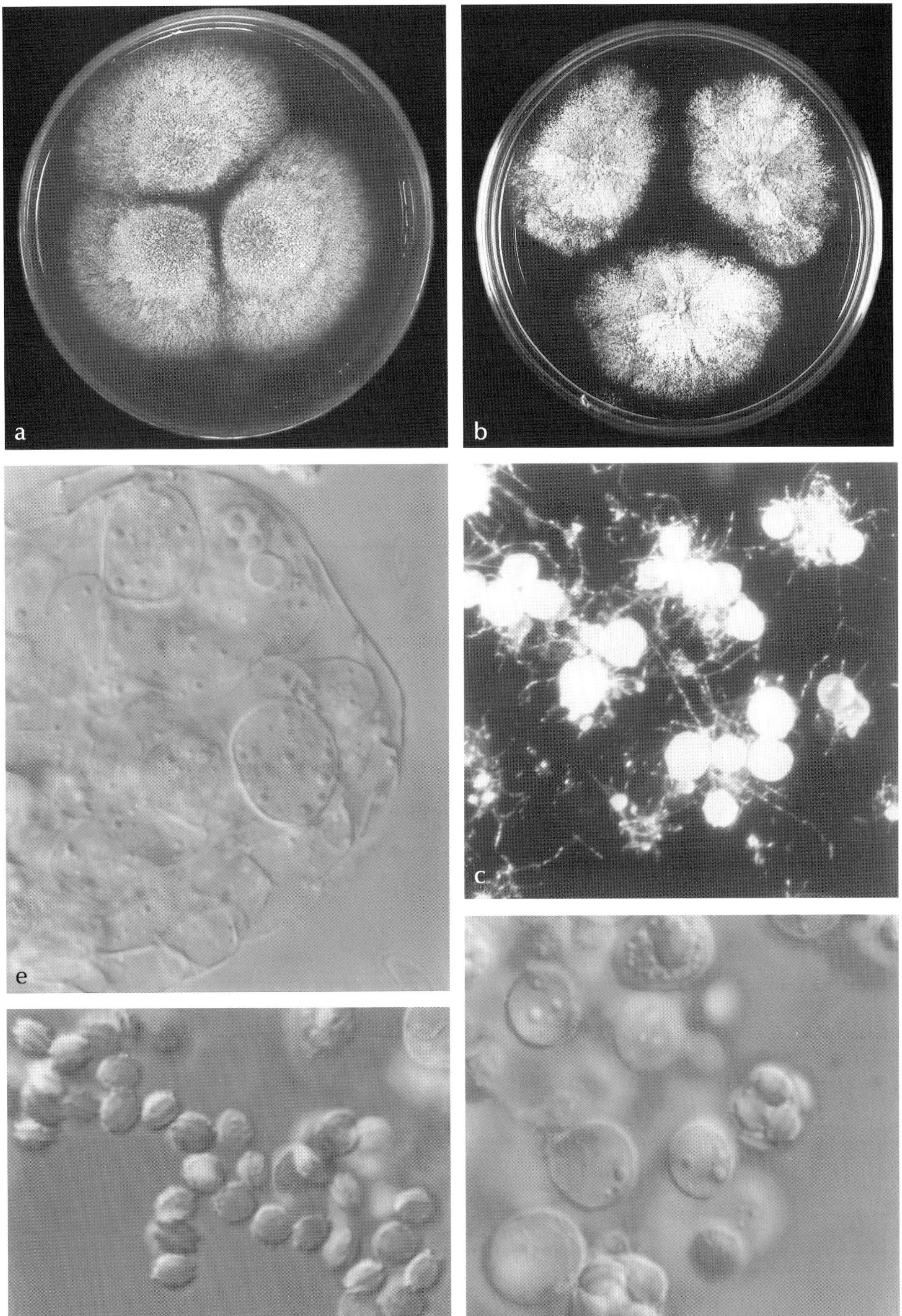
a
b
e
c
f
d

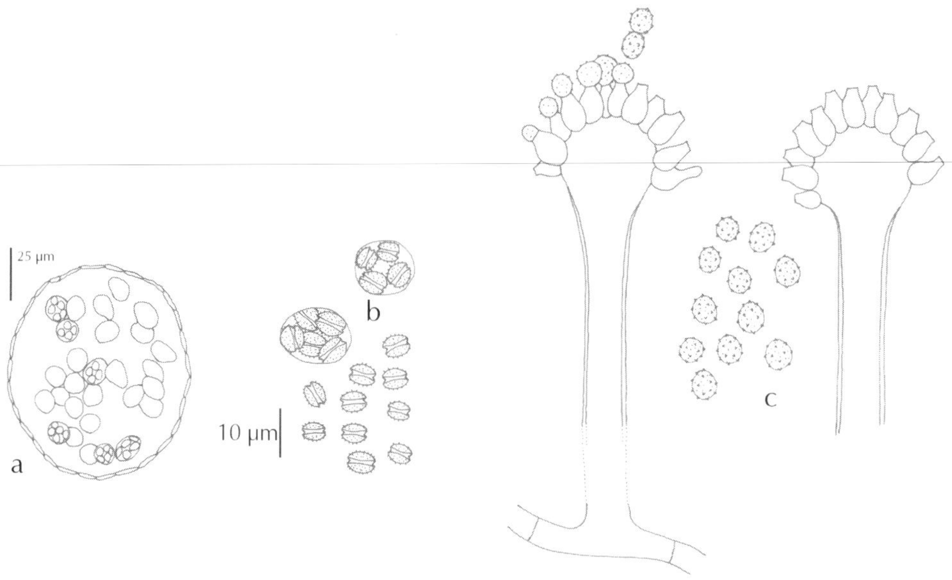

Fig. 14. a-c. *Eurotium amstelodami*. a. Ascoma with asci and ascospores; b. asci and ascospores; c. conidiophores and conidia of *Aspergillus vitis*.

Eurotium chevalieri L. Mangin

Anamorph: *Aspergillus chevalieri* L. Mangin

Colonies on Cz + 20% sucrose at 24-30°C spreading, attaining a diameter of 8 cm in 14 days, flat or somewhat wrinkled in the centre, margin irregular. Conidial heads in grey-green shades, usually projecting above a continuous layer of abundant yellow ascomata surrounded by orange-red hyphae. Reverse in brown shades, more intense in the centre. Ascomata abundantly spread over the entire colony, often more or less clustered, globose to subglobose, 8-10 µm in diam. Ascospores lenticular (lens-shaped), 4.5-5 x 3.5 µm, with smooth to slightly roughened surfaces and thin prominent flexuous equatorial crests. Conidial heads grey-green, radiate. Conidiophores 700-850 µm tall. Vesicles (sub)globose, 25-35 µm in diam. Conidia echinulate, ovate to ellipsoidal often with both ends more or less flattened, commonly 4.5-6 µm in length.

HABITAT: food, indoor

Mainly tropical and subtropical. Soils, seeds of wheat, sorghum, corn, rice and various foodstuffs (dried or milled), man. In indoor environments occurring world-wide.

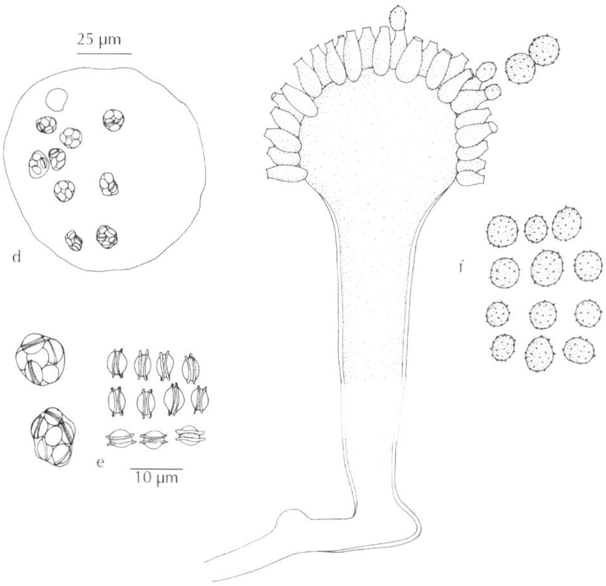

Fig. 14. d-f . *Eurotium chevalieri*. d. Ascoma with asci and ascospores; e. asci and ascospores; f. conidiophore and conidia of *Aspergillus chevalieri*.

Pl. 17. *Eurotium chevalieri*. Colonies after one week, a. on MEA + 20% sucrose; b. on Cz + 20% sucrose; c,d. ascomata, c. x 80, d. x 20; e. portion of ascoma with asci and ascospores, x 650; f. asci and ascospores, x 1750

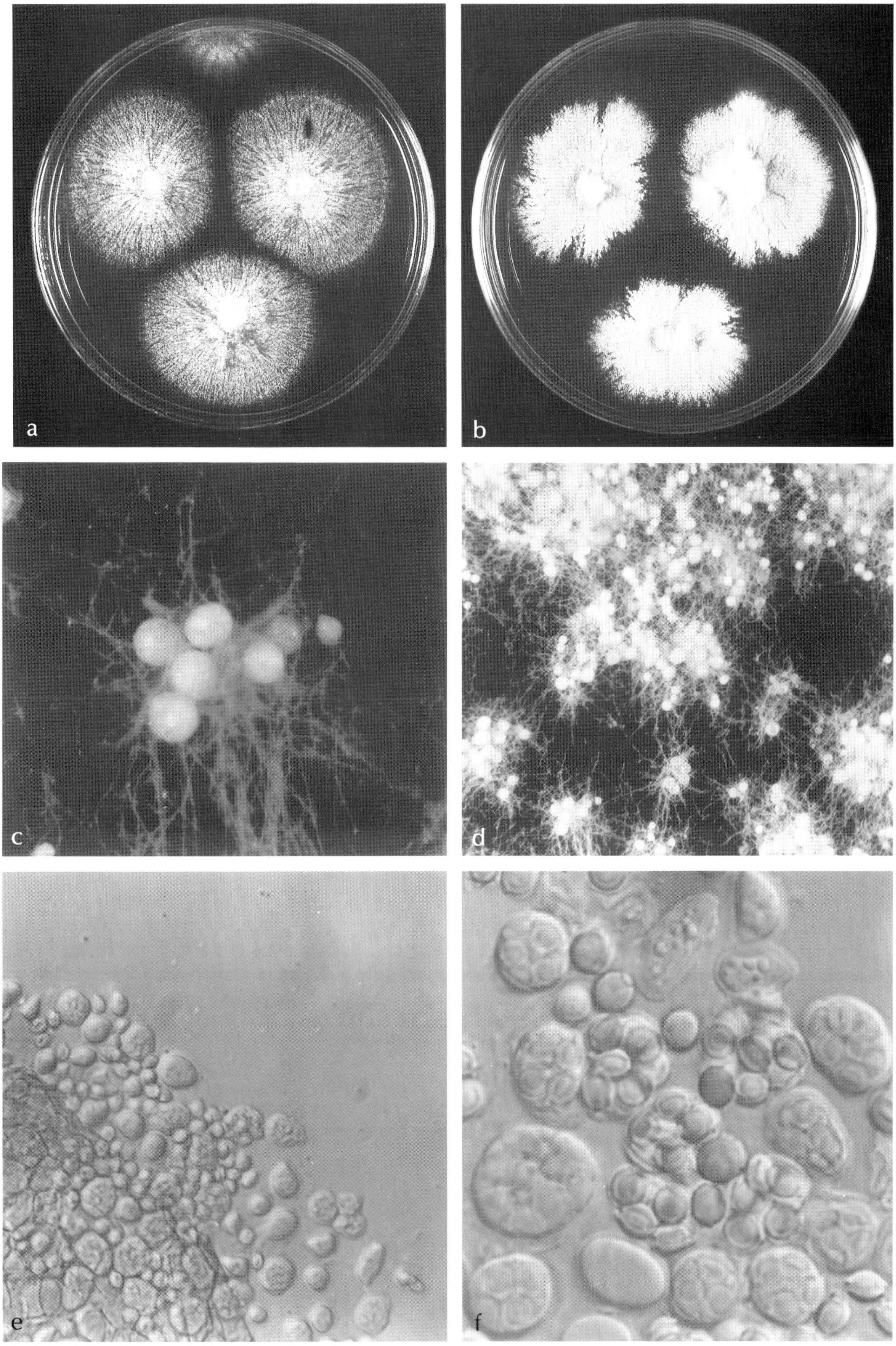
a
b
c
d
e
f

Eurotium herbariorum Link

Anamorph: *Aspergillus glaucus* Link

Colonies on Cz + 20% sucrose at 24°C spreading broadly and often irregularly, attaining a diameter of 5-7 cm in 14 days, flat, commonly characterized by broad zones or patches of dull green to grey-green conidial heads often alternating with more predominantly ascomatal orange-yellow areas. Reverse brown. Ascomata abundant, borne in loose networks of yellow to orange-red hyphae, yellow, globose to subglobose, mostly 75-100 µm, frequently up to 125 µm. Asci globose to subglobose, 10-12 µm in diam. Ascospores lenticular, 4.8-5.6 x 3.8-4.4 µm, smooth-walled without crests or ridges, sometimes with a trace of a furrow. Conidial heads in grey-green shades, radiate, or slightly pigmented to colourless, 500-1000 µm in diam. Conidia echinulate, ovate to (sub)globose, 4.5-7(8) µm in diam.

HABITAT: food, indoor

World-wide but predominant in tropical and subtropical areas: soils, grains, (e.g. wheat, corn), fruit and fruit-juice, peas, milled rice, dried food products, spices, meat products. May causes keratitis (Raper and Fennell, 1965) and indigestion in man. Contaminated food is toxic to rabbits. Commonly occurring in indoor environments, world-wide.

NOTE:

A similar species is *E. rubrum* with isolates producing an orange-red mycelium and finely roughened ascospores having a pronounced furrow (see also Kozakiewicz, 1989, Pitt, 1997).

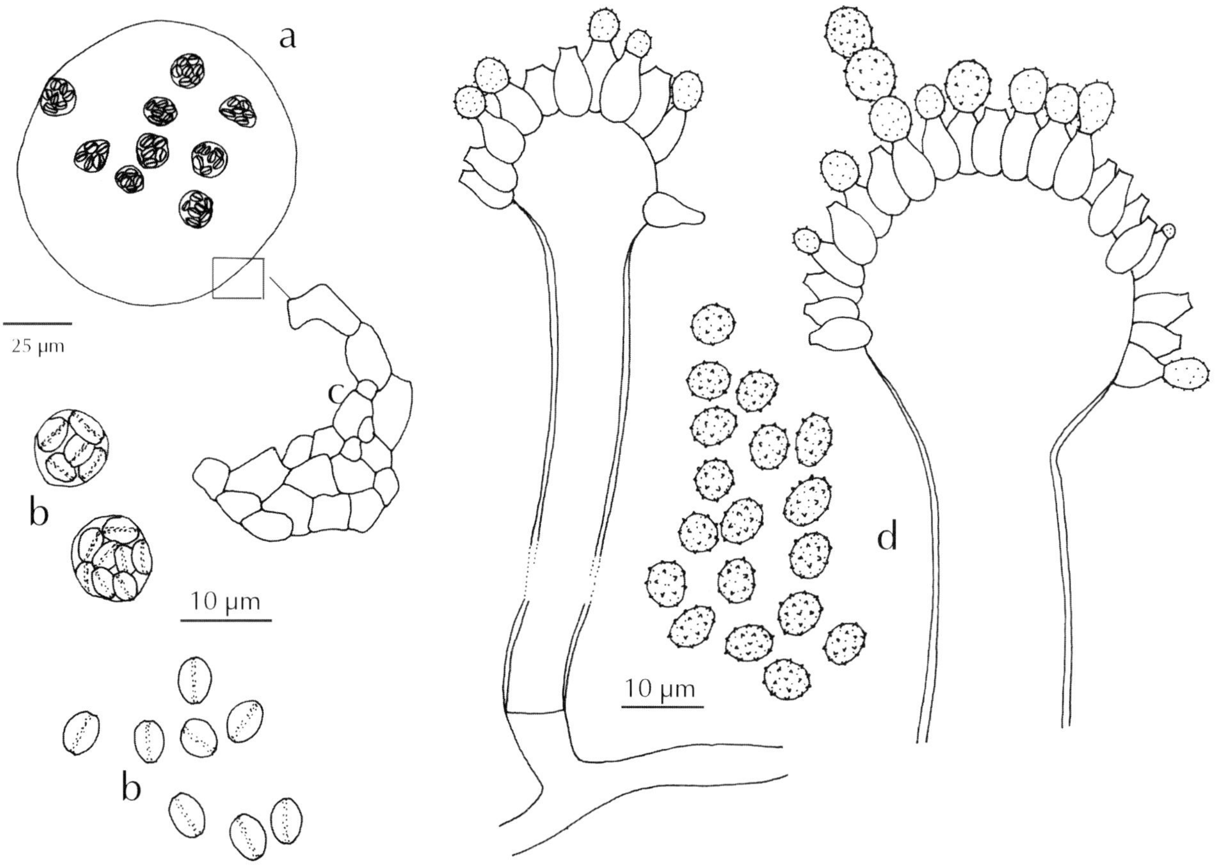

Fig. 15. a-c. *Eurotium herbariorum*. a. Ascoma with asci and ascospores; b. asci and ascospores; c. cells of ascoma wall; d. conidiophores and conidia of *Aspergillus glaucus*.

Pl. 18. *Eurotium herbariorum*. a. Colonies on MEA + 20% sucrose; b. conidiophore and conidia of *Aspergillus glaucus*, x 300; c. ascomata, x 20; d. ascoma, x 700; e. ascospores, x 1800; f. asci and ascospores, x 700.

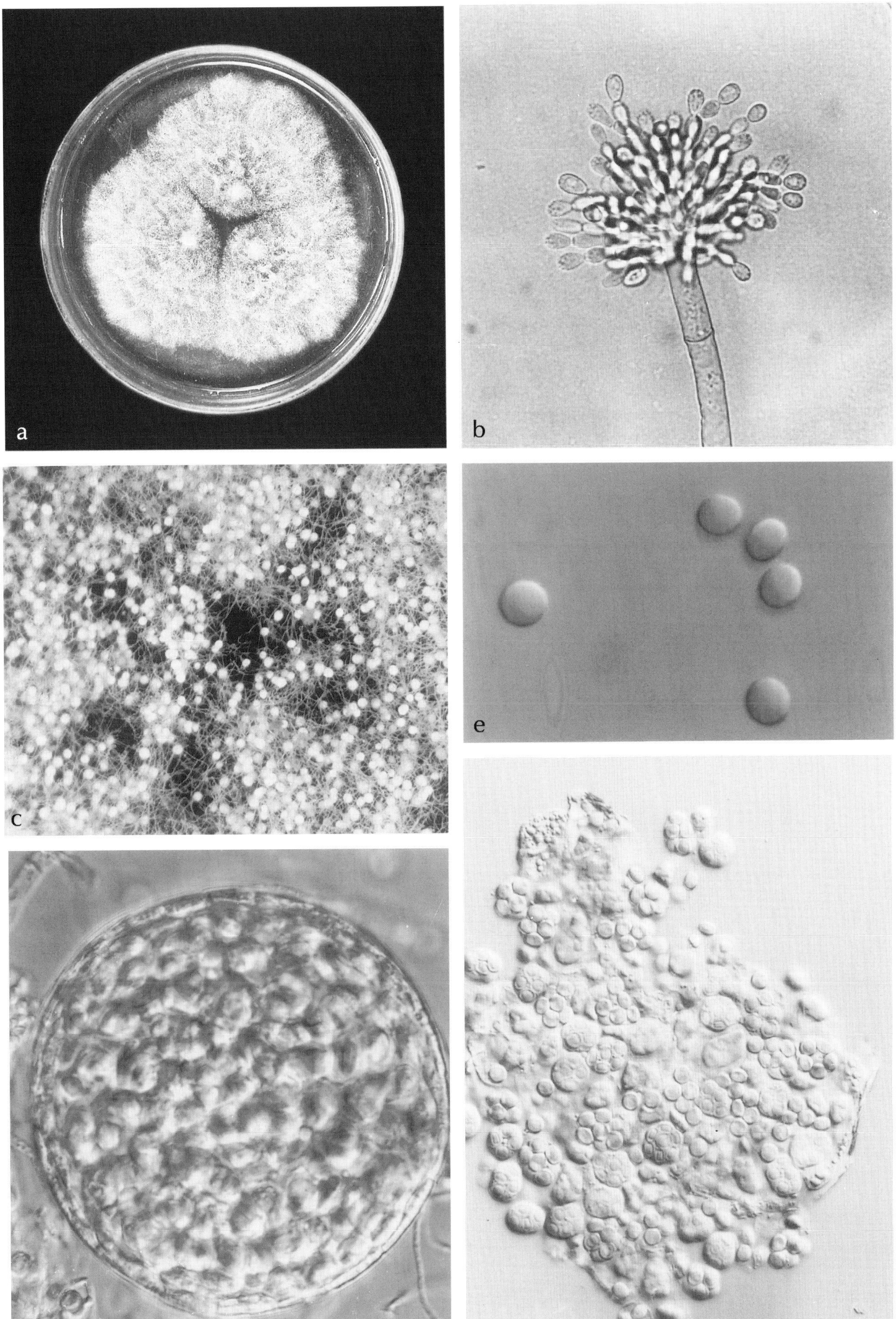
a
b
c
e
d
f

Monascus ruber v. Tieghem

=*Monascus purpureus* Went
Anamorph: *Basipetospora rubra* Cole & Kendrick

Colonies on MEA at 24°C attaining a diameter of 7 cm in 14 days, brownish with greyish-brown floccose mycelium in the centre. Growth on OA slower, attaining a diameter of 4.5 cm at 24°C, with a brick-red appearance. Reverse brown to red-brown, more intense near the centre. Ascomata borne terminally on a pedicel which is commonly up to 150 µm long (occasionally up to 1000 µm, wall consisting of hyphae, globose, 20-70 µm in diam, subhyaline to brownish-red. Asci globose to subglobose, 7.5-10 µm in diam. Ascospores yellowish ovate-ellipsoidal, often with a hyaline wall, (4)5-6 x 3-4 µm, smooth. Conidia borne in basipetal succession in chains, ovate to pyriform with a broadly truncate base, 6-8(11) x 5.6(-10) µm, mostly thin-walled, functioning as chlamydospores when thick-walled.

Temperature: growth between 18 and 40°C.

HABITAT: food
World-wide distribution: found in soil, cooked potatoes, rice, seeds of oat, soya, sorghum, tobacco, olives, silage.

NOTE:
Xeromyces (Monascus) bisporus is a related species often occurring on dry food products (e.g. dates) and raw materials. This species only grows on media with a high glucose content (MY50G) and is distinguished by the two-spored asci and *Fraseriella* anamorph, which looks similar to *Basipetospora*. (see p. 50)

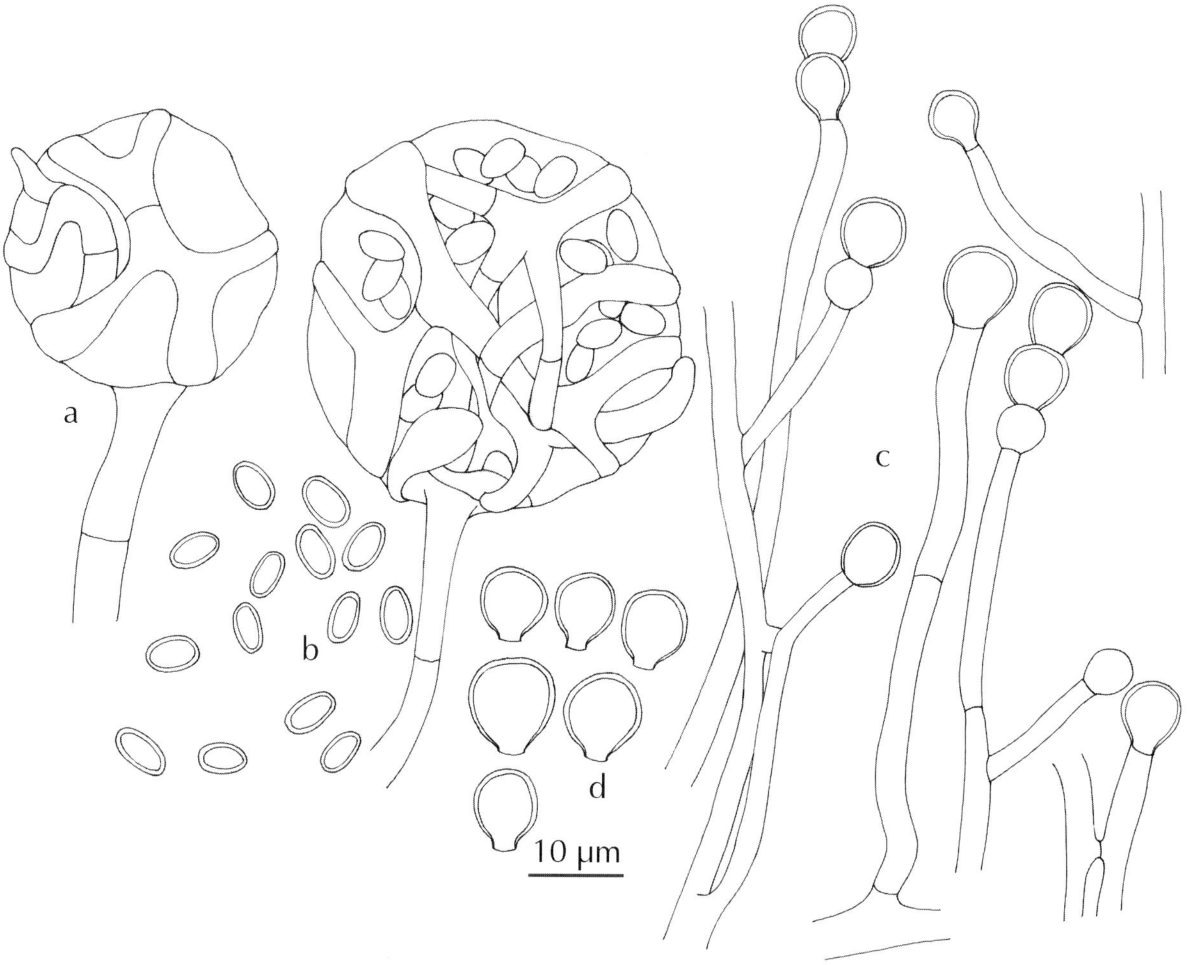

Fig. 16. *Monascus ruber*. a. Pedicillate ascomata; b. ascospores; c,d. conidial structures and conidia of *Basipetospora rubra*.
Pl. 19. *Monascus ruber*. a. Colony on MEA after 9 days; b-d. pedicellate ascomata with ascospores, b,c. x 450, d. x 500; e. ascospores, x 1200; f. conidial structures of *Basipetospora rubra*, x 750.

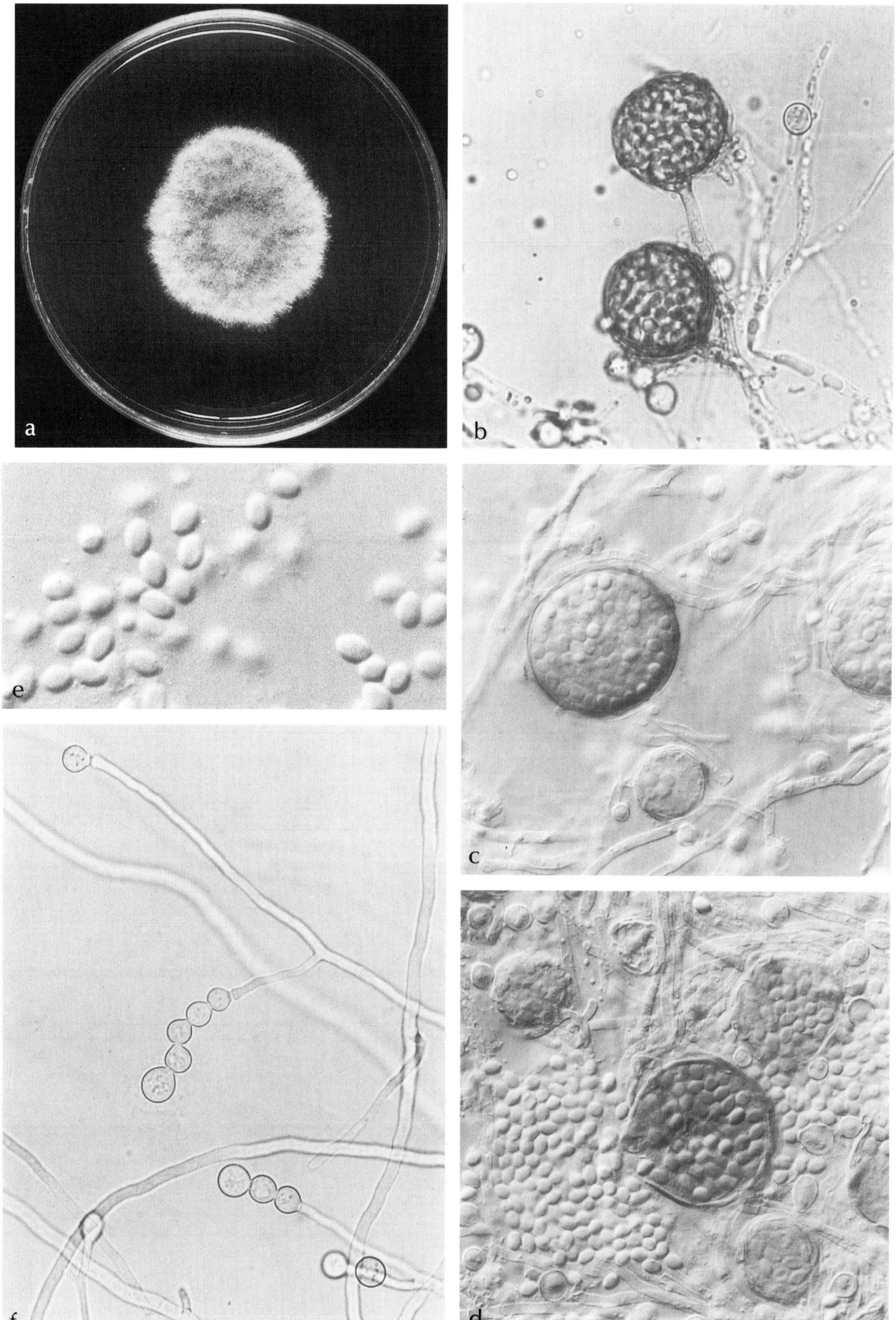
a
b
e
c
f
d

Neosartorya fischeri (Wehmer) Malloch & Cain

Anamorph: *Aspergillus fischerianus* Samson & W. Gams

Colonies on MEA at 24°C attaining a diameter of 6 cm in 14 days, white to creamish, granular appearance caused by a continuous layer of ascomata. Reverse uncoloured to creamish or slightly pinkish. Ascomata single or borne in small clusters within loose white-creamish hyphae, globose, 80-150(400) µm in diam, with a 2-3 layers thick wall of flattened cells. Asci 8-spored, globose to somewhat flattened, 8-10 x 10-12 µm. Ascospores biconvex, colourless, 6-7 x 4 µm (including crests), with two equatorial crests and irregular surface ridges. Conidial heads in olive-grey shades, columnar, uniseriate (see also *Aspergillus* p. 64). Conidiophores 300-500 µm tall, smooth-walled. Vesicles more or less elongated, 12-18 µm in diam. Conidia (sub)globose to ellipsoidal with slightly roughened walls, 2-2.5 µm in diam or up to 3 µm when globose.

Temperature: optimum 26-45°C; minimum 11-13°C; maximum 51-52°C. At 24°C the ascigerous state is predominant, at 37°C the conidial state is predominant.

Important (toxic) metabolites: verrucologen, fumitremorgin A & B.

HABITAT: food

World-wide distribution. Soil, (milled) rice, cotton, potatoes, groundnuts, leather, paper products. Sometimes pathogenic to animals and man (Wyllie *et al.* 1977/78, Raper and Fennell, 1965).

NOTE:

Raper and Fennell (1965) separated two other varieties differing by ascospore ornamentation. The var. *glaber* has smooth-walled ascospores, while the ascospores of the var. *spinulosus* are delicately spiny. Kozakiewicz (1989) raised the two varieties to species level. *Neosartorya* species are heat resistant and can cause spoilage of canned fruit products (see also Chapter 7).

In this guide four species (*N. fischeri, N. glabra, N. pseudofischeri and N. spinosa*) are treated. The key to distinguish the species is given on page 46. (Peterson, 1992)

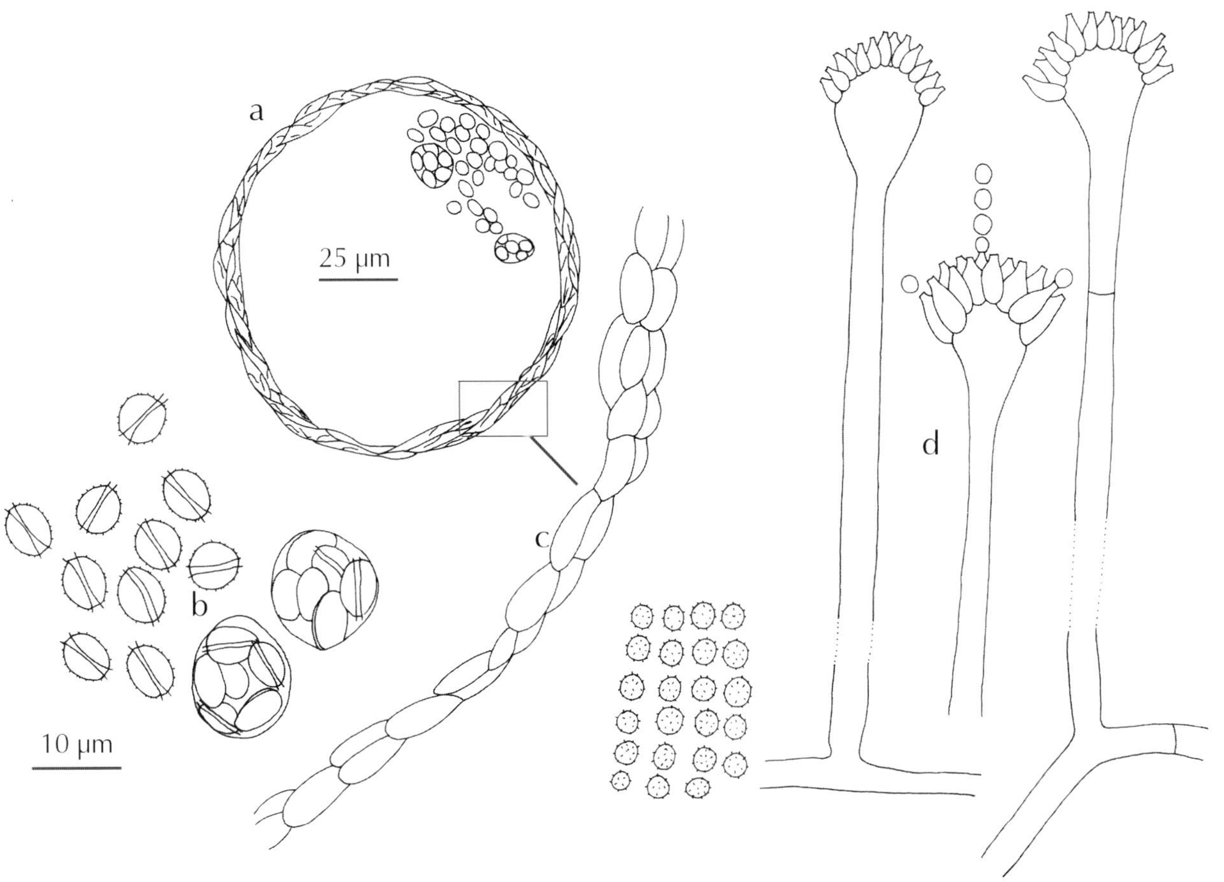

Fig. 17. *Neosartorya fischeri*. a. Ascoma with asci and ascospores; b. asci and ascospores; c. cells of ascoma wall; d. conidiophores and conidia of *Aspergillus fischerianus*.

Pl. 20. *Neosartorya fischeri*. a. Colonies on MEA after one week; b. conidiophores of *Aspergillus fischerianus*, x 450; c. ascomata covered by hyphae, x 80; d. ascomata with asci and ascospores, x 350; e. ascospores, x 2000; f. asci and ascospores, x 800.

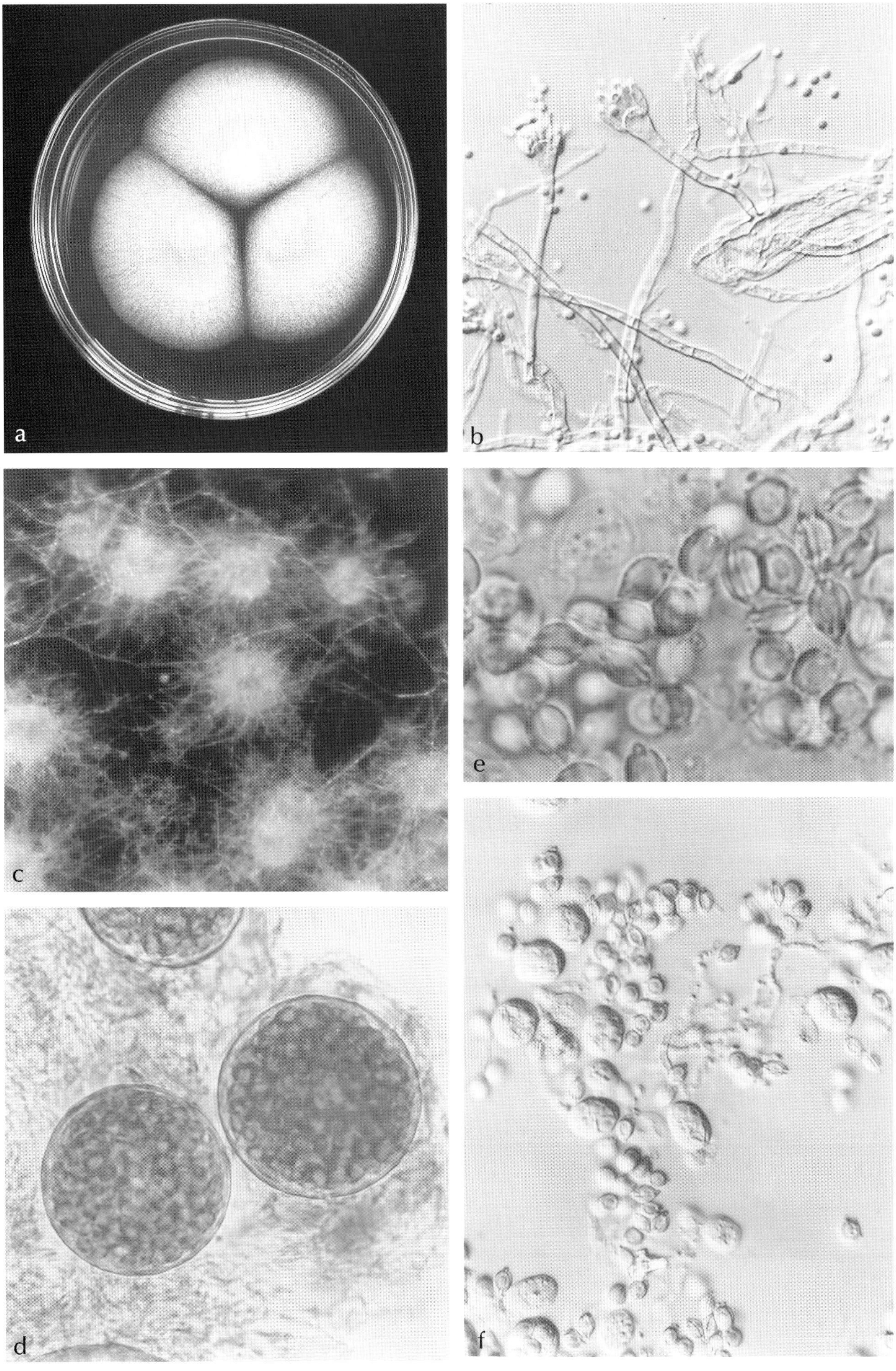
a
b
c
d
e
f

KEY TO THE DIFFERENT SPECIES (FORMERLY KNOWN AS VARIETIES OF *NEOSARTORYA FISCHERI*) BASED ON ASCOSPORE CHARACTERISTICS

1a. Ascospores smooth-walled (convex walls) in light microscopy .... *N. glabra*
1b. Ascospores ornamented 2

2a. Ascospores with spines on the convex walls .... *N. spinosa*
2b. Ascospores without spines 3

3a. Ascospores with a reticulate pattern of anastomosing ridges on the convex walls .... *N. fischeri*
3b. Ascospores without a reticulate pattern of anastomosing ridges, ornamentation is formed by triangular flaps of tissue .... *N. pseudofischeri*

Pl. 21. Ascopores of *Neosartorya*. a,b. *Neosartorya glabra*, x 1500; c,d. *N spinosa*, x 1875; e,f. *N. pseudofischeri*, x 1875

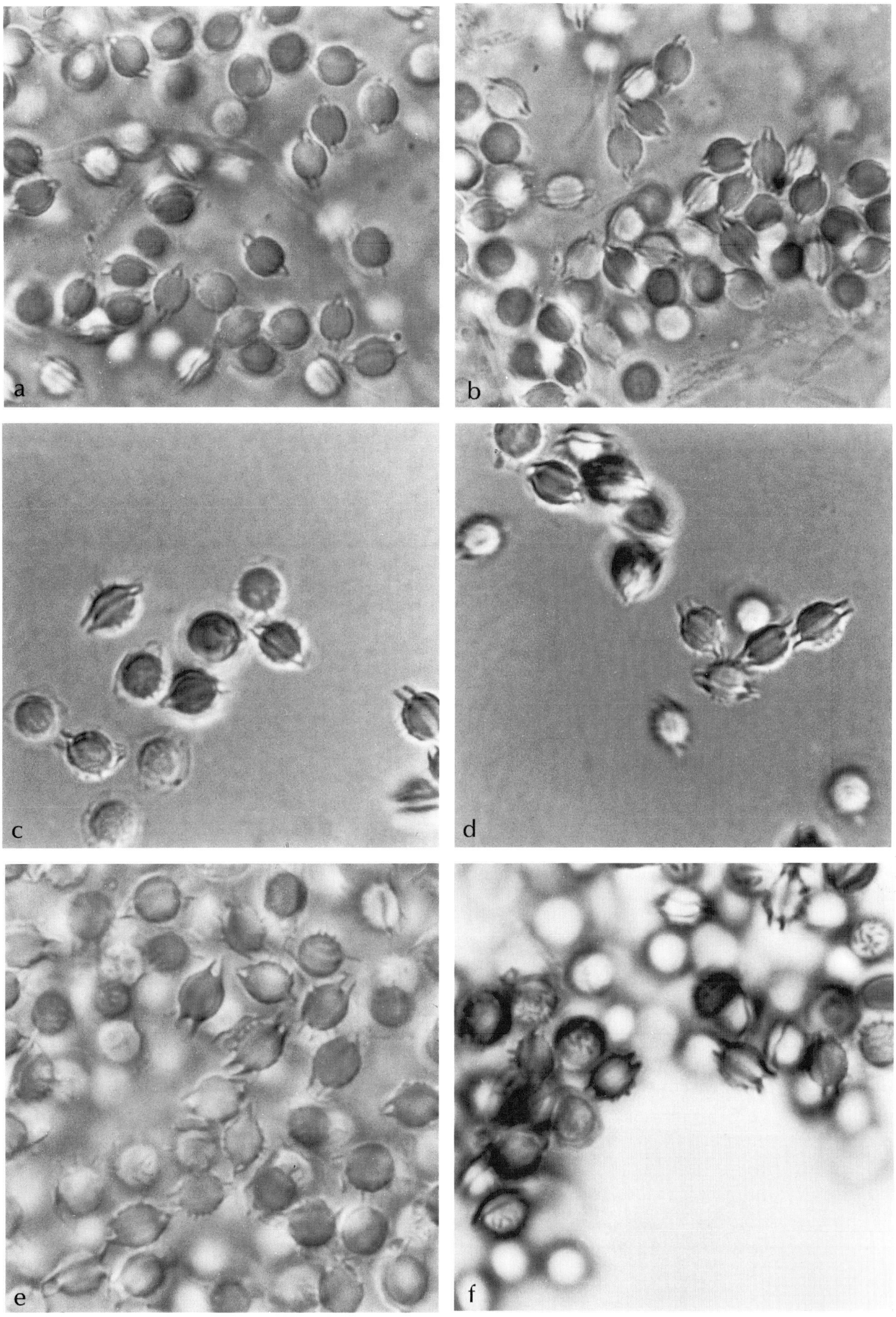
a
b
c
d
e
f

Talaromyces macrosporus (Stolk & Samson) Frisvad *et al.*

= *Talaromyces flavus* var. *macrosporus* Stolk & Samson
Anamorph: *Penicillium macrosporum* Frisvad *et al.*

Colonies on MEA at 25°C spreading broadly, attaining a diameter of 7 to 8 cm within 2 weeks, consisting of a basal felt in which numerous ascomata usually forming a continuous, thick yellow layer. Anamorph scanty, usually not affecting the colony appearance. Reverse ranging from orange red when young to orange brown in age, or purple red with the pigment diffusing in the agar. Ascomata yellow, or sometimes pinkish to purple-red, globose, 200-700 µm in diam., ripening within 2 weeks. Ascoma-covering consisting of a few layers of well developed networks of yellow pigmented hyphae, often heavily encrusted. Initials consisting of club-shaped ascogonia, around which thin antheridia coil tightly several times. Asci usually 8-spored, broadly ellipsoidal to subglobose, 8-11 x 7.5-9 µm. Ascospores yellow, sometimes reddish, more or less broadly ellipsoidal, 5-6.5 x 3.5-5.0 µm, thick-walled, spinulose. Conidiophores arising primarily from the substratum, especially in marginal areas, occasionally borne also as short branches from aerial hyphae overgrowing the ascomata, usually erect, 24-250 x 1.5-2.5 µm. Metulae 2 to 3 (-4) in the verticil, 10-15 x 1.7-2 µm, Phialides in whorls of 2 to 7, lanceolate, 8-12 x 1.7-2.5 µm. Conidia brownish green, ranging from subglobose to ellipsoidal, 2.2-3.5 x 2.0-2.5 µm.

Important (toxic) metabolites: **duclauxin**.

HABITAT: food

NOTE:
This species is heat-resistant and often occurs in (tropical) fruit-juices (see also Chapter 7). Ascospores may be soil-borne and thus contaminating the fruit. For the isolation of the species sometimes a heat shock is necessary to germinate the dormant ascospores (see also p. 339). *T. flavus* is a related species with smaller ascospores.

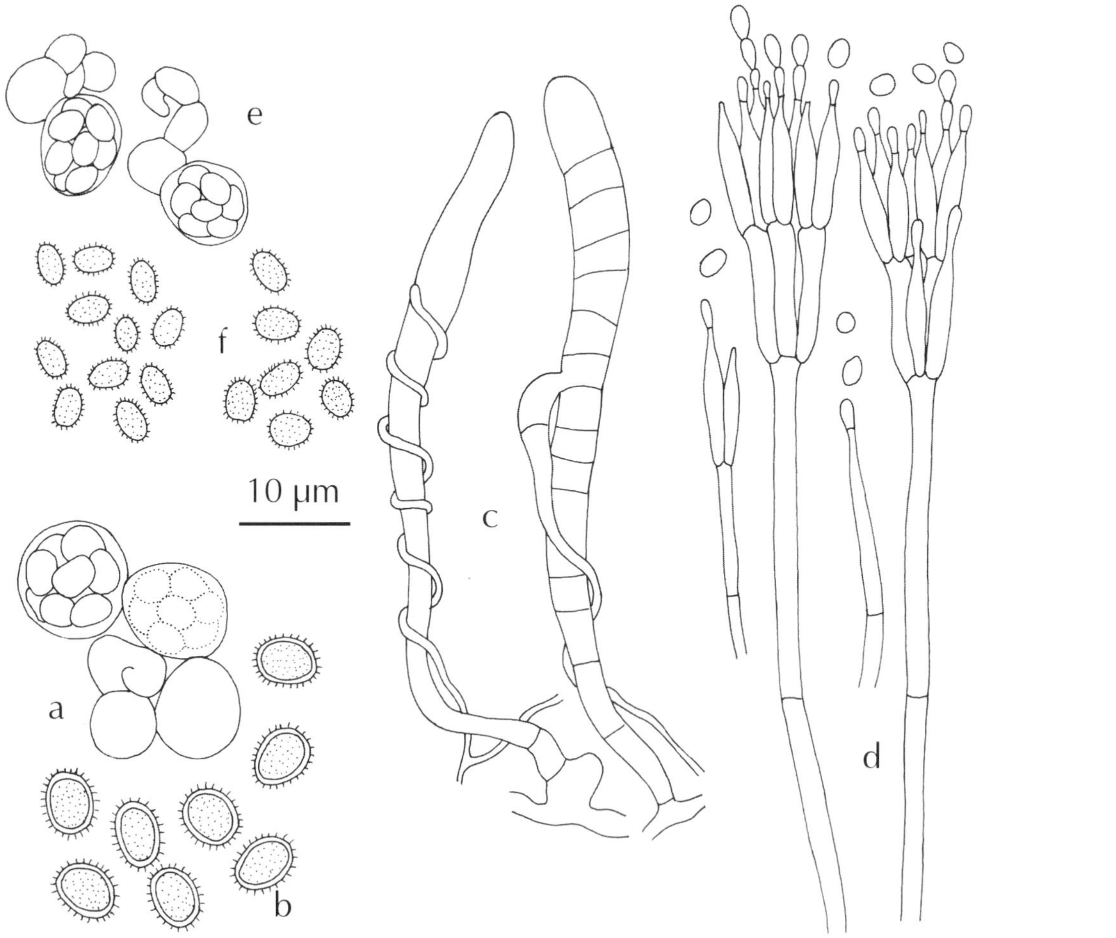

Fig. 18. *Talaromyces macrosporus.* a. Asci; b. ascospores; c. ascomatal intials; d. *Penicillium* anamorph , conidiophores and conidia; e,f. *Talaromyces flavus*; e. asci; f. ascospores (from Stolk and Samson, 1972).
Pl. 22. *Talaromyces macrosporus.* a. Colonies after two weeks on MEA; b. ascoma initial,s x 200; c-d. *Penicillium* anamorph, x 200; e. cleistothecium, x 75; f. asci, x 750; g, h. asci with ascospores of g. *T. flavus* and h. *T. macrosporus,* x 750.

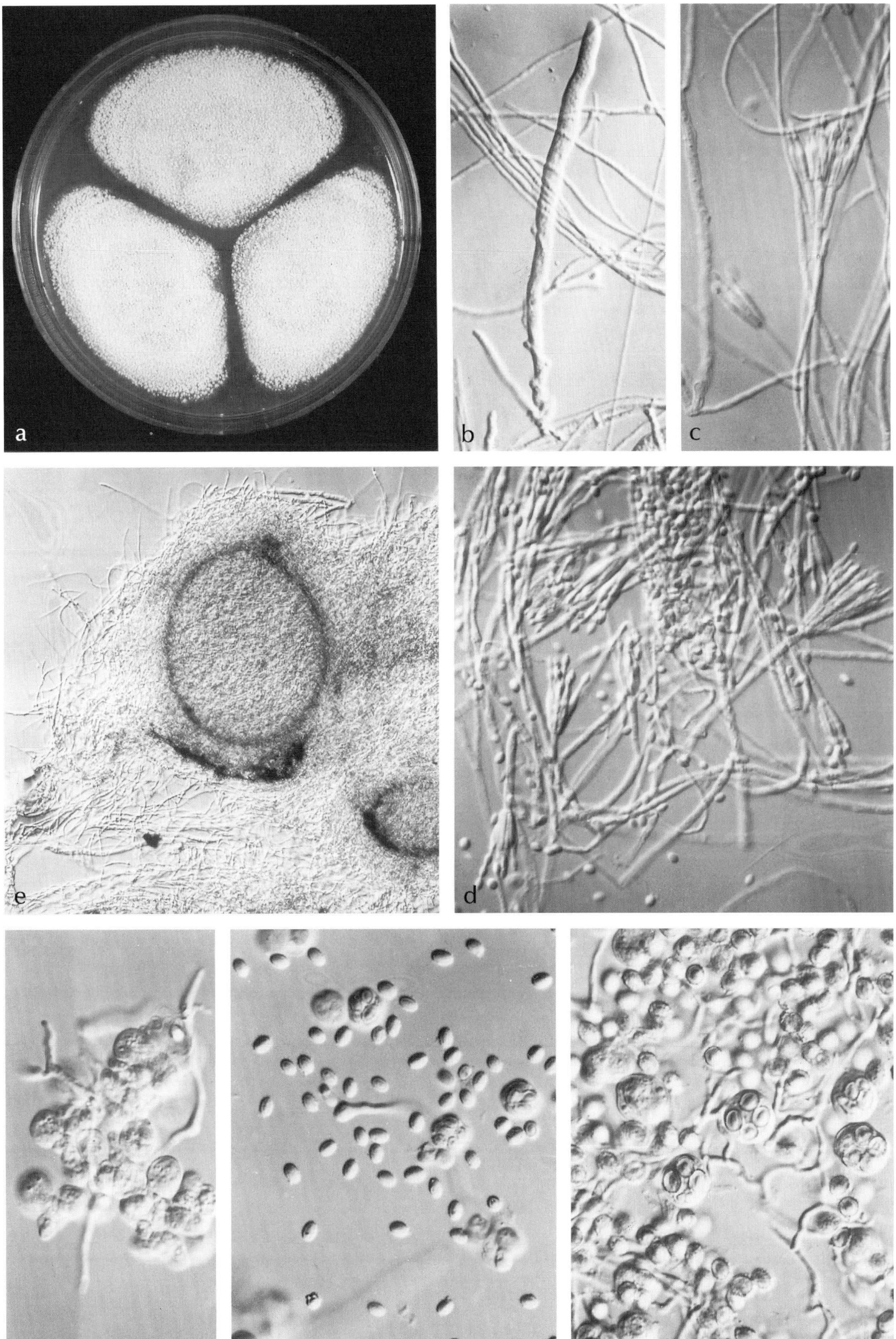
a
b
c
e
d
f
g
h

Xeromyces bisporus L.R. Fraser

= *Monascus bisporus* (Fraser) v. Arx
anamorph: *Fraseriella bisporus* Cif.& Corte

Colonies on Malt Yeast 50 % Glucose (MY50G) growing slowly reaching about 15 mm in diameter, after two weeks, low, whitish to translucent, more or less shiny. Development of cleistothecia after 2 weeks; cleistothecia (almost) sessile, maturing after a prolonged incubation period of 4-6 weeks, about 45-125 µm in diam, walls thin, hyaline; asci evanescent containing 2 ascospores. Ascospores more or less lunate, plano-convex, 10-12 x 4-5 µm. Aleurioconidia thick-walled, smooth, 15-20 x 12-15 µm, rounded at the top and with a flattened base.

Extremely xerophilic, is able to growth at a_w 0.61, with an optimum a_w for growth at 0.85 (Pitt & Hocking,1997). No growth on MEA or other high water activity media.

Temperature: good growth and sporulation at 25-30°C. Ascospores are heat resistant, with a $D_{82.2}$ of 2.3 min and a z value of 16°C (Pitt and Hocking, 1982, 1997).

HABITAT: food
Isolated from low wateractivity products, like tobacco, dried prunes, candied fruits, chocolate, licorice, spices. The fungus often will be visible after a long-term storage and on products with long shelf-lives (6 months and longer).

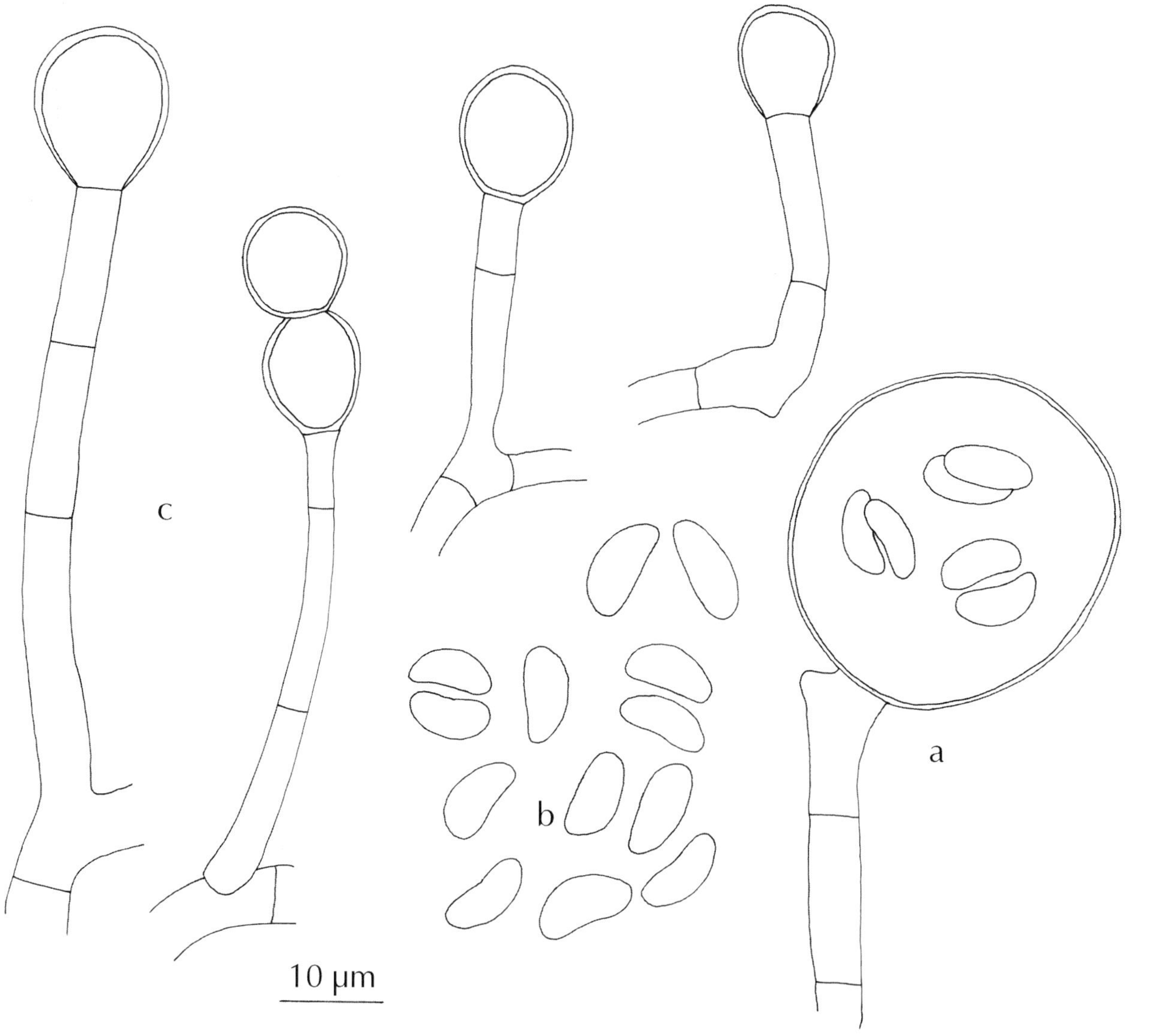

Fig. 19. *Xeromyces bisporus*. a. Ascoma with 2-spored asci; b. ascospores; c. aleurioconidia of anamorph *Fraseriella*.
Pl. 23. *Xeromyces bisporus*. a. Colony after prolonged incubation (2 months !) on MY50G; b,c. ascomata with ascospores, b. x 700; c. x 860; d. ascospores, x 650; e,f. Aleurioconidia of the anamorph *Fraseriella bisporus*, x700

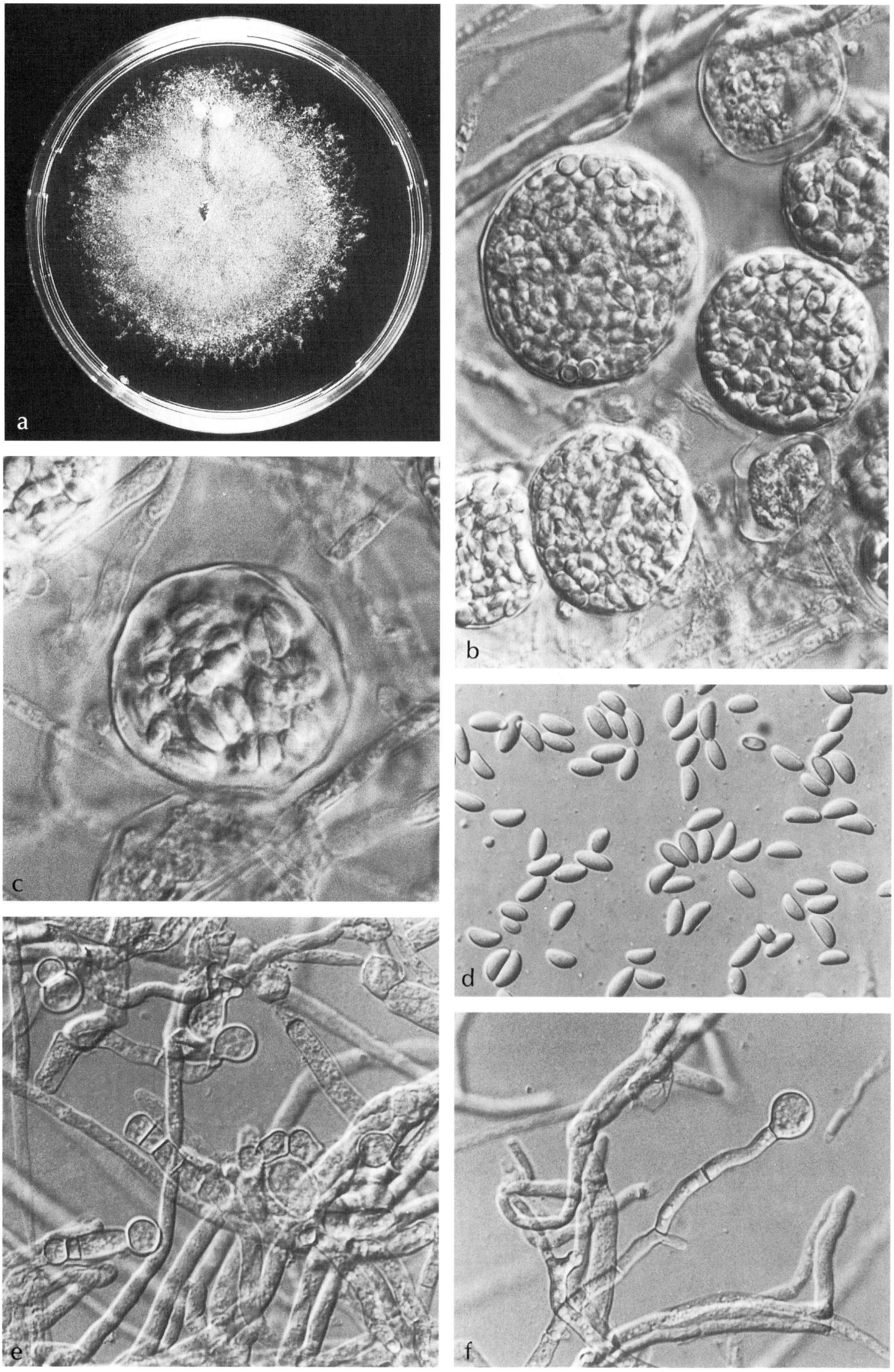
a
b
c
d
e
f

DEUTEROMYCETES

GENERAL CHARACTERISTICS

The Deuteromycetes or Fungi Imperfecti include important food- and airborne contaminants, many species being able to produce toxic metabolites. This class comprises "form-genera" and "form-species" for which only an **anamorph** (asexual or conidial state) is known. However, ascomycetous or basidiomycetous **teleomorphs** (sexual or perfect state) are connected with several species. The Deuteromycetes are usually divided on the basis of their conidium formation:

1. COELOMYCETES: conidiophores formed within fruit-bodies (conidiomata).

- **Melanconiales**: conidiophores in acervuli (flattened fructifications in the host plant, usually covered by a cuticle or epidermis).
- **Sphaeropsidales**: conidiophores in pycnidia (closed fruit-bodies usually with an apical opening).

2. HYPHOMYCETES or **MONILIALES**: conidiophores formed on simple or aggregated hyphae.

All species treated here, except *Phoma* (Sphaeropsidales) and *Epicoccum* (Melanconiales), belong to the Moniliales.

The main criterion to classify Deuteromycetes is based on their mode of conidium formation (**conidiogenesis**). Only the major types will be treated here. For a more detailed account the reader is referred to Cole and Samson (1979).

In the Deuteromycetes propagation occurs by means of **conidia** (specialized non-motile, asexual propagules, not formed by cleavage as in sporangiospores). The conidia are of various shape and colour and may arise solitarily, synchronously, in chains or in (slimy) heads. They are borne on a specialized cell (**conidiogenous cell**). The conidiogenous cells can be borne directly in or from a vegetative hypha or on differentiated supporting structures (stipe and branches). The entire system of fertile hyphae is called the **conidiophore**.

Conidia can be formed in:

- **acropetal** chains. One or more new conidiogenous loci (the places where conidia arise) are formed at the tip of each conidium. The youngest conidium is produced at the top.
- **basipetal** succession. The youngest conidium is formed at the base.
- **sympodial** succession. Each newly formed conidium moves into terminal position so that a geniculate, elongate or condensed rachis develops.

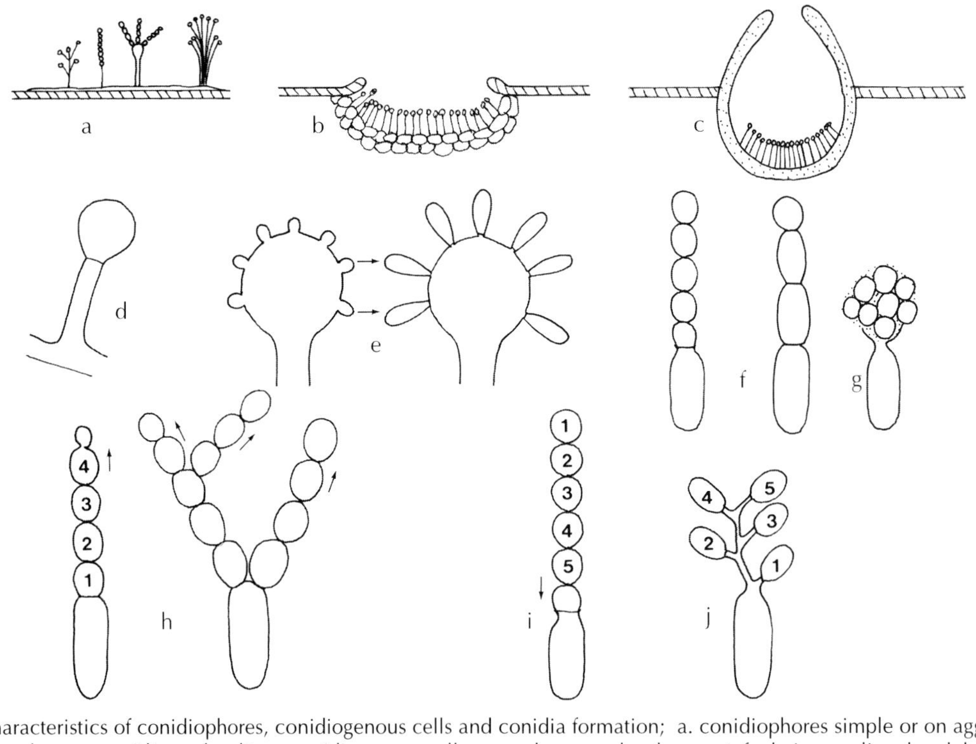

Fig. 20. a-j. Characteristics of conidiophores, conidiogenous cells and conidia formation; a. conidiophores simple or on aggregated hyphae; b. acervulus; c. pycnidium; d. solitary conidiogenous cell; e. synchronous development; f. chains; g. slimy head; h. acropetal succession; i. basipetal succession; j. sympodial succession.

The following major modes of conidiogenesis can be recognized:

THALLIC DEVELOPMENT (Fig. 21)

The conidia are formed solitarily or in chains (**arthroconidia**) from an undifferentiated part of the hypha. Cells are separated by septation and transformed into conidia, e.g. *Geotrichum*. In some genera (e.g. *Moniliella*) both thallic and blastic conidia are formed.

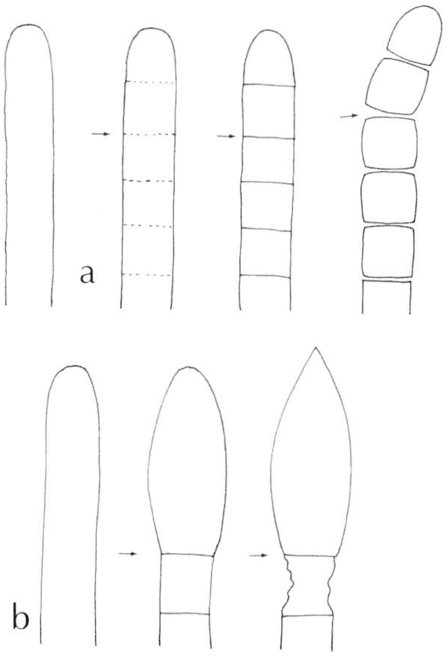

Fig. 21. a-b. Thallic development. a. in chains; b. solitarily

BLASTIC DEVELOPMENT (Fig. 22)

The wall of the conidiogenous cell becomes apically elastic and bulges out to form the conidial wall. The conidia can be produced solitarily, synchronously (e.g. *Botrytis, Aureobasidium*) or in acropetal chains (e.g. *Cladosporium*). They may have a narrow base (e.g. *Botrytis*) or a broad base (e.g. *Epicoccum*).

Some genera are characterized by **poroconidia** or tretic development. This mode of conidiogenesis is similar to blastic, but differs in that conidiogenous cells are dark and have an apical strongly pigmented wall-thickening through which the conidium is pierced. After conidium dehiscence a pore can be discerned (e.g. *Alternaria, Ulocladium*).

PHIALIDIC CONIDIOGENESIS (Fig. 23)

The conidia are produced in basipetal succession from an opening of a specialized cell (**phialide**). The phialide may be awl-shaped, flask-shaped or otherwise, or sometimes showing a collarette (cup-shaped structure at the apex). Conidia are produced in chains (e.g. *Penicillium, Aspergillus, Paecilomyces*) or aggregate in slimy or moist heads (e.g. *Trichoderma, Phialophora, Stachybotrys, Acremonium, Verticillium*).

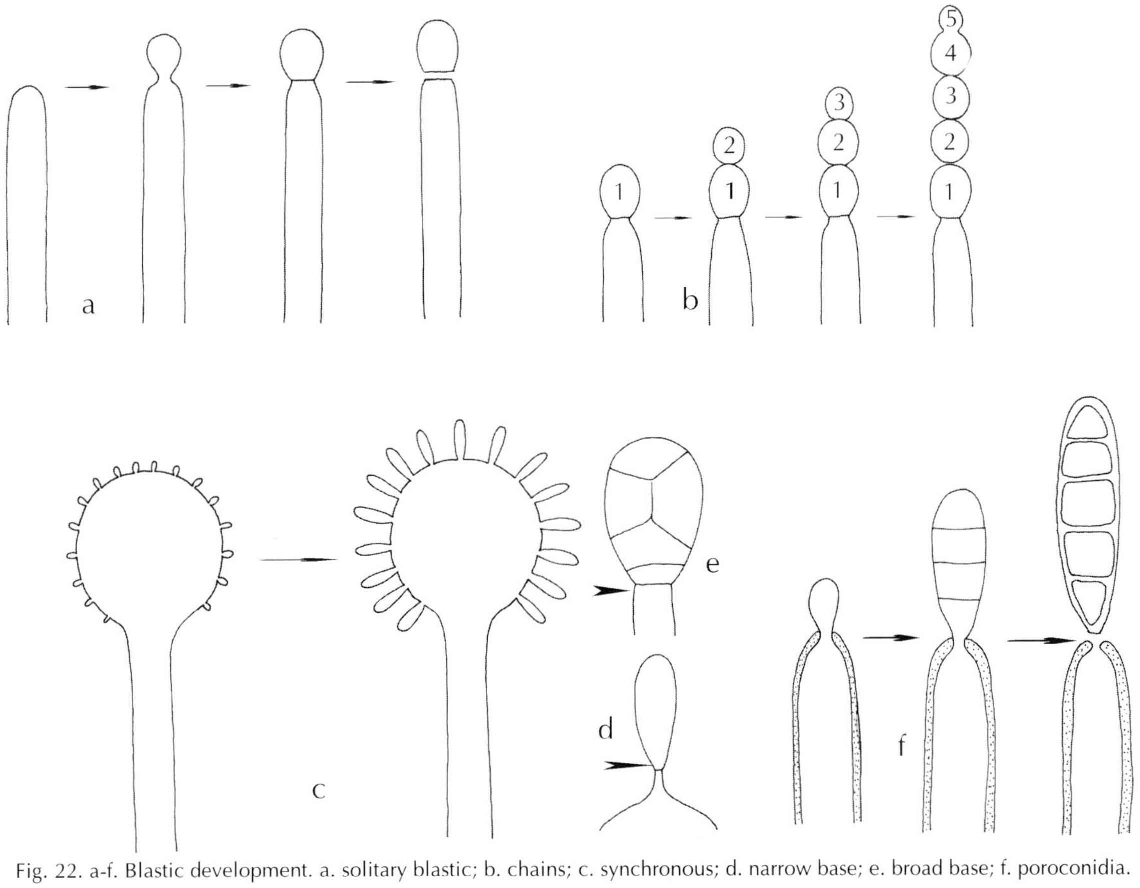

Fig. 22. a-f. Blastic development. a. solitary blastic; b. chains; c. synchronous; d. narrow base; e. broad base; f. poroconidia.

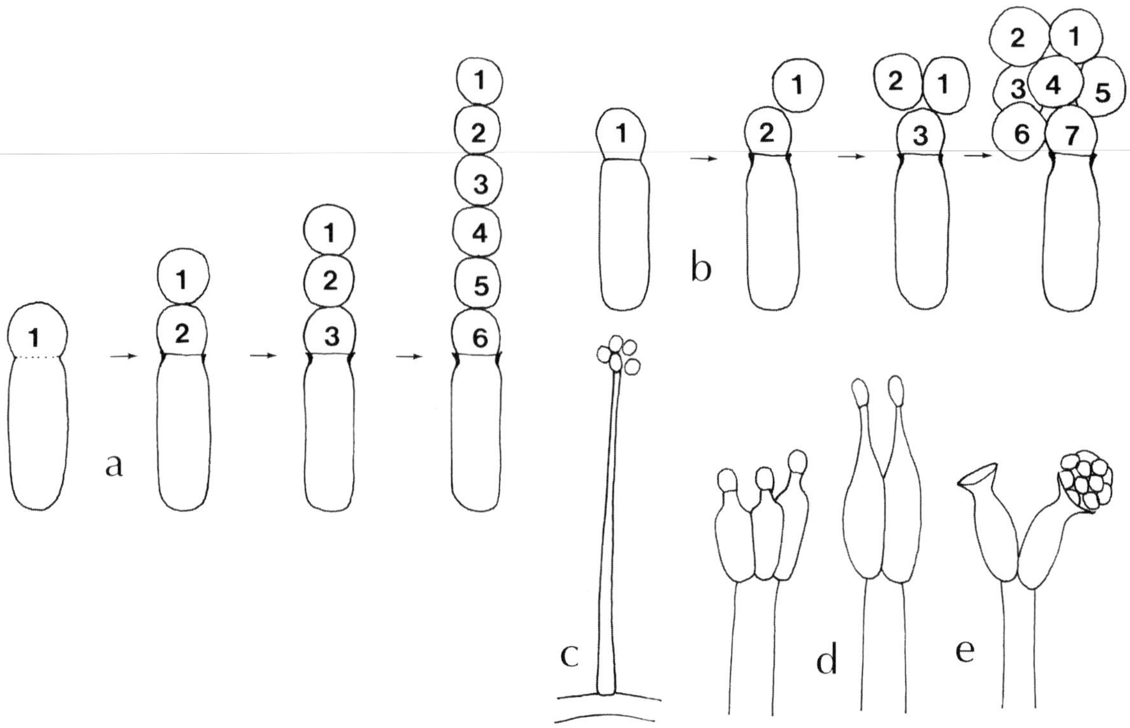

Fig. 23. a-e. Phialidic development. a. conidia in chains; b. conidia in heads; c. awl-shaped; d. flask-shaped; e. phialide with typical collarette.

ANNELLIDIC CONIDIOGENESIS (Fig. 24)

The conidia are formed from a series of short percurrent proliferations (annellations) on a conidiogenous cell (**annellide**). The annellide is often difficult to recognize with an ordinary light microscope, but can be distinguished by the increase in length of the conidiogenous apex (annellated zone during subsequent sporulation). A good microscopic character is also the broad truncate base of the conidia (e.g. *Scopulariopsis*).

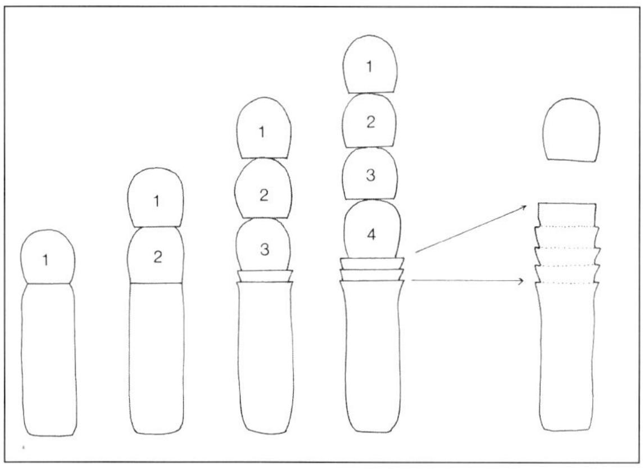

Fig. 24. Annellidic development, showing truncate conidia and elongation of the conidiogenous cell.

For general references on Deuteromycetes see Carmichael *et al.* (1980), Domsch *et al.* (1993), von Arx (1981). For special literature on monographic studies see Gams *et al.* (1987).

CULTIVATION FOR IDENTIFICATION

Most Deuteromycetes sporulate well on MEA (2%) and OA agar. For some genera e.g. *Fusarium*, SNA with a piece of sterile filter paper is very suitable for observing good sporulation. *Penicillium* and *Aspergillus* require Czapek or Czapek Yeast agar for correct identification (see also the description of each genus for culture methods).

Good microscopic mounts are required to examine the conidiogenesis. These are mostly made in lactic acid with some aniline or cotton blue or in water. For fragile structure such as in *Botrytis* and *Cladosporium* it is often helpful to prepare slides with the aid of a small piece of adhesive tape or cut out a square of transparent agar. The production of conidia in chains or slimy heads is best observed in the Petri-dish under low power with a light microscope. However, avoid inhalation of fungal products (volatiles) and conidia.

KEY TO THE COMMON FOOD- AND AIRBORNE GENERA OF THE DEUTEROMYCETES TREATED

1a. Conidia borne in pycnidia .... *Phoma* (p. 248)

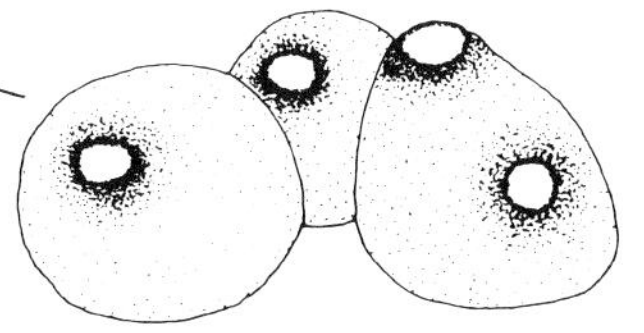

1b. Conidia not borne in pycnidia, but on hyphae, conidiophores, sporodochia or synnemata 2

2a. Conidia produced from special conidiogenous cells (phialides, annellides etc.) in chains or in heads, in basipetal succession 3

2b. Conidia not produced from special conidiogenous cells, but acropetally or by fragmentation of fertile hyphae, synchronously or single 14

3a. Conidia in dry chains 4

3b. Conidia in moist or slimy heads 10

4a. Conidia usually 2-celled, borne on filamentous conidiogenous cells, more or less obliquely inserted, arranged like a spike. Colonies pinkish .... *Trichothecium* (p. 264)

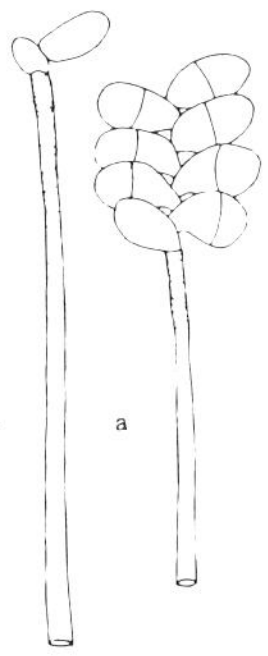

4b. Conidia always one-celled, borne on flask-shaped or cylindrical conidiogenous cells in straight chains. Colonies in various colours 5

5a. Colonies very restricted, reddish-brown. Conidia formed (in a quartet) by division of a cylindrical verrucose fertile hypha, cubic, becoming (sub)-globose. Xerophilic .... *Wallemia* (p. 268)

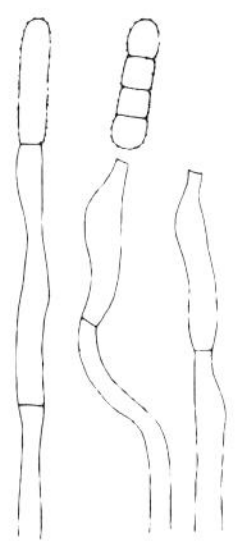

5b. Colonies usually not restricted (except xerophilic *Aspergillus* species). Conidia not formed after division of fertile hyphae 6

6a. Conidiophores with a typical apical swelling .... *Aspergillus* (p.64)

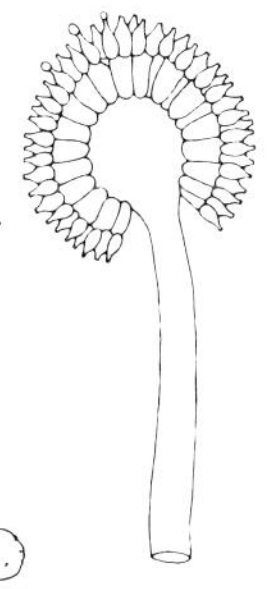

6b. Conidiophores without an apical swelling 7

7a. Conidiogenous cells annellidic. Conidia with a broad truncate base .... *Scopulariopsis* (p.252)

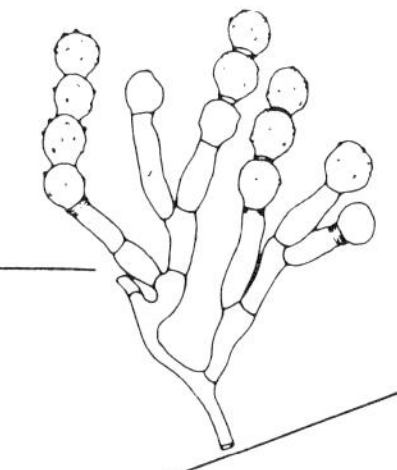

7b. Conidiogenous cells phialidic. Conidia without a broad truncate base 8

8a. Colonies dark grey to blackish. Phialides obovate (broadest at the top), conidia blackish. .... *Memnoniella* (p. 162)

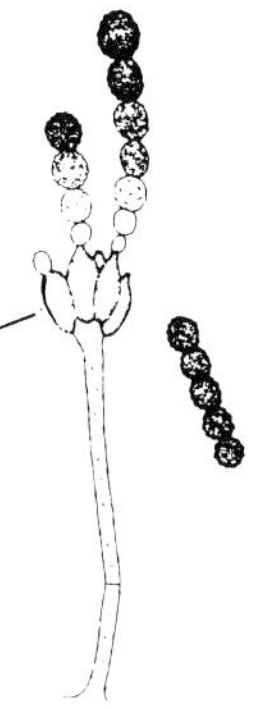

8b. Colonies not dark grey to blackish. Phialides flask-shaped or lanceolate (broadest at the base) 9

9a. Colonies yellow to brown. Phialides with a long neck .... *Paecilomyces* (p.170)

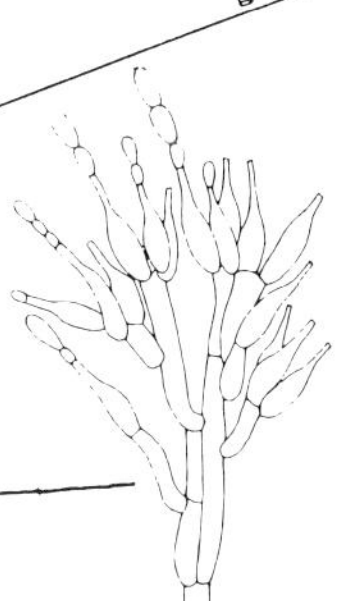

9b. Colonies often greenish (some species whitish). Phialides with a short neck .... *Penicillium* (p. 174)

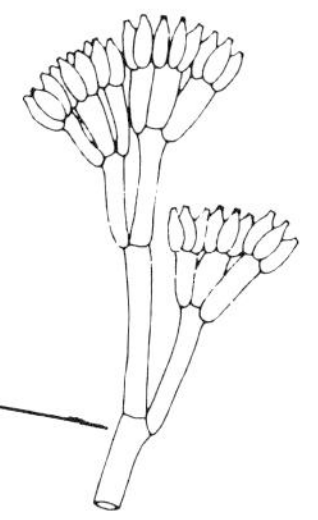

10a. Phialides long, awl-shaped, polyphialides absent .... *Acremonium* (p. 58)

10b. Phialides more or less flask-shaped and/or polyphialides present, or phialides reduced 11

11a. Colonies usually green (when grown in light) .... *Trichoderma* (p. 260)

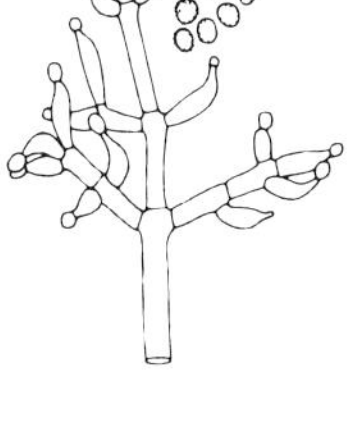

11b. Colonies whitish, yellow, purple, violet, pinkish, brown or blackish 12

12a. Colonies white, yellowish pinkish, purplish, sometimes greenish. Septate banana-shaped conidia usually present.................. *Fusarium* (p. 120)

12b. Colonies black, sometimes pinkish. Conidia not septate.. 13

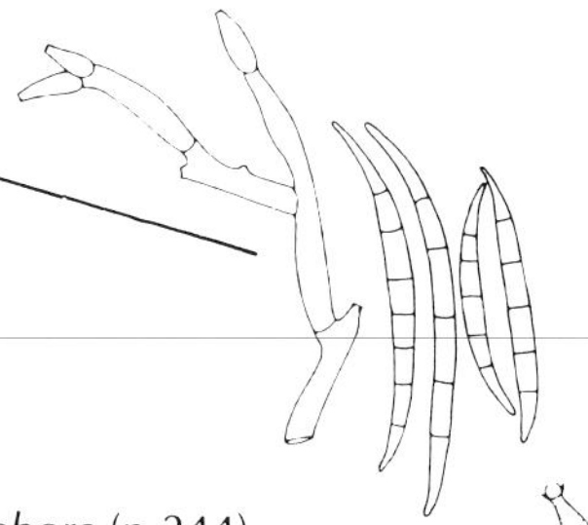

13a. Phialides solitary or in loose whorls, flask-shaped with a conspicuous collarette or phialides reduced. Conidiophores not distinct..................................... *Phialophora/Lecythophora* (p.244)

13b. Phialides in dense apical clusters, broadly clavate, widest near apex, without conspicuous collarette. Conidiophores distinct .... *Stachybotrys* (p.258)

14a. Colonies growing very fast, covering a 9 cm Petri-dish within a few days, loose, floccose, orange .... *Chrysonilia* (p.104)

14b. Colonies not orange and not covering a 9 cm Petri-dish within a few days 15

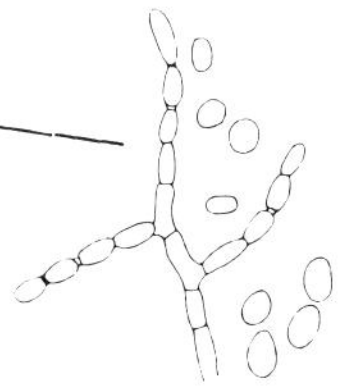

15a. Conidia only arthric 16

15b. Conidia both arthric and blastic or only blastic... 18

16a. Conidiophore stipes pigmented .... *Oidiodendron* (p. 168)

16b. Conidiophore stipes hyaline or absent 17

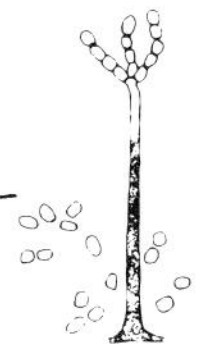

17a. Conidia hyaline, smooth-walled, more or less cylindrical. Intercalary conidia absent .... *Geotrichum* (p. 160)

17b. Conidia (sub)hyaline becoming faintly coloured, and rough-walled with age. Intercalary conidia present .... *Geomyces* (p. 158)

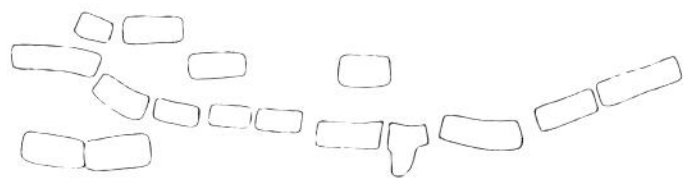

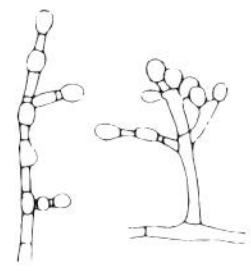

18a. Conidia formed in a quartet by division of a cylindrical fertile hypha .... *Wallemia* (p.268)

18b. Conidia not formed in a quartet............................. 19

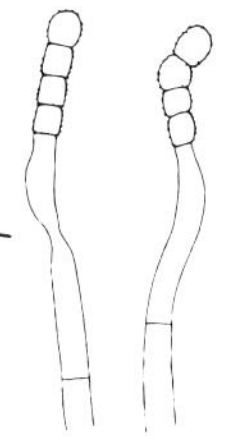

19a. Conidiogenous structure consisting of both arthroconidia and blastoconidia (compare also *Trichosporon* in yeasts and the thick-walled, brown arthroconidia-like hyphal cells in *Aureobasidium*)............................. *Moniliella* (p.164)

19b. Conidiogenous structures consisting of only blastoconidia... 20

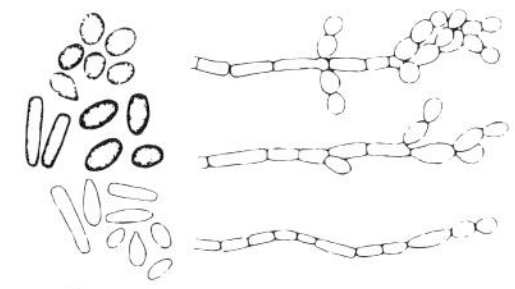

20a. Blastoconidia borne synchronously on hyphae, or from swollen cells or branches............. 21

20b. Blastoconidia not formed synchronously on hyphae or swollen cells or branches 22

21a. Conidia borne from denticles on terminally swollen conidiogenous cells. Conidiophores erect, apically branched (tree-like). Colonies thin, greyish-brown................. *Botrytis* (p.100)

21b. Conidia borne on hyphae or on almost completely swollen branches. Colonies yeast-like, creamish-yellow to light brown, pinkish-orange or blackish-green .... *Aureobasidium* (p.98)

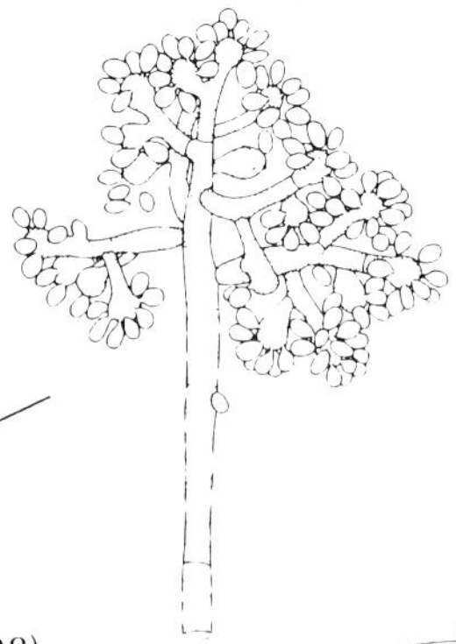

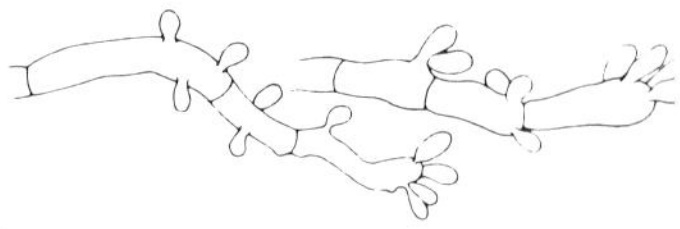

22a. Conidia formed singly on indistinct conidiophores, clustered and visible as black dots.. *Epicoccum* (p.118)

22b. Conidia formed singly or in chains, conidiophores distinct, not clustered...................... 23

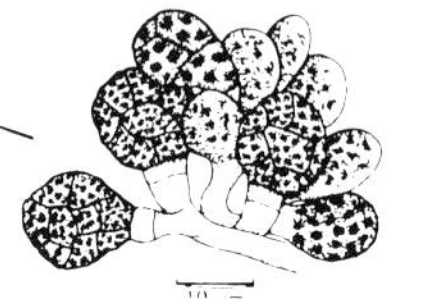

23a. Conidia in chains, smooth-walled. Colonies usually cream-coloured at first, darkening with age .... *Moniliella* (p.164)

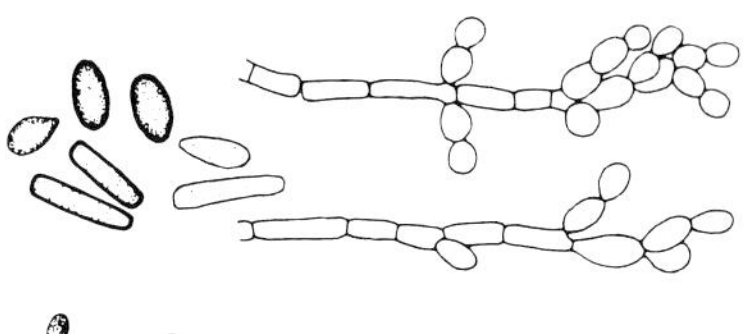

23b. Conidia in chains or single, rough-or smooth-walled. Colonies in greenish-black or greenish-brown shades 24

24a. Conidia rather thin-walled, mostly one-celled; basal conidia can be 2-celled, only with a transverse septum .... *Cladosporium* (p. 107)

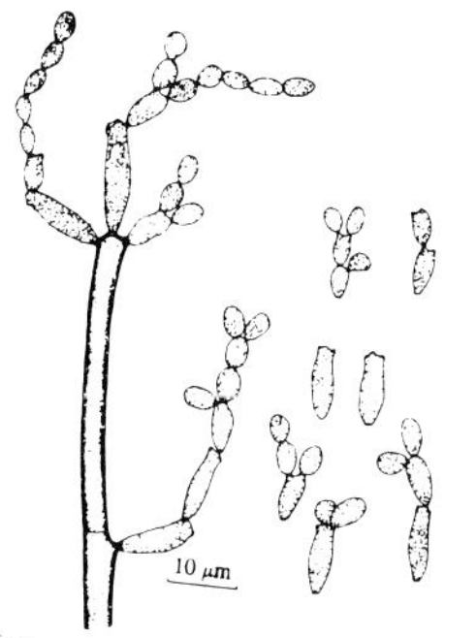

24b. Conidia septate with both transverse and longitudinal septa (muriform) or only with transverse septa 25

25a. Conidia smooth-walled, with transverse septa, slightly curved, end cells paler than central cell .... *Curvularia* (p.116)

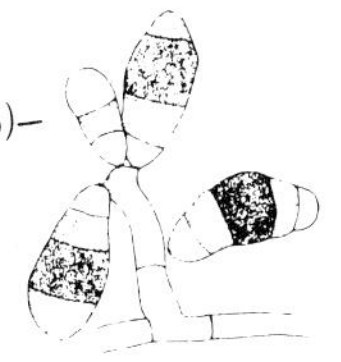

25b. Conidia rough-walled, with both transverse and longitudinal septa (muriform) 26

26a. Young conidia rounded at the base, mature conidia catenuate and/or rostrate .... *Alternaria* (p.62)

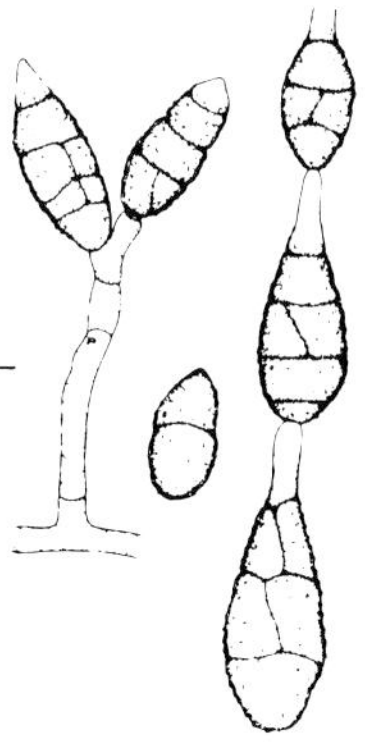

26b. Young conidia attenuated at the base, mature conidia single or in "false" short chains .... *Ulocladium* (p.266)

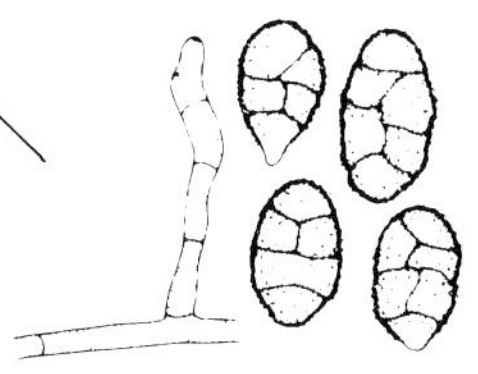

ACREMONIUM Link

Teleomorphs: *Emericellopsis* v. Beyma, *Mycoarachis* Malloch & Cain, *Nectria* (Fr.)Fr.

Colonies growing slowly, mostly less than 2.5 cm in diameter in 10 days. Vegetative mycelium mostly hyaline. Phialides mostly awl-shaped, simple, erect, arising from the substrate mycelium or from bundled aerial hyphae (fasciculate). If compound conidiophores occur, branches are only restricted to the basal part (basitonous). Conidia usually one-celled, hyaline or pigmented, mostly in slimy heads.

Most species are saprophytic and isolated from dead plant material and soil, some species are pathogenic to plants and humans and have world-wide distribution. For more detailed descriptions and keys see Gams (1971), Domsch *et al.* (1993).

CULTIVATION FOR IDENTIFICATION
Isolates are inoculated in streaks on MEA (2%) and incubated at 20°C in darkness for 10 days (for growth) and 2 days in daylight (for colour). Microscope slides are made in lactic acid with aniline blue. Observation with a low power microscope lens is helpful for identification.

KEY TO THE SPECIES TREATED

1a. Colonies yellowish-green to olive-green .... *A. butyri*
1b. Colonies whitish, often turning pale pink, pink or orange, but never in green shades 2

2a. Phialides arising from distinct bundled aerial hyphae, base not or hardly chromophilic, not granulose .... *A. strictum*
2b. Phialides often arising directly from substrate mycelium, or from mycelium pustules, base distinctly chromophilic, often granulose .... *A. charticola*

Acremonium butyri (v. Beyma) W. Gams

Teleomorph: *Nectria viridescens* C. Booth

Colonies on MEA at 20°C attaining a diameter of 1-2 cm in 10 days, more or less floccose in the center, yellowish-green to olive-green. Conidiophores usually branched and often showing chromophilic incrustations. Phialides arising from bundled aerial hyphae or from mycelium pustules, mostly with a visible collarette, 20-45 (-75) µm long. Conidia in slimy heads, slightly asymmetrical, elongate ellipsoidal to globose, 3.5-5.8 (-8.1) x 1.5-2.5 µm, hyaline, smooth-walled. Chlamydospores sometimes present.

HABITAT: food
World-wide distribution. Saprophytic. Reported from soil, leaf litter, dead wood, other fungi.

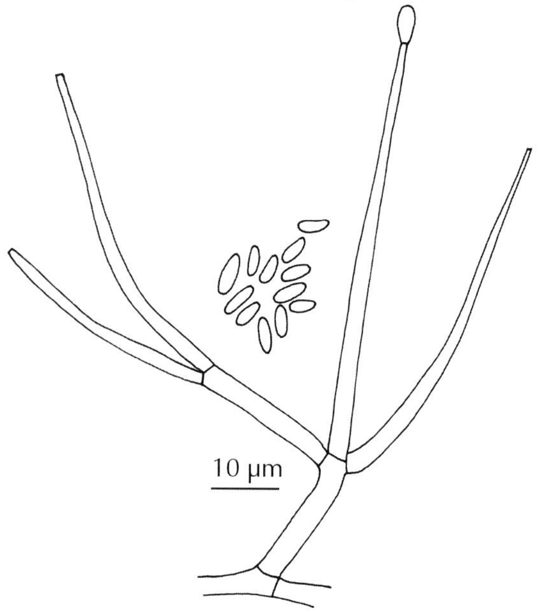

Fig. 25. *Acremonium butyri.* Conidiophore and conidia.

Pl. 24. a-d. *Acremonium butyri.* a. Colony on MEA after 10 days; b-d. conidiophores, conidia and phialides, x 2000; e-j. *Acremonium charticola.* e. colony on MEA after 10 days; f-j. phialides and conidia, x 2000

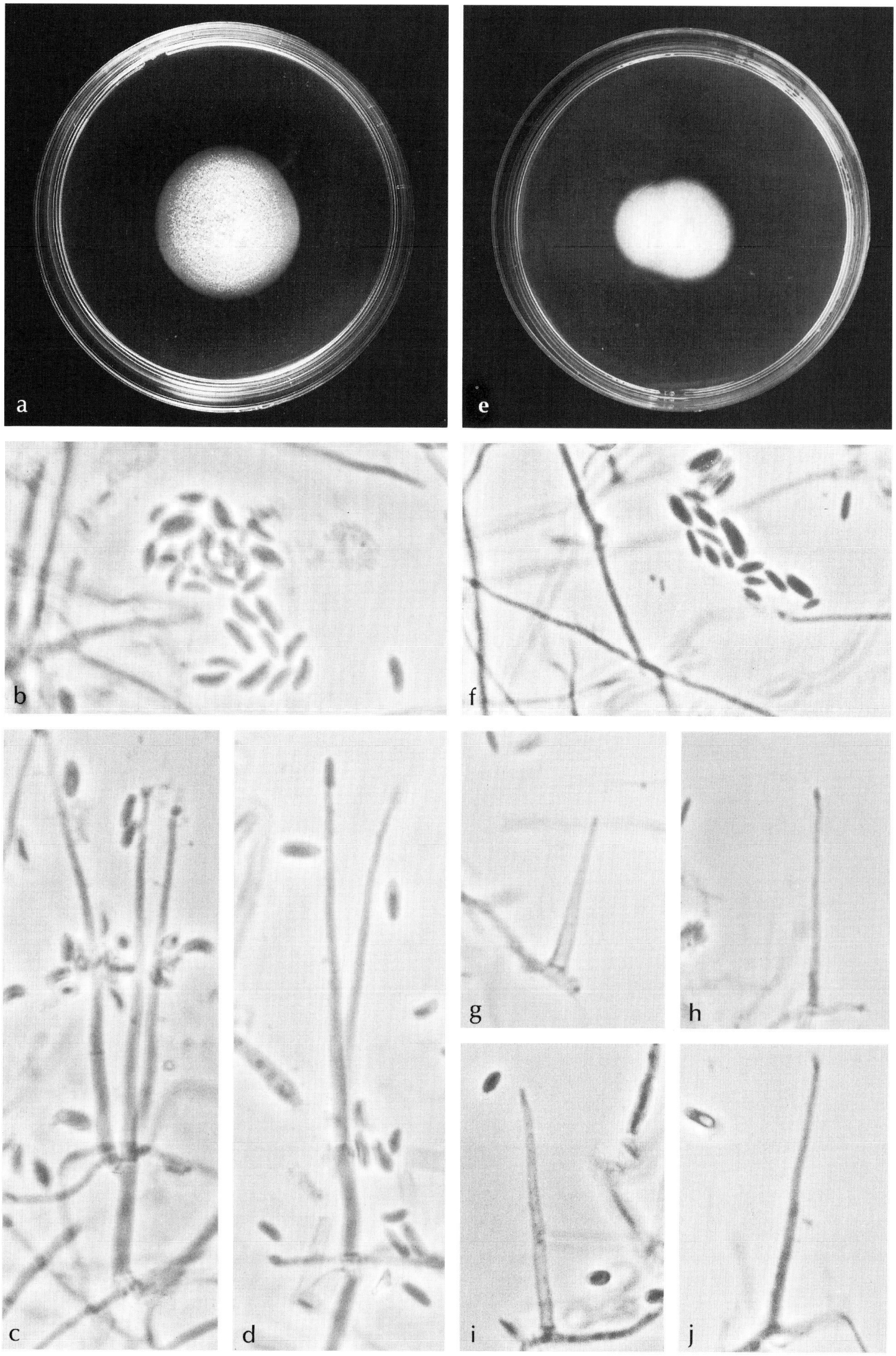
a
e
b
f
c
d
g
h
i
j

Acremonium charticola (Lindau) W. Gams.

Colonies on MEA at 20°C attaining a diameter of 1-1.5 cm in 10 days, often moist at first, becoming powdery and floccose with mycelium pustules, whitish or pink. Conidiophores often branched. Phialides often arising from substrate mycelium, but also from mycelium pustules, distinctly chromophilic at the base, often somewhat curved, without collarette, 15-45(-60) µm long. Conidia in slimy heads, ellipsoidal to short-cylindrical, 3.2-4.5 x 1.4-2.0 µm, hyaline, smooth-walled. Chlamydospores absent.

HABITAT; food, indoor
World-wide distribution. Probably causes apple- and pear-rot, sometimes pathogenic to humans. Isolated from moist cellar walls, mouldy carpets.

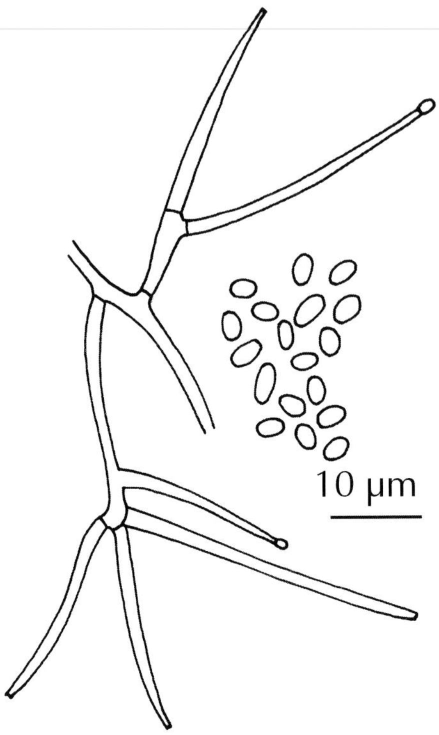

Fig. 26. *A. charticola.* Conidiophores and conidia.

Acremonium strictum W. Gams

Colonies on MEA at 20°C attaining a diameter of 1.5-2.5 cm in 10 days, floccose or velvety, especially in the center, or almost smooth and slimy, pink or orange. Conidiophores sometimes branched. Phialides usually arising from bundled aerial hyphae, erect, sometimes slightly chromophilic at the base, collarette rarely visible, 20-40(-65) µm long. Conidia in slimy heads, usually cylindrical or ellipsoidal and usually straight, 3.3-5.5(-7) x 0.9-1.8 µm, hyaline, smooth-walled. Chlamydospores absent.

HABITAT: food, indoor
World-wide distribution. Saprophytic. Isolated from soil, leaves of vascular plants, other fungi, hay. From mostly moist surfaces in indoor environments.

NOTE:
A. strictum is a very common and variable species.

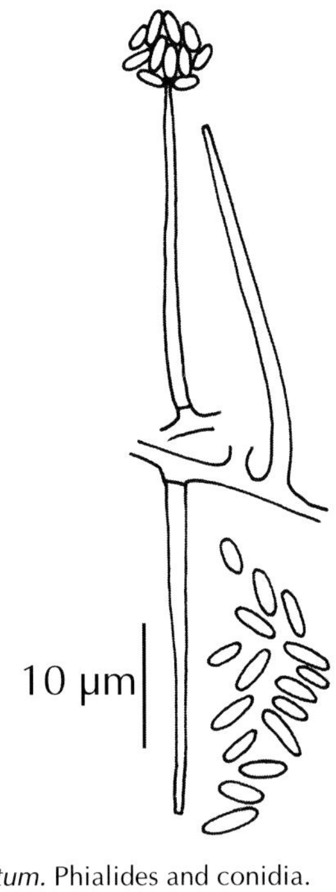

Fig. 27. *A. strictum.* Phialides and conidia.

Pl.25. *Acremonium strictum.* a. Colony on MEA after 10 days; b. conidiophores, x 600; c-f. conidia, conidiophore and phialides, x 2000

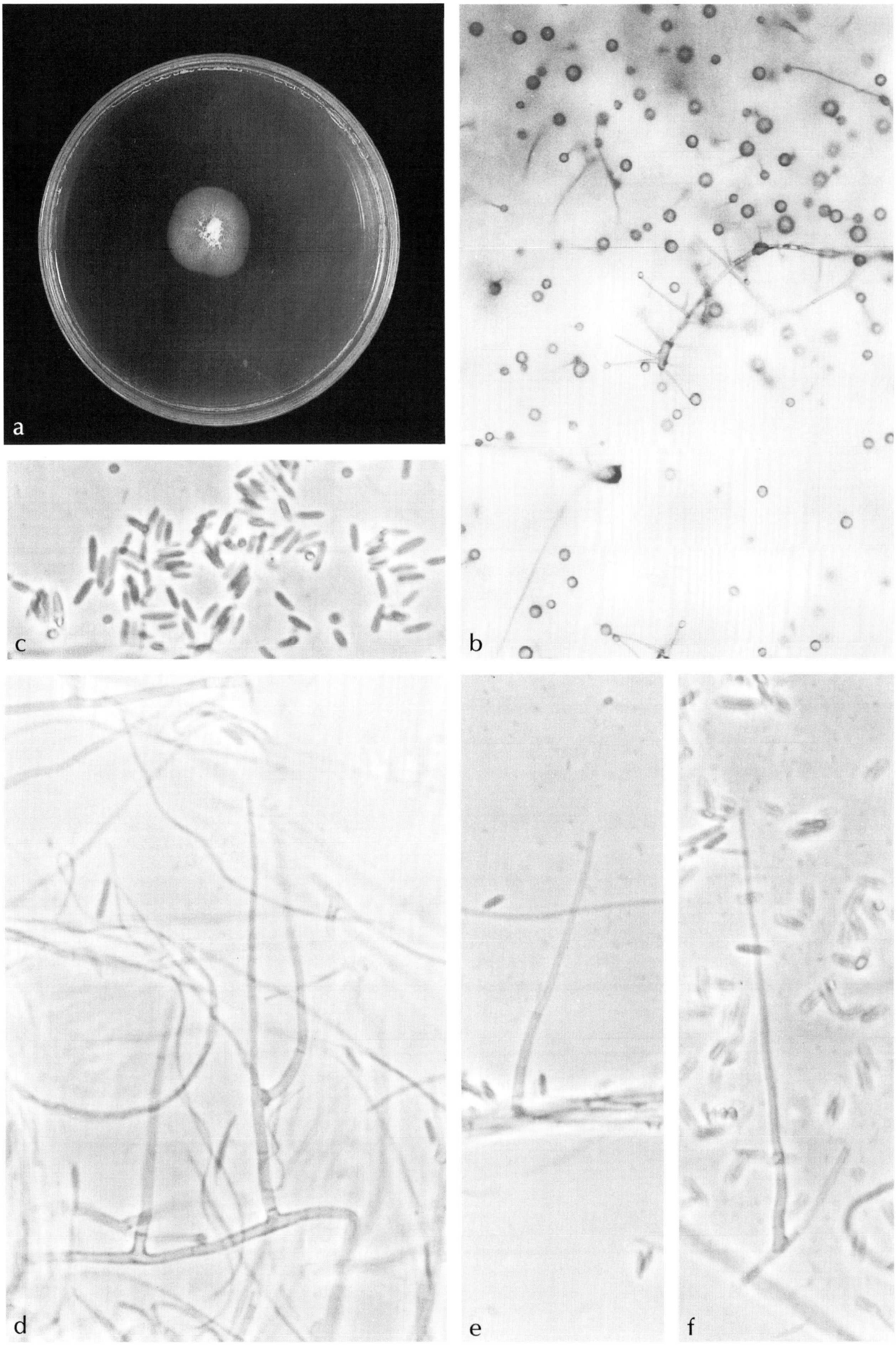
a
b
c
d
e
f

Alternaria alternata (Fr.) Keissler

= *Alternaria tenuis* Nees
Teleomorph: *Lewia* Simmons (for *Alternaria infectoria*)

Colonies on MEA at 25°C attaining a diameter of 6 cm in 7 days, black or olivaceous-black or greyish. Conidiophores 1-3 septate, simple or sometimes branched, straight or flexuous or sometimes geniculate, with one or several apical pores, up to 50 x 3-6 µm, medium brown, smooth-walled. Conidia in long often branched chains, obclavate, obpyriform, sometimes ovoid or ellipsoidal, often with a short conical or cylindrical beak, not exceeding one third of the conidial length, 18-83 x 7-18 µm, with up to 8 transverse and several longitudinal septa, pale to medium brown, smooth-walled or verrucose.

Temperature: optimum (22-) 25-28 (-30)°C; maximum 31-32°C; minimum 2.5-6.5°C (isolates from colder regions at 0°C to -2 or -5°C).

Important (toxic) metabolites: tenuazonic acid, alternariol, alternariol monomethylether, altertoxins.

HABITAT: food, indoor
World-wide distribution. A common saprophyte isolated from many kinds of plants, foodstuffs, soil, textiles and other (often moist) substrates and surfaces.

NOTE:
For more detailed descriptions see Neergaard (1945), Joly (1964), Simmons (1967), Ellis (1971), Domsch *et al.* (1993). Good sporulation can be obtained by growing cultures on potato-carrot or hay-infusion agar under near UV "black" light.

A. alternata may be less common than originally thought. *A. tenuissima* and *A. infectoria* are common in foods.

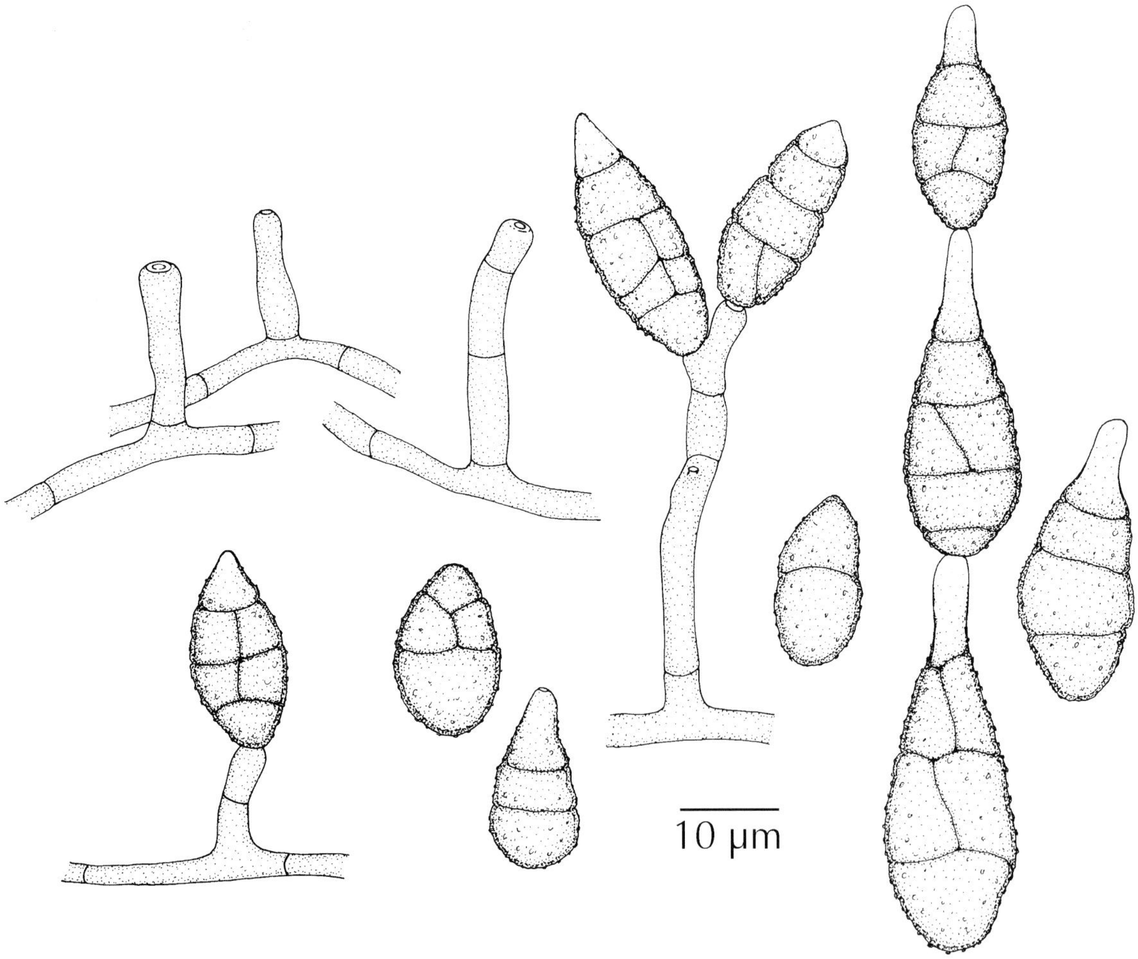

Fig.28. *Alternaria alternata*. Conidiophores and septate, often beaked conidia in chains.
Pl. 26. *Alternaria alternata*. a. Colony on PCA after one week; b-e. conidiophores and conidia, x 1000.

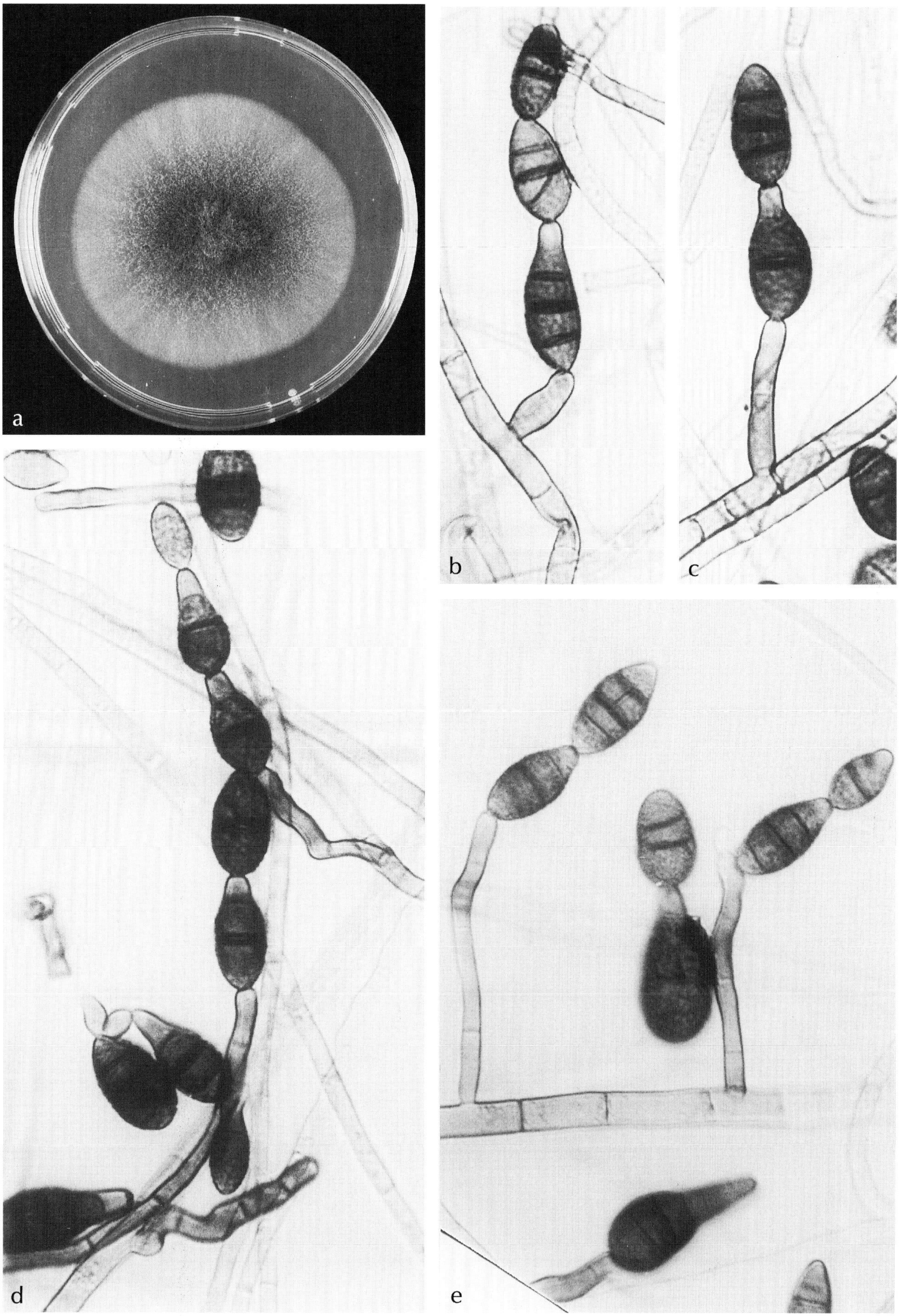
a
b
c
d
e

ASPERGILLUS Fr.: Fr.

Colonies usually growing rapidly, white, yellow, yellow-brown, brown to black or shades of green, mostly consisting of a dense felt of erect conidiophores. Conidiophores of an (usually aseptate) unbranched stipe, with a swollen apex (**vesicle**). Phialides borne directly on the vesicle (**uniseriate**) or on metulae (**biseriate**). Vesicle, phialides, metulae (if present) and conidia form the conidial head. Conidia in dry chains forming compact columns (**columnar**) or diverging (**radiate**), one-celled, smooth or ornamented, hyaline or pigmented. Species may produce **Hülle cells** (single or in chains, thick- and smooth-walled cells) or **sclerotia** (firm, usually globose, masses of hyphae).

Teleomorphs: *Eurotium, Emericella, Neosartorya* and other genera.

Aspergillus species are common contaminants of various substrates. In subtropical and tropical regions their occurrence is more common than the Penicillia. Several species have attracted attention as human and animal pathogens or because of their ability to produce toxic metabolites. Others are important for their role in fermentation of oriental food products or industrial application in the production of organic acids or enzymes.

The classification is mainly based on morphological characters. Raper and Fennell (1965) divided the genus into 18 groups and accepted 132 species with 18 varieties.

Samson (1979, 1992, 1994 a and b) provided a compilation of the species and varieties described since 1965 with a critical review of the validity of the published taxa. The genus now contains more than 180 species, with about 70 named teleomorphs. Typification of *Aspergillus* names has now been achieved because the majority of the *Aspergillus* names used by Raper and Fennell (1965) were typified by Samson and Gams (1986) and Kozakiewicz (1989).

In order to protect *Aspergillus* names in current use from being threatened or displaced by names that are no longer in use, and in order to eliminate uncertainties regarding their application spelling, gender, typification date and place or valid publication, Pitt and Samson (1993, 2000) proposed a list of species names in current use in the family *Trichocomaceae* (Fungi, *Eurotiales*). All published names and synonyms are listed in Samson and Pitt (2000).

Raper and Fennell (1965) subdivided *Aspergillus* into "Groups". This infrageneric classification has no nomenclatural status under the ICBN and Gams *et al.* (1986) replaced the group names by names of subgenera and sections (Table 1).

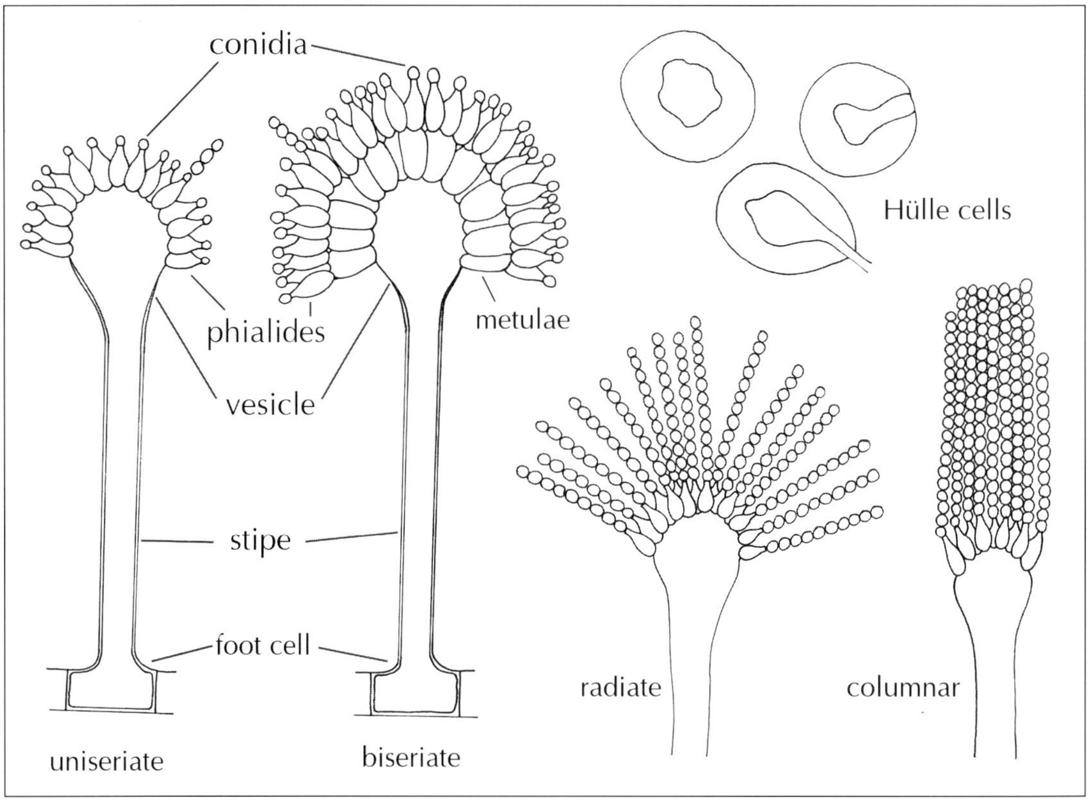

Fig. 29. Morphological structures in *Aspergillus*.

Table 1. Nomenclature of infrageneric taxa of the genus *Aspergillus*. Group-names following the classification of Raper and Fennell (1965) are given between brackets.

Subgenus	Section
Aspergillus	*Aspergillus (A. glaucus*-group)
	Restricti (A. restrictus-group)
Fumigati	*Fumigati (A. fumigatus*-group)
	Cervini (A. cervinus-group)
Ornati	*Ornati (A. ornatus*-group)
Clavati	*Clavati (A. clavatus*-group)
Nidulantes	*Nidulantes (A. nidulans*-group)
	Versicolor (A. versicolor-group)
	Usti (A. ustus-group)
	Terrei (A. terreus-group)
	Flavipedes (A. flavipes-group)
Circumdati	*Wentii (A. wentii*-group)
	Flavi (A. flavus-group)
	Nigri (A. niger-group)
	Circumdati (A. ochraceus-group)
	Candidi (A. candidus-group)
	Cremei (A. cremeus-group)
	Sparsi (A. sparsus-group)

Many *Aspergillus* species have an ascomycete teleomorph (Table 2); the teleomorph genera printed in bold contain food- and/or airborne species (see also under Ascomycetes (page 26).

Table 2. Ascomycetous genera with an *Aspergillus* anamorph.

Aspergillus anamorph	Teleomorph
Section *Aspergillus:*	***Eurotium*** Link:Fr., *Dichlaena* Mont. & Durieu.
Section *Fumigati:*	***Neosartorya*** Malloch & Cain.
Subgenus *Ornati:*	*Warcupiella* Subram., *Sclerocleista* Subram., *Hemicarpenteles* Sarbhoy & Elphick
Section *Nidulantes*:	***Emericella*** Berk. & Br.
Section *Flavipedes:*	*Fennellia* Wiley & Simmons
Section *Circumdati:*	*Neopetromyces* Frisvad & samson
Section *Flavi*	*Petromyces* Malloch & Cain
Section *Cremei:*	*Chaetosartorya* Subram.

Only the most common *Aspergillus* species are briefly described and keyed out in this guide, particularly those on food products and in indoor environments. However, for a more exact identify-cation of important isolates the taxonomic literature (Raper and Fennell, 1965; Samson, 1992, 1994 a and b) should be consulted. As already stated above Aspergilli are very common contaminants. Therefore we have not listed every substrate from which each species has been isolated. Data of habitats of many species are given by Domsch *et al.* (1993).

Practical laboratory guides for identification of common Aspergilli which are primarily based on morphology and colony characters have been published by Pitt and Hocking (1997), Klich and Pitt (1988a) and Tzean *et al.* (1990)

Identification of aflatoxigenic Aspergilli

Some specialized media are used for the rapid recognition of potentially aflatoxigenic Aspergilli. For the detection of *A. flavus* and *A. parasiticus* within 3 days Aspergillus Differential medium (Bothast and Fennell, 1974) or ***A****spergillus* ***F****lavus* and ***P****arasiticus* **A**gar (**AFPA**) (Pitt et al., 1983) can be used. Another medium containing the antibiotic bleomycin helps to distinguish *A. parasiticus* from *A. sojae* (Klich and Mullaney 1989). On this medium growth of both species is somewhat reduced, but *A. sojae* produces very restricted colonies while *A. parasiticus* colonies are at least 3 mm in diameter within six days. These selective media are especially appropriate for the rapid detection of mycotoxinogenic isolates in food commodities.

Profiles of secondary metabolites, including mycotoxins, are of great use in *Aspergillus* systematics. Klich and Pitt (1988b) examined the relationship between current species concepts and secondary metabolite production in sect. *Flavi* and found a good correlation between the production of aflatoxins B_1 + B_2, G_1 +G_2 and cyclopiazonic acid and the morphological characteristics of the species (see also under *A. flavus* and *A. oryzae*).

Table 3. Distinguishing characters of *Aspergillus flavus* and related spp. (from Samson & Frisvad, 1991).

Species	colony colour	conidia	AFPA	aflat B	aflat G	cyclop-ac
A. flavus	green	sm/fr	or	++	-	++
A. parasiticus	dark yellow green	r	or	++	++	-
A. nomius	green	r	or	++	++	-
A. oryzae	brown	sm/fr	cr	-	-	(+)
A. sojae	pale brown	r	or	-	-	-
A. tamarii	dark brown	r	br	-	-	++

Abbreviations: Conidia: sm/fr = smooth-walled to finely roughened; r = definitely roughened. AFPA = *Aspergillus flavus* and *parasiticus* agar (Pitt et al., 1983), reverse of the colony, or = orange , cr = cream, br = dark brown; aflat B = Aflatoxin B_1 and B_2, aflat G = Aflatoxin G_1 and G_2; cyclop-acid = production of cyclopiazonic acid.

CULTIVATION FOR IDENTIFICATION

Isolates are inoculated at three points on Czapek and MEA 2% and incubated at 25°C. Cultivation on CREA may be helpful. For xerophilic species such as *A. penicillioides* and *Aspergillus* species with *Eurotium* teleomorphs, Czapek (Yeast autolysate) and MEA with 20-40% sucrose should be used. For the three point-inoculated cultures glass Petri-dishes are preferred to plastic Petri-dishes. Most species sporulate within 7 days. Colour and structure of the conidial head (columnar or radiating) are best observed with a dissecting microscope. Microscopic mounts are made in lactic acid with or without aniline blue and a drop of alcohol is added (70-90%) to remove air bubbles and excess of conidia. As a few Aspergilli are pathogenic to man (e.g. *A. fumigatus*), inhalation of conidia must be avoided.

KEY TO THE SPECIES TREATED

1a. Colonies white, black or in yellow, brown or grey colours 2
1b. Colonies in some shade of green 8

2a. Conidial heads white, often wet .... *A. candidus*
2b. Conidial heads yellow, some shade of brown or black 3

3a. Conidial heads dark brown to black .... *A. niger*
3b. Conidial heads not dark brown to black, but olive, yellow-brown or other shades of brown 4

4a. Conidial heads columnar, often cinnamon-brown to pinkish-brown .... *A. terreus*
4b. Conidial heads not columnar, colour yellow or brown 5

5a. Conidial heads olive to light brown; stipe brown. Hülle cells often produced .... *A. ustus*
5b. Conidial heads not olive; stipe hyaline or yellowish. Hülle cells absent 6

6a. Conidial heads pure yellow, conidia smooth to finely roughened....................... *A. ochraceus*
6b. Conidial heads yellow-brown, conidia ornamented 7

7a. Conidia conspicuously ornamented with warts and tubercles, outer and inner wall can be distinguished .... *A. tamarii*
7b. Conidia mostly roughened, outer and inner wall can not be distinguished .... *A. wentii*

8a. Conidiophores typically brown, Hülle cells and *Emericella* teleomorph mostly present (see also p. 34) .... *A. nidulans*
8b. Conidiophores not typically brown, *Emericella* teleomorph absent 9

9a. Colonies on Czapek or MEA mostly restricted (colony diameter usually less than 1.5 cm within one week) see also *Eurotium* (11) 10
9b. Colonies growing faster with a diameter usually larger than 1.5 cm 11

10a. Colonies variably coloured, conidial heads biseriate, sometimes Hülle cells present .... *A. versicolor*
10b. Colonies grey green, conidial heads uniseriate, on MEA or Czapek growing very restricted with poor sporulation, on low water activity media showing better development, Hülle cells not formed .... *A. penicillioides*

11a. Yellow *Eurotium* teleomorph produced in old cultures or on low water activity media, colonies spreading on low water activity media .... *A. glaucus* (compare also *Eurotium*, p. 36)
11b. Yellow *Eurotium* teleomorph absent 12

12a. Conidial heads yellow-green to dark yellow green 13
12b. Conidial heads blue to dark blue green or strikingly blue green ("Delft blue").... 15

13a. Conidial heads predominantly uniseriate, conidia dark yellow green, conspicuously echinulate .... *A. parasiticus*
13b. Conidial heads uni- and biseriate 14

14a. Conidia minutely echinulate, yellow green .... *A. flavus*
14b. Conidia irregularly roughened or smooth, greenish olive .... *A. oryzae*

15a. Conidial heads biseriate, colonies "Delft blue green" .... *A. sydowii*
15b. Conidial heads uniseriate 16

16a. Conidial heads columnar, vesicles broadly clavate, conidia rough to echinulate .... *A. fumigatus*
16b. Conidial heads radiate, splitting into several columns with age, vesicles narrowly clavate, smooth-walled .... *A. clavatus*

Aspergillus candidus Link

Colonies on Czapek agar at 25°C attaining a diameter of 1.0-1.5 cm within 7 days, usually thin with some aerial mycelium, intermixed with conidiophores arising from the agar or from the aerial mycelium. Conidial heads white, later cream, in fresh isolates, often wet. Conidiophore stipe hyaline to slightly yellowish, smooth-walled. Small reduced conidial heads often present. Vesicles globose to subglobose, 10-50 µm diam. Phialides sometimes borne directly on the vesicle, but mostly on metulae, 5-8 x 2.5-3.5 µm. Metulae 5-8 x 2-3 µm. Conidia globose to subglobose, 2.5-4.0 µm in diam, hyaline, thin- and smooth-walled. Sclerotia sometimes present, reddish purple to black.
Colonies on MEA growing faster and sporulating more densely. Poor growth on CREA.

Important (toxic) metabolites: terphenyllin, xanthoascin.

HABITAT: food
This species is often found as a storage fungus e.g. on cereals. Fresh isolates often have wet heads which is unusual in Aspergillus. However, with subculturing this character will disappear.

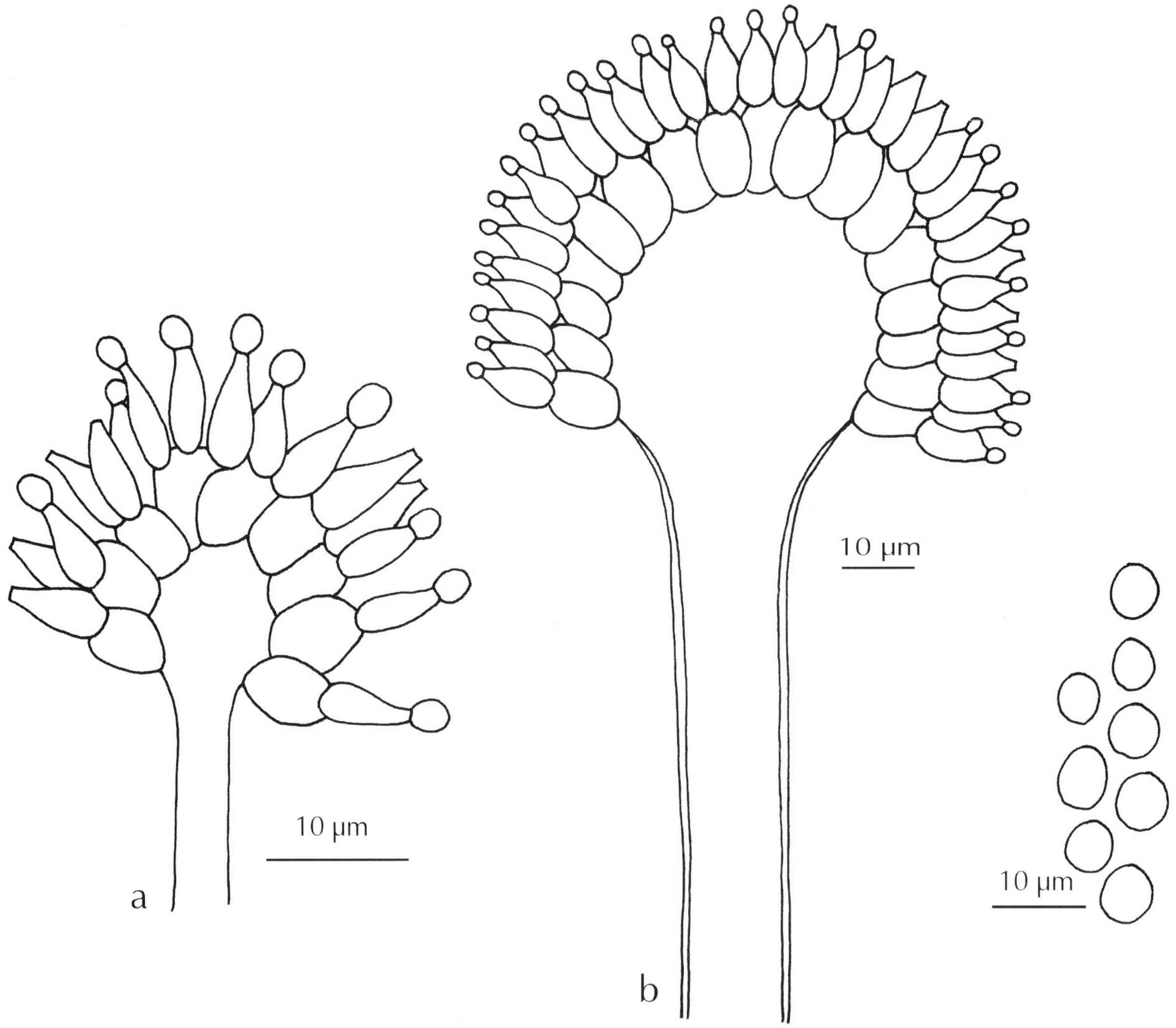

Fig. 30. *Aspergillus candidus*. Conidiophores and conidia. a. reduced head, b. well developed head.
Pl. 27. *Aspergillus candidus*. Colonies after one week, a. on Cz, b. on MEA; c. conidial head, x 1000; d. conidia, x 1250; e. atypical reduced conidial head, x 1000.

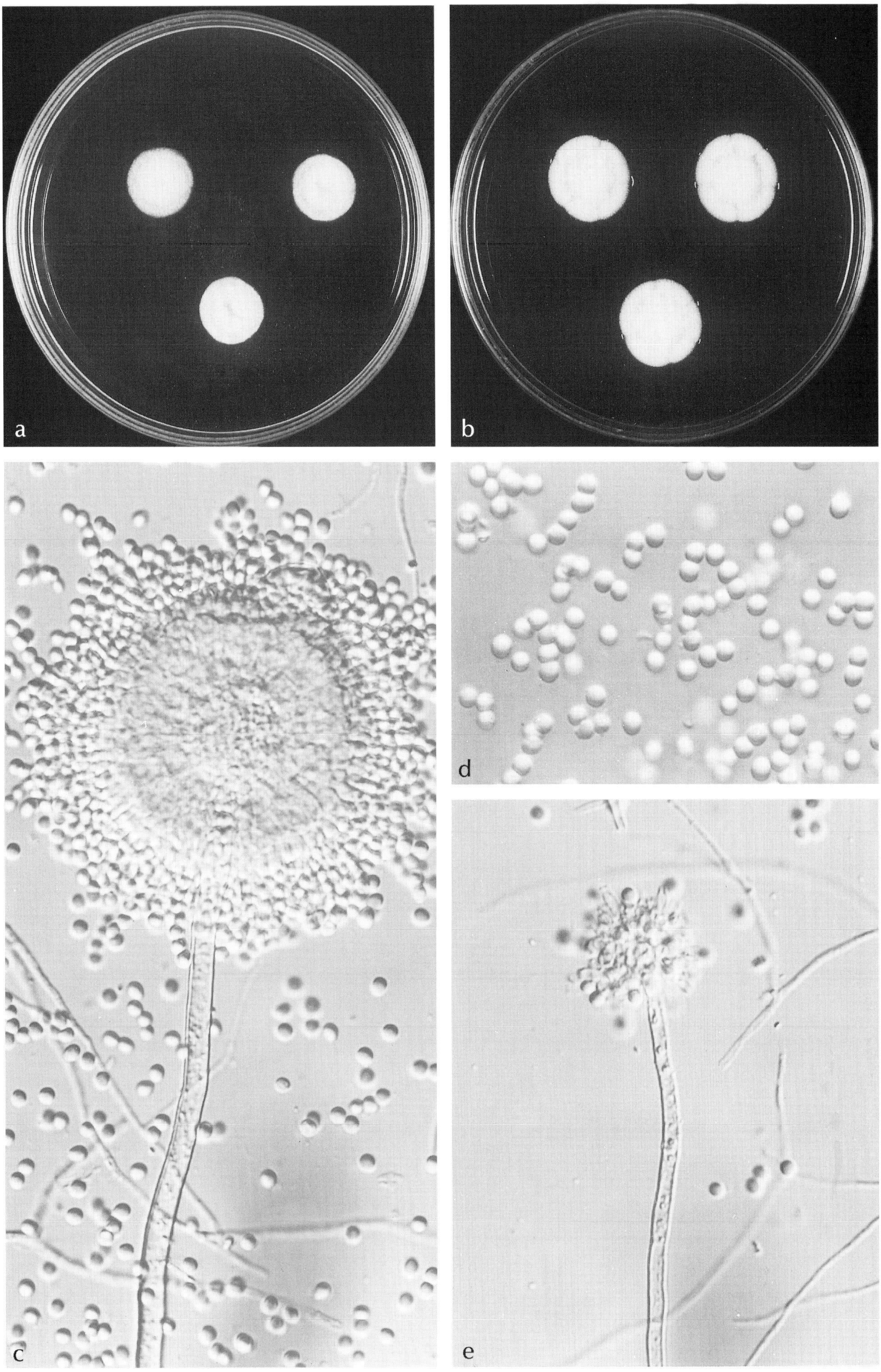

Aspergillus clavatus Desm.

Colonies on Czapek agar at 25°C attaining a diameter of 2.5-3.5 cm within 7 days, mostly consisting of a dense felt of blue-green, usually long conidiophores. Conidial heads clavate usually splitting into several divergent columns. Conidiophores 1.5-3.0 mm in length, hyaline, smooth-walled. Vesicles typically clavate, mostly 40-60 µm wide. Phialides borne directly on the vesicle, varying in size, (2.5)7-(3.5)8 x 2-3 µm. Conidia ellipsoid, 3.0-4.5 x 2.5-4.5 µm, greenish, smooth-walled.
Colonies on MEA show similar growth, but conidiophores often less abundant.
Good growth on CREA.

Important (toxic) metabolites: **patulin**, ascladiol, **cytochalasin E**, tryptoquivalins.

HABITAT : food,(indoor)
Especially common in malt(ing) houses, breweries and compost.

NOTE:
The length of the conidiophore is an important criterion when distinguishing *A. clavatus* from similar species. However, light (daylight or near UV) can influence this character.

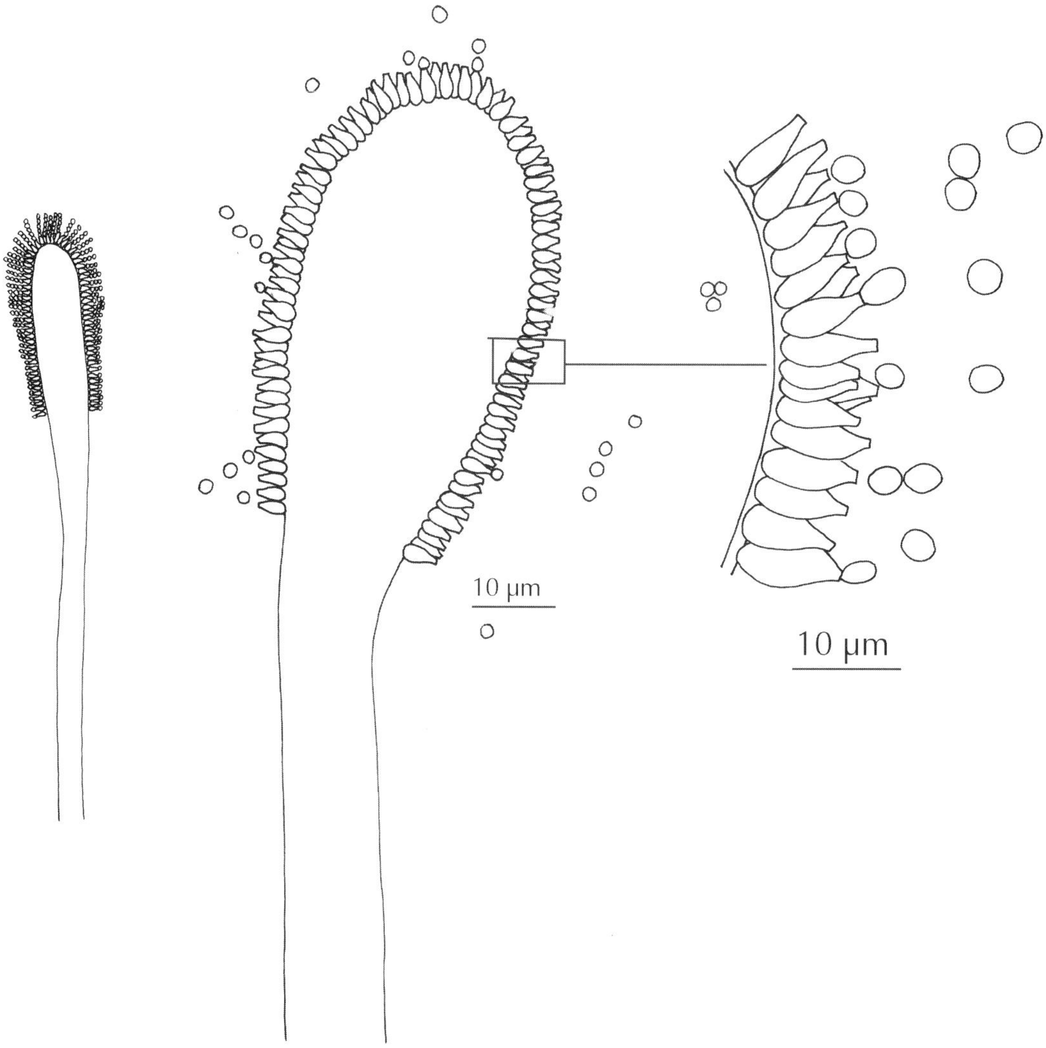

Fig. 31. *Aspergillus clavatus.* Conidiophores and conidia.
Pl. 28. *Aspergillus clavatus.* Colonies after one week, a. on Cz, b. on MEA; c,d, conidiophores with conidial heads and detail, c. x 300, d. x 700; e. conidia, x 1250.

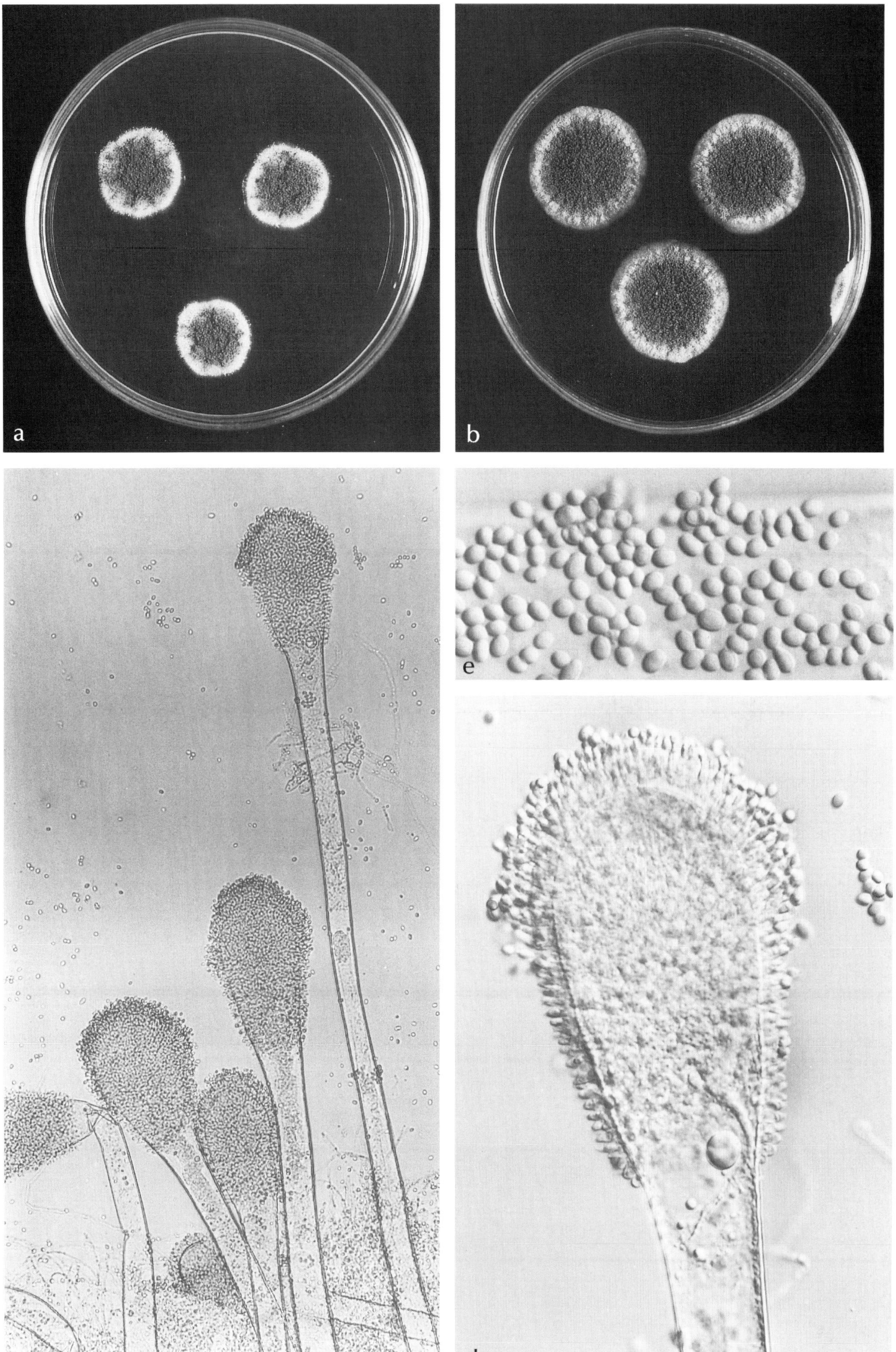
a
b
c
e
d

Aspergillus flavus Link

Colonies on Czapek agar at 25°C attaining a diameter of 3-5 cm within 7 days, usually consisting of a dense felt of yellow-green conidiophores. Conidial heads typically radiate, later splitting into several loose columns, yellow-green becoming dark yellow-green. Conidiophores hyaline, coarsely roughened, up to 1.0 mm (some isolates up to 2.5 mm) in length. Vesicles globose to subglobose, 25-45 µm in diam. Phialides borne directly on the vesicle or on the metulae, 6-10 x 4.0-5.5 µm. Metulae 6.5-10 x 3-5 µm. Conidia globose to subglobose, 3.6 µm in diam, pale green, echinulate. Sclerotia often produced in fresh isolates, variable in shape and dimension, often brown to black.

Colonies on MEA growing faster, otherwise similar. Poor growth on CREA, orange reverse on AFPA.

Important (toxic) metabolites: kojic acid, **3-nitropropionic acid, cyclopiazonic acid, aflatoxin** $\mathbf{B_1}$, aspergillic acid.

HABITAT: food
Common in (ground)nuts, spices, oil seeds, cereals, occasionally in dried fruits (e.g. figs).

NOTE:
The distinction between *A. flavus* and *A. parasiticus*, *A. oryzae* and other similar species is often difficult to observe because many intergrading strains occur. See also page 65 (Table 3) and note under *A. oryzae*.

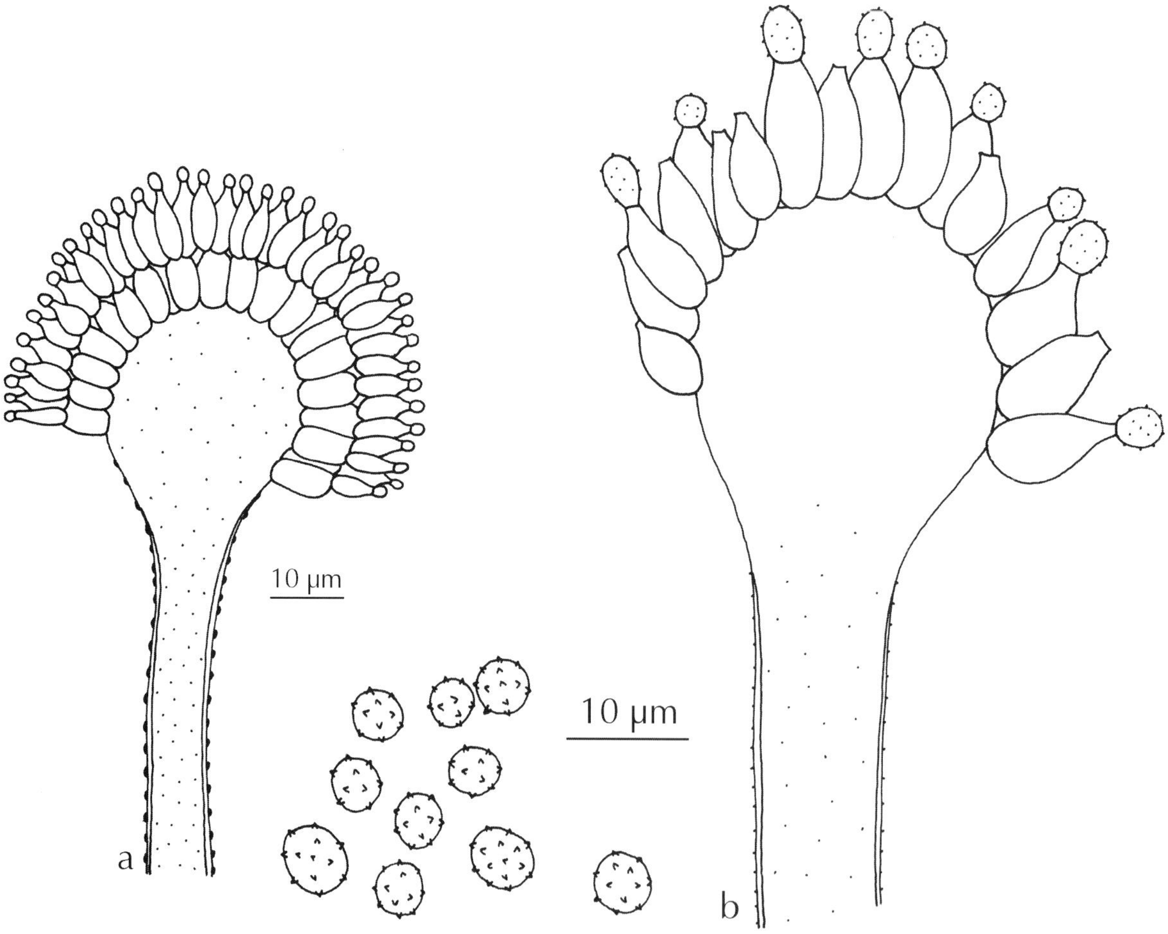

Fig. 32. *Aspergillus flavus.* Conidiophores and conidia. a. mature biseriate head, b. young uniseriate head.
Pl. 29. *Aspergillus flavus.* Colonies after one week, a. on Cz, b. on MEA; c, d. conidial head and tip of conidiophore, x 1000; e. conidia, x 1000.

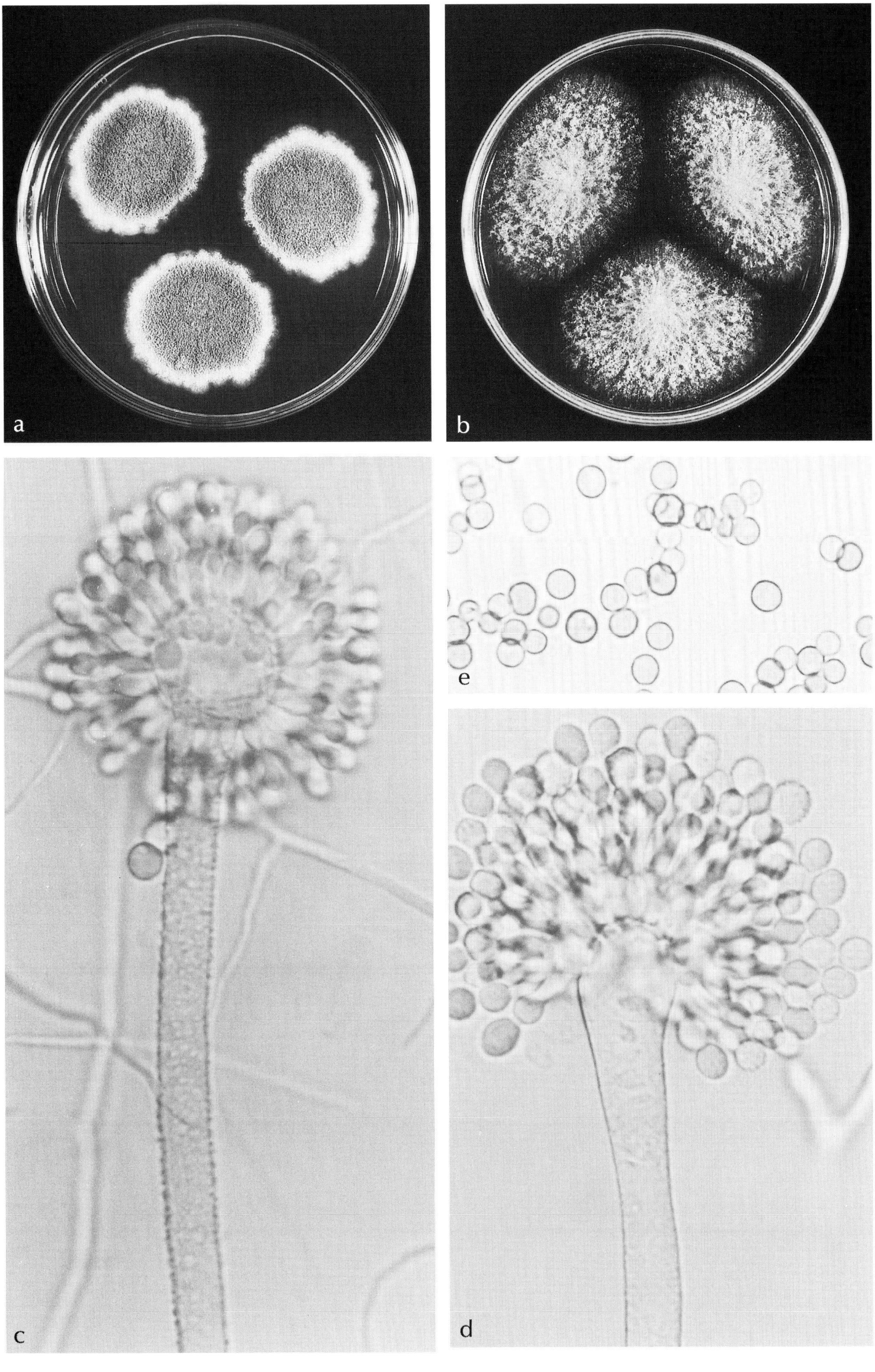
a
b
c
e
d

Aspergillus fumigatus Fres.

Colonies on Czapek agar at 25°C attaining a diameter of 3-5 cm within 7 days, consisting of a dense felt of dark green conidiophores intermixed with aerial hyphae bearing conidiophores. Conidial heads typically columnar. Conidiophores short, smooth-walled, green, particularly in the upper part. Vesicles broadly clavate, 20-30 µm in diam. Phialides directly borne on the vesicle, often greenish pigmented, 6-8 x 2-3 µm. Conidia globose to subglobose, 2.5-3.0 µ in diam, green, rough-walled to echinulate.
Colonies on MEA growing faster and sporulating heavier. Poor growth on CREA.

Important (toxic) metabolites: **gliotoxin, verrucologen, fumitremorgin A & B**, fumitoxins, tryptoquivalins.

HABITAT : food, indoor
Common in heated cereals, garbage, house dust, compost, humidifier systems, HVAC etc.

NOTE:
This species grows at high temperatures (up to 55°C) and low oxygen tensions and is a common contaminant. Because of its pathogenicity heavy sporulating cultures should be handled with care. Raper and Fennell (1965) placed isolates with smooth ellipsoidal conidia in a separate variety *ellipticus*. Isolates obtained from clinical samples (e.g. sputum, lung tissue) often show poor sporulation and atypical degenerated conidiophores. For the detection of *A. fumigatus* cultures can be best incubated at 35-40°C. Colonies usually will develop faster than when incubated at 25°C.

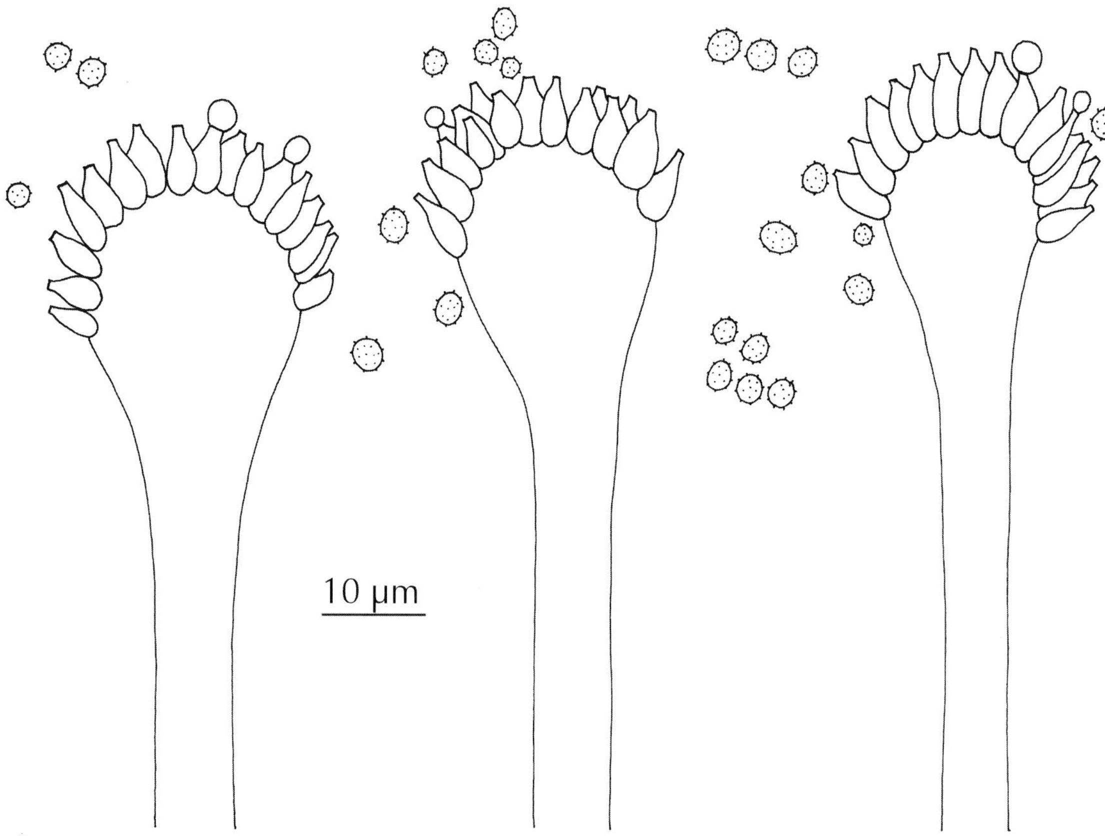

Fig. 33. *Aspergillus fumigatus*. Conidiophores and conidia.
Pl. 30. *Aspergillus fumigatus*. Colonies after one week, a. on Cz, b. on MEA; c,d. conidial head and tip of conidiophore, x 900; e. conidia, x 1500.

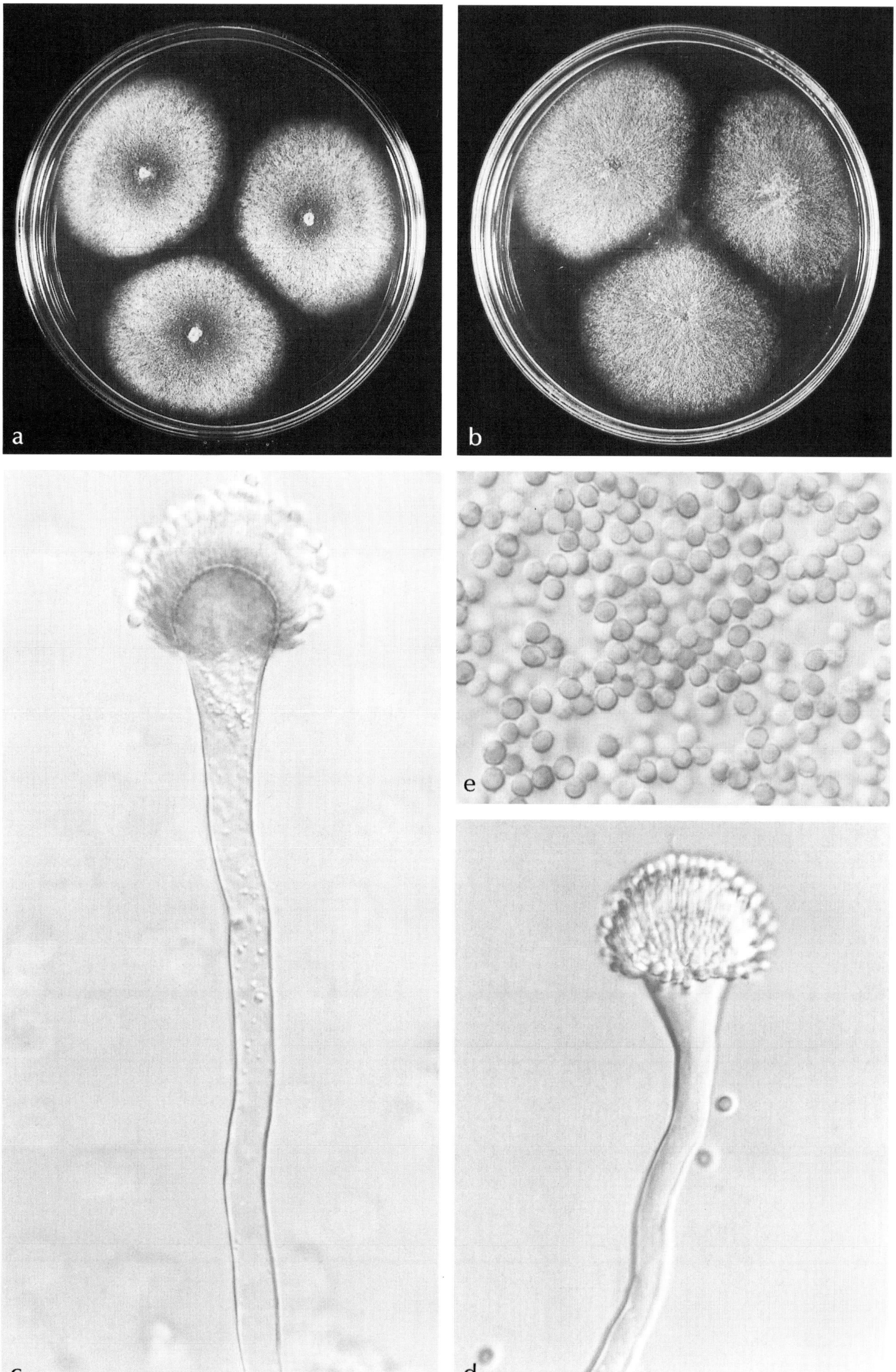
a
b
c
e
d

Aspergillus niger van Tieghem

Colonies on Czapek agar at 25°C attaining a diameter of 4-5 cm within 7 days, consisting of a compact white or yellow basal felt with a dense layer of dark brown to black conidiophores. Conidial heads, black, radiate, tending to split into loose columns with age. Conidiophore stipes smooth-walled, hyaline but also in brown colours. Vesicles globose to subglobose, 50-100 µm in diam. Phialides borne on metulae, 7.0-9.5 x 3-4 µm. Metulae hyaline to brown, often septate, 15-25 x 4.5-6.0 µm. Conidia globose to subglobose, 3.5-5 µm in diam, brown, ornamented with irregular warts, spines and ridges.
Colonies on MEA thinner but sporulating densely. Poor growth on CREA.

Important (toxic) metabolites: naphtho-γ-pyrones, malformins, ochratoxin A (few isolates).

HABITAT : food, indoor

This species is a common contaminant on various substrates (sun dried foods and spices).

NOTE:
Distinction from species such as *A. phoenicis* and *A. awamori* is difficult and was usually based on conidial characters. Several studies have shown that the biseriate black Aspergilli belong to the *A. niger* complex. Using molecular techniques two taxa can be differentiated within this complex: *A. niger* and *A. tubigensis*. However, both species can not be separated using characters of colony growth and morphology (see Samson, 1992, 1994 a and b). For the separation of taxa within the *A. niger* aggregate see Varga *et al.* (2000) and Parenicova *et al.* (2000).

A. carbonarius is common on grapes and may be the species responsible for ochratoxin A content in wine.

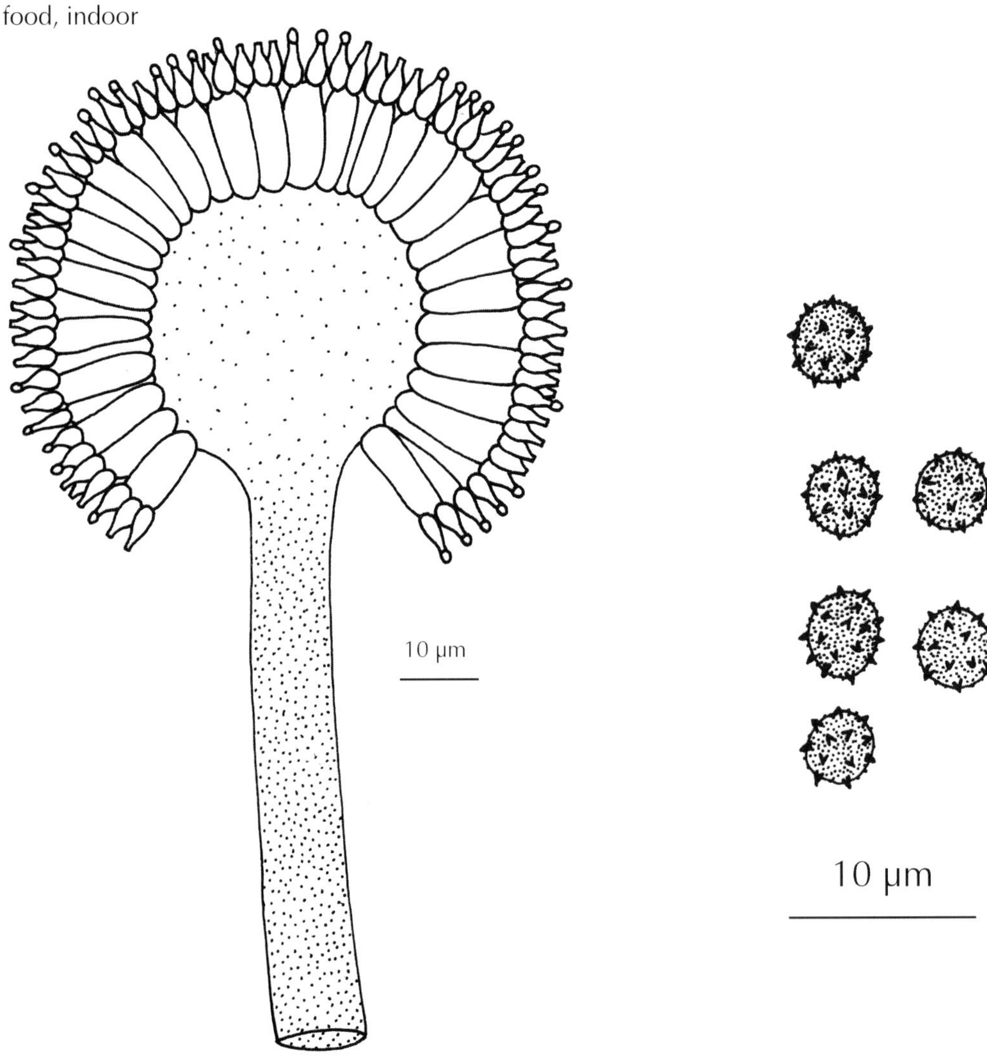

Fig. 34. *Aspergillus niger.* Conidiophores and conidia.
Pl. 31. *Aspergillus niger.* Colonies after one week, a. on Cz, b. on MEA; c,d. conidial heads, c. x 300, d. x 750; e. conidia, x 1200.

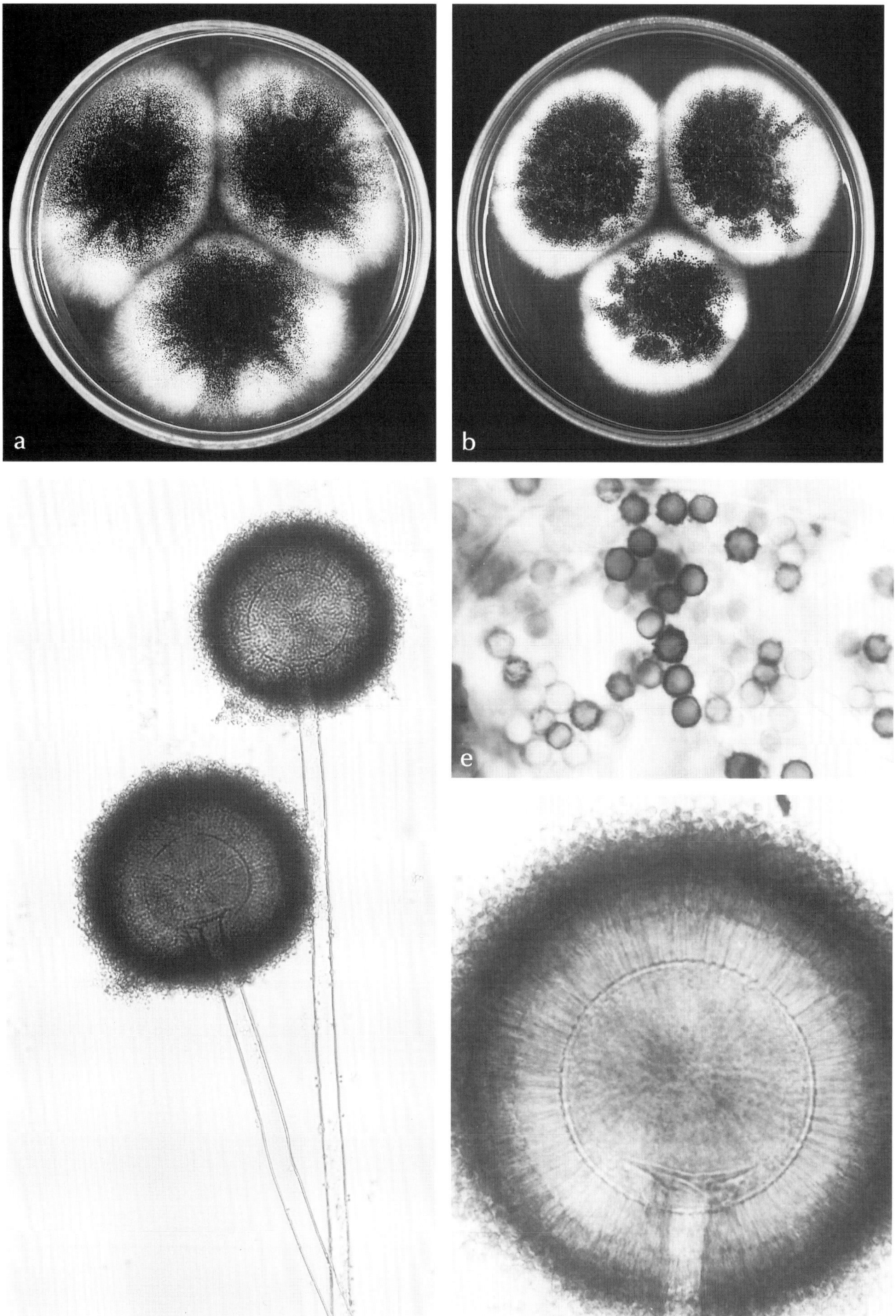
a
b
c
e
d

Aspergillus ochraceus Wilhelm

Colonies on Czapek at 25°C attaining a diameter of 2.5-3.5 cm within 7 days, usually consisting of a dense felt of yellow conidiophores. Conidial heads yellow, globose when young, later splitting into two or more compact columns. Conidiophore stipe up to 1.5 mm in length, yellow to pale brown, rough-walled. Vesicles globose, hyaline, 35-50 µm in diam. Phialides borne on metulae, 7-11 x 2.0-3.5 µm. Metulae 15-20 x 5-6 µm. Conidia globose to subglobose, 2.5-3.0 µm in diam, hyaline, finely rough- or smooth-walled. Sclerotia usually present, first white, later lavender to purple, irregular in shape.
Colonies on MEA growing faster with less crowded conidiophores and reduced sclerotium production.

Poor growth on CREA.

Important (toxic) metabolites: **penicillic acid, ochratoxin A**, xanthomegnin, viomellein, vioxanthin.

HABITAT: food, indoor
A. ochraceus is common on coffee beans, spices.

NOTE:
This species can be confused with other yellow Aspergilli e.g. *A. melleus, A. sclerotiorum, A. ostianus* and *A. alliaceus. A. melleus* Yakawa can be distinguished by the often subglobose to ellipsoidal conidia. The species also differs by having yellow to brown sclerotia, but these are often absent.

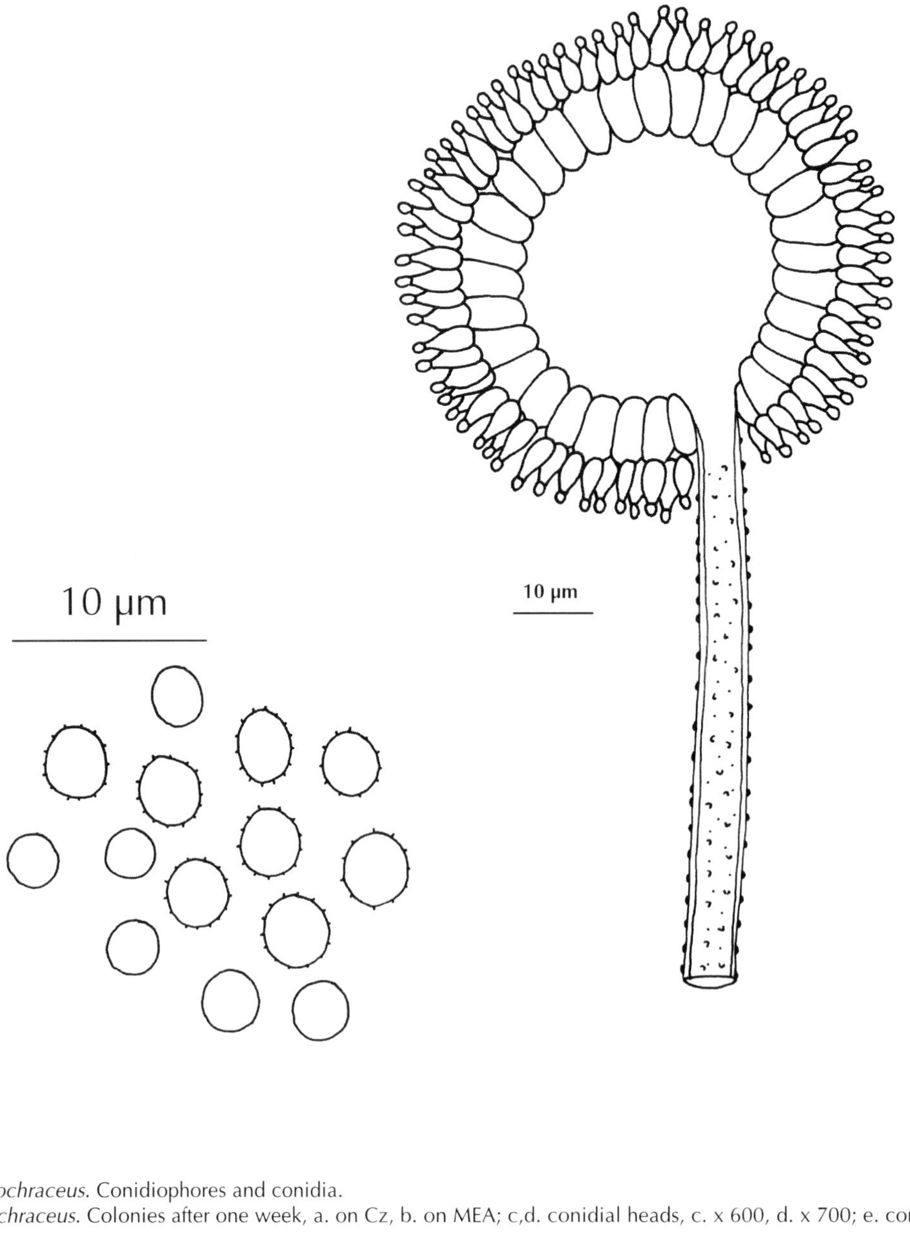

Fig. 35. *Aspergillus ochraceus*. Conidiophores and conidia.
Pl. 32. *Aspergillus ochraceus*. Colonies after one week, a. on Cz, b. on MEA; c,d. conidial heads, c. x 600, d. x 700; e. conidia, x 1000.

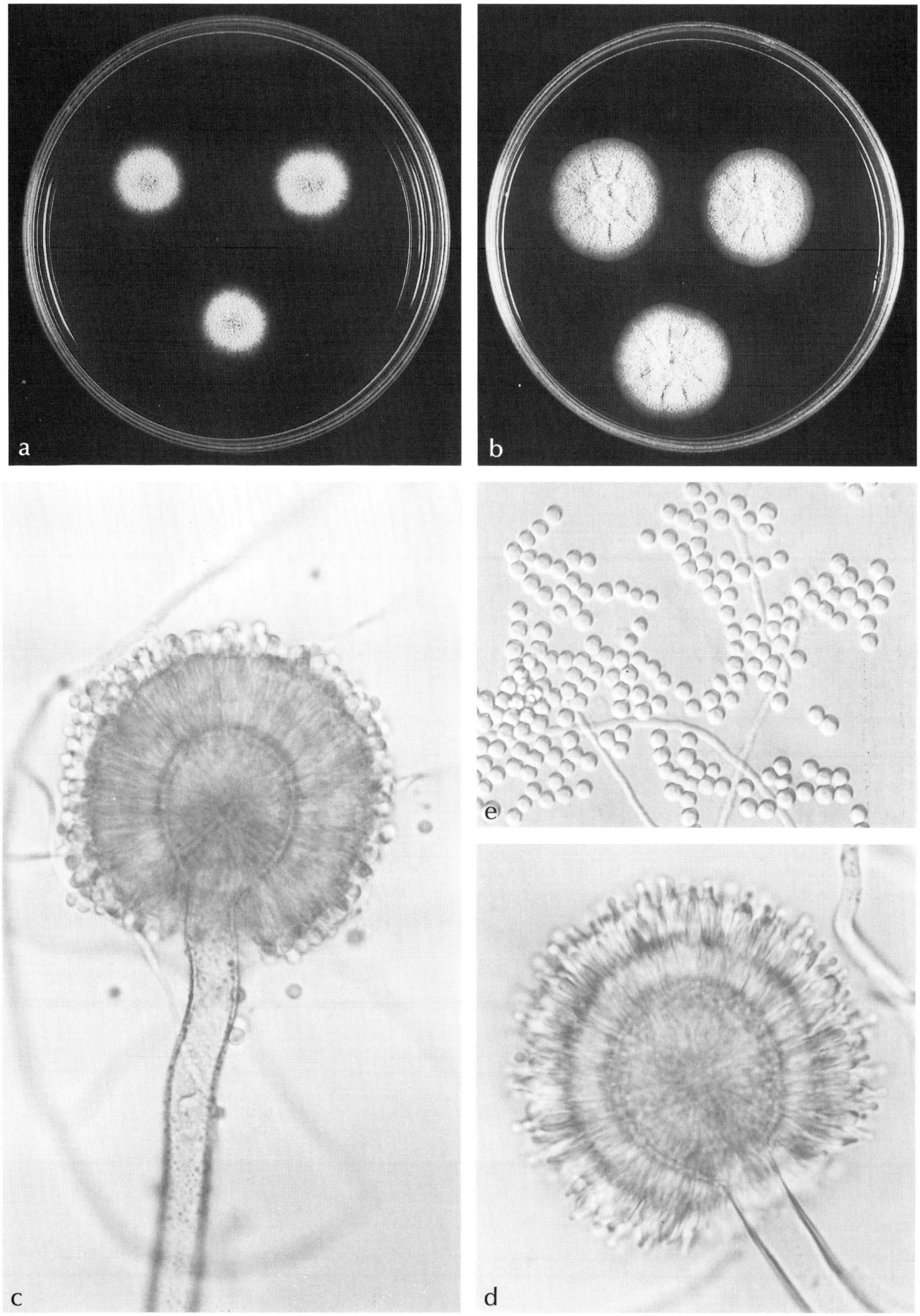
a
b
c
e
d

Aspergillus oryzae (Ahlburg) Cohn

Colonies on Czapek agar at 25°C attaining a diameter of 4-5 cm within 7 days, consisting of a felt of long conidiophores often intermixed with aerial mycelium. Conidial heads radiate, pale greenish yellow, later becoming light to dull brown. Conidiophores hyaline, up to 4-5 mm in length, mostly rough-walled. Vesicles subglobose, 40-80 µm in diam. Phialides often directly borne on the vesicle or on metulae, usually measuring 10-15 x 3-5 µm. Metulae 8-12 x 4-5 µm. Conidia ellipsoidal when young, globose to subglobose when mature, 4.5-8 µm in diam, green, smooth to finely rough-walled.
Colonies on MEA growing faster though somewhat thinner. Poor growth on CREA, cream coloured reverse on AFPA.

Important (toxic) metabolites: kojic acid, **cyclopiazonic acid, 3-nitropropionic acid**.

HABITAT: food
The occurrence of *A. oryzae* is mostly restricted to fermented food products or industrial environments. Therefore it is normally not expected in food samples of in the indoor environment.

NOTE:
The distinction between this species and *A. flavus* is often difficult because of the many intergrading strains. Raper and Fennell (1965) considered colour and roughness of the conidia important characters, but these are often difficult to recognize. Typical *A. oryzae*-isolates differ from *A. flavus* by lighter green colonies and larger, less ornamented conidia, which are usually not equal in size. See also page 65 (Table 3). The occurrence of *A. oryzae* is restricted to food and industrial environments.

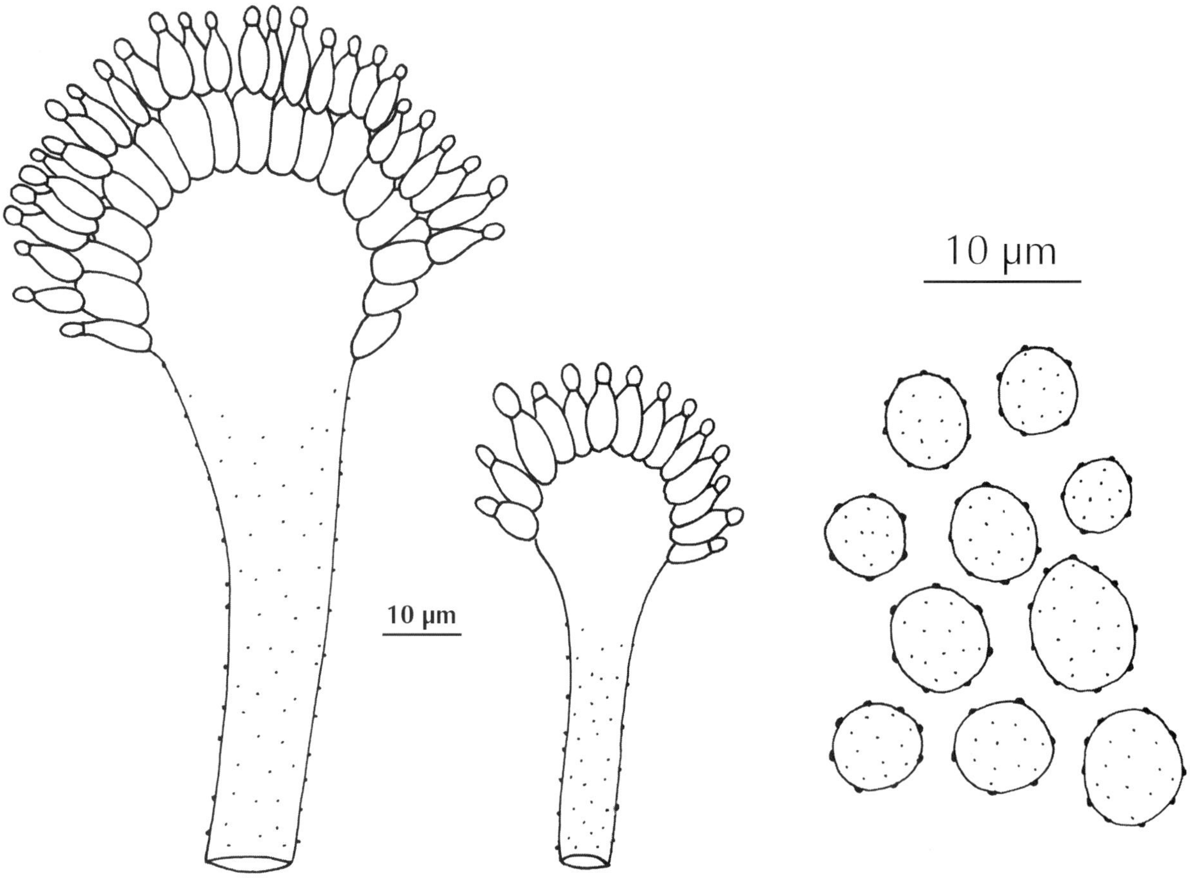

Fig. 36. *Aspergillus oryzae*. Conidiophores and conidia.
Pl. 33. *Aspergillus oryzae*. Colonies after one week, a. on Cz, b. on MEA; c,d. conidial heads, x 500; e. conidia, x 1000.

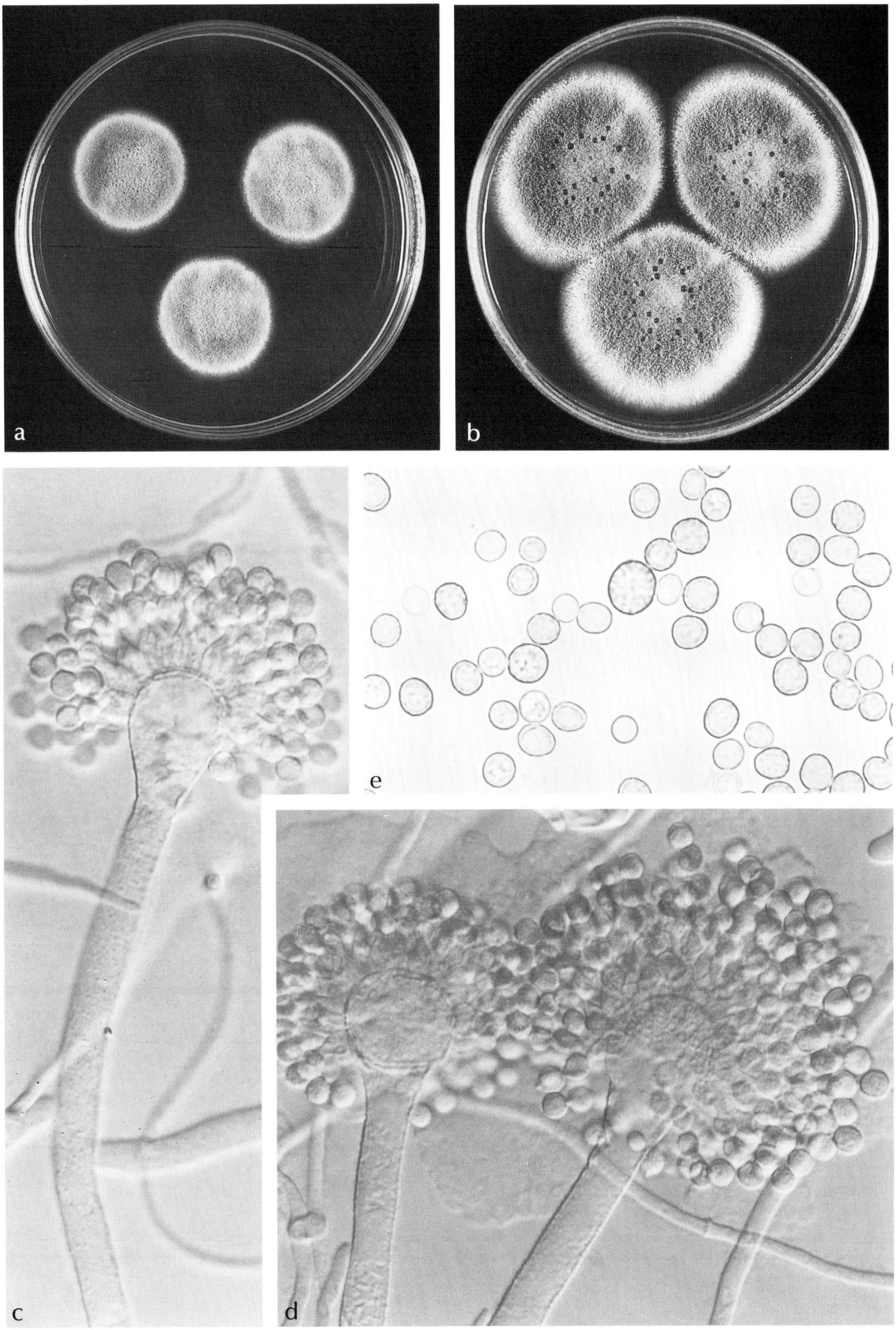
a
b
c
d
e

Aspergillus parasiticus Speare

Colonies on Czapek agar at 25°C attaining a diameter of 2.5-3.5 cm within 7 days, usually consisting of a dense felt of green conidiophores. Conidial heads green, radiate. Conidiophores mostly 300-700 µm long, hyaline, rough-walled. Vesicles subglobose, 20-35 µm in diam. Phialides usually borne directly on the vesicle, 7-9 x 3-4 µm, hyaline to pale green. Conidia globose, 3.5-5.5 µm in diam, yellow-green, conspicuously rough-walled. Colonies on MEA growing faster, otherwise similar. Poor growth on CREA.

Important (toxic) metabolites: kojic acid, aspergillic acid, **aflatoxin B_1, B_2, G_1, G_2**.

HABITAT: food
Occurring in cereals, (ground)nuts, soil.

NOTE:
A. parasiticus is characterized by mainly uniseriate radiate conidial heads and conspicuously rough-walled conidia. For the distinction between the related species *A. flavus, A. sojae* and *A. oryzae* (see also page 65, Table 3).

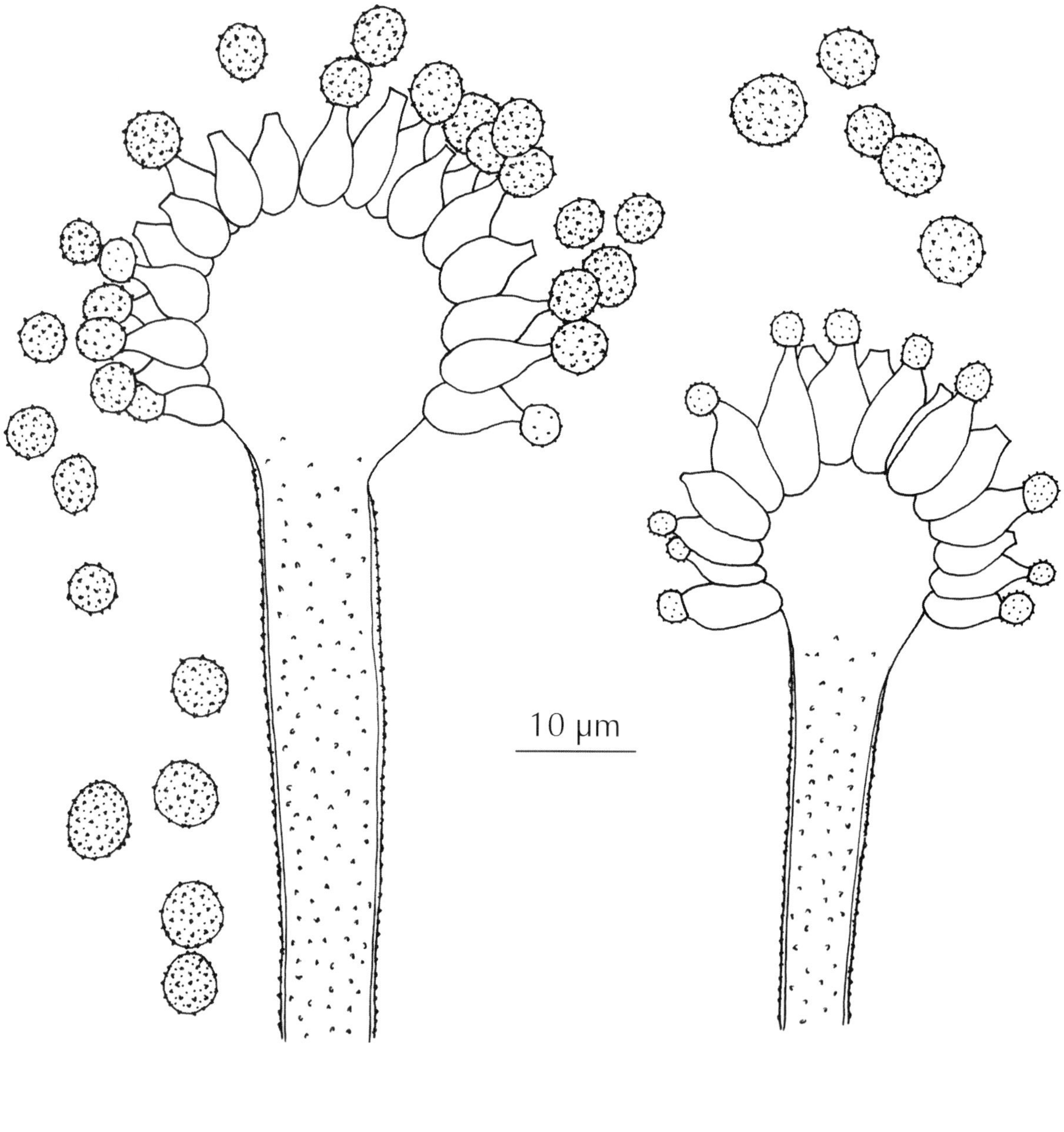

Fig. 37. *Aspergillus parasiticus.* Conidiophores and conidia.
Pl. 34. *Aspergillus parasiticus.* Colonies after one week, a. on Cz, b. on MEA; c,e. conidial heads, x 1400; d. conidia, x 1400.

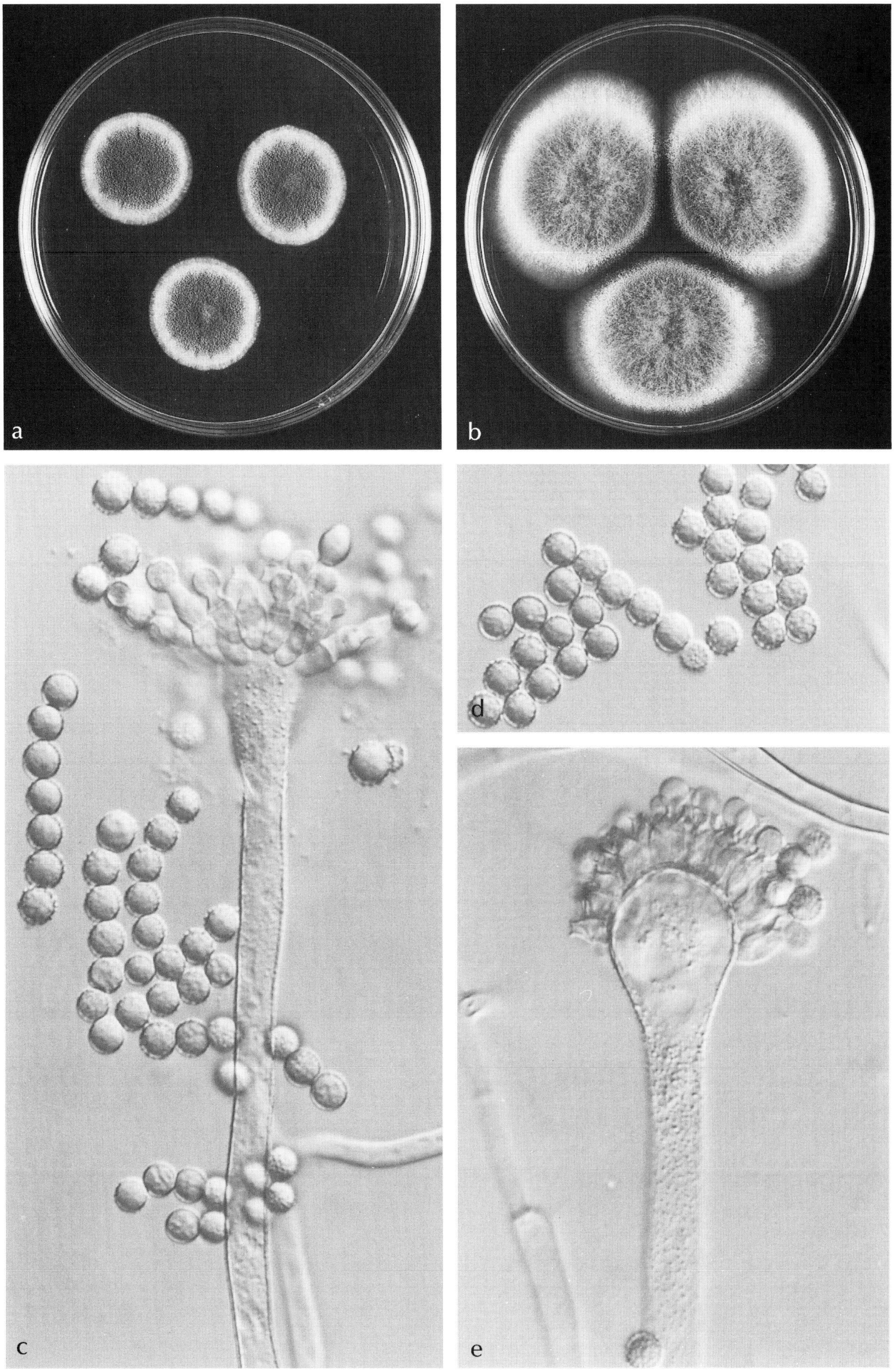
a
b
c
d
e

Aspergillus penicillioides Speg.

Colonies on Czapek agar at 25°C showing no or poor growth. Colonies on Czapek agar with 20-40% sucrose developing slowly, sometimes sporulating after 14 days. On MEA with 20-40% sucrose growing and sporulating well, consisting of a dense felt of dark to dull green conidiophores, arising from the agar or from the scanty aerial mycelium. Conidiophore stalks hyaline, smooth-walled. Conidial heads radiate when young, becoming loosely columnar. Phialides borne directly on the vesicle, 6-9 x 2.5-3.0 µm, covering more than half of the vesicle. Conidia first ellipsoidal often with flattened ends, becoming ellipsoidal to subglobose with age, 3-4(5) µm in diam., conspicuously echinulate.

HABITAT: food, indoor
This species is common on products with a low wateractivity such as spices, dried fruit, (ground)nuts, bakery products. It is also common in musea, archives and other indoor environments e.g. furniture, carpets and in house dust (Samson and Hoekstra, 1994).

NOTE:
A closely related species also often occurring in indoor environments is *Aspergillus restrictus* G. Smith. This species is distinguished by dark green colonies. The phialides are situated only on the upper half of the vesicles which gives the conidial head a typical columnar feature. The conidia are borne cylindrical to barrel-shaped, whereas the conidia of *A. penicillioides* are more eillipsoidal in the beginning. Both species are often overlooked when isolating fungi by means of standard media, as it only appears on old plates or slant cultures. Detection, isolation and cultivation on low water activity media (e.g. DG18) is therefore, strongly recommended.

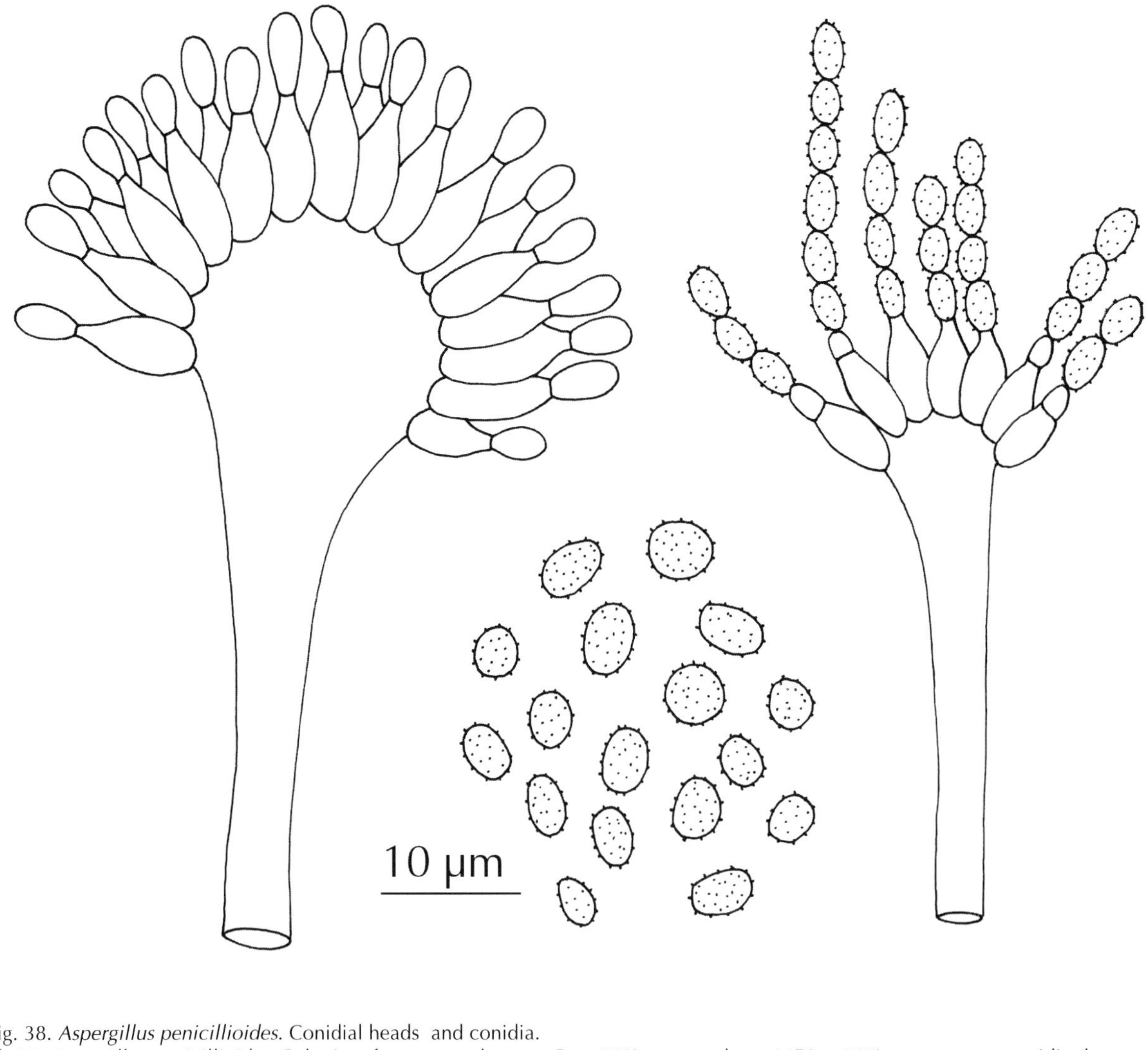

Fig. 38. *Aspergillus penicillioides.* Conidial heads and conidia.
Pl. 35. *Aspergillus penicillioides.* Colonies after one week, a. on Cz + 20% sucrose, b. on MEA + 20% sucrose; c-e. conidiophores and conidia, c. x 900, d. x 1500, e. x 600.

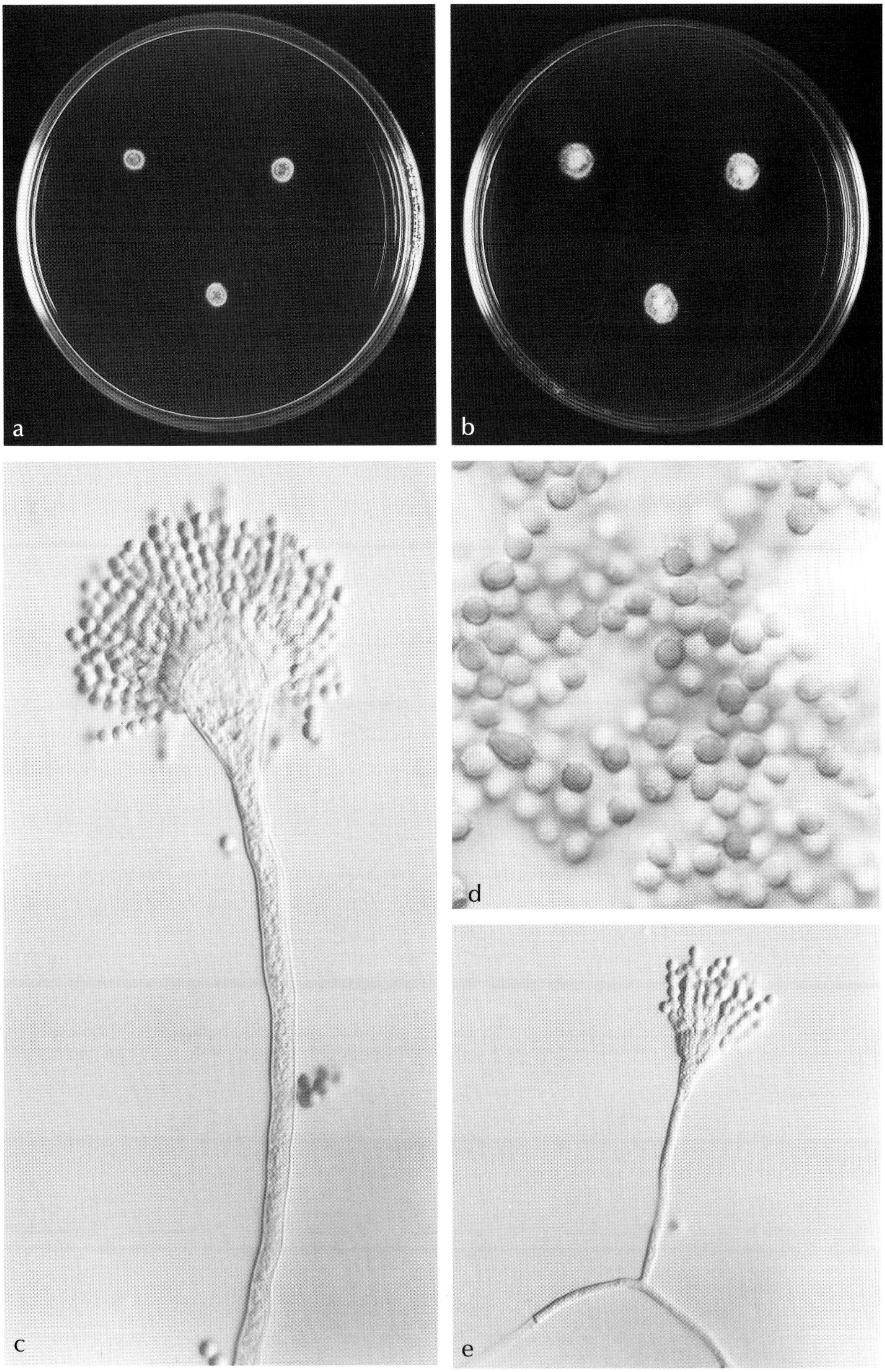
a
b
c
d
e

Aspergillus sydowii (Bain. & Sart.) Thom & Church

Colonies on Czapek at 25°C attaining a diameter of 3-4 cm in 2 weeks, velvety sometimes somewhat floccose, usually consisting of a dense felt of intensely blue green ("Delft blue") conidiophores. Conidial heads blue green, radiate; exudate reddish brown often abundant. Reverse mostly in shades of reddish brown to dark red (maroon). Conidiophore stipe up to 500 µm in length, hyaline sometimes slightly coloured, smooth-walled. Heads biseriate, vesicles globose, spathulate to subclavate, hyaline up to 20 µm in diam. Phialides borne on metulae, 5-7(-10) x 2.0-3.5 µm. Metulae 4-6(-7) x 2-3.5 µm. Conidia globose, (2.5)3-4 µm in diam, hyaline, conspicuously rough-walled. Reduced conidial heads on short stipes present in the aerial mycelium. On MEA colonies reaching 1.6-2.5 cm in diam after one week and 4-5 cm in 2 weeks, the bright blue green colour is less pronounced compared to Czapek. Good growth on CREA.

HABITAT: food, indoor
World-wide distribution, often isolated from soil, cotton, beans, nuts, straw, cotton fabrics, leather, paper, archives; cases of invasive aspergillosis are known.

NOTE:
Aspergillus sydowii is very close to *A. versicolor* and can be recognized by the blue green colonies, the larger and and more rough-walled conidia. Also the reduced heads are commonly occurring in *A. sydowii. A. sydowii* is encountered in the temperate zones more often in indoor environments and less in food, but is commonly occurring in food in S. Asia.

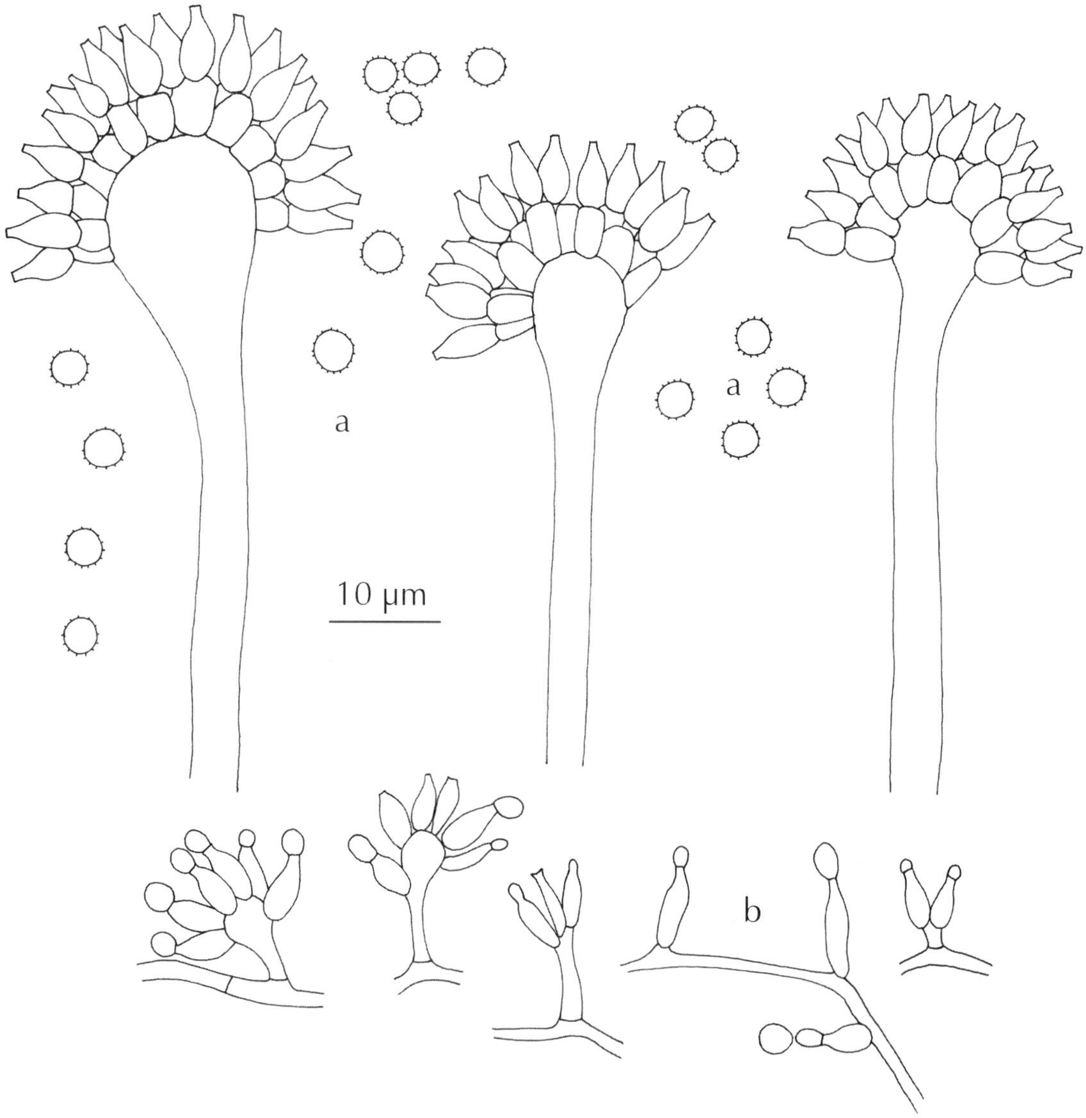

Fig 39. *Aspergillus sydowii*. a. Conidiophores and conidia ; b. reduced conidial heads and phialides.
Pl. 36. *Aspergillus sydowii*. Colonies after one week, a. on CYA, b. on MEA; c,f. conidiophores, c. x 750, f. x 1750; d. reduced conidal head; e. conidia x 1700.

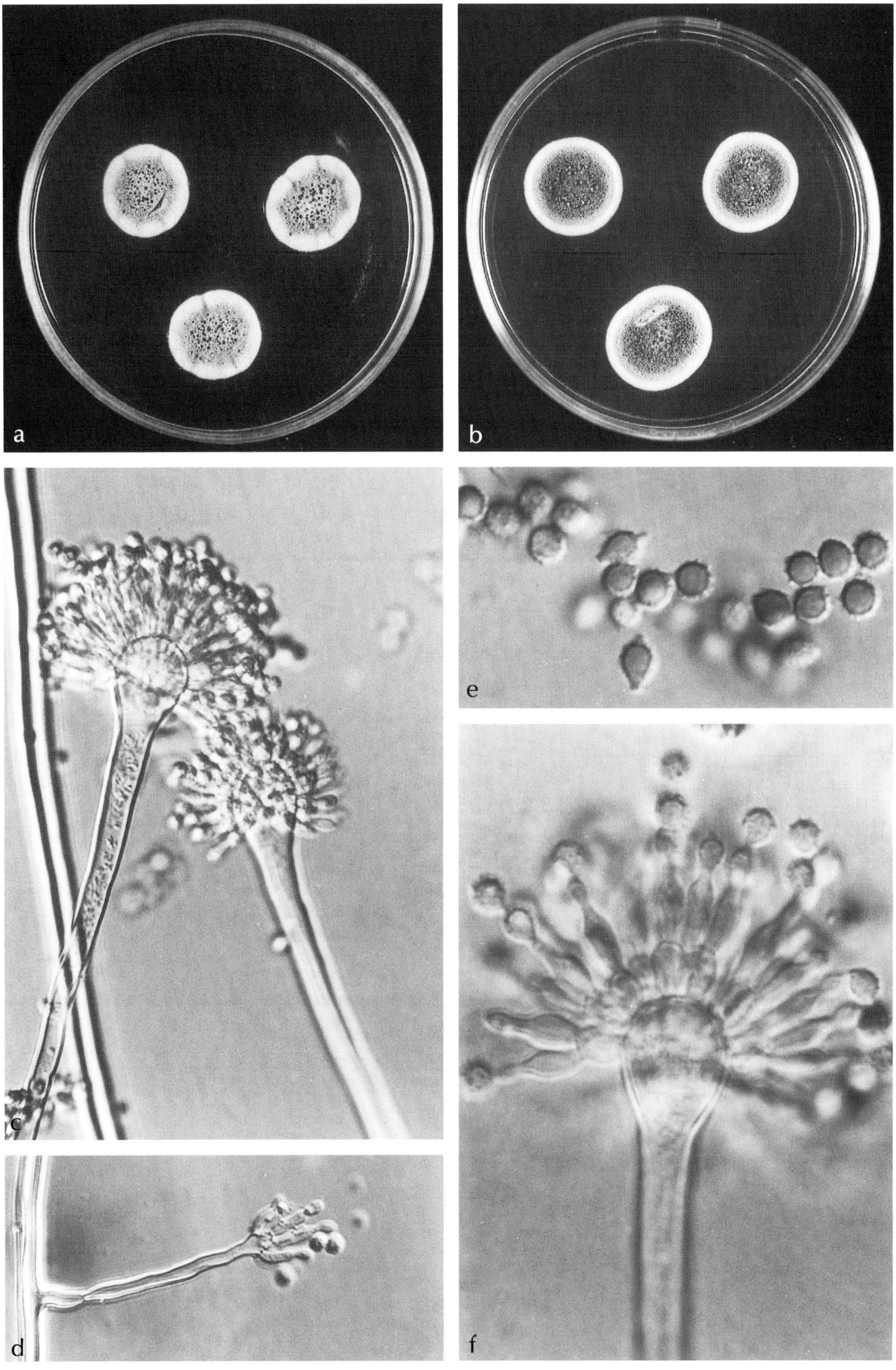
a
b
c
d
e
f

Aspergillus tamarii Kita

Colonies on Czapek agar at 25°C attaining a diameter of 4-5 cm within 7 days, mostly sporulating densely and initially forming a dense felt of yellow-brown conidiophores, which rapidly turn dark green-brown. Conidiophore stalk hyaline, mostly conspicuously rough-walled. Conidial heads radiate, splitting into separate columns. Vesicles globose to subglobose, 25-50 µm in diam. Phialides borne directly on the vesicle or on metulae (mostly on large heads), 10-15 x 4-8 µm. Metulae 7-10 x 4-6 µm. Conidia globose to subglobose, 5-6.5(8) µm in diam, brownish yellow, conspicuously ornamented with tubercles and warts, the outer and inner conidial wall visible.

Poor growth on CREA.

Important (toxic) metabolites: **cyclopiazonic acid**, fumigaclavines.

HABITAT: food
common on tropical commodities, species, corn.

NOTE:
Molecular genetic study showed that *A. tamarii* strains producing aflatoxins were variants of *A. caelatus* and not *A. tamarii* (Peterson, 2000).

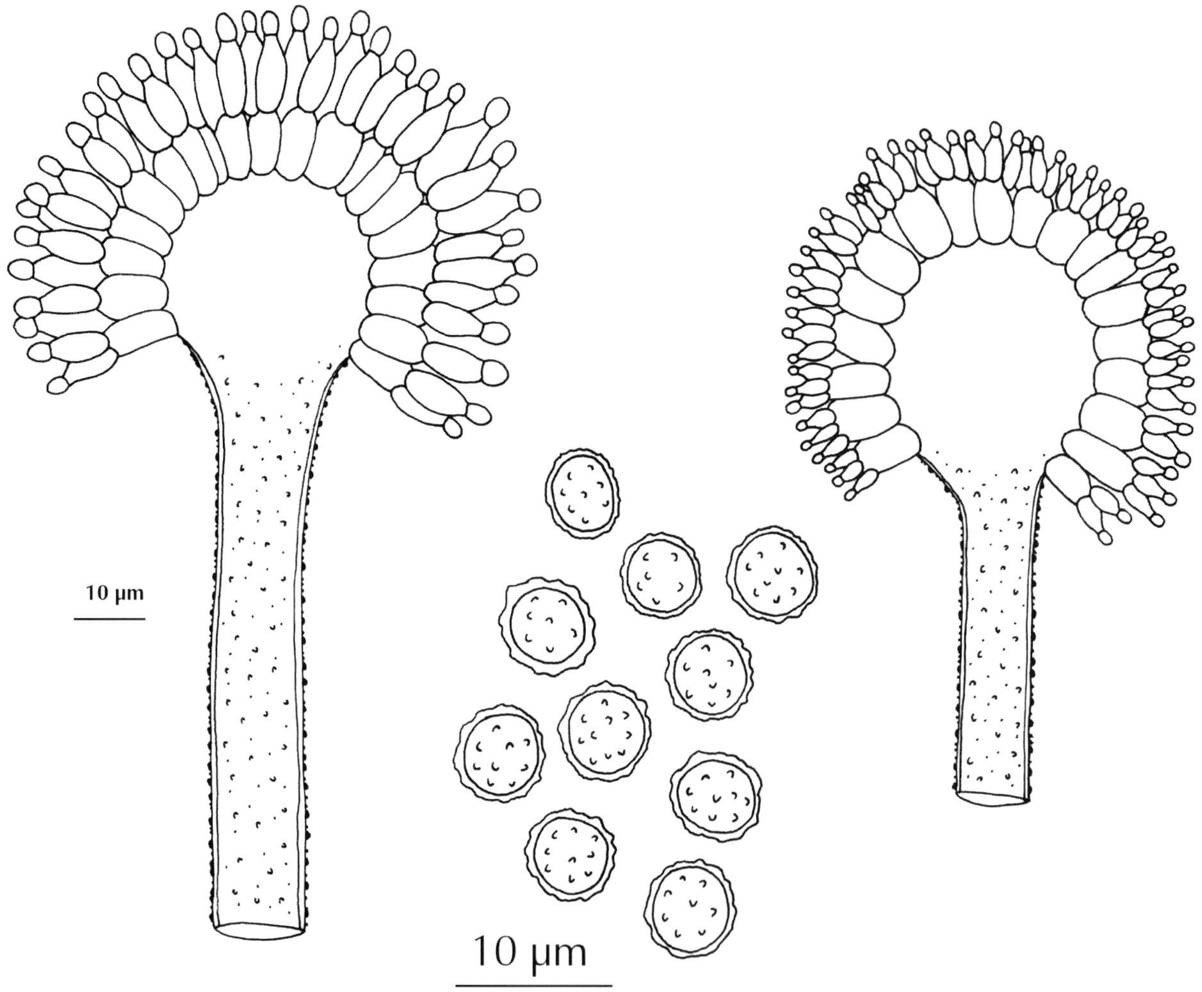

Fig. 40. *Aspergillus tamarii.* Conidiophores and conidia.
Pl. 37. *Aspergillus tamarii.* Colonies after one week, a. on Cz, b. on MEA; c-e. conidial heads and conidia, c,d. x 950, e. x 1250.

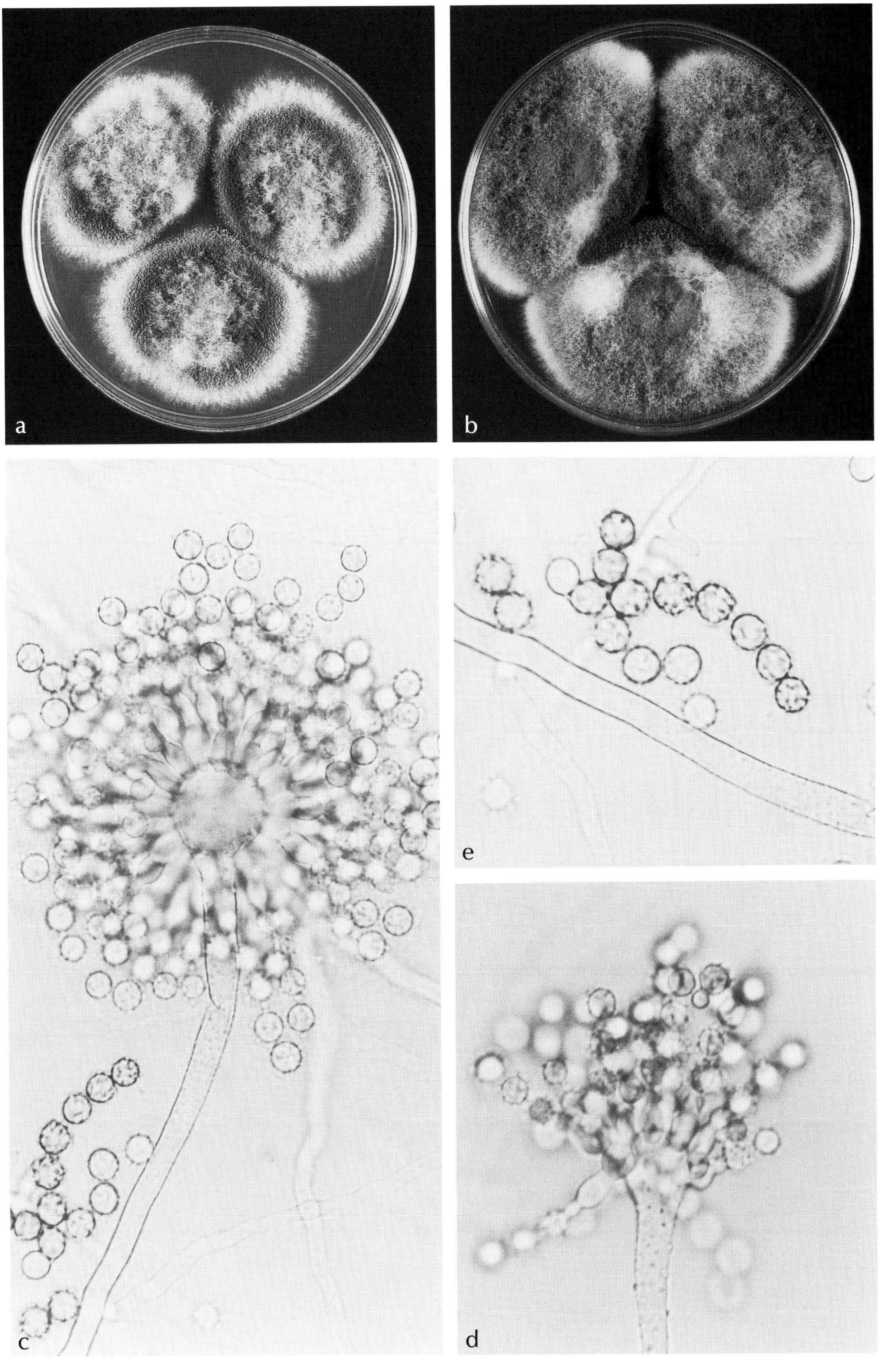
a
b
e
c
d

Aspergillus terreus Thom

Colonies on Czapek agar at 25°C attaining a diameter of 3.5-5.0 cm within 7 days, mostly consisting of a dense felt of yellow-brown conidiophores, becoming darker with age. Conidial heads yellow brown, compact, columnar, mostly 150-500 x 30-50 µm. Conidiophores hyaline, smooth-walled. Vesicles subglobose, 10-20 µm in diam. Phialides borne on metulae, 5-7 x 1.5-2.0 µm. Metulae 5-7 x 2.0-2.5 µm. Conidia globose to ellipsoidal, 1.5-2.5 µm in diam, hyaline to slightly yellow, smooth.
Colonies on MEA similar but growing faster and sporulating more densely. Poor growth on CREA.

Important (toxic) metabolites: terrein, **patulin, citrinin, citreoviridin**.

HABITAT : food, indoor
Common in soil, spices, silage.

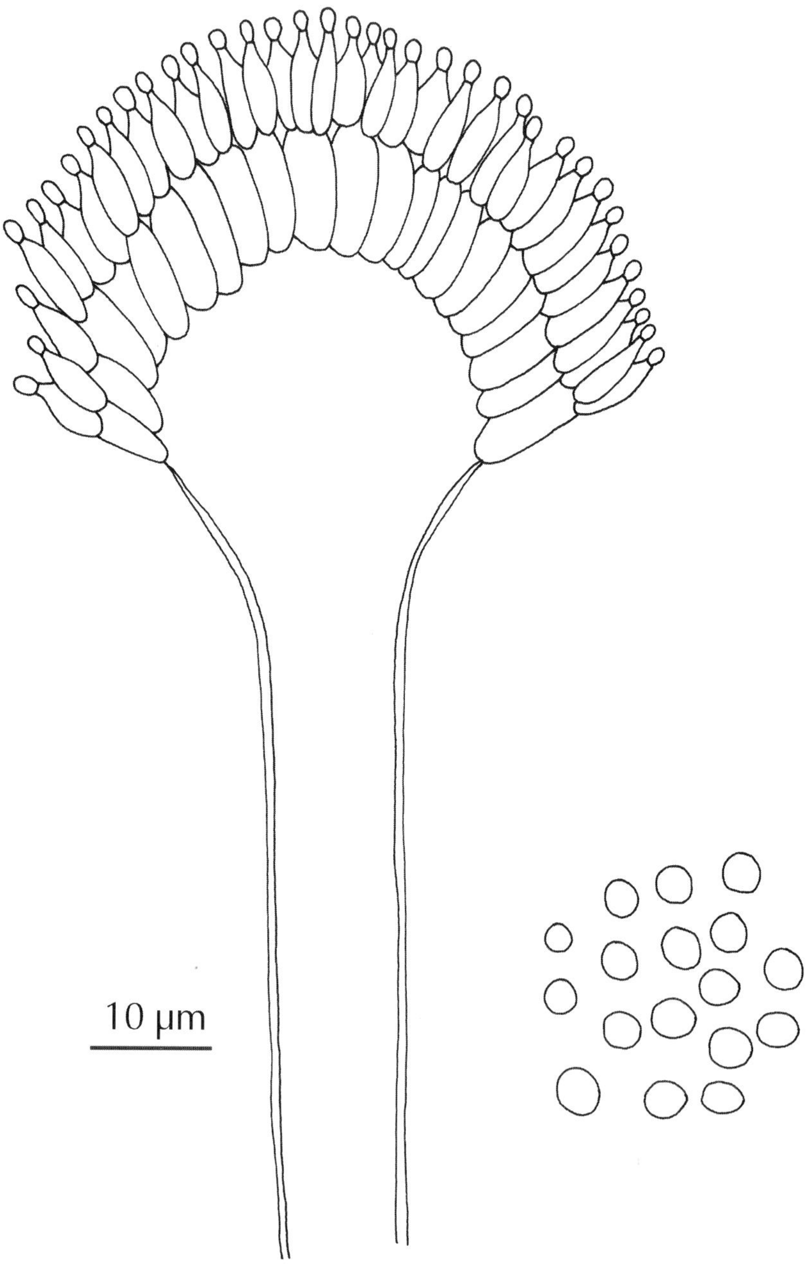

Fig. 41. *Aspergillus terreus*. Conidiophores and conidia.
Pl. 38. *Aspergillus terreus*. Colonies after one week, a. on Cz, b. on MEA; c-e. conidiophores and conidia, c. x 1000, d. x 300, e. x 900.

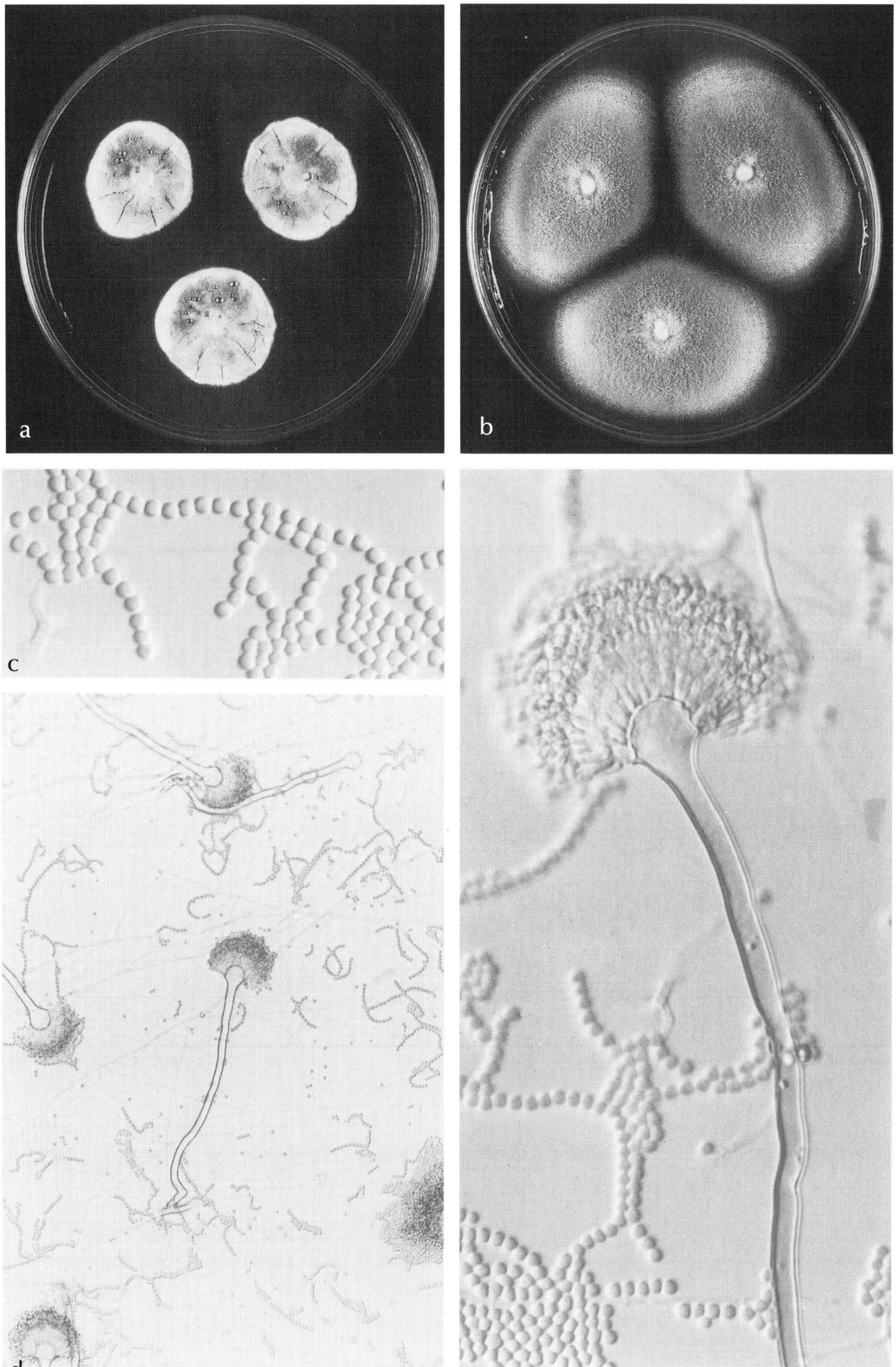
a
b
c
d
e

Aspergillus ustus (Bain.) Thom and Church

Colonies on Czapek agar at 25°C attaining a diameter of 4.5-6.0 cm within 7 days, at first white or cream to pale yellowish, later light to dark drab or olive-grey. Conidial heads radiate, usually splitting into more or less well-defined columns in age, variable in size, 100-125 µm in diam. Conidiophores arising from submerged hyphae, up to 400 µm long by 3.0 to 6.0 µm, often short conidiophores borne on aerial mycelium, stipe brown; vesicles hemispherical to subglobose, 7.0 to 15 µm in diam. Phialides on metulae 4-7 x 3-4.0 µm, metulae 5-7 x 2.5-3.0 µm.

Conidia globose, 3.0-4.5 µm, roughened, echinulate to marked with conspicuous colour bars. Hülle cells often produced, irregular in size, serpentine to helicoid.
Good growth on CREA.

Important (toxic) metabolites: **austamide, austdiol, austins, austocystins.**

HABITAT : food, indoor
Common in soil, cereals and groundnuts.

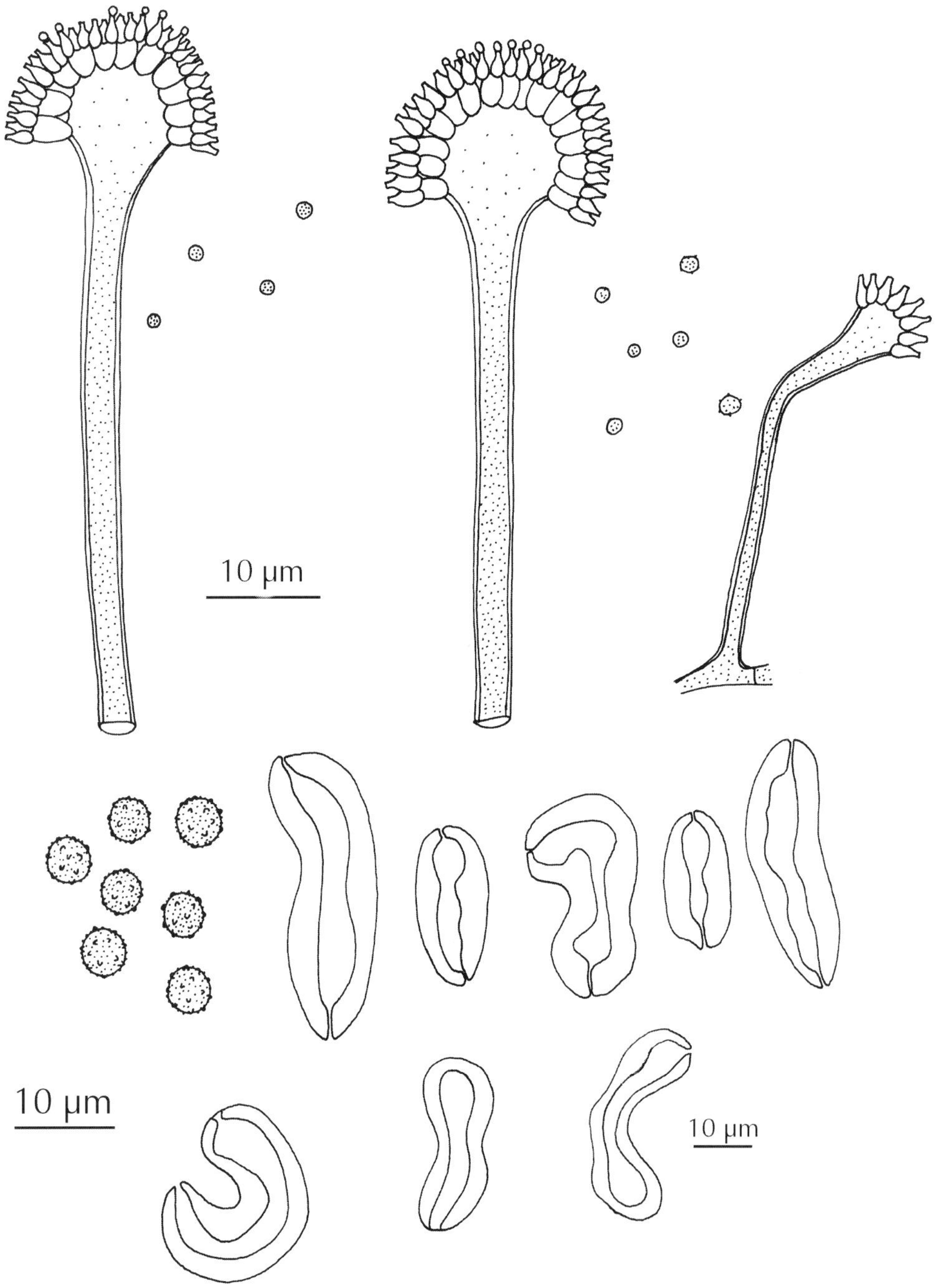

Fig. 42. *Aspergillus ustus*. Conidiophores, conidia and Hülle cells.
Pl. 39. *Aspergillus ustus*. Colonies after one week, a. on Cz, b. on MEA ; c,d. conidiophores, c. x 750, d. x 200; e. conidia, x 1500; f. Hülle cells x 750.

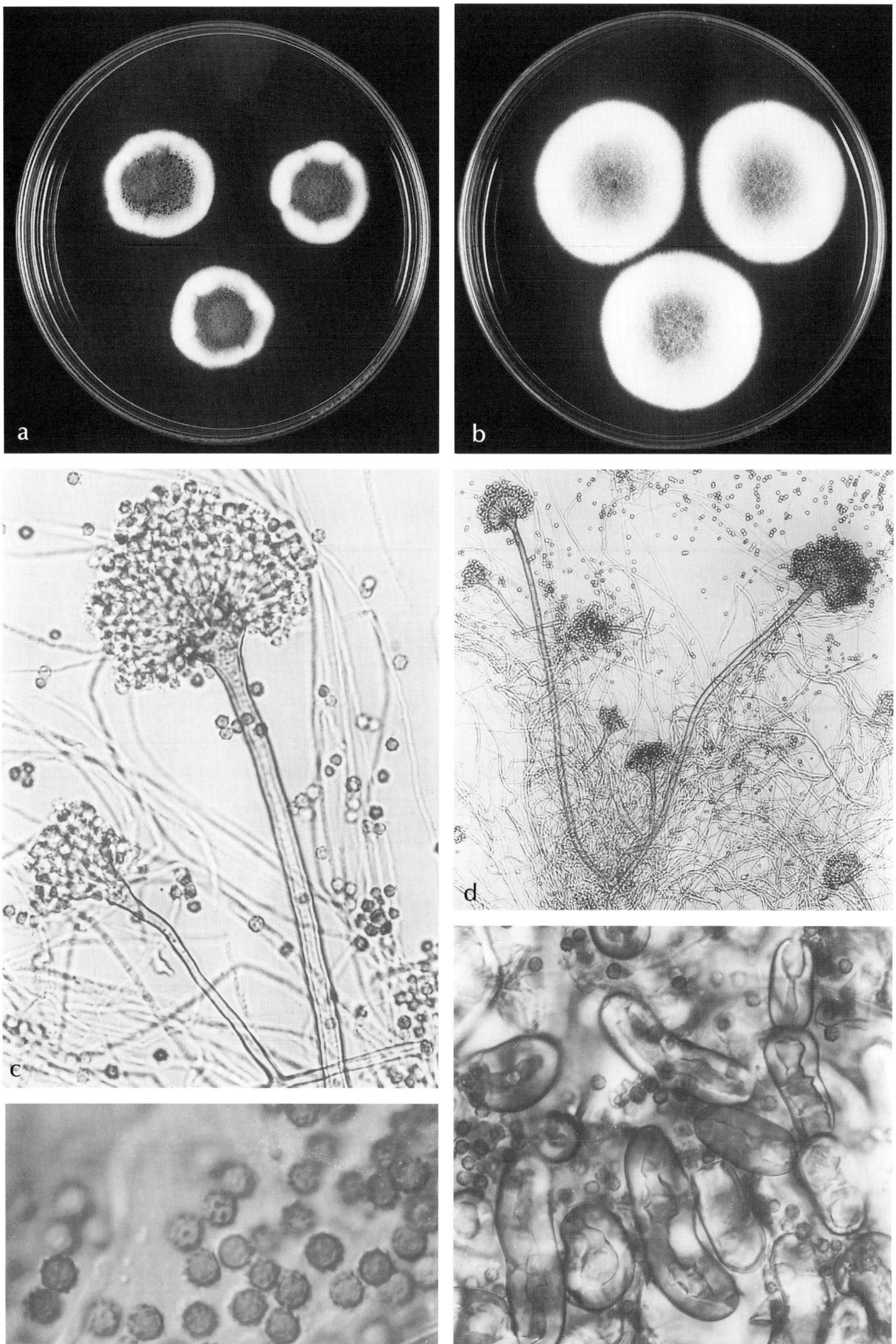
a
b
c
d
e
f

Aspergillus versicolor (Vuill.) Tiraboschi

Colonies on Czapek agar at 25°C attaining a diameter of 1.0-1.5 cm within 7 days, often poorly sporulating after one week but richer after 14 days, either consisting of a dense felt of conidiophores or of aerial and closely interwoven mycelium bearing the conidiophores. Colour at first white, then changing to yellow, orange-yellow to yellow-green, often intermixed with flesh to pink colours. Conidiophores hyaline or slightly pigmented, smooth-walled. Vesicles subglobose to ellipsoidal, 12-16 µm in diam. Phialides borne on metulae, 5.5-8.0 x 2.5-3.0 µm. Metulae 5.0-7.5 x 2.0-2.5 µm.
Conidia globose, 2-3.5 µm in diam, echinulate. Hülle cells sometimes present.

Colonies on MEA growing faster and with less aerial mycelium. Colour mostly in darker green shades.

Good growth on CREA.

Important (toxic) metabolites: **sterigmatocystin**, nidulotoxin.

HABITAT : food, indoor
Common on cheese, cereals, soil, (ground)nuts, spices, dry meat products. The species is very common in indoor environments (Samson and Hoekstra, 1994). Isolates may produce pronounced earthy volatiles. The colony pattern and diameter of indoor isolates may vary, but the micromorphology is often similar.

NOTE:
this species can be confused with *A. sydowii* (see p. 86).

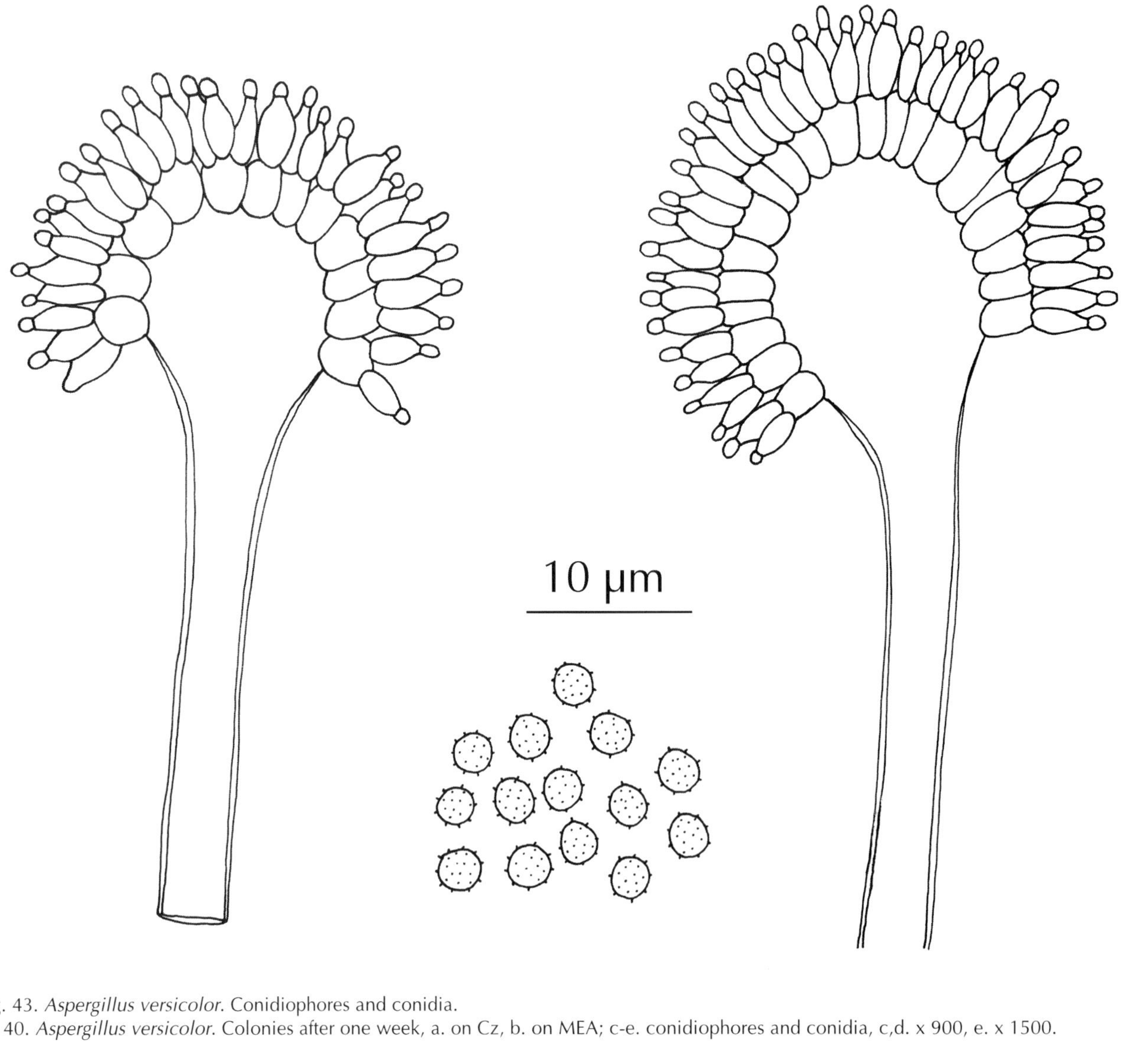

Fig. 43. *Aspergillus versicolor.* Conidiophores and conidia.
Pl. 40. *Aspergillus versicolor.* Colonies after one week, a. on Cz, b. on MEA; c-e. conidiophores and conidia, c,d. x 900, e. x 1500.

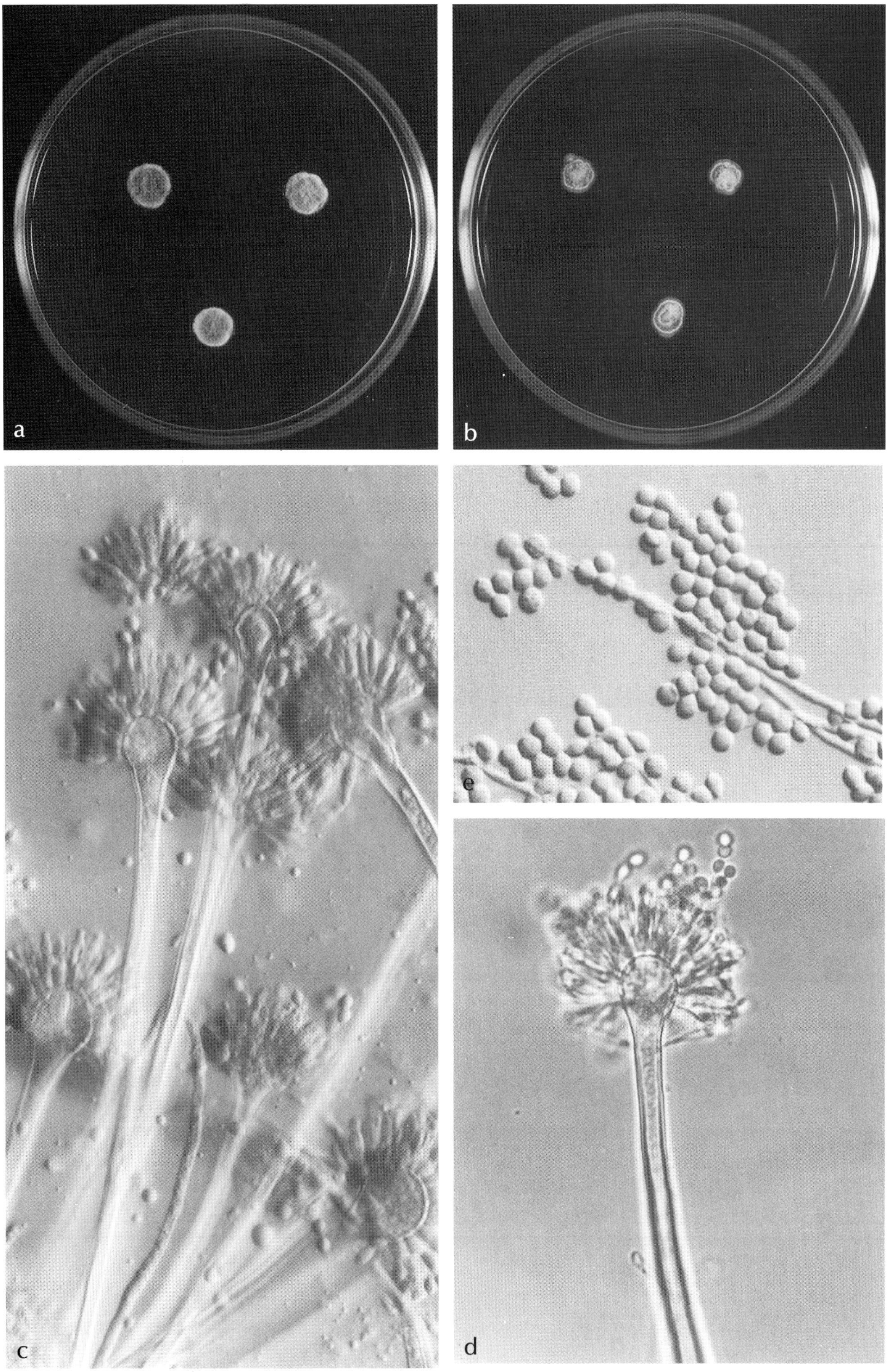
a
b
c
e
d

Aspergillus wentii Wehmer

Colonies on Czapek agar at 25°C, growing restrictedly, attaining a diameter of 2-3.5 cm in 7 days, usually somewhat floccose, conidial heads yellow-brown to coffee brown in age. Conidial heads large, globose, becoming radiate and ranging up to 500 µm in diameter; conidiophores up to several millimetres in length by 10 to 25 µm in diam., colourless; vesicles globose up to 80 µm in diam., fertile over the entire surface. Phialides borne on metulae, 10-20 x 3-5 µm, metulae 6-8 x 2.5-3.0 µm. Conidia first ellipsoidal, then subglobose to globose, yellow-brown, 4.5-5.0(6.0) µm, varying from smooth- to rough-walled and often showing a network of colour bars and ridges, without double wall characteristic as in *A. tamarii*. Poor growth on CREA.

Important (toxic) metabolites: emodin, wentilacton.

HABITAT: food
Common in (sub)tropical commodities: corn, (ground)nuts, tobacco, cereals.

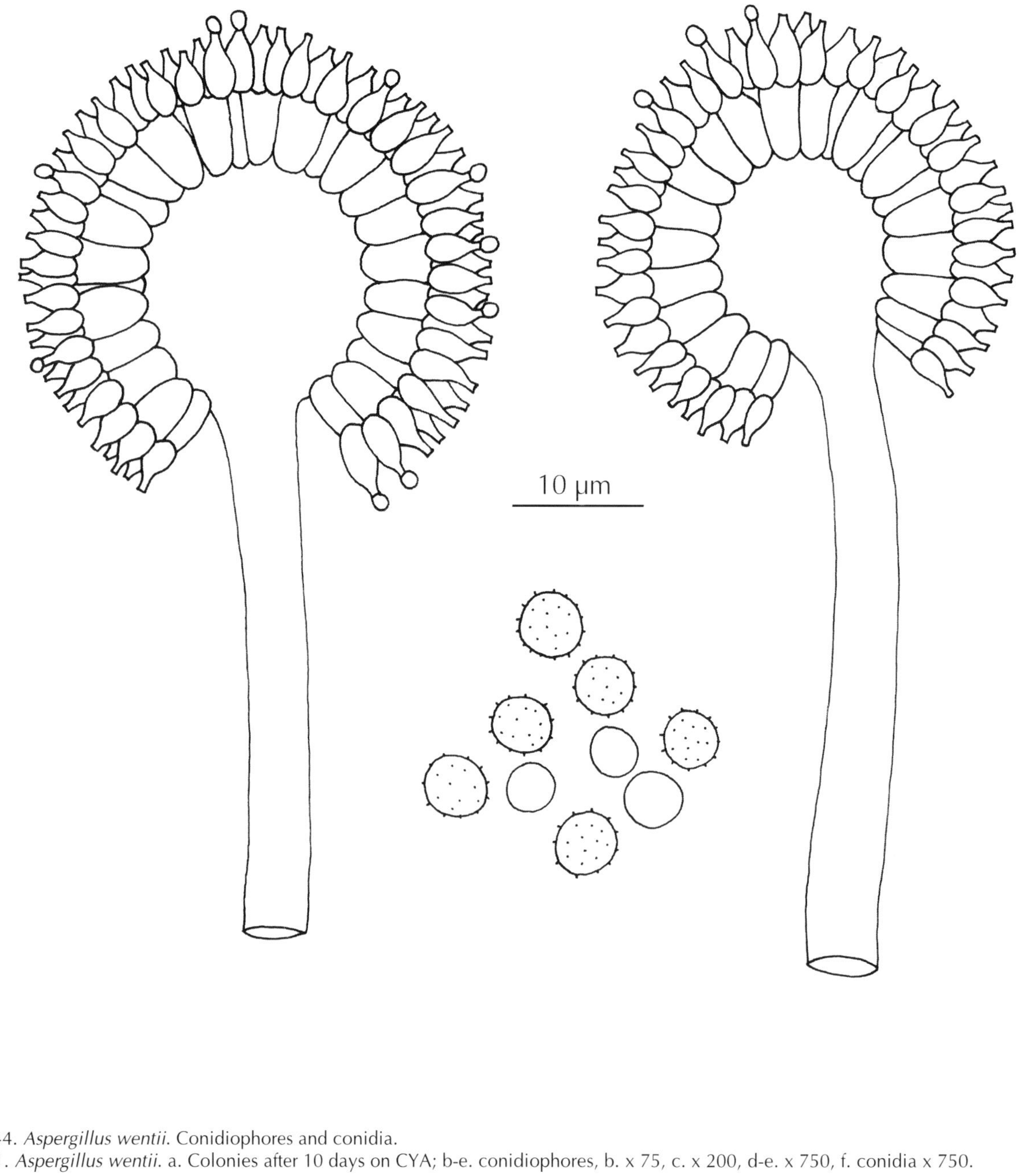

Fig. 44. *Aspergillus wentii*. Conidiophores and conidia.
Pl. 41. *Aspergillus wentii*. a. Colonies after 10 days on CYA; b-e. conidiophores, b. x 75, c. x 200, d-e. x 750, f. conidia x 750.

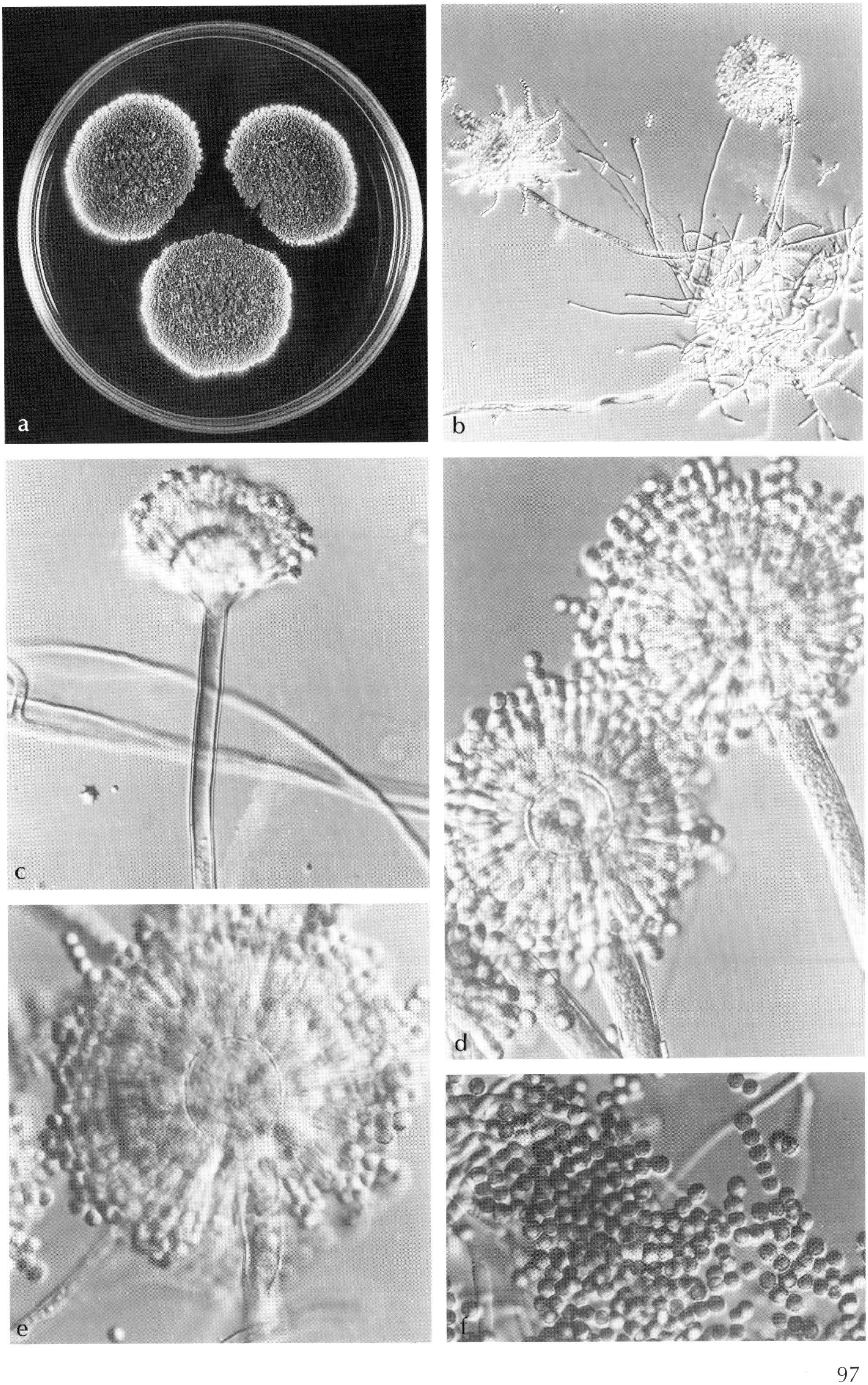
a
b
c
d
e
f

Aureobasidium pullulans (De Bary) Arnaud

Colonies on MEA at 24°C attaining a diameter of 4 cm in 7 days, smooth, soon covered with a slimy mass of spores, yellow, cream, pink, brown or black. Aerial mycelium sometimes formed, scanty, thinly floccose. Hyphae 2-16 µm in diam, hyaline, smooth- and thin-walled. In older cultures dark-brown, thick-walled hyphae may be formed. These thick-walled hyphae may act as chlamydospores or often fall apart into separate cells (arthroconidia). Conidiogenous cells undifferentiated, intercalary or terminal on subhyaline hyphae or arising as short lateral branches. Blastic conidia produced synchronously in dense groups from indistinct scars or denticles (they can be produced all over the cell surface). Conidia one-celled, (7.5)9-11(-16) x (3.5) 4-5.5(-7) µm, often with an indistinct hilum (= a mark or scar at the point of attachment), hyaline, smooth-walled. Secondary smaller conidia often produced. Endoconidia not common, formed by an intercalary cell and released into a neighbouring empty cell.

Temperature range for growth 2-35°C; optimum 25°C; maximum 35°C (higher in some human pathogenic isolates) see Hermanides-Nijhof (1977), Domsch *et al.* (1993).

HABITAT: food, indoor
Saprophyte with a world-wide distribution. Isolated from soil, leaf surfaces of plants (e.g. oak, beech, maple, poplar). It has been recorded on seeds of wheat, barley, oats, on flour, tomato, fresh harvested pecans, fruit (e.g. pears, grapes), fruit-drinks, frozen fruit-cake, human skin and nails, common in indoor environments in humid places (e.g. bathrooms, window frames).

CULTIVATION FOR IDENTIFICATION:
To observe conidial development the isolate is inoculated in a streak and incubated in either darkness or diffuse daylight at 24°C for 2-3 days.

NOTE:
This species has two varieties: *A. pullulans* (de Bary) Arnaud var. *pullulans*, with a colony which remains pink, light brown, or yellow for at least three weeks, and *A. pullulans* (de Bary) Arnaud var. *melanogenum* Hermanides-Nijhof which soon becomes black or greenish-black due to dark hyphae which often fall apart into separate cells.

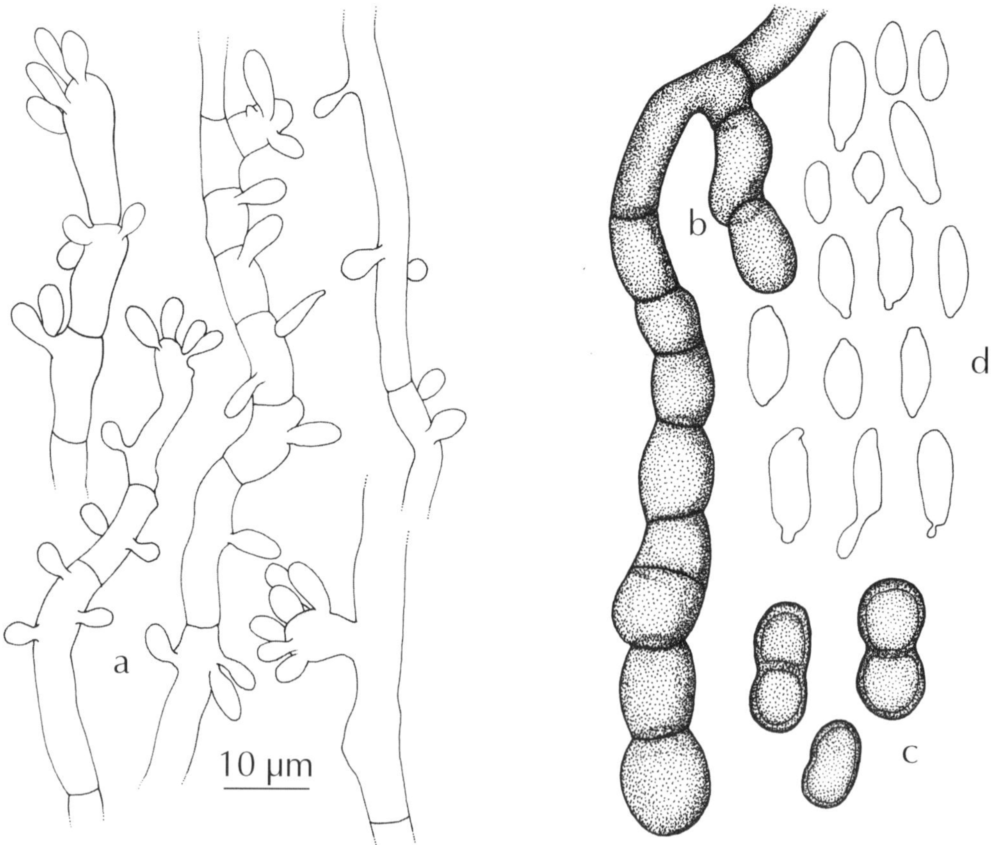

Fig. 45. *Aureobasidium pullulans*. a. Conidiogenous cells producing (blasto)conidia synchronously; b-c. thick-walled, dark hypha, this hypha may fall apart into separate cells (arthroconidia); d. conidia.
Pl. 42. *Aureobasidium pullulans*. a. Colonies on MEA after one week; b,c. dark thick-walled hyphae, which can fall apart into separate cells, x 600; d-g. conidiogenous cells and conidia, x 500.

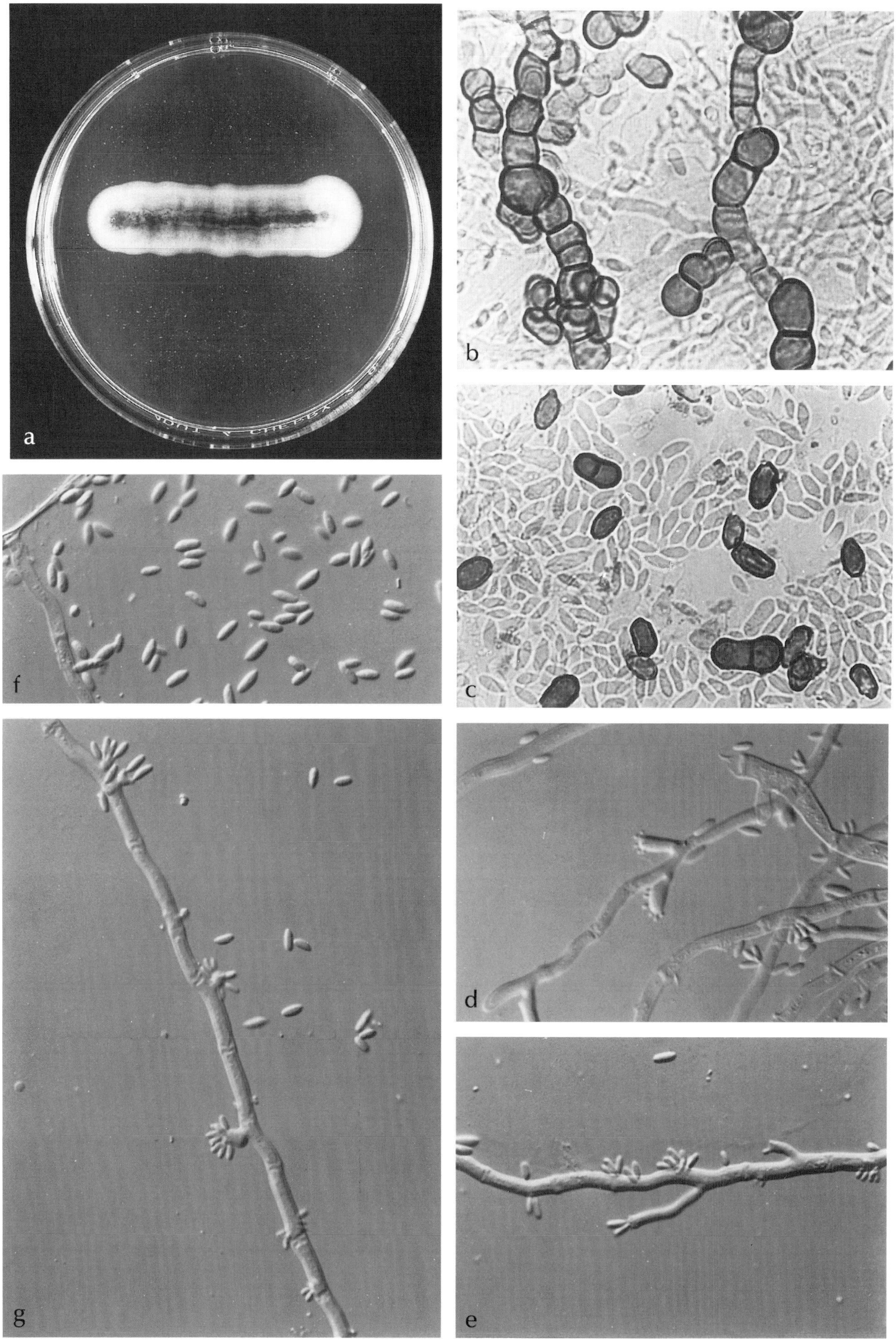
a
b
c
d
e
f
g

BOTRYTIS Mich. ex Pers.

Teleomorph: *Botryotinia* Whetzel (= *Sclerotinia* Fuckel pro parte)

Colonies broadly spreading, hyaline at first, becoming light grey to dark brown. Sclerotia frequently formed, both on natural substrate and in culture. Conidiophores erect, brown, mononematous, solitary or in groups, branched with branches mostly restricted to the apical region, with terminal swollen conidiogenous cells, producing numerous blastoconidia simultaneously on short denticles. Conidia one-celled (occasionally 2-3-celled), pale brown, globose to ovate or ellipsoidal, smooth-walled or almost so. There is often also a phialidic state belonging to the genus *Myrioconium*, with small spherical or subspherical colourless microconidia.

HABITAT:
The genus includes important plant pathogens with a world-wide distribution. For more detailed descriptions see Ellis (1971), Jarvis (1977), Domsch *et al.* (1993).

CULTIVATION FOR IDENTIFICATION
Descriptions are made after cultivation for 10 days at 20°C, in diffuse daylight in Petri-dishes. Conidium production can be stimulated by near-UV ("black") light and growing the cultures on hay-infusion or potato agar.

KEY TO THE SPECIES TREATED

1a. Conidiophores up to 1 mm long. Conidia narrowly ellipsoidal, (5)7-11 x (3)5-6(8) µm .... *B. aclada*
1b. Conidiophores up to 2 mm long or more. Conidia obovoid, 8-14 x 6-9 µm........ *B. cinerea*

Botrytis aclada Fres.

= *Botrytis allii* Munn

Colonies at 20°C attaining a diameter of about 6 cm or more in 10 days, hyaline becoming grey to greyish-brown. Sclerotia 1-5 mm in diam, frequently found on natural substrates, but less so in culture. Conidiophores abundant, rather short, about 1 mm long. Conidia narrowly ellipsoidal, sometimes pyriform or cuneiform, occasionally septate in culture, (5)7-11 x (3)5-6(8) µm.

Temperature: optimum 20-25°C; minimum below 5°C; maximum about 35°C.

HABITAT: food
World-wide distribution. Occurring on various species of *Allium*, causing neck rot in onions and shallots during storage.

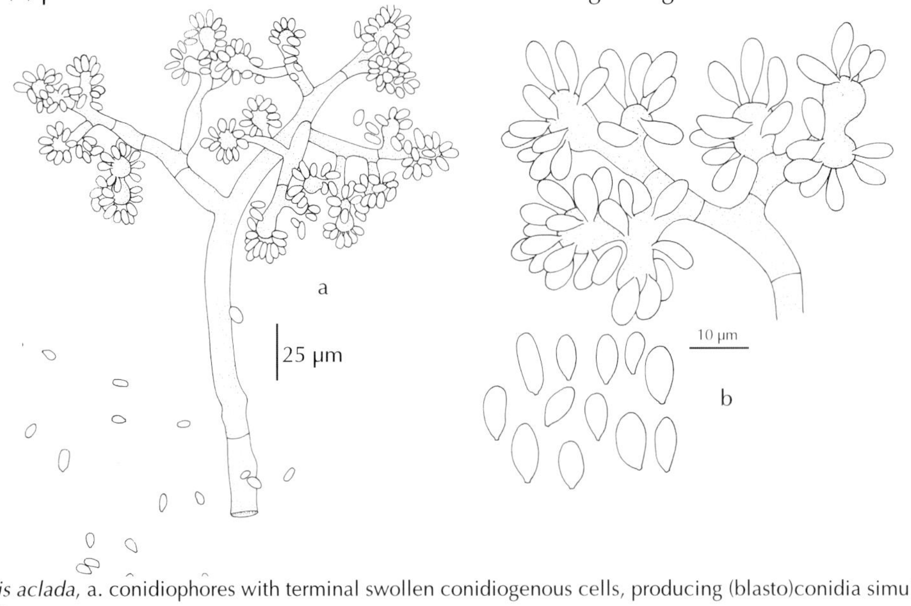

Fig. 46 a-b. *Botrytis aclada*, a. conidiophores with terminal swollen conidiogenous cells, producing (blasto)conidia simultaneously on short denticles, b. conidia.
Pl. 43. *Botrytis aclada*. a. Colony after 10 days; b-d. young and older conidiophores producing conidia simultaneously from swollen conidiogenous cells, b. x 200, c. x 500, d. x 750; e. conidia, x 1250.

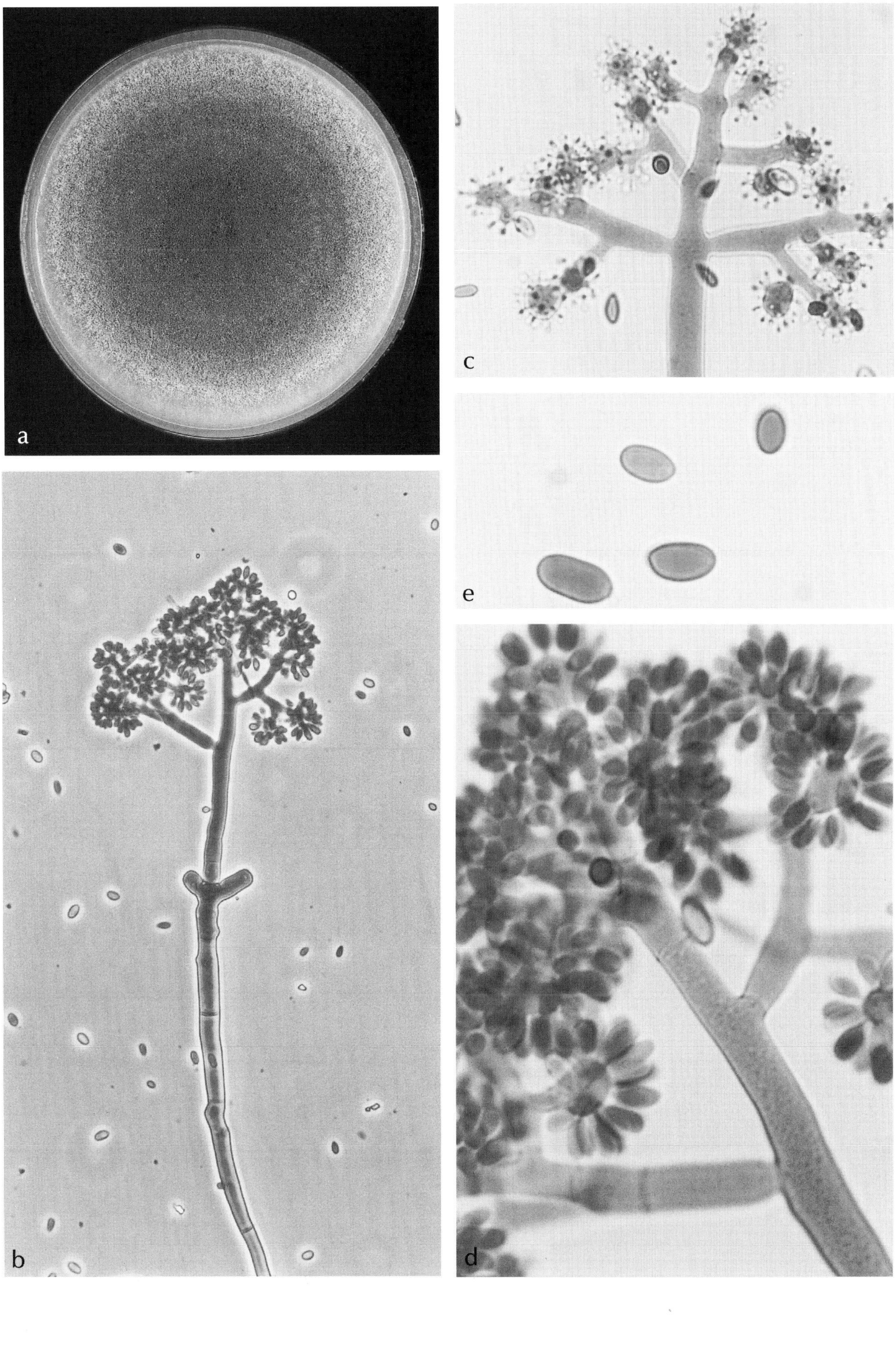
a
b
c
e
d

Botrytis cinerea Pers.

Colonies at 20°C attaining a diameter of 6 cm or more in 10 days, hyaline at first, later becoming grey to greyish-brown. Sclerotia black, consisting of a medulla and a dark brown to black cortical layer of cells, extremely variable in shape and size. Conidiophores frequently up to 2 mm and 16-30 µm wide, brown below, smooth-walled, with an apical head of alternate branches.
Conidia obovoid, 8-14 x 6-9 µm, usually with a protuberant hilum, pale brown, hydrophobic, smooth-walled. Microconidia produced on sporodochia (spermodochia), flask-shaped phialides often present, globose, 2.5-3 µm in diam.

Temperature: optimum (21)22-25(30)°C; minimum (-2)5-12°C; maximum (28)33-35°C.

HABITAT: food, indoor (often on plants, fruits)
World-wide distribution but mainly occurring in humid temperate and subtropical regions. It can be facultatively parasitic on a wide range of plants. This so-called "grey-mould" can damage flowers, leaves, stems and causes fruit- and leaf-rots in grapes, strawberries, cabbage and lettuce.
Also isolated from soils as well as stored and transported fruit and vegetables, contamination probably occurred already in the field.

NOTE:
Botrytis cinerea is a species-complex belonging to several distinct teleomorphs.

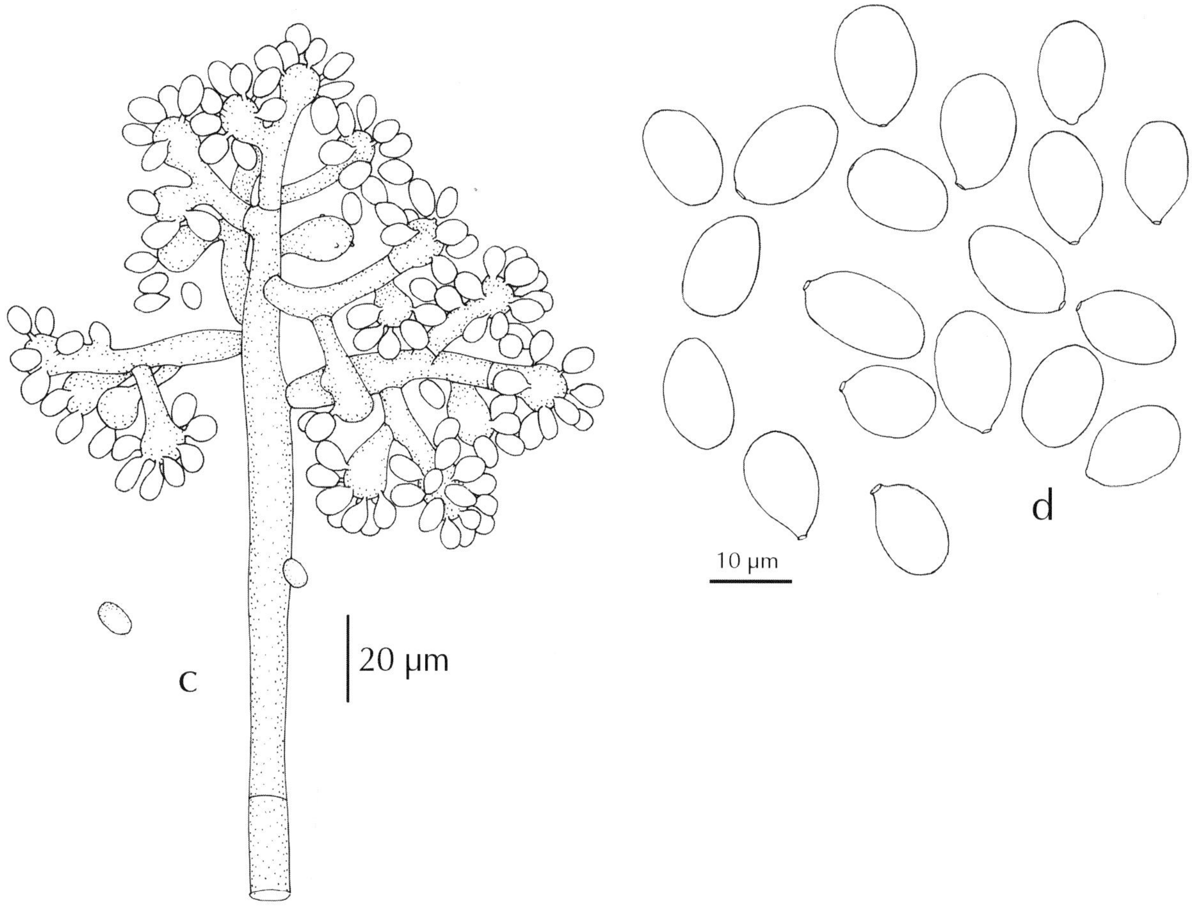

Fig. 46. *Botrytis cinerea*. c. Conidiophore with terminal swollen conidiogenous cells, producing (blasto)conidia simultaneously on short denticles; d. conidia.
Pl. 44. *Botrytis cinerea*. a. Colony after 10 days; b,c. conidiophores producing conidia simultaneously from swollen conidiogenous cells, b. x 200, c. x 750; d. conidia, x 750.

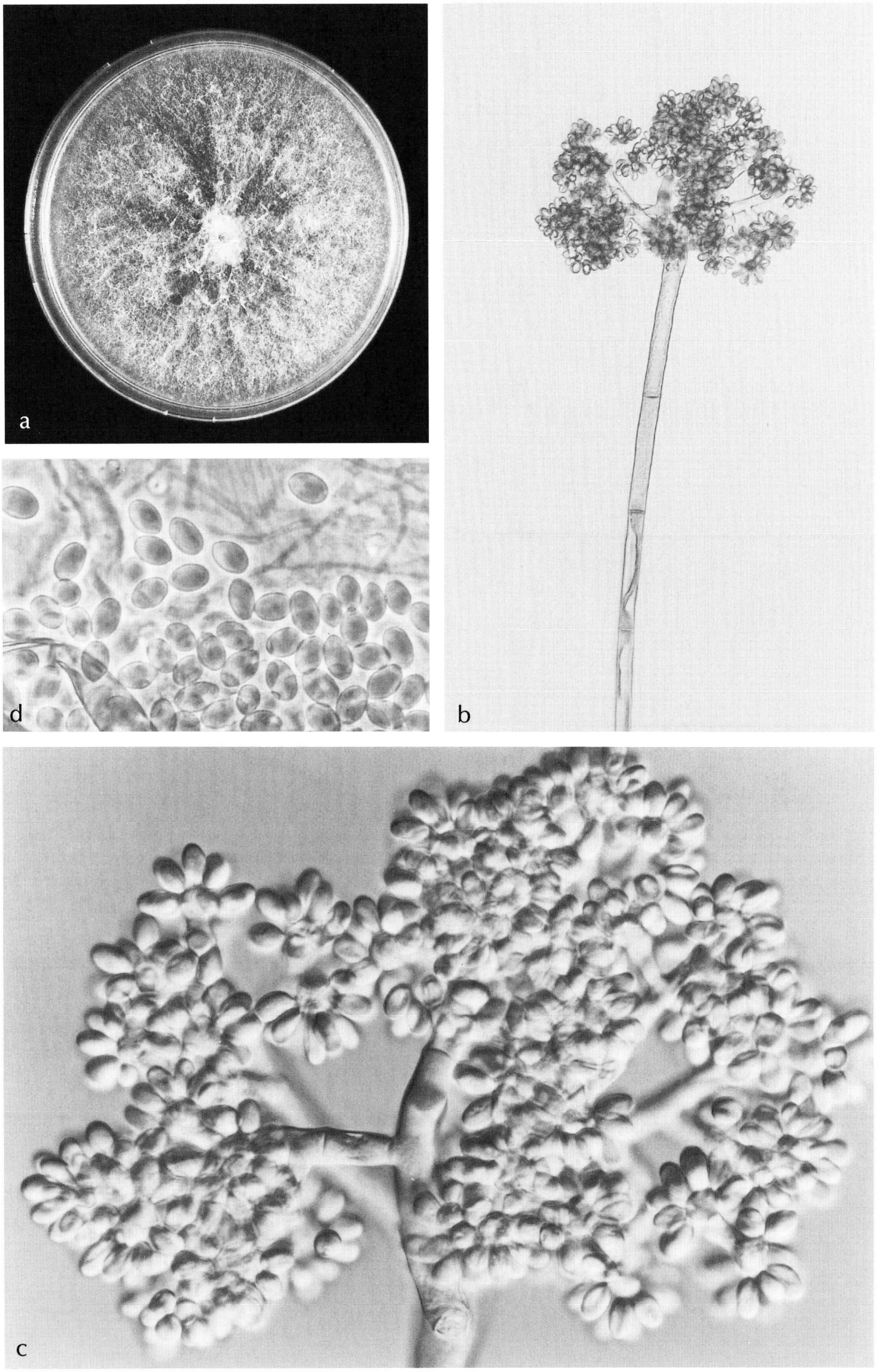
a
d
b
c

Chrysonilia sitophila (Mont.) v. Arx

= *Monilia sitophila* (Mont.) Sacc.
= *Penicillium sitophilum* Mont.

Teleomorph: *Neurospora sitophila* Shear & Dodge.

Colonies at 24°C growing fast, attaining a diameter of 2.5 cm in one day, with irregular tufts, especially at the margin of the Petri-dish, initially colourless, becoming pinkish to orange. Conidiogenous hyphae more or less ascending, smooth-walled, septate, with lateral branches which form chains of conidia. Conidia one-celled, in chains, connected by a narrow hyaline strand, becoming separated quickly, dispersed as a powdery mass under dry conditions, ellipsoidal, or more or less cylindrical or globose to subglobose (often also irregular in shape and size), (7)10-15 x 5-10(11)µm, hyaline, pinkish to orange in mass, smooth-walled.

HABITAT: food, indoor
Reported from Europe, U.S.A., Japan, Surinam, Indonesia. Known as the "red bread mould", occurring on bread and related products, silage, meat, sometimes in connection with field, transportation and storage rot of fruit (e.g. apples, strawberries, raspberries). Sometimes occurring as a contaminant in indoor environments.

NOTE:
For more detailed descriptions see Shear & Dodge (1927) Hashmi *et al.* (1972), von Arx (1981). Mycelium, conidiogenous hyphae and conidia often grow out of the Petri-dish and are able to lift the lid. Because of the explosive growth of this fungus, spoilage can occur within a few days. This fungus is able to grow into slants or Petri-dishes of other cultures so that contamination in the laboratory and hospitals easily can occur.

Chrysonilia crassa (Shear & Dodge) v. Arx

= *Monilia crassa* Shear & Dodge

Teleomorph: *Neurospora crassa* Shear & Dodge.

The species mainly differs from *Chrysonilia sitophila* by the smaller conidia measuring (5)6-7(8) x 4-6 µm.

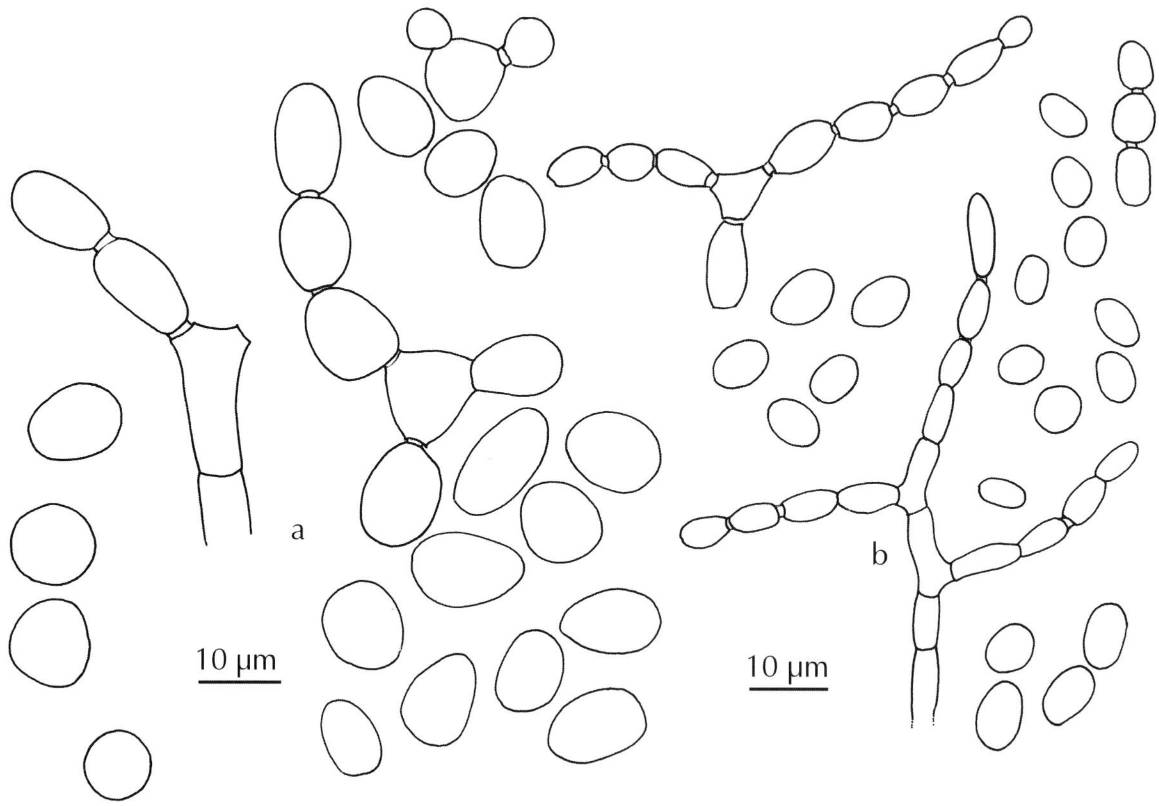

Fig. 47. a. *Chrysonilia sitophila.* Branched conidiophores forming chains of conidia; b. *Chrysonilia crassa.* Branched conidiophores forming chains of conidia.
Pl. 45. a-d. *Chrysonilia sitophila.* a. Colony after 3 days; b-d. conidiophores branched, each branch forming chains of conidia, b. x 500; c,d. x 1000; e. *Chrysonilia crassa*, x 500.

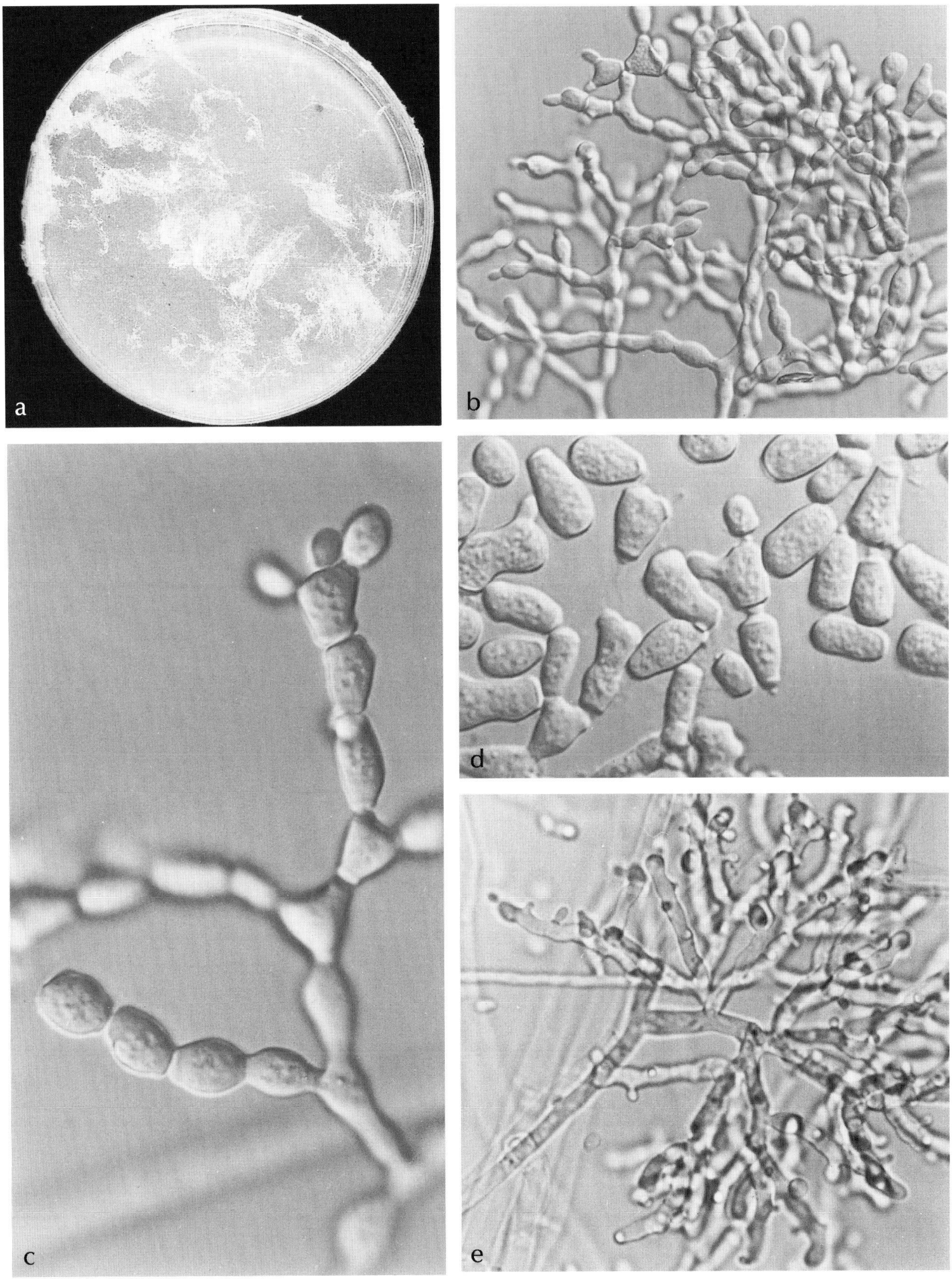
a
b
c
d
e

CLADOSPORIUM Link

Teleomorph: *Mycosphaerella* Johanson.

Colonies mostly olivaceous-brown to blackish-brown or with a greyish-olive appearance, velvety or floccose becoming powdery due to abundant conidia, rather slow-growing. Vegetative hyphae, conidiophores and conidia pigmented.

Conidiophores more or less distinct from vegetative hyphae, erect, straight or flexuous, unbranched or branched only in the apical region, with geniculate sympodial elongation in some species. Conidia in branched acropetal chains formed synchronously or in succession from multiple conidiogenous loci, lower conidia often larger and septate (so-called "ramoconidia") upper conidia one-celled, ellipsoidal, fusiform, ovoid, (sub)globose, often with distinct scars, pale to dark olivaceous-brown, smooth-walled, verrucose or echinulate; (blasto)conidia mostly formed on denticles in groups of 1-3 at the apex of the conidiophore, subapically below a septum or on the tip of previously formed conidia. Stroma sometimes present. Two types of conidial structures can be distinguished, the *Cladosporium* and the *Hormodendrum* type (or form). The *Cladosporium* type is characterized by the presence of geniculate and sympodially elongated conidiophores which give the upper portion of the conidiophore a zig-zag appearance; the *Hormoden-drum* type is characterized by racemosely branched conidiophores with acropetal conidial chains arising at several levels. However the two types cannot always be sharply defined.

HABITAT: food, indoor

A genus with a world-wide distribution. Several species are plant pathogens or saprophytic and more or less host-specific on old or dead plant material.

NOTE:

For more detailed descriptions and keys see De Vries (1952), Ellis (1971), Domsch *et al.* (1993).

CULTIVATION FOR IDENTIFICATION:

Colonies are inoculated on MEA and incubated in diffuse daylight at 18-20°C for 7-10 days. Due to the fragile conidiophore structure it is often difficult to obtain good microscope slides. It is helpful to use transparent adhesive tape and lightly touch the colonies. Put a drop of lactic acid on a slide and place the tape over the slide. (see also p. 2)

KEY TO THE SPECIES TREATED

1a. Conidiophores not elongating sympodially; conidia usually not exceeding 4.5 µm in width, smooth or slightly roughened 2
1b. Conidiophores commonly elongating sympodially; conidia usually exceeding 5.0 µm in width, distinctly verrucose 3

2a. Most one-celled conidia (sub)globose, 3-4.5 µm in diameter, finely roughened .... *C. sphaerospermum*
2b. Most one-celled conidia elongate, 3-7(-11) x 2-4(5) µm, smooth-walled or nearly so .... *C. cladosporioides*

3a. One-celled conidia mostly 5.5-13 x 4-6 µm, 2-3-celled conidia also present, somewhat larger .... *C. herbarum*
3b. One-celled conidia mostly 7-17 x 5-8 µm, 2-3-celled conidia common, considerably larger .... *C. macrocarpum*

Cladosporium cladosporioides (Fres.) de Vries

Colonies on MEA at 18-20°C attaining a diameter of (1.5)3-4 cm in 10 days, velvety becoming powdery due to abundant conidia, olivaceous-brown or greyish-green. Reverse greenish-black.

Conidiophores arising laterally or sometimes terminally from the hyphae, up to 350 µm long, but generally much shorter and 2-6 µm wide, without sympodial elongations and swellings, bearing branched conidial chains, pale to mid-olivaceous-brown, smooth-walled or verruculose. Ramoconidia towards the base of chain, 0-1(-2)-septate, more or less cylindrical, up to 30 x 3-5 µm, brown or greenish-brown, smooth-walled or sometimes minutely verruculose. Conidia in acropetal branched chains, mostly one-celled, ellipsoidal to lemon-shaped, 3-7(-11) x 2-4(5) µm, brown or greenish-brown, mostly smooth-walled, sometimes verruculose.

Temperature: optimum 20-28°C; minimum -3 to -10°C.

HABITAT: food, indoor

A very common species with a world-wide distribution and occurring on plant material and in soil. Isolated from air, soil, textiles, foodstuffs, seeds, crops etc. Also common in indoor environments.

NOTE:

C. cladosporioides is closely related to *C. sphaerospermum*

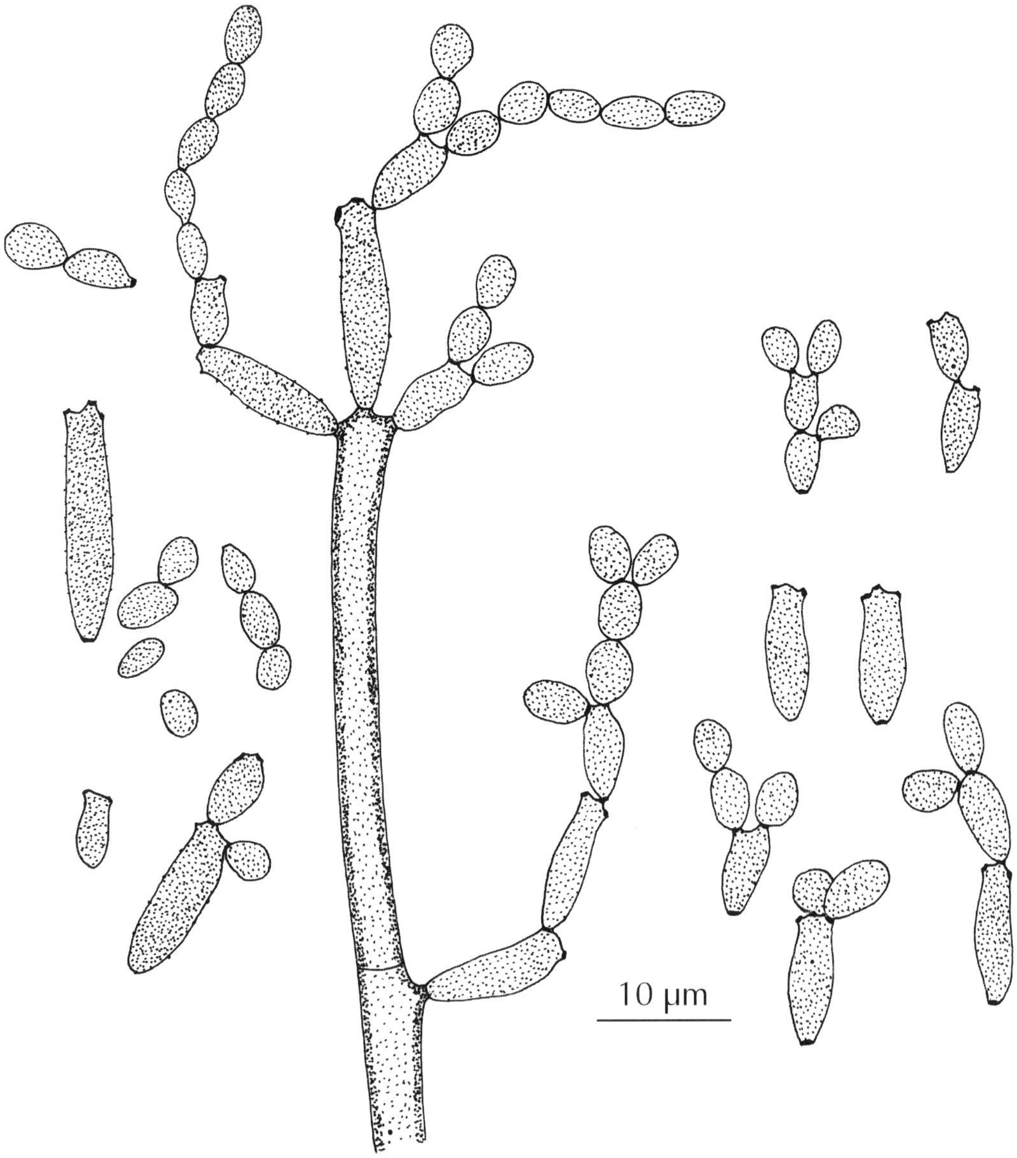

Fig. 48. *Cladosporium cladosporioides.* Conidiophore (without swellings) and conidia.

Pl. 46. *Cladosporium cladosporioides.* a. Colony on MEA after one week; b, c, e. branched conidial chains, b,e, x 800, c x 725, d. conidia, x 1275.

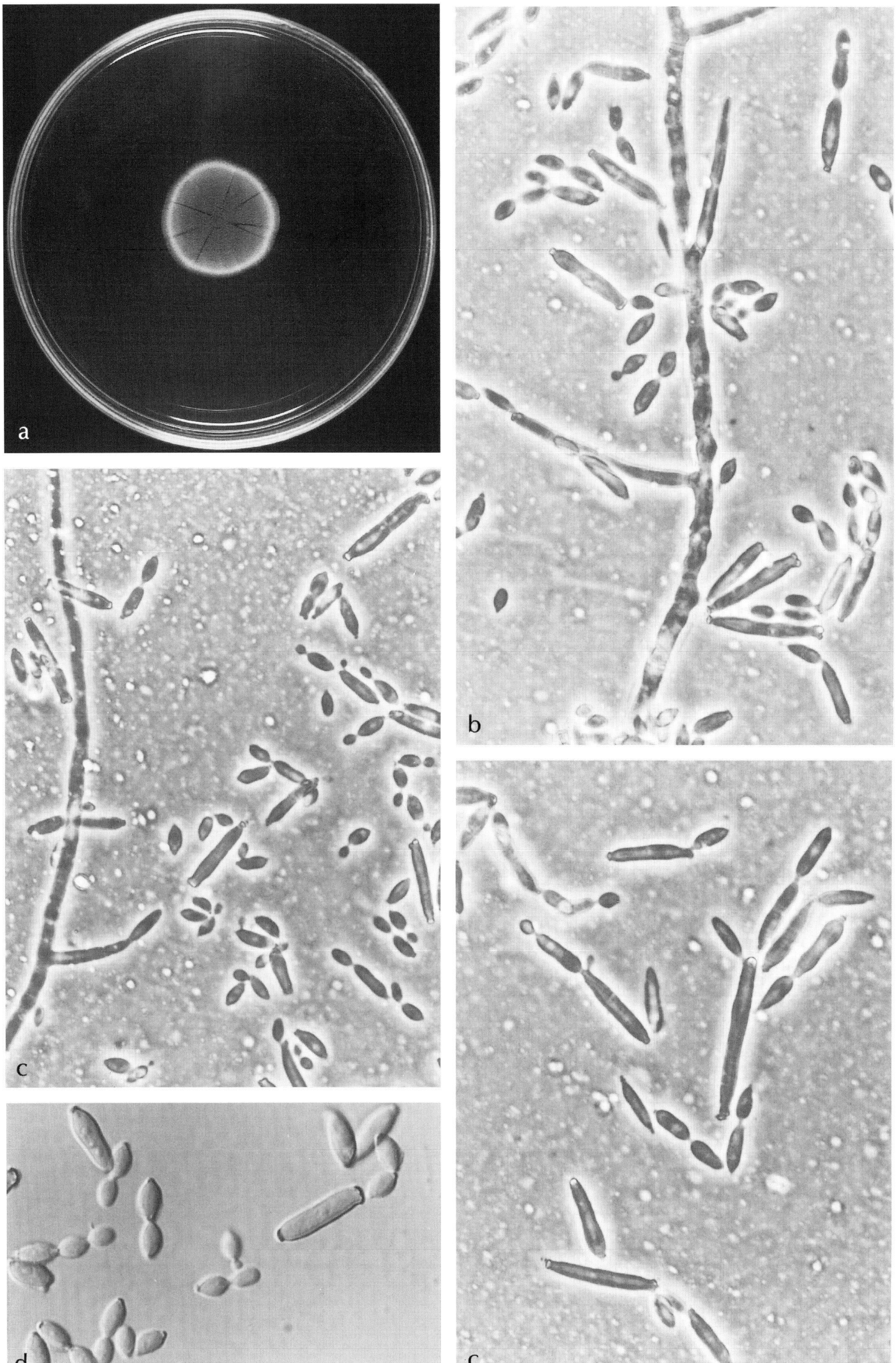
a
b
c
c
d

Cladosporium herbarum (Pers.) Link

Teleomorph: *Mycosphaerella tassiana* (de Not.) Johanson

Colonies on MEA at 18-20°C attaining a diameter of (2.5)3-3.7 cm in 10 days, velvety, locally powdery due to conidia, olivaceous-green to olivaceous-brown. Reverse greenish-black. Conidiophores mostly arising laterally or sometimes terminally from the hyphae, up to 250 μm long and 3-6 μm wide, with terminal and intercalary swellings (7-9 μm in diam) and geniculate elongations, pale to mid-olivaceous-brown or brown, smooth-walled. Conidia in long, often branched chains, ellipsoidal to cylindrical with rounded ends, rather frequently 2-or more-celled, one-celled 5.5-13 x 4-6 μm, golden brown, distinctly verruculose, with more or less protuberant scars.

Temperature: optimum 18-28°C; maximum 28-32°C; minimum -6°C.

HABITAT: food, indoor
Common species with a world-wide distribution, especially abundant in temperate regions on dead or dying plant substrates and other organic matter. Isolated from air, soil, foodstuffs, stored fruit, cereal grains (especially wheat), groundnuts, paint, textiles etc.

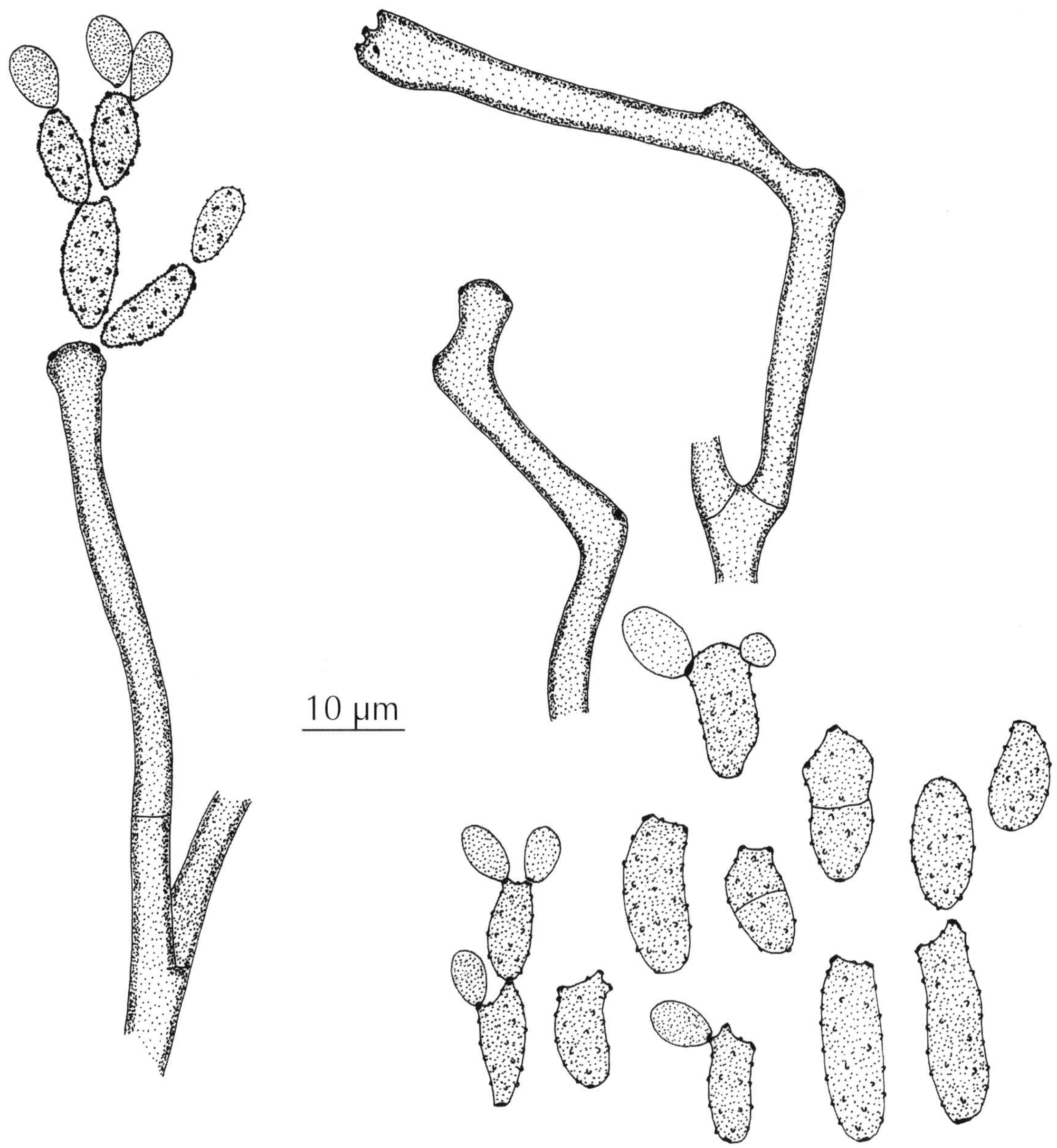

Fig. 49. *Cladosporium herbarum.* Conidiophores (often with swellings) and conidia.
Pl. 47. *Cladosporium herbarum.* a. Colony on PDA after one week; b-e. conidiophores and conidia, b. x 1015 c,d, x 720, e. x 1045.

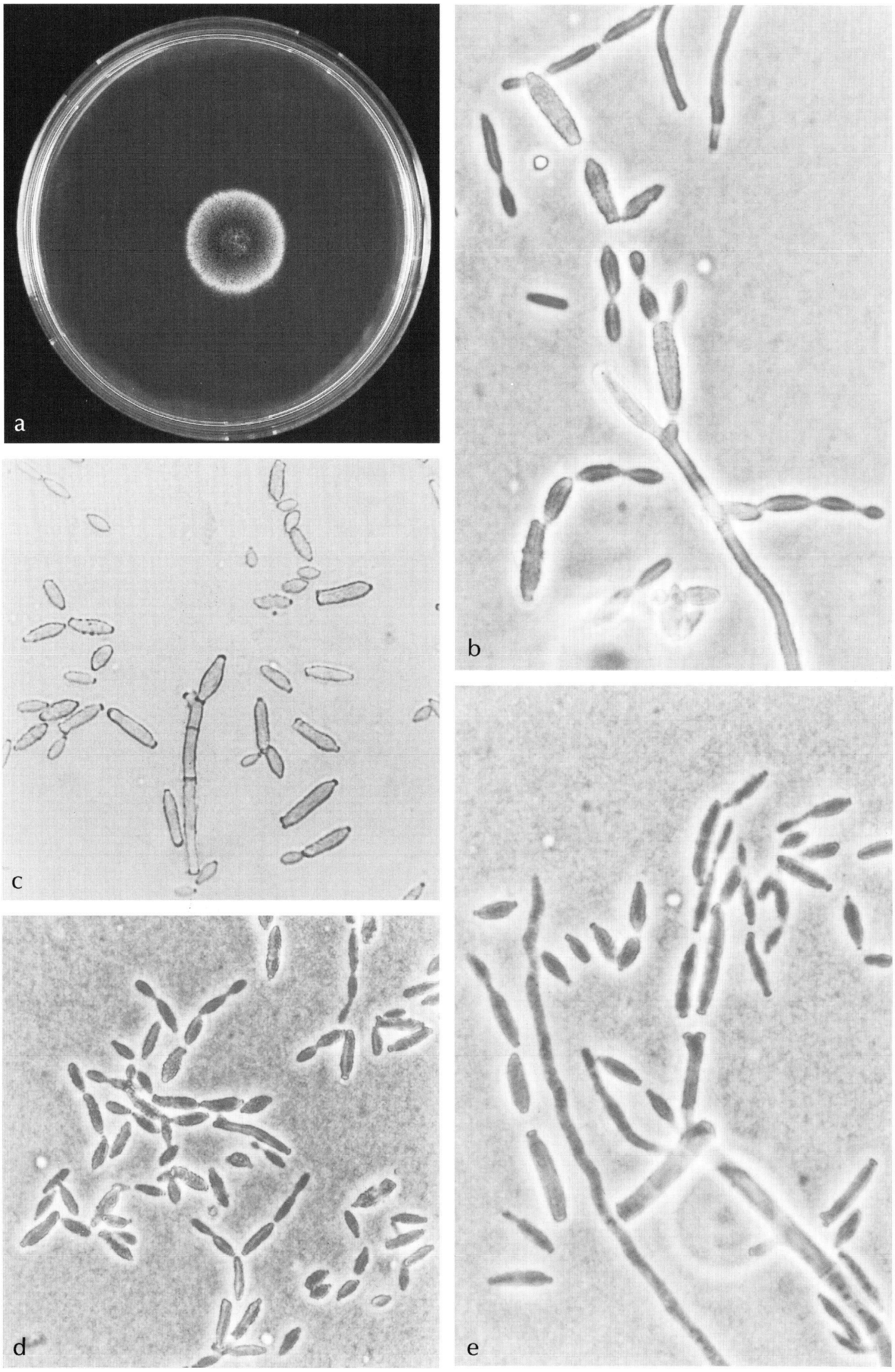
a
b
c
d
e

Cladosporium macrocarpum Preuss

Colonies on MEA at 18-20°C attaining a diameter of 2.5-3.5 cm in 10 days, velvety, often covered with greyish aerial mycelium, olive-green. Reverse greenish-black. Conidiophores arising laterally from the hyphae, up to 300 µm long and 4-8 µm wide, often becoming geniculate and nodose by sympodial elongation, terminal and intercalary swellings, 9-11 µm in diam (if present), pale to mid-brown, or olivaceous-brown, smooth or partly verruculose. Conidia usually in rather short chains, 0-3-septate, ellipsoidal; one-celled mostly, 7-17 x 5-8 µm, pale to mid-brown or olivaceous brown, densely verruculose.

HABITAT: food (not common), indoor
World-wide distribution, common in temperate regions, particularly on dead plants, but less common than *C. herbarum* and *C. cladosporioides*. Isolated from soil, seeds of cereals, feathers of free-living birds.

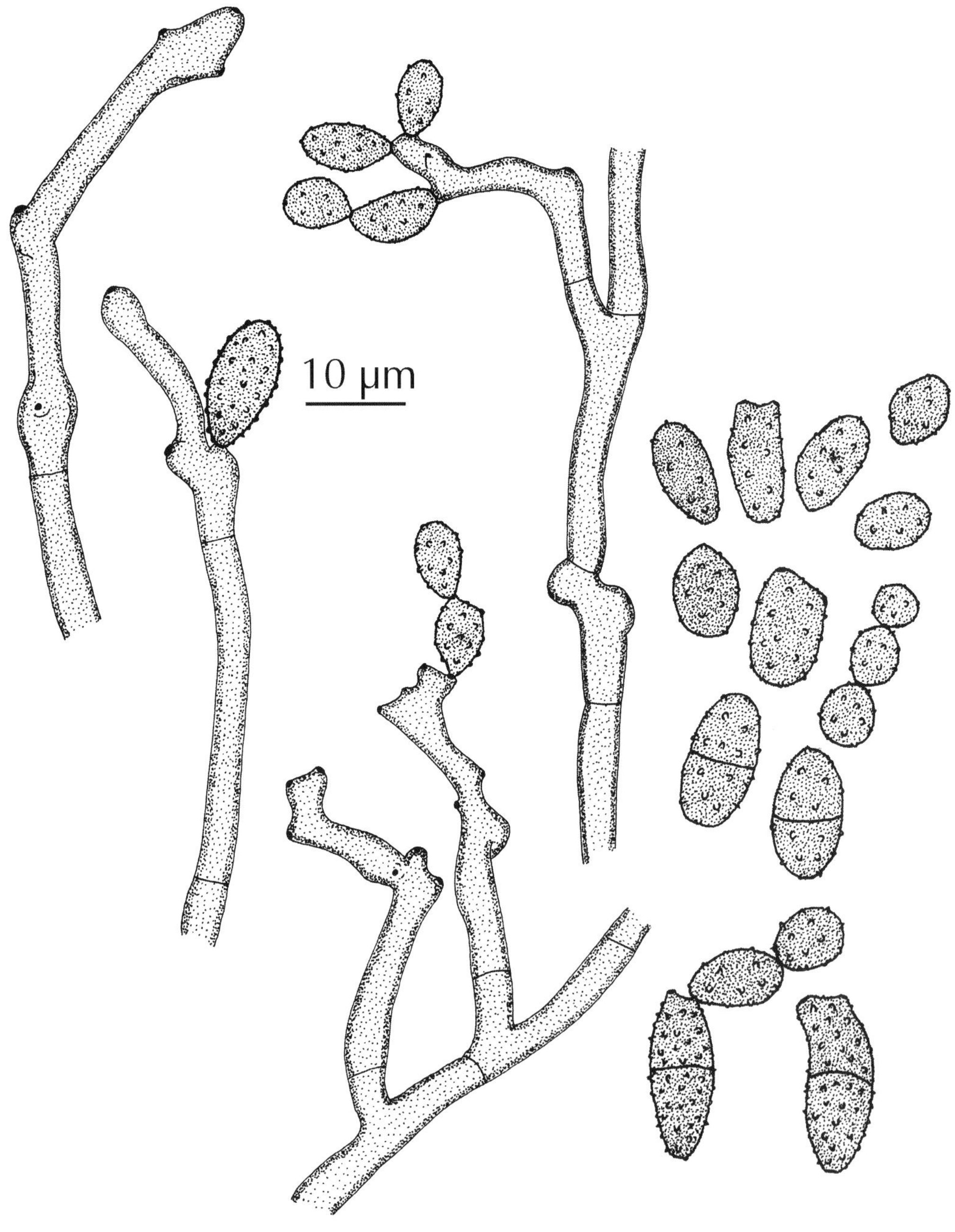

Fig. 50. *Cladosporium macrocarpum*. Conidiophores (often becoming geniculate or nodose) and conidia.
Pl. 48. *Cladosporium macrocarpum*. a. Colony on MEA after one week; b-f. conidiophores and conidia, b,c,e,f. x 600, d. x 1000.

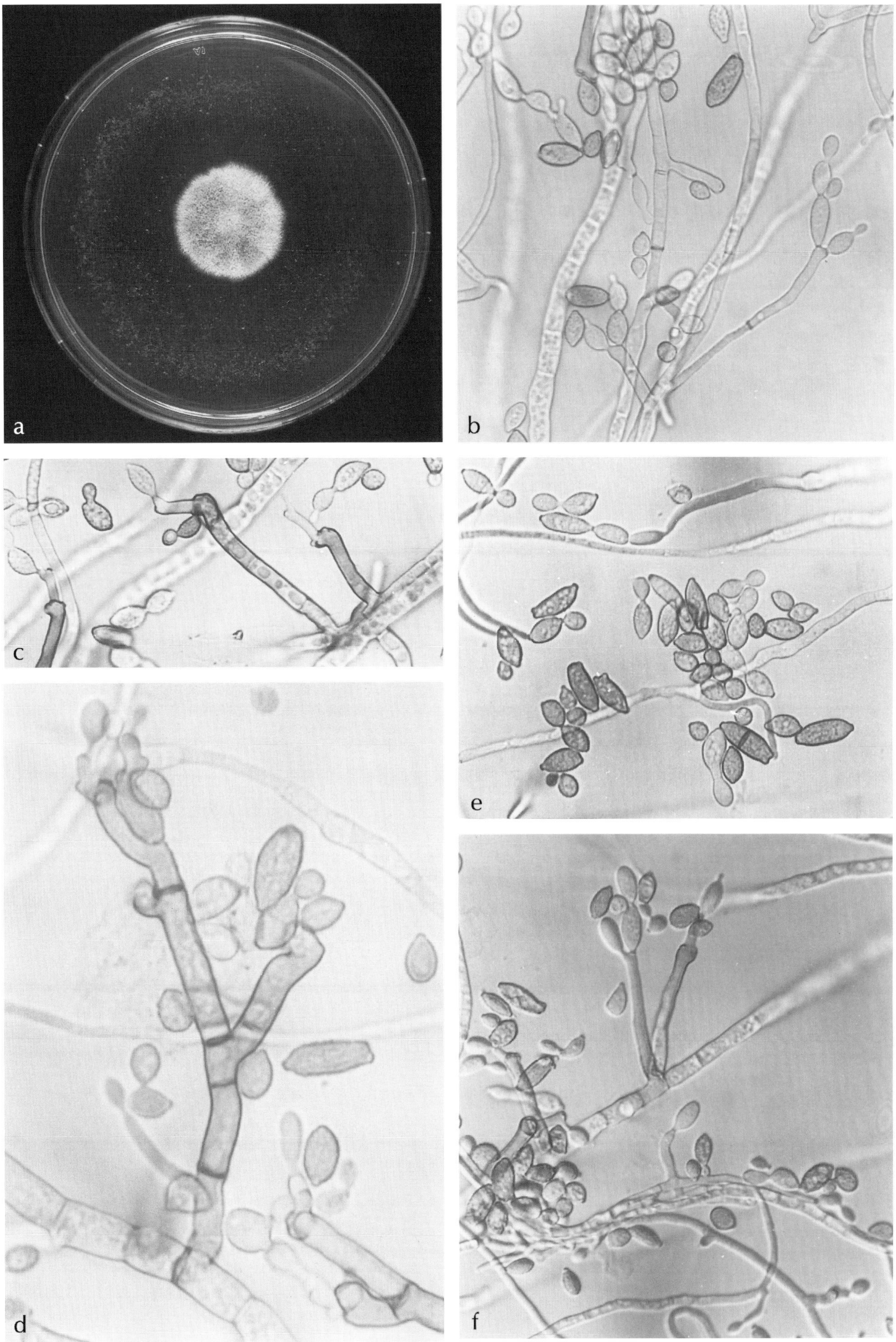
a
b
c
d
e
f

Cladosporium sphaerospermum Penzig

Colonies on MEA at 18-20°C attaining a diameter of 1.5-2 cm in 10 days, velvety or powdery, olive-green to olivaceous-brown. Reverse greenish-black. Conidiophores arising laterally and terminally from the hyphae, up to 300 µm long, but generally much shorter and 3-5 µm wide, not geniculate, bearing several branched conidial chains, pale to dark-olivaceous-brown, smooth-walled or verruculose. Ramoconidia towards the base of the chain, 0-3-septate, more or less elongate, up to 33 x 3-5 µm, smooth-walled or verruculose. Conidia in acropetal chains, mostly one-celled, (sub)globose, 3-4(-7) µm in diam, brown or greenish-brown, minutely verruculose.

HABITAT: food, indoor
A common species with a world-wide distribution and occurring as a secondary invader on many different plants. Isolated from air, soil, foodstuffs, seeds, paint, textiles etc. and occasionally from man and animals.

NOTE:
The difference between this species and *C. cladosporioides* is not always sharp as the former can also have relatively short conidia.

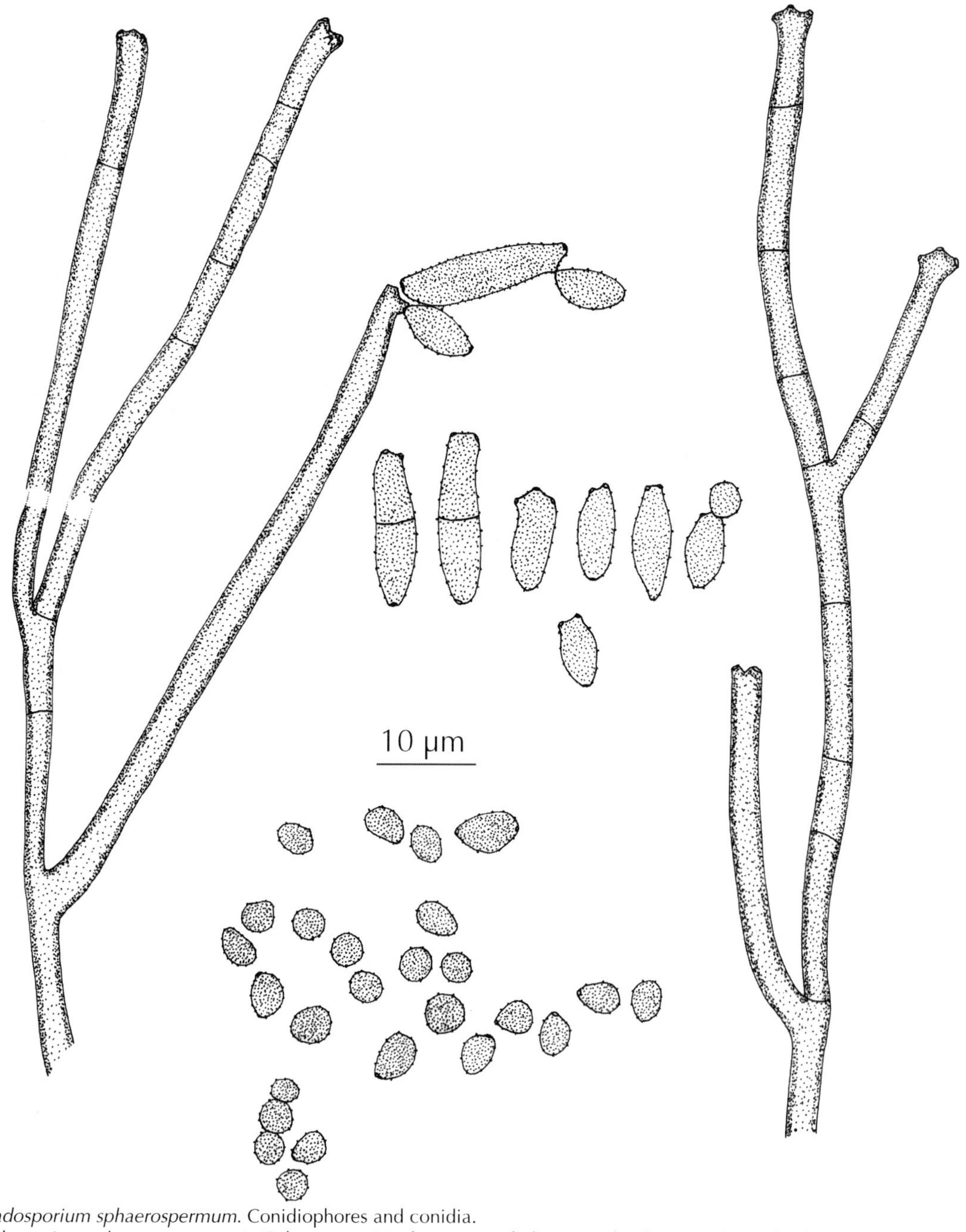

Fig. 51. *Cladosporium sphaerospermum.* Conidiophores and conidia.
Pl. 49. *Cladosporium sphaerospermum.* A. Colony on MEA after one week; b-e. conidiophores and conidia, b. x 500, c-e. x 1000.

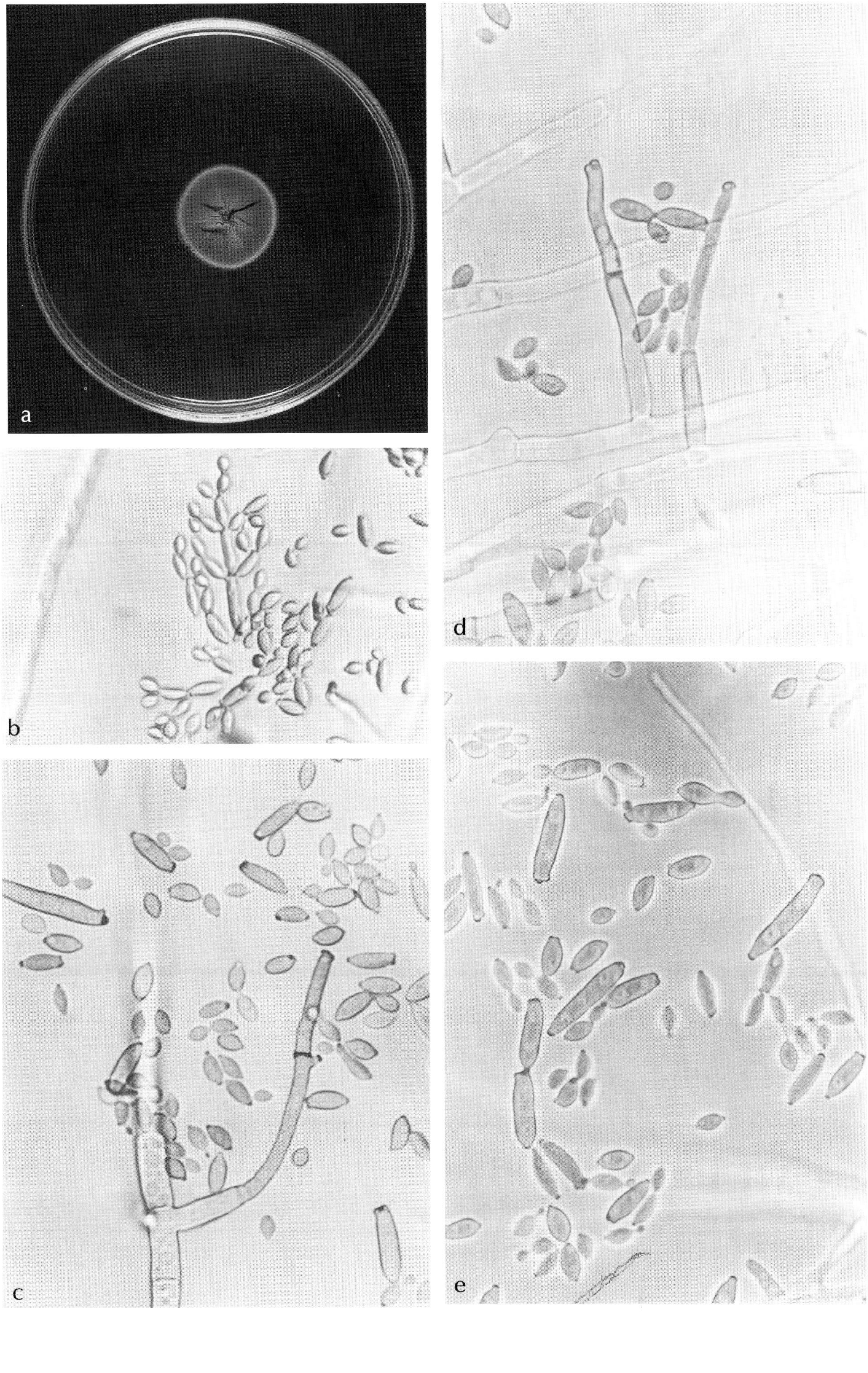
a
b
c
d
e

Curvularia geniculata (Tracy & Earle) Boedijn

Teleomorph: *Cochliobolus geniculatus* Nelson

Colonies, PCA and OA moderately fast growing, reaching a diameter of 3-4 cm within a week at 25°C, brownish-black, velvety sometimes becoming somewhat cottony. Reverse brown. Conidiophores erect, pigmented becoming paler near the tip, sometimes geniculate up to 600 µm long, septate. (Poro)conidia usually 4-septate, mostly curved unilaterally at the broadest and most pigmented central cell, adjacent cells less pigmented and the end cells subhyaline, 18-35(37) x 8-14 µm. Heterothallic.

Temperature: optimum temperature range 24-30°C. Sporulation is enhanced by incubation under near-UV light.

HABITAT: food, indoor
World-wide: isolated from soil and plant materials with a preference for the tropics and warmer climates. Frequently isolated from cellulosic materials e.g. paper, archives.

NOTE:
The species *Curvularia lunata* is more commonly occurring on food and is distinguished by having 3-septate conidia.

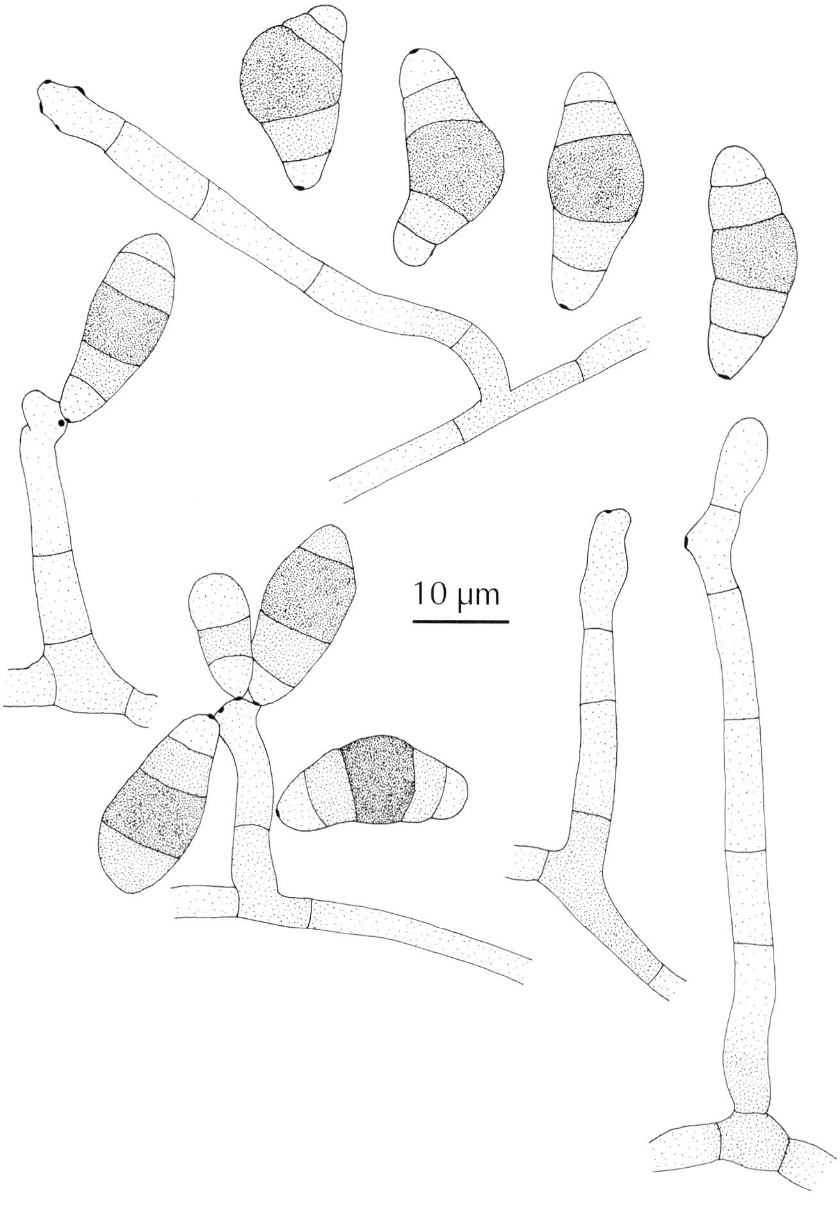

Fig. 52. *Curvularia geniculata*. Conidiophores and conidia.
Pl. 50. *Curvularia geniculata*. a. Colony on OA after one week; b-d. conidiophores ad conidia, b. x 775, c,d. x 1000

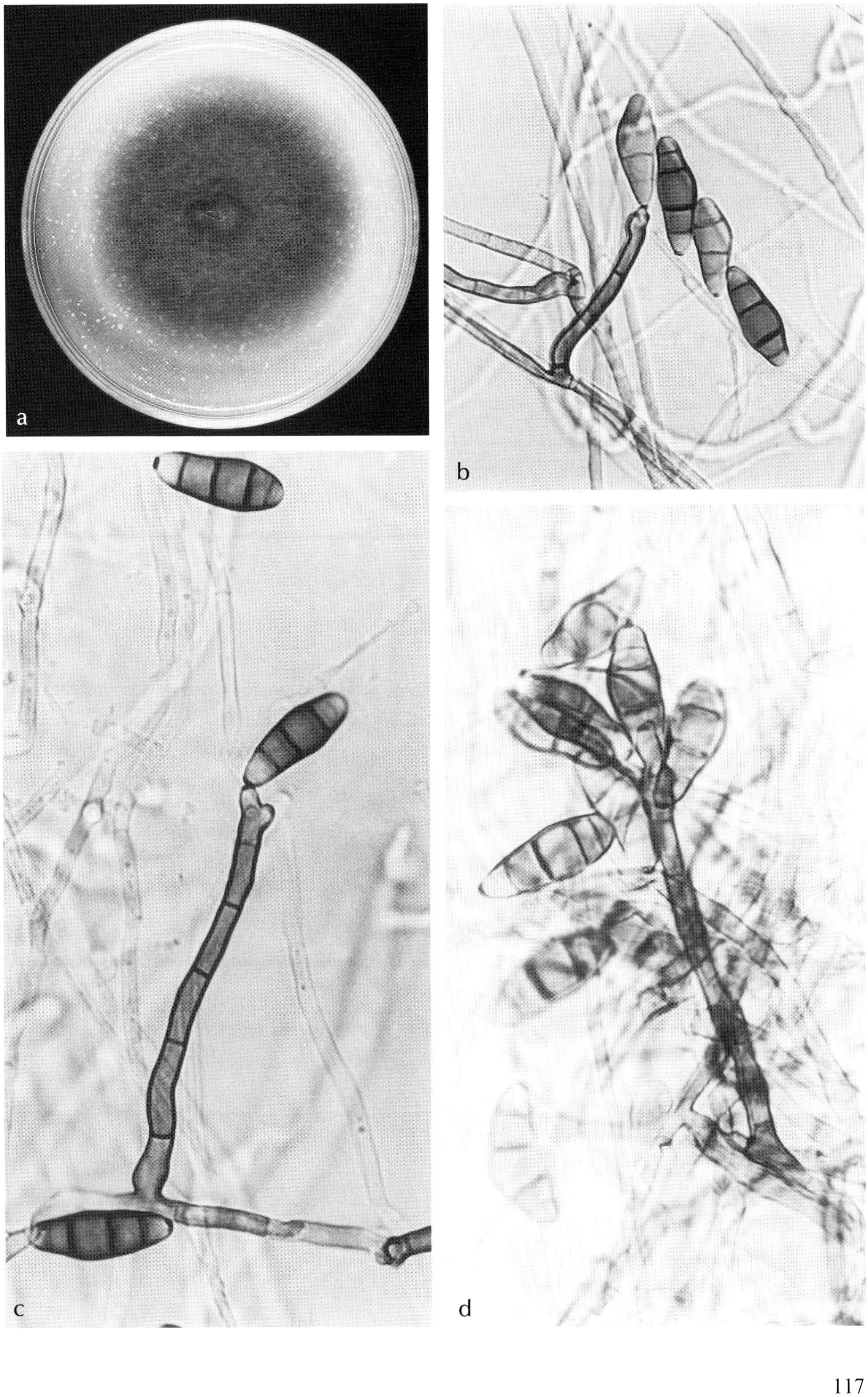
a
b
c
d

Epicoccum nigrum Link

= *Epicoccum purpurascens* Ehrenb.

Colonies on MEA or OA at 20°C attaining a diameter of at least 6 cm in 10 days, lanose to felty, yellow-orange, red or brown, sometimes greenish. Reverse similar, often more intensely coloured. Pulvinate sporodochia appearing in 8-10 days, visible as black spots, 100-2000 µm in diam. Conidiophores in clusters, straight or more or less flexuous, terminating in more or less isodiametric conidiogenous cells, colourless to pale brown, short, 5-15 x 3-6 µm. (Blasto)conidia formed singly, globose to pyriform, commonly 15-25 µm in diam, with funnel-shaped base and broad attachment scar, often with pale protuberant basal stalk cell, dark golden brown, verrucose, obscuring the septa which divide the conidia into up to 15 cells.

Temperature: optimum 23-28°C; minimum (-3-)4°C; maximum 45°C.

HABITAT: food, indoor

World-wide distribution, a common secondary invader of dead parts of diverse plants; soil, seeds (e.g. corn, barley, oats, wheat), beans, mouldy paper, textile, insects, human skin and sputum.

NOTE:

For more detailed descriptions the reader is referred to Schol-Schwarz (1959), Ellis (1971), Domsch *et al.* (1993). Cultures often remaining sterile, sporulation, can be induced by exposure to near-UV ("black") light.

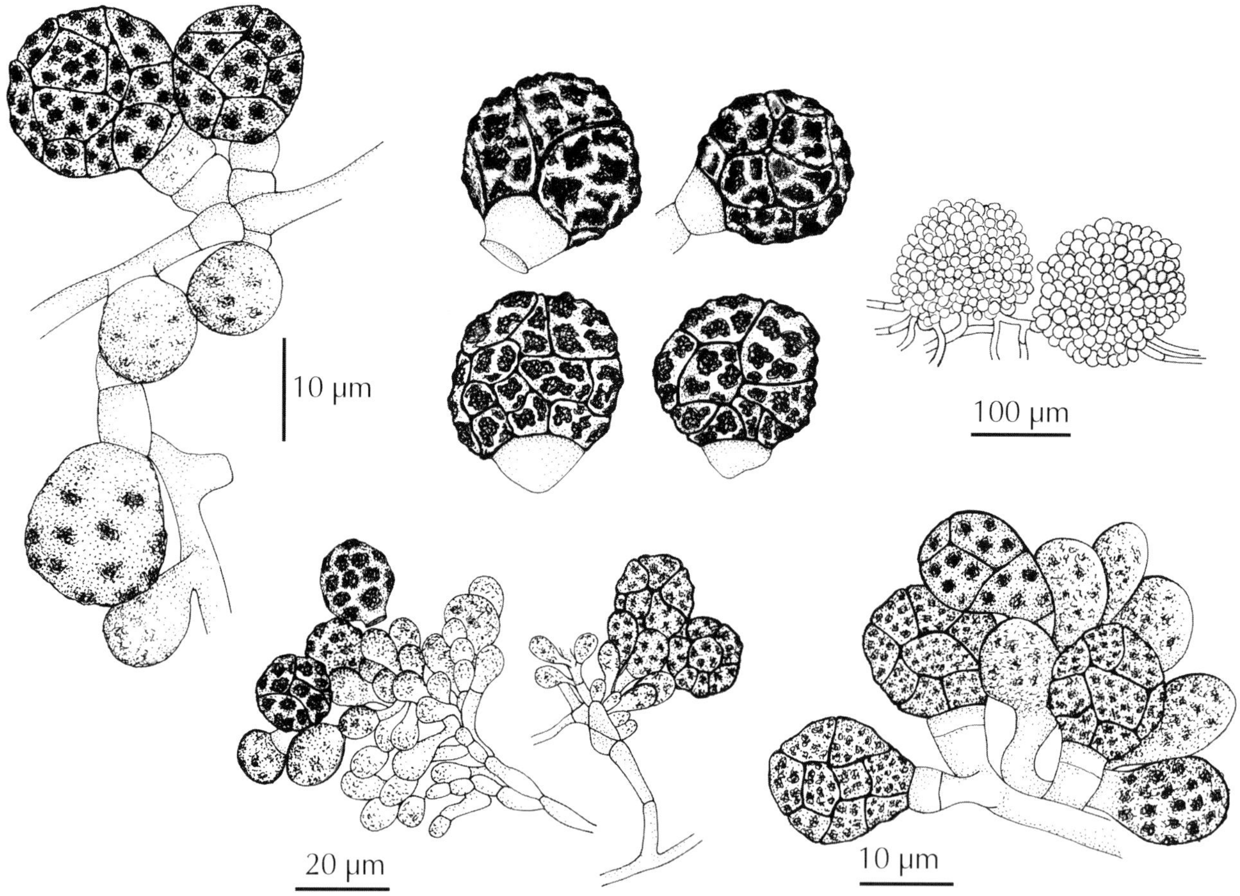

Fig. 53. *Epicoccum nigrum*. Sporodochia, clustered conidiophores with multi-celled conidia.

Pl. 51. *Epicoccum nigrum*. a. Colony on OA after one week; b. sporodochia visible as black dots, in Petri-dish, x 40; c. multi-celled conidia, borne from short conidiogenous cells, x 1000; e. multi-celled conidia with truncate bases, x 500; d,f. young sporodochia, x 500.

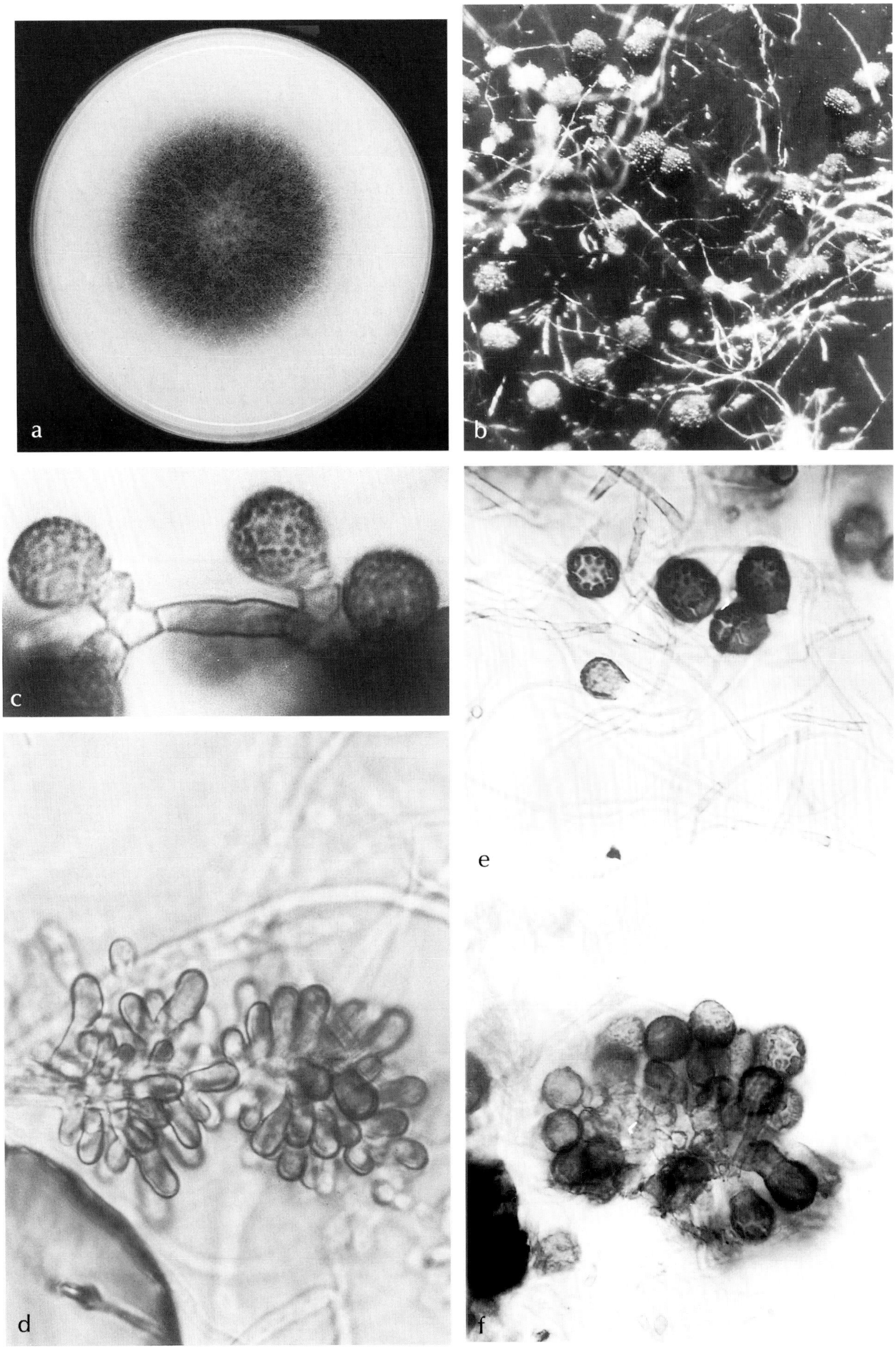
a
b
c
d
e
f

FUSARIUM Link: Fr.

in collaboration with U. Thrane and K. A. Seifert
BioCentrum – DTU, Technical University of Denmark, DK-2800 Lyngby, Denmark and Agriculture and Agri-Food Canada, Ottawa, Canada.

Colonies on PDA or PSA usually growing fast, white, cream-coloured, yellowish, brownish, pink, reddish or violet. Aerial mycelium felty, cottony, usually abundant in fresh isolates, but sometimes reduced or absent in profusely sporulating strains. Colonies on SNA or Carnation Leaf Agar (CLA, see also under Cultivation for identification) usually unpigmented, or with pink, orange, violet or red pigmentation on or near the filter paper or leaves, with sparse aerial mycelium.

Conidiophores sometimes of two types.
Type 1. Complexly branched conidiophores aggregate into slimy pustules (**sporodochia**) or forming continuous slimy masses of spores (**pionnotes**) (Fig. 54). Phialides slender and tapering, with one fertile opening (**monophialide**), on highly branched conidiophores. Producing banana-shaped, septate macroconidia, usually with a pedicellate basal cell (foot cell). **Macroconidia** are produced in clear, yellow, orange, blueish or brownish slime. Suboptimally developed macroconidia are also found in the aerial mycelium of some strains.
Type 2. Unbranched or sparingly branched conidiophores arising from the aerial mycelium, or directly from the agar surface. Phialides slender and tapering, short in some species, longer in others, with one fertile opening (**monophialide**) or several fertile openings (**polyphialide**). One or two-celled **micro-conidia** are produced in chains or wet heads. Microconidia can be ellipsoidal, ovoid, allantoid, globose, pyriform or citriform; species with micro-conidia usually have a characteristic mixture of these shapes. In some species, solitary, dry, septate, fusiform conidia are produced from polyphialides in the aerial mycelium; these are called **mesoconidia** by some authors.

Chlamydospores may be present or absent, occurring within hyphae (intercalary) or terminal, sometimes in macroconidia, single, in pairs, in chains, or in clusters, unpigmented or yellowish to brownish. Dark sclerotia are produced by some strains.

Teleomorphs: *Albonectria* Rossman & Samuels, *Cosmospora* Rabenhorst, *Haematonectria* Samuels & Nirenberg, *Gibberella* Sacc. *Fusarium*-like anamorphs are also produced by some species of *Plectosphaerella* Kleb. and *Monographella* Petr. Perithecia are produced in culture by strains of only a few species; for other species, crossing of opposite mating types is possible with specialized media and incubation conditions. Here, the descriptions refer only to the anamorphs, but the teleomorph is mentioned if known.

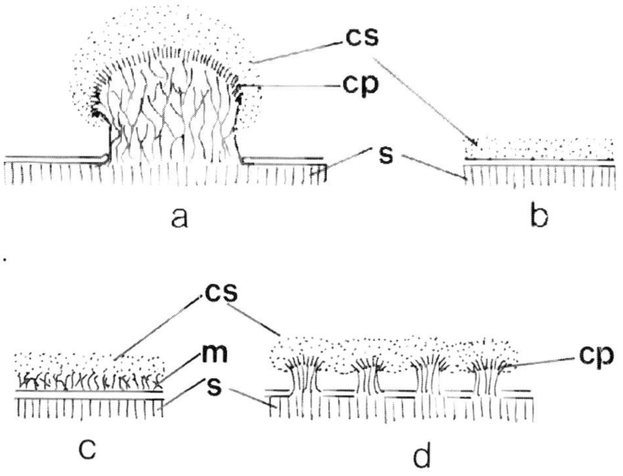

Fig. 54. Types of colony textures in *Fusarium*. a. sporodochium, b-d. pionnotes, conidial slime (cs) directly on substrate (s), covering mycelium (m), covering sporodochia (in d). s = substrate, cs = conidial slime, cp = conidiophores, m = mycelium

Many *Fusarium* species have a world-wide distribution, including plant pathogens of crops widely grown around the globe (eg. wheat, maize). Many species are poorly known, and apparently of restricted geographical distribution. The plant pathogenic species cause root and stem rots, vascular wilts, fruit rot, and infect seeds. Three of the five internationally regulated mycotoxins are produced by *Fusarium* species. Some species are opportunistic pathogens of man or livestock.

For more detailed descriptions and keys, the reader is referred to Booth (1971), Wollenweber and Reinking (1935), Domsch *et al.* (1993), Gerlach and Nirenberg (1982), Burgess *et al.* (1994), Nelson *et al.* (1983), Joffe (1986), and Nirenberg (1989; 1990). For mycotoxin producing *Fusarium* species see Wyllie and Morehouse (1977/78), Marasas *et al.* (1984), and Thrane (1989), Rossman *et al.*, (1999). Following the desciption for each species, references to the publications mentioned above are given in codes. B = Burgess *et al.* (1994); GN = Gerlach and Nirenberg (1982); NTM = Nelson *et al.* (1983) - followed by a page number. A computerized key to common *Fusarium* species operates over the internet at the following URL http://res.agr.ca./brd/fusarium/

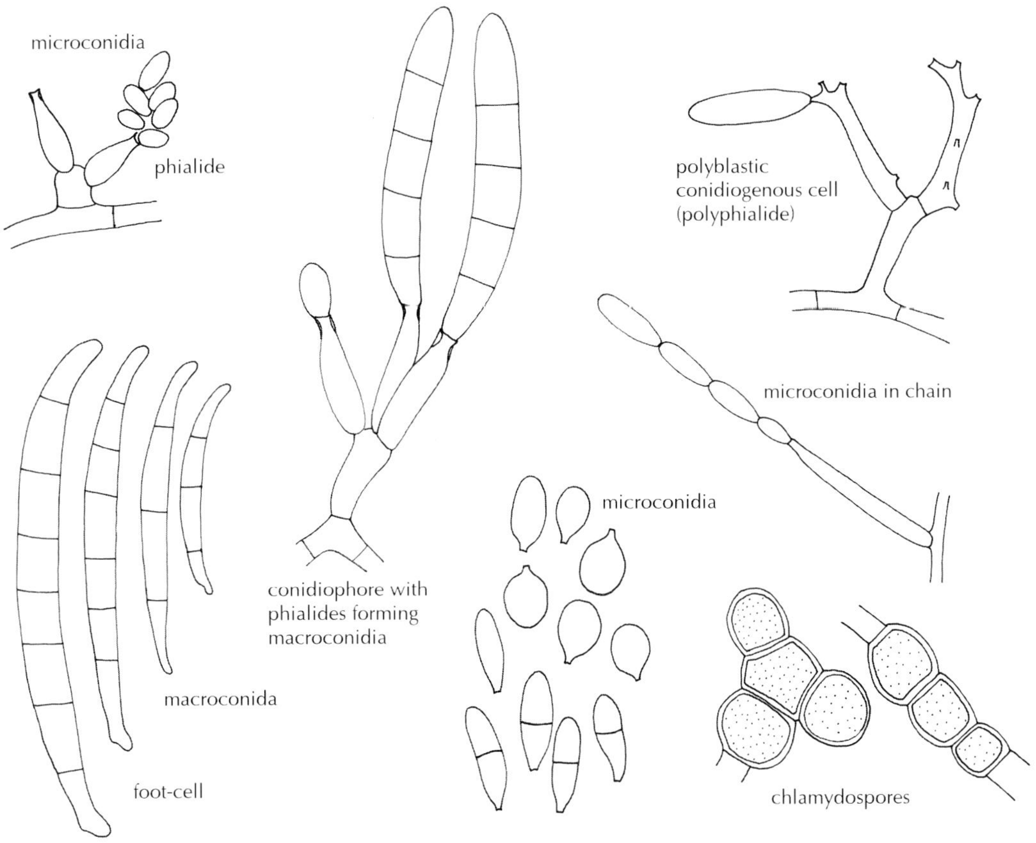

Fig. 55. Morphological structures in *Fusarium*.

CULTIVATION FOR IDENTIFICATION
Cultivation on two media is recommended, one for measuring colony growth rates and optimizing expression of diagnostic pigments, the second for optimizing development of microscopic structures necessary for identification.

Medium 1: Potato Dextrose Agar (PDA) or Potato Sucrose Agar (PSA) are mostly widely used. Incubation at 25°C for 7-14 days in diffuse daylight or in the dark. Formulations made with potatoes, rather than powders, are recommended by some authors. Similar pigments and colony profiles develop tmeal Agar, but colony diameters can be difficult to measure. MEA has not been widely used in *Fusarium*.

Medium 2. Low nutrient media such as SNA (Synthetischer Nährstoffarmer agar = Synthetic Nutrient Agar: Nirenberg, 1976) with a piece of sterile filter paper or CLA (Carnation Leaf Agar, Fischer *et al.*, 1982) are most widely used. Incubation at 25°C for 7-14 days under mixed black light (near-UV, emission ca. 310-360 nm) and artificial daylight (cool white tubes), on a 12 hr light: 12 hr dark cycle, is necessary to maximize production of sporodochia. Burgess et al. (1988) show the design of one incubation setup. Incubation of cultures in the dark is necessary for the critical identification of some species in section Liseola.
Microscopical slides should be mounted in lactic acid or Shear's mounting medium. A slide can also be made using a SNA block (1 x 1 cm) with the sporulating fungus transferred to a slide, add a drop of 0.25% triton-X (in water), and a clover slip. These slides can even be observed with oil immersion. Note that in microscopical slides made in water the conidia will swell up after 5 to 10 minutes.

Identification of species of *Fusarium* can be difficult because of the need to differentiate rather similar shapes of macroconidia, which can be difficult without experience. Characters of the microconidia (shapes, arrangement into wet heads or chains, presence or absence of polyphialides) are usually easier to interpret for those species that produce them. *Fusarium* cultures tend to degenerate if they are isolated or maintained on rich media. Certain species, such as *F. oxysporum* and *F. equiseti* are almost impossible to maintain in their wild-type state.

KEY TO THE SPECIES TREATED

1a. Aseptate microconidia produced abundantly, especially in the aerial mycelium............... 2
1b. Aseptate microconidia rare or absent in the aerial mycelium.. 15

2a. Some round, pyriform or citriform microconidia present, especially in the aerial mycelium (see Fig. 56 a-c) 3
2b. No round, pyriform or citriform microconidia present, aseptate microconidia ellipsoidal, reniform, clavate or fusiform (see Fig. 56 d-g)... 5

3a. Most or many conidiogenous cells producing microconidia with more than one opening (polyphialides, or polyblastic).. *F. sporotrichioides*
3b. Conidiogenous cells producing microconidia with only one opening (monophialide) 4

4a. Monophialides producing microconidia plump, most microconidia round, sporodochia and macroconidia sometimes not produced, macroconidia variable in shape.................. *F. poae*
4b. Monophialides producing microconidia relatively long and slender, microconidia a mixture of citriform, ellipsoidal and reniform, sporodochia and macroconidia usually produced, macroconidia relatively uniformly falcate (type A) .... *F. tricinctum*

5a. Polyphialides present in the old or new parts of the aerial mycelium, intermixed with monophialides 6
5b. Only monophialides present in the aerial mycelium.. 9

6a. Microconidia produced in short, dry chains in the aerial mycelium .... *F. proliferatum*
6b. Microconidia produced in wet heads in the aerial mycelium, or if dry not in chains......... 7

7a. Colonies on PDA or PSA with at least some red pigment; macroconidia (if produced) curved, slender to medium in stature (type A or D).. 8
7b. Colonies on PDA or PSA with at least some purple or violet colour, macroconidia (if produced) more or less straight, slender, with parallel walls (type B)............................ *F. subglutinans*

8a. Polyphialides often with more than 3 openings producing conidia, macroconidia falcate (type A), colonies on PDA growing more than 60 mm in 4 days .... *F. sporotrichioides*
8b. Polyphialides usually with 2-3 openings producing conidia, macroconidia slender with more or less parallel walls (type D), colonies on PDA growing less than 60 mm in 4 days .... *F. avenaceum*

9a. Microconidia produced in long, dry chains in the aerial mycelium.................................. *F. verticillioides*
9b. Microconidia in wet heads in aerial mycelium 10

10a. Macroconidia conspicuously curved, sometimes with an elongated apical cell, usually abundantly produced.. 11
10b. Macroconidia straight to slightly curved, apical cell usually not elongated, macroconidia sometimes not abundantly produced... 13

11a. Macroconidia slender (type D), multiseptate fusiform conidia lacking a foot-shaped basal cell present, no chlamydospores produced.. *F. avenaceum*
11b. Macroconidia broader, fusiform, multiseptate fusiform conidia lacking a foot-shaped basal cell not produced, chlamydospores produced 12

12a. PDA colonies with no red pigments, usually brown... *F. equiseti*
12b. PDA colonies with red pigments .... *F. acuminatum*

13a. Macroconidia widest above the middle (wedge-shaped), apical cell shaped like a dolphin nose, chlamydospores produced in chains or clumps .... *F. sambucinum*
13b. Macroconidia with more or less parallel walls, widest near centre, chlamydospores produced singly or in pairs, refractile 14

14a. Monophialides in aerial mycelium long, microconidia usually ellipsoidal to fusiform, or clavate, macroconidia with a rounded apical cell and often lacking a conspicuous foot cell, sometimes sparsely produced... *F. solani*
14b. Monophialides in aerial mycelium short, comma-shaped microconidia usually present, macroconidia with a pointed apical cell and relatively conspicuous foot cell ..... *F. oxysporum*

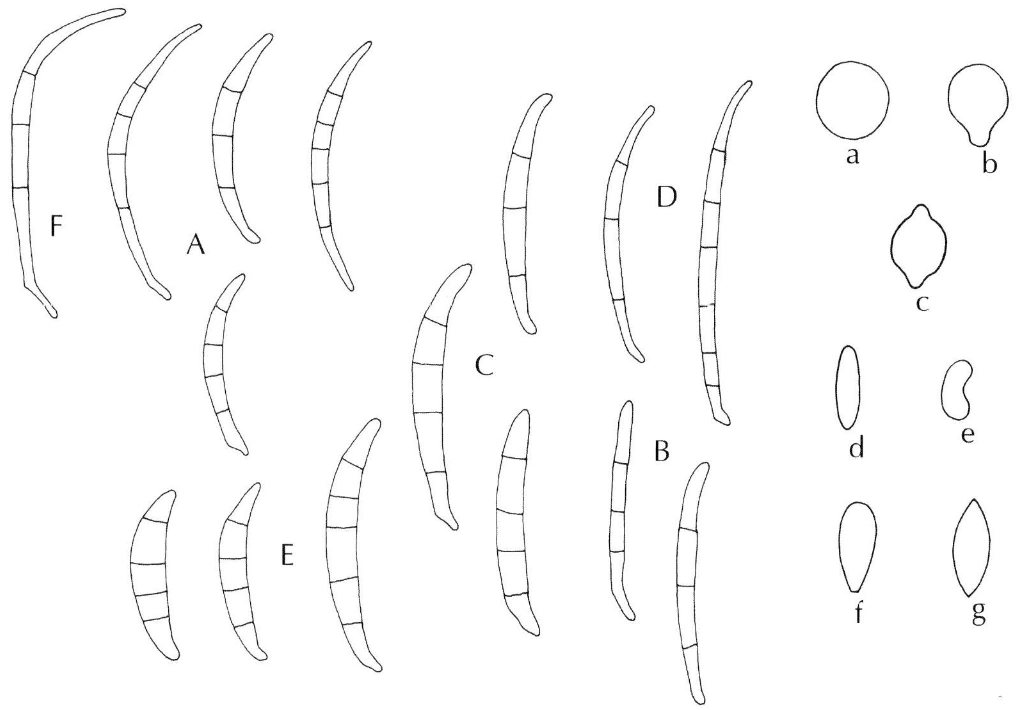

Fig. 56. Shape of micro- and macroconidia of *Fusarium*

15a. Macroconidia widest above the middle, wedge-shaped 16
15b. Macroconidia generally widest near centre, not wedge-shaped 17

16a. Macroconidia with a blunt apical cell, often lacking a well-defined foot-cell, more than 6 µm wide, sporodochia orange or brown, typically occurring on grains .... *F. culmorum*
16b. Macroconidia often with an apical cell resembling a dolphin's snout, foot cell relatively distinct, less than 6 µm wide, sporodochia yellow or orange, typically occurring on potatoes .... *F. sambucinum*

17a. Producing fusiform, multiseptate conidia lacking a foot cell 18
17b. Producing only conidia with relatively typical foot cells 21

18a. Colonies on PDA lacking red pigments 19
18b. Colonies on PDA with red pigments 20

19a. Most phialides polyphialides or polyblastic, conidia dry, sporodochia not produced .... *F. semitectum*
19b. Only monophialides produced, conidia wet, macroconidia often with an extended apical cell and elongated foot cell (type F), sporodochia orange .... *F. equiseti*

20a. Macroconidia falcate, of medium stature (type A), conidia wet, chlamydospores produced .... *F. acuminatum*
20b. Macroconidia slender with more or less parallel walls (type D), dry conidia produced from polyphialides at least in young colonies .... *F. avenaceum*

21a. Walls of macroconidia more or less parallel, macroconidia straight or gently curved (type C), sporodochia lightly coloured, not brown, colonies on PDA growing more than 60 mm in 4 days .... *F. graminearum*
21b. Macroconidia broadest near the middle, appearing swollen, often with a conspicuous foot cell almost perpendicular to the main body of the conidium, sporodochia brownish, colonies on PDA growing less than 60 mm in 4 days .... *F. crookwellense*

Table 4. Colony and morphological characters of the *Fusarium* species treated.

Species	Reverse on PDA	Sporo-dochial colour	Macroconidia shape	Micro-conidia shape	Microconidia phialides	Microconidia groups	chlamydospores	substrates
F. acuminatum	red	orange	A, widest below middle	d, e, f	mono	slimy heads	single, chains or clumps	grains, seeds, fruits vegetables
F. avenaceum	red	orange	D	d, f, g	mono + poly (especially at colony margin)	mostly dry	none	grains seeds
F. crookwellense	red	brownish	E, widest in middle	-	-	-	single, chains or clumps	grains seeds potatoes
F. culmorum	red yellow	orange brownish	E, widest above middle	(d, f)	(mono)	(slimy heads)	single, chains or clumps	grains
F. equiseti	brown	orange	A, F	d, e, f	mono	slimy heads	single, chains or clumps, brown	dead plants
F. graminearum	red yellow	colourless reddish	C	-	-	-	single, chains or clumps	grains seeds
F. oxysporum	colourless blue orange	yellow orange blue	C	d, e, f	mono	slimy heads	single or pairs	roots soil
F. poae	red	colourless orange	E	a, b	mono plump	slimy heads	single, chains or clumps	grains
F. proliferatum	violet	colourless	B	d, f	poly + mono	short dry chains	none	maize
F. sambucinum	colourless yellow, red	orange	E, widest above middle	d, f	mono	slimy heads	single, chains or clumps	potatoes
F. semitectum	colourless brown	none	A	g	polyblastic	mostly dry	single, chains or clumps	fruits vegetables
F. solani	yellow red	yellow colourless blue	C	d, f, g	mono long	slimy heads	single or pairs	roots soil
F. sporo-trichioides	red	orange colourless	A	a, b, d, f, g	poly + mono	mixture of dry and slimy heads	single, chains or clumps	grains seeds
F. subglutinans	violet	colourless	B	d, g	poly	slimy heads	none	maize
F. tricinctum	red	orange	A	c, d, e	mono	slimy heads	single, chains or clumps	grains seeds
F. verticillioides	violet	colourless	B	d, f	mono	long dry chains	none	maize

SYNOPTIC IDENTIFICATION KEY FOR COMMON *FUSARIUM* SPECIES.

1. *F. acuminatum*
2. *F. avenaceum*
3. *F. crookwellense*
4. *F. culmorum*
5. *F. equiseti*
6. *F. graminearum*
7. *F. oxysporum*
8. *F. poae*
9. *F. proliferatum*
10. *F. sambucinum*
11. *F. semitectum*
12. *F. solani*
13. *F. sporotrichioides*
14. *F. subglutinans*
15. *F. tricinctum*
16. *F. verticillioides*

Colony diameter on PSA after 4 days at 25 C less than 60 mm: 1, 2, 5, 7, 8, 9, 10, 11, 12, 14, 15, 16
Colony diameter on PSA after 4 days at 25 C more than 60 mm: 3, 4, (5), 6, 8, 13
Growth on PSA at 37 C (colony diameter > 2 mm): (5), 7, 9, 11, 12, 14, 16
Reverse on PSA after 7-10 days at 25 C being rose, red, blue, violet: 1, 2, 3, 4, 6, 7, 8, 9, 10, 12, 13, 14, 15, 16
Reverse on PSA after 7-10 days at 25 C being light, yellow, brown (no red or blue tinge): (4), 5, (6), 7, (8), 10, 11, 12, (14), (16)
Growth on TAN at 25°C (colony diameter > 2 mm):1, 2, 5, 7, 9, 10, 11, 12, 13, 14, 15, 16
Only multiseptate fusiform conidia present: (1), (2), 3, 4, (5), 6, (10), 11
Aseptate conidia produced abundantly: (2), 7, 8, 9, 12, 13, 14, 15, 16
Aseptate conidia present, round, pyriform or citriform: 8, 13, 15
Aseptate conidia present, ellipsoidal, reniform, clavate, or fusiform: 1, (2), (4), 5, 7, 9, 10, 12, 13, 14, 15, 16
Aseptate conidia produced in dry chains: 9, 16
Polyphialides or polyblastic cells present: 2, 9, 11, 13, 14
Multiseptate fusiform conidia without distinct foot-cell present:(1), 2, 4, (5), (10), 11, 12
Chlamydospores present: 1, 3, 4, 5, 6, 7, 8, 10, 11, 12, 13, 15

SECONDARY METABOLITES:

Antibiotic Y: 1, 2, 15
Aurofusarin: 1, 2, 3, 4, 6, (8), 10, 13, 15
Beauvericin: 9, 14
Butenolide: 3, 4, 6, 8, 10, 13, 15
Chlamydosporol: 1, 2, (11), 15
Chrysogine: 3, 4, 5, 6, 11, 15
Culmorin: 3, 4, 6,
Cyclonerodiols: 2, 3, 4, 6, 8,
Deoxynivalenol and derivatives: 4, 6,
Diacetoxyscirpenol: 5, 8, 10, (11)
Enniatins: 1, 2, 10, 15
Equisetin: 5, 11
Fumonisins: 9, 14, 16
Fusaproliferin: 9, 14
Fusapyrone: 9, 11, 14
Fusaric acid: 3, 7, 9, 12, 14, 16
Fusarin C: 2, 3, 4, 6, 8, 9, (11), 13, 15, 16
Fusarochromanone: 5
Gamma-lactones: 8
Gibepyrone A: 7, 16
Moniliformin: 1, 2, 7, 9, 14, 16
Naphthoquinone pigments: 7, 9, 12, 14, 16
Nectriafurone: 7, (12)
Nivalenol and derivatives: 3, 4, 5, 6, 8
T-2 toxin and derivatives: (1), 8, (11), 13
Visoltricin: 15
Zearalenone: 3, 4, 5, 6, 11

Fusarium acuminatum Ellis & Everhart

Teleomorph: *Gibberella acuminata* Wollenw.

Colonies on PDA or PSA at 25°C, attaining a diameter of 2.3-5.9 cm in 4 days, normally strongly carmine red, peach-beige and finally brownish floccose aerial mycelium ochraceous-brownish. Sporodochia and pionnotal layers salmon to bright orange. Reverse red-brownish coloured. Conidiophores: at first lateral phialides in the aerial mycelium later branching, densely branched when forming sporodochia, monophialidic. Microconidia absent or occasionally sparsely produced in the aerial mycelium. Macroconidia strongly curved, sickle-shaped, slender 3-5 septate, mostly 24-54 x 3-4.5 µm, with pedicellate basal cell. Chlamydospores intercalary and in chains. (B102, GN187, NTM100).

Important (toxic) metabolites: 4-acetamido-4-hydroxy-2-butenoic acid-þ-lactone ("butenolide"), antibiotic Y, chlamydosporol, enniatins, monilifor-min, **trichothecenes**.

HABITAT: food
World-wide distribution in temperate as well as tropical countries. Isolated from soil, plants, seeds of cereals, citrus, bananas, can cause stem rot of maize, foot and root rot of legumes.

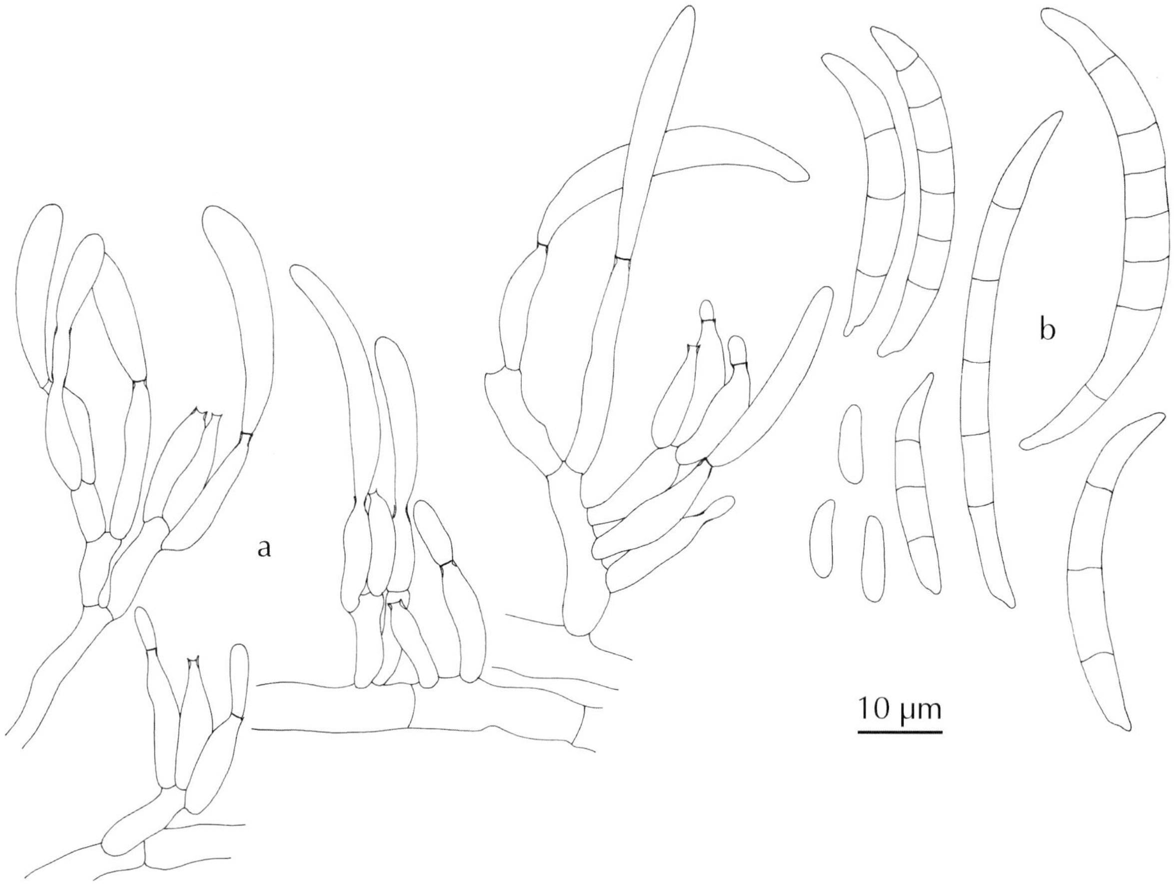

Fig. 57. *Fusarium acuminatum*. a. Conidiophores and phialides; b. macro- and micro- conidia, microconidia are sparsely formed in the aerial mycelium.
Pl. 52. *Fusarium acuminatum*. a. Colony on PDA after one week; b, d-f. conidiophores, lateral phialides and macroconidia, x 800; c. macroconidia, x 800.

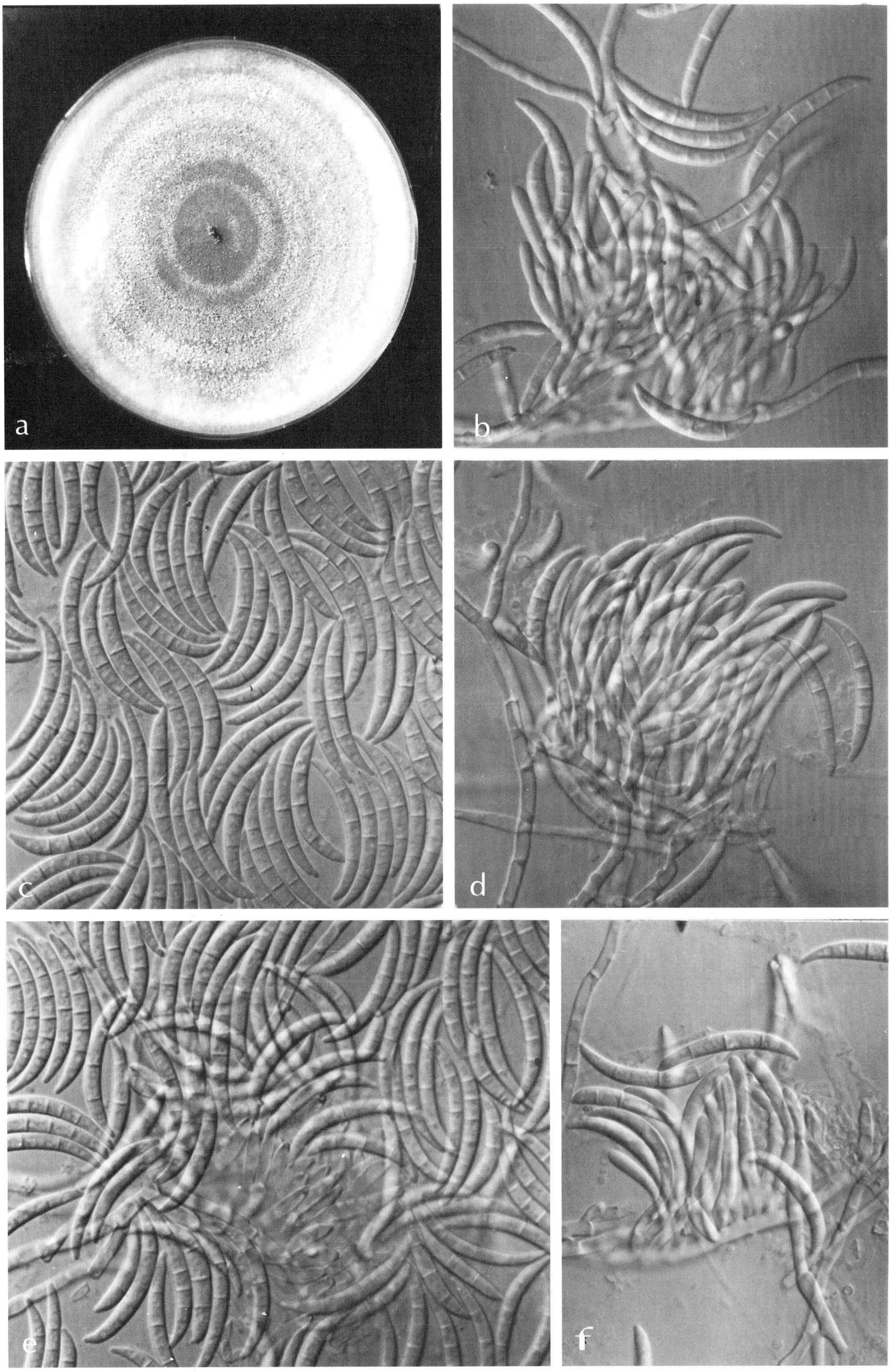
a
b
c
d
e
f

Fusarium avenaceum (Corda: Fr.) Sacc.

Teleomorph: *Gibberella avenacea* R.J. Cook

Colonies on PDA or PSA at 25°C attaining a diameter of 3-5.9 cm in 4 days, yellow or reddish with a whitish floccose overgrowth of aerial mycelium, the latter becoming peach-coloured or reddish-brown and felty with age. Reverse in yellowish or brown to reddish shades. Conidiophores simple or more or less branched, arising from the aerial mycelium or in orange sporodochia with monophialides or polyphialides. Microconidia from polyblastic conidiogenous cells, fusiform, 0-3-septate, variable in size, 6-30 x 2.5-4.5(6) µm. Macroconidia borne on phialides, narrowly fusiform or slender, sickle-shaped to almost straight, narrowing at both ends, 4-7 septate with an elongated apical cell and a distinct foot cell, mostly 35-89(90) x 3.5-4(6) µm, in mass more or less orange coloured. Chlamydospores absent in mycelium, sometimes present in conidia. (B96, GN139, NTM80).

Temperature: optimum 25°C; minimum -3°C; maximum 31°C.

Important (toxic) metabolites: antibiotic Y, chlamydosporol, enniatins, fusarin C, moniliformin.

HABITAT: food
World-wide distribution, mainly in temperate zones, but also occurring in the (sub)tropics: soil, plant material. This species causes root rots in wheat, rye, maize, clover, lucerne and damaged vegetables, fruits peaches, apples, pears, seeds of oats, barley, and wheat.

NOTE:
This species can be extremely variable in appearance and colour, commonly peach, reddish or brownish-yellow.

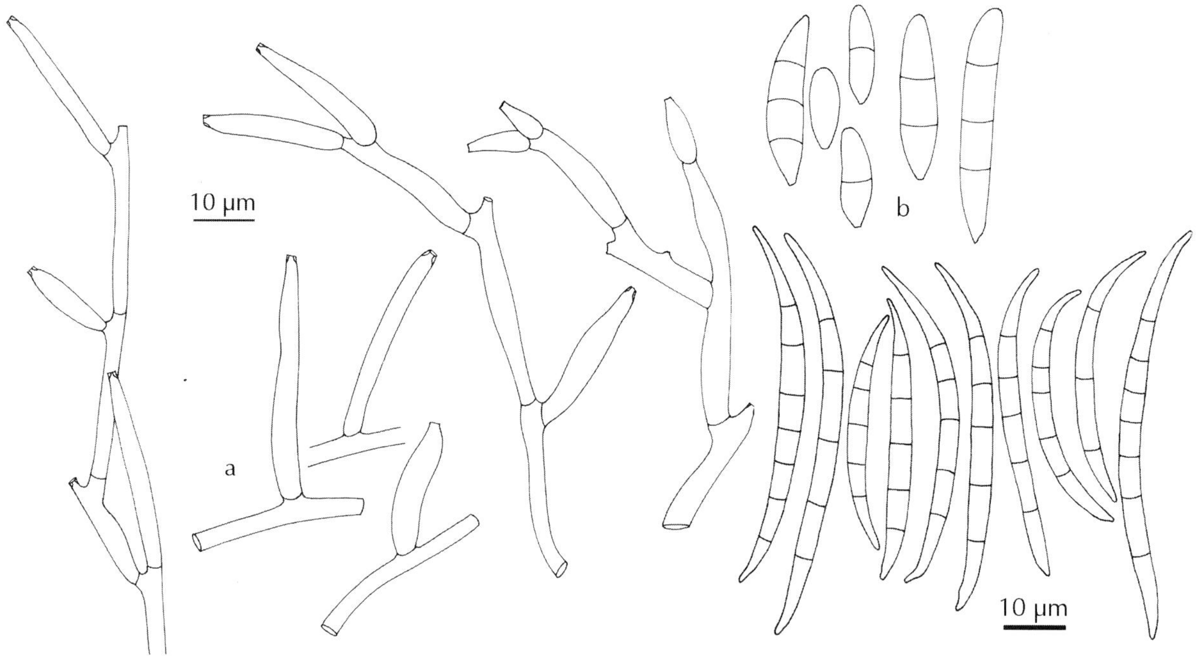

Fig. 58. *Fusarium avenaceum*. a. Conidiophores with simple and polyblastic conidiogenous cells; b. microconidia, 1-3 septate, more or less fusiform; macroconidia, 4-7-septate, slender, sickle-shaped to almost straight (orig. partly after E.J. Hermanides-Nijhof).
Pl. 53. *Fusarium avenaceum*. a. Colony on PDA after one week; b, c. conidiophores with monophialides, x 1000; d-f. sporodochial macroconidia, long, slender, curved to almost straight, 4-7-septate, x 1000.

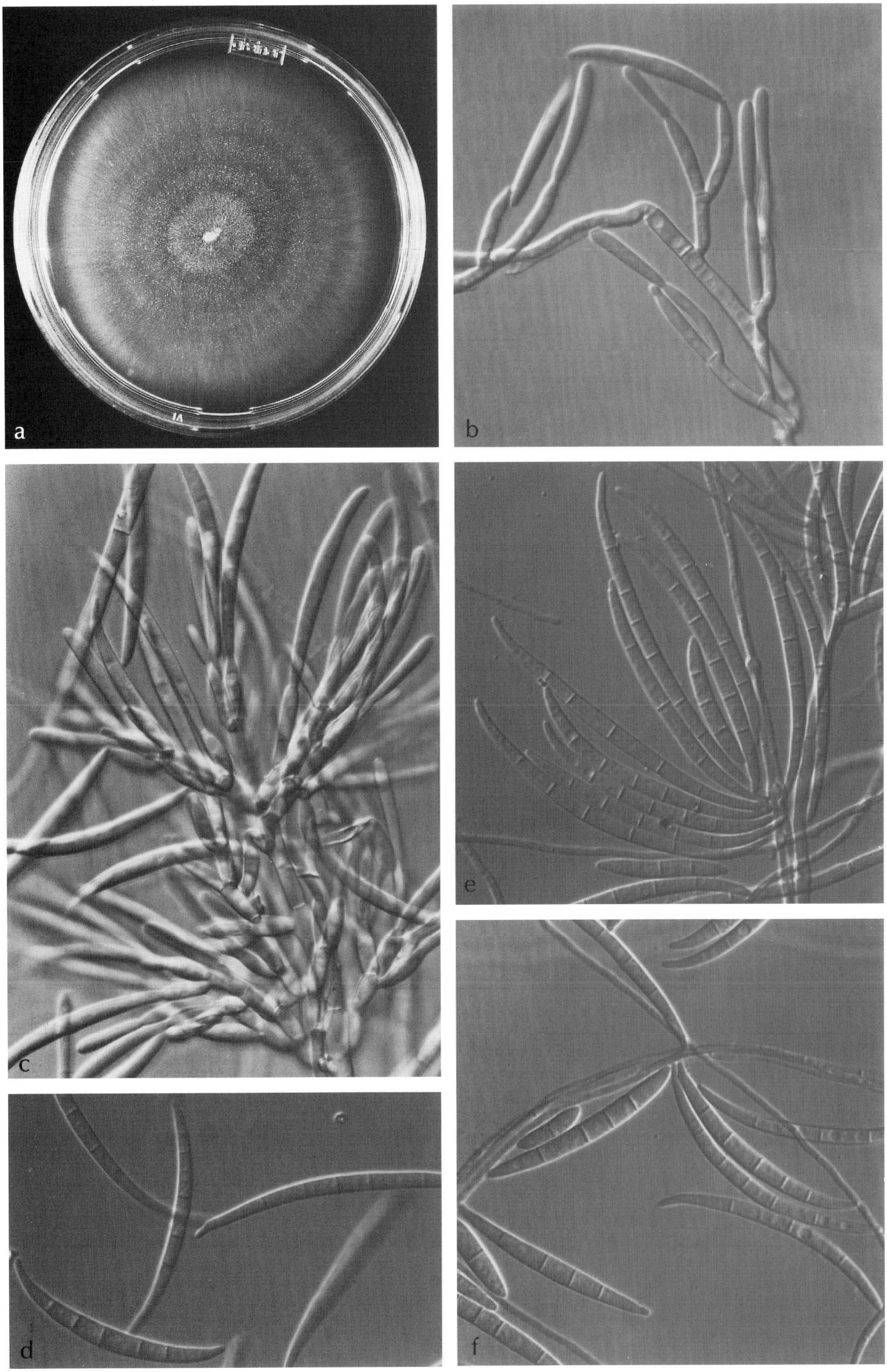
a
b
c
d
e
f

Fusarium crookwellense Burgess, Nelson & Toussoun

= *Fusarium cerealis* (Cooke) Sacc.

Teleomorph: unknown

Colonies on PDA and PSA at 25°C attaining a diameter of 7.5-9 cm in 4 days. Aerial mycelium floccose and whitish to yellowish or pinkish at first, becoming ochraceous to brownish-red and felty with age. Sporodochia orange to red-brown produced in the central part of older cultures.
Reverse and agar surface reddish to purple or brownish.
Conidiophores branched or unbranched bearing monophialides. Microconidia absent. Macroconidia dorsally curved to sickle-shaped, with a distinctly curved apical cell and distinct foot cell, commonly 5-septate: 35-40(50) x 5-6.3(7.5) µm.
Chlamydospores occurring in mycelium or in conidia. (B87: *F. crookwellense*, NTM121: *F. crookwellense*).

Important (toxic) metabolites: 4-acetamido-4-hydroxy-2-butenoic acid þ-lactone ("butenolide"), culmorin, fusarin C, **trichothecenes, zearalenone**.

HABITAT: food
World-wide distribution: soil, seeds of cereals. Causes foot and root rot and head blight in cereals. Causes rot of potatoes.

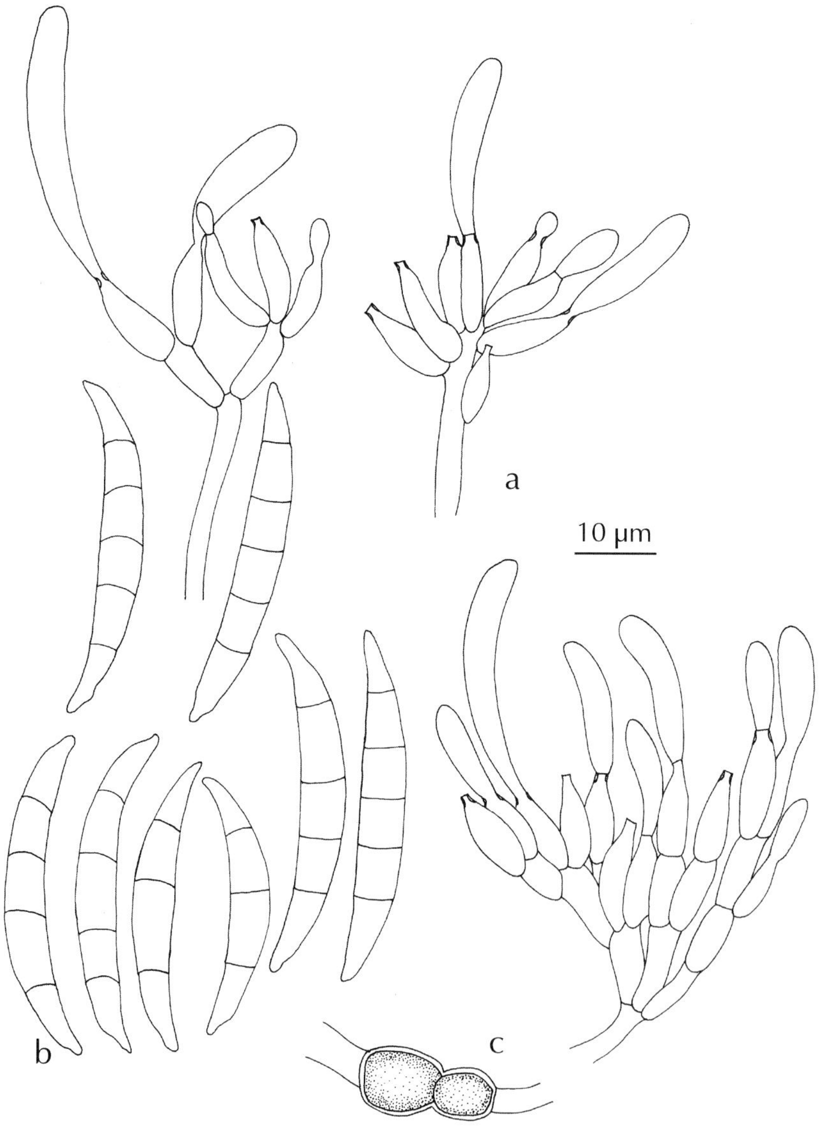

Fig. 59. *Fusarium crookwellense*. a. Conidiophores with monophialides; b. macroconidia; c. chlamydospores.
Pl. 54. *Fusarium crookwellense*. a. Colony on PDA after one week; b,d,e. conidiophores and macroconidia, b. x 960, d. x 940, e. x 1065; c. macroconidia, x 670.

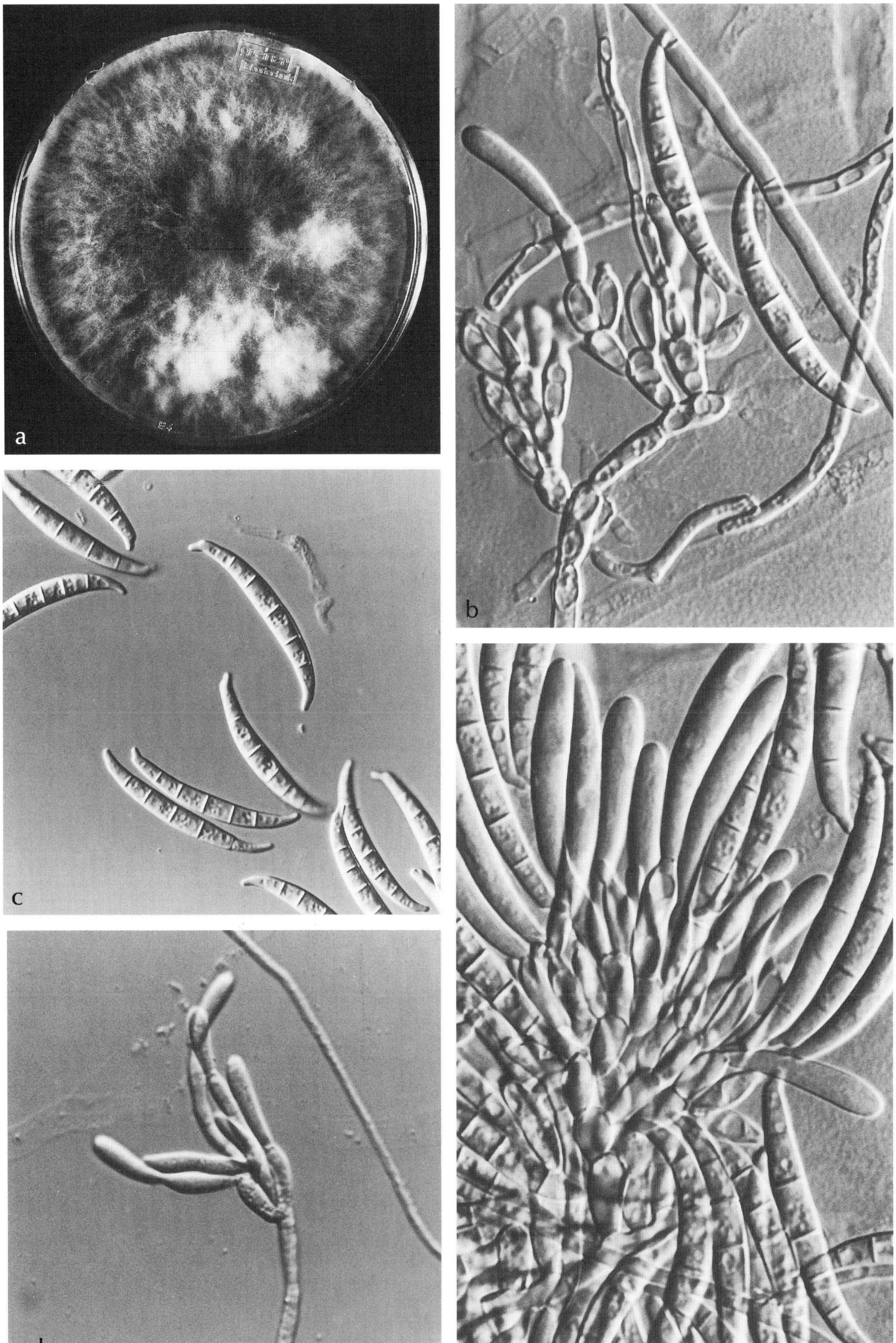
a
b
c
d
e

Fusarium culmorum (W.G. Smith) Sacc.

Teleomorph: unknown

Colonies on PDA and PSA at 25°C attaining a diameter of 7.5-9 cm in 4 days. Aerial mycelium floccose and whitish to yellowish or pinkish at first, becoming ochraceous to brownish-red and felty with age. Reverse and agar surface reddish to purple or brownish. Conidiophores usually branched with short and wide phialides from which rather thick-walled macroconidia are borne in the aerial mycelium, or in sporodochia or in pionnotes with a red-brown slime. Microconidia absent. Macroconidia fusiform to sickle-shaped, usually 5-septate (sometimes 3-4- or 6-8-septate), with a more or less pointed apex and a distinct footcell, 3-septate: 26-36 x 4-6 µm, 4-septate: 30-46 x 5-7 µm, 5-septate: 34-50 x 5-7 µm. Chlamydospores brownish, smooth to rough-walled, present in mycelium or in conidia, intercalary or sometimes terminal, in chains, in clumps, solitary or absent. (B82, GN225, NTM115).

Temperature: optimum 25°C; minimum 0°C; maximum 31°C.

Important (toxic) metabolites: 4-acetamido-4-hydroxy-2-butenoic acid þ-lactone ("butenolide"), culmorin, fusarin C, **trichothecenes, zearalenone**.

HABITAT: food, indoor
World-wide distribution with a slight preference for the temperate zones: soil, seeds of oats, barley, wheat, and other cereals. Causes foot and root rot and head blight in cereals. Attacks maize. Causes storage rot of apples, potatoes, sugar beet.

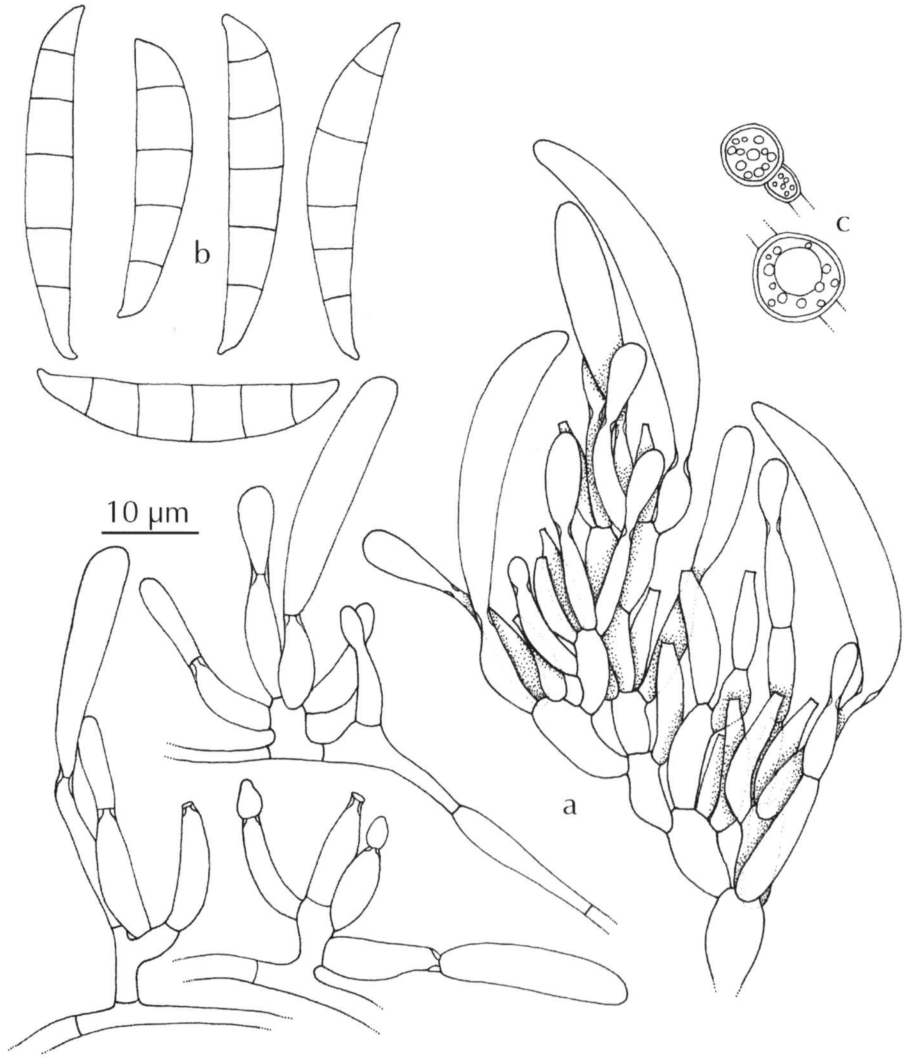

Fig. 60 *Fusarium culmorum*. a. Conidiophores forming only one type of conidium; b. conidia; c. chlamydospores. (orig. partly after E.J. Hermanides-Nijhof).
Pl. 55. *Fusarium culmorum*. a. Colony after 12 days; b-d. conidiophores and macroconidia, b. x 200, c-d. x 800; e, f. macroconidia, x 800.

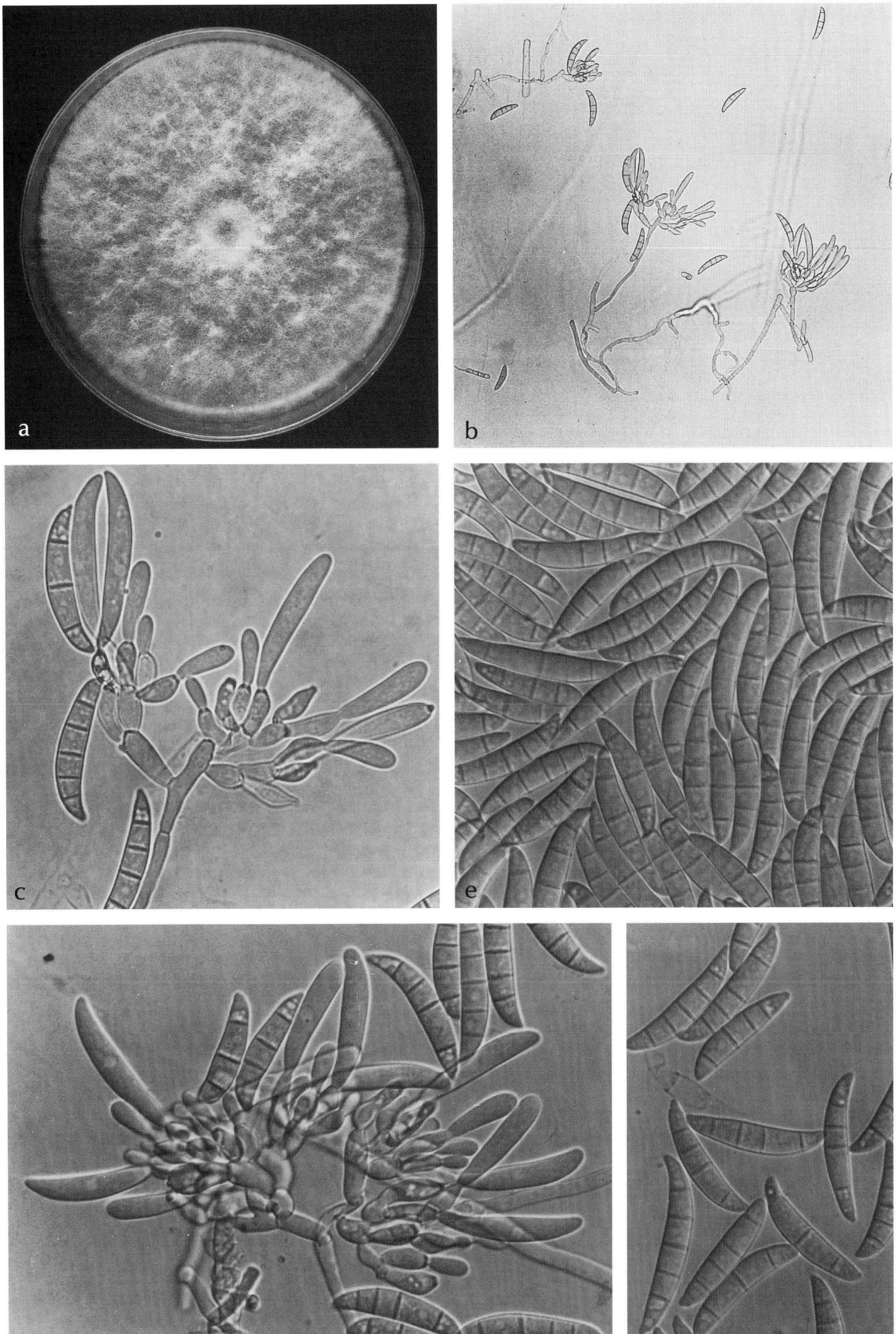
a
b
c
e
d
f

Fusarium equiseti (Corda) Sacc.

Teleomorph: *Gibberella intricans* Wollenw.

Colonies on PDA or PSA at 25°C attaining a diameter of 4.5-6.9 cm in 4 days, with floccose aerial tufts, becoming creamish to yellowish-brown with age. A tendency towards pionnotes and sporodochia can be observed in some isolates. Reverse peach-coloured, changing to light-brown or dark-brown. Conidiophores branched or unbranched, monophialidic. Microconidia absent or rare, spindle-shaped or ovoid, 0-2-septate, 6-24 x 2.5-4 µm. Macroconidia sparse at first and borne from simple lateral phialides on the aerial mycelium. After about 14 days branched conidiophores are also produced. Macroconidia sickle-shaped with a distinct foot cell and in some isolates an elongated curved apical cell, 3-5(7)-septate, typically 5-septate. 3-septate: (15)22-45 x (2.5)3-5 µm, 5-septate: (30)40-58 x (3)3.7-5 µm, 7-septate: 42-60 x 4-5.9 µm. Chlamydospores abundant, pale brown at maturity, smooth- or rough-walled, intercalary, solitary, in chains or clumps, in the hyphae or in conidia. (B109, GN177, NTM89).

Temperature: optimum 21(30)°C; minimum -3°C; maximum depending of the isolate about 35°C.

Important (toxic) metabolites: equisetin, fusarochromanone, **trichothecenes, zearalenone**.

HABITAT: food
World-wide distribution in soils and damaged plant tissues. Reported from seeds and fruits which have been in contact with soil.

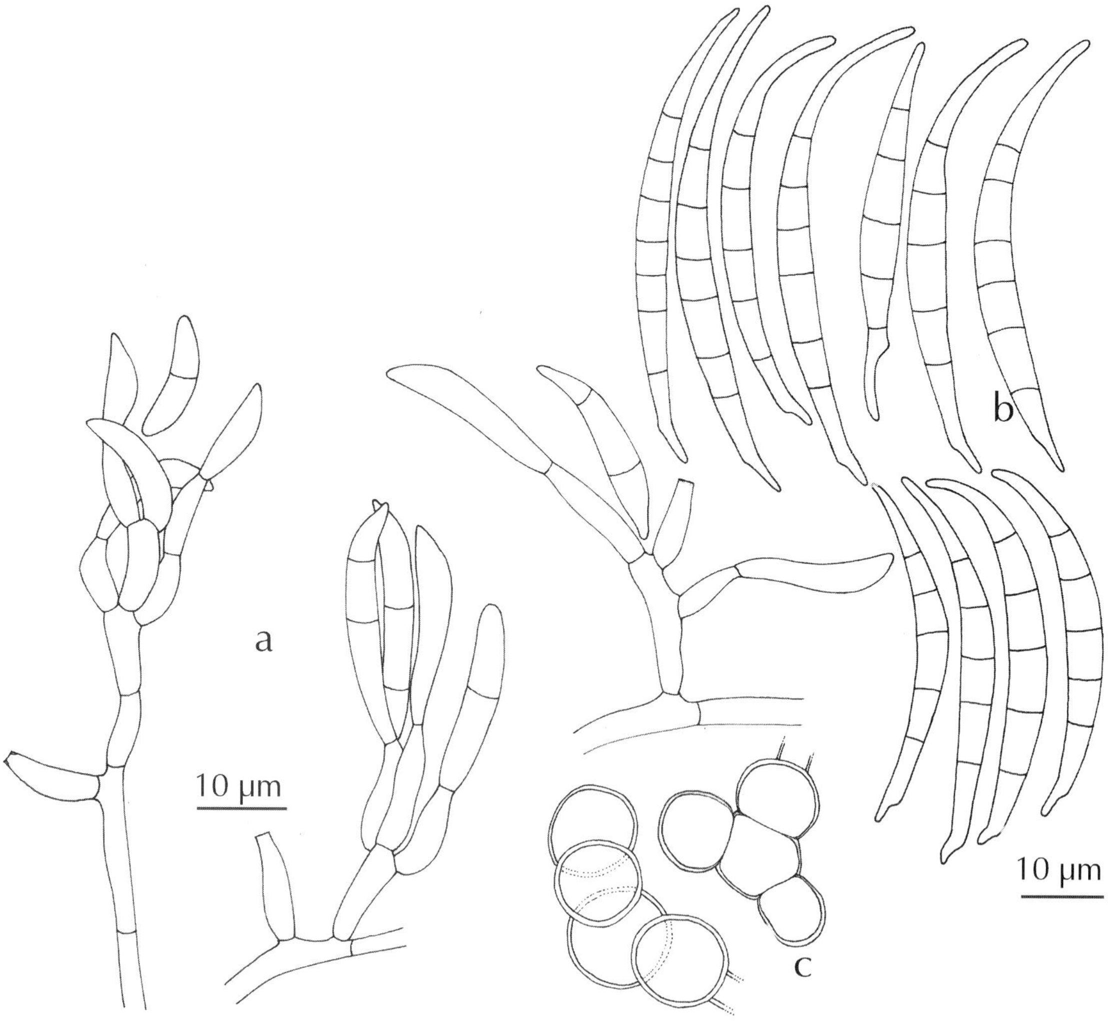

Fig. 61. *Fusarium equiseti*. a. Conidiophores; b. macroconidia; c. chlamydospores. (orig. partly after E.J. Hermanides-Nijhof)
Pl. 56. *Fusarium equiseti*. a. Colony on PDA after one week; b. conidia with elongated apical cells and distinct foot cells, x 1000; c, e, f. conidiophores with monophialides and conidia, x 1000; d. chlamydospores, x 750.

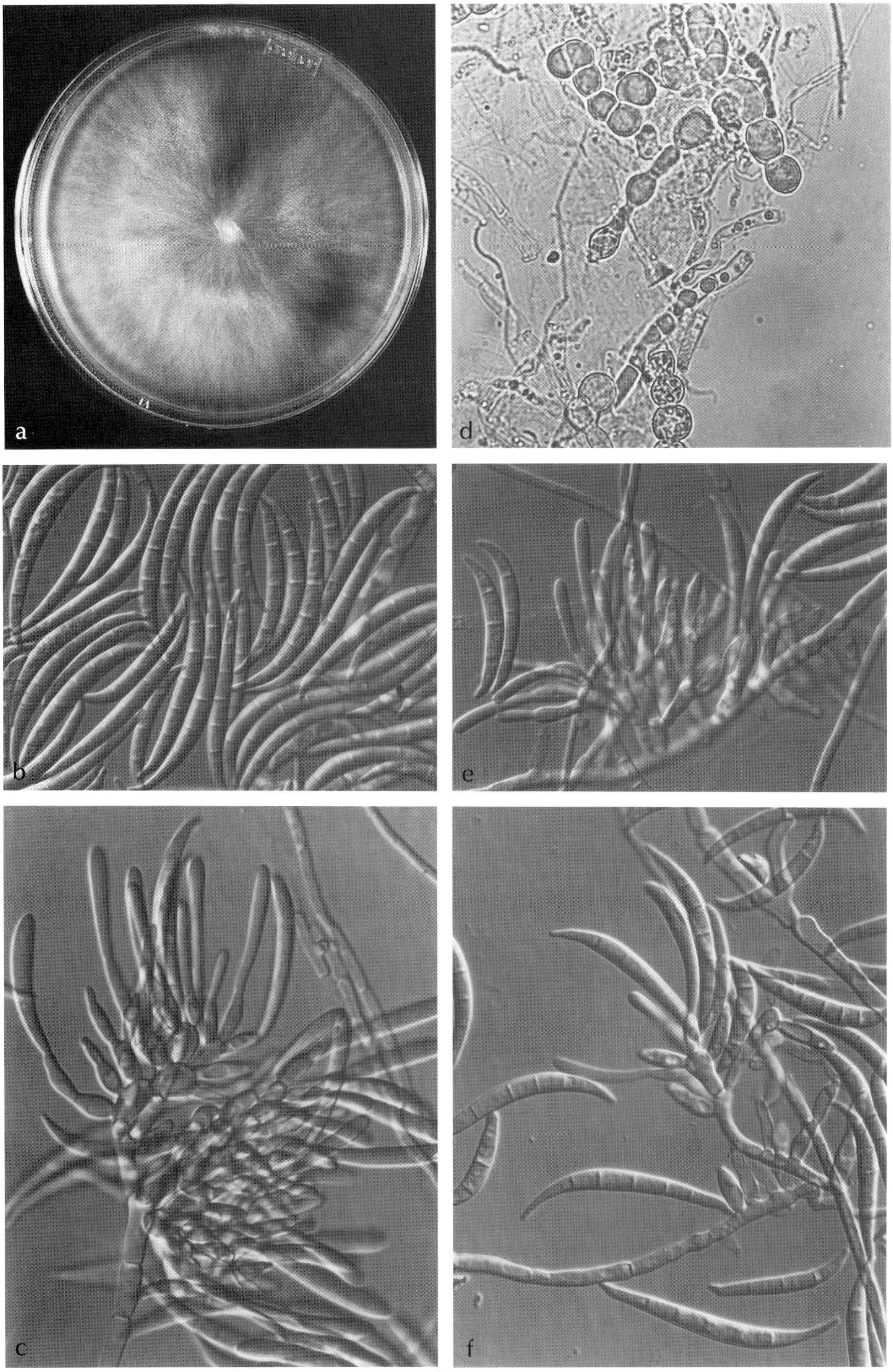
a
d
b
e
c
f

Fusarium graminearum Schwabe

Teleomorph: *Gibberella zeae* (Schw.:Fr.) Petch

Colonies on PDA or PSA at 25°C growing fast, attaining a diameter of 7.5-9 cm in 4 days, greyish rose to red, often vinaceous becoming brownish at age. Aerial mycelium floccose, whitish becoming brownish to rose. Reverse in red to reddish brown shades. Conidiophores at first as simple monophialides on the hyphae later strongly branched, sometimes percurrently but not sympodially proliferating. Microconidia absent. Macroconidia slender, sickle-shaped, curved with pointed and curved apical and pedicellate cells, mostly (3)5-6(9)-septate, 41-60 (80) x 4-5.5 µm. Chlamydospores sometimes present, mostly intercalary and in chains, often absent. (B89, GN241, NTM118).

Important (toxic) metabolites: 4-acetamido-4-hydroxy-2-butenoic acid γ-lactone ("butenolide"), culmorin, fusarin C, **trichothecenes, zearalenone**.

HABITAT: food
World-wide distribution, colonizer and sometimes a parasite of mainly graminaceous plants. Reported from soil, seeds of cereals (particularly barley and corn) and grasses.

NOTE:
Sporulation often sparse (induction by near UV-light). Perithecia are often formed of SNA or CLA, especially when the medium begins to dry out. They are about 150-250 µm diam., dark purple to black, turning yellow in acid, with 3-septate ascospores ca. 20-29 x 3-5 µm.

A similar species, *F. pseudograminearum* Aoki & O'Donnell has recently been recognized for the cause of crown rot of wheat in arid environments, formerly known as *F. graminearum* Group I (Aoki & O'Donnell, 1999).

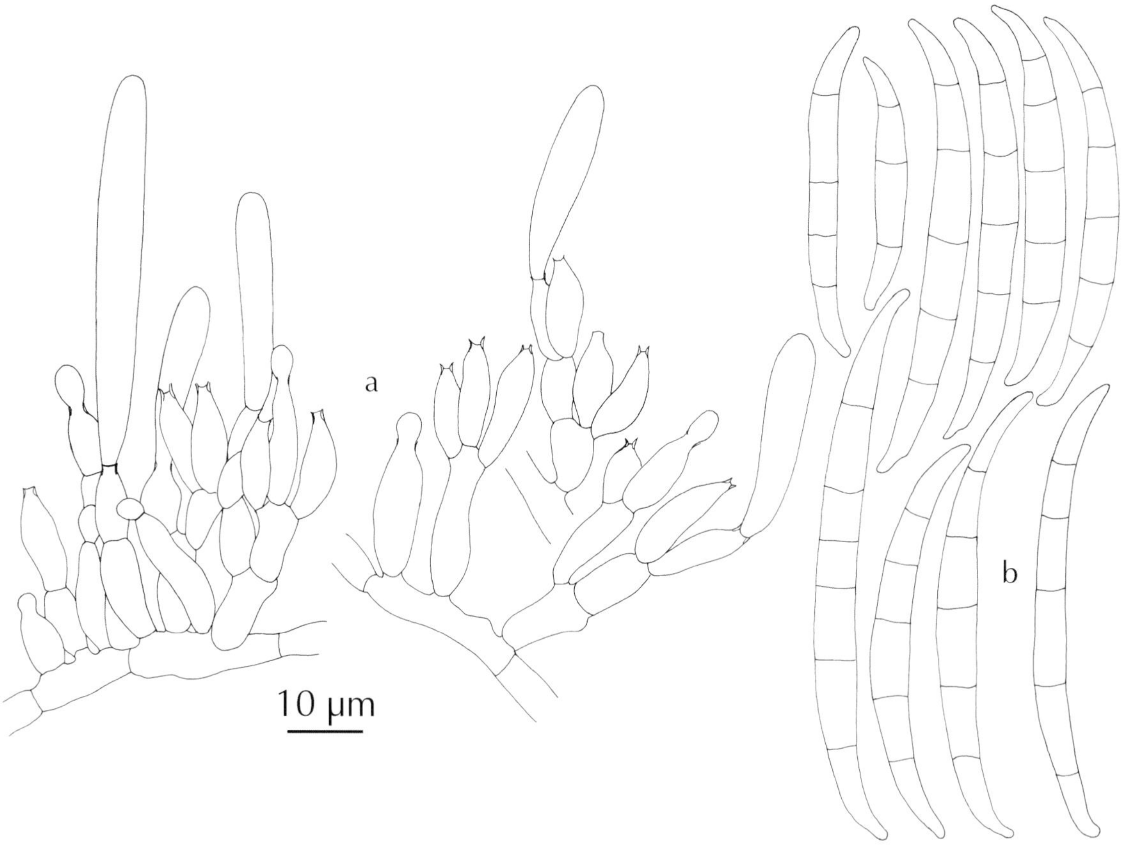

Fig. 62. *Fusarium graminearum*. a. Conidiophores and phialides; b. macroconidia.
Pl. 57. *Fusarium graminearum*. a. Colony on PDA after one week; b,e. macroconidia, x 800; c,d,f. conidiophores and macroconidia, x 800.

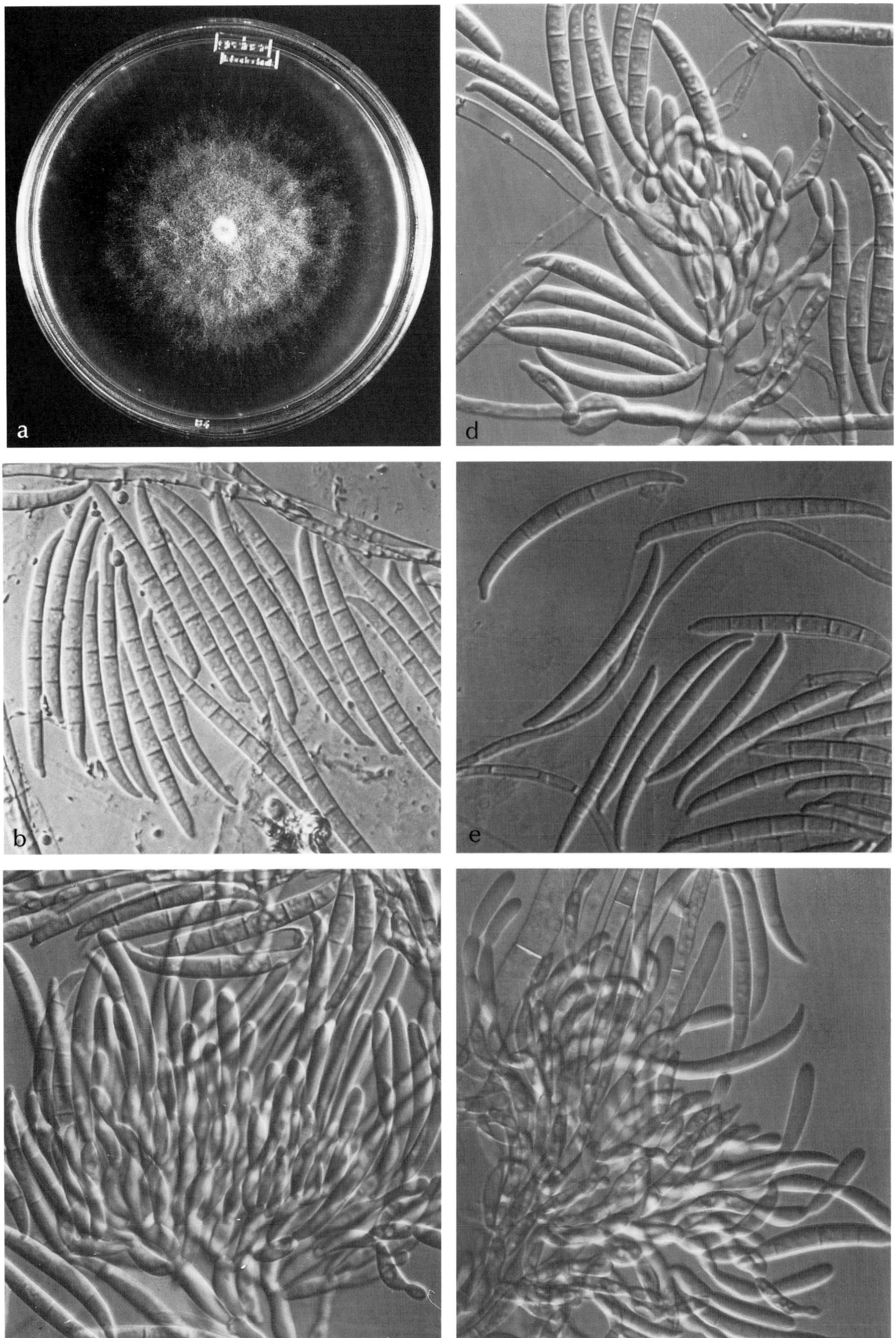
a
b
c
d
e
f

Fusarium oxysporum Schlecht.:Fr.

Teleomorph: unknown

Colonies on PDA or PSA at 25°C attaining a diameter of 3-5.5 cm in 4 days. Aerial mycelium sparse or floccose, becoming felty, whitish or peach, usually with a purple tinge, more intense near the medium surface. In some strains orange sporodochia may occur. Reverse yellowish or in purple shades. Conidiophores branched or unbranched bearing monophialides. Microconidia 0(-2)-septate, borne on laterial, simple (often reduced) phialides or on phialides on short branched conidiophores, generally abundant, in false heads, variable in shape and size, ovoid-ellipsoidal to cylindrical, straight or slightly curved, 5-12 x 2.2-3.5 µm. Macroconidia sparse in some strains, borne on phialides on branched conidiophores or in sporodochia, 3(-5)- septate, fusiform, more or less curved, pointed at both ends with a pedicellate basal cell, usually 3-septate, (20)27-46(50) x 3-4.5(5) µm. Chlamydospores in hyphae or in conidia, hyaline, smooth- or rough-walled, (sub)globose, 5-15 µm in diam, terminal or intercalary, in chains, in pairs or single. (B74, GN345, NTM142).

Temperature: optimum 25-30°C; minimum 5°C; maximum at 37°C or below.

Important (toxic) metabolites: fusaric acid, moniliformin, naphthoquinone pigments, nectriafurone.

HABITAT: food, indoor

World-wide distribution. Occurring chiefly as a soil saprophyte but can be pathogenic to numerous plants, isolated from cereal grains, flax, groundnuts, soya, peas, beans, cotton, bananas, onion bulbs, potatoes, oranges, apples, beet. Causes storage rot (e.g. in maize, beet).

NOTE:

F. oxysporum is one of the most economically important species within the genus, but also the most variable. The species can grow under anaerobic conditions.

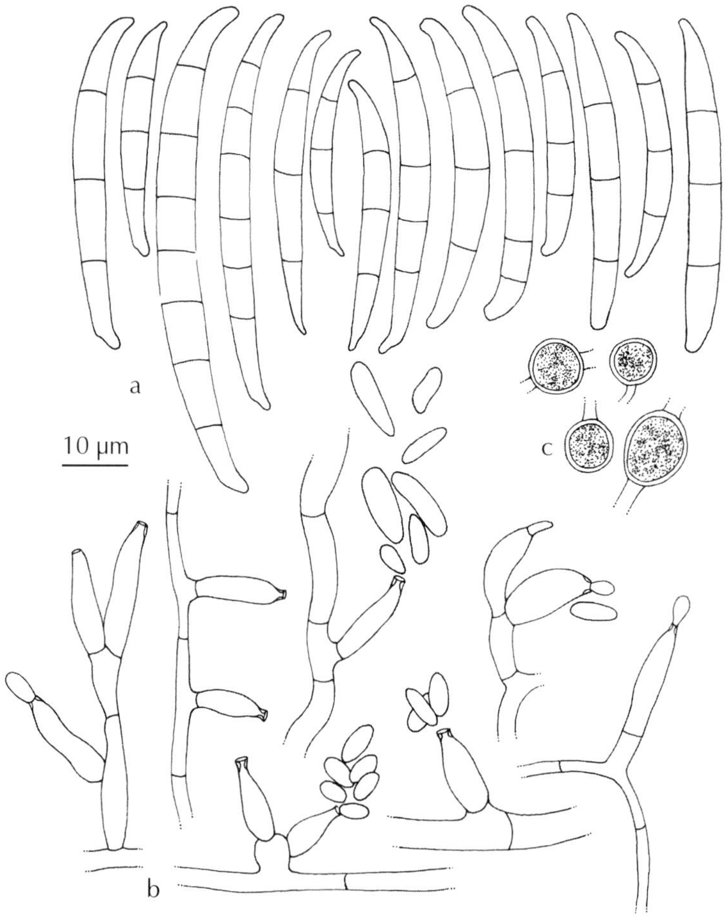

Fig. 63. *Fusarium oxysporum*. a. Macroconidia; b. conidiophores and phialides with microconidia; c. chlamydospores.
Pl. 58. *Fusarium oxysporum*. a. Colony on PDA after one week; b. conidiophore with microconidia, x 900; c,d. micro- and macroconidia, c. x 700, d. x 1500; e. phialides with microconidia and chlamydospore, x 1500; f. phialides with microconidia, x 1500;

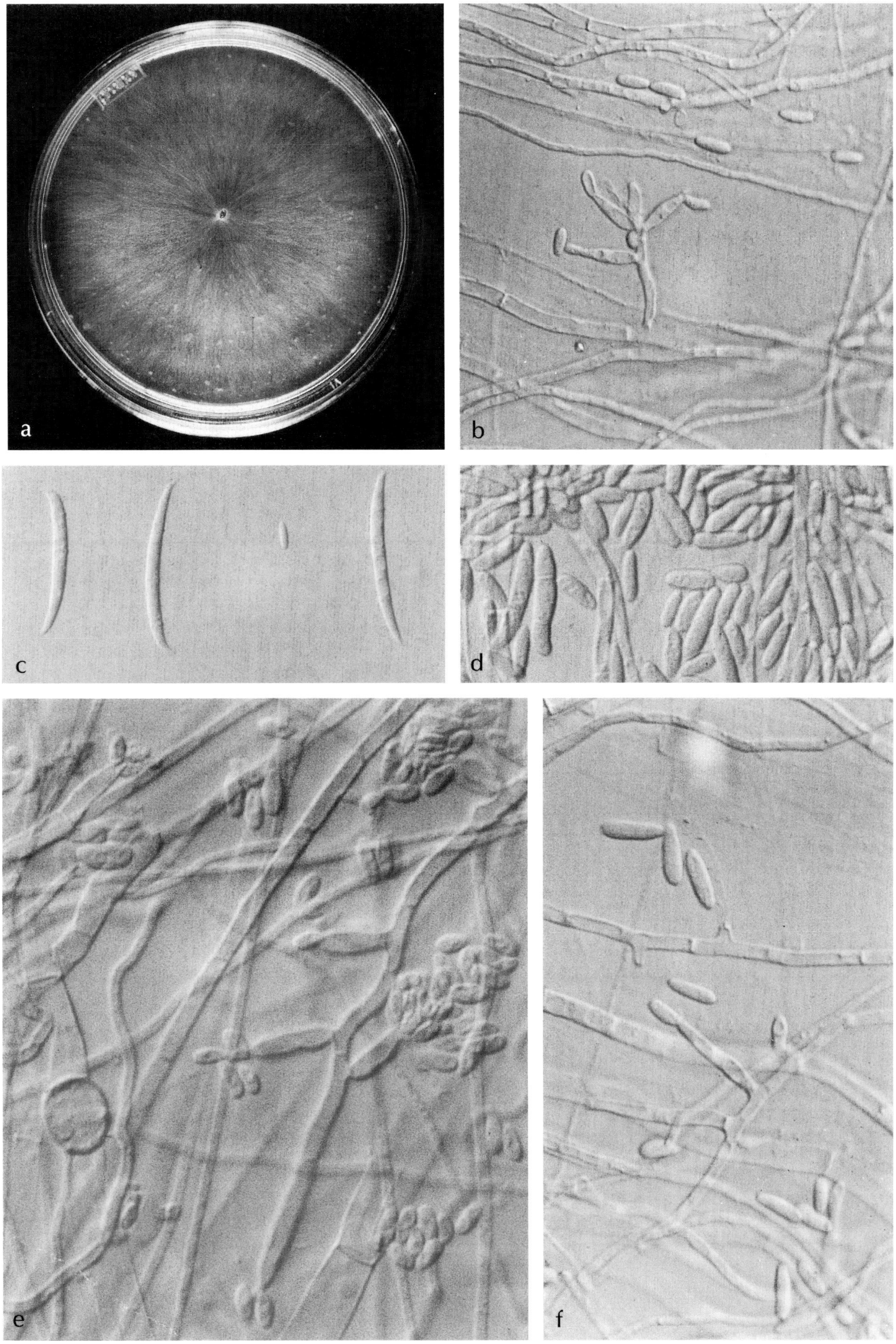
a
b
c
d
e
f

Fusarium poae (Peck) Wollenw.

Teleomorph: unknown

Colonies on PDA or PSA at 25 °C, attaining a diameter of 5.5-8.8 cm in 4 days. Aerial mycelium cottony, white or pale pinkish, peach, near the agar surface red to violet. Odour fruity (peach-like). Reverse in yellowish to reddish shades. Conidiophores at first unbranched later usually branched with short and broad monophialides, not in sporodochia. Microconidia, abundantly produced, mostly one-celled, napiform, pyriform, 6-10 x 5.5-7.5 µm. Macroconidia sparsely produced, slightly curved, mostly 2-3- septate, 18-38 x 3.8-7 µm, sometimes 5- septate and up to 56 µm long.

Chlamydospores not produced, swollen hyphal portions sometimes observed. (B52, GN115, NTM64).

Temperature: optimum 22.5-27.5°C; minimum 2.5 to 9°C; maximum 32-33°C.

Important (toxic) metabolites: 4-acetamido-4-hydroxy-2-butenoic acid-γ-lactone ("butenolide"), fusarin C, **trichothecenes**.

HABITAT: food

Mainly in temperate regions. Common on gramineous plants, in soil. Isolated from cereal seeds.

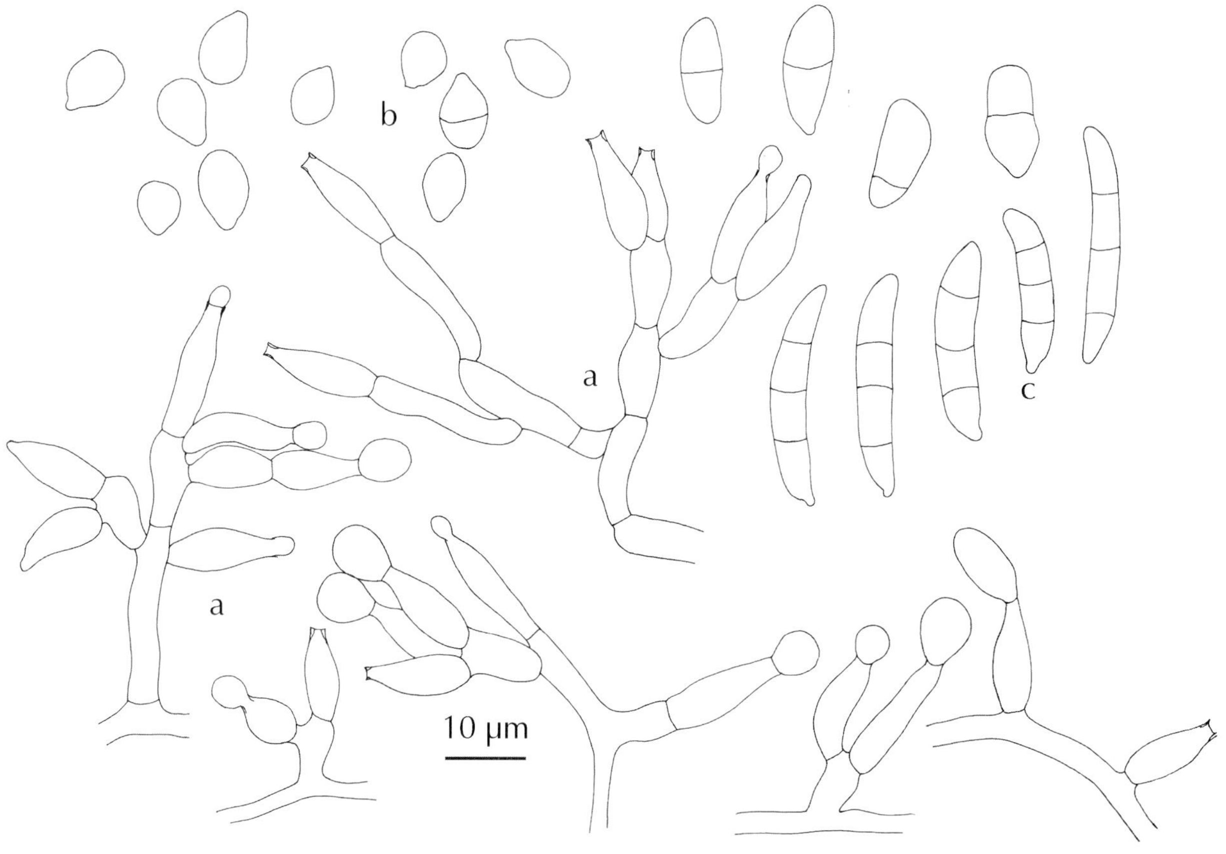

Fig. 64. *Fusarium poae.* a. Conidiophores and phialides; b. microconidia; c. macroconidia.

Pl. 59. *Fusarium poae.* a. Colony on PDA after one week; b-d, f. conidiophores, phialides and predominantly microconidia, x 750; e. micro- and macroconidia, x 750; g. phialide and microconidia, x 2000.

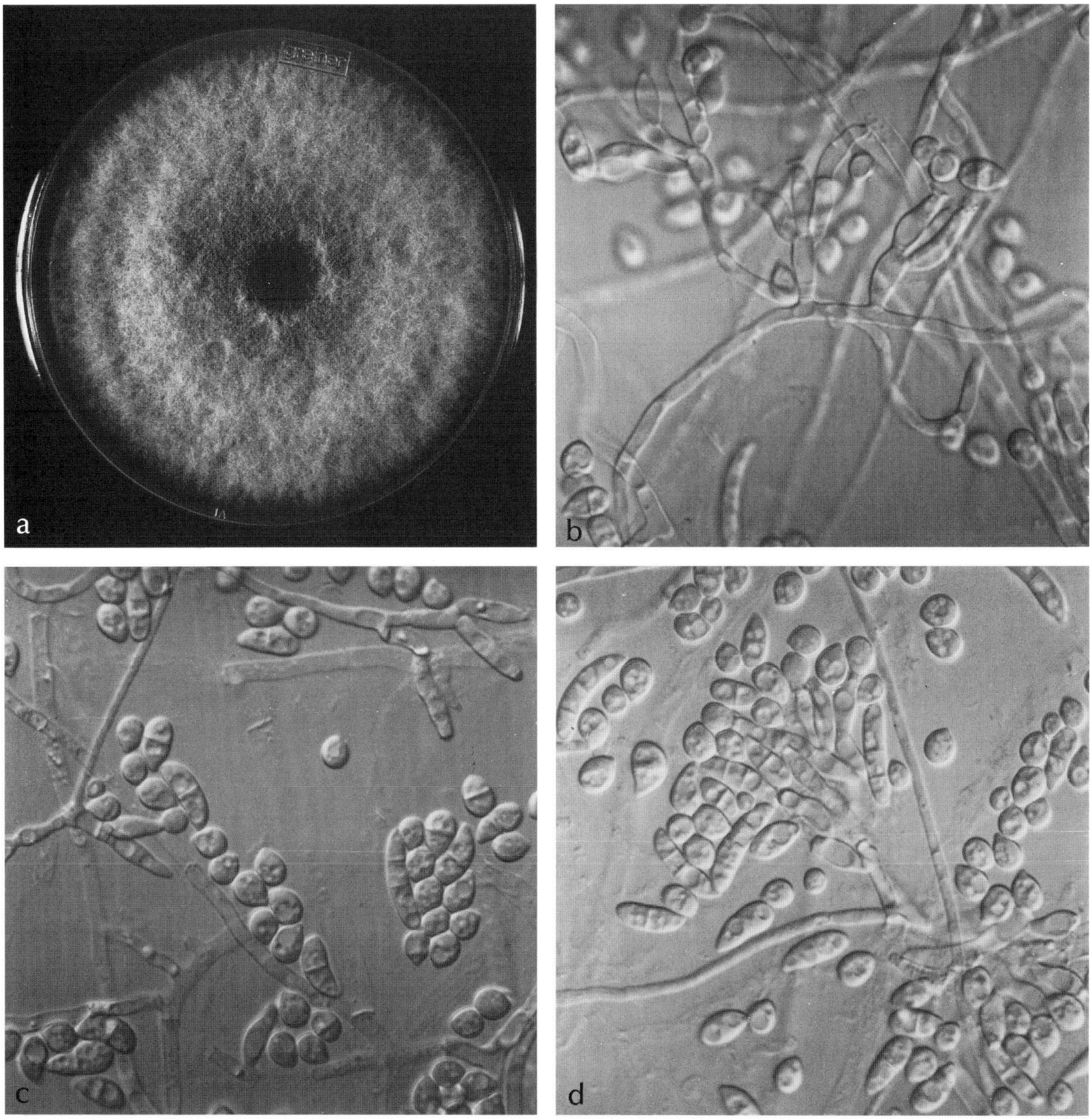
a
b
c
d
e
f
g

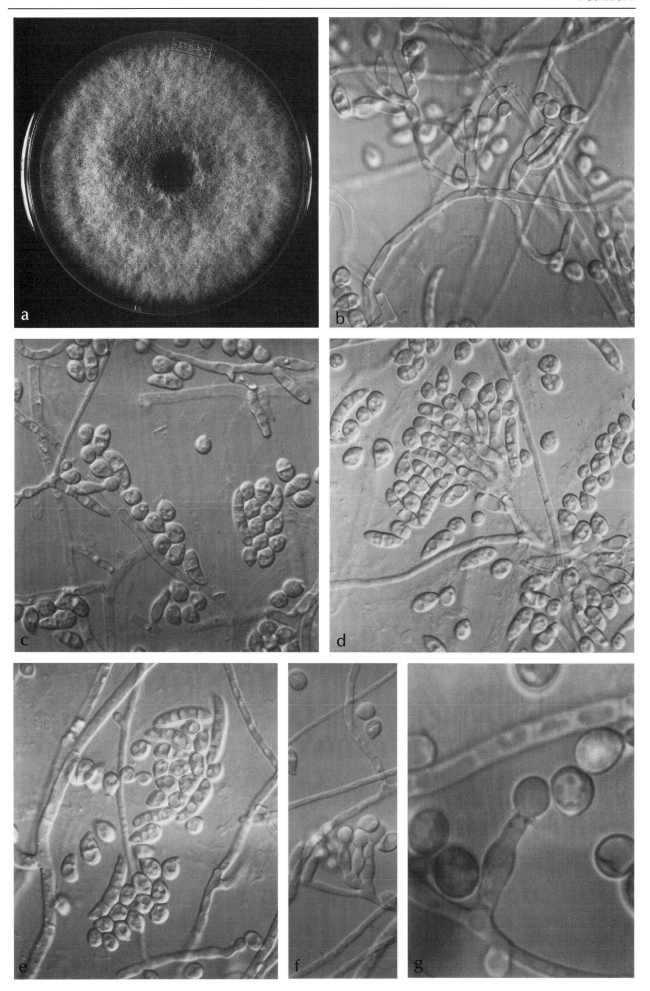

Fusarium proliferatum (Matsushima)Nirenberg

Teleomorph: unknown

Colonies on PDA or PSA at 25°C attaining a diameter of 3.5-5.5 cm in 4 days. Aerial mycelium floccose, white, pale pinkish or greyish violet. Reverse ranging from light coloured to greyish (dark) violet. Black sclerotia may occur. Conidiophores unbranched at first, later branched with mono- and polyphialides. Microconidia, usually one-celled, produced abundantly starting in the aerial mycelium, in long chains and in false heads, clavate with a flattened base, mostly 7-9 x 2.2-3.2 µm, also pyriform Microconidia may be present, measuring mostly 7-11 x 4.7-7.7 µm. Macroconidia, rarely produced, usually 3- or 5-septate, slightly sickle-shaped to rather straight, 3-septate, mostly 30-46 x 3.3-4.1 µm, 5-septate, mostly 47-58 x 3.4-4.4 µm. Chlamydospores absent. (B62, GN309, NTM132).

Important (toxic) metabolites: beauvericin, **fumonisins**, fusaric acid, fusarin C, fusaproliferin, fusapyrone, moniliformin, naphthoquinone pigments.

HABITAT: food
Occurring in tropical and subtropical countries. Isolated from orchids, figs, rice, corn and other cereals. Responsible for foot and fruit rots and leaf spots.

NOTE:
This a species complex. with similar morphology to *F. proliferatum* which have recently been described based mainly on DNA characters. They are mostly host specific and/or of limited geographical distribution (Nirenberg and O'Donnell, 1998).

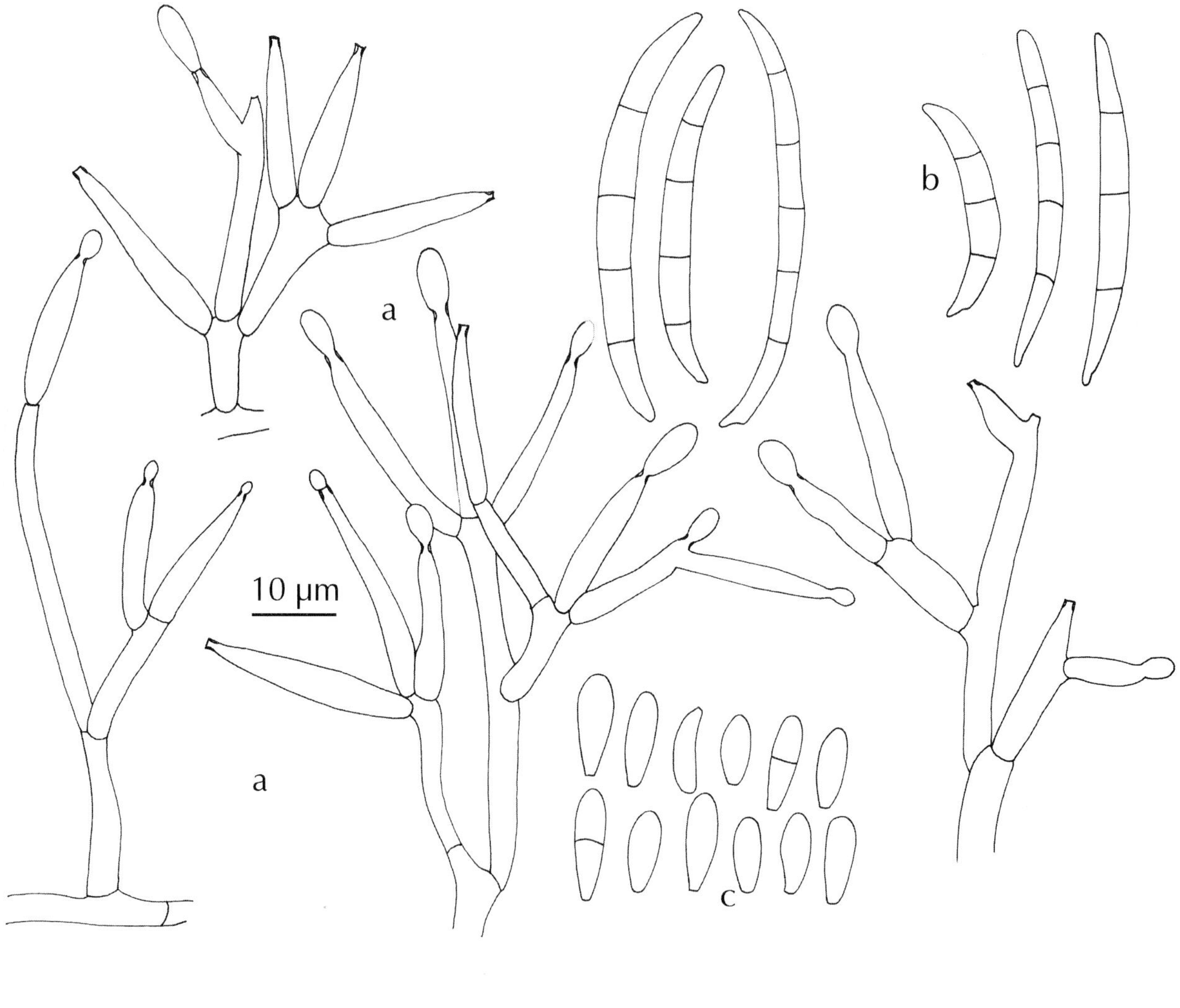

Fig. 65. *Fusarium proliferatum*. a. Conidiophores with mono- and polyphialides; b. macroconidia; c. microconidia.
Pl. 60. *Fusarium proliferatum*. a. Colony on PDA after one week; b,d. conidiophores and microconidia, x 680; e. polyphialide and microconidia, x 1250; c. macroconidia, x 700; f. microconidia, x 1250.

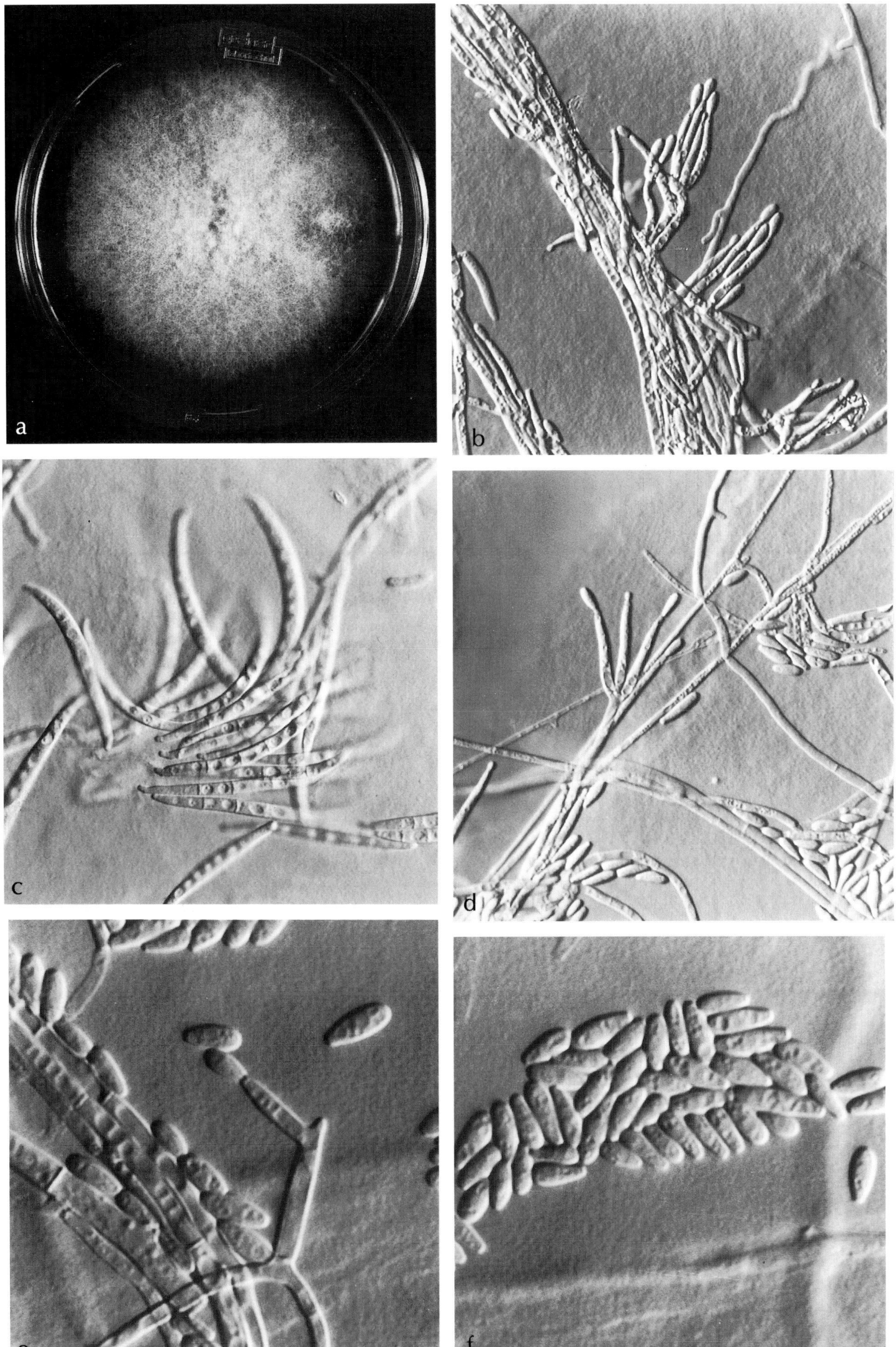
a
b
c
d
e
f

Fusarium sambucinum Fuckel

= *F. sulphureum* Schlecht *sensu* Wolenw.

Teleomorph: *Gibberella pulicaris* (Fr.:Fr.) Sacc.

Colonies on PDA or PSA at 25°C attaining a diameter of 3.4-5.9 cm in 4 days. Aerial mycelium floccose, white to greyish rose. Reverse white, yellowish, rose, greyish red. Dark red cultures are rarely seen. Conidiophores arising as single lateral monophialides, later branching sparsely, and verticillately branched in sporodochia or pionnotes. Microconidia not produced. Macroconidia uniform in type and size, sickle-shaped, strongly curved dorsiventrally, thick-walled, stout, distinctly septate, 3-5(-7)-septate; 3-septate measuring mostly 22-35 x 4.0-5.2 µm, 4-5-septate, mostly 26-44 x 4.0-5.6 µm, 6-7-septate, mostly 37-50 x 4.5-5.6 µm. Chlamydospores may be present, single, in chains and in clumps. (B85, GN202, NTM111).

Important (toxic) metabolites: 4-acetamido-4-hydroxy-2-butenoic acid-γ-lactone ("butenolide"), enniatins, **trichothecenes**.

HABITAT: food

Occurring world-wide on cereals, potatoes, grasses, bark of trees, also reported from soils.

NOTE:

The name *F. sulphureum*, frequently used for the potato pathogen in North America, is a synonym. Two related species have recently been recognized, including *F. venetatum*, the fungus grown commercially as the single cell protein Quorn™.

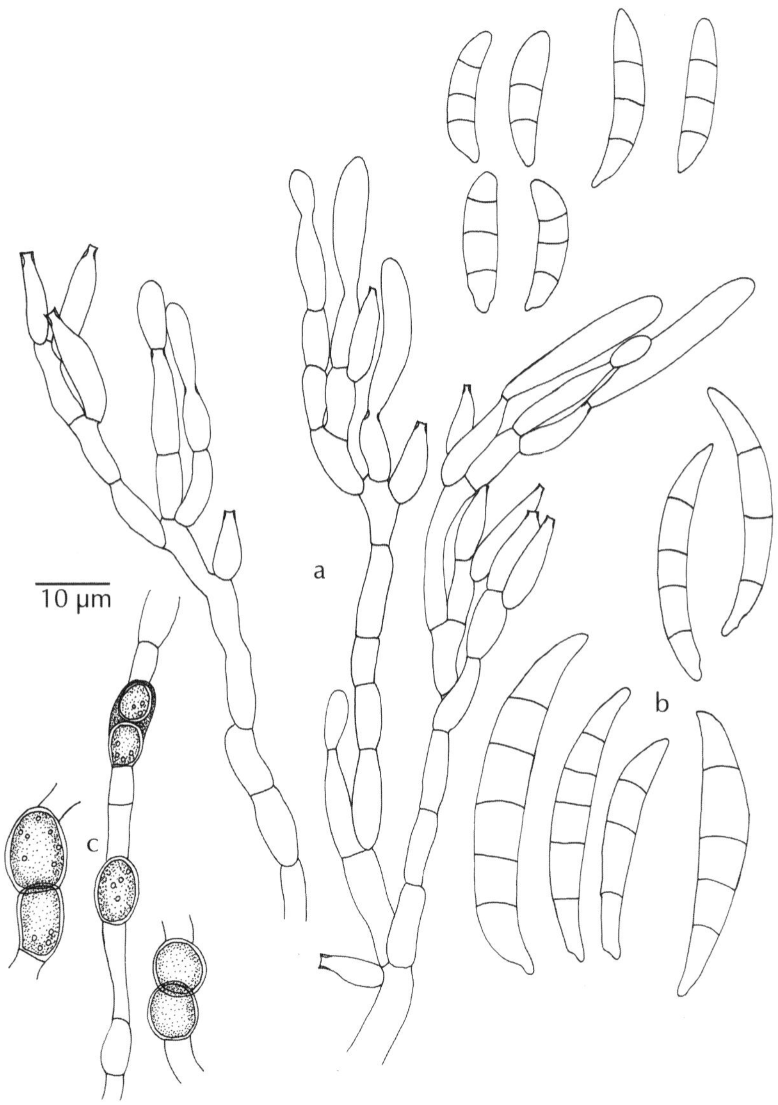

Fig. 66. *Fusarium sambucinum* a. Conidiophores with monophialides; b. macroconidia; c. chlamydospores.

Pl. 61. *Fusarium sambucinum*. a. Colony on PDA after one week. b-e. conidiophores with monophialides and macroconidia, b and d. x 715, c. x 850, e. x 665; f. macroconidia x 800.

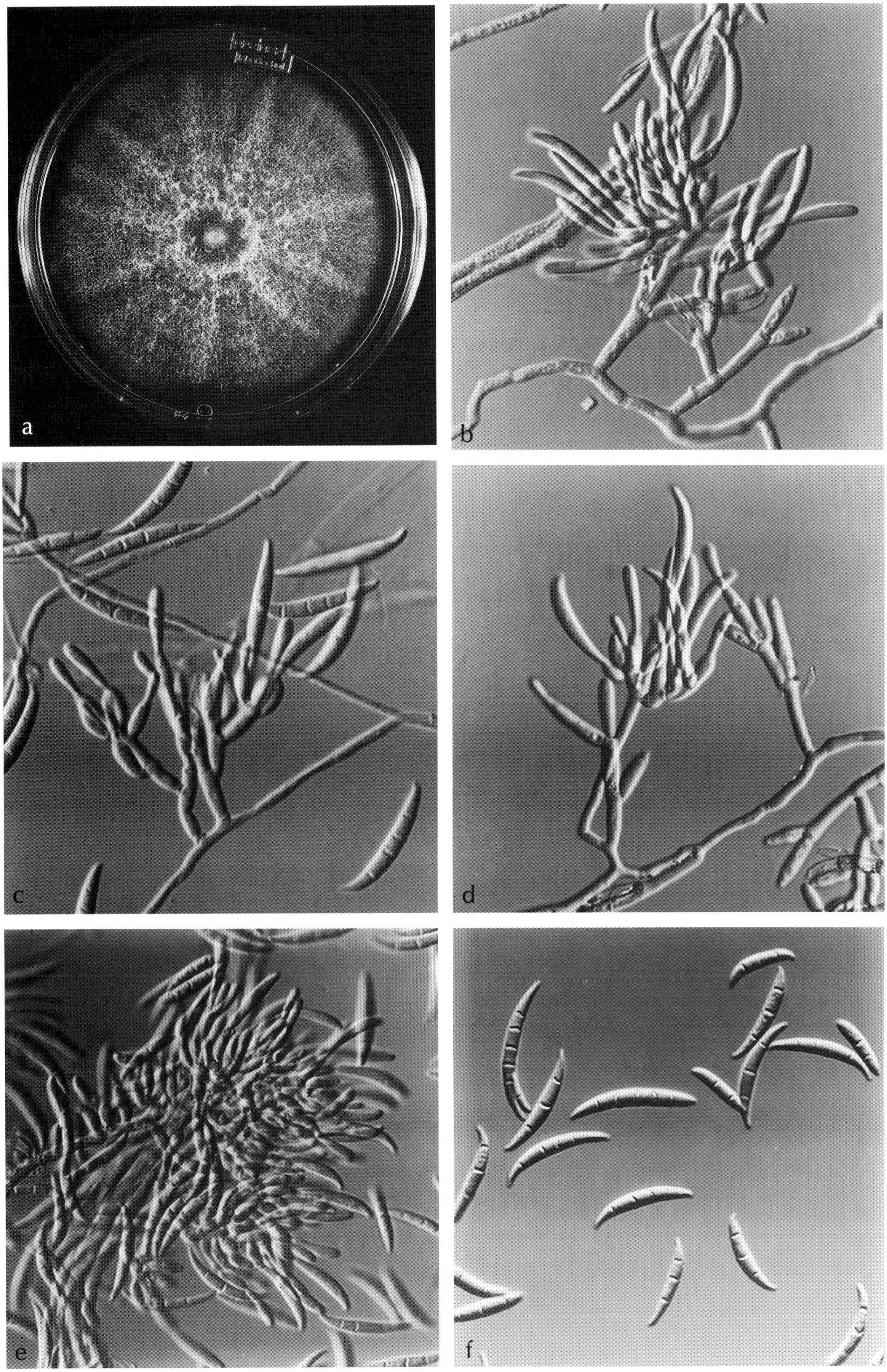
a
b
c
d
e
f

Fusarium semitectum sensu Wollenw.

= *Fusarium incarnatum* (Roberge) Sacc. pro parte
= *Fusarium pallidoroseum* (Cooke) Saccardo

Teleomorph: unknown

Colonies on PDA or PSA at 25°C attaining a diameter of 4.5-6.9 cm in 4 days, white, often with a peach tinge. Aerial mycelium floccose, or in tufts, whitish or peach, often changing to brownish (14-21 days). Reverse becoming peach to ochraceus brown, never red. Sporodochia absent. No distinct division in micro-and macroconidia. (Macro)conidia borne from loosely branched conidiophores or from short lateral phialides in young aerial mycelium. Conidiogenous cells at first produce conidia from single apical pores, later becoming polyblastic sympodial ("polyphialide") cells. Conidia 3-5(7)-septate, fusiform, straight or somewhat curved, foot cell not pedicellate but wedge-shaped, apical cell beaked, 0-2-septate conidia may be present. 3-septate: 17-28 x 2.5-4(5)µm; 5-septate: 22-40(55) x 3-4.5 µm. Chlamydospores often sparse, hyaline, smooth-walled, globose, 5-10 µm in diam, intercalary, single or in chains, in hyphae and conidia. (B116, GN155+159, NTM84).

Temperature: optimum 25°C; minimum 3(or15)°C; maximum 34-37°C.

Important (toxic) metabolites: equisetin, fusapyrone, **zearalenone**.

HABITAT: food
Distribution common, especially in tropical and subtropical countries. Often causing storage rot in groundnuts, bananas, citrus, potatoes, tomatoes, melons, cucumber, seldom pathogenic to living plants. Isolated from soil from both tropical and temperate zones. Often imported into Europe with fruit.

NOTE:
In older cultures (especially after mass transfer) a greater variation in conidial shape and size may be observed with additional 1-septate pyriform to obovate conidia (10-12 x 2.5-3.5 µm). This species can be distinguished from *F. equiseti* by the presence of polyblastic conidiogenous cells and the lack of pedicillate foot cells. The *F. semitectum*-complex contains several taxa, which need further taxonomic investigation. Some authors use *Fusarium incarnatum* as the correct name.

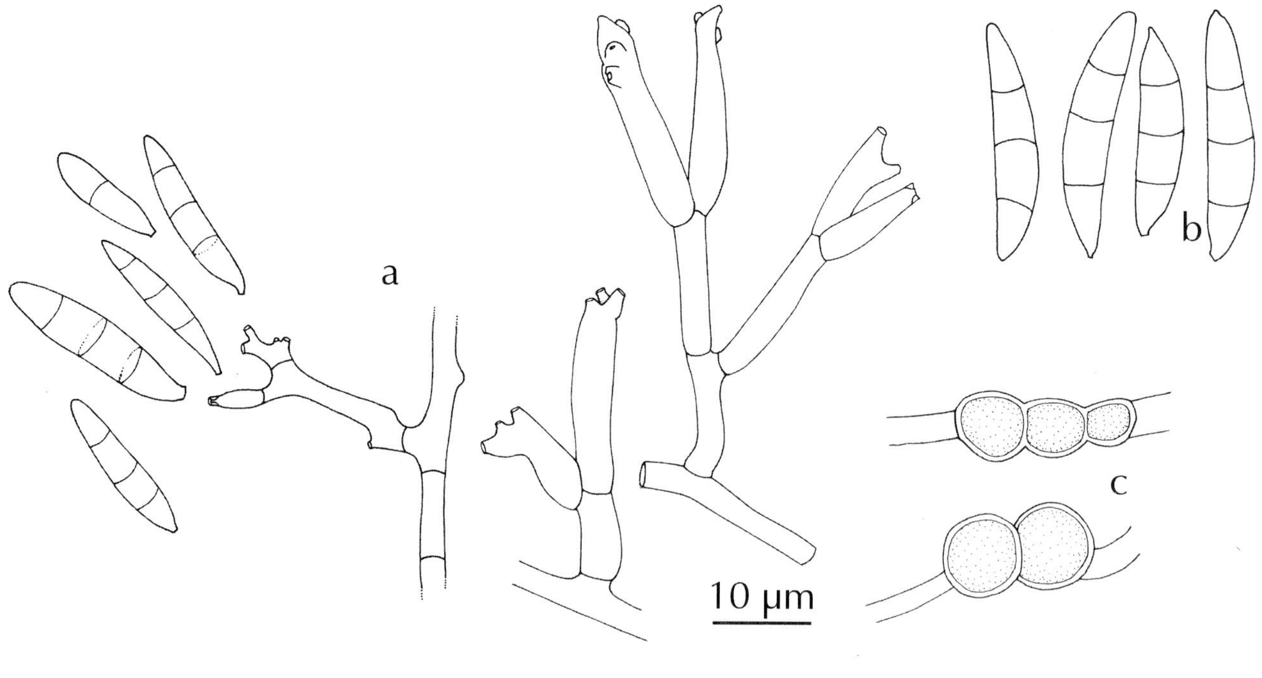

Fig. 67. *Fusarium semitectum.* a. Polyblastic conidiogenous cells; b. conidia; c. chlamydospores.
Pl. 62. *Fusarium semitectum.* Colony on PDA after one week; b-f. polyblastic conidiogenous cells and conidia, x 1250.

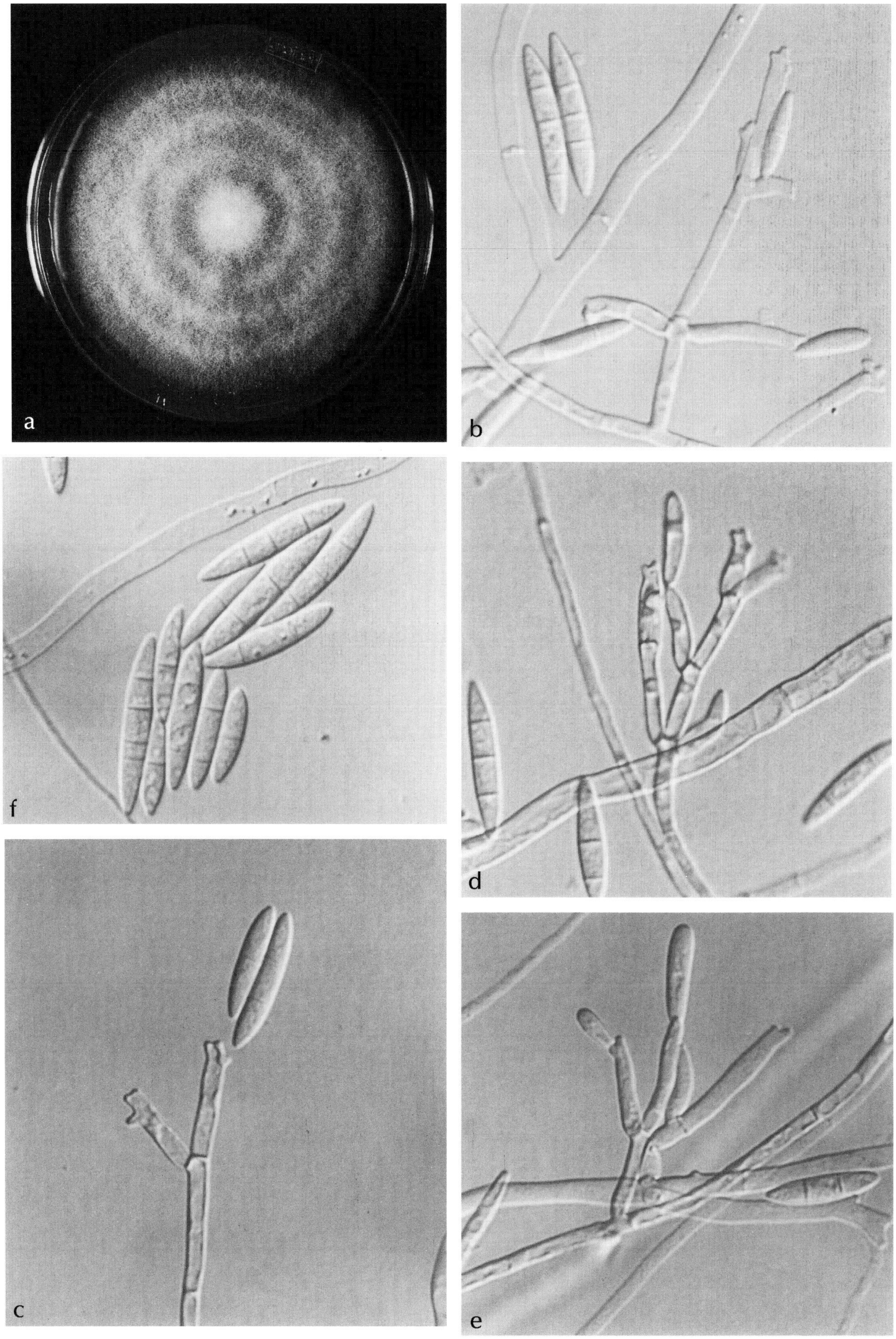
a
b
f
d
c
e

Fusarium solani (Mart.) Sacc.

Teleomorph: *Haematonectria* sp.

Colonies on PDA or PSA at 25°C attaining a diameter of 2.5-5 cm in 4 days. Aerial mycelium sparse or dense and floccose, sometimes leathery, greyish-white, cream to buff, conidial slime formed in sporodochia or pionnotes. Agar surface sometimes green to bluish-brown. Reverse often in greenish or brownish shades. Conidiophores unbranched or branched bearing monophialides. Microconidia usually abundant, ovoid or oblong, 0-1-septate, 8-16(24) x 2-4(5) µm, formed from elongated and sometimes verticillate conidiophores. Macroconidia formed after 4-7 days from short multi-branched conidiophores which may form sporodochia, 3-5(-7)-septate (usually 3-septate), fusiform, cylindrical, often moderately curved, 27-52(65) x 4.4-6.8 µm, with an indistinctly pedicellate footcell and a short blunt apical cell. Chlamydospores hyaline, smooth-or rough-walled, globose to ovoid, in hyphae or in conidial cells, either terminal, on lateral branches, intercalary or in chains. (B76, GN364, NTM146).

Temperature: optimum 27-31°C, depending on a tropic or temperate origin of the isolate; growth still possible at 37°C.

Important (toxic) metabolites: fusaric acid, naphthoquinone pigments.

HABITAT: food, indoor
World-wide distribution: soil, wide range of plants and animals. Causes keratitis in man, root and stem rot and spoilage. Often found in wounds and damaged organisms (secondary infections).

NOTE:
This is a species complex. Molecular studies indicate that more than 40 biological and phylogenetic species are presently grouped under the name *F. solani*. Critical morphological studies and proposals of formal names for these species have yet to be published.

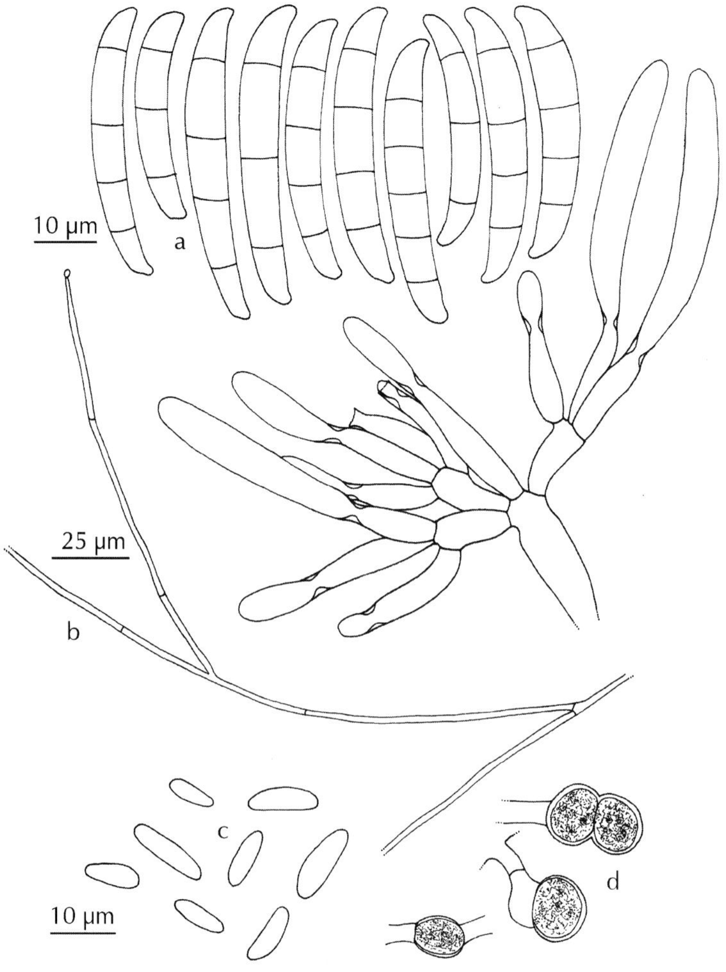

Fig. 68. *Fusarium solani*. a. Macroconidia and conidiophore; b-c. microconidia formed from elongated conidiophores; d. chlamydospores (orig. partly after E.J. Hermanides-Nijhof)
Pl. 63. *Fusarium solani*. a. Colony on PDA after one week; b,d. elongated conidiophores and phialides forming microconidia, b. x 400, d. x 450; c. microconidia, x 1000; f. conidiophores forming micro- and macroconidia, x 450; e. micro- and macroconidia, x 400.

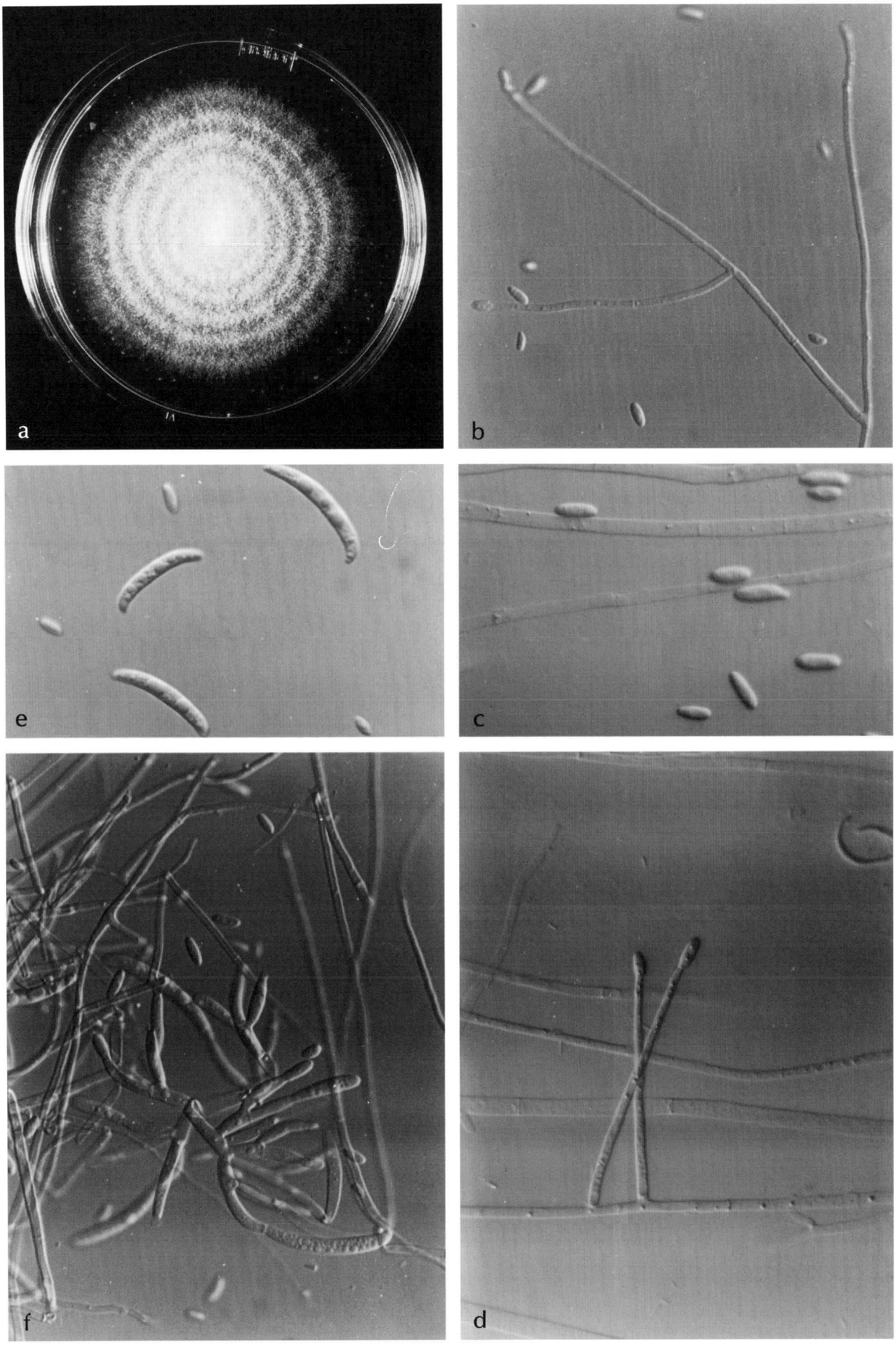
a
b
e
c
f
d

Fusarium sporotrichioides Sherb.

Teleomorph: unknown

Colonies on PDA or PSA at 25 °C attaining a diameter of 6.5-8.8 cm in 4 days. Aerial mycelium floccose or cottony, whitish, later becoming yellow, pink-red to purple or brownish. Reverse in red and yellow to brownish shades. Sporodochia red-brown, abundant or absent. Conidiophores unbranched and branched with mono- and polyphialides. Microconidia abundantly produced in the aerial mycelium, 0-1(-2) septate, ovoid, pyriform, napiform, fusoid, variable in size; pyriform mostly 6-14 x 5.3-7.1 µm, fusoid mostly 8-16 x 3.0-4.2 µm. Macroconidia produced in the aerial mycelium or in sporodochia 3-to 5-septate, sickle-shaped, moderately curved, without a distinct foot-cell, often with a bent apical cell, 3-septate mostly 21-36 x 3.6-4.8 µm, 5-septate mostly 40-47 x 4.1-5.3 µm. Chlamydospores numerous, brownish when mature, globose, smooth-walled. Sclerotial plectenchymatous bodies dark red to brown, often produced on rice or wheat grains. (B54, GN129, NTM70).

Temperature: optimum 22.5-27.5(-30)°C; minimum 2.5-7°C; maximum 35°C.

Important (toxic) metabolites: 4-acetamido-4-hydroxy-2-butenoic acid-γ-lactone ("butenolide"), fusarin C, **trichothecenes**.

HABITAT: food, indoor
World-wide distribution, in temperature and tropical zones: soil, wide variety of plants, especially cereal crops (barley, wheat, oats), cucumber seeds, pome fruit.

NOTE:
A similar species, *F. chlamydosporum*, which lacks globose conidia in the aerial mycelium and tends to produce brown colonies on PDA, is sometimes encountered in indoor air, particularly in air conditioning systems.

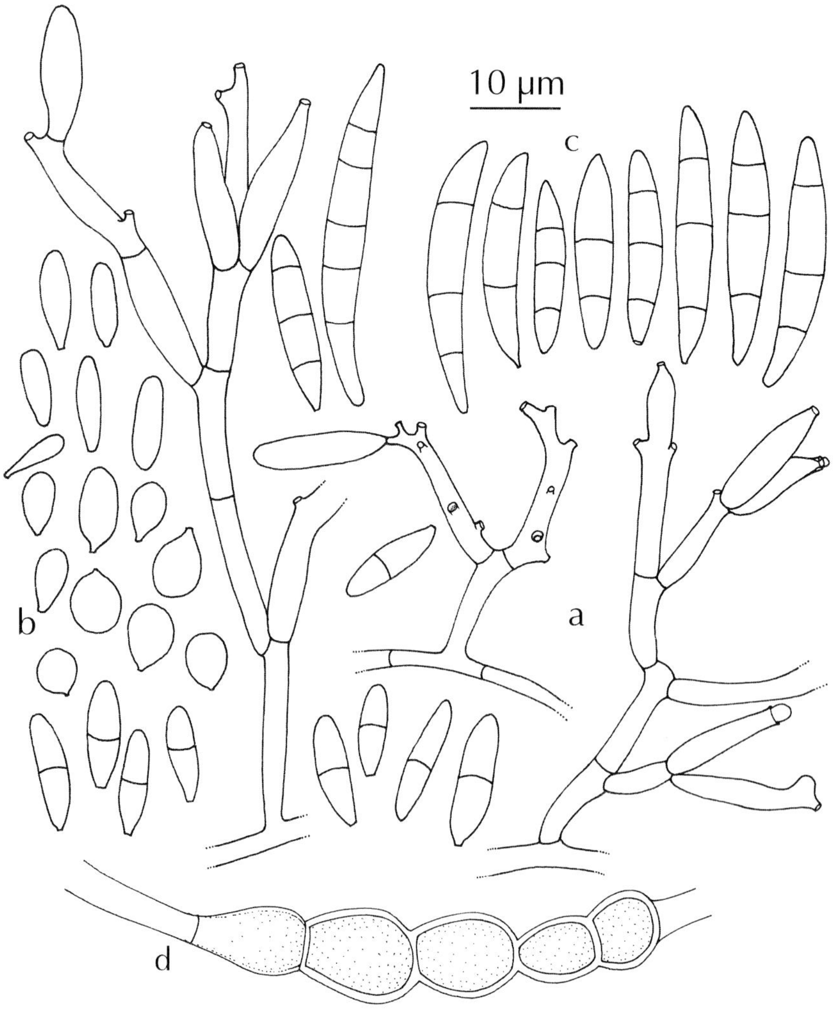

Fig. 69. *Fusarium sporotrichioides*. a. Conidiophores with polyphialides; b. microconidia; c.macroconidia; d. chlamydospores.
Pl. 64. *Fusarium sporotrichioides*. a. Colony on PDA after one week; b-f. conidiophores, micro- and macroconidia, b,f. x 700, c-e. x 1000.

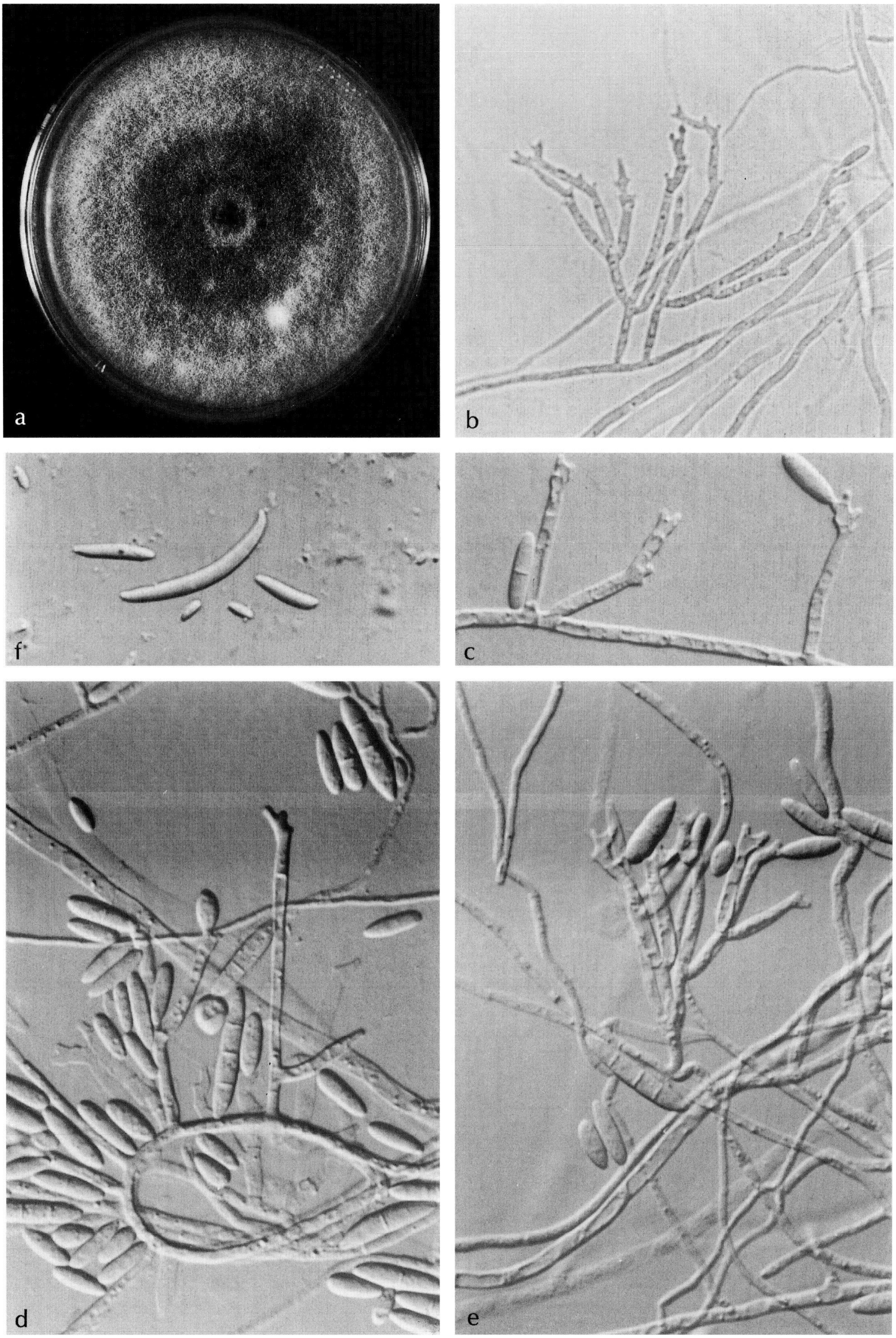
a
b
f
c
d
e

Fusarium subglutinans (Wollenw. & Reinking) Nelson, Toussoun & Marasas
= *Fusarium sacchari* (Butler) W. Gams var. *subglutinans* (Wollenw. & Reinking) Nirenberg

Teleomorph: *Gibberella subglutinans* (Edwards) Nelson, Toussoun & Marasas
= *Gibberella fujikuroi* (Sawada) Wollenw. var. *subglutinans* Edwards

Colonies on PDA or PSA at 25°C attaining a diameter of 3.5-5.5 cm in 4 days. Aerial mycelium floccose, white, pale pinkish or greyish ochre, violet shades may occur. Reverse ranging from light coloured to greyish (dark) violet. Black sclerotia may occur. Conidiophores unbranched at first, later branched with mono- and polyphialides. Microconidia usually one-celled, produced in false heads in the aerial mycelium (starting after two days) oval, ellipsoid or allantoid, mostly 9-12 x 2.5-3.5 µm. Subdeveloped macroconidia ("mesoconidia") appear before the macroconidia formation. Macroconidia 3-5-septate, slightly sickle-shaped, rather straight, 3-septate, mostly 27-54 x 3.4-4.2 µm, 5-septate, mostly 53-63 x 3.5-4.5 µm. Chlamydospores absent. (B65, GN325, NTM135).

Important (toxic) metabolites: beauvericin, fumonisins, fusaproliferin, fusapyrone, fusaric acid, fusarin C, moniliformin, naphthoquinone pigments.

HABITAT: food
Common in tropical and subtropical areas on wide range of plants. Responsible for stalk rot and cob rot of corn.

NOTE:
This a species complex. Ten phylogenetic species with similar morphology to *F. subglutinans* have recently been described based mainly on DNA characters. They are mostly host specific and/or of limited geographical distribution (Nirenberg & O'Donnell, 1998).

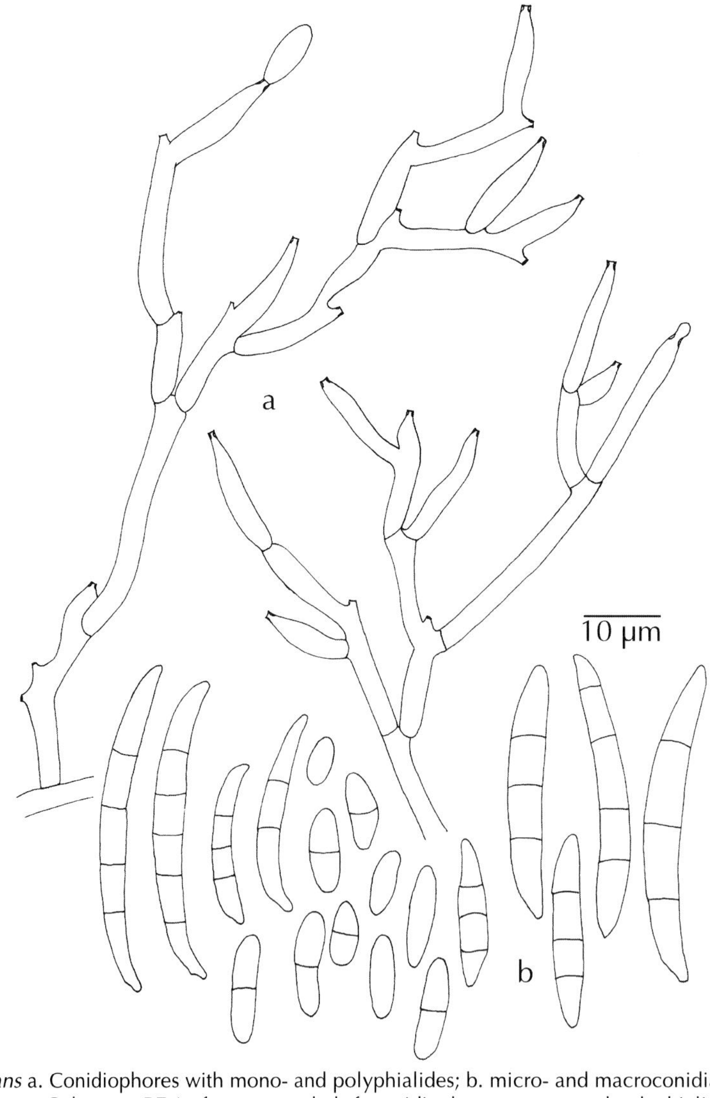

Fig. 70. *Fusarium subglutinans* a. Conidiophores with mono- and polyphialides; b. micro- and macroconidia.
Pl. 65. *Fusarium subglutinans.* a. Colony on PDA after one week; b-f. conidiophores, mono- and polyphialides with micro- and macroconidia, b. x 770, c,d x 840, e. x 1085, f x 1160.

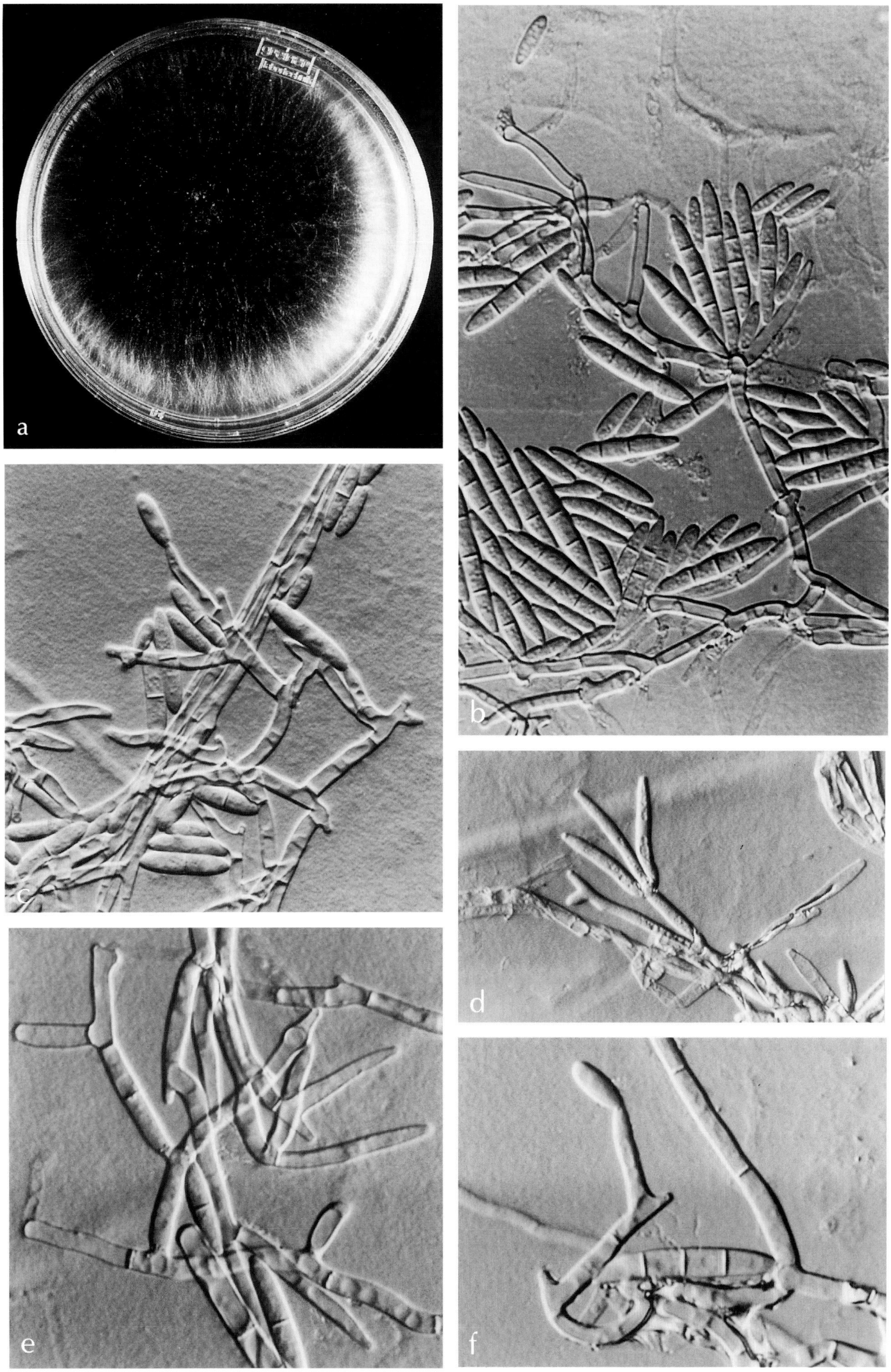
a
b
c
d
e
f

Fusarium tricinctum (Corda) Sacc.

Teleomorph: *Gibberella tricincta* El-Gholl, McRitchie, Schoulties & Ridings

Colonies on PDA or PSA reaching 3.2-5.5 cm diam. in four days at 25 °C. Aerial mycelium forming a compact cushion, red to vinaceous or purple, partly also white or ochraceous. Reverse red to vinaceous, but yellow cultures may occur. Conidiophores mostly richly branched, bearing slender monophialides 10-30 x 2-3 µm not sympodially proliferating. Microconidia, one-celled rarely 2-celled, scattered in the aerial mycelium as a cream powder, often lemon-shaped (citriform, sometimes pyriform, napiform, ellipsoidal or fusiform, mostly 8-11(-14) x 4.5-7.5 (-9.0) µm. Macroconidia mostly produced in sporodochia, moderately curved 3-5- septate, mostly 24-46 x 3.2-4.1 µm, 5-septate mostly 33-50 x 3.6-4.6 µm. Chlamydospores not common. (B53, GN125, NTM67).

Temperature: optimum at 22-23°C; minimum at 0-10°C; maximum at 31-32.5°C.

Important (toxic) metabolites: 4-acetamido-4-hydroxy-2-butenoic acid-γ-lactone ("butenolide"), antibiotic Y, chlamydosporol, enniatins, fusarin C.

HABITAT: food
World-wide distribution, isolated e.g. from cultivated soils, cereals, red clover, carnation.

NOTE:
Strains sometimes lose the ability to produce citriform microconidia after several transfers.

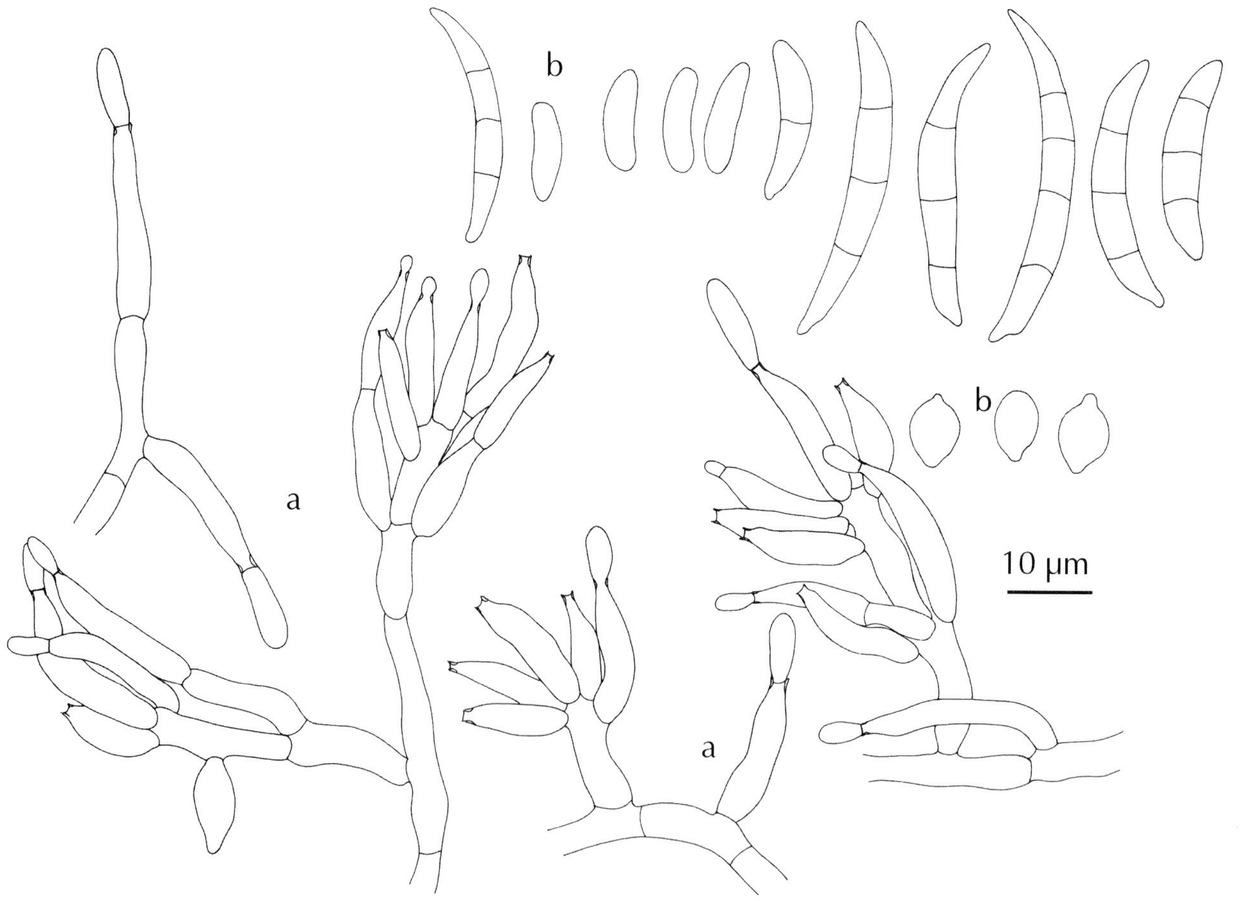

Fig. 71. *Fusarium tricinctum*. a. Conidiophores. b. micro- and macroconidia.
Pl. 66. *Fusarium tricinctum*. a. Colony on PDA after one week; b, d. conidiophores and macroconidia, x 750; c. macroconidia and microconidia, x 750; e, f. slender monophialides producing microconidia, x 750 .

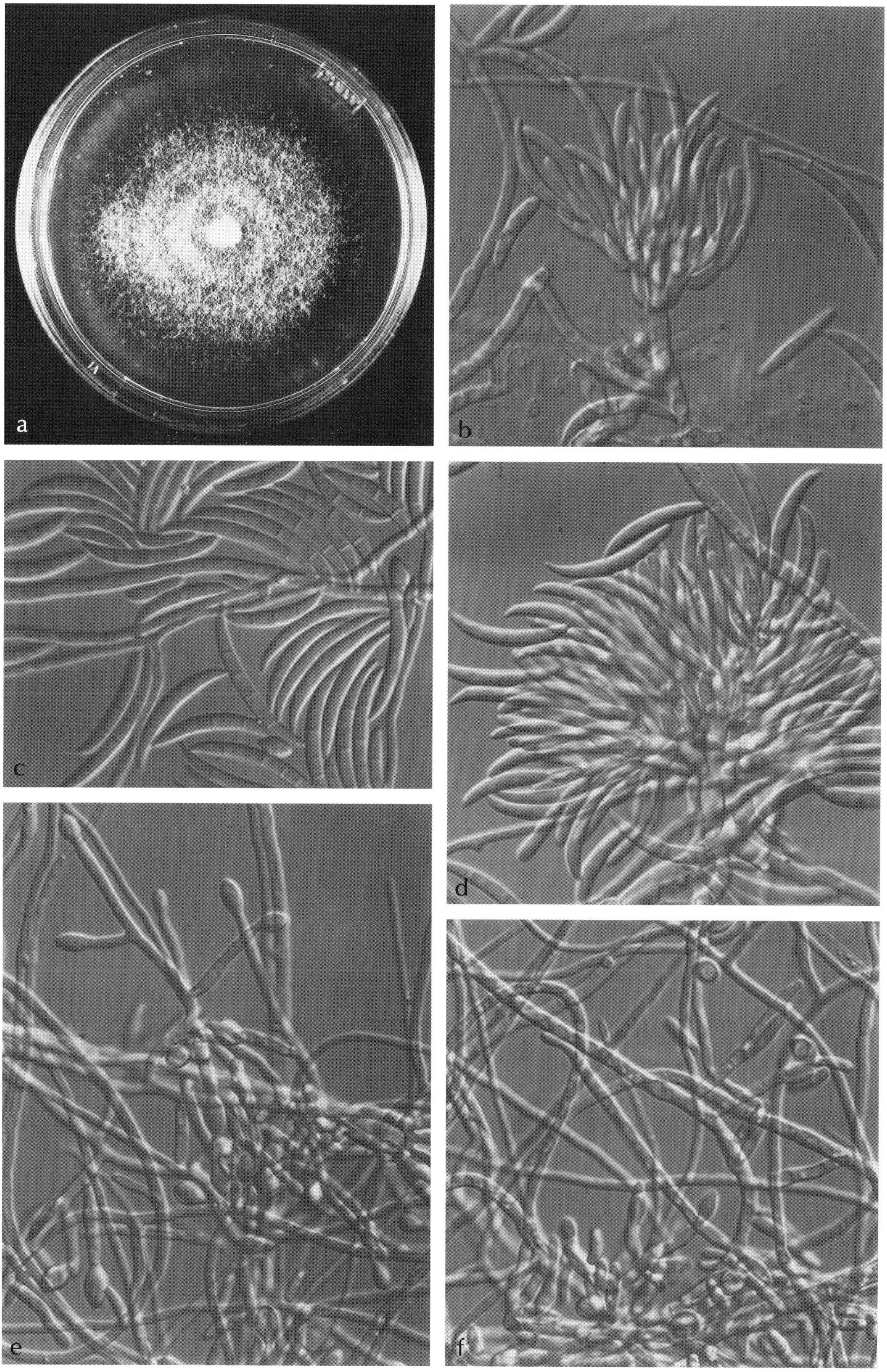
a
b
c
d
e
f

Fusarium verticillioides (Sacc.) Nirenb.

=*Fusarium moniliforme* Sheld. *s. str.*

Teleomorph: *Gibberella moniliformis* Winel.

Colonies on PDA or PSA at 25°C, attaining a diameter of 3.5-5.5 cm in 4 days (6.2 cm in diam on SNA in 10 days, Nirenberg, 1976), peach, pale cream, violet to lilac. Aerial mycelium dense, delicately floccosse to felty or with a powdery appearance due to micro-conidium production, whitish at first becoming rose-whitish (pale flesh-coloured) to vinaceous. Sporodochia rarely formed, brownish-rose. Reverse in dark violet, lilac, vinaceous or creamish shades. (Micro)conidiophores mostly unbranched, formed on aerial mycelium, with 1-3 simple elongated phialides. Microconidia in long chains (can be observed in Petri-dish under low power), rarely in false heads, 0-(sometimes 1-2)-septate, clavate with truncate base, in mass rose-beige, 4.3-19 x 1.5-4.5 µm. Macro-conidium formation in many strains rare. (Macro)conidiophores formed as lateral branches on hyphae and may consist of a single basal cell, bearing 2-3 phialides or consisting of 2-3 metulae. Macroconidia slender, 3-7-septate, (mostly 3-or 5-septate), straight or slightly curved, fusiform, thin-walled with elongated often curved apical cell and pedicellate basal cell, 3-septate mostly 30-46 x 2.7-3.6 µm; 5-septate mostly 47-58 - 3.1-3.6 µm. Chlamydospores absent. Sclerotia rarely present, blackish-blue, probably perithecial initials.

Perithecia of the teleomorph *Gibberella moniliforme* Winel. usually only occurring on dead plant material. (B60: *F. moniliforme*, GN301, NTM128: *F. moniliforme*).

Temperature: optimum 22.5-27.5(-35)°C; minimum 2.5-5°C; maximum (32) 37.5°C.

Important (toxic) metabolites: **fumonisins**, fusaric acid, fusarin C, moniliformin, naphthoquinone pigments.

HABITAT: food

Distribution: tropics and subtropics, rarely temperate zones in Europe (except for the greenhouses): corn, rice, sugar-cane, bananas, asparagus, cotton.

NOTE:

Causes keratitis in man, toxic to animals fed with contaminated foodstuff. The species can grow under anaerobic conditions and tolerates >15% NaCl in the medium. Production of macroconidia may be induced by "black light". A similar species, with yellow colony pigmentation on PDA, has been described from *Sorghum* sp. as *F. thapsinum*.Klittich *et al.*

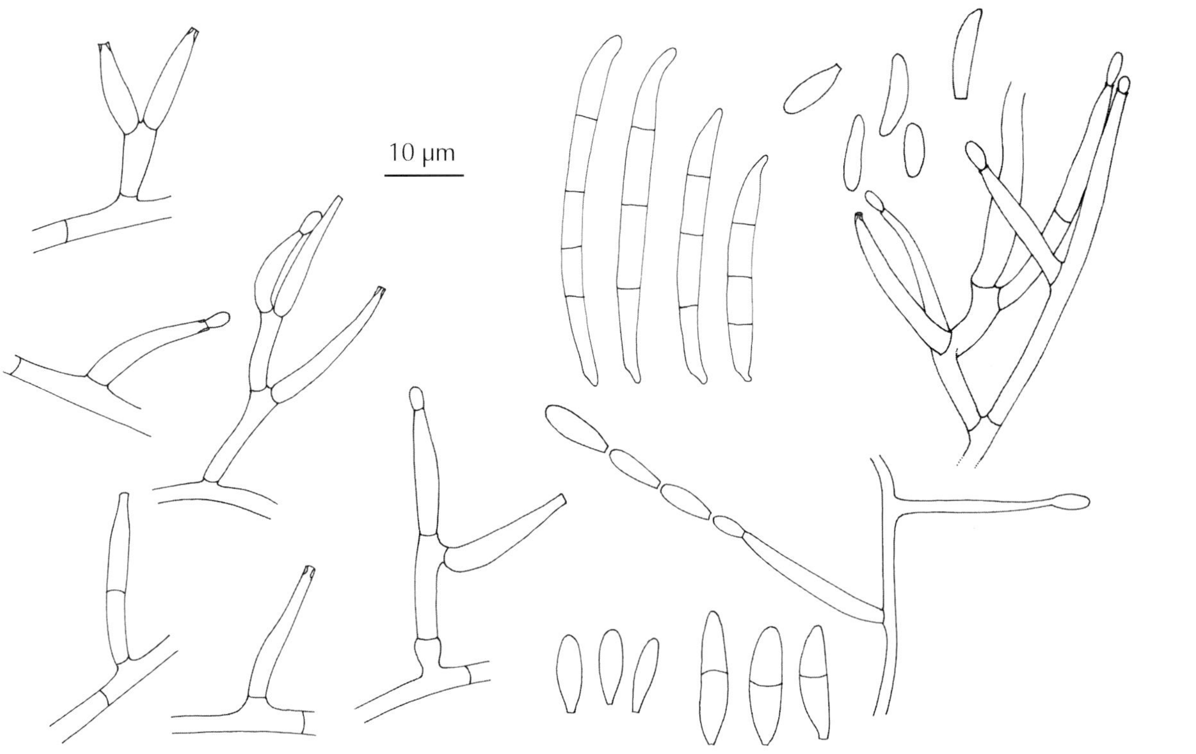

Fig. 72. *Fusarium verticillioides* . Conidiophores, phialides, macro- and microconidia, the latter formed in long chains from phialides. (orig. partly after E.J. Hermanides-Nijhof)

Pl. 67. *Fusarium verticillioides*. a. Colony on PDA after one week; b. catenulate microconidia, in Petri-dish, x 150; c, e. micro- and macroconidia, x 800; d,f, conidiophores, x 800.

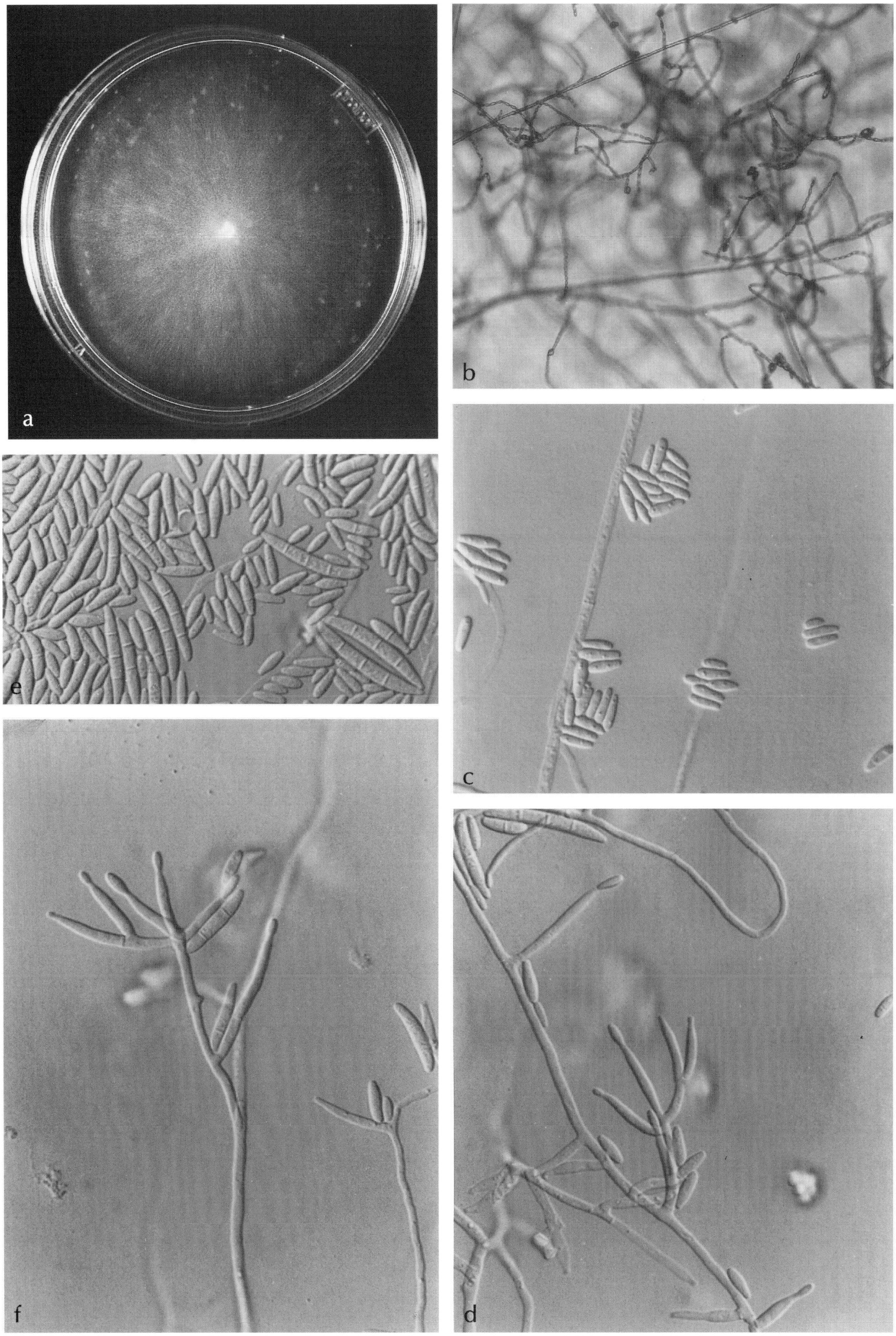
a
b
c
d
e
f

Geomyces pannorum (Link) Sigler & Carmichael

Colonies slow growing, initially white becoming greyish, buff or brownish with age, velvety often tufted. Reverse creamish, yellowish with a reddish brown center. Alternate arthroconidia produced from fertile hyaline hyphae, in short chains of 2-4, in terminal or lateral position, barrel-shaped, obovoid, cuneiform, often with a truncate base, 3-6 x 2-4 µm subhyaline, initially thin- and smooth-walled, becoming rough-walled and thicker walled with age. Intercalary conidia, barrel-shaped, 3-6 x 2-5 µm, smooth-walled at first becoming rough-walled.

Temperature: good growth and sporulation at 20-25°C .

HABITAT: indoor
World-wide distribution, soil, compost, archives.

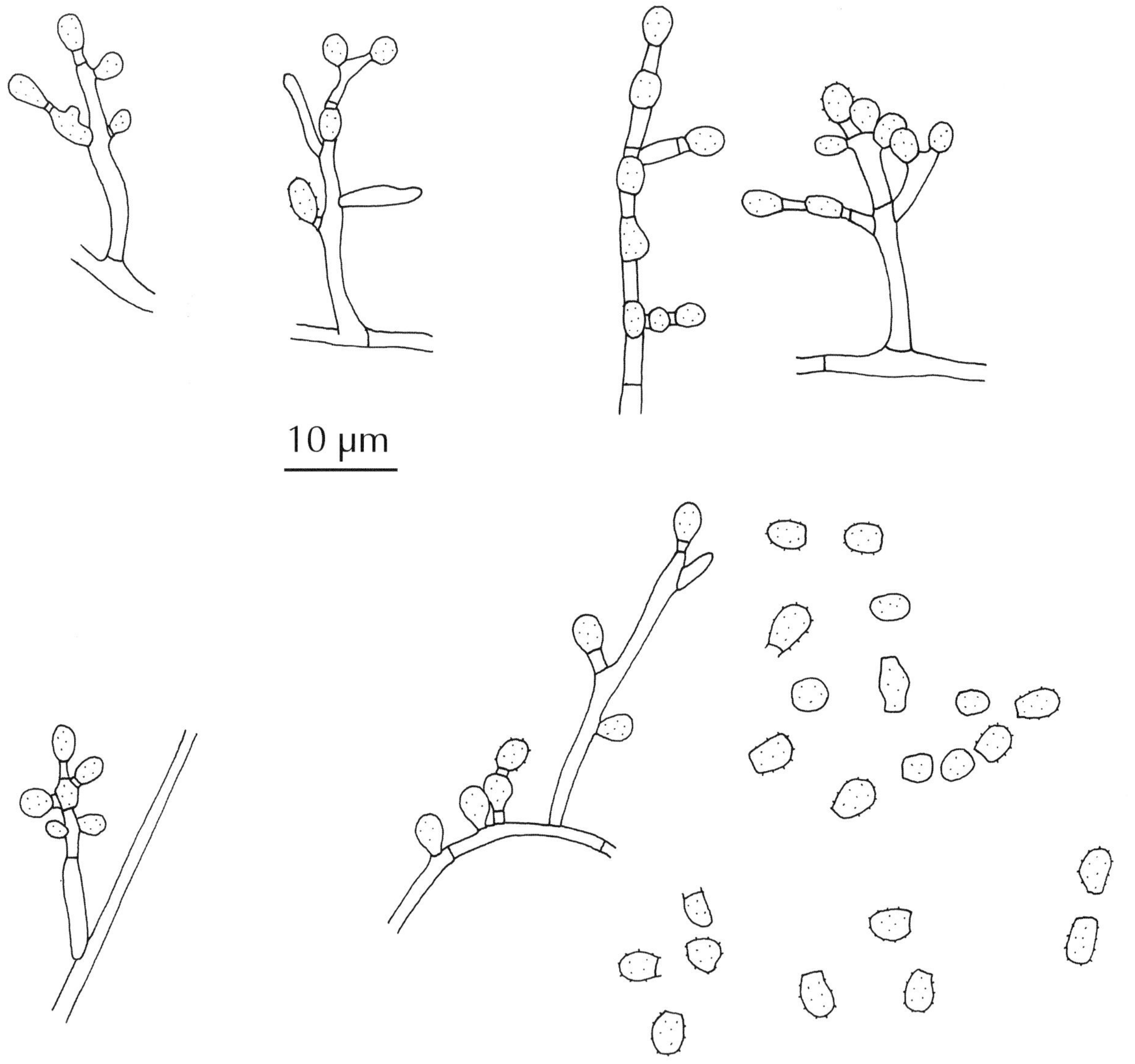

Fig. 73. *Geomyces pannorum.* Conidiophores and conidia.
Pl. 68. *Geomyces pannorum.* a. Colonies on Cherry decoction agar after four weeks; b-f. conidiophores and conidia, x 700

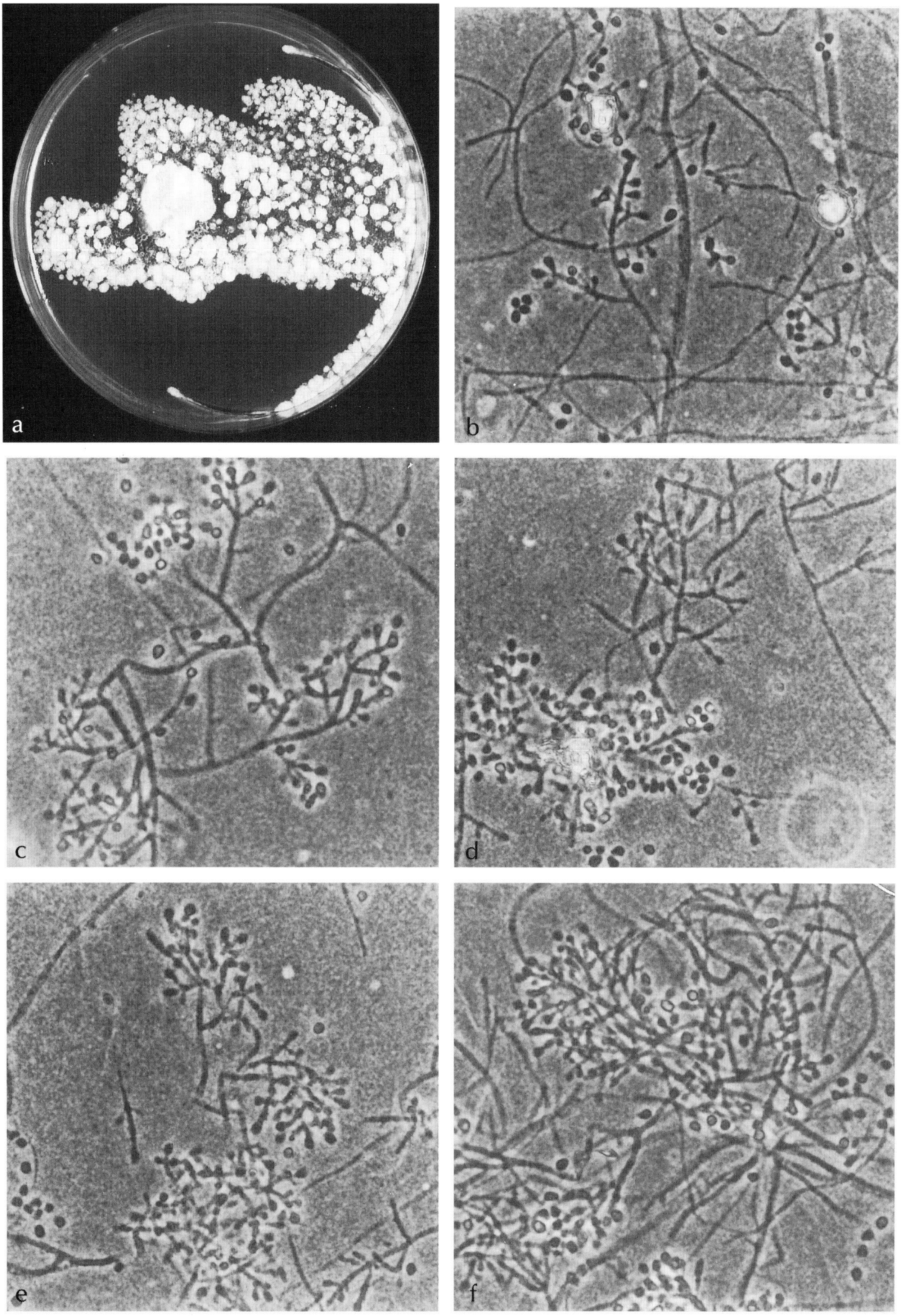
a
b
c
d
e
f

Geotrichum candidum Link

Teleomorph: *Galactomyces geotrichum* (Butler & Petersen) Redhead & Malloch

Colonies on MEA at 24°C attaining a diameter of 7 cm in 7 days, white, smooth, often butyrous or membranous, odour often sweet. Advancing hyphae septate, dichotomously branched (forked), 7-11 µm wide. Conidia cylindrical, barrel-shaped or ellipsoidal, mostly 6-12(-20) x 3-6(-9) µm, formed by breaking up fertile hyphae, chains mostly aerial, erect or decumbent.
Temperature: optimum 25-27°C (when isolated from plants) and 30-31°C.

HABITAT: food, indoor
World-wide distribution: soil, water, air, corn and other cereals, rice, grain, rotting paper, textiles, grapes, citrus, bananas, tomatoes, cucumber, frozen fruit-cake, fruit-juices, bread, milk and milk products, animals and man.

NOTE:
For more detailed descriptions see Domsch *et al.* (1993) and de Hoog *et al.* (1986). This species is very common. Morphology and colony development resembles that of typical yeasts, except in the production of distinct mycelium. Ascospores of the teleomorph, *Galactomyces geotrichum* are rarely produced.

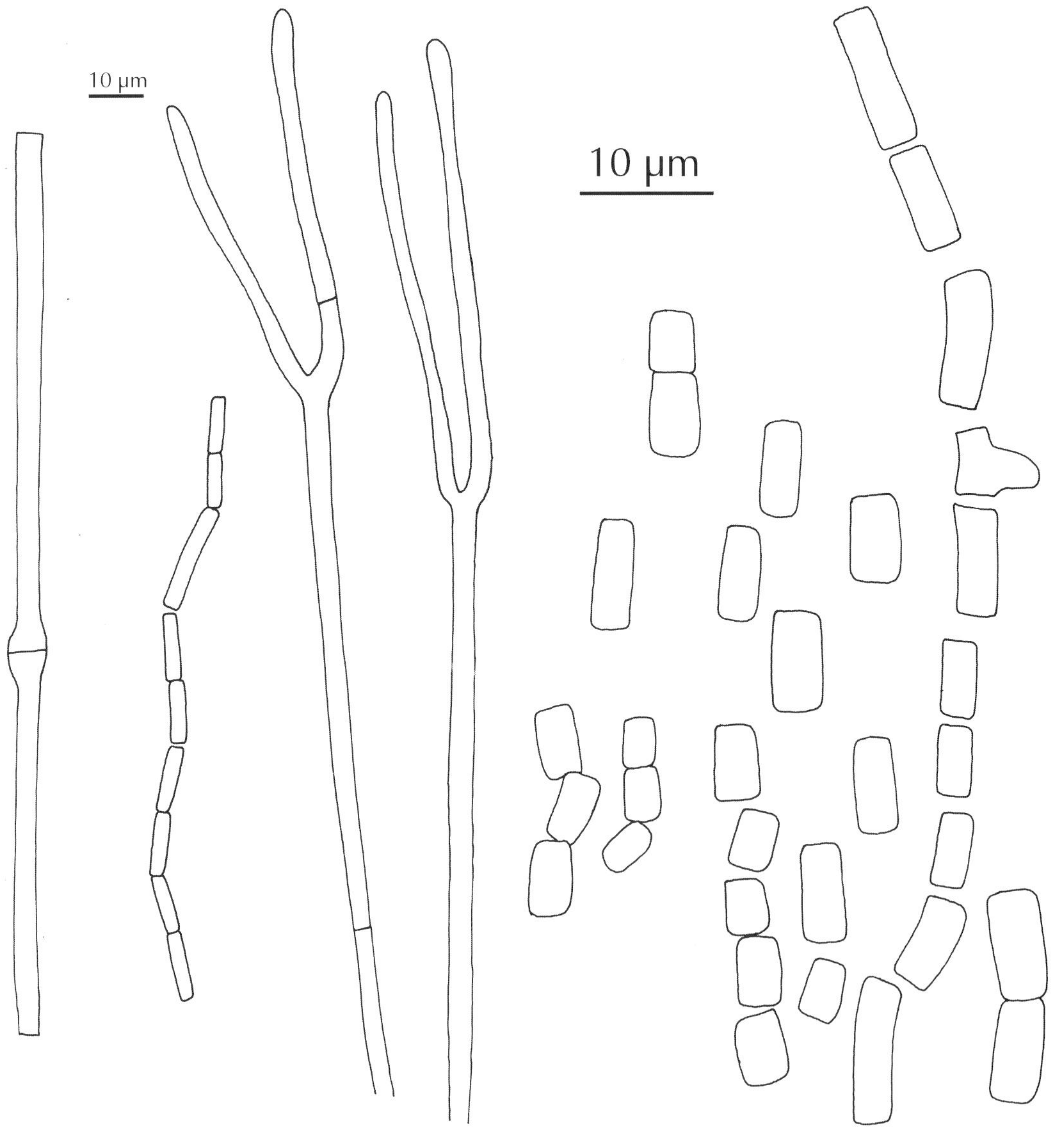

Fig. 74. *Geotrichum candidum.* Hyphae dichotomously branched, conidia formed by breaking up fertile, mostly aerial hyphae.
Pl. 69. *Geotrichum candidum.* a. Colony on MEA after one week; b. margin of colony with dichotomously branched (forked) hyphae, x 175; c,d. disarticulation of fertile hyphae into (arthro)conidia, x 1000; e,f. conidia, x 1000.

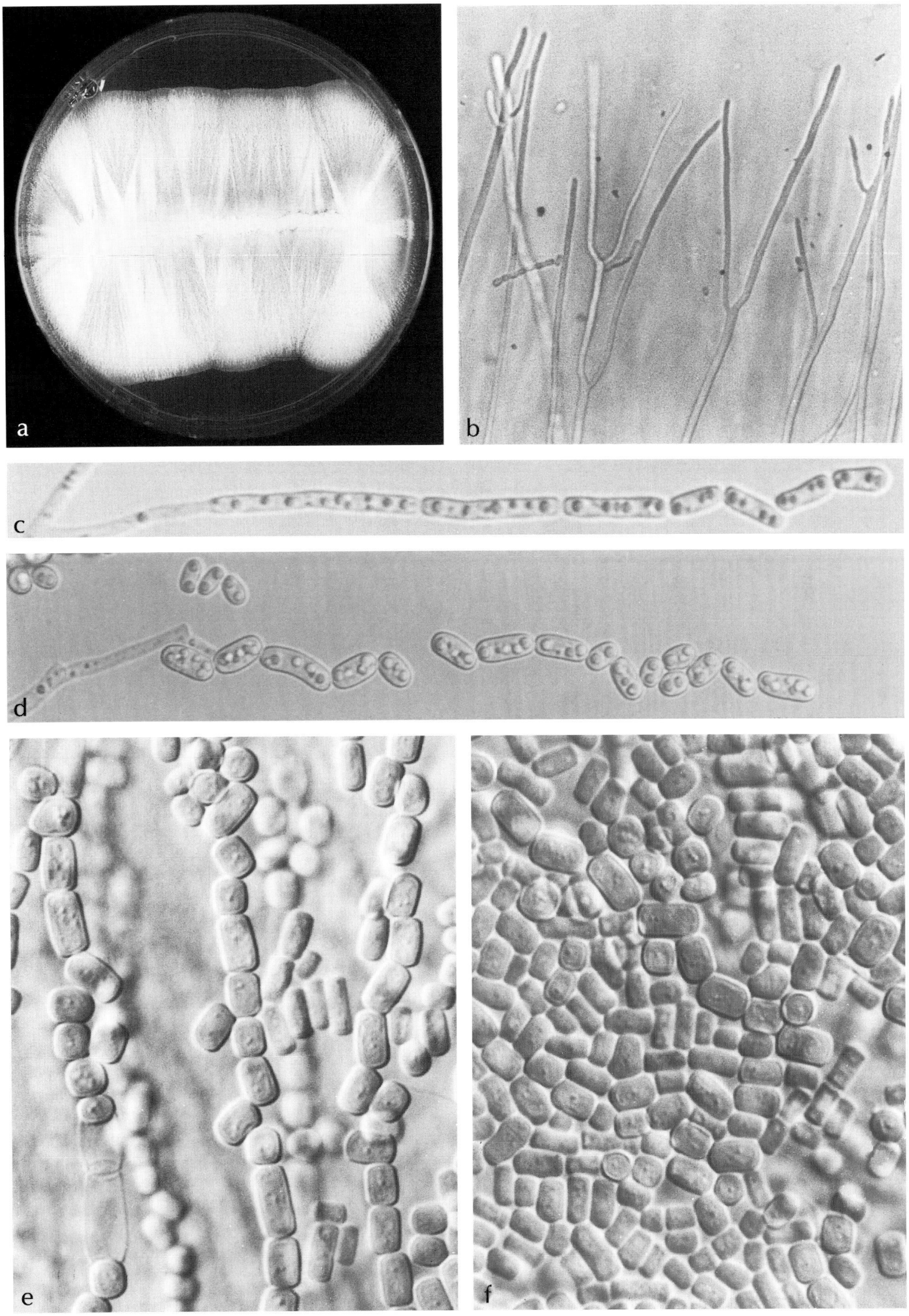
a
b
c
d
e
f

Memnoniella echinata (Riv.) Galloway

= *Stachybotrys echinata* (Riv.) G. Smith

Colonies on OA reaching 1,5 cm within a week at 25°C, on MEA attaining a diameter of about 2-3 cm, appearing blackish to blackish-green and powdery due to the conidial masses. Reverse yellowish brown to grayish. Conidiophores simple (occasionally forked), about 100 µm long and 3-6 µm wide, hyaline or greyish at first, later becoming olivaceous brown to black and roughened, bearing clusters of 4-8 phialides. Phialides obovate or ellipsoidal, 7-10 x 3-4 µm, hyaline at first, becoming olivaceous-grey, smooth-walled. Conidia in chains, (sub)globose to somewhat compressed, 3.5-6 µm in diameter, hyaline at first becoming dark olive-grey to brownish-black and coarsely rough-walled to warted with age. Temperature range for good growth: 15-30 °C (with a maximum temperature for conidium germination of 37°C.

Important (toxic) metabolites: Griseofulvins, the profiles of toxic compounds are similar to *Stachybotrys chartarum* (Jarvis *et al.*, 1998)

HABITAT: indoor
Although the fungus has a world-wide distribution it has been isolated mainly from soils in tropical countries. Also isolated from cellulosic materials (e.g. paper, wallpaper), textiles, dead plant material.

NOTE:
For detailed descriptions and keys see Ellis (1971), Domsch *et al.* (1993). Exposure to this fungus on natural substrates or in culture should be avoided. Recent molecular studies using PCR sequencing showed that *Memnoniella echinata* and *Stachybotrys chartarum* are closely related (Haugland and Heckman, 1998).

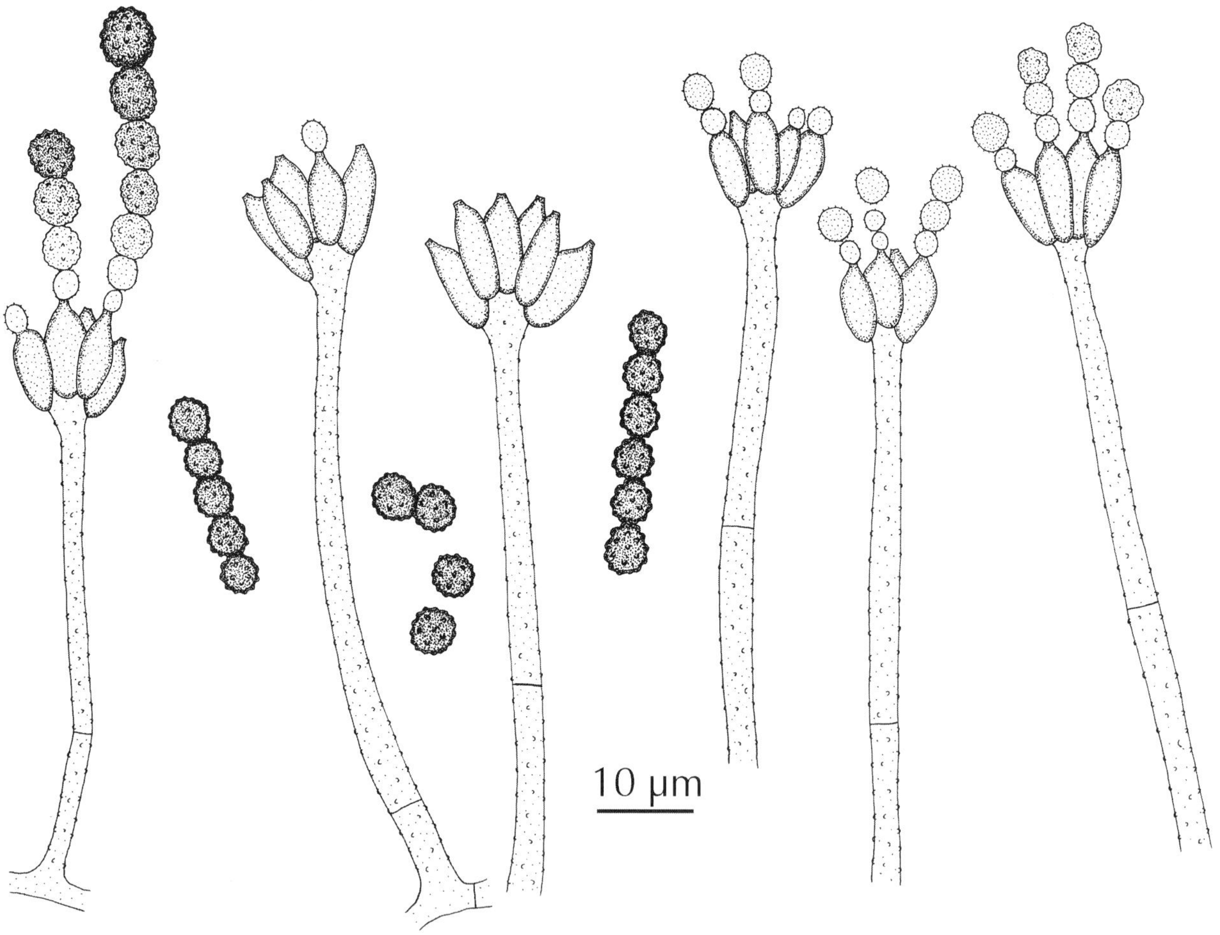

Fig. 75. *Memnoniella echinata.* Conidiophores and conidia.
Pl. 70. *Memnoniella echinata.* a. Colony after 10 days on OA; b-d.,f. conidiophores and conidia in chains, b,c x 675, d,f. x 1835; e. conidia in chains, x 1690.

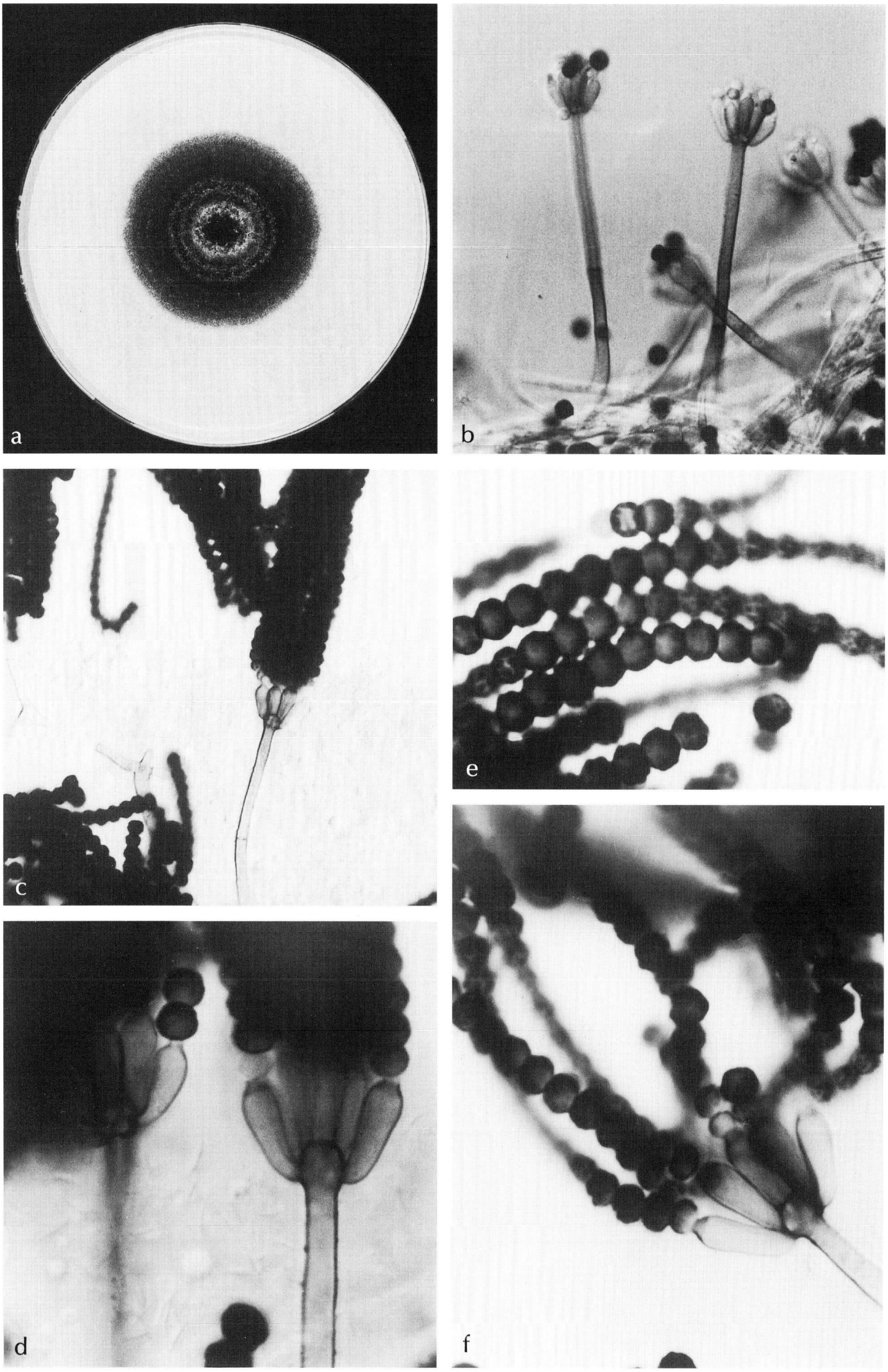
a
b
c
e
d
f

MONILIELLA Stolk & Dakin

Colonies usually restricted, smooth, velvety or cerebriform, cream-coloured at first, darkening to pale olivaceous or blackish-brown with age. Budding cells often present, ellipsoidal to subcylindrical, frequently composing a pseudomycelium (as in yeasts). Hyphae often branched, hyaline or coloured, septate, with a strong tendency to fragment. (Blasto)conidia usually one-celled, arising in acropetal chains from undifferentiated hyphae. (Arthro)conidia formed by fragmentation of hyphae, large, often becoming brown and thick-walled after detachment. Chlamydospores sometimes present, one-celled, subglobose, thick-walled, dark-brown. For more detailed descriptions see Stolk and Dakin (1966), Dakin and Stolk (1968), de Hoog (1979), for conidiogenesis Martinez and de Hoog (1979), Cole and Samson (1979).

CULTIVATION FOR IDENTIFICATION
Isolates are inoculated on MEA (2%) (or PCA, cherry agar) and cultivated at 20-24°C for 7-14 days in diffuse daylight.

KEY TO THE SPECIES TREATED

1a. Differentiated, dark chlamydospores present in hyaline hyphae ....*M. acetoabutens*
1b. Differentiated chlamydospores absent; conidia often becoming dark ....*M. suaveolens*

Moniliella acetoabutens Stolk & Dakin

Colonies on MEA at 24°C attaining a diameter of 3-4 cm in 7 days, growth rate varying per strain, velvety to more or less lanose, white to brownish-grey, narrowly zonate. Reverse blackish-brown or greenish-brown, especially in the central part. Odour yeast-like. Hyphae hyaline, septate, branched, tending to fragment into long thick-walled hyphal fragments (up to 150 µm long) or into thin-walled fragments (arthroconidia). (Blasto)conidia arising terminally or laterally in short chains from undifferentiated hyphae, apical conidia broadly ellipsoidal, 4.5-9 x 3.5-6 µm with indistinct scars, lower conidia cylindrical with obtuse or slightly rounded ends, usually about 10 µm long. Chlamydospores intercalary or terminal, solitary or in short chains, globose, 8.5-11 µm in diam, dark brown, smooth- and thick-walled (particularly on MEA).

HABITAT: food
Acidophilic, quite commonly occurring in pickles and vinegar, also recorded from fruit-juices, syrups and mint sauces, reported from Europe, U.S.A.

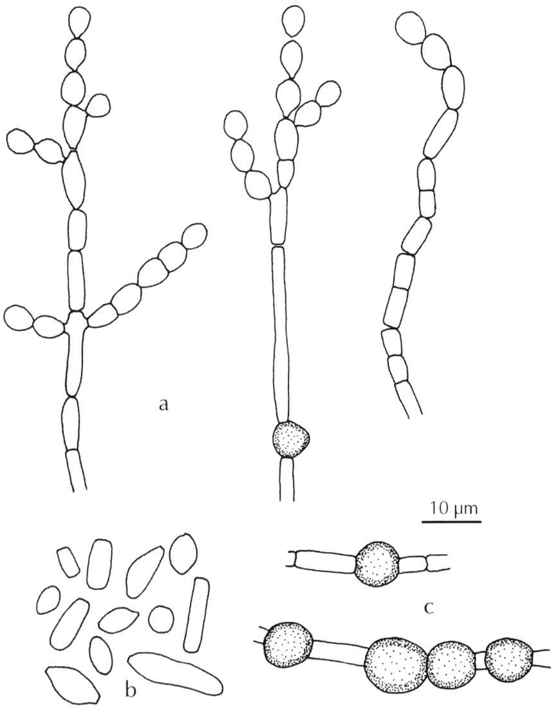

Fig. 76. a-c. *Moniliella acetoabutens.* a. Acropetal chains of (blasto)conidia arising from disarticulated hyphae; b. conidia; c. chlamydospores;
Pl. 71. *Moniliella acetoabutens.* a. Colony on MEA after one week; b-e. (blasto)conidia arising in acropetal chains from disarticulated hyphae, x 1000.

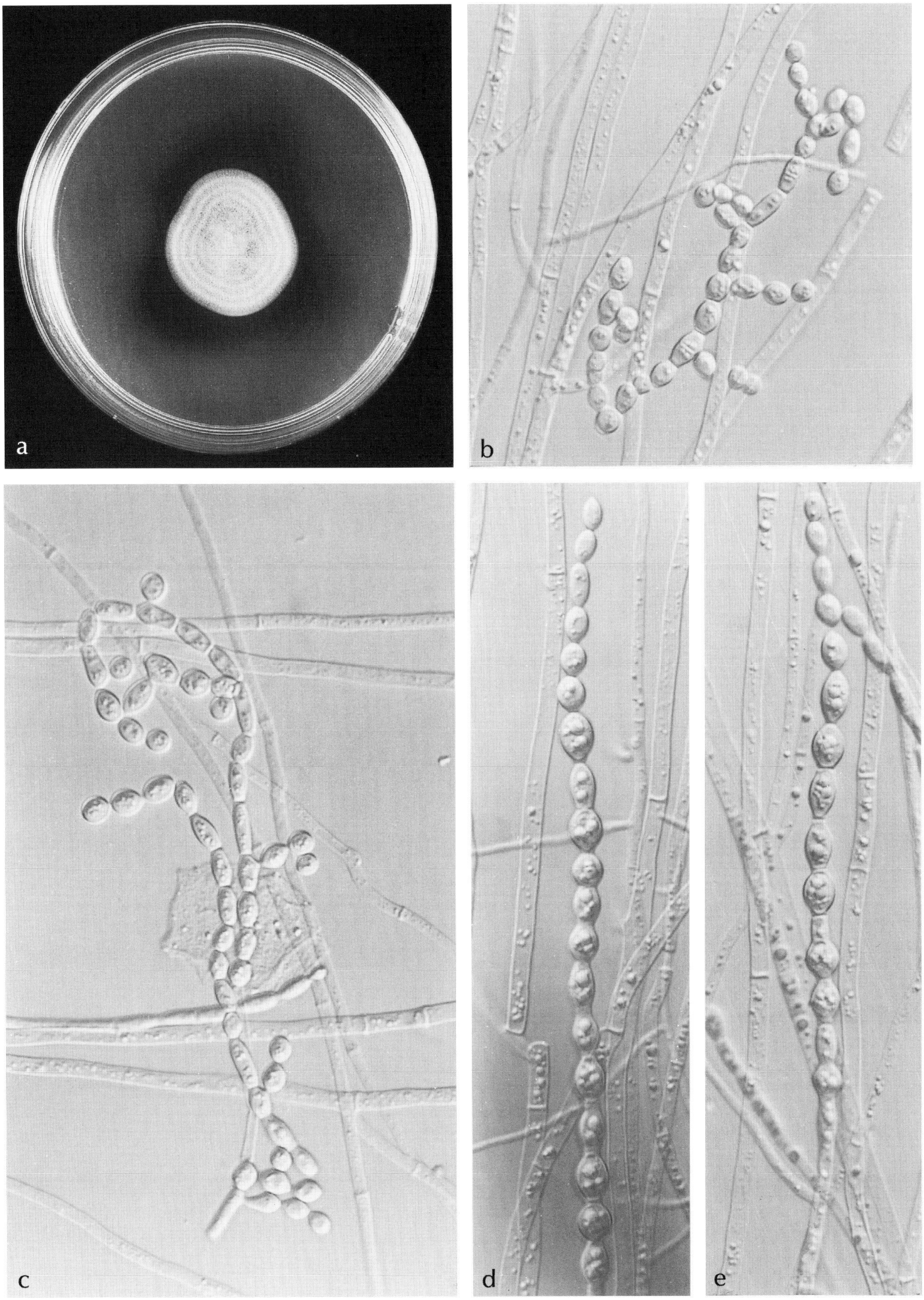

Moniliella suaveolens (Lindner) v. Arx

Colonies on MEA at 24°C attaining a diameter of 2.5-3 cm in 7 days, smooth with indistinct radial furrows, in the central part somewhat butter-like (on PCA more lanose in the centre with white to greyish aerial mycelium), whitish to cream-coloured or greyish-brown. Reverse pale yellow to greyish-brown or dark-brown in the centre. Odour faintly yeast-like. Hyphae hyaline, usually gradually changing over into conidial chains at the apex, often with lateral branches arising just below the septa, often with a basal septum when mature. Lower hyphal parts disarticulate into separate cells which often have subapical protruding scars. Conidiogenous cells terminal or intercalary, undifferentiated, cylindrical with distinct scars at their distal parts, often converted into ramoconidia, subhyaline. (Blasto)conidia arising acropetally, terminal or lateral, in unbranched or once-branched chains, fusiform, nearly cylindrical, saturnoid or ellipsoidal, 12-27(-40) x 4-5.8 µm, truncate at both ends, sometimes 1-septate, apical conidia distally rounded, scars distinct, (sub)hyaline to brown.

HABITAT: food

Lipo- and osmophilic. Isolated from fatty substances such as margarine and dairy products, e.g. causing black spots on cheese, as well as from sweet substances and phyllospheres e.g. leaf surface of tobacco.

NOTE:

Two varieties are distinguished (de Hoog, 1979): var. *suaveolens* with colonies spreading, usually pale and conidia elongate, (sub)hyaline; var. nigra with colonies restricted, usually dark and conidia ellipsoidal, usually becoming darker brown.

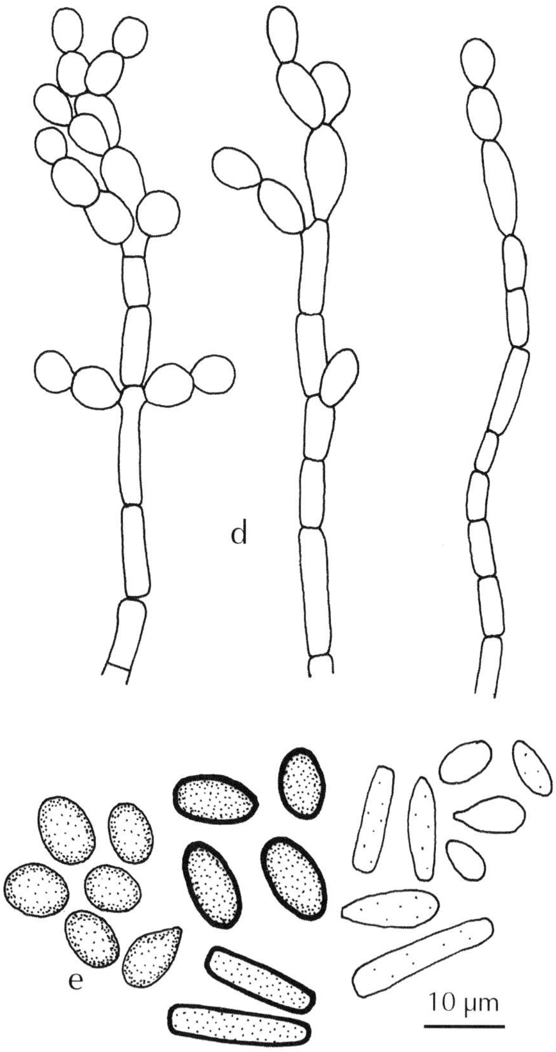

Fig. 76 d-e. *Moniliella suaveolens*; d. acropetal chains of (blasto)conidia; e. conidia. (orig. G.S. de Hoog).
Pl. 72. *Moniliella suaveolens*. a. Colony on MEA after one week; b,c. acropetal chains of (blasto)conidia and lower hyphal part breaking up into long cells, x 1500.

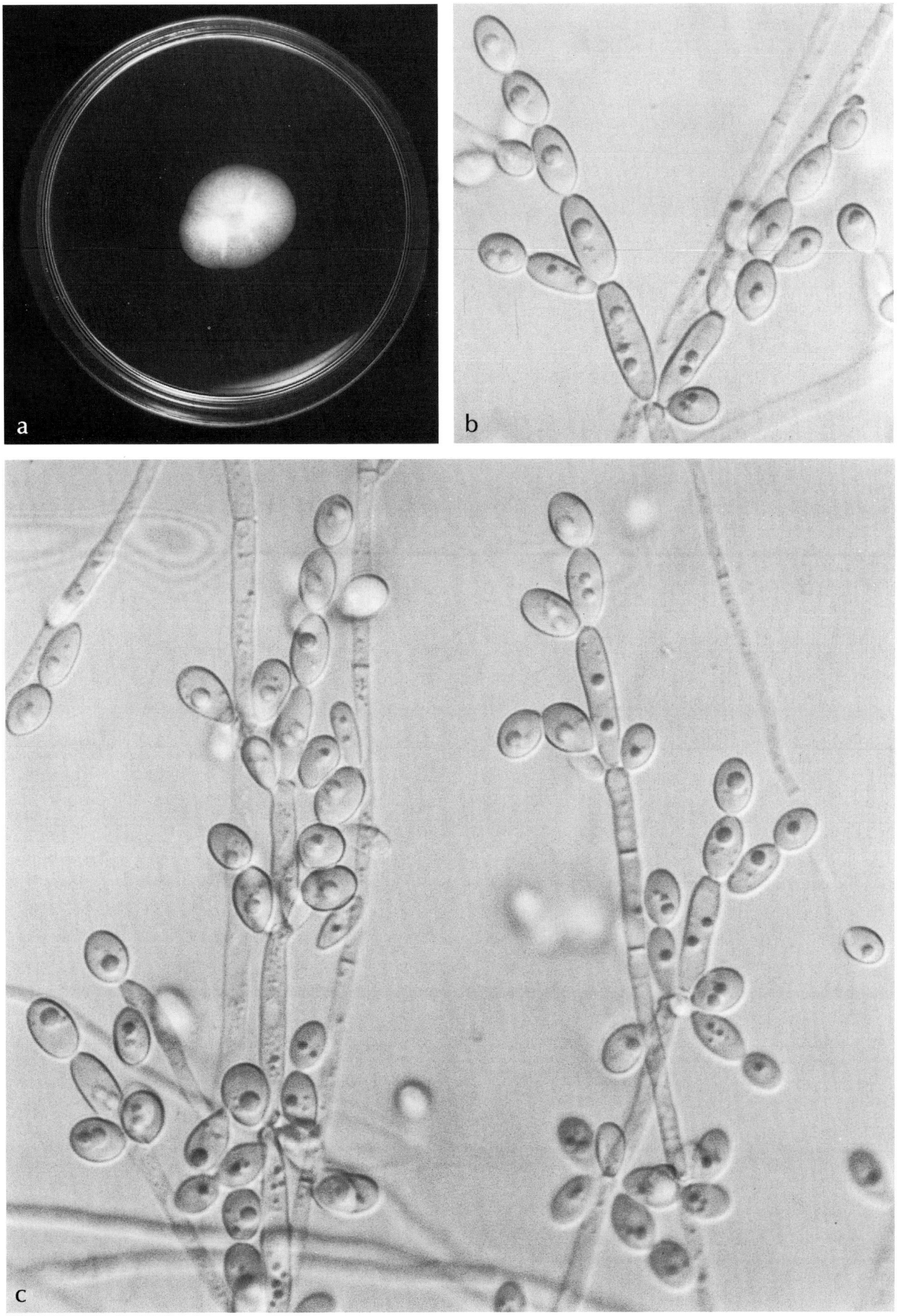
a
b
c

Oidiodendron griseum Robak

Colonies on hay agar greyish to olivaceous brown, velvety often with tufts towards the central part of the colonies, slow- growing reaching about up to 1.5 cm in 14 days. Reverse dark brown. Conidiophores erect, pigmented, up to 100 µm long and 1.5-2.0 µm wide, apically branched. Branches, consisting of chains of arthroconidia. The arthroconidia may also be produced from undifferentiated hyphae. Between the arthroconidia which are produced basipetally often empty cells are present. Arthroconidia pale grey-green in mass, smooth-walled or nearly so, 2.0-3.5 x 1.5-2.0 µm, ovoid to short cylindrical.

Temperature: good growth and sporulation at 25-30°C

HABITAT: indoor
Isolated from soils in forests in Scandinavia, Britain, France, Africa, New Zealand, North America and Canada, Trinidad. Also from humus, garden compost, decaying wood, pine litter, cedar and peat bogs, hay. Frequently isolated from paper, archives, and timber.

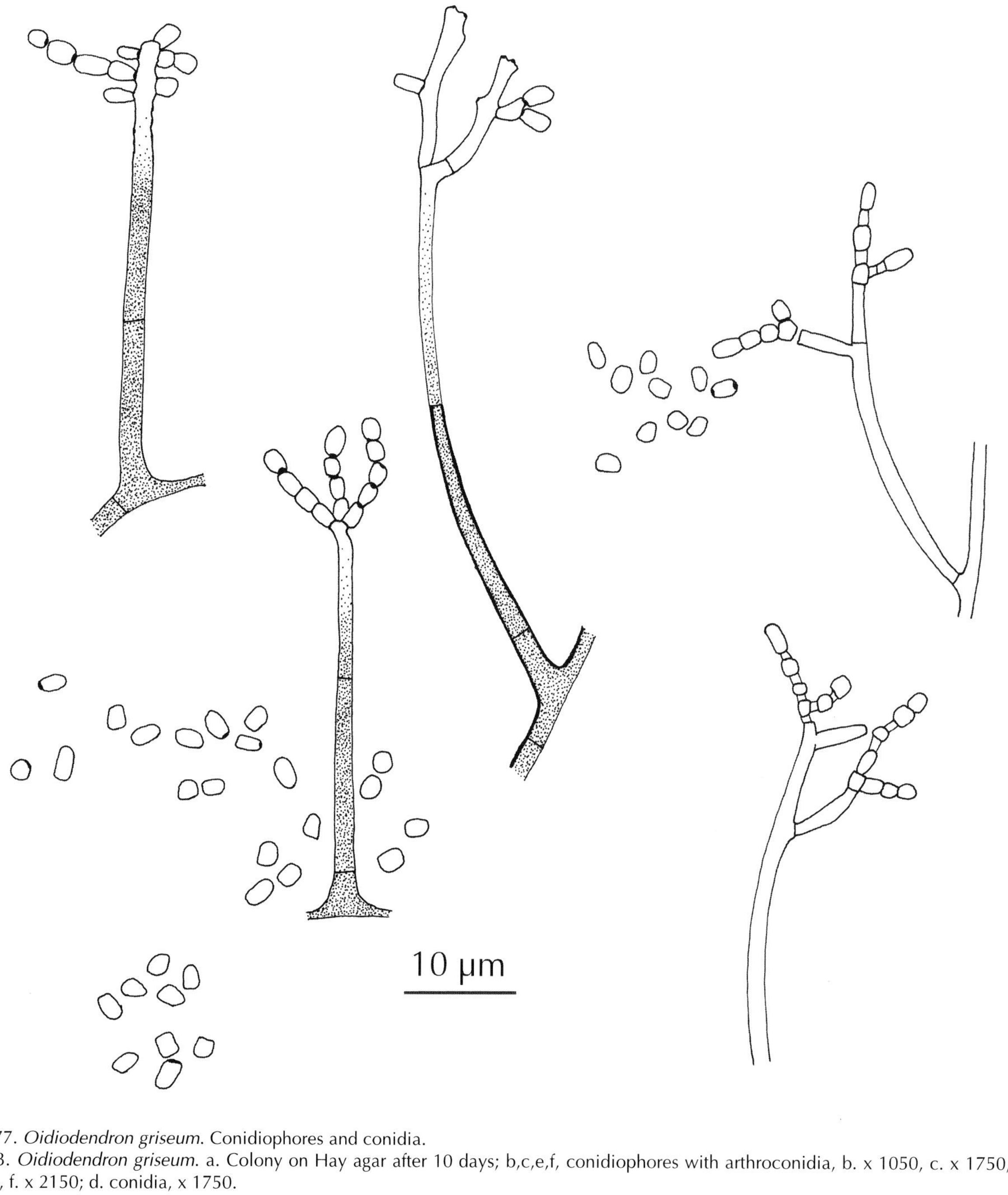

Fig. 77. *Oidiodendron griseum*. Conidiophores and conidia.
Pl. 73. *Oidiodendron griseum*. a. Colony on Hay agar after 10 days; b,c,e,f, conidiophores with arthroconidia, b. x 1050, c. x 1750, e. x 1500, f. x 2150; d. conidia, x 1750.

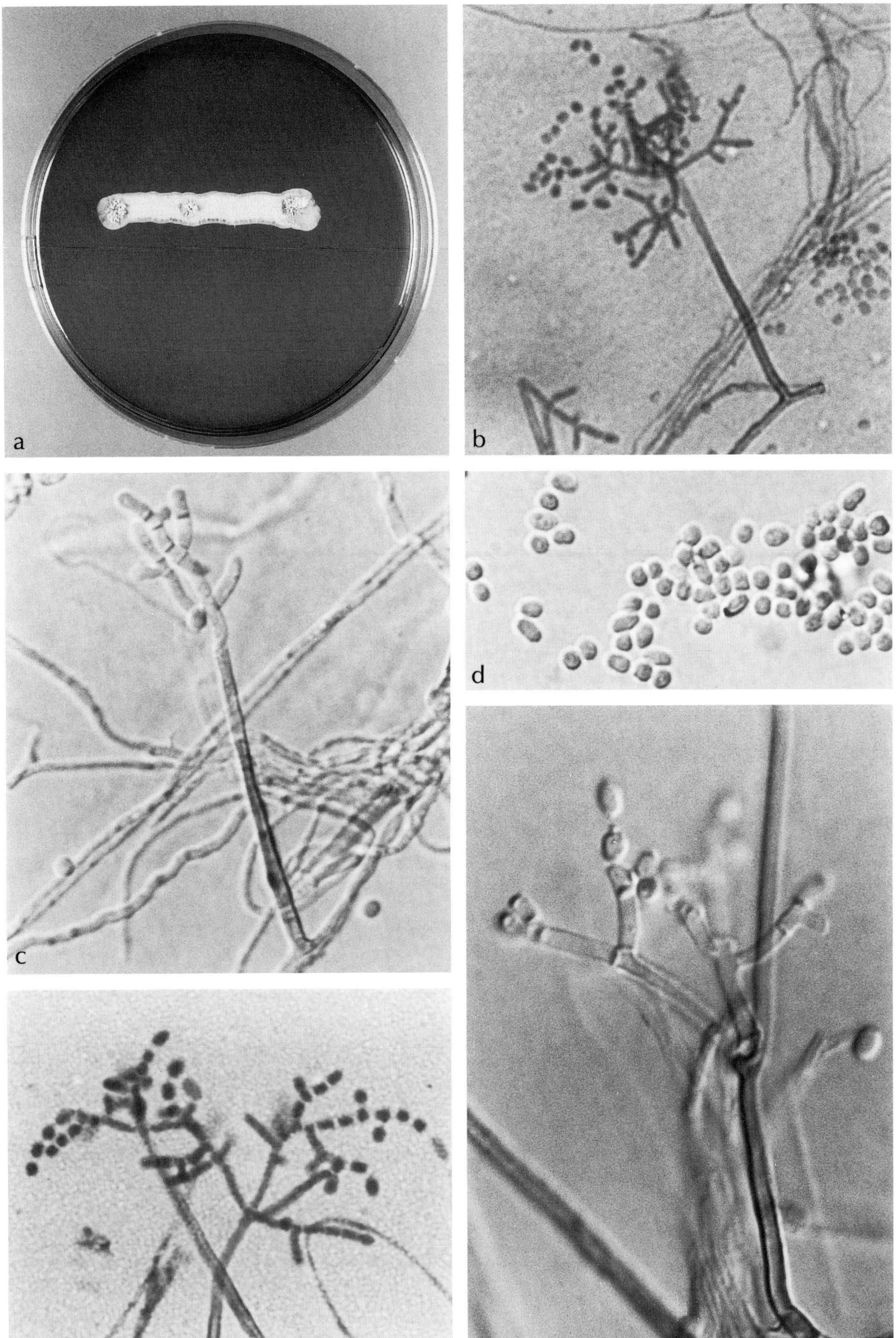
a
b
c
d
e
f

Paecilomyces variotii Bain.

Colonies on MEA at 25°C, growing rapidly, attaining a diameter of 3-5 cm within 7 days, consisting of a dense felt of numerous conidiophores giving the colonies a powdery appearance, yellow-brown. Odour sweet aromatic. Conidiophores consisting of dense whorls of verticillately arranged branches, each bearing 2 to 7 phialides. Phialides solitary or in whorls, variable in size but mostly 12-20 x 2.5-5.0 µm, consisting of a cylindrical to ellipsoidal basal portion, tapering abruptly to a long, cylindrical, 1-2 µm wide neck. Conidia in long divergent chains, one-celled, hyaline to yellow, smooth-walled, subglobose, ellipsoidal to fusiform, variable in size but mostly 3.0-5.0 x 2-4 µm. Chlamydospores usually present, single or in short chains, brown to dark brown, smooth to slightly rough, subglobose to pyriform, thick-walled, 4-8 µm.

Important (toxic) metabolites: patulin, viriditoxin.

HABITAT: food, indoor
P. variotii is a common contaminant in air and often isolated from substrates originating from higher temperatures e.g. compost. The species is thermophilic, growing at a maximum of about 50°C (60°C).

NOTE:
The anamorphs of *Byssochlamys* and *Thermoascus crustaceus* (Samson, 1974) are similar. *P. variotii* has been reported as the causative agent of mycosis, and therefore isolates should be handled with caution. *P. variotii* can grow in extreme, often thermophilic environments and also has recently been reported to be resistant against preservatives (e.g. sorbic, benzoic and propionic acid). See also page 172.

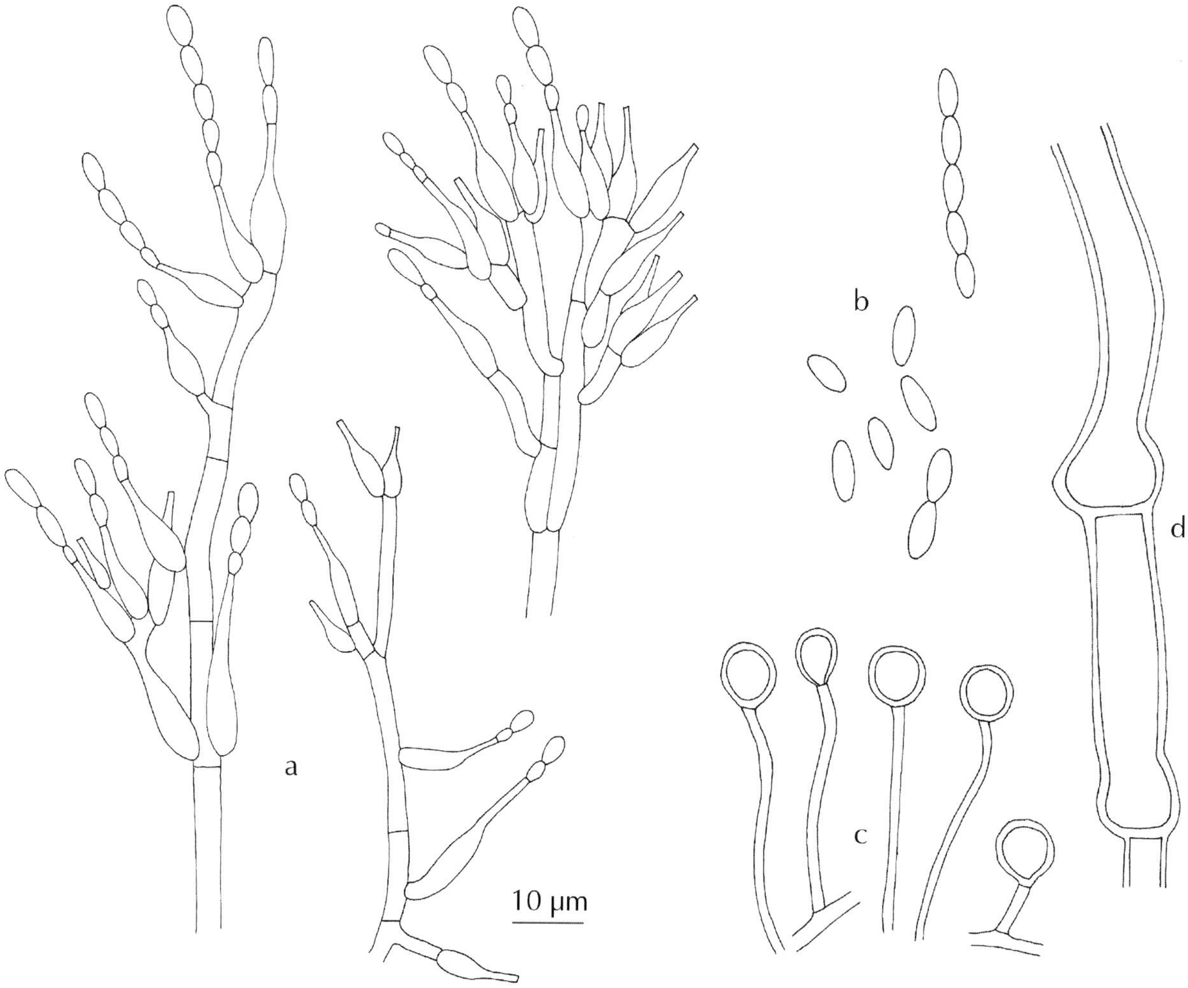

Fig. 78. *Paecilomyces variotii*. a. Conidiophores; b. conidia; c. chlamydospores ; d. thick-walled hyphal element.
Pl. 74. *Paecilomyces variotii*. a. Colonies on MEA after 10 days; b-d. conidiophores, x 1000; e. conidia, x 1000

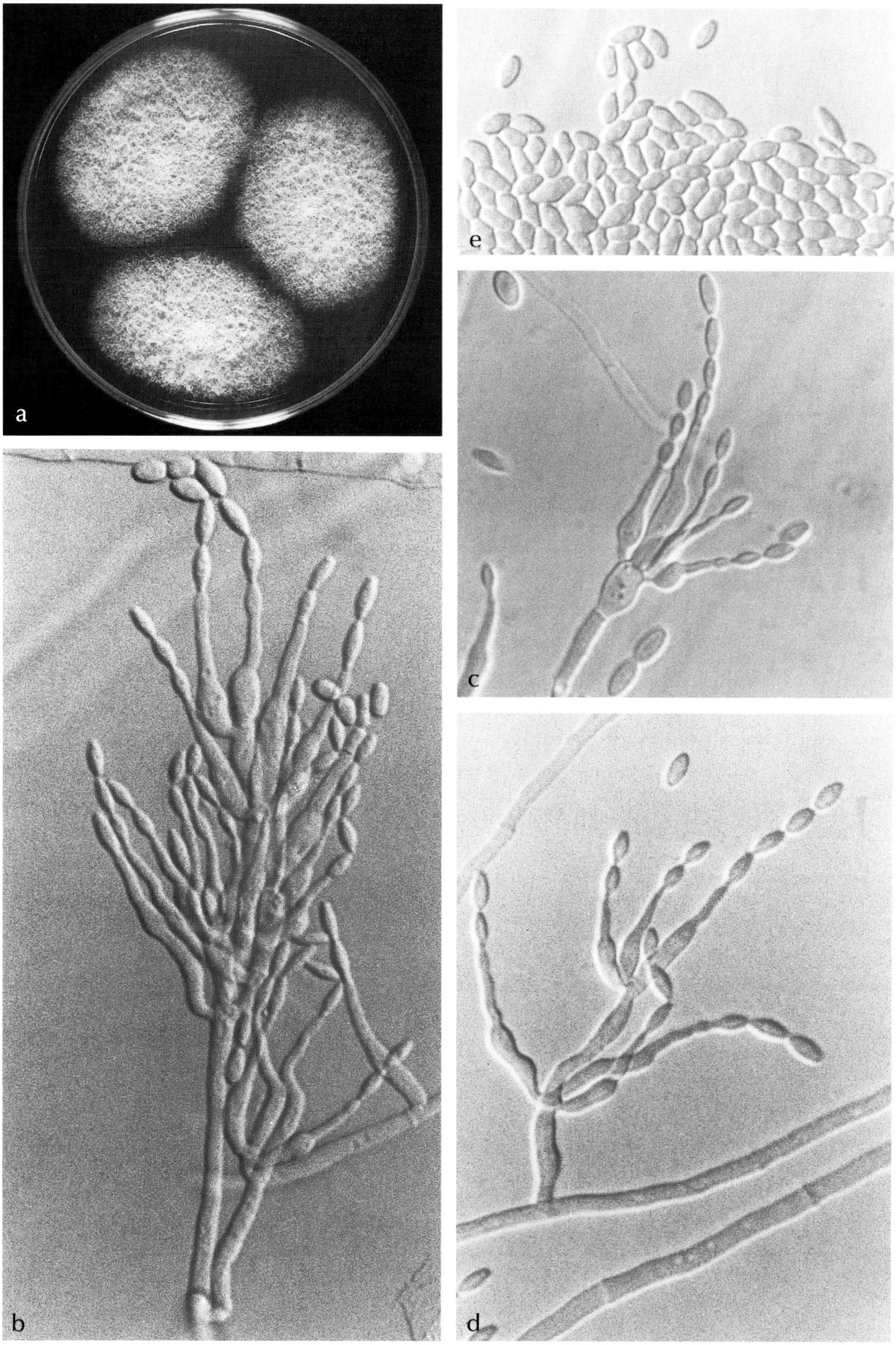
a
b
e
c
d

Paecilomyces variotii Bain. (continuation)

P. variotii represents a complex of different taxa and a molecular taxonomic study is urgently needed to reveal the delimitation of the taxa. Strains of *Paecilomyces variotii* sometimes produce thick-walled hyphae and abundant chlamydospores (Plate 75 and Figure 78). These strains proved to be heat resistant surviving treatments of 97°C.

These structures are also found in *Paecilomyces spectabilis* the anamorph of *Talaromyces spectabilis* Udagawa and Suzuki (1994). *T. spectabilis,* which is very similar to *Byssochlamys,* was found in heat processed fruit beverage in Japan and soil in Nepal. In culture, ascospores of *T. spectabilis* are rarely found and the cultures have to be incubated for at least 2 weeks on Oatmeal agar or malt extract agar.

Pl. 75. *Paecilomyces variotii.* a,b. Conidiophores, x 1000; c. conidia x 950. d-g. thick-walled hyphal elements and chlamydospores, x 1000

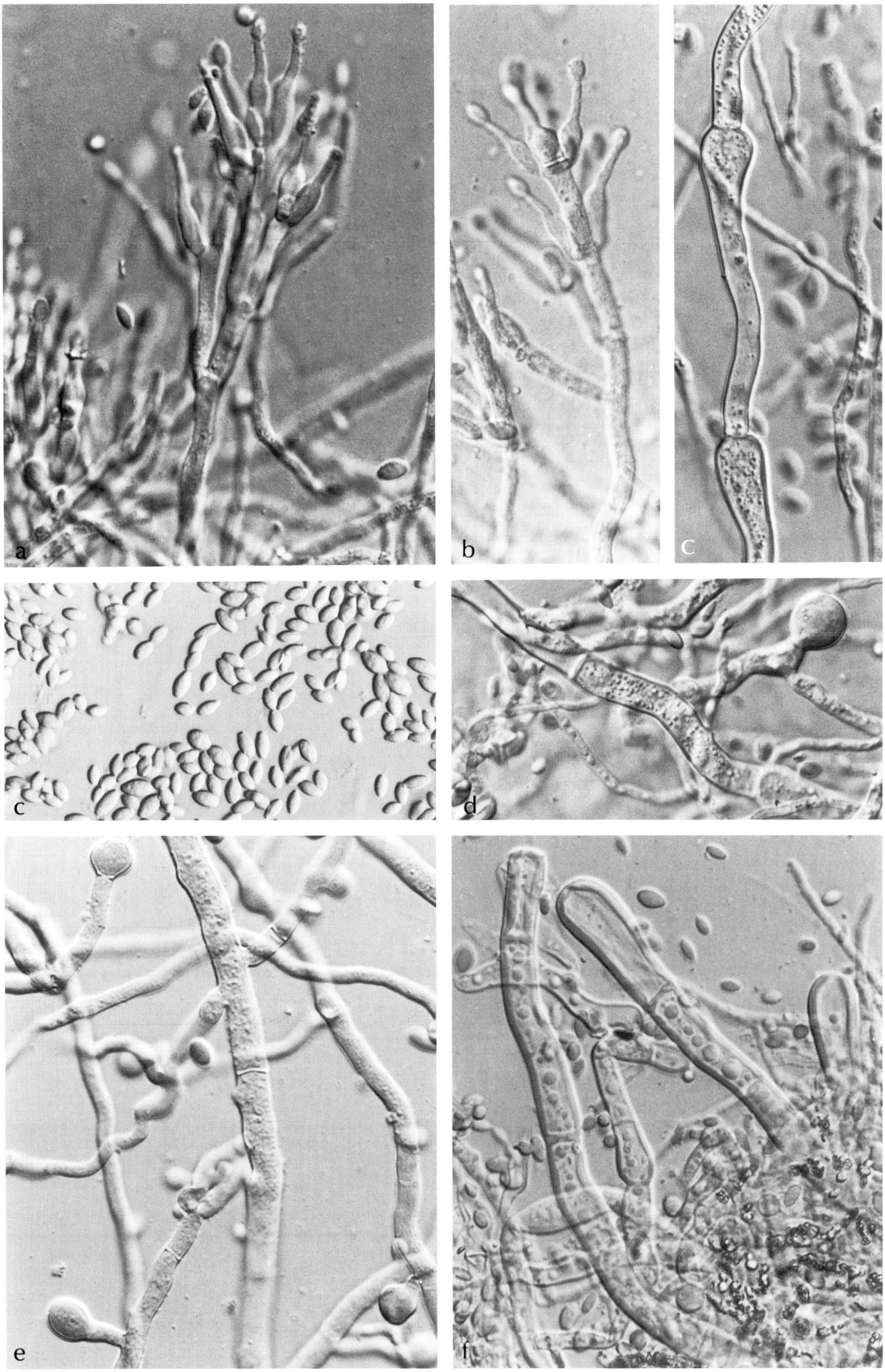
a
b
c
c
d
e
f

PENICILLIUM Link: Fr.

Teleomorphs: *Eupenicillium, Talaromyces, Hamigera*

In the collaboration with Jens C. Frisvad.
BioCentrum-DTU, Technical University of Denmark, DK-2800 Lyngby, Denmark

Colonies usually growing rapidly, usually in shades of green, sometimes white, mostly consisting of a dense felt of conidiophores. The conidiophore may arise from the substrate (**velvety**), from aerial hyphae (**lanose**), from prostrate bundled hyphae (**funiculose**) or from erect loosely or compactly bundled hyphae (**fasciculate** or **synnematous**) (Fig. 80 f-k). Conidiophores hyaline, smooth- or rough-walled, they can be single (**mononematous**) or bundled (synnematous), consisting of a single stipe terminating in either a whorl of phialides (**monoverticillate**) or in a penicillus (Fig. 79). The penicillus existing of branches and metulae (penultimate branches which bear a whorl of phialides). All cells between metulae and stipe are referred to as branches.

Branching pattern either **one-stage branched** (biverticillate), **two-stage branched** (terverticillate) or **three-** (quaterverticillate) to **more-stage branched** (Figs 80 and 81). Phialides usually flask-shaped consisting of a cylindrical basal part and a distinct neck, or lanceolate (= acerose: more or less narrow basal part tapering to a somewhat pointed apex). Conidia in long dry chains, divergent or in columns, globose, ellipsoidal, cylindrical or fusiform, hyaline or greenish, smooth- or rough-walled. **Sclerotia** are produced by some species.

Many species of *Penicillum* are common contaminants on various substrates and are known as potential mycotoxin producers (see Chapters 5 and 6). Correct identification is therefore important when studying possible *Penicillium* contamination.

The classification of *Penicillium* species by their cultural characters were emphasized by Raper and Thom (1949) but these have proved to be very variable. Pure morphological characteristics can be used for the classification, in addition with features of growth rate and water activity with different media and temperatures (Pitt, 1979). The use of profiles of secondary metabolites are also proven to be valuable (Frisvad, 1981, 1985; Frisvad and Filtenborg, 1983, Lund and Frisvad, 1994 and see Chapter 2). For more recent detailed descriptions and keys of Penicillia see also Pitt (2000).

For some species odour and exudate production will help to recognize the taxa, but it should be stressed that inhalation of conidia and volatiles may cause health effects. However, human pathogenic species as in *Aspergillus* are rare in *Penicillium* and only limited to *P. marneffei*.

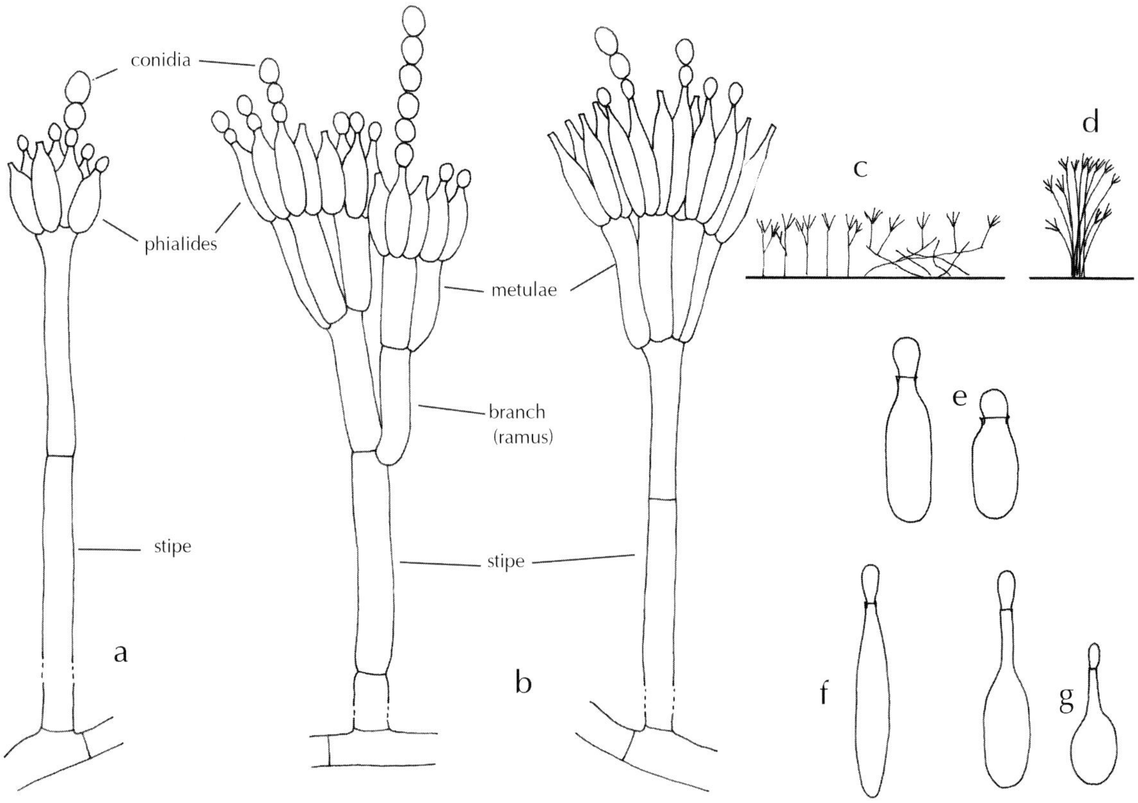

Fig. 79. a-g. Morphological structures in *Penicillium*. a-b. Conidiophore structure; c. mononematous; d. synnematous; e. flask-shaped; f. lanceolate (= acerose); g. *Paecilomyces*-type.

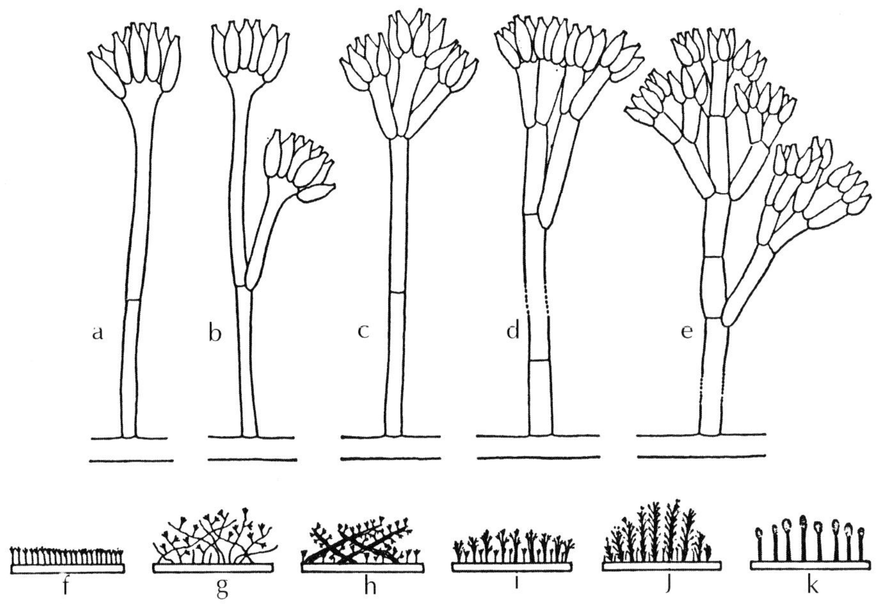

Fig. 80. a-e. Types of conidiophore branching in *Penicillium*. a,b. simple (=monoverticillate); c. one-stage branched (=biverticillate); d. two-stage branched (= terverticillate); e. three-stage branched (= quaterverticillate); f-k. colony types in *Penicillium*. f. velvety; g. lanose; h. funiculose; i-k. fasciculate.

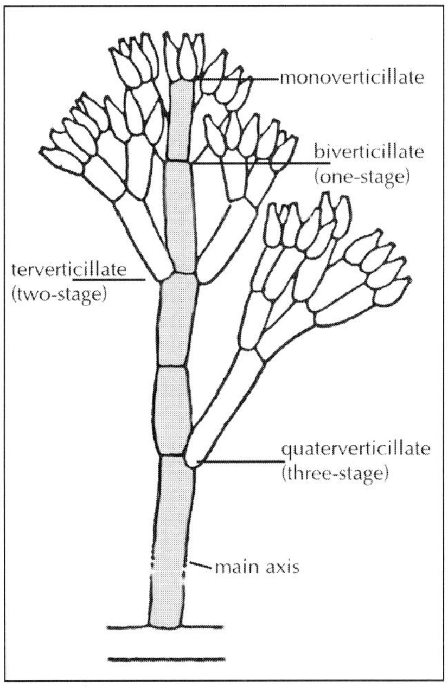

Fig. 81. Terminology of the different branching patterns of a penicillus. Each level of branching is counted from the top of the main axis. Note, that the phialides are not counted when using the one-, two- and three- staged terminology.

CULTIVATION FOR IDENTIFICATION

Samson and Pitt (1985) recommended that isolates should be inoculated at three points on Czapek Yeast autolysate agar (CYA) and MEA (2%) and incubated at 25° C, either in dark or in light. However, the cultivation on Czapek agar (Cz) is also appropriate. Note, that atypical growth and pigmentation of Penicillia on CYA and Cz may occur, often due to a lack of trace elements, particularly cupper. Difficulties can be avoided by adding a copper/zinc solution (see also Appendix). Glass Petri-dishes give better results than plastic Petri-dishes when inoculated with a dry inoculum. Most species sporulate within 5-7 days. Microscopic mounts are made in lactic acid, with or without aniline or cotton blue, and a drop of alcohol is added to remove air bubbles and excess of conidia. For some species in the terverticillate *Penicillium* group also Creatine sucrose agar (CREA) is very useful for distinguishing closely related species (see also Table 5 on page 177 and Chapter 2). On CREA characteristics of colony growth, production of acids (turning the medium from purple to yellow) and base production can be used as diagnostic features of the species. Lund (1995) described a method to differentiate *Penicillium* species by the detection of indole metabolites using the filterpaper method. Several *Penicillium* species are known to produce indole secondary metabolites (chaetoglobosin C, cyclo-piazonic acid, isofumigaclavin A and rugulovasine A and B).

Procedure: an agar plug is cut out from the central part of the colony and put into a the lid of a Petri dish with the mycelium side up. A piece of filter paper is dipped in **Ehrlich reagent** and placed on the mycelium. Within about 10 minutes a violet ring (zone) will appear in case of indole metabolite production.

Ehrlich reagent: 4-Dimethylaminobenzaldehyde (2 g) (Merck) is dissolved in 96 % ethanol (85 ml) and 37 % hydrochloric acid (15 ml) (Merck) is added.

In the following dichotomous key the most common Penicillia are keyed out. However, there are several other species which can occur in food or in indoor environments. Many species are very similar and identification is not very simple. Related species are given as note and the descriptions are given in a smaller font. The terverticillate Penicillia belonging to the *P. aurantiogriseum/verrucosum* complex are difficult to identify solely on the basis of their morphological structures. On pages 178-181 we have included two synoptic keys. One for the Penicillia with rough-walled stipes and one for species with smooth-walled stipes. In these keys more species are given than in the dichotomous key. Please note that for recognizing the ornamentation of the stipe microscopical slides should be examined with oil immersion. In addition it is strongly advised that when doubt exists several preparations should be made preferably from malt extract agar and that ornamentation may occur when colonies are older than seven days.

KEY TO THE SPECIES TREATED

1a. Colonies white (or very pale greyish-green).. 2
1b. Colonies in some shade of green, sometimes with yellow aerial mycelium. 3

2a. Conidiophore with rough stipe.. *P. camemberti* (1)
2b. Conidiophore with smooth stipe .... *P. nalgiovense*

3a. Colonies on Czapek growing and sporulating poorly (on CYA good growth occur), conidiophores short with distinct large phialides (15-20 µm long) and ellipsoidal to cylindrical conidia; responsible for citrus rot........................ *P. digitatum*
3b. Colonies on Czapek and CYA growing and sporulating well, conidiophore with distinct long stipe and usually smaller phialides (5-12 µm)........................ 4

4a. Conidiophores monoverticillate (simple, unbranched) .... *P. glabrum*
4b. Conidiophores branched.. 5

5a. Phialides acerose (lanceolate), conidiophores predominantly with terminal whorl of metulae and phialides (biverticillate), occasionally also terverticillate........................ 6
5b. Phialides flask-shaped, conidiophores biverticillate, terverticillate to quaterverticillate. 9

6a. Colonies growing fast, diameter more than 1.5 cm within one week on MEA 7
6b. Colonies restricted, diameter less than 1.5 cm within one week on MEA 8

7a. Colonies funiculose, conidia subglobose-ellipsoidal, 2.5-3.5 µm long .... *P. funiculosum*
7b. Colonies velvety, conidia large, ellipsoidal, 3.5-5 (7) µm long .... *P. oxalicum*

8a. Conidia rough, ellipsoidal.. *P. rugulosum*
8b. Conidia smooth or rough, often fusiform.. *P. variabile*

9a. Conidiophores predominantly biverticillate 10
9b. Conidiophores terverticillate to quaterverticillate.. 11

10a. Colonies restricted on MEA, mostly growing less than 1.5 cm diam within one week; reverse yellow; metulae 3-5, equal in length .... *P. citrinum*
10b. Colonies growing more than 1.5 cm diam on MEA within one week, reverse dark green to blackish after 7 days; metulae 2-4, unequal in length .... *P. corylophilum*

11a. Conidiophore stipe smooth-walled on Czapek agar and MEA, occasionally roughened on MEA.. 12
11b. Conidiophore stipe both on Czapek and MEA finely to distinctly rough or warted 17

12a. Conidiophores large up to 500-2000 µm tall, compact with 4-6 µm wide stipes 13
12b. Conidiophores smaller with stipes of 2.5-4.0 µm wide 14

13a. Stipes up to 800 µm, conidia globose to subglobose, colonies restricted .... *P. brevicompactum*
13b. Stipes up to 2000 µm, conidia ellipsoidal, colonies more spreading .... *P. olsonii*

14a. Colonies velvety, often with yellow exudate and reverse; conidia globose to ellipsoid.............. *P. chrysogenum*
14b. Colonies with aggregated conidiophores (fasciculate), yellow exudate mostly lacking; conidia subglobose, ellipsoid to cylindrical 15

15a Phialides short, less than 6.5 µm long, conidiophores often quaterverticillate ....*P. griseofulvum*
15b. Phialides mostly longer than 6.5 µm, conidiophores terverticillate.. 16

16a. Colonies 4-5 cm in diam within 14 days; conidia subglobose to ellipsoid; responsible for rot of pomaceous fruit ....*P. expansum*
16b. Colonies 2.0-2.5 cm in diam within 14 days; conidia ellipsoid to cylindrical; responsible for citrus rot ....*P. italicum* (2)

17a. Conidia echinulate .... *P. echinulatum* (3)
17b. Conidia smooth to finely rough 18

18a. Conidiophore stipe conspicuously warted, conidia globose, 4-5 µm in diam, colonies velvety without odour, reverse typically dark green .... *P. roqueforti* (4)
18b. Conidiophore stipe rough, but usually not warted, conidia globose to ellipsoidal, 3-4.5 µm in diam, colonies fasciculate, often with pronounced odour, reverse not dark green...19
(*P.aurantiogriseum*-complex, see also Table 5, p. 177)

19a. Colonies with yellow mycelium and orange brown exudate .... *P. hirsutum* (5)
19b. Colonies without yellow mycelium and orange brown exudate 20

20a. Colonies on common media Cz and MEA at 25°C in 7 days not exceeding 10 mm in diam, yellow green, conidia (2.5-)2.8-3.2(-3.5) µm diam, rarely larger .... *P. verrucosum* (6)
20b. Colonies on Cz and MEA at 25°C in 7 days usually exceeding 10 mm in diam 21

21a. Conidia relatively small (2.5-)2.8-3.2(-3.5) µm in diam, when ellipsoidal up to 3.5(-4) µm in length; (weak) growth on CREA, acid production but no base production 22
21b. Conidia relatively large, 3-4(-5) µm in diam., when ellipsoidal up to 4.5-6 µm in length; good growth on CREA, usually both acid and base production 23

22a. Colonies grey green on MEA, conidia subglobose to ellipsoid, poor growth on CREA .... *P. aurantiogriseum* (7)
22b. Colonies blue green on MEA, fast rate and good sporulation on all media, conidia subglobose to broadly ellipsoid, growth on CREA .... *P. polonicum* (8)

23a. Conidial areas dull green to greyish green, in fresh isolates forming crusts mostly after 10 days .... *P. crustosum*
23b. Conidial areas in various blue-green or green shades, colonies not forming crusts 24

24a. Conidial areas in fresh isolates dark blue-green or dark green; reverse on Cz colourless, yellowish or brownish; conidia (sub)globose .... *P. solitum*
24b. Conidial areas in fresh isolates pale blue-green, yellow-green or greyish green; reverse on Cz colourless or yellowish conidia subglobose to ellipsoidal .... *P. commune* (9)

Note: 1: related species is *P. caseifulvum* (p. 192)
2: related species is *P. ulaiense* (p. 220)
3: related species is *P. discolor* (p. 208)
4: related species *are P. carneum* and *P. paneum* (p. 232)
5: related species *are P. allii, P. hordei, P. albocoremium* (p. 218)
6: related species is *P. nordicum* (240)
7: related species see *P. cyclopium, P. aurantiocandidum, P. freii* (p. 204) and *P. viridicatum* (p. 242)
8: related species is *P. melanoconidium* (p. 228)
9: related species is *P. palitans* (p. 198)

Table 5. Diagnostic differences between important food- and airborne two-stage branched *Penicillium* species.

	Ehrlich reaction	Colony Colour CYA	Colony Colour MEA	Reverse Colour CYA	Creatine Growth	Creatine Acid	Conidium shape	Conidium size	Conidium Ornamentation
P. aurantiogriseum		bg	bg	br-c-or	w	+	(s)-e	3-3.5 µm	s
P. echinulatum	NC	dg	dg	cr(br)	++	+ba	g-s	3.4-4 µm	r
P. commune	violet	ggr	ggr	cr(br)	++	+ba	(s)-e	3.5-5 µm	s
P. cyclopium		ggr	bg	br-c-o	w	+	g-s	3-3.5 µm	s
P. crustosum	yellow/NC	ggr	ggr	cr-y	++	+ba	g-s	3.5-4 µm	s
P. discolor	violet	dg	dg	cr(br)	++	+ba	g-s	3.5-4 µm	r
P. hirsutum		yg	yg	cr-br	++	+	g	2.5-3 µm	s
P. palitans	violet	g	g	Cr(br)	++	+ba	g-s	3.5-4 µm	s
P. polonicum		bg	bg	y-p-br	+	+	s-e	3-4.0 µm	s
P. solitum	NC	dg	dg	cr(br)	++	+ba	g-s	3.5-5 µm	s(fr)
P. verrucosum		yg	yg	cr(br)	w/-	-	g-s	3-3.5 µm	s
P. viridicatum		g	g	p-br-or	w	+	g-s	3-3.5 µm	fr

Conidial colour: bg: blue-green, g: green, ggr: grey-green, yg: yellow green, dg: dark green. Reverse colour: or: orange, br: brown, c: curry, p: pinkish, cr: cream, y: yellow. Growth on CREA: w: weak, ba: base production. Conidial shape: g: globose, e: ellipsoid, s: subglobose. Ornamentation: s: smooth, fr: finely rough; Ehrlich reaction: NC: no colour

SYNOPTIC KEY TO COMMON FOOD- AND AIRBORNE PENICILLIA WITH *SMOOTH-WALLED* STIPES:

Species list:
1. *P. atramentosum*
2. *P. brevicompactum*
3. *P. caseifulvum*
4. *P. chrysogenum*
5. *P. citrinum*
6. *P. corylophilum*
7. *P. digitatum*
8. *P. expansum*
9. *P. funiculosum*
10. *P. glabrum*
11. *P. griseofulvum*
12. *P. italicum*
13. *P. nalgiovense*
14. *P. olsonii*
15. *P. oxalicum*
16. *P. purpurogenum*
17. *P. rugulosum*
18. *P. sclerotigenum*
19. *P. ulaiense*
20. *P. variabile*

Conidial shape:
ellipsoidal: 1, (4), (7), 8, 9, 11, (12), 15, 16, 17, 18, (19), 20
globose to subglobose: 2, 3, 4, 5, 6, 10, (11), 13, 14
cylindrical: 7, 12, 19

Conidial ornamentation:
rough: 17, (20)
finely roughened: 2, 10, 14, 20
smooth: 1, 2, 3, 4, 5, 6, 7, 8, 9, 10, 11, 12, 13, 14, 15, 16, 18, 19, (20)

Conidial size (average):
>3.5 um diam: 1, 2, 3, 4, 7, 12, 13, 14, 15, 18, 19
<3.5 um diam: 4, 5, 6, 8, 9, 10, 11, 13, 16, 17, 20

Phialide shape:
Lanceolate: 9, 16, 17, 20
Ampulliform: 1, 2, 3, 4, 5, 6, 7, 8, 10, 11, 12, 13, 14, 15, (17), 18, 19

Phialide length:
< 6.5 um: 11
7<x<9.5 um: 1, 2, 4, 5, 6, 8, 10, 12, 13, 14, 18
10<x<12 um: (1), 3, (4), (5), 6, 8, 9, 10, 12, (13), (14), 15, 16, 17, 18, 19, 20
> 13 um: (3), 7, (12), (15), (19)

Penicilli:
Metulae divergent: (1), 4, 5, 6, 7, 11, 13
Multiramulate: (1), (2), 14
One asymmetric ramus: 1, 2, 3, 4, 8, (11), 12, (13), (18)
Symmetric, metulae adpressed: (5), 9, (14), 16, 17, 20
Asymmetric, two- stage branched (= terverticillate): 1, 2, 3, 4, 8, 11, 12, 13, (14), (15),18, 19
Simple, no metulae, no rami: (6), 10
One- stage branched (= biverticillate): (4), 5, 6, 7, 9, (13), 15, 16, 17, (18), 20

Production of secondary metabolites:
Botryodiploidin: 2
Brevianamide A: 2
Chaetoglobosin A, B, C: 8
Chrysogine: 4, 13
Citreoisocoumarinol: 6
Citrinin: 5, (8)
Communesin A & B: 8
Cyclopiazonic acid: 11
Cyclopeptin: (3)
Deoxybrevianamide E: 12,13
Dehydrodeoxybrevianamide E: 12, 19
Dichlorodiaportin, diaportinol & diaportinic acid: 13
Dipodazin: (13)
Expansolide: 8
Fulvic acid: 10, 11
Citromycetin: 10
Griseofulvin: 11, 18
2-(4-hydroxyphenyl)-2-oxo acetaldehyde oxime: 14
Italicic and italinic acid: 12
Meleagrin: 1, 4
Met O: 2, 14
Mitorubrin: 16, (17), (20)
Mycochromenic acid: 2
Mycophenolic acid: 2
Nalgiovensin & nalgiolaxin: (13)
Oxaline: 1, 15
Patulin: 8, 11, 18
Penicillin: 4, 13
Pebrolides: 2
Phomenone: (6)
PI-3: 11
PR-toxin: (4?)
Raistrick phenols: 2
Roquefortine C: (1), 4, 8, 11, (15), 18
Rugulosin: 17, 20
Rugulovasine A & B: 1, 3
Sclerotigenin: 18
Secalonic acid D & F: 9, 15
Skyrin: 17, 20
Tanzawaic acid: 5
Tryptoquivalins: 7
Verrucolone (arabenoic acid): 12, 14

Growth on creatine-sucrose agar:
Very good (uninhibited): 1, 3, 8
Rather good, but inhibited: (4), ((8))
Poor growth, strongly inhibited: 2, 4, 5, 6, 7, 9, 10, 11, 12, 13, 14, 15, 16, 17, 18, 19, 20

Acid production on creatine sucrose agar:
Poor or nil: 1, 2, 4, 5, 6, 7, 9, 11, 12, 13, 14, 15, 16, 17, 18, 19, 20
Strong: 2, 3, 4, 8, 10, 15

Base production following acid production on creatine sucrose agar:
Obvious after 7-9 days of growth: 3, 8
No base production: 2, 4, 10, 13, 15
Medium staying violet: 1, 2, 4, 5, 6, 7, 9, 11, 12, 13,1 4, 15, 16, 17, 18, 19, 20

Growth at 37°C:
Strong growth: 9, 15
Some growth: (4), 5, 16, (17), (20)
No growth: 1, 2, 3, 4, 6, 7, 8, 10, 11, 12, 13, 14, 17, 18, 19, 20

Conidium colour *en masse* on CYA:
Dull green: 2, 4, 6, 8, 9, 10, 14, 15, 16, 17, 18
Grey-green: 3, 6, 9, 11, 12, 14, 15, 18, 19, 20
Blue-green: (4), 5, (8), (12), (16)
Pure green to yellow green: (4), (8), (16)
Dark green: 1, (4), (10), (13), (17)
Olive green: 7
White: (13)

Reverse colour on CYA:
Yellow: 3, 4, 5, (8), 10, 13, 15, (18)
Creamish beige to pale: 2, 3, 4, (5), 6, 7, 8, 9, 10, 11, 12, 13, 14, 15, 16, 17, 18, 19, 20
Dark brown to dark yellow or orange brown: 1, (2), 8, (9), 10, 11, 12, 13, (18), 20
Brown (yellowish, orange or pinkish): (1), (2), (4), (5), 8, 9, 10, 13, 17, (18), 20
Dark greenish to brownish black centre: (6)
Blood red centre: 16

Reverse colour on YES:
Orange, yellowish orange: 3, (4), 8, (10), (12), 13, (18)
Yellow: 4, 5, (8), 10, 13, 15, 17, 18, 20
Yellow brown: 1, 5, (6), 8, (10), 11, (12), 13, 14, 19, 20
Cream yellow: 2, 4, 5, 6, 7, (8), (9), 10, 11, (13), 14, 15, 17, 18, 19
Beige to brown: 1, 2, 7,
Dark brown: 1
Red brown: (8), 12, 16
Rose to carmine red: (2)
Red: 12, 16
Blackish green: (6)

Degree of sporulation on YES agar:
Strong: 1, 2, 3, 4, 5, 7, 8, 10, 11, 12, (13), 14, 15, 16, 17, 19, 20
Weak: 6, (7), 8, 9, (12), 13, (15)

Exudate droplets on CYA:
Clear exudate droplets: 1, 2, 3, 4, 8, (9), (10), 11, (12), 13, (16)
Yellow to brown exudate droplets: 1, 2, 4, 5, 8, (10), 11, (13),

Colony texture (on MEA):
Fasciculate: 8, 11, 12, 19
Mealy: 8, 11, 12, 19
Velutinous: 1, 2, 4, 5, 6, 7, (8), 10, (11), (12), 13, 14, 15, (16), 17, 18, 20
Funiculose: 9, 16
Floccose: 3, 13
Crustose: 12, 15, 18, 19

Colony diameter (CYA, after one week at 25°C):
< 15 mm: (7), 17, 19, 20
15-19 mm: 2, 3, 7, (11), 13, (14), 19, 20
20-24 mm: (1), 2, 3, (4), 5, 6, 7, 11, (12), 13, (14), 19
25-29 mm: 1, 2, (3), 4, 5, 6, 7, 8, 9, 11, 12, 13, 14, 15, 16, 19
30-34 mm: 1, 4, 5, (6), 7, 8, 9, 10, 11, 12, 13, 14, 15, 16
35-39 mm: 1, 4, 7, 8, 9, 10, 12, 14, 15, 16
40-44 mm: (1), 4, 7, 8, (9), 10, 12, (14), 15, (16), 18
45-49 mm: 7, 8, 10, 12, 15, 18
50-54 mm: 7, (10), 15, 18
>55 mm: 7, 15, 18

Colony diameter (MEA, after one week at 25°C):
< 15 mm: 2, (5), (8), (13), (14), 17, 19, 20
15-19 mm: 2, 3, 5, 8, 11, 13, (14), (15), 17, 19, 20
20-24 mm: 1, 2, 3, (4), 5, 8, 11, 13, 14, 15, 19, 20
25-29 mm: 1, 2, 3, 4, 7, 8, 9, 11, 12, 13, 14, 15
30-34 mm: 1, 4, 6, 7, 8, 9, (11), 12, (13), 14, 15, 16
35-39 mm: 1, 4, 6, 7, 8, 9, 12, 14, 15, 16
40-44 mm: 4, 6, 7, 8, 9, 10, 12, 15, 16, 18
45-49 mm: 4, (6), 7, (8), 10, 12, 15, (16), 18
50-54 mm: 4, 7, 10, (12), 18
> 54 mm: 7, (10), 18

Colony diameter (YES, after one week at 25°C):
< 19 mm: 17, 20
20-24 mm: 2, (19), 20
25-29 mm: 2, 5, 6, (7), 9, 11, 13, 16, 19
30-34 mm: 1, 2, 3, 5, 6, 7, (8), 9, 11, 13, 15, 16, 19
35-39 mm: 1, 3, 5, (6), 7, 8, 9, 10, 11, (12), 13, 14, 15, 16, 19
40-44 mm: 1, 3, (4), 7, 8, (9), 10, (11), 12, 13, 14, 15, (16)
45-49 mm: 1, 4, 7, 8, 10, 12, (13), 14, 15
50-54 mm: 1, 4, 7, 8, 10, 12, 14, 15, 18
55-60 mm: 1, 4, 7, 8, 10, 12, 15, 18
> 60 mm: 1, 4, 7, 8, 12, 15, 18

Sclerotia present: (14), 18

Habitat:
Cereals with low lipid content: 2, 9, 11, 20
Cereals with high lipid content: 2, 9, 11, 15, 20
Rye bread: 6
Onions: 10
Citrus fruits: 7, 12, 19
Pomaceous fruits: 8, (20)
Yams: 18
Tomatoes: 2, 14
Cucumbers: 15
Cabbage: 5
Pate: 2
Cheese: 1, 3, (4), 13
Salami, dried ham: 2, 4, (8), (11), 13, 17
Nuts: 4, 6
Coffee: 2, 5

Indoor environments: 2, 4, (5), (9), 10, 14, (17), (20)

Geographical distribution:
Temperate: 1, 2, 3, 4, (5), 6, 8, (9), 10, (11), 13, 14, 17, 20
Subtropical: 2, 4, 5, 6, 7, 8, 9, 10, 11, 12, 13, 14, 15, 16, 17, 18, 19, 20
Tropical: 2, 4, 5, 9, 14, 15, 16, (17), 18, (20)

SYNOPTIC KEY TO COMMON FOOD AND AIR-BORNE PENICILLIA WITH *ROUGH-WALLED* STIPES

Species list:
1. *P. aethiopicum*
2. *P. albocoremium*
3. *P. allii*
4. *P. aurantiogriseum*
5. *P. aurantiocandidum*
6. *P. camemberti*
7. *P. carneum*
8. *P. commune*
9. *P. crustosum*
10. *P. cyclopium*
11. *P. discolor*
12. *P. echinulatum*
13. *P. freii*
14. *P. hirsutum*
15. *P. hordei*
16. *P. melanoconidium*
17. *P. nordicum*
18. *P. palitans*
19. *P. paneum*
20. *P. polonicum*
21. *P. roqueforti*
22. *P. solitum*
23. *P. tricolor*
24. *P. verrucosum*
25. *P. viridicatum*

Conidial shape:
ellipsoidal: 1, (8)
globose to subglobose: 2, 3, 4, 5, 6, 7, 8, 9, 10, 11, 12, 13, 14, 15, 16, 17, 18, 19, 20, 21, 22, 23, 24, 25

Conidial ornamentation:
rough: 11, 12
finely roughened: 15, 25
smooth: 1, 2, 3, 4, 5, 6, 7, 8, 9, 10,13, 14, 16, 17, 18, 19, 20, 21, 22, 23, 24, (25)

Conidial size (average):
>3.5 um diam: 1, 6, 7, 8, 9, 11, 12, (17), 18, 19, 20, 21, 22
<3.5 um diam: 2, 3, 4, 5, 10, 13, 14, 15, 16, 17, 23, 24, 25

Production of secondary metabolites:
Anacine: 4, 17, 20
Asteltoxin: 23
Auranthine: 4
Aurantiamine: 4, 13
Brevianamide A: 25
Chaetoglobosin A, B, C: 11
Chrysogine: (2), 3, (14)
Citrinin: 24
Compactins: 14, 22
Cyclopaldic acid: (8)
Cyclopiazonic acid: 8, 18
Cyclopenol, cyclopenin, cyclopeptin, dehydrocyclopeptin: 2, 3, 5, (8), 9, 10, 11, 12, 13, 14, (15), (18), 20, 22
Fulvic acid: (3)
Griseofulvin: 1
Marcfortine: 19
Meleagrin: 2, 3, 14, 16
3-methoxyviridicatin: (2), 5, 10, 13, 20
Mycophenolic acid: 7, 21
Ochratoxin A: 17, 24
Oxaline: (2), 3, 16
Palitantin: 8, 11, 12, (18)
Patulin: 7, 19
Penitrem A: 9, 16
PI-3: 3
PR-toxin: 21
Puberuline, verrucofortin: 5, 10, 20, 23
Puberulonic acid: 5
Roquefortine C: 2, 3, 7, 9, 14, (15), 16, 19, 21
Rugulovasine A & B: (8)
Sclerotigenin: 16
Solistatin: 22
Terrestric acid: 4, 9, 14, 15, 23
Territrems: 12
Tryptoquivalins: 1
Verrucine: 24
Verrucosidin: 4, 16, 20
Verrucolone (arabenoic acid): 17, 24
Viridicatin: (8), 9, 11, 12, (18), 22
Viridicatol: (2), 5, (8), 9, 10, 11, 12, 13, 14, (15), (18), 20, 22
Viridicatumtoxin: 1
Xanthomegnin, viomellein and vioxanthin: 10, 13, (16), 23, 25

Growth on 0.5 % acetic acid:
Strong growth: 7, 19, 21
No growth: 1, 2, 3, 4, 5, 6, 8, 9, 10, 11, 12, 13, 14, 15, 16, 17, 18, 20, 22, 23, 24, 25

Growth on creatine-sucrose agar:
Very good (uninhibited): 6, 7, 8, 9, 11, 12, 18, 19, 21, 22
Rather good, but inhibited: 1, (3), 14, 15, 20
Poor growth, strongly inhibited: 2, 3, 4, 5, 10, 13, 16, 17, 23, 24, 25

Acid production on creatine-sucrose agar:
Poor or nil: (2), 3, 7, 17, 19, 21, 24
Strong: 1, 2, (3), 4, 5, 6, 8, 9, 10, 11, 12, 13, 14, 15, 16, 18, 20, 22, 23, 25

Base production following acid production on creatine-sucrose agar:
Obvious after 7-9 days of growth: 6, 8, 9, 11, 12, 18, 22
No base production: 1, 2, 3, 4, 5, 10, 13, 14, 15, 16, 17, 20, 23, 24, 25
Medium remaining violet: 7, 19, 21

Growth on nitrite-sucrose agar:
Growth: 6, (8), 7, 17, 19 ,21, 24
No or very thin growth: 1, 2, 3, 4, 5, 8, 9, 10, 11, 12, 13, 14, 15, 16, 18, 20, 22, 23, 25

Growth at 37°C:
Some growth: 1
No growth: 2, 3, 4, 5, 6, 7, 8, 9, 10, 11, 12, 13, 14, 15, 16, 17 18, 19, 20, 21 22, 23, 24, 25

Conidium colour *en masse* on CYA:
Dull green: 1, (6), 7, 8, 9, 15, 17, 19, 21
Grey-green: 8, (9), 10, 23
Blue-green: 1, 2, 4, 5, (8), (9), 13, (14), (15), 20, (22)
Pure green to yellow green: 3, 14, 17, (18), 24, 25

Dark green: 11, 12, 16, 18, 22
White: 6, (17)

Conidium colour *en masse* on MEA:
Blue green: 2, 4, 5, 10, 13, (16), 20
Grey-green to dull green: 1, (2), 3, (6), 7, 8, 9, 11, 12, 14, 15, 16, 19, 21, 22, 23
Green to yellow green: 11, 12, (14), 15, 16, 17, 18, 24, 25

Reverse colour on CYA:
Blackish green: 21
Yellow: 1, (2), (4), (5), (10), (13), (14), 16, (25)
Cream to light brown: (2), (4), (5), 6, 7, 8, 9, (10), 11, 12, (13), (15), 16, 17, 18, 19, (20), 22, 24, (25)
Dark brown to yellowish dark brown: 3, 23
Yellow, orange or pinkish brown: 4, 5, (9), 10, 13, 14, 15, 20, 25

Reverse colour on YES:
Brilliant orange to red: 11
Orange, yellowish orange: 11, 12, (15), 22
Yellow: 1, (2), 4, 5, (8), 9, 10, 11, 12, 13, 15, 16, 17, 18, (19), 20, 22, 25
Yellow brown: 1, 2, 3, 7, 14, 19, 23
Cream yellow: (1), 6, 7, 8, 15, 17, 19
Red violet, red brown, terracotta: 24
Rose to carmine red: (7), (19)
Blackish green: 21

Degree of sporulation on YES agar:
Strong: 1, (2), 3, (4), 7, (8), 9, 11, 12, (14), 16, 17, 18, 19, 20, 21, (22), (25)
Weak: 2, 4, 5, 6, 8, 10, 13, 14, 15, (17), 22, 23, 24, 25

Exudate droplets on CYA:
Clear exudate droplets: 1, 2, 4, 5, (8), (9), 10, 13, (16), 17, (18), (20), (22), (24), 25
Yellow to brown exudate droplets: (3), (8), 14, 15, (17), (20), (22), 23, (24), (25)

Colony texture (on MEA):
Fasciculate: (1), 2, (3), (4), (8), (9), (10), 11, (12), (13), 14, 15, (20), (25)
Mealy: 1, 2, 3, 4, 5, 8, 9, 10, 11, 12, 13, 14, 15, 16, 17, 18, 20, 22, 23, 24, 25
Velutinous: 1, 3, 4, 5, 7, 8, 9,10, 11, 12, 13, (14),16, 17, 18, 19, 20, 21, 22, 23, 24, 25
Floccose: 2, 6, 15
Crustose: (1), (4), (8), 9, (13)

Colony diameter (CYA, after one week at 25°C):
< 14 mm: (4), (17), (24)
15-19 mm: (4), (5), (8), (10), (13), (16), 17, (22), 24, (25)
20-24 mm: 4, 5, 6, (7), 8, 10, (11), 13, 16, 17, 18, (20), 22, 23, 24, 25
25-29 mm: 1, (2), (3), 4, 5, 6, (7), 8, 10, 11, 12, 13, (14), (15), 16, 18, 20, (21), 22, 23, 25
30-34 mm: 1, 2, 3, (4), (5), (6), (7), 8, 9, (10), (11), 12, (13), 14, 15, 18, 20, (21), (22), (23), (25)
35-39 mm: 1, 2, 3, 7, (8), 9, (11), 12, 14, 15, (19), 20, 21
40-44 mm: (1), (2), (3), 7, 9, 19, 21
> 45 mm: 7, 19, 21

Colony diameter (MEA, after one week at 25°C):
< 14 mm: (2), 17, 24
15-19 mm: (2), (5), (8), (13), 17, (21), 22, 24, 25
20-24 mm: (2), (3), 4, 5, 6, (7), 8, 10, 11, 12, 13, (15), 16, (17), 18, (20), (21), 22, 23, (24), 25
25-29 mm: 1, (2), 3, 4, 5, 6, (7), 8, (9), 10, 11, 12, 13, 15, 16, 18, 20, (21), 22, 23, 25
30-34 mm: 1, 2, 3, 4, 5, (6), (7), 8, 9, 10, 11, 12, 13, 14, 15, 16, 20, (21),22, 23, 25
35-39 mm: 1, 2, 3, 4, 7, (8), 9, (10), 14, 15, 20, (21), (25)
40-44 mm: 1, 2, 7, 9, 14, 15, 19, 20, 21
> 45 mm: (1), (2), 7, 14, 19, 21

Colony diameter (YES, after one week at 25°C):
< 19 mm: (17), ((21)), (24)
20-24 mm: 13, 16, 17, 24, 25
25-29 mm: 4, 5, 6, 8, 10, 13, 16, 17, 24, 25
30-34 mm: 1, 4, 5, 6, 8, 10, 11, 12, 13, 16, 17, 20, 21, 22, 23, 24, 25
35-39 mm: 1, 2, 3, 4, 5, 6, 8, 9, 10, 11, 12, 13, 14, 15, 16, 20, 22, 23, 25
40-44 mm: 1, 2, 3, 4, 5, 6, 8, 9, 10, 11, 12, 14, 15, 18, 20, (21), 22, 25
45-49 mm: 1, 2, 3, 7, 8, 9, 11, 12, 14, 15, 18, 20, (21), 22
50-54 mm: 1, 2, 3, 7, 9, 11, 12, 19, 21
55-60 mm: 1, 2, 7, 9, 19, 21
> 61 mm: 2, 7, 19, 21

Habitat:
Cereals with low lipid content: 4, 5, 10, 13, 15, 16, 20, (22), 23, 24, 25
Cereals with high lipid content: 1, 4, 5, (9), 10, 13, (15), 16, 20, (22), 24, 25
Silage: 7, 19, 21
Rye bread: 7, 19, 21
Onions: 2, 3
Garlic: 3
Ginger: 2, (20)
Pomaceous fruits: 9, 22
Margarine: 8, (9),11, 12
Cheese: 6, 8, (9), 11, (12), 17, 21, (22), (24)
Salami, dried ham: (1), (2), (4), 7, 8, (9), (12), 17, 20, (22)
Nuts: 8, 9, 11, (12), (17), (22)

Indoor environments: 4, 8, 18, 20

Geographical distribution:
Polar: 17, 22
Temperate: 2, (3), 4, 5, 6, 7, 8, 9, 10, 11, 12, 13, 14, 15, 16, 17, 18, 19, 20, 21, 22, 23, 24, 25
Subtropical: 1, 2, 3, 4, (5), (6), 7, 8, 9, 10, 13, 16, 17, (18), 19, 20, 21, (22), 25
Tropical: 1, 4, (8), 9, (10), (16), 20, 25

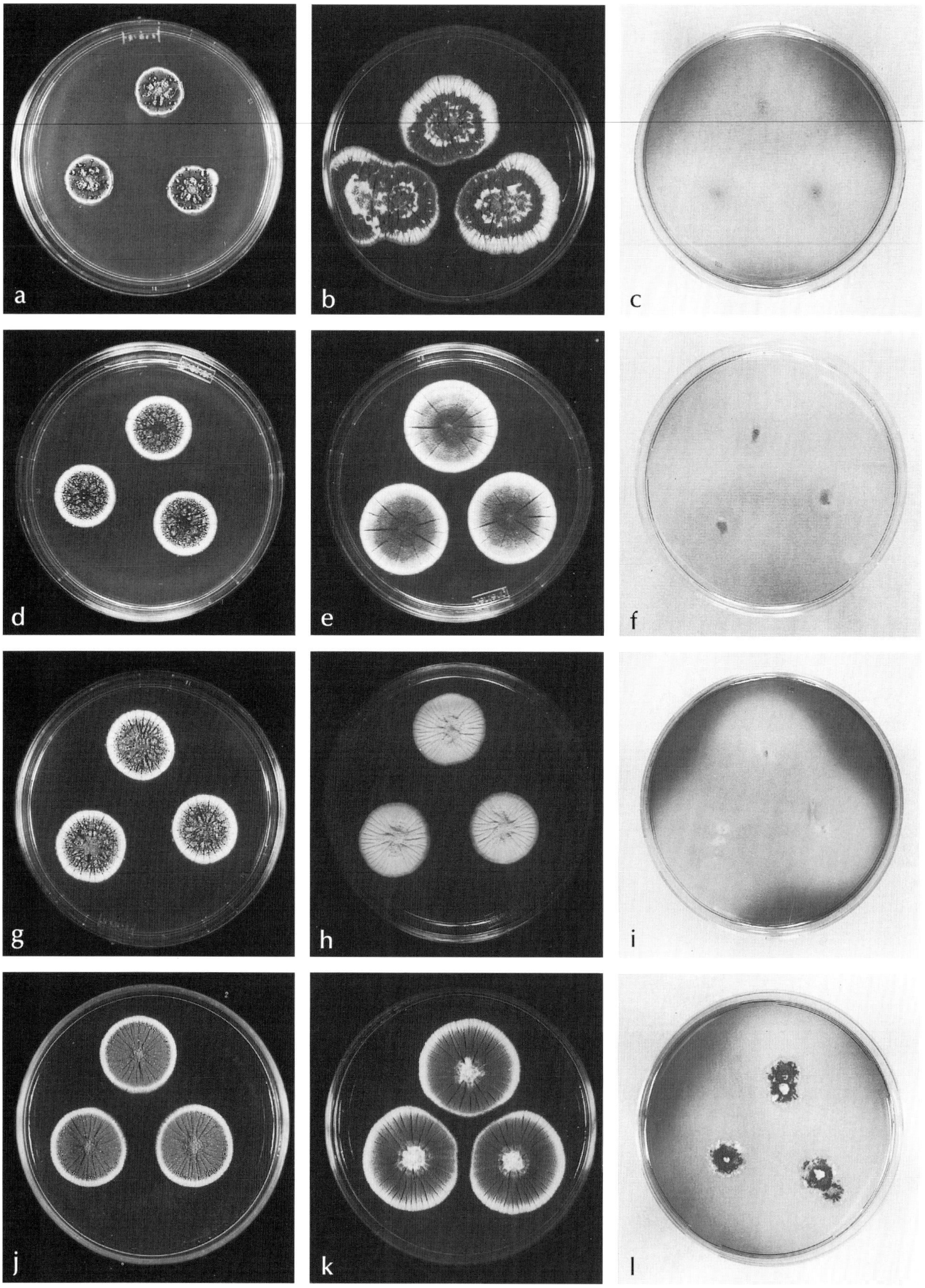

Pl. 76. Colonies after 7 days on CYA, YES and CREA. *P. aurantiogriseum* a. CYA, b, YES, c. CREA; *P. viridicatum* d. CYA, e, YES, f. CREA; *P. cyclopium* g. CYA, h, YES, i. CREA; *P. polonicum* j. CYA, k, YES, l. CREA.

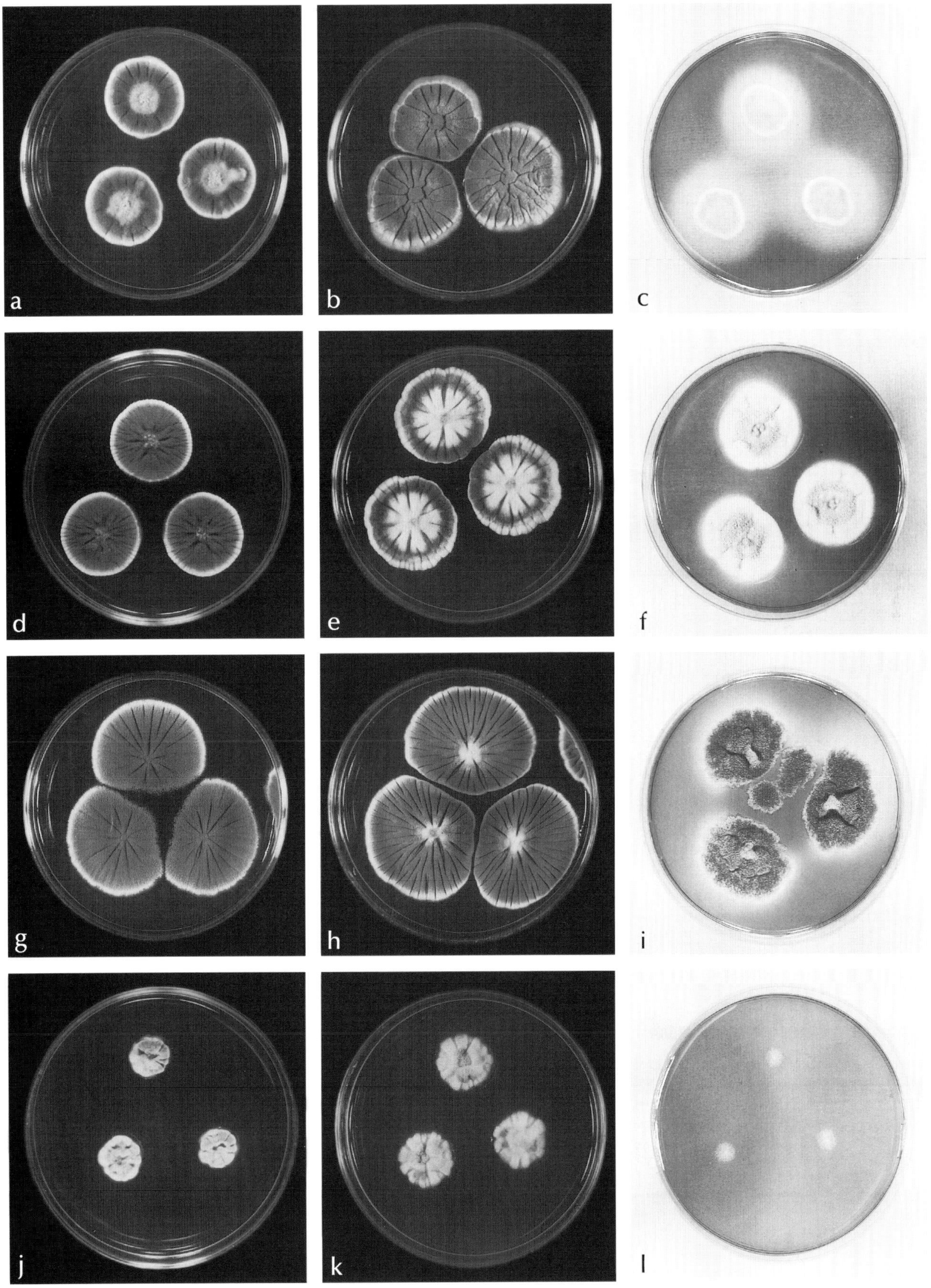

Pl. 77. Colonies after 7 days on CYA, YES and CREA. *P. solitum* a. CYA, b, YES, c. CREA; *P. commune* d. CYA, e, YES, f. CREA; *P. crustosum* g. CYA, h, YES, i. CREA; *P. verrucosum* j. CYA, k, YES, l. CREA.

Penicillium aethiopicum Frisvad

Colonies on Czapek agar and CYA at 25°C growing rather rapidly producing dull green conidia with a granular to fasciculate colony surface, always with large clear to beige exudate droplets. The colony reverse is more sulcate than is usual in *Penicillium* species and coloured golden yellow. On MEA the conidia are also dull green with a pale or dull yellow brown reverse. On YES agar strong sporulation, reverse coloured golden yellow to curry. On CREA weak to moderate growth and strong acid production. On CYA at 37°C nearly always growth, diameter after 7 days (0-) 3-9 mm. Conidia smooth and ellipsoidal, conidiophores two-stage branched (terverticillate) with all elements adpressed, stipes smooth to finely roughened.

Important toxic metabolites: viridicatumtoxin, tryptoquivalins and tryptoquivalons

Secondary metabolites with unknown toxicity: griseofulvin, dechlorogriseofulvin, lichexanthone.

HABITAT:
Cereals and oils seeds, pantropical.

Pl. 78. *Penicillium aethiopicum.* Colonies after one week, a. on Cz; b. on MEA; c-e. conidiophores, x 950.

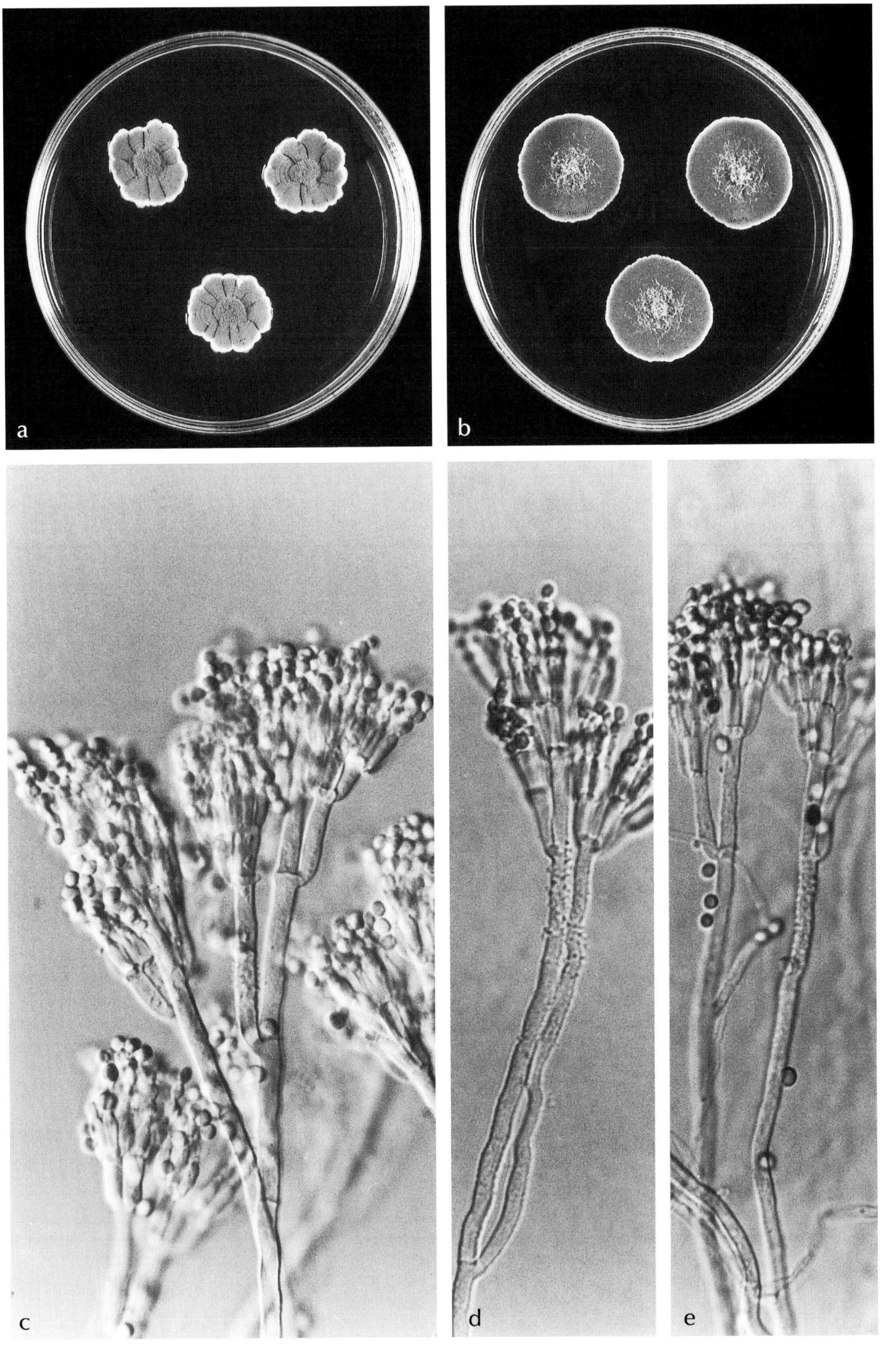
a
b
c
d
e

Penicillium atramentosum Thom

Colonies on Czapek agar and CYA at 25°C growing moderately fast producing dull to dark green conidia and velutinous colonies, with dark brown exudate droplets. The colony reverse is sulcate to plicate, coloured brownish orange to dark brown. On MEA the conidia are dull green with a pale to greish orange reverse. On YES agar good sporulation, reverse coloured beige light brown. On CREA strong growth and no acid production. Conidia smooth and subglobose conidiophores two-stage branched (terverticillate) with elements divergent but occasionally adpressed, stipes smooth.

Important toxic metabolites: roquefortine C

Secondary metabolites with unknown toxicity: meleagrin, oxaline, rugulovasine A & B.

HABITAT: Cheese

Pl 79. *Penicillium atramentosum.* Colonies after one week, a. on Cz; b. on MEA; c,d. conidiophores, x 950.

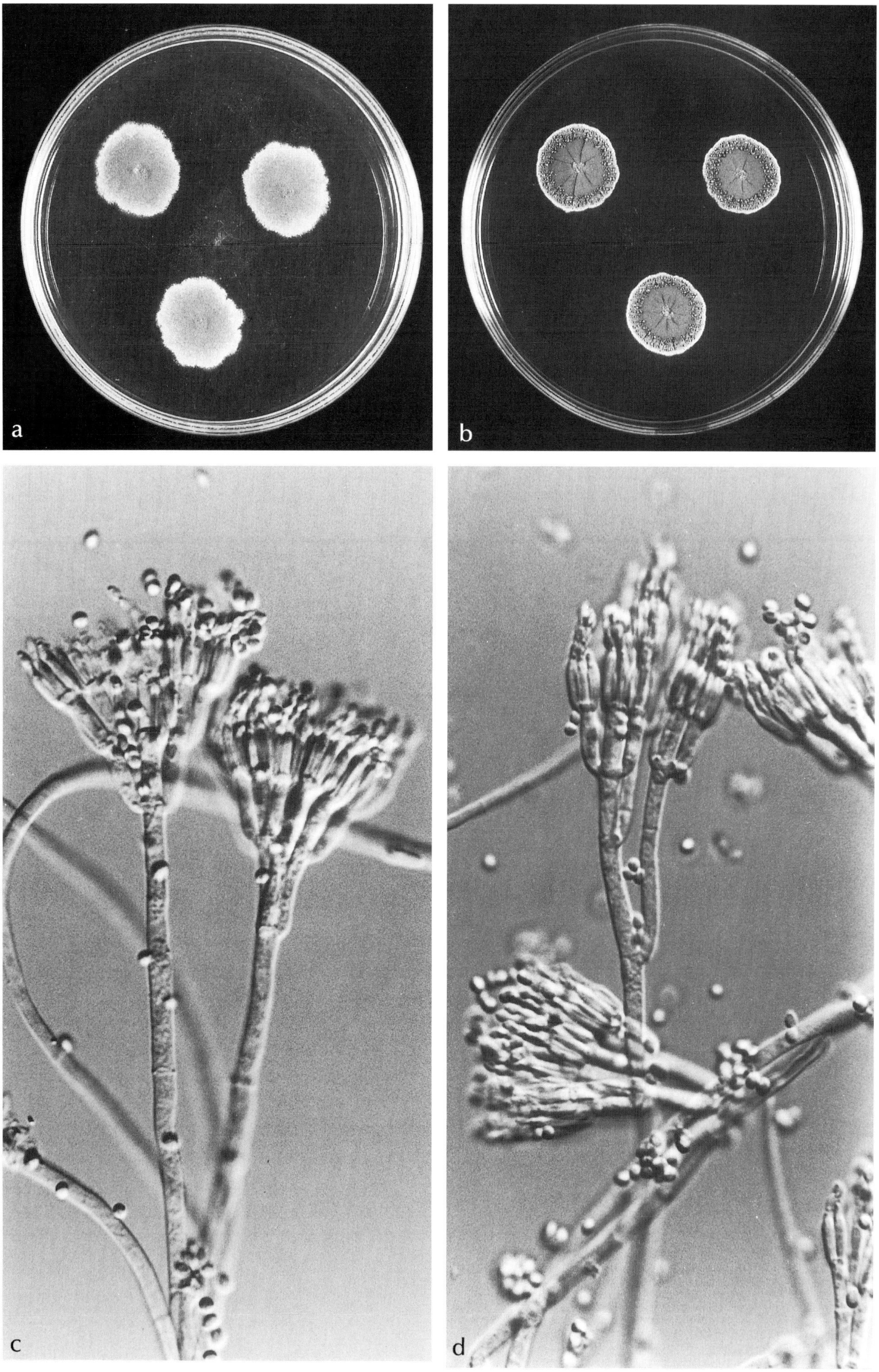
a
b
c
d

Penicillium aurantiogriseum Dierckx

Colonies on Czapek agar and CYA at 25° C growing restrictedly producing grey to dark blue green conidia with a granular to fasciculate colony surface, often with exudate droplets. The colony reverse is orange brown with the colour often diffusing into the agar medium. Strains which are maintained for a long time on agar, the reverse colour may be yellow to orange. On MEA the conidia are blue green with a strong blue element and colonies have a distinct yellow reverse, often with the yellow colour diffusing into the medium. On YES agar the degree of sporulation variable, reverse colour distinct yellow. On CREA weak growth but strong acid production (few isolates may have a poor acid production). Conidiophores two-stage branched (terverticillate) with all elements adpressed, stipes rough-walled, conidia smooth and globose to subglobose

Important toxic metabolites: nephrotoxic glycopeptides, verrucosidin, penicillic acid, terrestric acid

Secondary metabolites with unknown toxicity: aurantiamin, auranthine, anacine.

HABITAT: food, indoor
The primary habitat is cereals.

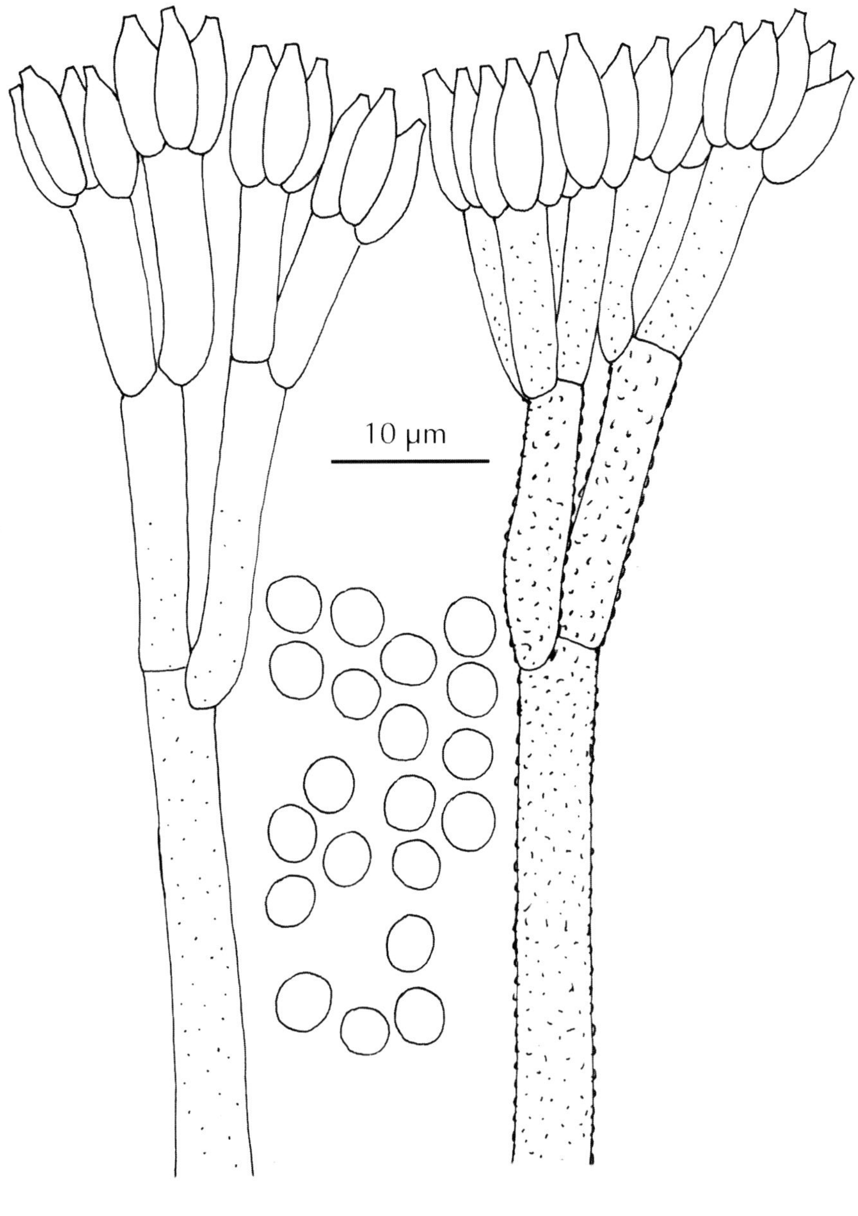

Fig. 82 *Penicillium aurantiogriseum.* Conidophores and conidia.
Pl. 80. *Penicillium aurantiogriseum.* Colonies after one week, a. on Cz, b. on MEA; c,d. conidiophores, x 950; e. conidia, x 1750.

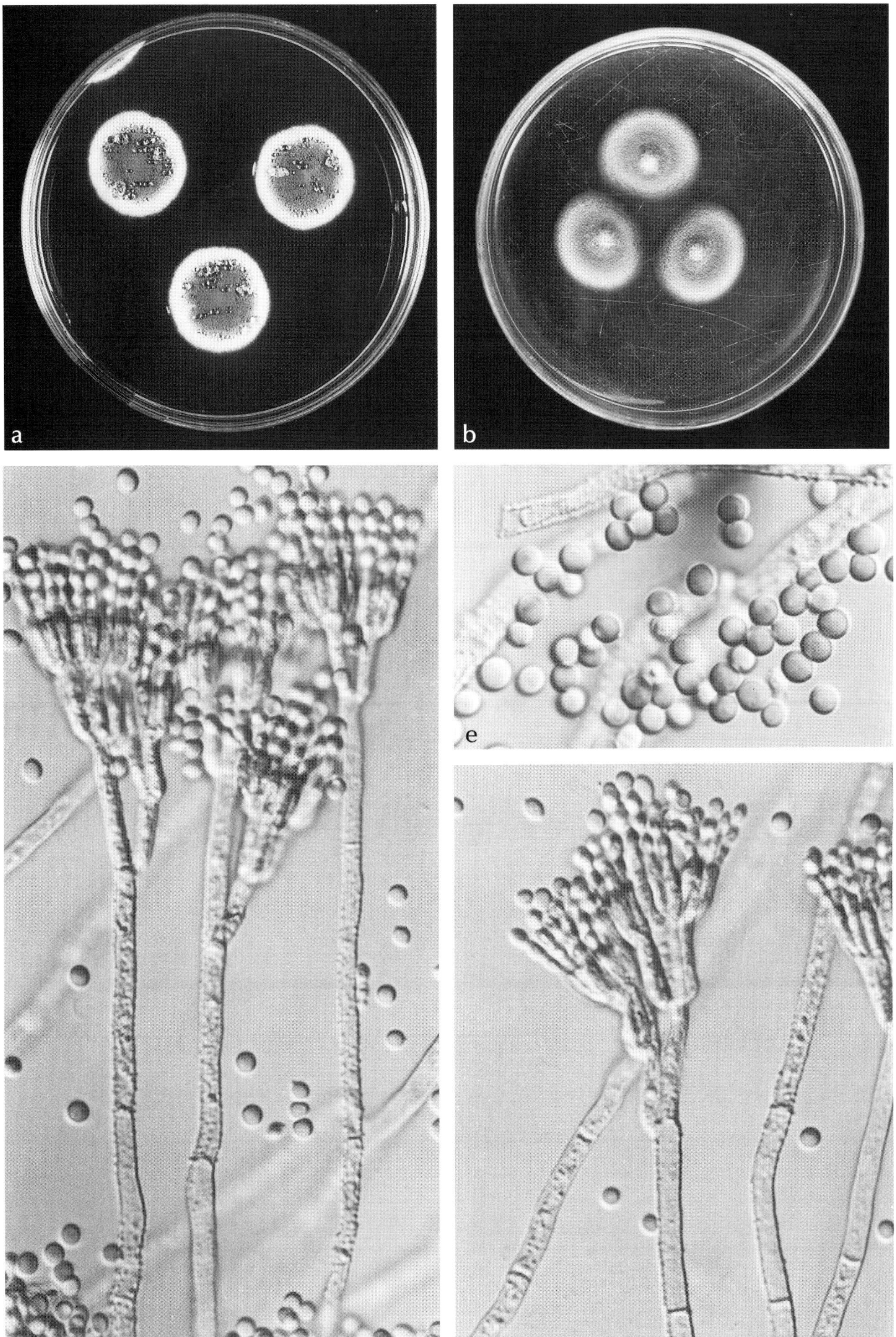
a
b
c
e
d

Penicillium brevicompactum Dierckx

Colonies on Czapek agar at 25°C, mostly attaining a diameter of 1-1.5 cm within 7 days, consisting of a dense felt of large, compact conidiophores. Colour grey-green to yellow-green. Conidiophores 300-500 µm long, consisting of a smooth, large, 4-6 µm wide stipe terminating in a compact penicillus, terverticillate to quaterverticillate. Branches and metulae usually inflated, metulae 9-12 x 4-7µm. Phialides flask-shaped, 7-10 x 3.5 µm. Conidia globose to subglobose, smooth or slightly roughened, 3-4.5 µm in diameter.
Colonies on MEA grow faster and less dense. On CREA poor growth.

Important (toxic) metabolites: botryodiploidin, mycophenolic acid, brevianamide A, met O.

HABITAT: food, indoor
In soil, groundnuts, fruits and corn, fruit-juices, commonly occurring in indoor environments.

NOTE:
This species is characterized by its restricted growth and the large compact penicilli. Using the dissecting microscope the conidiophores resemble those of *Aspergillus*.

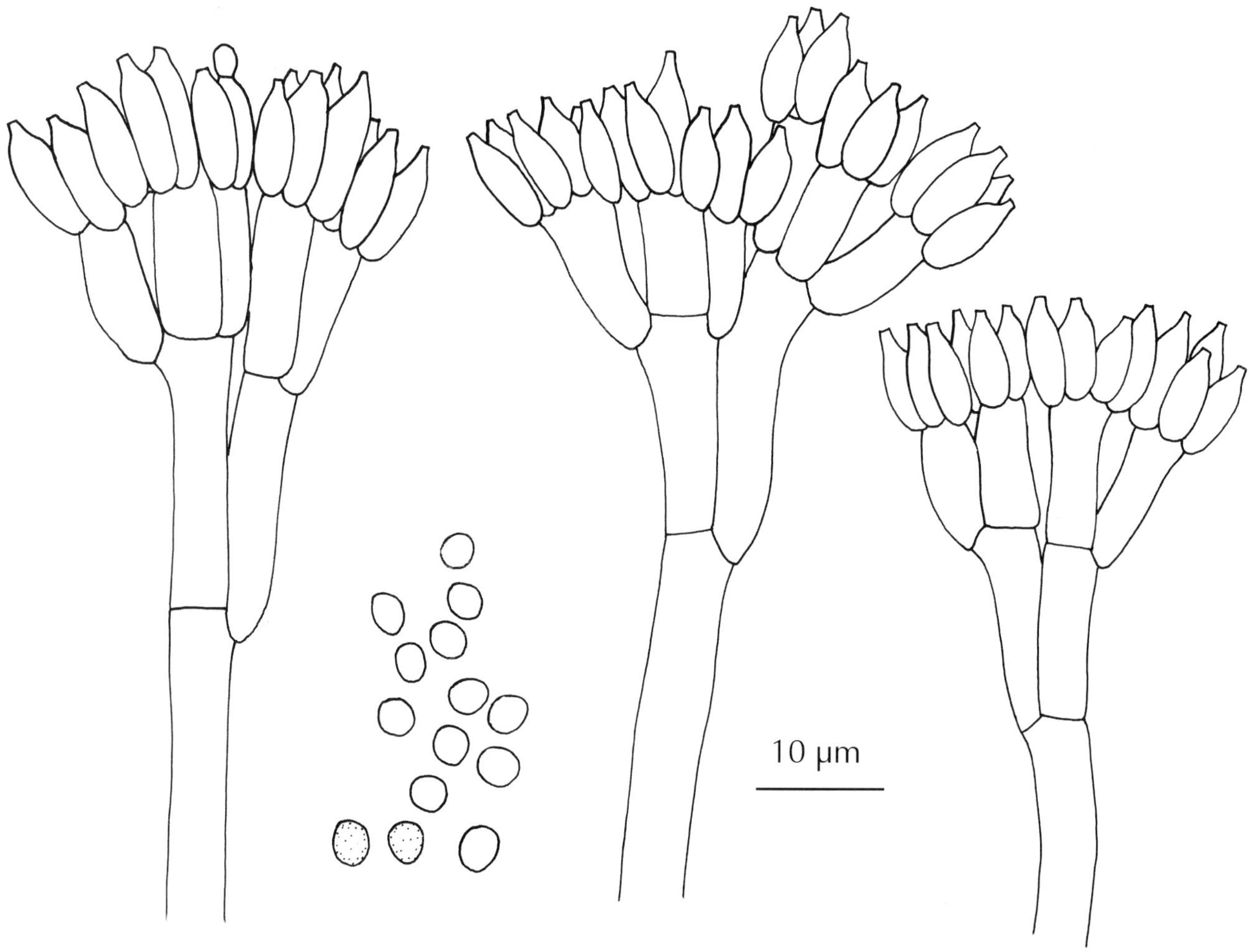

Fig. 83. *Penicillium brevicompactum*. Conidiophores and conidia.
Pl. 81. *Penicillium brevicompactum*. Colonies after one week, a. on Cz; b. on MEA; c-d. conidiophores, c. x 1350, d. x 1500; e. conidia, x 800.

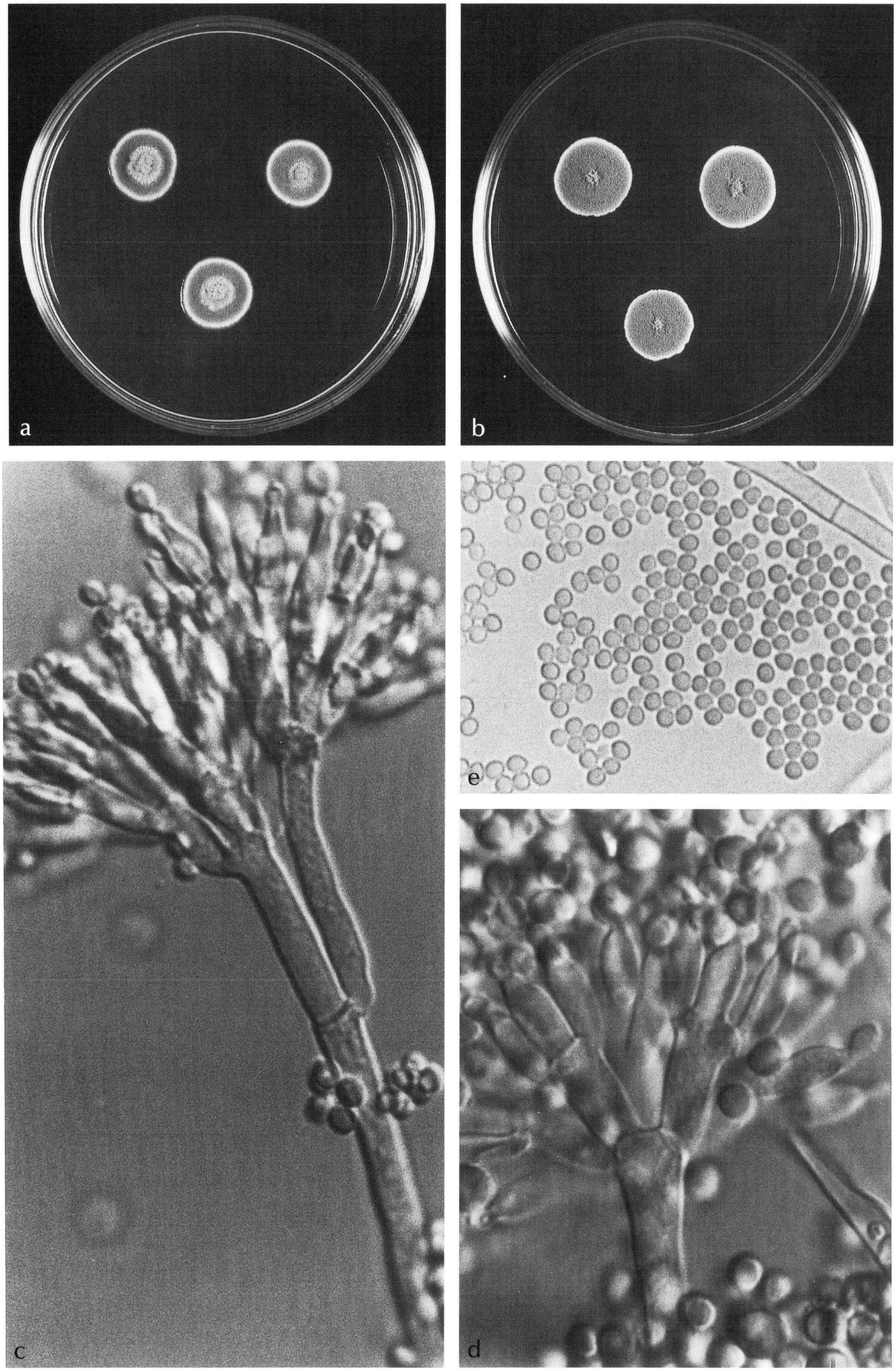
a
b
c
e
d

Penicillium camemberti Thom

Colonies on Czapek agar at 25° C growing slowly, attaining a diameter of 2-3.5 cm within two weeks consisting of a raised floccose aerial mycelium usually up to 1 cm high, at first white, remaining so or changing to yellowish, pinkish (rare) or greenish grey, the latter shade usually appearing rather late. Odour mouldy or not pronounced. Exudate rarely present as colourless drops. Conidiogenous structures arising from submerged hyphae, occasionally from aerial hyphae. Conidiophores up to 500 µm long and 2.5-4.0 µm wide, bi- to terverticillate; conidiophore stipe rough-walled, rarely smooth-walled, sometimes becoming ornamented. Metulae 8-14 x 2.5-3.0 µm, giving rise to 3 to 6 phialides. Phialides flask-shaped with short necks, 10-13 x 2.5 µm. Conidia in tangled chains, globose to subglobose, or broadly ellipsoidal, 4.0-5.0 x 3.0-4.5 µm, hyaline or slightly greenish.
Colonies on MEA, conidiophores more abundant and more roughened than on Czapek agar. Good growth on CREA.

Important (toxic) metabolites: cyclopiazonic acid.

HABITAT: food
Soft cheeses and its surrounding environments; occasionally isolated from meat.

Penicillium caseifulvum F. Lund, Filt. & Frisvad

Colonies on Czapek agar and CYA at 25° C growing restrictedly producing grey green conidia and deep floccose colonies, with small clear exudate droplets. The colony reverse is coloured creamish yellow to brown yellow. On MEA the conidia are grey green colonies have a light yellow reverse and are floccose. On YES agar rather good sporulation with blue green conidia, reverse vividly orange. On CREA strong growth and strong acid production and subsequent base production. Conidiophores two-stage branched (terverticillate) with all elements adpressed, stipes smooth, conidia smooth and globose to subglobose

Important toxic metabolites: rugulovasine A & B
secondary metabolites with unknown toxicity: cyclopeptin (Lund *et al.*, 1998).

HABITAT: food
Surface of blue cheeses and occasionally other cheeses.

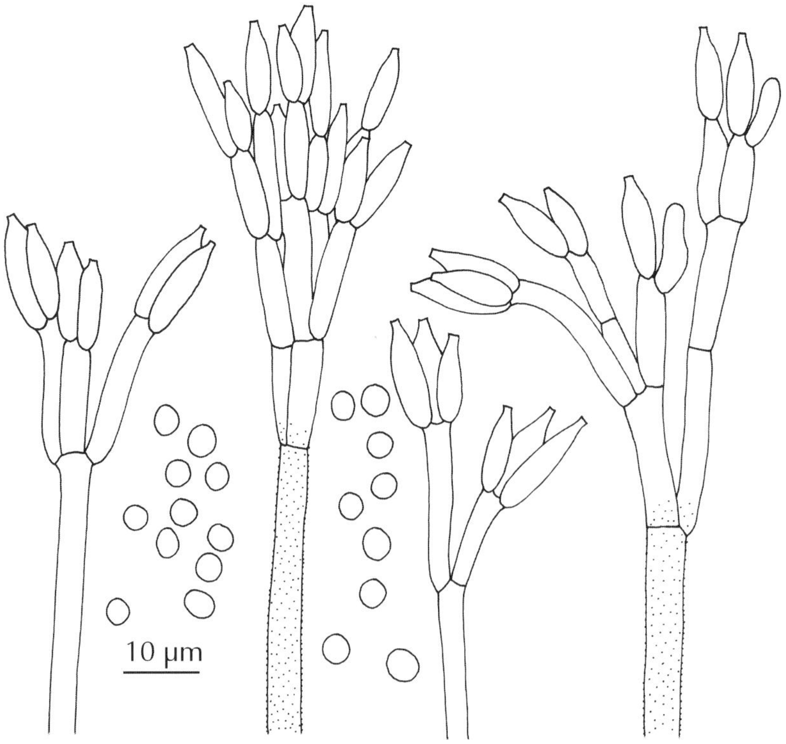

Fig. 84. *Penicillium camemberti.* Conidiophores and conidia.
Pl. 82. *Penicillium camemberti.* a. Colonies on Cz after one week; b. Camembert cheese; c-e. conidiophores with smooth- to rough-walled stipes, x 1000; f. conidia, x 1250.

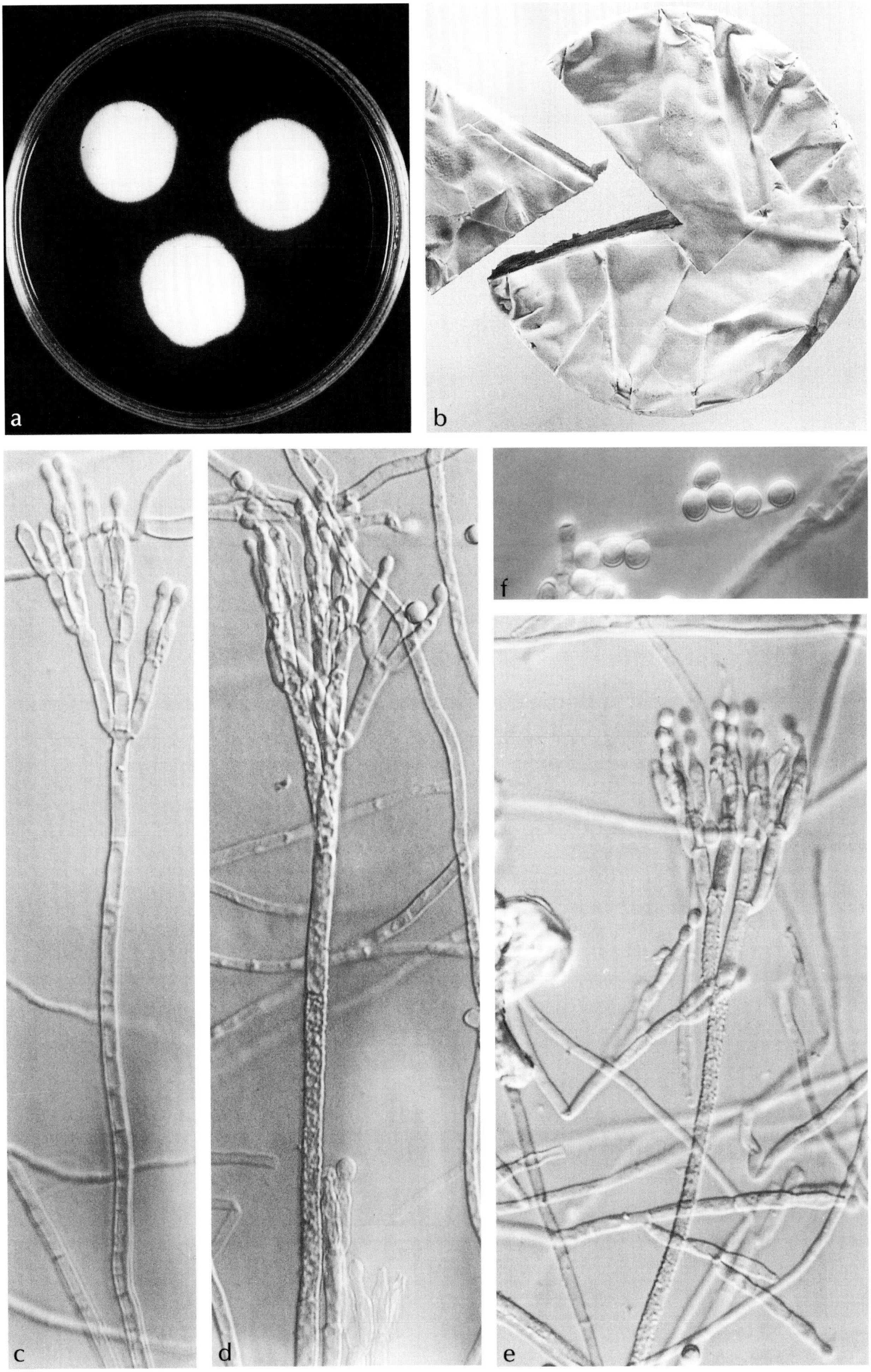
a
b
c
d
f
e

Penicillium chrysogenum Thom

Colonies on Czapek agar at 25°C growing rapidly, attaining a diameter of 4 to 5 cm within 10 days; sometimes more restricted, velvety to somewhat floccose, yellow-green or pale green-blue, becoming darker at age. Exudate typically produced as yellow drops, sometimes hyaline or absent. Odour usually aromatic, fruity. Conidiophores arising from the substrate, mononematous, usually ter- to quaterverticillate, in some strains varying from bi- to terverticillate to more complex branched; stipes 250-500 x 2.5-3.5 (4.0) µm, smooth-walled; branches somewhat divergent; metulae 8-15 x 2.0-2.3 µm more or less cylindrical, smooth-walled, bearing 3 to 6 phialides. Phialides flask-shaped, often with a thickened wall, mostly measuring 7-10 x 2.0-2.5 µm. Conidia at first subglobose to ellipsoidal, remaining so or later becoming globose, 3.0-4.0 x 2.8-3.8 µm, hyaline or slightly greenish, smooth-walled, usually produced in loose columns.

Colonies on MEA mostly growing faster, flat, velvety, occasionally slightly floccose; exudate limited or absent. On CREA poor growth and acid production.

Important (toxic) metabolites: roquefortine C, meleagrin, penicillin.

HABITAT: food, indoor

Very common on various food products, also products with low a_W. Also frequently occurring in indoor environments.

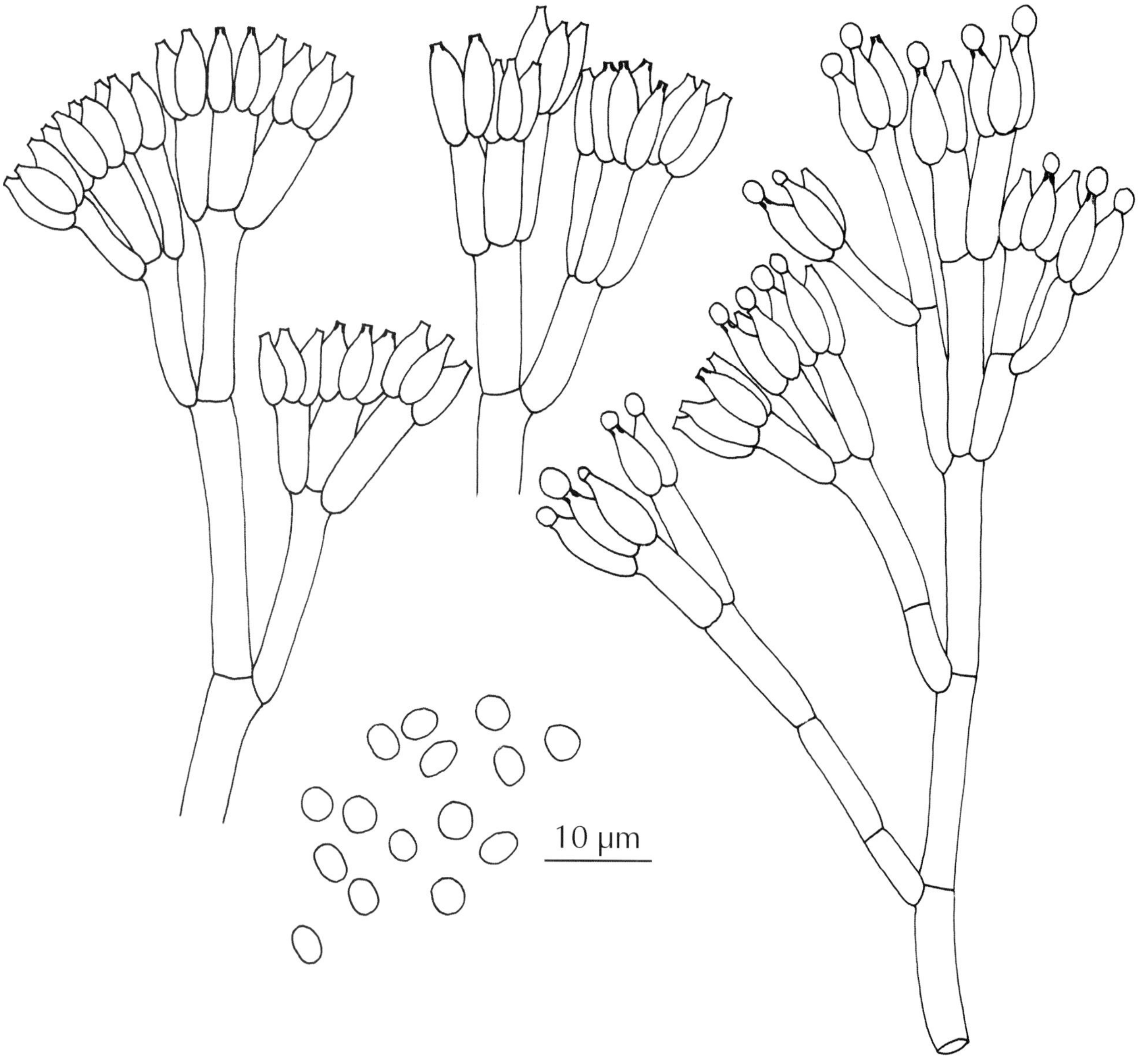

Fig. 85. *Penicillium chrysogenum*. Conidiophores and conidia.

Pl. 83. *Penicillium chrysogenum*. Colonies after one week, a. on Cz, b. on MEA; c,d. conidiophores, x 1000; e. conidia, x 1000.

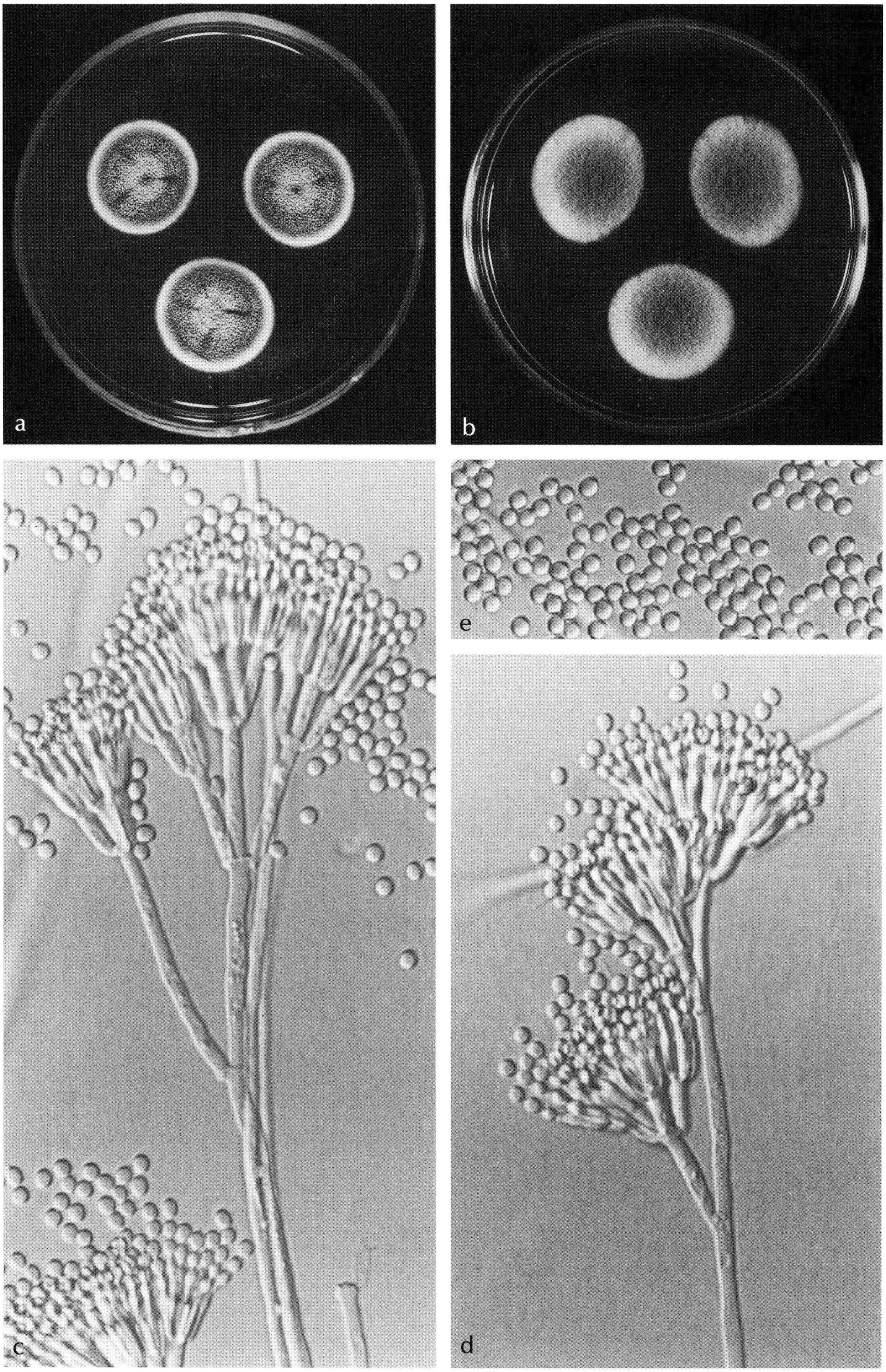
a
b
c
d
e

Penicillium citrinum Thom

Colonies on Czapek agar at 25° C, growing restrictedly, attaining a diameter of 1-1.5 cm within 7 days, consisting of a dense felt of conidiophores, sometimes appearing leathery, blue green. Reverse normally yellow to orange. Conidiophores 50-200 x 2-3 µm, smooth-walled with 3-5 divergent metulae in a whorl. Metulae 12-20 x 2-3 µm each, bearing 6-10 phialides. Phialides flask-shaped, 8-10 x 2-2.5 µm. Conidia produced in columns, globose to subglobose, smooth-walled, or finely rough, hyaline to greenish, 2.5-3.0 µm.

Colonies on MEA usually growing faster and less dense.

Important (toxic) metabolites: citrinin.

HABITAT: food, (indoor)
Common in (tropical) cereals and spices, but with a world-wide distribution, sometimes also occurring in indoor environments.

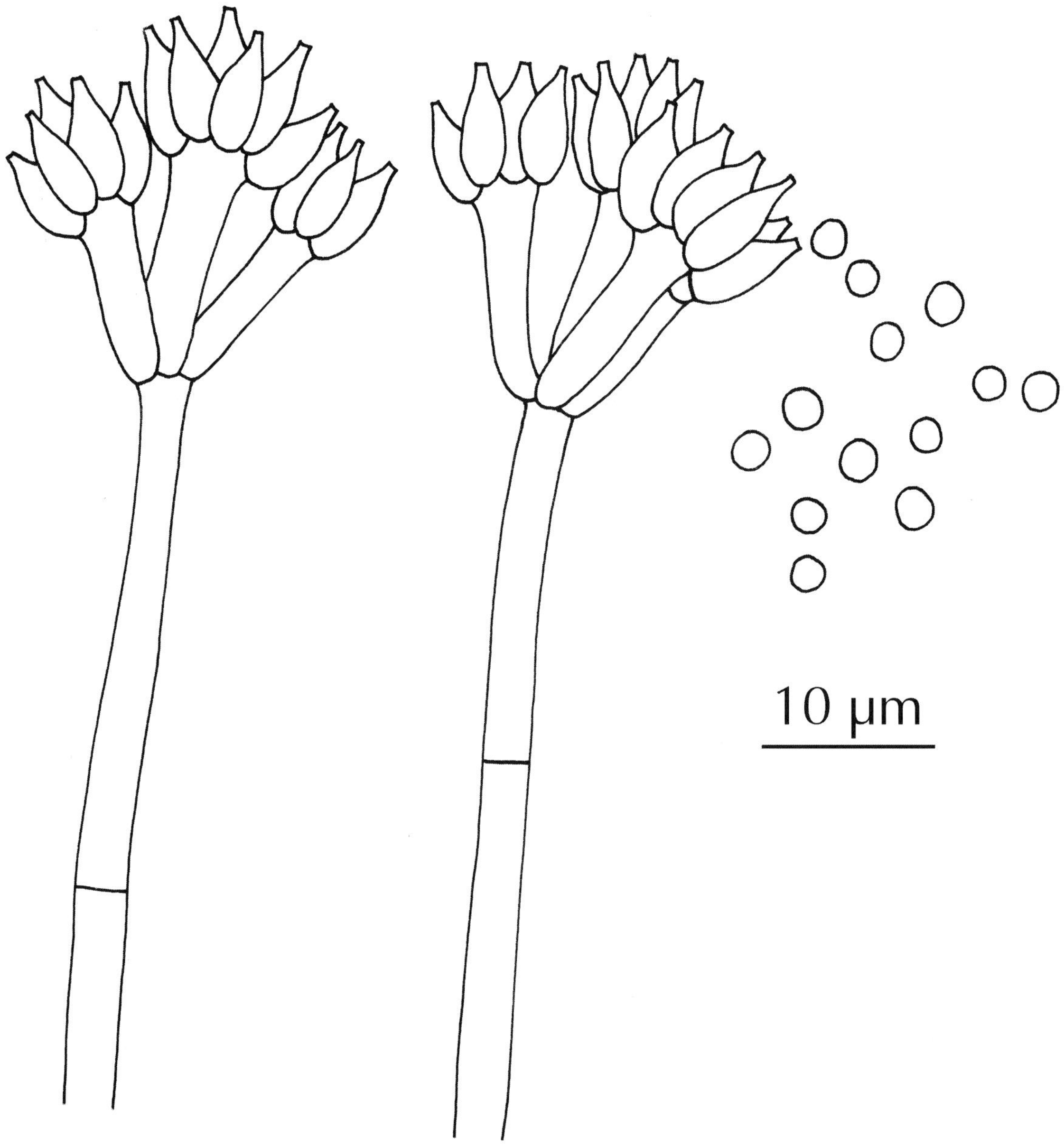

Fig. 86. *Penicillium citrinum*. Conidiophores and conidia.
Pl. 84. *Penicillium citrinum*. Colonies after one week, a. on Cz, b. on MEA; c,d. conidiophores with conidia, x 950.

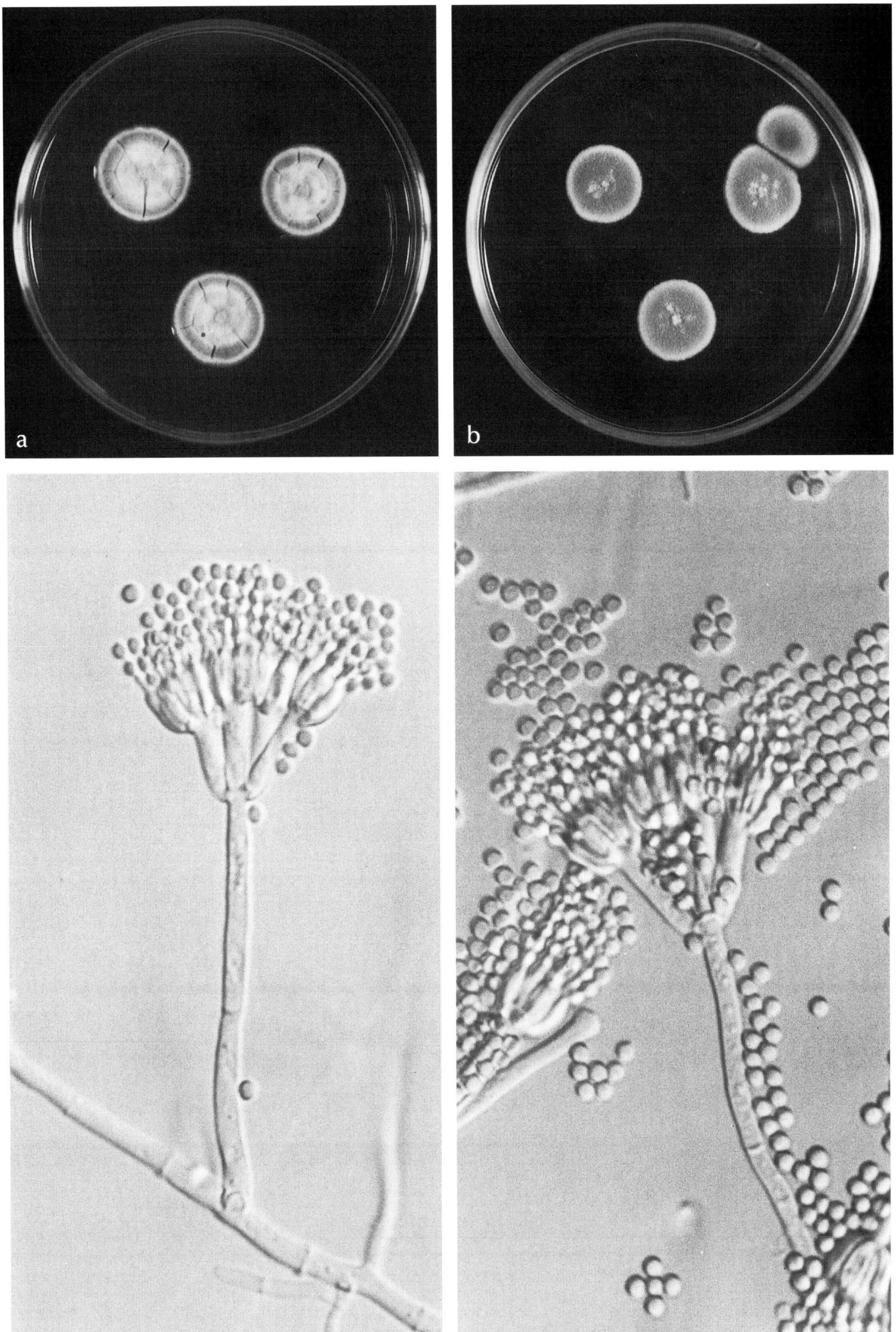
a
b
c
d

Penicillium commune Thom

Colonies on Czapek agar and CYA at 25° C growing rather restrictedly, attaining a diameter of 2.5-3 cm within 7 days, producing grey green to greyish turquoise conidia and velutinous to slightly floccose colonies, with clear to beige exudate droplets. The colony reverse is coloured creamish dull yellow to brown yellow. On MEA growing faster, the conidia are grey green, colonies have a pale or light yellow to sun yellow reverse velutinous to granular, rarely fasciculate. On YES agar poor to rather good sporulation with blue green conidia, reverse creamish yellow to yellow. On CREA strong growth and strong acid production and subsequent base production.
Conidiophores two-stage branched (terverticillate) with all elements adpressed, stipes rough-walled. Conidia smooth and subglobose to ellipsoidal, 3.5-5 µm in diam.

Important toxic metabolites: cyclopiazonic acid, rugulovasine A & B

Secondary metabolites with unknown toxicity: cyclopenin, cyclopenol, dehydrocyclopeptin, cyclopeptin, viridicatol, viridicatin, cyclopaldic and cyclopolic acid. Except for cyclopiazonic acid, the other secondary metabolites are apparently not produced by all isolates of the species.

HABITAT: food, indoor
The most common contaminant on hard and soft cheeses, but also common on meat products.

Penicillium palitans Westling

Colonies on Czapek agar and CYA at 25° C growing rather restrictedly producing green to dark green conidia and velutinous to slightly granular colonies, with clear to beige exudate droplets. The colony reverse is coloured creamish dull yellow to brown yellow often with a brown center. On MEA the conidia are green, colonies have a pale or light yellow to sun yellow reverse colonies velutinous. On YES agar good sporulation with green conidia, reverse yellow. On CREA strong growth and strong acid production and subsequent base production. Conidiophores two-stage branched (terverticillate) with all elements adpressed, stipes rough-walled. Conidia smooth and subglobose. The conidium colour and texture of this species on YES or CYA is so characteristic that it can be used to differentiate *P. palitans* from *P. solitum* and related species from series *Viridicata*, based on image analysis (Dörge *et al.* 2000).

Important toxic metabolites: cyclopiazonic acid, fumigaclavine A & B

Secondary metabolites with unknown toxicity: cyclopenin, cyclopenol, dehydrocyclopeptin, cyclopeptin, viridicatol, viridicatin, palitantin.

HABITAT: food, indoor
Hard cheeses, wooden surfaces, indoor air, occasional on meat products.

Pl. 85. *Penicilium commune*. Colonies after one week, a. on Cz., b. on MEA; c,d. conidiophores with conidia, x 950; e. conidia, x 1750.

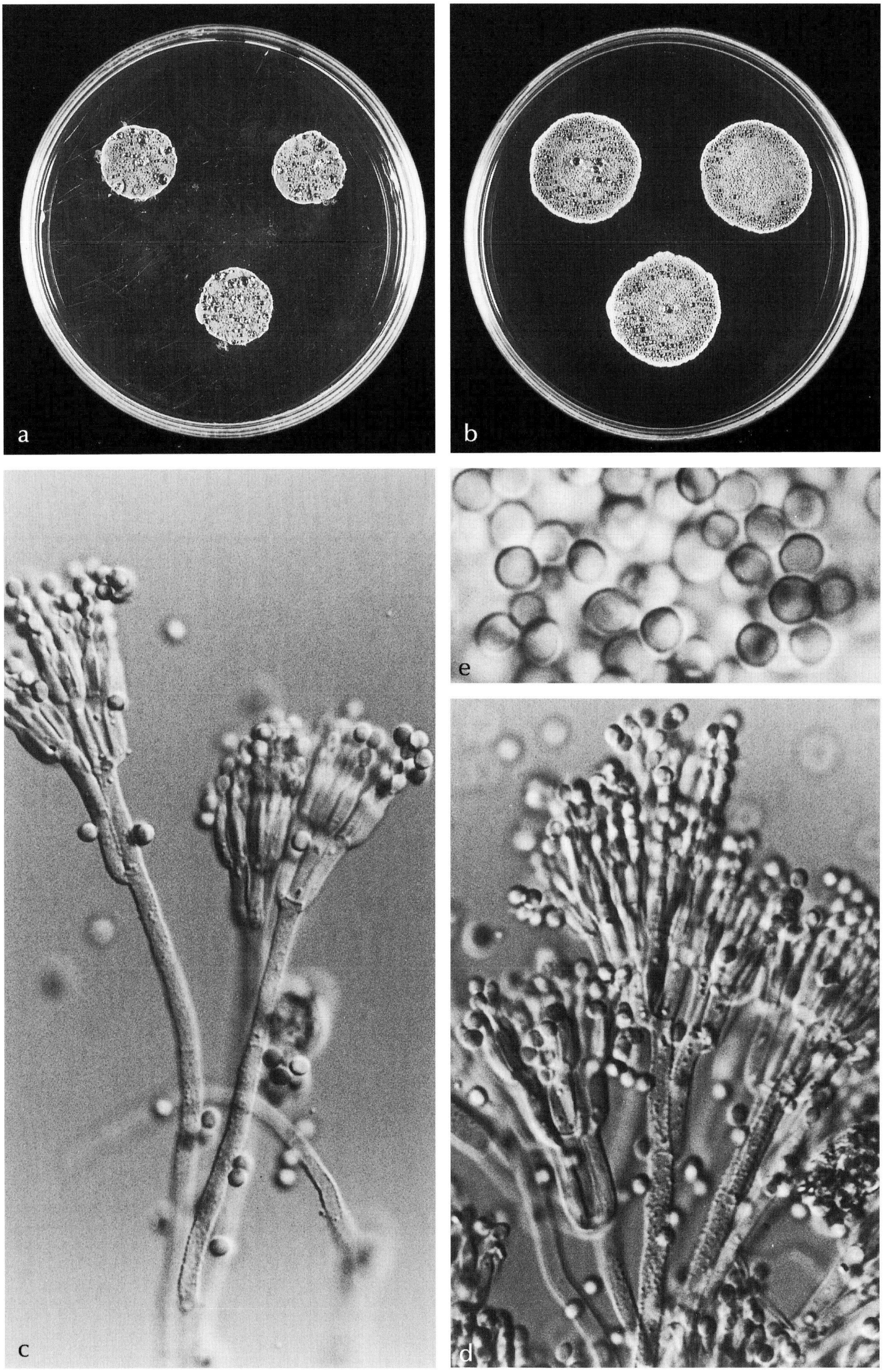
a
b
e
c
d

Penicillium corylophilum Dierckx

Colonies on Czapek agar at 25°C, mostly growing fast, attaining a diameter of 2-3 cm within 7 days. Consisting of a velvety layer of conidiophores. Colour blue-green soon becoming grey-green. Reverse in typical isolates green at first, soon becoming dark green to blackish; often arranged in a radial pattern. Conidiophores in young colonies often simple, unbranched, but soon terminating in a whorl of 2-4 metulae, smooth-walled. Metulae often of unequal length, mostly 12-20 x 2-3 µm. Phialides flask-shaped, 8-12 x 2.0-2.5 µm. Conidia subglobose to ellipsoidal, smooth, 2.5-3.2 x 2.5-3.0 µm.

Colonies on MEA similar but less dense and growing somewhat faster. Poor growth and no acid production on CREA.

HABITAT: food, indoor
Isolated from various food products e.g. cereals, frozen fruit cakes, acid liquids.

NOTE:
Penicillium corylophilum can be recognized by its one-stage branched conidiophores, the usually variable length of metulae and the dark green reverse.

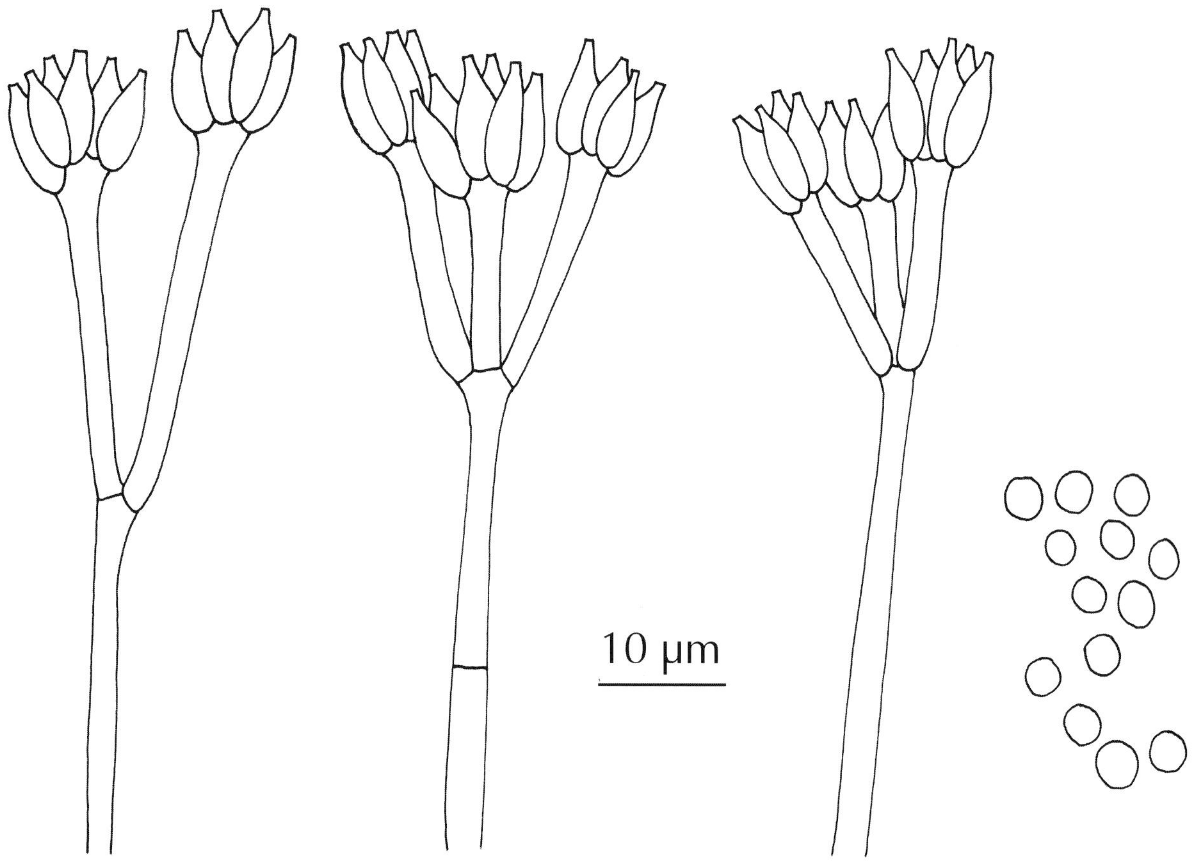

Fig. 87. *Penicillium corylophilum.* Conidiophores and conidia.
Pl. 86. *Penicillium corylophilum.* Colonies after one week, a. on Cz, b. on MEA; c,d. conidiophores with conidia, x 1350.

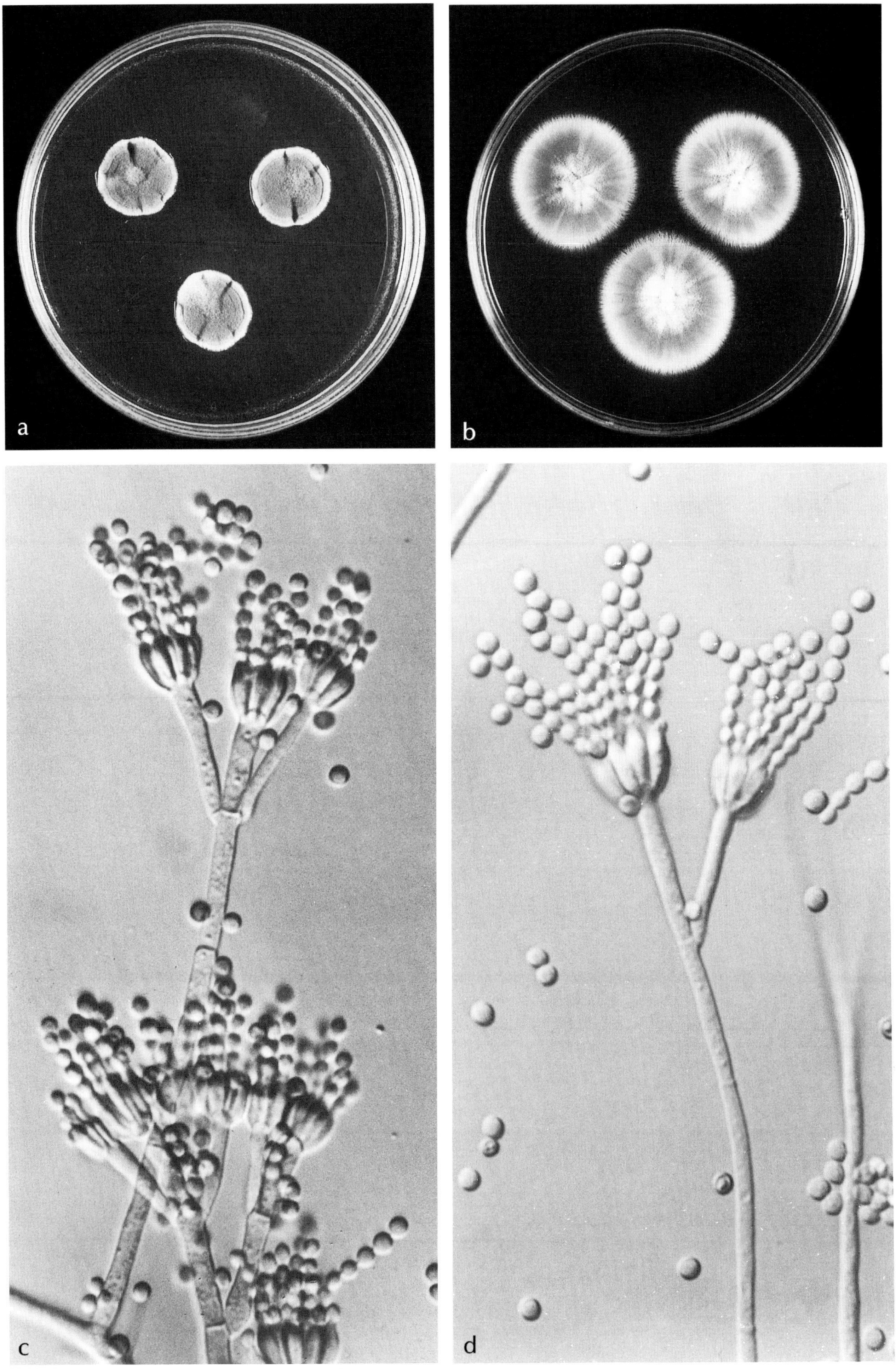
a
b
c
d

Penicillium crustosum Thom

Colonies on Czapek agar and CYA growing fast, attaining a diameter of 3-4 cm within 7 days, producing grey green to dull green (but turquoise near the colony margin) producing a crustose layer of conidia with a granular to fasciculate texture, often with clear to pale brown exudate. The colony reverse is cream to pale beige or yellow brown. On MEA the conidia are grey green to full green and the conidial layer is crustose and granular to fasciculate and colonie have a pale to yellow beige reverse. On YES agar the colonies are heavily sporulating and colonies have a bright yellow reverse. On CREA the growth is very good with acid and subsequent base production. Conidiophores two-stage branched (terverticillate) with all elements adpressed, stipes rough-walled. Conidia smooth-walled, globose to subglobose, 3.5-4 μm in diam.

Important toxic metabolites: penitrem A-F, terrestric acid, roquefortine C

Secondary metabolites of unknown toxicity: cyclopenin, cyclopenol, dehydrocyclopeptin, cyclopeptin, viridicatol, viridicatin, styrene, 2-methylisoborneol, geosmin, dimethyl-disulphide.

HABITAT: food
Rather common in oil seeds and nuts, occasional in cheese and meat products. *P. crustosum* can produce a rot in apples.

Pl. 87. *Penicillium crustosum*. Colonies after one week, a. on Cz, b. on MEA; c-e. conidiophores and conidia, x 950.

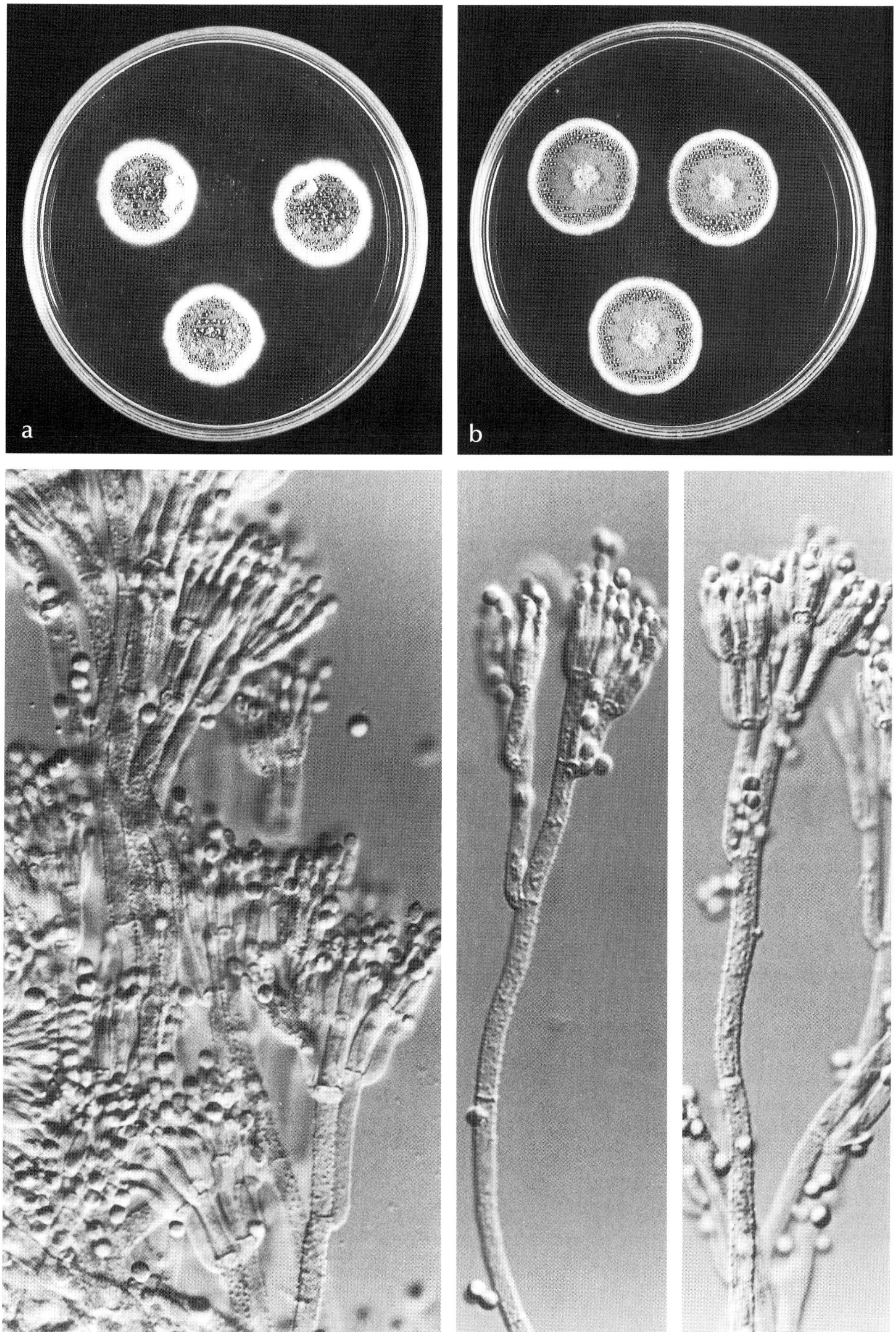
a
b
c
d
e

Penicillium cyclopium Westling

Colonies on Czapek agar and CYA at 25°C growing restrictedly producing grey green conidia with a granular to fasciculate colony surface, often with exudate droplets. The colony reverse is orange to red or pinkish brown with the colour often diffusing into the agar medium or more rarely creamish yellow. On MEA the conidia are blue green with a strong blue element and colonies have a distinct yellow reverse, often with the yellow colour diffusing into the medium. On YES agar there is no sporulation and the colony mycelium if often strongly yellow, reverse colour distinct yellow. On CREA weak growth but strong acid production. Conidiophores two-stage branched (terverticillate) with all elements adpressed, stipes rough-walled. Conidia smooth-walled, globose to subglobose, 3-3.5 µm in diam.

Important toxic metabolites: xanthomegnin, viomellein,vioxanthin, penicillic acid

Secondary metabolites with unknown toxicity: cyclopenin, cyclopenol, dehydrocyclopeptin, cyclopeptin, viridicatol, 3-methoxyviridicatin, verrucofortine (=verrucosine), puberuline, rugulosuvine, leucyltryptophanyldiketopiperazine.

HABITAT:
The primary habitat is cereals.

Penicillium aurantiocandidum Dierckx

Colonies on Czapek agar and CYA at 25°C growing restrictedly producing grey to blue green conidia with a granular colony surface, often with exudate droplets. The colony reverse is orange to pinkish brown with the colour often diffusing into the agar medium. On MEA the conidia are blue green with a strong blue element and colonies have a distinct yellow reverse, often with the yellow colour diffusing into the medium. Sporulation is less pronounced than in other species in *Viridicata*, often with a non-spurulating floccose margin. On YES agar there is no sporulation, reverse colour distinct yellow. On CREA weak growth but strong acid production. This species is close to *P. cyclopium* and differs by its ability to produce puberulonic acid and inability to produce xanthomegnin, viomellein and vioxanthin. Conidiophores two-stage branched (terverticillate) with all elements adpressed, stipes rough-walled. Conidia smooth-walled, globose to subglobose.

Important toxic metabolites: penicillic acid.

Secondary metabolites with unknown toxicity: puberulonic acid, puberulic acid, cyclopenin, cyclopenol, dehydrocyclopeptin, cyclopeptin, viridicatol, 3-methoxyviridicatin, verrucofortine (=verrucosine), puberuline, verrucosinol, demethylverrucosine, dehydroverrucofortine, rugulosuvine, leucyltryptophanyldiketopiperazine.

HABITAT:
The primary habitat is cereals.

Penicillium freii Frisvad & Samson

Colonies on Czapek agar and CYA at 25°C growing restrictedly producing blue to turquoise conidia with a fasciculate and crustose colony surface, with large clear exudate droplets. The colony reverse is cream yellow to yellow or curry, rarely yellow to pinkish brown. On malt agar the conidia are blue green with a strong blue element and colonies have a distinct yellow reverse, often with the yellow colour diffusing into the medium. On YES agar there is no sporulation or rarely some sporulation in the center of the colony, reverse colour distinct yellow. On CREA weak growth but strong acid production. Conidiophores two-stage branched (terverticillate) with all elements adpressed, stipes rough-walled. Conidia smooth-walled, globose to subglobose.

Important toxic metabolites: xanthomegnin, viomellein, vioxanthin, (penicillic acid).

Secondary metabolites with unknown toxicity: aurantiamin, cyclopenin, cyclopenol, dehydrocyclopeptin, cyclopeptin, viridicatol, 3-methoxy-viridicatin.

HABITAT:
The primary habitat is cereals. Most common in temperate regions of the world.

Pl. 88. *Penicillium cyclopium.* Colonies after one week, a. on Cz, b. on MEA; c,d. conidiophores, x 950.

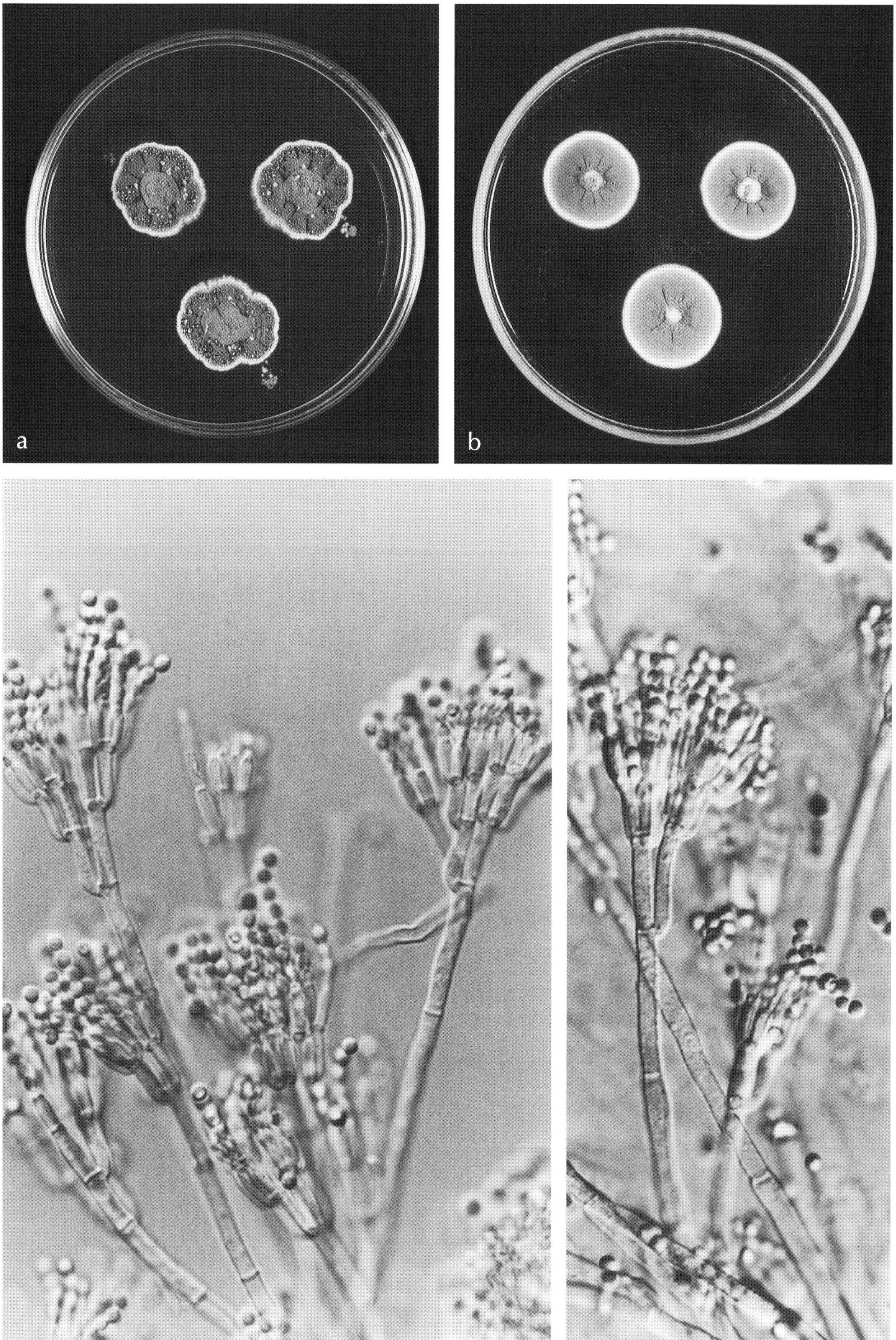
a
b
c
d

Penicillium digitatum Sacc.

Colonies on Czapek agar at 25°C, growing and sporulating poorly, attaining a diameter of about 1 cm within 14 days. On MEA growth is rapid, colonies attaining a diameter of 4-6 cm within 7 days, velvety, yellow- to brown-green. Conidiophores irregularly branched, consisting of short stipes with few metulae and branches terminating in whorls of 3-6 phialides, smooth-walled. Phialides often solitary, cylindrical with a short neck, variable in size, 15-30 x 3.5-5.0 µm. Conidia ellipsoidal to cylindrical, smooth, olive-green in mass, variable in size, but mostly 3.5-8.0 x 3.0-4.0 µm.
Poor growth and no acid production on CREA.

Important (toxic) metabolites: tryptoquivalins.

HABITAT: food
Cause of rot of various citrus fruit, sometimes isolated from other substrates e.g. corn, rice, meat, fruit-juices.

NOTE:
The shape of the phialide and the conidiophore structure are atypical for *Penicillium. P. digitatum* can be considered as an intermediate between *Penicillium* and *Paecilomyces. P. italicum* and *P. digitatum* are common on citrus fruit. The species differs from *P. italicum* by its irregular conidiophores, the large olive-green conidia and poor growth on Czapek agar.

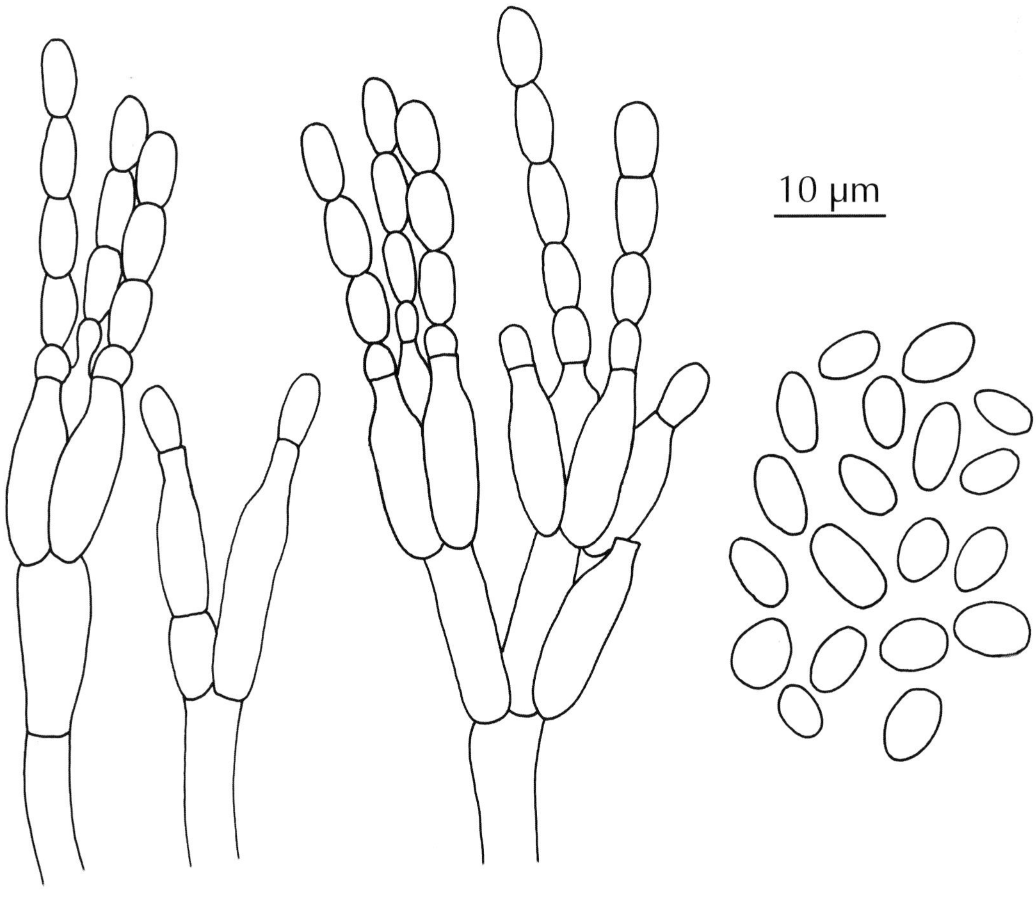

Fig. 88. *Penicillium digitatum*. Conidiophores and conidia.
Pl. 89. *Penicillium digitatum*. a. Colonies on MEA after one week; b. natural infection of orange; c. conidia, x 1000; d,e. conidiophores, x 1000.

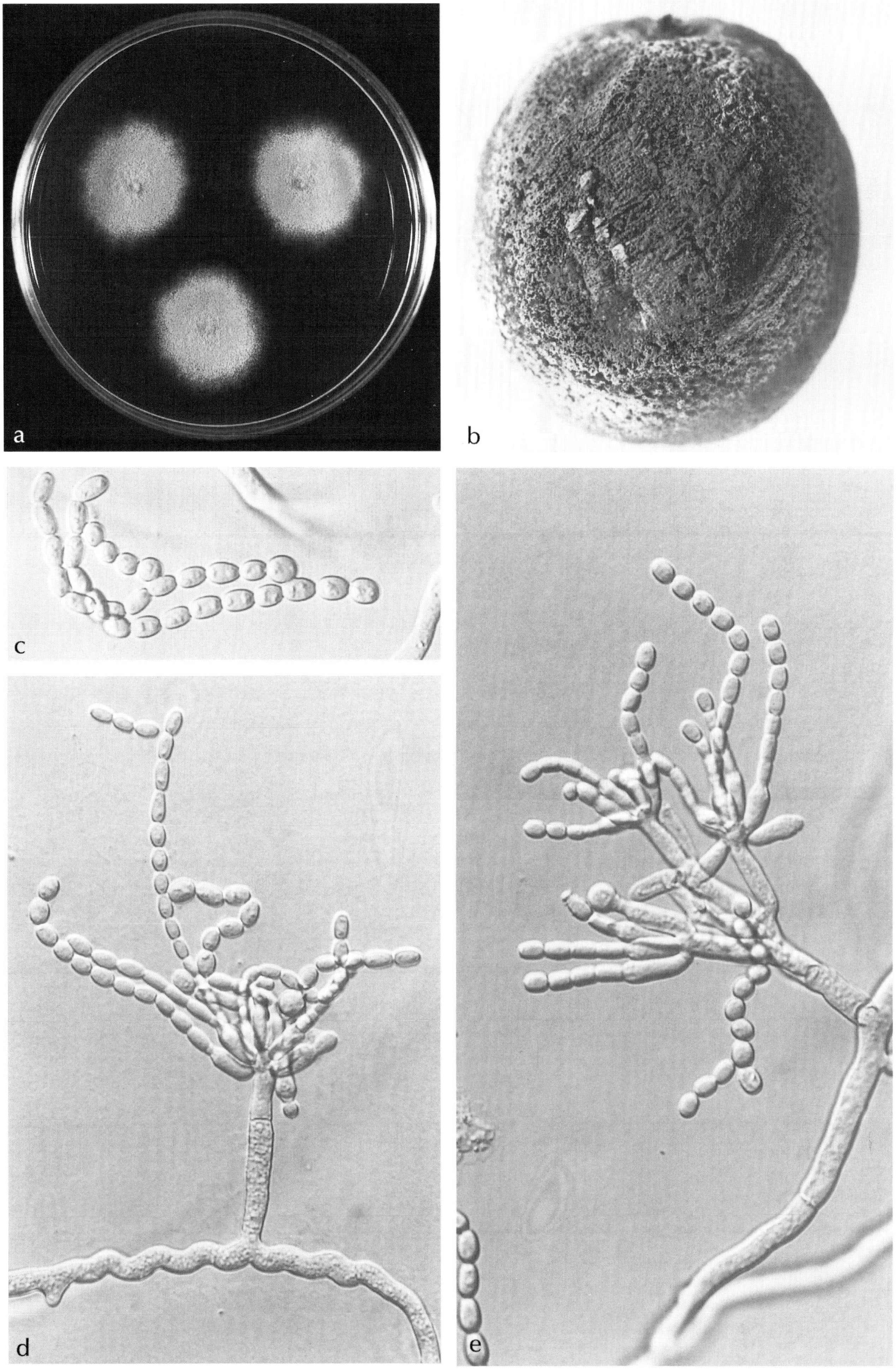
a
b
c
d
e

Penicillium echinulatum Fassatiová

Colonies on Czapek agar at 25°C growing rapidly, attaining a diameter of 2.8-3.5 cm within 7 days (4.5-5.0 cm within 14 days). Colour grey-green to dark green. Odour pronounced, earthy. Reverse uncoloured to yellow. Exudate absent or as colourless droplets. Conidiophores mononematous but loosely synnematous in marginal areas, hyaline, rough-walled, terverticillate, stipes 200-400 x 3.0-3.5 µm; branches appressed to the main axis. Metulae cylindrical, smooth-walled or nearly so, 10-15 x 2.5-3.3 µm, bearing 4 to 8 phialides. Phialides flask-shaped with a short but distinct neck, 7-10 x 2.2-2.8 µm. Conidia globose to subglobose, 3.5-4.5 µm, greenish, rough-walled to echinulate. On CREA good growth and acid production.

Important (toxic) metabolites: territrems.

HABITAT: food
Common on margarine, cheese and other lipid containing substrates.

NOTE:
A species which is morphologically similar to *P. echinulatum is P. discolor*. These taxa can be distinguished by the following criteria:

Colony dark green, conidia 3.5-4.5 µm
diam. Ehrlich test: no colour....... ***P. echinulatum***
Colony dark blue green, conidia 4-5 µm
diam. Ehrlich test: violet .... ***P. discolor***

Penicillium discolor Frisvad & Samson

Colonies on Czapek agar and CYA at 25°C growing restrictedly producing dark glaucous grey to dark green conidia with a floccose to velutionous colony surface, with clear to orange exudate droplets. The reverse is cream to light brown with a dark brown center. On MEA the colonies are granular to strongly fasciculate with dark green conidia, colony reverse pale or yellow. On YES agar the colonies are sporulating well and the reverse is yellow to orange, with a bright orange red to red soluble pigment being produced after 6-12 days. On CREA the growth is very good with acid and subsequent base production.

Important toxic metabolites: chaetoglobosin A, B & C

Secondary metabolites with unknown toxicity: cyclopenin, cyclopenol, dehydrocyclopeptin, cyclo - peptin, viridicatol, viridicatin, palitantin, geosmin, 2-methylisoborneol.

HABITAT: food
P. discolor have been associated to cultivated plant roots, walnuts, black walnuts, pecans, acorns, hazelnuts and natamycin treated cheeses.

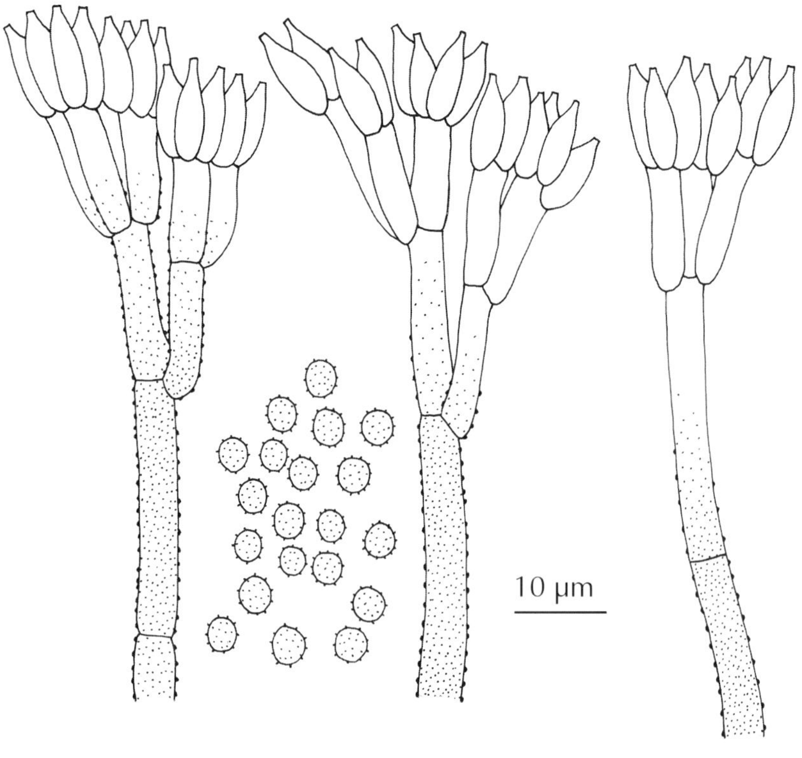

Fig. 89. *Penicillium echinulatum.* Conidiophores and conidia.
Pl. 90. *Penicillium echinulatum.* Colonies after one week, a. on Cz, b. on MEA; c-e. conidiophores, x 1000; f. conidia, x 1500.

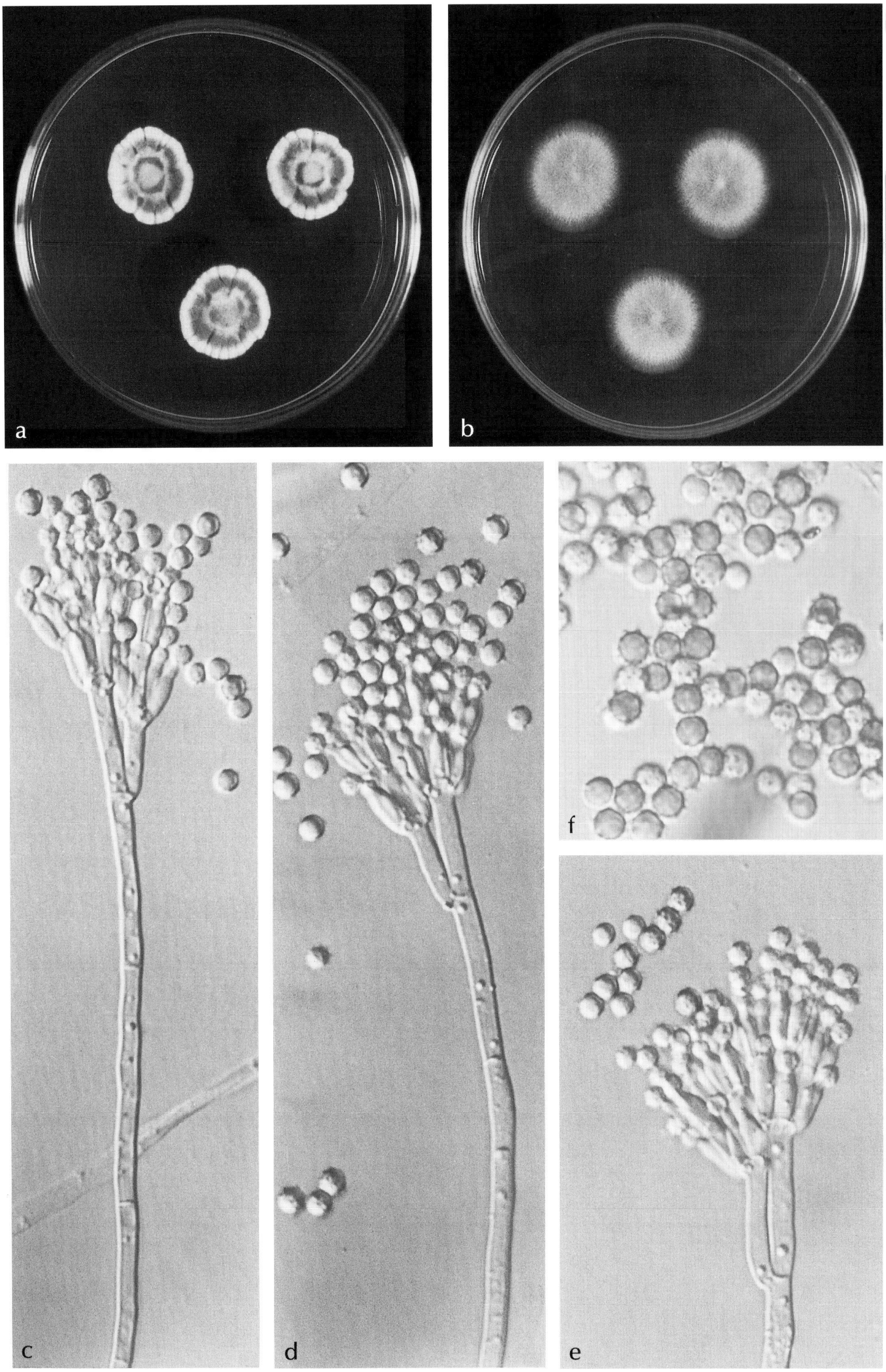
a
b
c
d
f
e

Penicillium expansum Link

Colonies on Czapek and MEA at 25°C growing rapidly, attaining a diameter of 4-5 cm within 14 days, colour yellow- to blue-green. Odour pronounced, aromatic-fruity, suggesting apples. Exudate absent or as hyaline drops. Reverse uncoloured to yellowish or yellow-brown. Conidiophores mononematous but in fresh isolates predominantly loosely synnematous, especially in the marginal areas, hyaline, ter- to quaterverticillate with the branches usually appressed against the main axis; stipes usually smooth-walled, sometimes very finely roughened, this latter feature more pronounced on MEA. Metulae more or less cylindrical, 10-15 x 2.2-3.0 µm, bearing 5 to 8 phialides. Phialides cylindrical with a short but distinct neck, 8-12 x 2.0-3.5 µm. Conidia subglobose to ellipsoidal, 3.0-3.5 x 2.5-3.0 µm, greenish, smooth-walled.
Good growth and acid production on CREA.

Important (toxic) metabolites: roquefortine C, patulin, citrinin, communesins, chaetoglobosin C.

HABITAT : food
The species is responsible for rapid rot of pomaceous fruit. Also common on nuts.

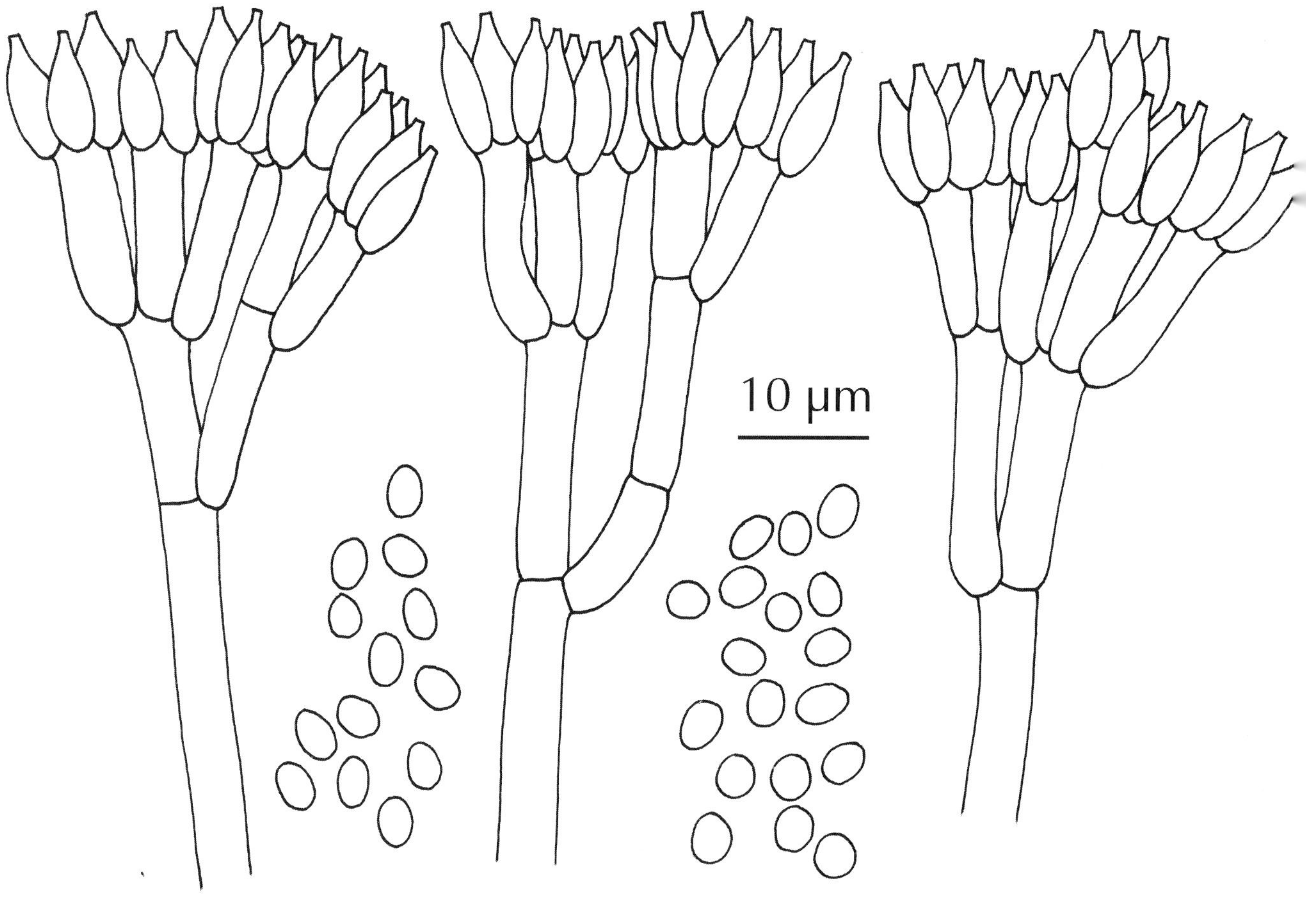

Fig. 90. *Penicillium expansum*. Conidiophores and conidia.
Pl. 91. *Penicillium expansum*. Colonies on Cz after one week; b. apple infected with *P. expansum*, this species is responsible for rapid rot of pomaceous fruit; c,d. conidiophores, x 1000; e. conidia, x 1000.

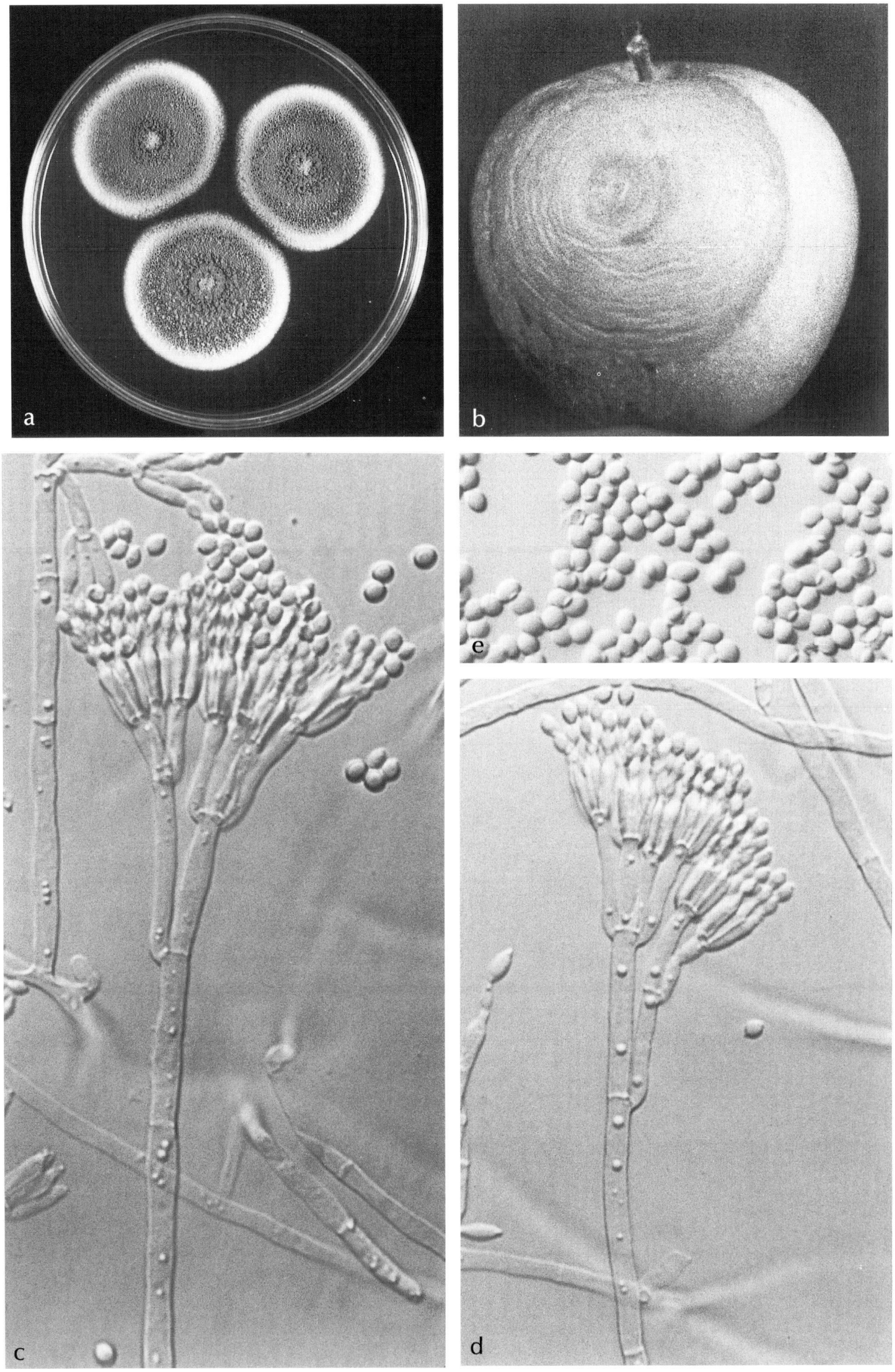
a
b
e
c
d

Penicillium funiculosum Thom

Colonies on Czapek agar at 25°C growing rapidly attaining a diameter of 3-4 cm within 7 days, consisting of a dense felt of often yellow vegetative mycelium intermixed with grey to yellow-grey conidiophores. Reverse pink to deep red, or orange-brown. Odour often pronounced in old cultures, earthy. Conidiophores arising from the agar surface or from bundled aerial hyphae, biverticillate, consisting of a smooth-walled stipe, terminating in a whorl of 5-8 metulae. Metulae 10-13 x 2-3 µm, terminating in a whorl of 3-6 phialides. Phialides lanceolate, 10-12 x 1.5-2.5 µm. Conidia subglobose to ellipsoidal, smooth to finely roughened, 2.5-3.5 x 2-2.5 µm. Colonies on MEA similar, but with richer sporulation. Poor growth and acid production on CREA.

HABITAT: food, indoor
Common on tropical cereals and fruits, nuts. Occasionally occurring in indoor environments.

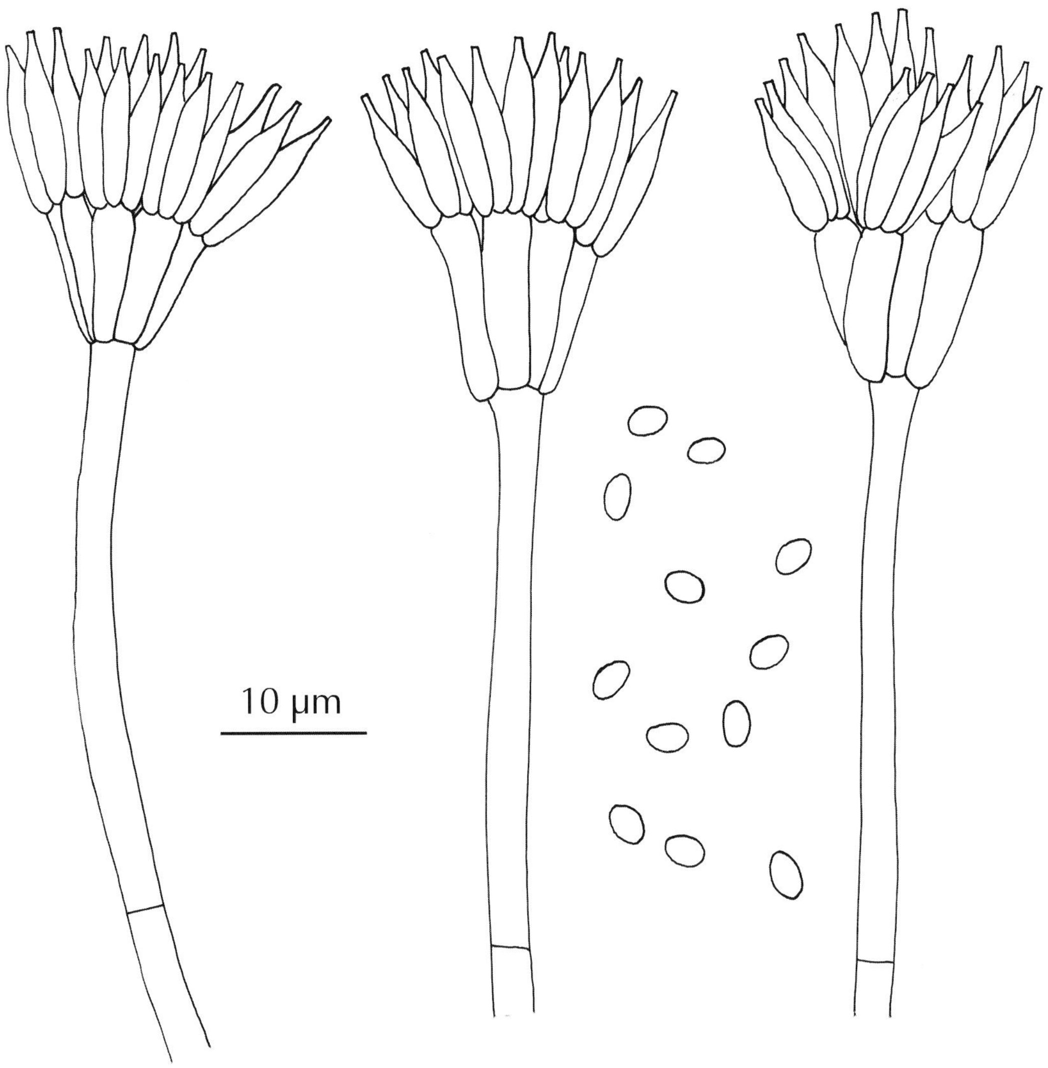

Fig. 91. *Penicillium funiculosum.* Conidiophores and conidia.
Pl. 92. *Penicillium funiculosum.* Colonies after 14 days, a. on Cz, b. on MEA; c,d. conidiophores, c,d. x 1000; e. detail of penicillus, x 2500; f. conidia, x 1800.

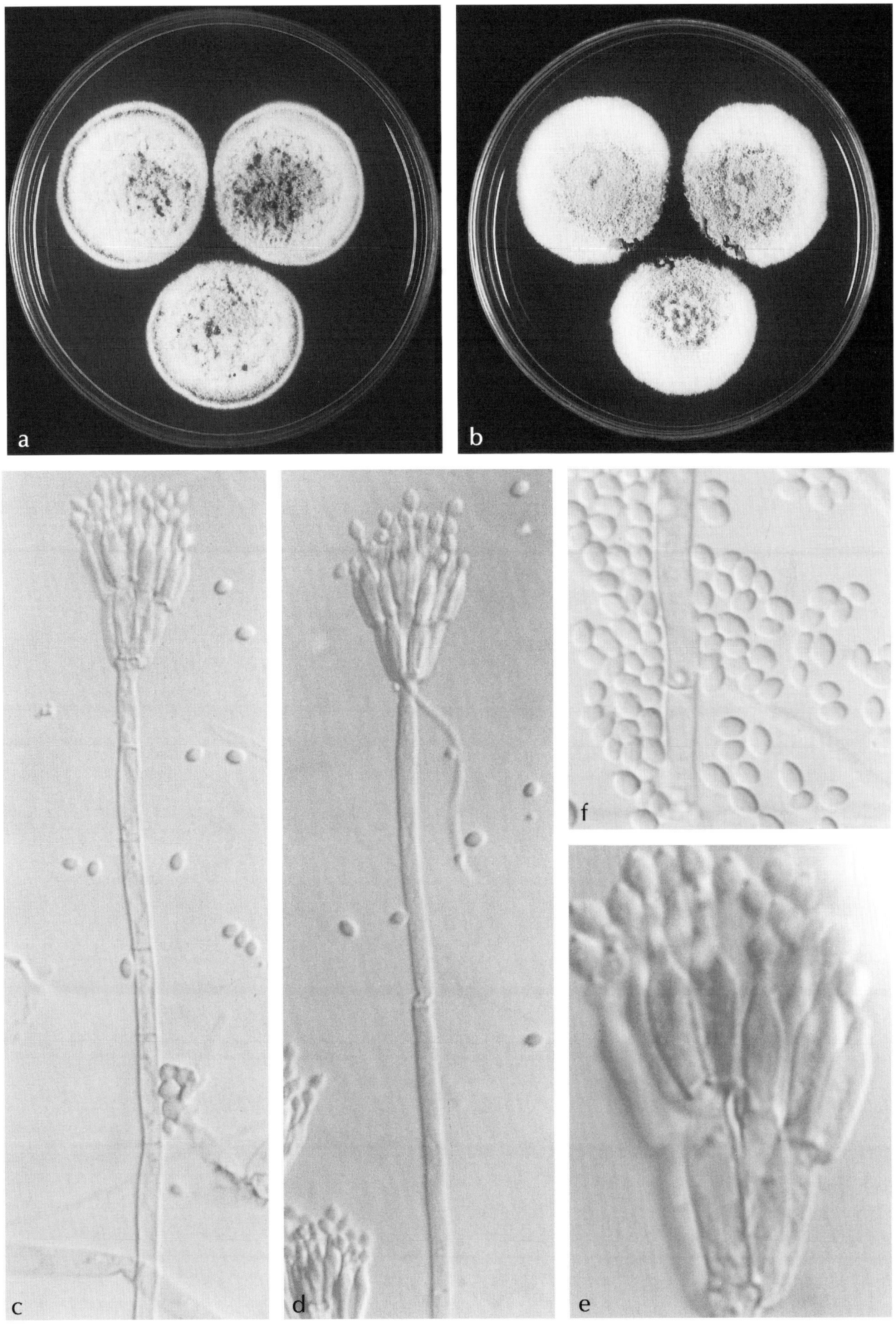
a
b
c
d
f
e

Penicillium glabrum (Wehmer) Westling

= *Penicillium frequentans* Westling

Colonies on Czapek agar and MEA at 25°C growing rapidly, attaining a diameter of 4-5 cm within 7 days, consisting of a dense felt of erect conidiophores, velvety, grey-green. Reverse usually yellow to yellow-orange. Conidiophores monoverticillate, consisting of an unbranched stipe, smooth to finely roughened, terminating in a whorl of 10-12 phialides. Phialides flask-shaped, 8-12 x 3-3.5 µm. Conidia produced in typical long columns, globose to subglobose, smooth to finely roughened, 3-3.5 µm in diam.

Important (toxic) metabolites: citromycetin.

HABITAT: food, indoor
Isolated from (dried) fruits, nuts, frozen cakes, fruit-juices, cereals, but also from compost and occurring in indoor environments.

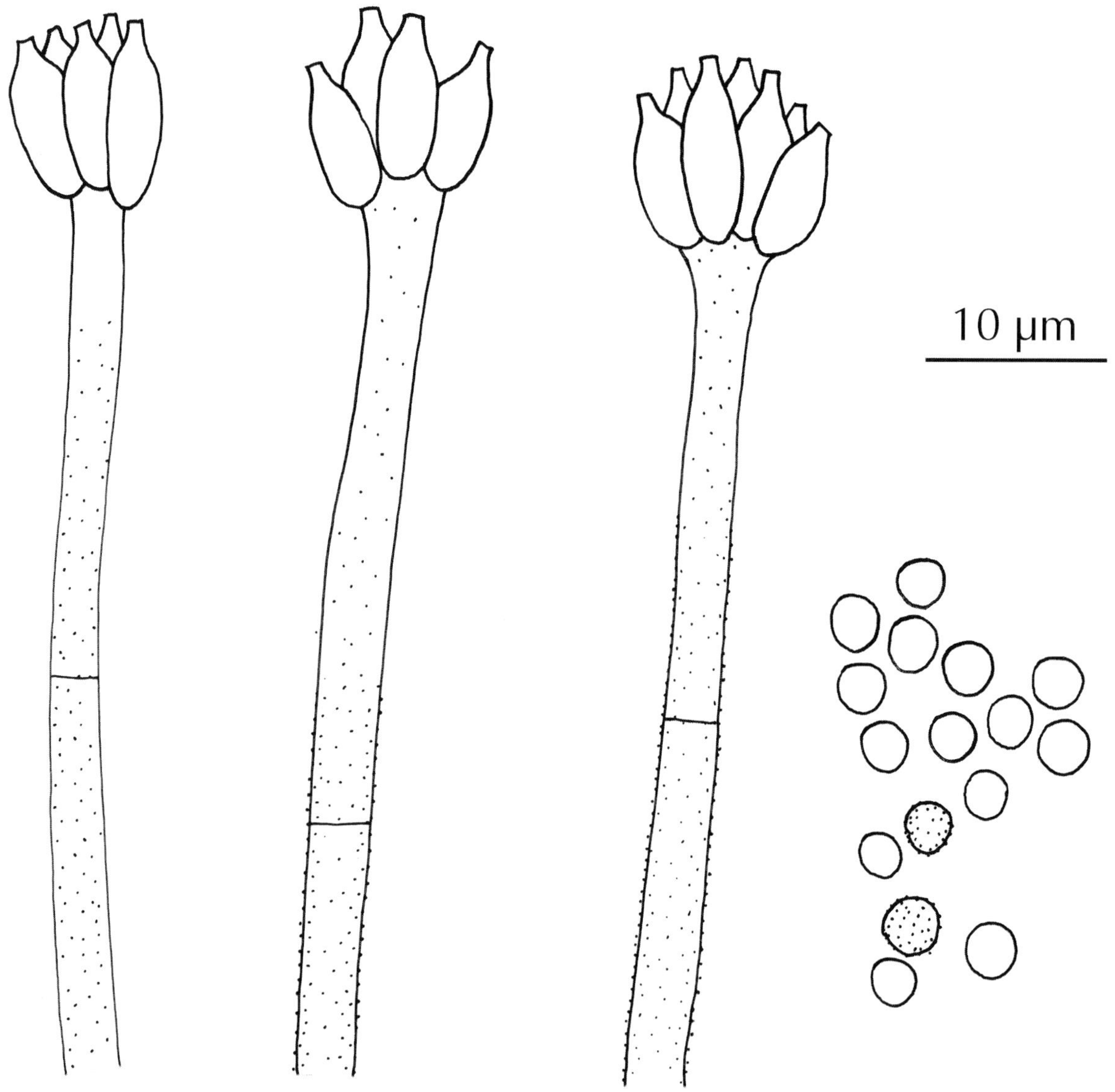

Fig. 92. *Penicillium glabrum.* Conidiophores and conidia.
Pl. 93. *Penicillium glabrum.* Colonies after one week, a. on Cz, b. on MEA; c,d. conidiophores, x 1000; e. conidia, x 1000.

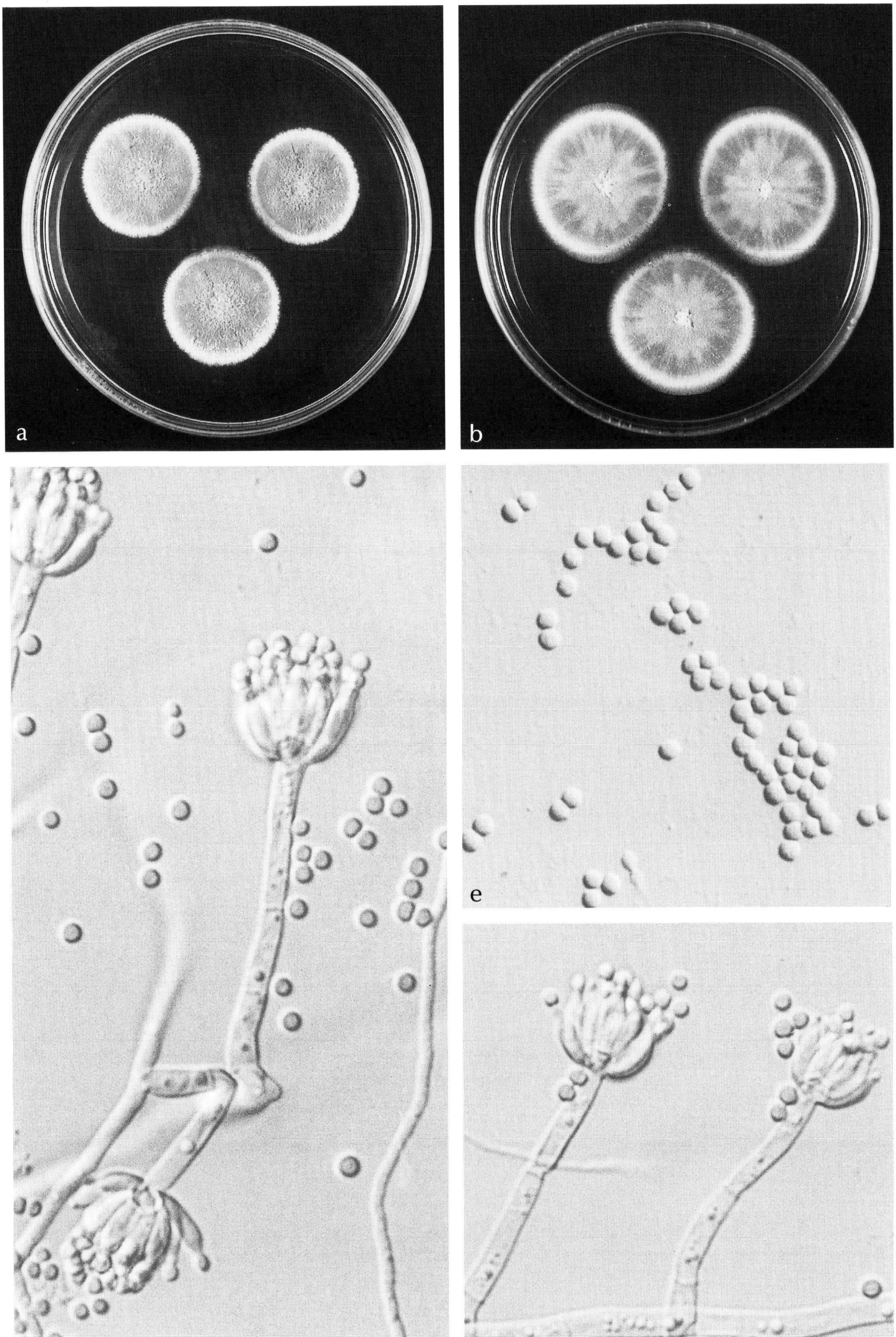
a
b
c
e
d

Penicillium griseofulvum Dierckx

= *Penicillium patulum* Bain.
= *Penicillium urticae* Bain.

Colonies on Czapek agar at 25°C restricted, attaining a diameter of 1.8-2.0 cm within 7 days (2.0-2.5 cm within 14 days). Colour variable, but mostly grey-green to yellow-green. Odour aromatic, usually not pronounced. Exudate usually absent or hyaline when present. Reverse yellowish to orange-brown to red-brown. Conidiophores mononematous or (loosely) synnematous, especially in the marginal areas, smooth-walled, hyaline, irregular ter- to quaterverticillate, with branches strongly divergent. Stipes undulate, 400-500 x 3.0-4.0 µm. Phialides more or less cylindrical with a very short, inconspicuous neck, 4.5-6.5 x 2.2-2.5 µm. Conidia ellipsoidal, sometimes subglobose, 2.5-3.5 x 2.2-2.5 µm, hyaline to greenish, smooth-walled.

Growth on MEA faster, 2.0-2.4 cm in diameter within 7 days, usually yellow-green due to a richer sporulation.

Important (toxic) metabolites: roquefortine C, cyclopiazonic acid, patulin, griseofulvin.

HABITAT: food/feed
World-wide distribution. Isolated from soil, common on cereals, feed.

NOTE:
The short phialides with an inconspicuous neck are the typical features of this species.

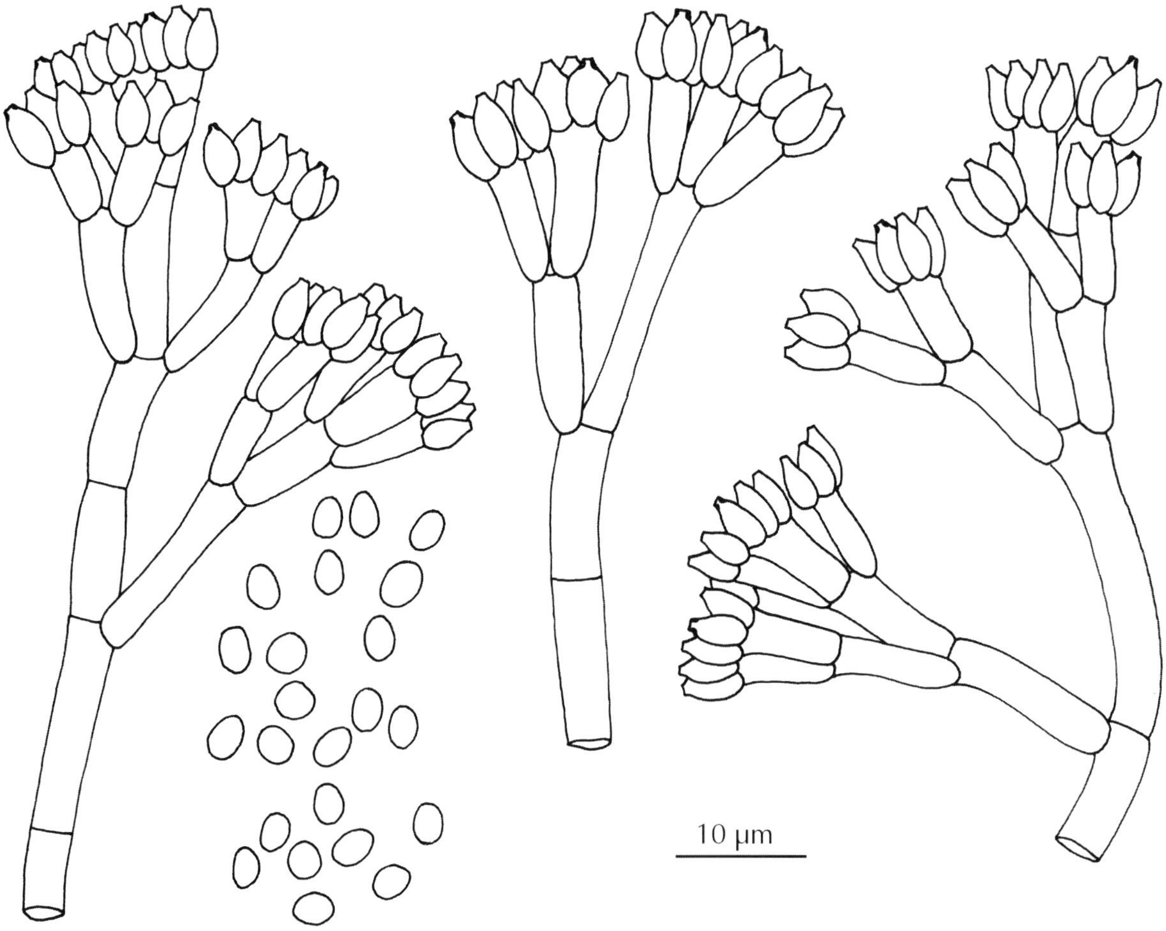

Fig. 93. *Penicillium griseofulvum*. Conidiophores and conidia.
Pl. 94. *Penicillium griseofulvum*. Colonies after one week, a. on Cz, b. on MEA; c,d. conidiophores, c. x 1000, d. x 1700; e. conidia, x 1700.

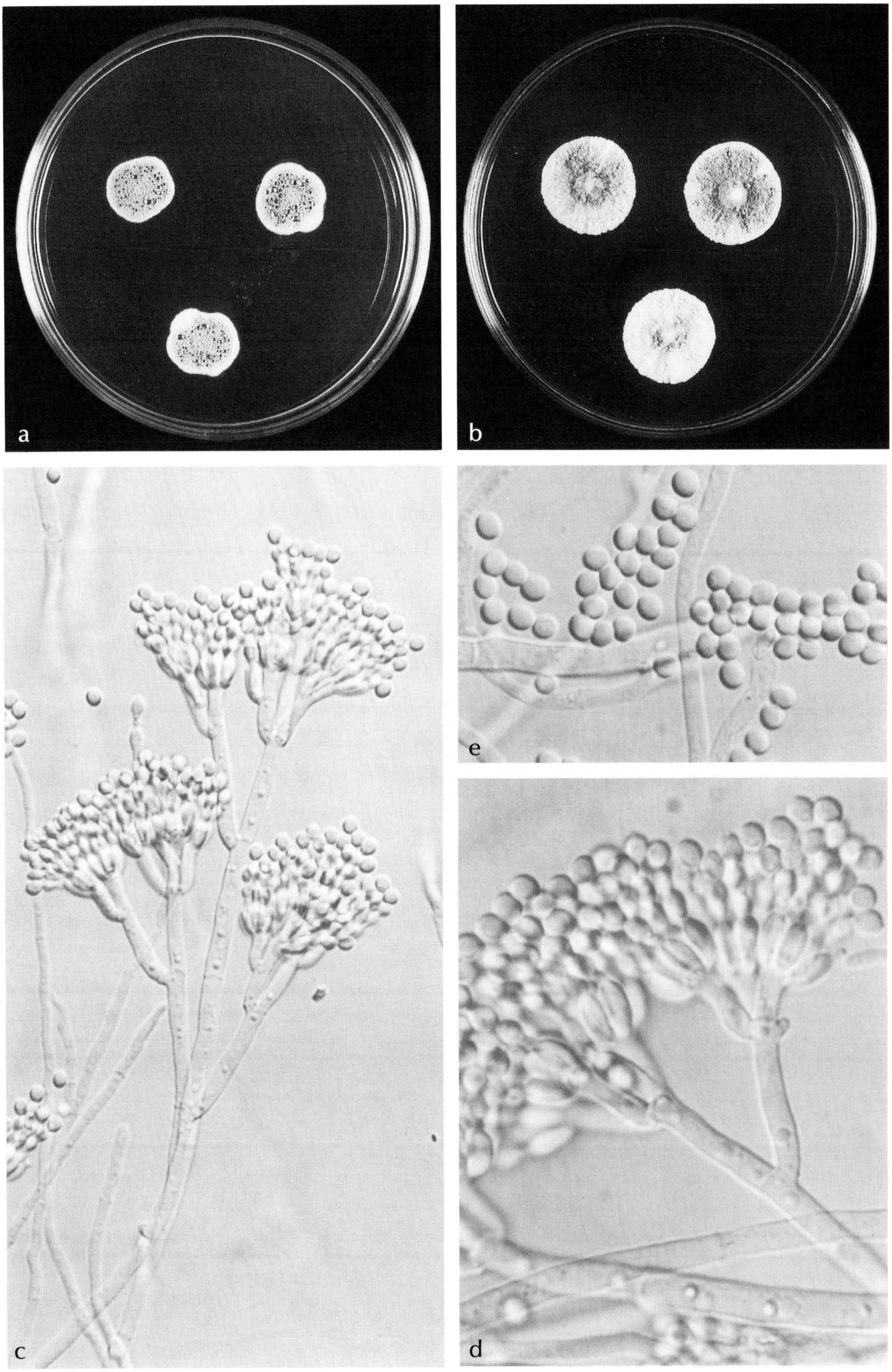
a
b
c
e
d

Penicillium hirsutum Dierckx

=*Penicillium corymbiferum* Westling
=*Penicillium verrucosum* Dierckx var. *corymbiferum* (Westling) Samson *et al.*

Colonies on Czapek agar at 25°C growing rapidly, attaining a diameter of 2-3 cm within 7 days (3-4 within 14 days), usually zonate in marginal areas, strongly fasciculate especially in fresh isolates, after prolonged subculturing becoming more or less velvety. Colour yellow-green, intermixed with yellow vegetative mycelium. Exudate in fresh isolates abundantly produced as brown drops. Reverse yellow-brown to reddish-brown. Odour pungent.

Important (toxic) metabolites: roquefortine C, terrestric acid.

NOTE:
P. hirsutum is known as the causal agent of *Penicillium* rot of liliaceous bulbs. Other related species are *P. allii* (on garlic), *P. albocoremium* (on onions) and *P. hordei* (on cereals).

Penicillium albocoremium (Frisvad) Frisvad

Colonies on Czapek agar and CYA at 25°C growing rapidly producing blue green conidia and deep floccose colonies, with large clear to yellow exudate droplets. The colony reverse is coloured light brown to yellow brown. On MEA the conidia are blue green with a pale or weakly yellow reverse, fasciculate often producing feathery synnemata without a distinct capitulum. On YES agar poor to no sporulation, reverse coloured cream to light brown. On CREA weak to moderate growth and strong acid production. Conidiophores two-stage branched (terverticillate) with all elements adpressed, stipes roughened. Conidia smooth and globose to subglobose.

Important toxic metabolites: roquefortine C

Secondary metabolites with unknown toxicity: cyclopenin, cyclopenol, dehydrocyclopeptin, cyclopeptin, viridicatol, 3-methoxyviridicatin, meleagrin, chrysogine.

HABITAT:
Onions, potatoes, carrots, ginger and other plant roots.

Penicillium allii Vincent & Pitt

Colonies on Czapek agar and CYA at 25°C growing rapidly producing dull to greyish green conidia and granular to fasciculate colonies, with redbown to terracotta exudate droplets. The colony reverse has little or no sulcation, coloured yellow brown to dark brown. On MEA the conidia are dull green with a yellow to yellow brown reverse, granular to fasciculate. On YES agar good sporulation, reverse coloured beige to warm yellow brown. On CREA weak to moderate growth and no, weak or moderate acid production. Conidiophores two-stage branched (terverticillate) with all elements adpressed, stipes roughened. Conidia smooth and globose to subglobose.

Important toxic metabolites: roquefortine C

Secondary metabolites with unknown toxicity: meleagrin, PI-3, dehydrofulvic acid.

HABITAT:
Garlic and onions. *P. allii* produces a destructive rot in garlic.

Penicillium hordei Stolk

Colonies on Czapek agar and CYA at 25°C growing restrictedly producing green conidia with a highly floccose colony surface, often with yellow brown to dark brown exudate droplets. The colony reverse is creamish yellow to yellow with a dark brown center. On MEA the colonies are fasciculate to synnematous (yellow synnemata), the conidia are green and colonies have a distinct yellow reverse. Sporulation is rather poor. On YES agar there is no sporulation, reverse colour yellow to orange. On CREA rather good growth but inhibited and strong acid production. Conidiophores two-stage branched (terverticillate) with all elements adpressed, stipes rough-walled. Conidia finely rough and globose to subglobose.

Important toxic metabolites: roquefortine C, terrestric acid.

HABITAT:
The primary habitat is cereals, especially barley.

Pl. 95. Colonies after one week. a,b. *P. hirsutum*, a. on Cz, b. on MEA; c,d. P. allii, c. on Cz, d. on MEA; e,f. *P. hordei*, e. on Cz, f. on MEA.

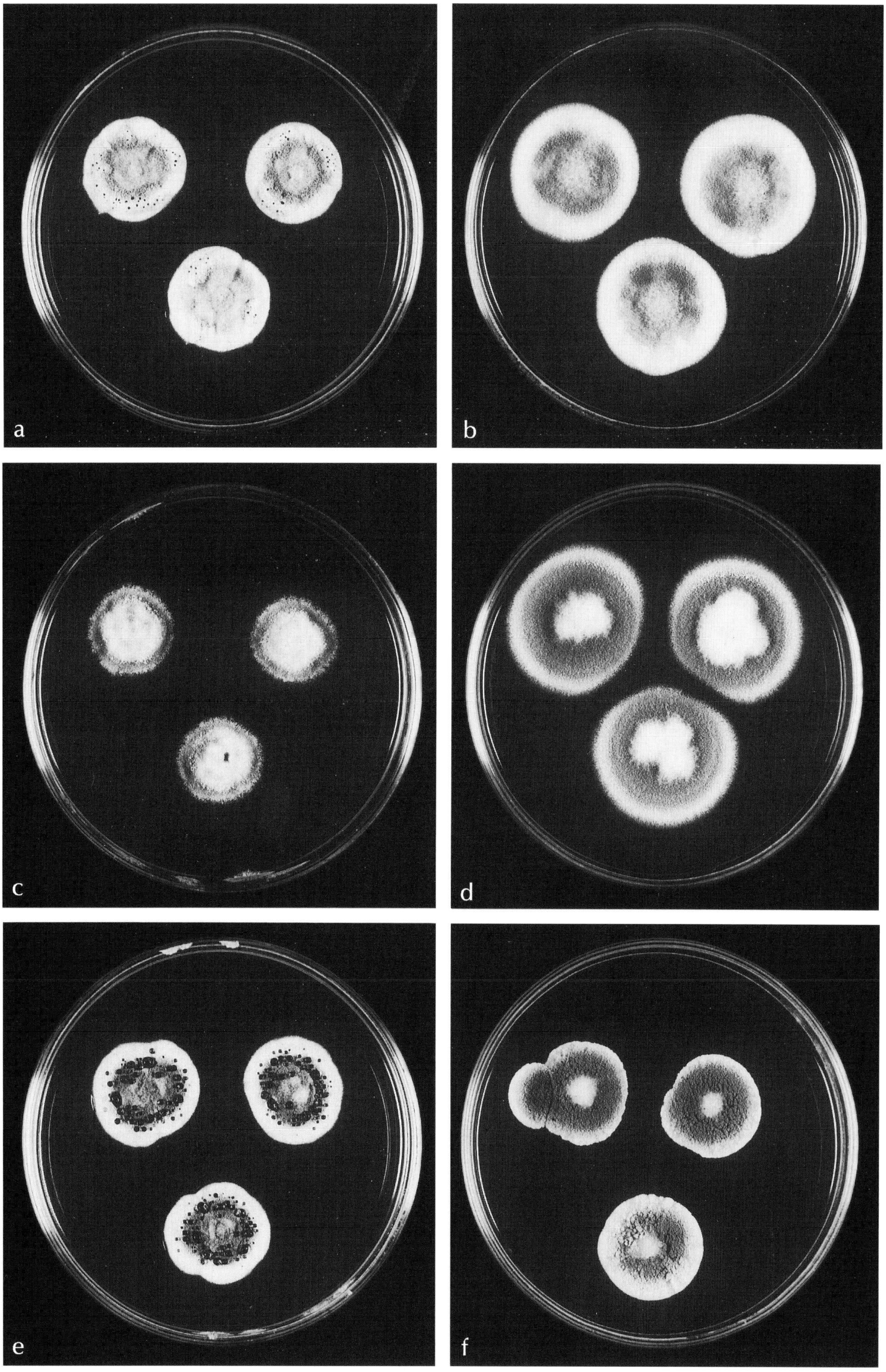
a
b
c
d
e
f

Penicillium italicum Wehmer

Colonies on Czapek agar at 25°C restricted, attaining a diameter of 2.0-2.5 cm within 14 days; more spreading on MEA. Colour grey-green. Odour aromatic, suggesting perfume. Exudate mostly absent, collecting in hyaline drops when present. Reverse uncoloured to yellow-brown. Conidiophores sometimes mononematous, more commonly loosely synnematous, especially in the marginal areas; the synnemata sometimes developing from bundles of the vegetative hyphae submerged in the agar; conidiophores smooth-walled, hyaline, tervertcillate, usually with the branches appressed. Stipes 100-250 x 3.5-5.0 µm. Metulae more or less cylindrical, smooth-walled, 15-20 x 3.5-4.0 µm, bearing 3 to 6 phialides each. Phialides slender, cylindrical with short but distinct necks, 8-15 x 2.0-5.0 µm. Conidia cylindrical, but often becoming ellipsoidal to subglobose, 4.0-5.0 x 2.5-3.5 µm, greenish, smooth-walled. Colonies on MEA growing faster, 5-6 cm in diameter within 14 days, usually less fasciculate.

HABITAT:
This species causes fruitrot in citrus and can also be found in fruit-juices and soil.

Penicillium ulaiense Hsieh, Su & Tzean

Colonies on Czapek agar and CYA at 25°C growing restrictedly producing greyish green conidia with a fasciculate to strongly synnematous colony surface that appears crustose because of abundant conidium production, occasionally with clear exudate droplets. Synnemata with white stipes and green sporulating capitula. The colony reverse is pale or beige. On MEA the conidia are greyish green and colonies are fasciculate to synnematous with a pale to light yellow reverse. On YES agar there is strong sporulation, reverse colour creamish yellow occasionally with a brown center. On CREA weak growth and no acid production. Conidiophores one-stage branched (asymmetrically biverticillate) with metulae adpressed, stipes smooth-walled. Conidia large, smooth and cylindrical to ellipsoidal.

Important metabolites: deoxybrevianamide E.

HABITAT:
The primary habitat is citrus fruits treated with fungicides (Holmes *et al.*, 1994)

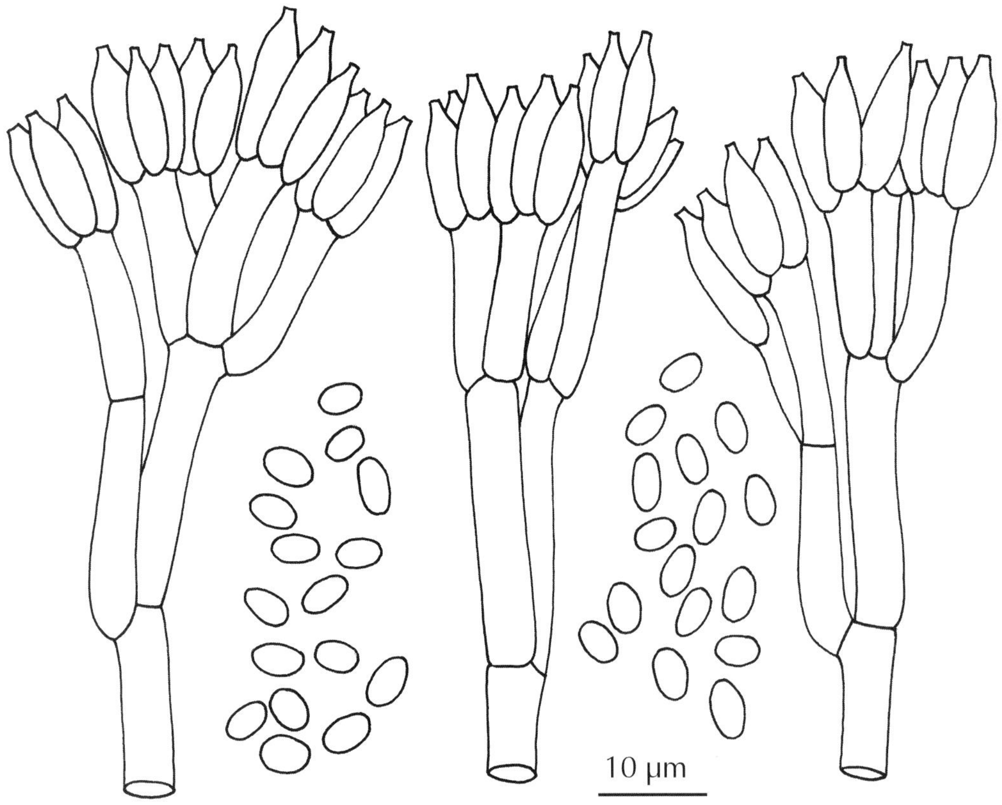

Fig. 94. *Penicillium italicum.* Conidiophores and conidia.
Pl. 96. *Penicillium italicum.* Colonies after one week, a. on Cz, b. on MEA; c,d. conidiophores, x 1200; e. conidia, x 1300.

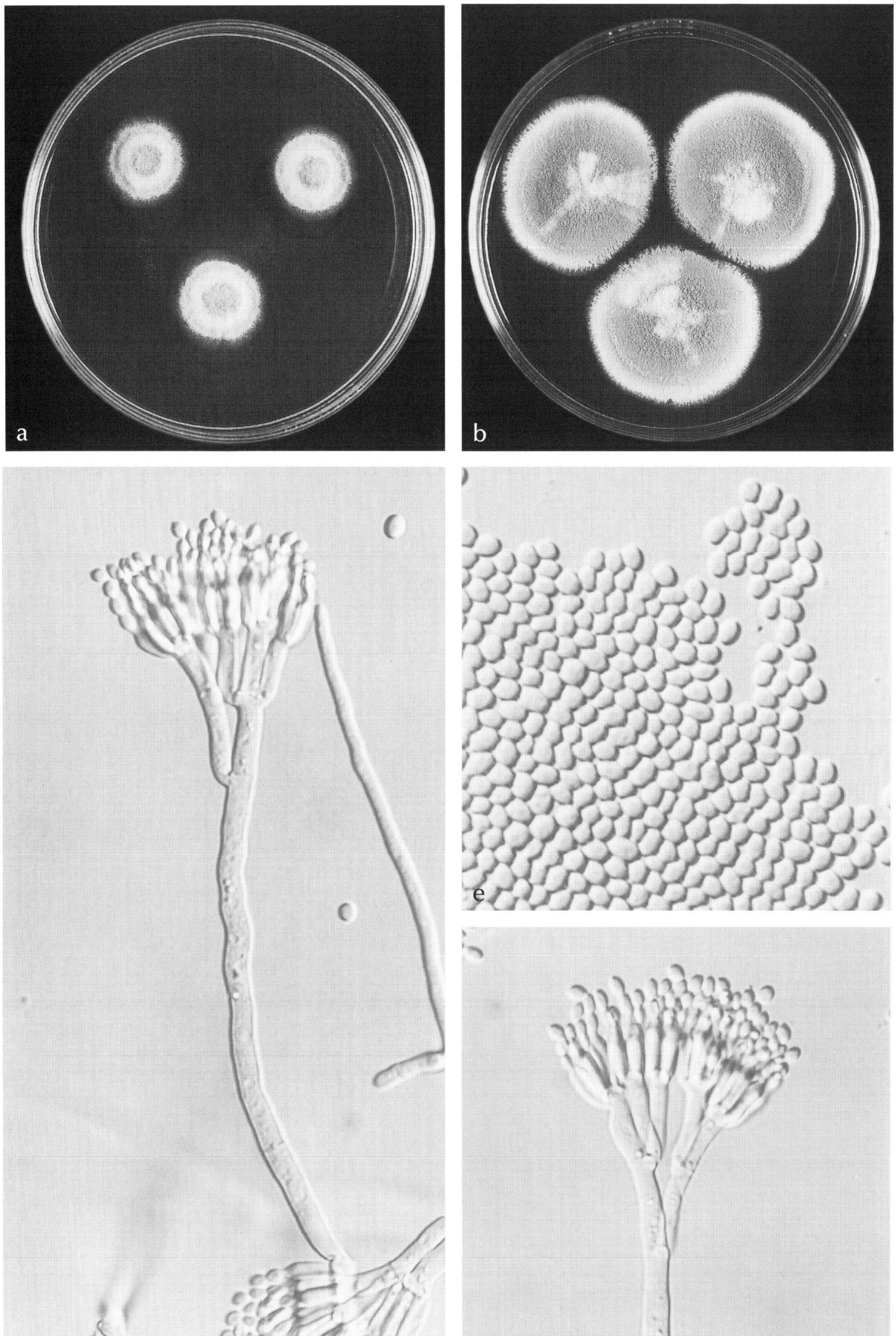
a
b
c
e
d

Penicillium nalgiovense Laxa

Colonies on Czapek agar at 25°C attaining a diameter of 2.5-3 cm within 7 days, white sometimes becoming pale green with age, consisting of a dense felt of conidiophores. Reverse usually yellow. Conidiophores ter- to quaterverticillate or more-stage branched, smooth-walled, hyaline. Metulae 7-15 x 2.5-3 µm, terminating in a whorl of 2-6 phialides. Phialides flask-shaped, often with an inconspicuous but wide neck, 8-10 x 2-2.5 µm. Conidia globose to subglobose, smooth, hyaline, 3-4 µm in diam.
Colonies on MEA as on Czapek agar but thinner.

Important (toxic) metabolites: penicillin.

HABITAT: food
P. nalgiovense was originally described from cheese, but the species has been recently isolated from salami-sausages. It is now used as a starter culture for the fermentation of certain types of salami-sausages. These starter cultures and isolates from meat products show a morphological resemblance to *P. chrysogenum*.

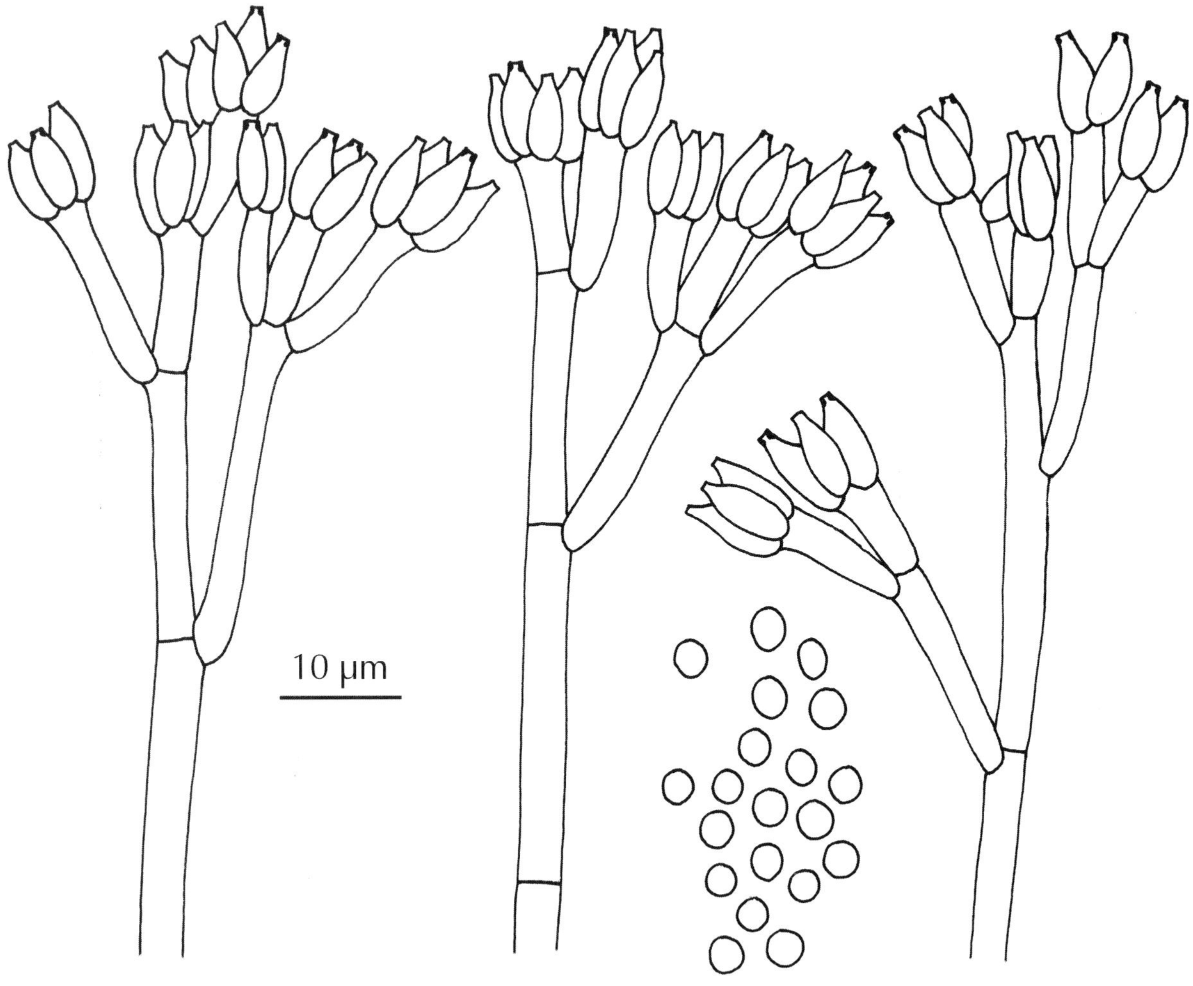

Fig. 95. *Penicillium nalgiovense*. Conidiophores and conidia.
Pl. 97. *Penicillium nalgiovense*. Colonies after one week, a. on Cz, b. on MEA; c,d. conidiophores, c. x 1000, d. x 1250.

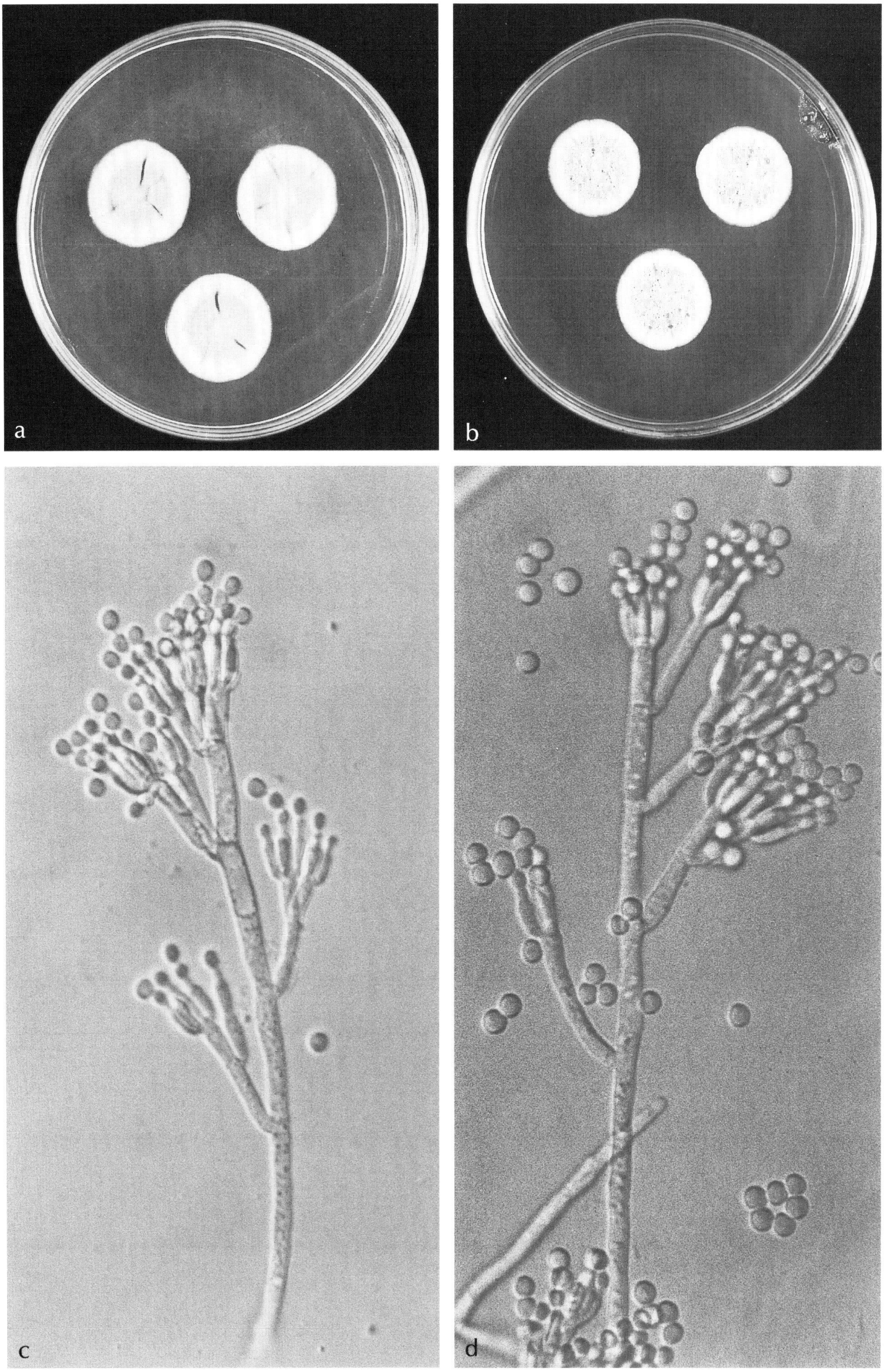

Penicillium olsonii Bain. & Sartory

Colonies on Czapek agar and CYA at 25°C growing rather fast, grey-green to greenish-glaucus. On Czapek colonies attaining a diameter of 2 cm after 7 days, velvety and occasionally with clear exudate droplets. Reverse cream yellow to pale brown. On MEA colonies attaining a diameter of 2.5-3.5 cm after 7 days, velvety, the same colour as on Czapek. Reverse pale. On YES agar there is strong sporulation, reverse colour creamish yellow. On CREA weak growth and no acid production.
Conidiophores two-stage branched (terverticillate), often multiramulate, branches and metulae adpressed, penicillus compact and arranged towards the apex, tall, 800-1000 µm or up to 2000 µm long; stipes broad typically 4-6 µm, smooth-walled. Metulae in verticals of 3-5, 10-12 x 3-4 µm. Phialides flask-shaped with a short neck, 9-10 x 2-2.5 µm. Conidia subglobose to ellipsoidal, 3-4 x 2.5-3.0 µm, almost smooth to finely roughened.

Important metabolites: verrucolone, 2-(4-hydroxyphenyl)-2-oxoacetaldehydeoxime, bis(2-ethylhexyl)phthalate.

HABITAT: food, indoor
The primary habitats are tomatoes, mould ripened salami, and beans. Common in soils of ornamental plants e.g. in greenhouses. But also occurring indoors in dwellings.

Note:
The penicillus superficially resembles an *Aspergillus* head in the stereomicroscope. This species resembles *P. brevicompactum.*

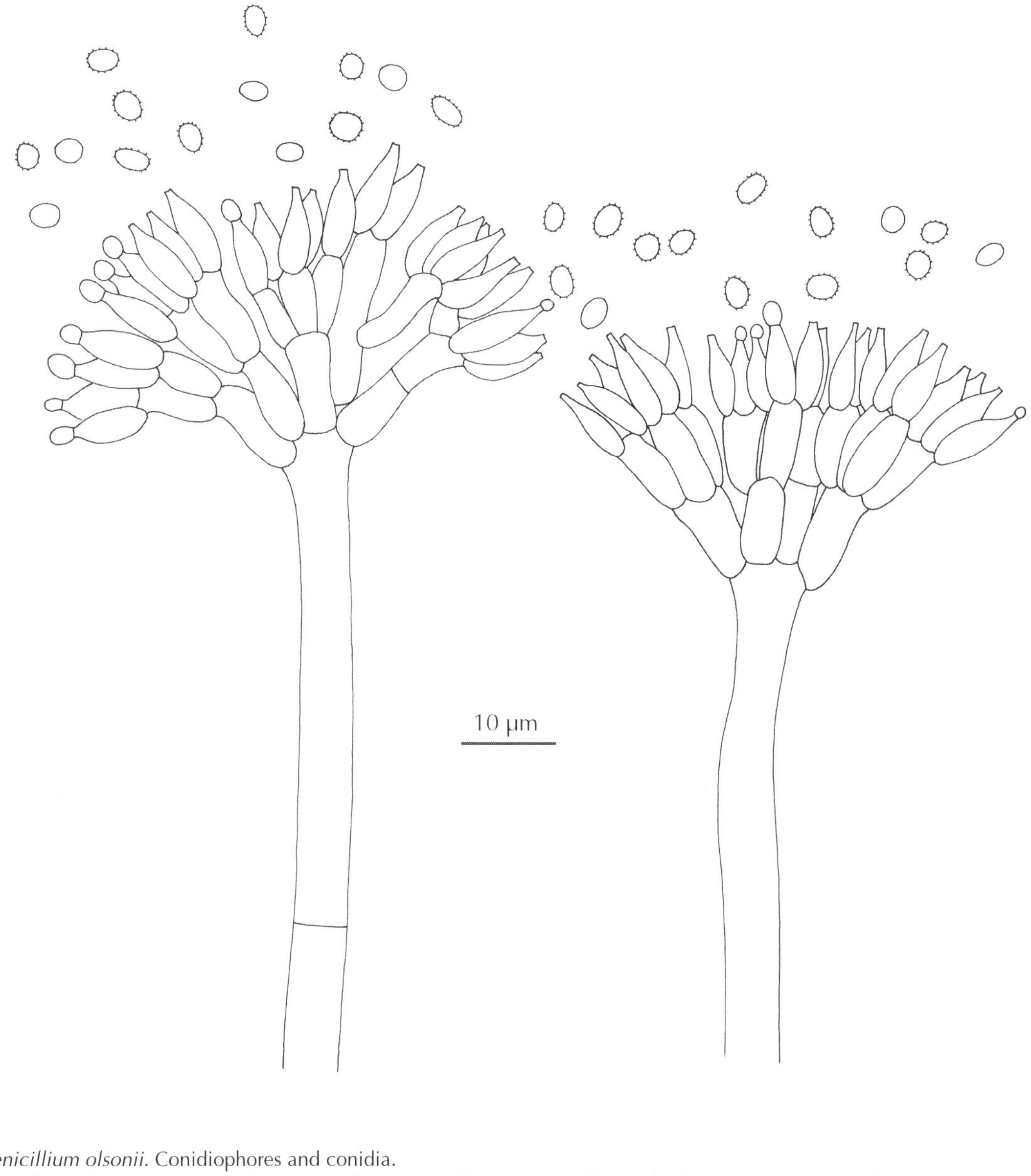

Fig. 96. *Penicillium olsonii.* Conidiophores and conidia.
Pl. 98. *Penicillium olsonii* . Colonies after one week, a. on Cz, b. on MEA; c-d. conidiophores, x 950; e. conidia, x 2000

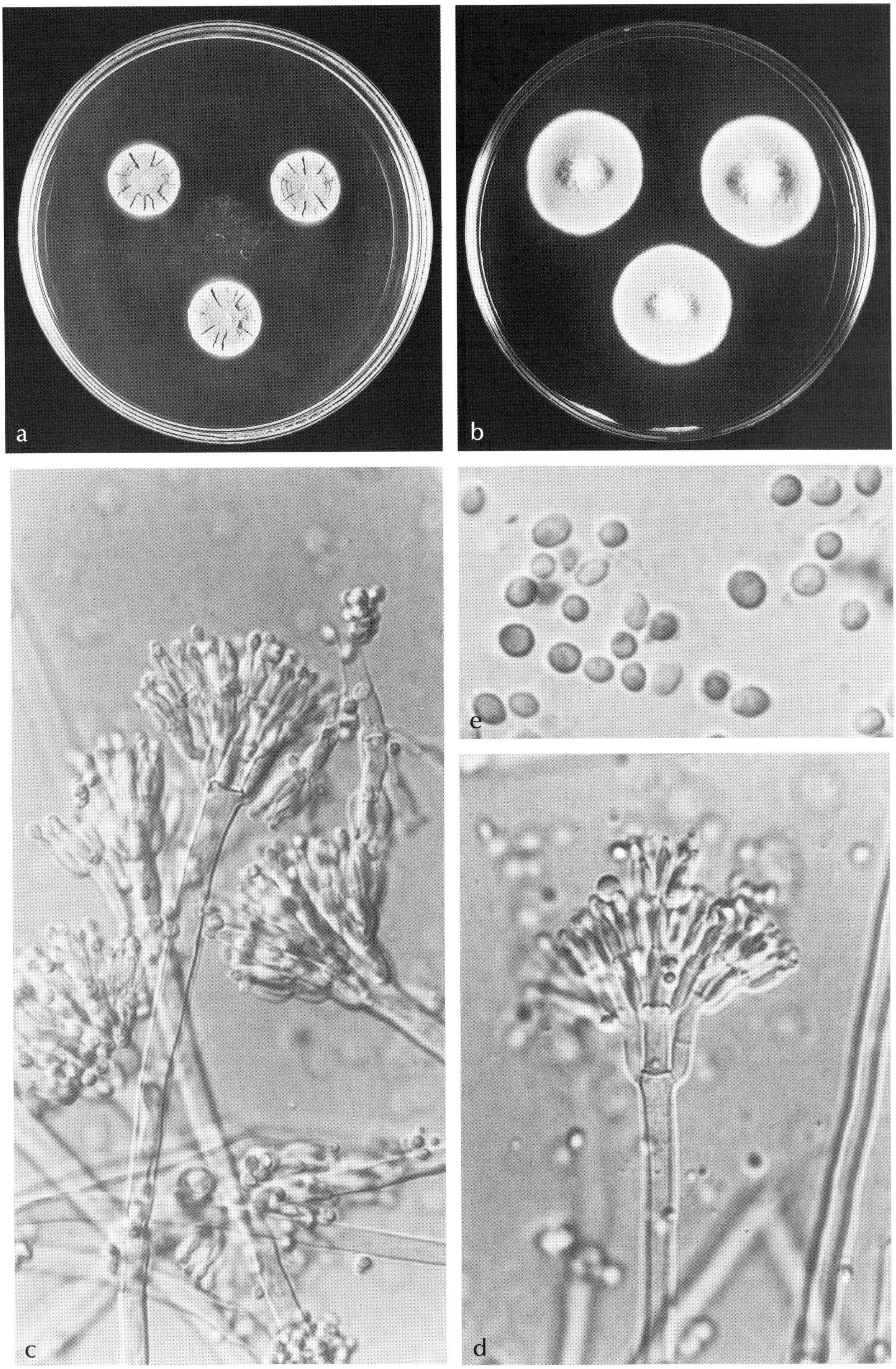
a
b
e
c
d

Penicillum oxalicum Currie & Thom

Colonies on Czapek agar and CYA at 25°C growing fast attaining a diameter of 3.5-6 cm in 10 days, producing dark greyish green conidia with a velvety colony surface that appears crustose because of abundant conidium production, occasionally with clear exudate droplets and a shiny, silky appearance under low power magnification. The colony reverse is cream yellow to yellow, curry or pinkish. On MEA the conidia are green and colonies are velutinous and have a yellow reverse. On YES agar there is strong sporulation, reverse colour distinct yellow. On CREA weak growth and variable acid production. Conidiophores mostly one-stage branched (asymetrically biverticillate), sometimes with additional branch, stipes smooth-walled; metulae 2-4, adpressed, 15-25 (30) µm long. Phialides cylindrical to lanceolate with a narrow tip, 10-15(20) µm long. Conidia large, ellipsoidal, 3.5-5(7) µm, smooth-walled (sometimes finely roughened).
On CYA at 37°C growth is fast.

Important toxic metabolites: secalonic acid D & F, roquefortine C

Secondary metabolites with unknown toxicity: meleagrin, oxaline, anthglutin, oxalicine, oxalic acid.

HABITAT: food
The primary habitat is oil seeds and cereals, especially maize, yams, greenhouse cucumbers and other tropical and subtropical products, soil.

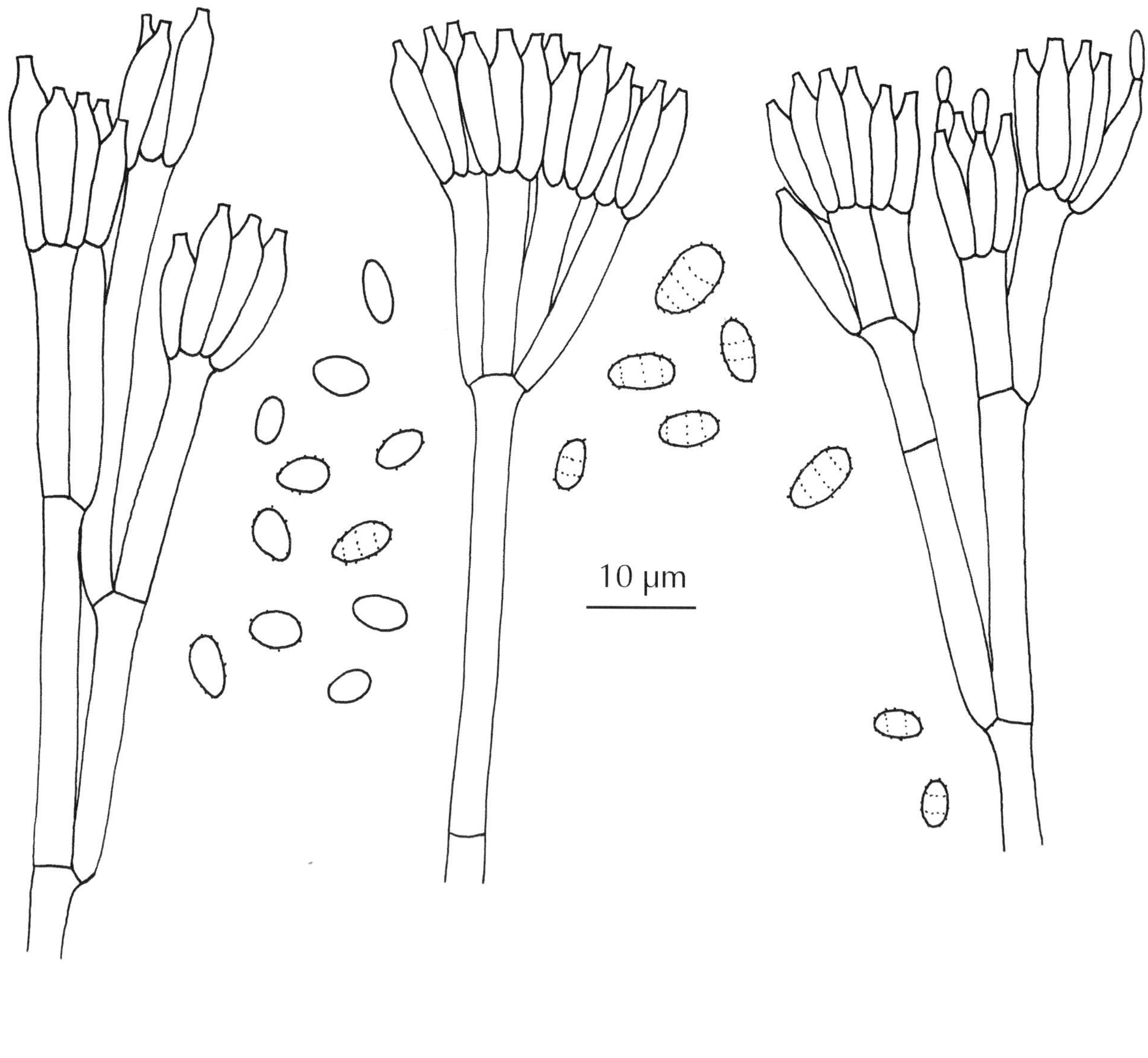

Fig. 97. *Penicillium oxalicum*. Conidophores and conidia.
Pl. 99. *Penicillium oxalicum*. Colonies after one week, a. on Cz, b. on MEA; c-d, f. conidiophores, x 950; e. conidia, x 2300.

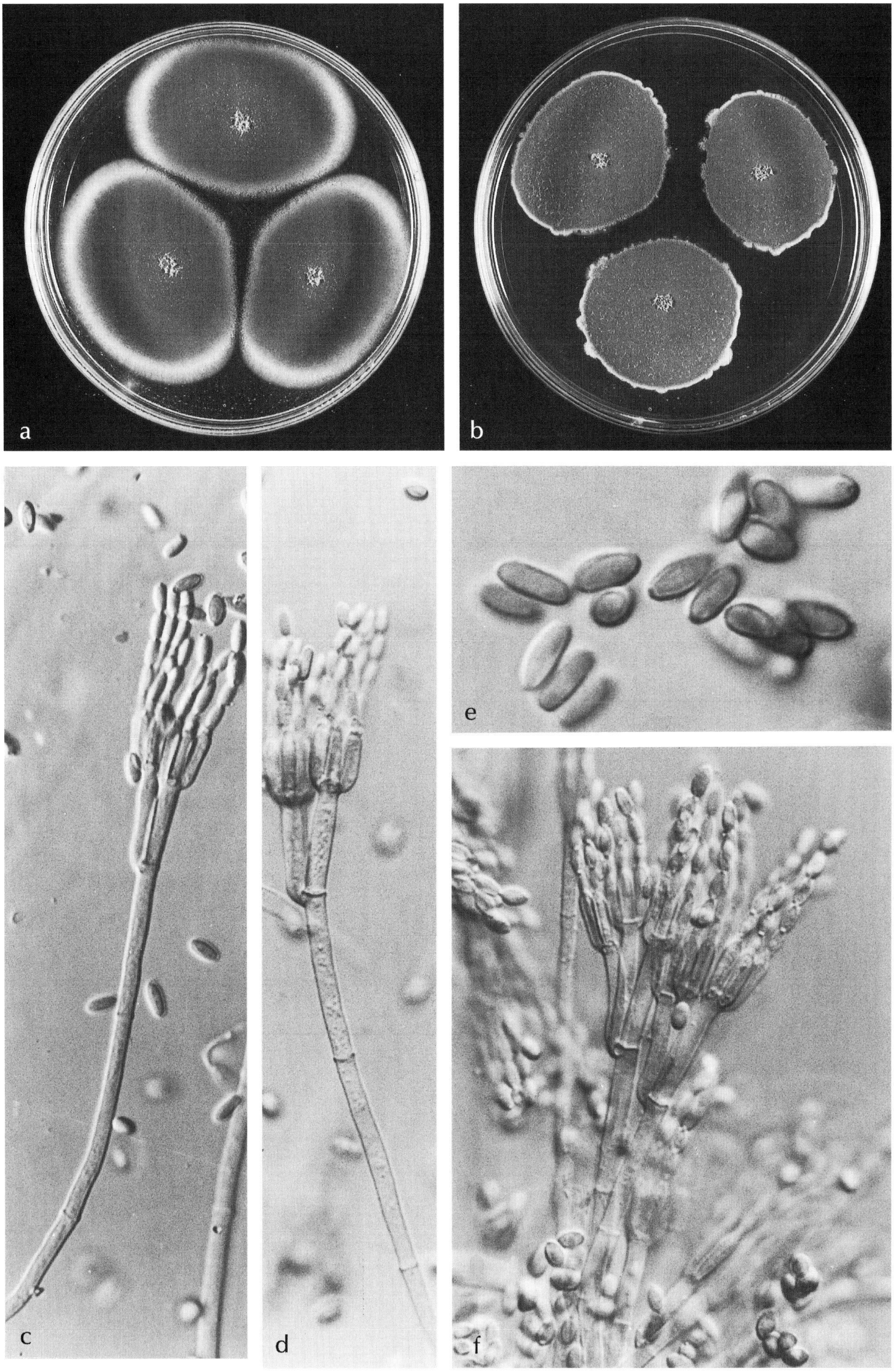
a
b
c
d
e
f

Penicillium polonicum Westling

Colonies on Czapek agar and CYA at 25°C growing fast producing (dark) blue green conidia with a granular colony surface, often with exudate droplets. The colony reverse is cream yellow, orange to pinkish brown with the colour often diffusing into the agar medium. On MEA the conidia are blue green with a strong blue element and colonies have a distinct yellow reverse, often with the yellow colour diffusing into the medium. On YES agar there is strong sporulation, reverse colour distinct yellow. On CREA rather good growth and strong acid production. Unlike species growing really well on CREA, *P. polonicum* does not produce base after acid production. Conidiophores two-stage branched (terverticillate) with all elements adpressed, stipes rough-walled. Conidia smooth and globose to subglobose, 3-4 μm in diam.

Important toxic metabolites: nephrotoxic glycopeptides, penicillic acid

Secondary metabolites with unknown toxicity: cyclopenin, cyclopenol, dehydrocyclopeptin, cyclopeptin, viridicatol, 3-methoxyviridicatin, verrucofortine (=verrucosine), puberuline, rugulosuvine, leucyltryptophanyldiketopiperazine, aspterric acid, anacine,methyl-4-[2-(2R)-hydroxyl-3-butynyl-oxy]benzoate, pseurotins, γ-elemene.

HABITAT: food, indoor
The primary habitat is cereals and meat products.

Penicillium melanoconidium (Frisvad) Frisvad & Samson

Colonies on Czapek agar and CYA at 25°C growing restrictedly producing dark green conidia with a velvety to weakly granular colony surface, with clear exudate droplets. The colony reverse is cream yellow to yellow or curry. On MEA the conidia are green and colonies have a yellow reverse. On YES agar there is strong sporulation, reverse colour distinct yellow. On CREA weak growth but strong acid production. Conidiophores two-stage branched (terverticillate) with all elements adpressed, stipes rough-walled. Conidia smooth and globose to subglobose.

Important toxic metabolites: penitrem A, verrucosidin, penicillic acid, roquefortine C, (xanthomegnin, viomellein, vioxanthin)

Secondary metabolites with unknown toxicity: meleagrin, sclerotigenin.

HABITAT:
The primary habitat is cereals.

Pl. 100. *Penicillium polonicum.* Colonies after one week, a. on Cz, b. on MEA; c-e. conidiophores and conidia, x 950.

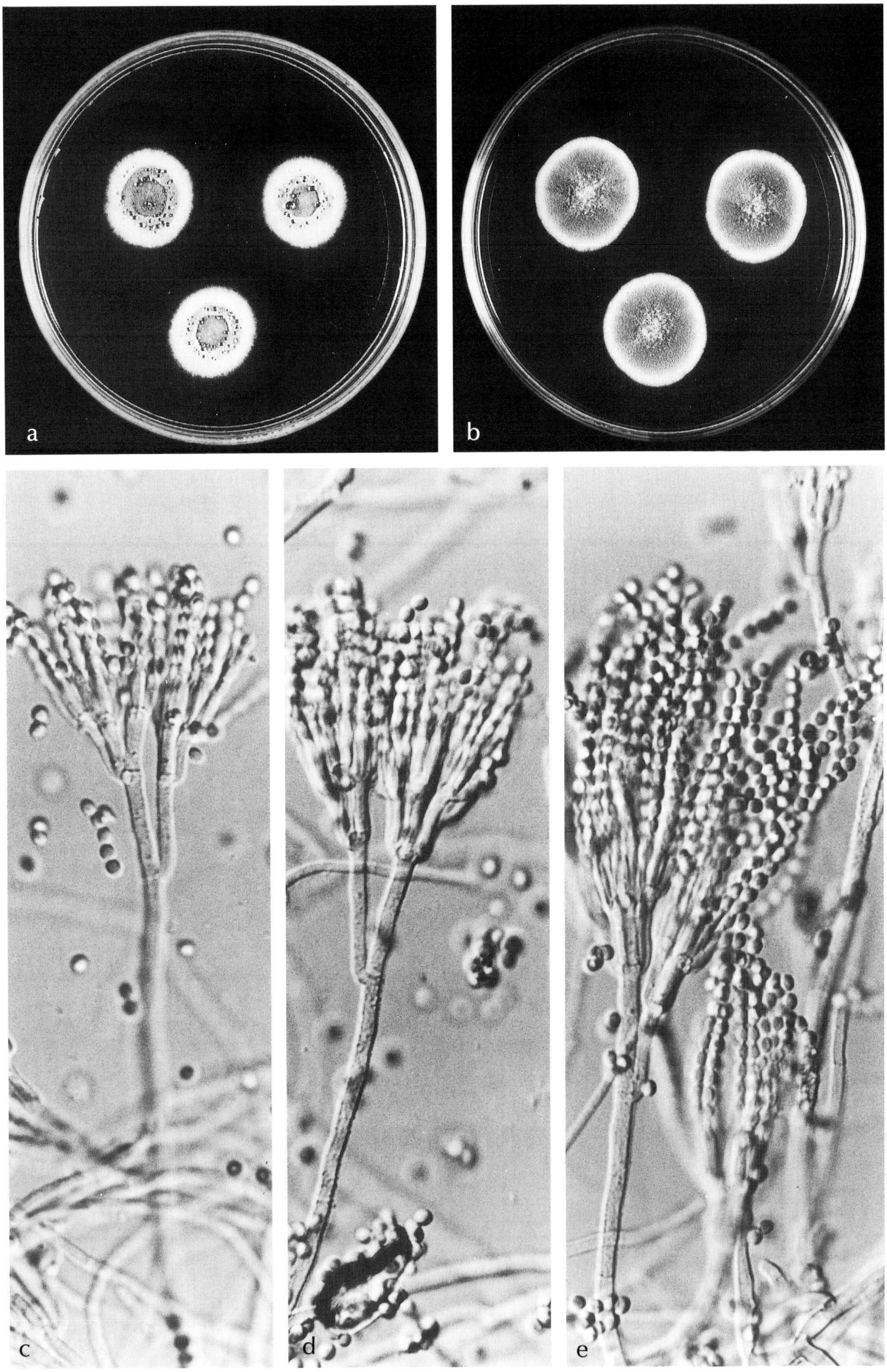
a
b
c
d
e

Penicillium roqueforti Thom

Colonies on Czapek agar and MEA at 25°C growing rapidly, attaining a diameter of 4-5 cm within 14 days, consisting of a dense felt of erect conidiophores, velvety; in other cultures becoming more lanose with production of aerial vegetative mycelium. Colour blue-green, later becoming darker. Exudate in fresh isolates often present as hyaline droplets. Odour mostly absent or not pronounced. Reverse greenish, often changing to darker shades of green to black. Conidiophores ter- to quaterverticillate, 100-200 x 4.0-6.5 µm, the stipes typically ornamented with conspicuous warts, some strains less roughened at the apex of conidiophore stipes, but usually ornamented at the base. Metulae 10-15 x 3-4.5 µm, rough-walled, giving rise to clusters of 4-7 phialides. Phialides flask-shaped with a short neck, 8-12 x 3.0-3.5 µm. Conidia in loose columns, globose to subglobose, greenish, smooth-walled, mostly 4-6 µm, occasionally up to 8 µm.

Reduced, often smooth-walled conidiogenous structures occur submerged in the agar. Conidia are mostly larger than those produced by aerial conidiophores. Sclerotia can be formed in old cultures.
Good growth, but no acid production on CREA.

Important (toxic) metabolites: roquefortine C, isofumigaclavine A & B, PR-toxin, mycophenolic acid.

HABITAT: food
Widely used in fermentation of various types of blue cheeses, but also common as spoilage agent in refrigerated stored foods, meat and meat products, rye-breads, silage.

NOTE:
Common on substrates with high levels of acetic or propionic acid or high concentration of CO_2 or ethylacetate or low levels of oxygen.

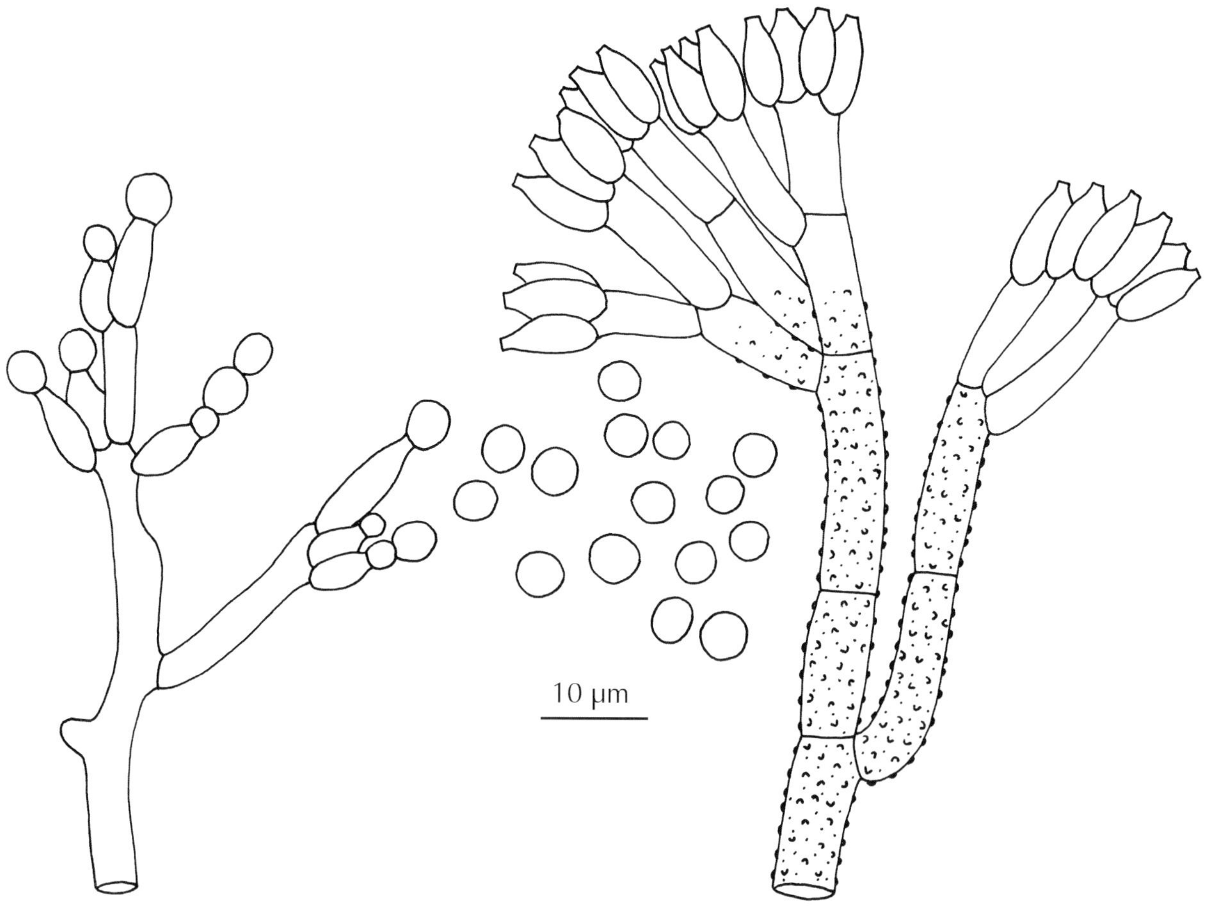

Fig. 97. *Penicillium roqueforti.* Conidiophores and conidia.
Pl. 101. *Penicillium roqueforti.* a. Colonies after one week on MEA; b. piece of Roquefort cheese in which the mould is visible as dark coloured (green) veins in the cheese; c,d. conidiophores and conidia, x 1000; e. conidial structures in submerged agar, x 400.

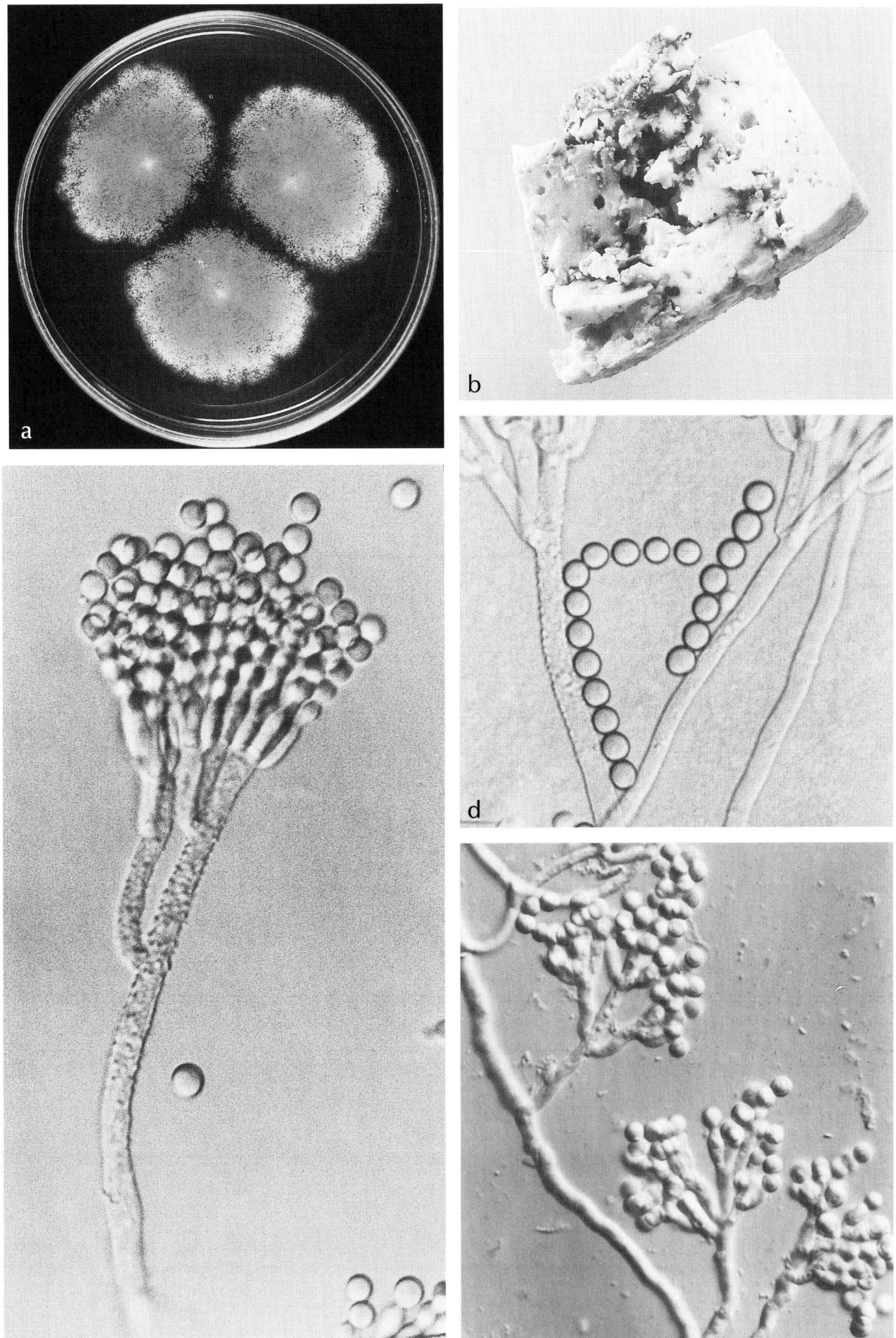
a
b
c
d
e

Penicillium carneum (Frisvad) Frisvad

Colonies on Czapek agar and CYA at 25°C growing rapidly producing dull green conidia with a velvety colony surface, with few small clear exudate droplets. The colony reverse not sulcate and coloured pale to beige. On MEA the conidia are also dull green with a pale reverse. On YES agar strong sporulation, reverse coloured beige to avellaneous. On CREA strong growth and no acid production, occasionally weak acid production under the colony. Conidiophores two-stage branched (terverticillate) with all elements adpressed, stipes rough to warted. Conidia smooth, relatively large and globose.

Important toxic metabolites: penitrem A, patulin, roquefortine C, isofumigaclavine A & B, occasionally penicillic acid

Secondary metabolites with unknown toxicity: cyclopaldic acid, mycophenolic acid, geosmin (Frisvad & Filtenborg, 1989; Boysen *et al.*, 1996).

HABITAT: food
Salami, rye bread, silage.

Penicillium paneum Frisvad

Colonies on Czapek agar and CYA at 25°C growing rapidly producing dull green to dark blue green conidia with a velvety colony surface, with few small clear exudate droplets. The colony reverse not sulcate and coloured cream to beige. On MEA the conidia are also dull green with a pale reverse. On YES agar strong sporulation, reverse coloured creamish yellow to yellow or pink. On CREA strong growth and no acid production, occasionally weak acid production under the colony though. Conidiophores two-stage branched (terverticillate) with all elements adpressed, stipes rough to warted. Conidia smooth, relatively large and globose.

Important toxic metabolites: patulin, roquefortine C, botryodiploidin

Secondary metabolites with unknown toxicity: marcfortines A, B and C (Boysen *et al.*, 1996).

HABITAT: food
Rye bread, silage.

Pl. 102. Colonies after one week a,b. *Penicillium roqueforti*, a. on Cz, b. on MEA; c,d. *P. paneum*; c. on Cz, d. on MEA; e,f. *P. carneum*, e on Cz, f. on MEA.

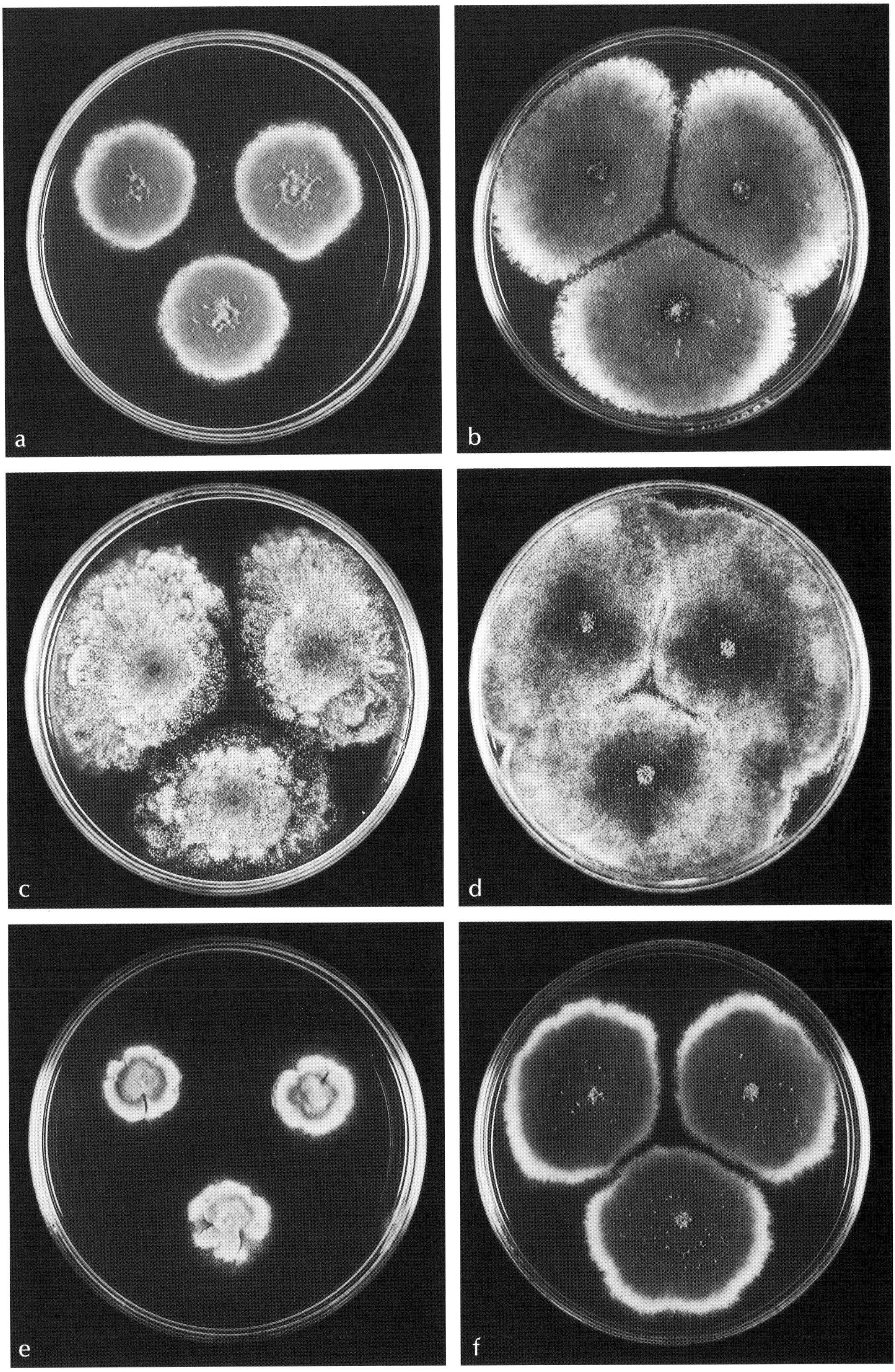
a
b
c
d
e
f

Penicillium rugulosum Thom

Colonies on Czapek agar and MEA at 25°C growing restrictedly, attaining a diameter of about 1 cm within 7 days, usually consisting of a compact and dense felt of conidiophores, velvety, yellowish-green to dark green. Conidiophores often biverticillate. Metulae in whorls of 5 to 7, 9-12 x 2-2.5 µm. Phialides cylindrical with a conspicuous neck or sometmes lanceolate, 10-12 x 1.5-2.5 µm. Conidia ellipsoidal, conspicuously roughened, 3-3.5 x 2.5-3 µm.

Important (toxic) metabolites: rugulosin.

HABITAT: food, indoor
Soil, but also isolated from meat, corn.

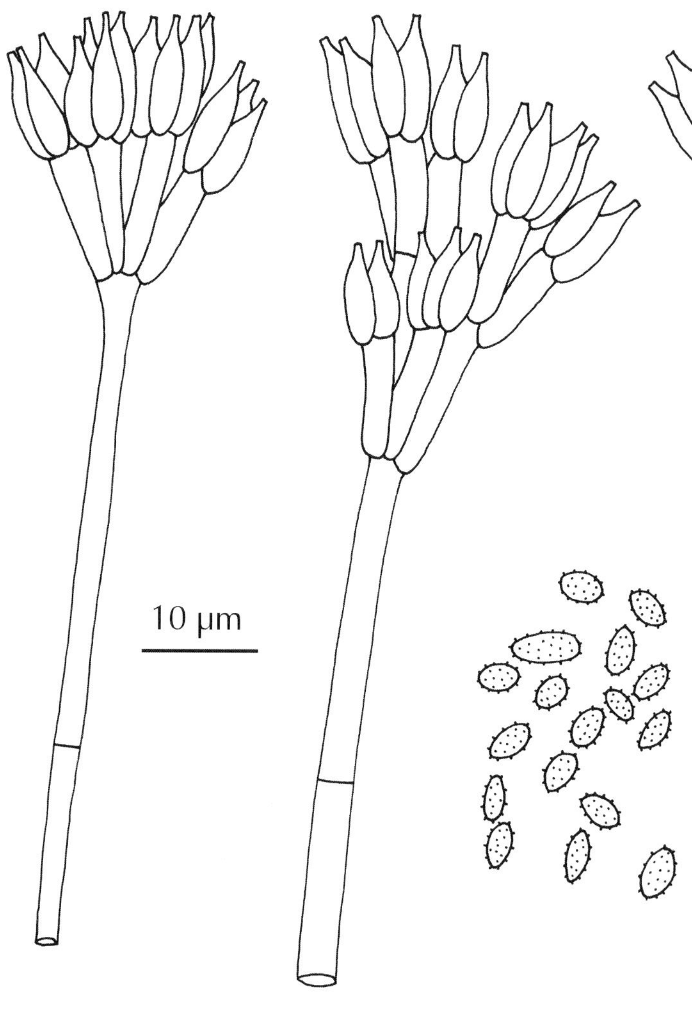

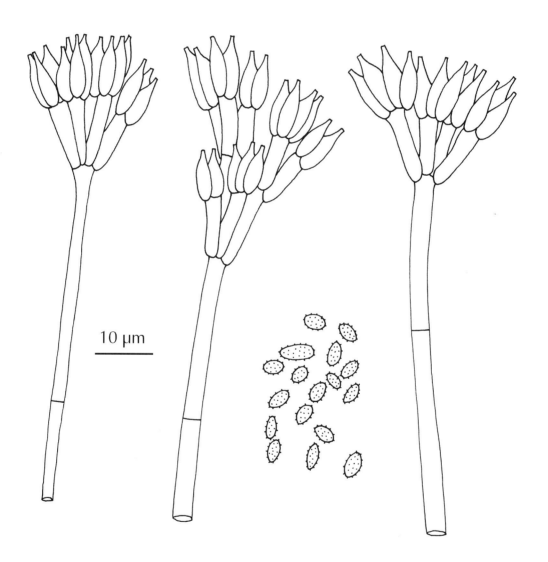

Fig. 99. *Penicillium rugulosum.* Conidiophores and conidia.
Pl. 103. *Penicillium rugulosum.* Colonies after one week, a. on Cz, b. on MEA; c-e, conidiophores and conidia, x 1250.

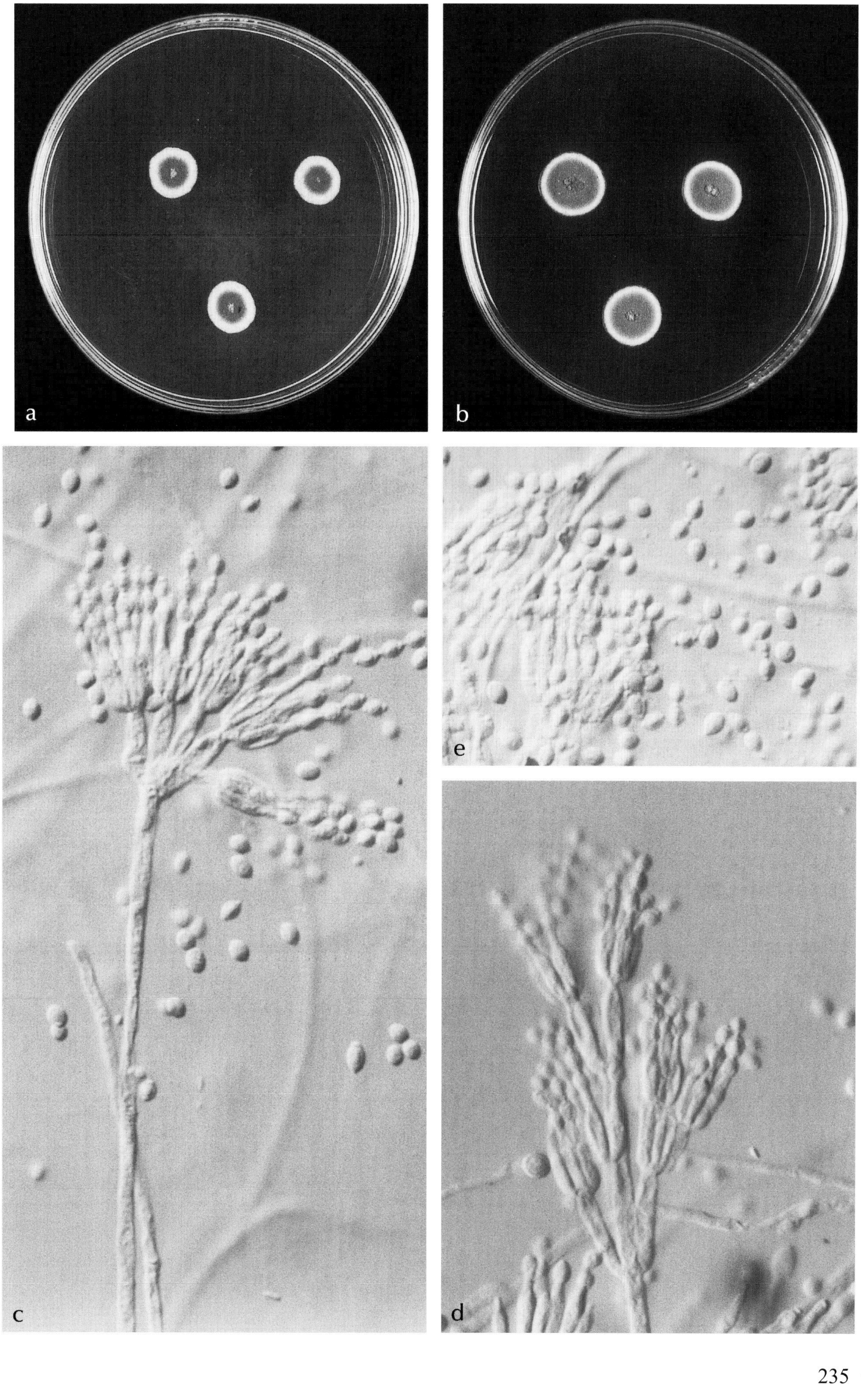
a
b
c
e
d

Penicillium solitum Westling

Colonies on Czapek agar and CYA at 25°C growing rather restrictedly producing dark green to dark blueish green conidia and velutinous to slightly granular colonies, with clear to dark brown exudate droplets. The colony reverse is coloured creamish dull yellow to brown yellow often with a brown centre. On MEA the conidia are green, colonies have a pale or sun yellow reverse, colonies velutinous. On YES agar good sporulation with green conidia, reverse yellowish orange. On CREA strong growth and strong acid production and often subsequent base production. Conidiophores two-stage branched (terverticillate) with all elements adpressed, stipes rough-walled. rough-walled. Conidia smooth and subglobose, 3.5-5 µm in diam.

Important metabolites: cyclopenin, cyclopenol, dehydrocyclopeptin, cyclopeptin, viridicatol, viridicatin, compactin, dehydrocompactin, solistatin (Frisvad and Filtenborg, 1989; Sørensen *et al.*, 1999).

HABITAT: food
Contaminant on hard cheeses, liver paté and other meat products. Common on air dried fish and meat products from Faroe Island. Can be weakly pathogenic to apples (Frisvad, 1981; Pitt *et al.*, 1991).

Pl. 104. *Penicillium solitum*. Colonies after one week, a. on Cz, b. on MEA; c,d. conidiophores and conidia, x 950.

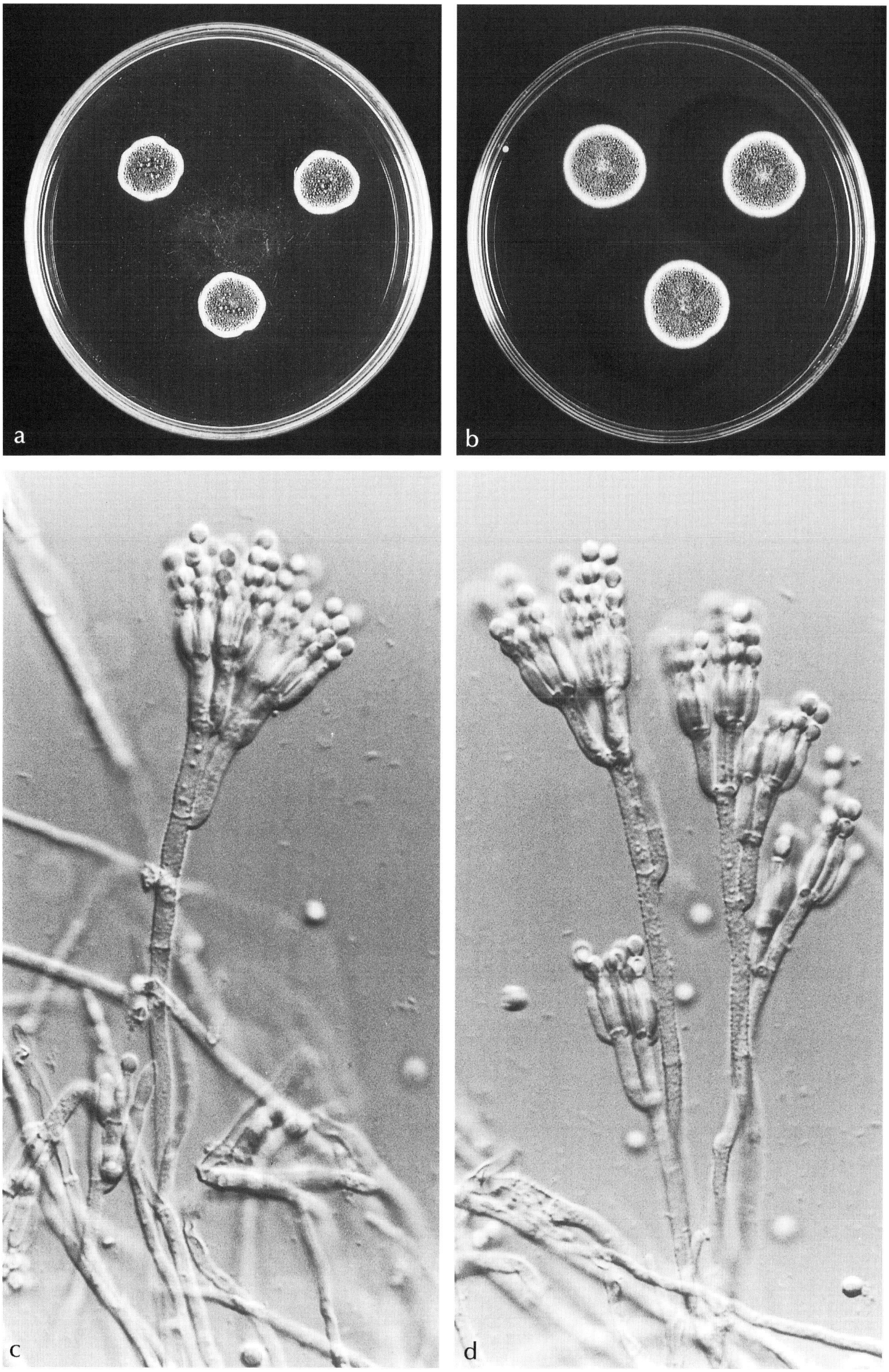
a
b
c
d

Penicillium variabile Sopp

Colonies on Czapek agar at 25°C growing restrictedly, attaining a diameter of 1-1.5 cm within 7 days, consisting of a yellow mycelial felt with conidiophores often concentrated in the central part. Reverse yellow to orange-brown. Conidiophores often arising from aerial hyphae, biverticillate, consisting of a smooth-walled stipe, terminal in a whorl of 5-7 metulae. Metulae 7.5-12 x 2.5-3.0 µm. Phialides lanceolate, 10-12 x 2.0-2.5 µm. Conidia ellipsoidal to fusiform, smooth to irregular roughened, often with striations, variable in size, but mostly 3.0-3.5 x 2-2.5 µm.
Colonies on MEA growing somewhat faster and with less yellow mycelium.

Important (toxic) metabolites: rugulosin.

HABITAT: food, indoor
Isolated from fruits, cereals, rice, corn, fruit-juices, groundnuts.

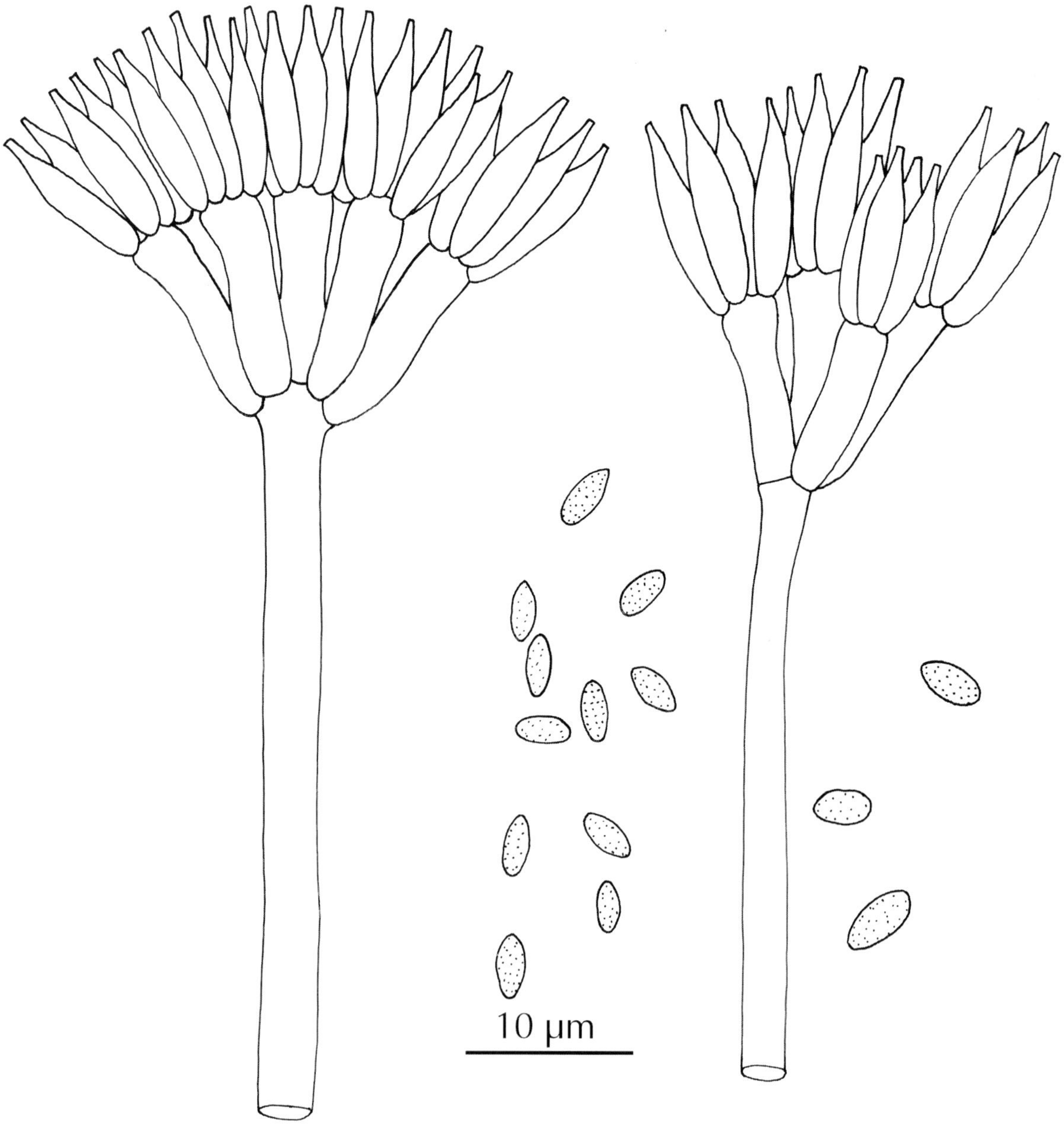

Fig. 100. *Penicillium variabile*. Conidiophores and conidia.
Pl. 105. *Penicillium variabile*. Colonies after one week, a. on Cz, b. on MEA, c,d. conidiophores, c. x 1250, d. x 1800; e. conidia, x 2000.

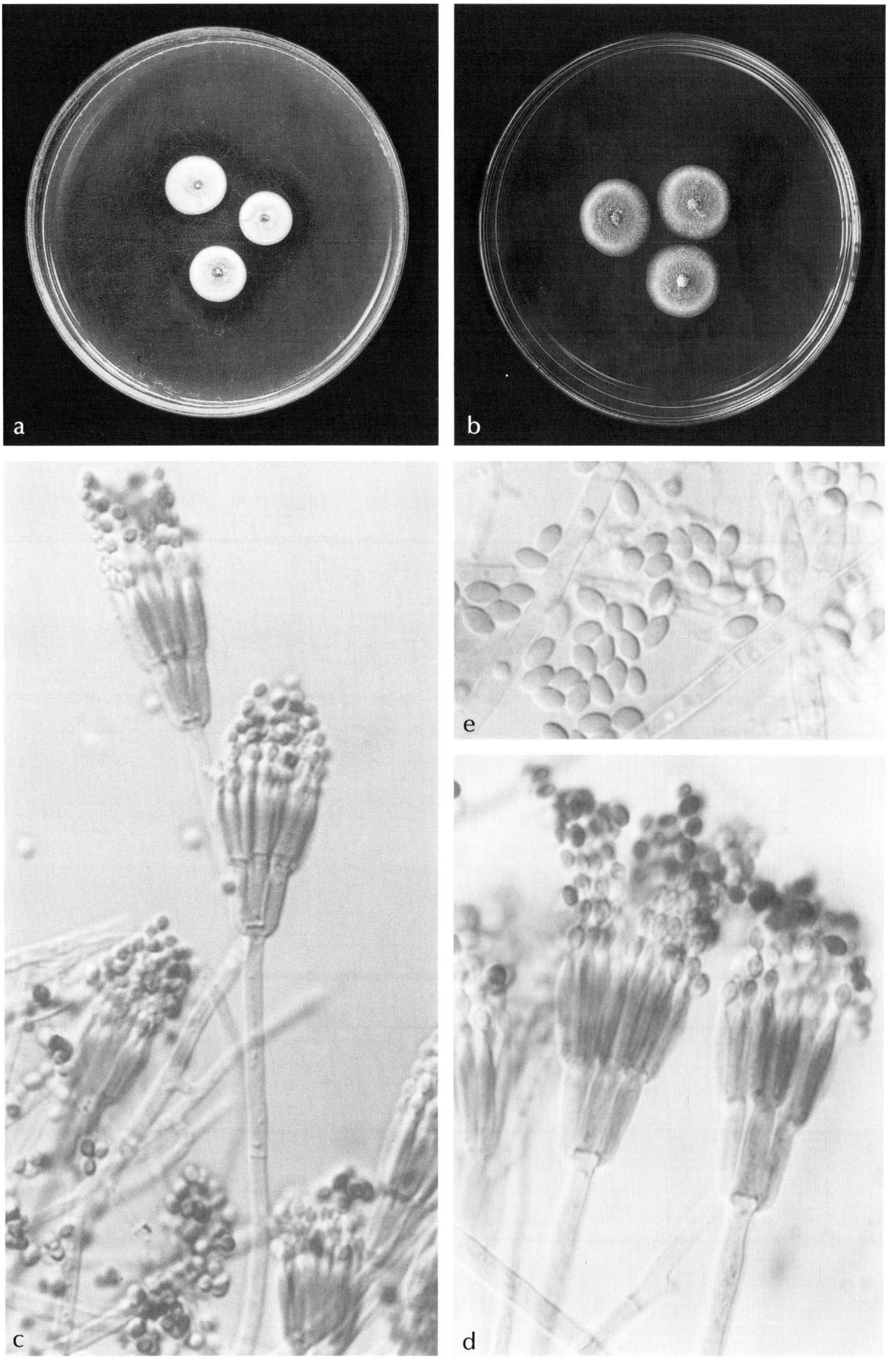
a
b
c
e
d

Penicillium verrucosum Dierckx

Colonies on Czapek agar and CYA at 25°C growing restrictedly, attaining a diameter of less than 1 cm within 7 days (on Czapek), producing green conidia with a velvety to weakly granular colony surface, with clear exudate droplets. The colony reverse is cream yellow often with a brown center. On MEA colonies are granular to fasciculate and the conidia are green. On YES agar there is often no or weak sporulation, and the colonies have a unique terracotta to violet brown reverse. On CREA weak growth and no acid production. Conidiophores two-stage branched (terverticillate) with all elements adpressed, stipes rough-walled. Conidia smooth-walled, globose to subglobose, 3-3.5 µm in diam.

Important toxic metabolites: ochratoxin A, citrinin secondary metabolites with unknown toxicity: verrucolone (= arabenoic acid) and verrucines (Frisvad and Filtenborg, 1989; Larsen *et al.*, 1998).

HABITAT: food
The primary habitat is cereals in temperate regions of the world, occasionally found on cheese.

Penicillium nordicum Dragoni & Cantoni ex Ramírez

Colonies on Czapek agar and CYA at 25°C growing restrictedly producing white or green to grey green conidia with a velvety to weakly granular colony surface, with clear exudate droplets. The colony reverse is cream yellow often with a brown center. On MEA the conidia are green and colonies have a yellow reverse. On YES agar there is often strong sporulation, reverse colour creamish yellow. On CREA weak growth and no acid production. Conidiophores two-stage branched (terverticillate) with all elements adpressed, stipes rough-walled. Conidia smooth-walled, globose to subglobose.

Important toxic metabolites: ochratoxin A
secondary metabolites with unknown toxicity: verrucolone (= arabenoic acid), anacine.

HABITAT: food
The primary habitat is meat products (salami, ham), occasionally found on cheese.

Pl. 106. *Penicillium verrucosum* Colonies after one week a. on Cz, b. on MEA; c, d. conidiophores, x 1000; e. conidia, x 1100.

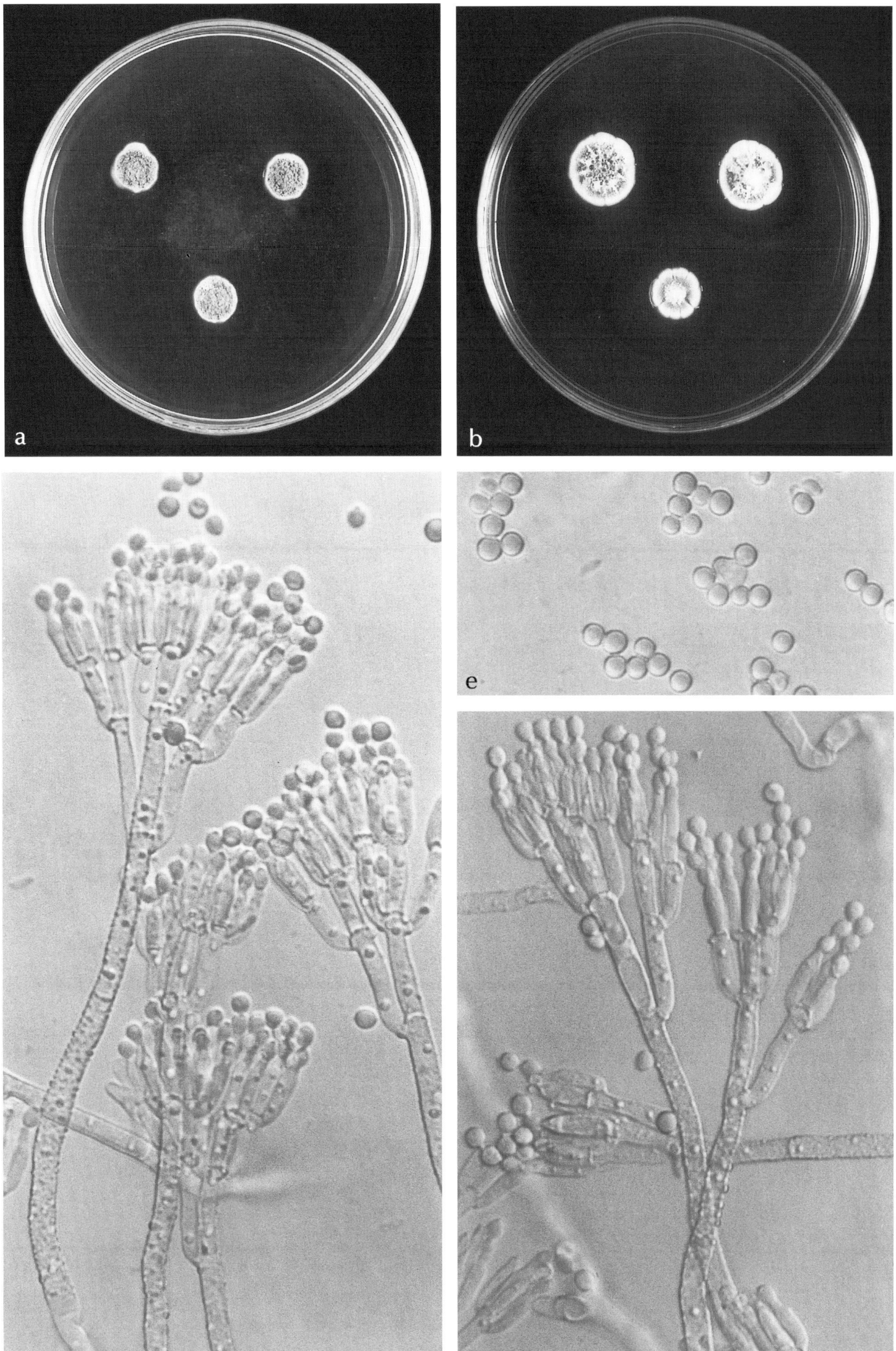
a
b
c
e
d

Penicillium viridicatum Westling

Colonies on Czapek agar and CYA at 25°C growing restrictedly producing yellow green to green conidia with a granular to fasciculate colony surface, often with exudate droplets. The colony reverse is orange to red or pinkish brown with the colour often diffusing into the agar medium or more rarely creamish yellow. On MEA the conidia are yellow green to green, colonies are granular to fasciculate have a distinct yellow reverse, often with the yellow colour diffusing into the medium. On YES agar there is no sporulation and the colony mycelium if often strongly yellow, reverse colour distinct yellow. On CREA weak to moderate growth and strong acid production. Conidiophores two-stage branched (terverticillate) with all elements adpressed, stipes rough-walled. Conidia smooth-walled, globose to subglobose, 3-3.5 µm in diam.

Important toxic metabolites: xanthomegnin, viomellein, vioxanthin, xanthoviridicatin D & G, penicillic acid, viridic acid

Secondary metabolites with unknown toxicity: brevianamide A, viridamine (Frisvad & Filtenborg, 1989).

HABITAT: food
The primary habitat is cereals.

Penicillium tricolor Frisvad, Seifert, Samson & Mills

Colonies on Czapek agar and CYA at 25°C growing restrictedly producing grey to dull green conidia with a velutinous colony surface, with clear to reddish brown exudate droplets. The colony reverse is brownish orange to yellowish brown with a brown centre. On MEA the conidia are greyish green with a pale to yellow reverse.
On YES agar there is strong sporulation, reverse colour curry to yellow brown. On CREA poor growth but strong acid production.
Conidiophore stipes very rough-walled or warted (tuberculate). Conidia smooth and globose to subglobose.

Important toxic metabolites: xanthomegnin, viomellein, vioxanthin, asteltoxin, terrestric acid

Secondary metabolites with unknown toxicity: verrucofortine (=verrucosine), puberuline, rugulosuvine, leucyltryptophanyldiketopiperazine, styrene (Frisvad *et al.*, 1994).

HABITAT:
The primary habitat is cereals.

Pl. 107. *Penicillium viridicatum*. Colonies after one week, a. on Cz, b. on MEA; c-e. conidiophores and conidia, x 950.

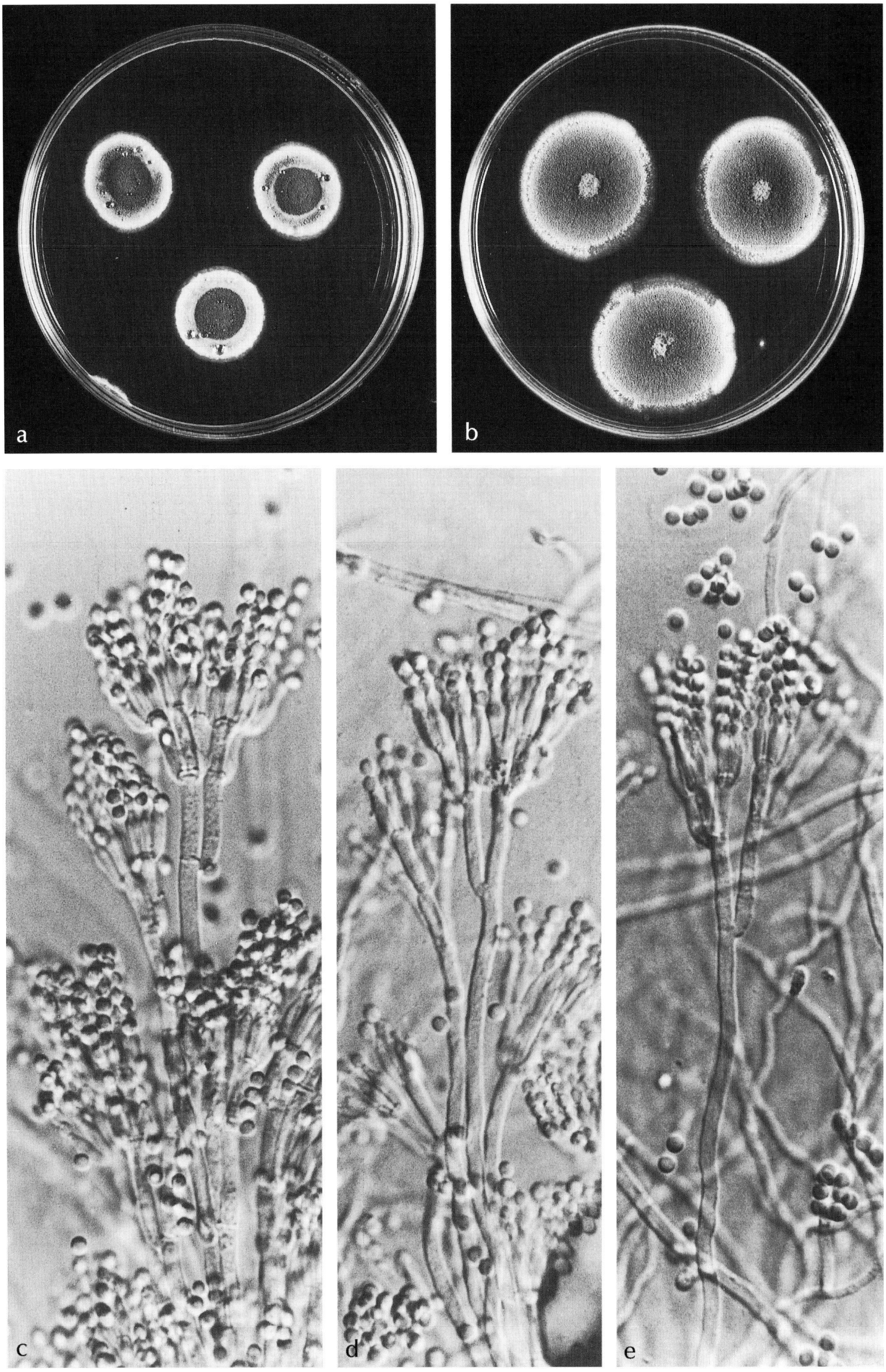
a
b
c
d
e

PHIALOPHORA Medlar

Teleomorphs: *Pyrenopeziza* Fuckel, *Mollisia* (Fr.) Karst., *Ascocoryne* Groves & Wilson, *Coniochaeta* (Sacc.) Massee, *Gaeumannomyces* v. Arx & Olivier.

Colonies usually growing slowly, olivaceous-black, sometimes hyaline to pinkish or becoming brown with age, with or without aerial mycelium. Phialides commonly with a distinct collarette, flask-shaped, cylindrical or reduced (without a basal septum), borne on branched conidiophores, often in clusters or arising solitarily from hyphae. Conidia one-celled, globose to ellipsoidal, sometimes slightly curved, more or less hyaline in slimy heads or chains.

HABITAT:
Phialophora species have been isolated from decaying wood, foodstuffs (e.g. butter, margarine), apples, soil, diseased human and animal tissue, also occurring as parasites or saprophytes in plant material.

NOTE:
For more detailed descriptions and keys see Schol-Schwarz (1970), Domsch *et al.* (1993). *Phialophora* is a heterogeneous genus, but can be more or less grouped according to the teleomorph connections. The former *Phialophora hoffmannii* with cream to orange-pink and slimy colonies and reduced phialides is placed in the genus *Lecythophora* Nannf. with the teleomorph *Coniochaeta*.

CULTIVATION FOR IDENTIFICATION
Isolates are inoculated in a streak on MEA or OA and incubated for 10-12 days at 20°C in diffuse daylight.

KEY TO THE SPECIES TREATED

1a. Colony remaining pinkish .... *Lecythophora hoffmannii*
1b. Colony dark olive-brown .... *Phialophora fastigiata*

Lecythophora hoffmannii (van Beyma) W. Gams
= *Phialophora hoffmannii* (van Beyma) Schol-Schwarz

Colonies on MEA at 20°C attaining a diameter of 2-3 cm in 10 days, smooth, flat (often with a leathery appearance), with a central tuft, pinkish. Reverse pinkish buff. Hyphae more or less hyaline, smooth- and thin-walled, often fasciculate. Phialides either single on hyphae or in dense clusters on short lateral branches, or intercalary along undifferentiated hyphae, hyaline, cylindrical to flask-shaped, sometimes elongate, 7-20 µm long, or subglobose to ovoid, 3-7 x 2-3 µm; intercalary phialides cylindrical with one or several scattered collarettes. Conidia ellipsoidal to cylindrical, sometimes somewhat curved, commonly 3-3.5 x 1.5-2.5 µm, subhyaline, smooth- and thin-walled.

HABITAT: food (indoor)
Reported from Switzerland, Scotland; isolated from rotten wood, foodstuffs (e.g. butter), soil, humans and animals.

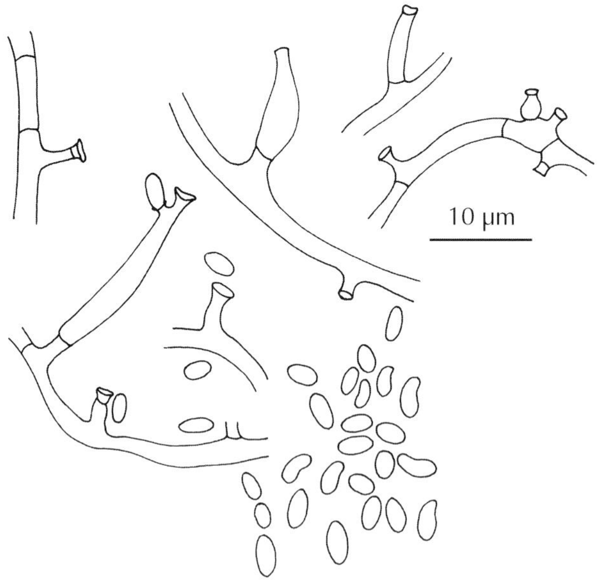

Fig. 100 a. *Lecythophora hoffmannii*. Phialides with collarettes and conidia.

Pl. 108. a-d. *Phialophora fastigiata*. a. Colony on MEA after 10 days; b. conidia, x 1000; c,d. conidiophores, phialides with collarettes, c. x 1000, d. x 2000; e-g. *Lecythophora hoffmannii*. e. colony on MEA after 10 days; f. conidia, x 1000; g. phialides, x 1000.

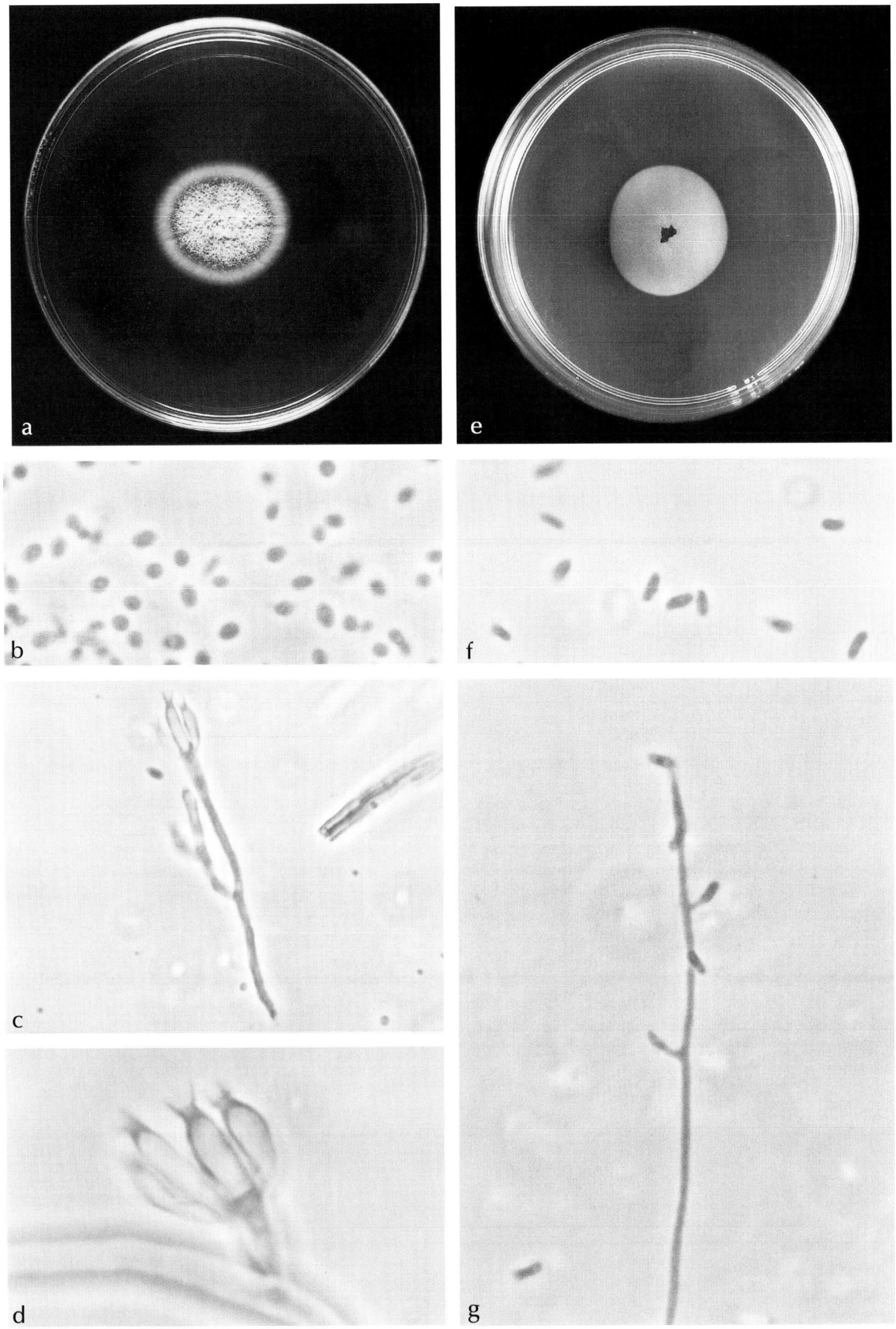
a
b
c
d
e
f
g

Phialophora fastigiata (Lagerb. & Melin) Conant

Colonies on MEA at 20° C attaining a diameter of 2.3-2.5(3) cm in 10 days, more or less velvety, often with an almost hyaline margin, olivaceous-brown. Reverse dark brown to black. Hyphae brownish, walls usually becoming distinctly encrusted with age. Phialides borne singly, laterally or terminally on the hyphae or in clusters, flask-shaped, brownish with a collarette, 6-10(-30) x 2.2-3(-4.2) µm. Conidia subglobose, obovoid to ellipsoidal, often apiculate at base, sometimes slightly curved, 3-6(-7) x (1.5)2-3(-3.8) µm, subhyaline to light brown.

HABITAT: food, indoor
Reported throughout Europe, N. America, Uganda, the USSR, New Zealand and Australia. Isolated from wood and wood pulp, soil, wheat, seeds of cultivated grasses.

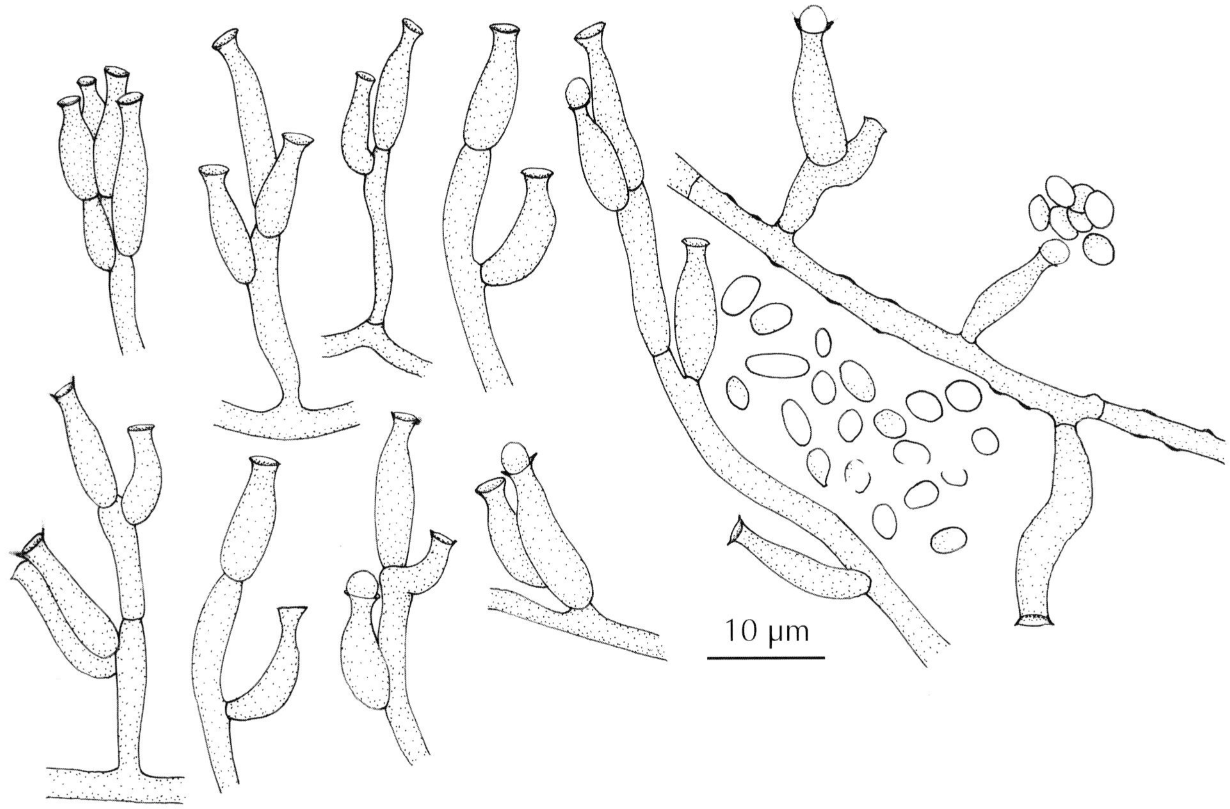

Fig. 101 b. *Phialophora fastigiata*. Phialides with collarettes and conidia.

PHOMA Sacc.

Colonies with pigmented separate pycnidia (or pycnidial conidiomata), usually each with single ostioles, sometimes multi-ostiolate. Conidiogenous cells with a narrow phialidic opening, more or less the same structure as inner-cells of pycnidia, producing abundant slimy conidia. Conidia usually one-celled, ellipsoidal to cylindrical, hyaline or coloured. Chlamydospores in some species formed in single chains or aggregated in "dictyochlamydospores" (resembling the conidia of *Alternaria*).

NOTE:
Phoma is the largest (more than 2000 species) and most widely distributed genus of the Sphaeropsidales, with species reported from soil, as saprophytes on various plants and on dead plant material, and occasionally pathogenic to plants and humans. For more detailed descriptions and keys see Boerema *et al.* (several publications since 1960), Dorenbosch (1970), Sutton (1980); for soil-borne species see Domsch *et al.* (1993). A list of references is also given in Gams et al. 1998.

KEY TO THE SPECIES TREATED

1a. Catenulate dictyochlamydospores typically present. Colonies without a red tinge .*Ph. glomerata*
1b. Catenulate dictyochlamydospores absent. Colonies with a reddish appearance....*Ph. macrostoma*

Phoma glomerata (Corda) Wollenweber & Hochapfel

Colonies on OA at 20-22°C attaining a diameter of 5-5.7 cm in 7 days, aerial mycelium usually sparse, (sometimes in greyish woolly sectors), with abundant pycnidia, olivacous buff to dull green, becoming darker due to dictyochlamydospore production. Pycnidia superficial on or immersed in agar, usually with one ostiole (sometimes 2-3), more or less globose (20)30-180(300) x (40)60-200(600) µm, light-coloured to black or blackish-brown (pycnidia often coalesce to form a complex with many ostioles). Conidia one-celled (occasionally 1-septate) mostly ovoid to ellipsoidal, slightly curved or straight, 5-9 x 2.5-3(3.5) µm, hyaline to light-olivaceous, with two or more guttules. Dictyochlamydospores in branched or unbranched chains of 2-20 (or more).

CULTIVATION FOR IDENTIFICATION
Isolates are inoculated centrally in a plastic Petri-dish in one point position on OA, CMA or PCA, incubated at 20-22°C in darkness for 7 days, followed by 7 days alternate light/darkness. Characteristic pigments often are produced on MEA. Sporulation is stimulated by near UV-light (12h near UV/12h darkness).

HABITAT: food, indoor
World-wide distribution; isolated from a wide variety of plants and plant material, from soil, butter, rice grain, cement, litter, paint, paper and wool. The fungus can be pathogenic to humans and attack different kinds of fruit, e.g. as the cause of tomato-rot.

NOTE:
Colonies can show variation in formation of pycnidia, aerial mycelium and chlamydospores; isolates often degenerate in culture.

Pl. 109. *Phoma glomerata*. a. Colony on OA after one week; b. pycnidia with ostioles, in Petri-dish, x 40; c,e. pycnidia with ostioles, x 100; d. dictyochlamydospores, often in chains (resembling the conidia of *Alternaria*), x 500; f. conidiogenous cells, x 1970; g. conidia, x 1000.

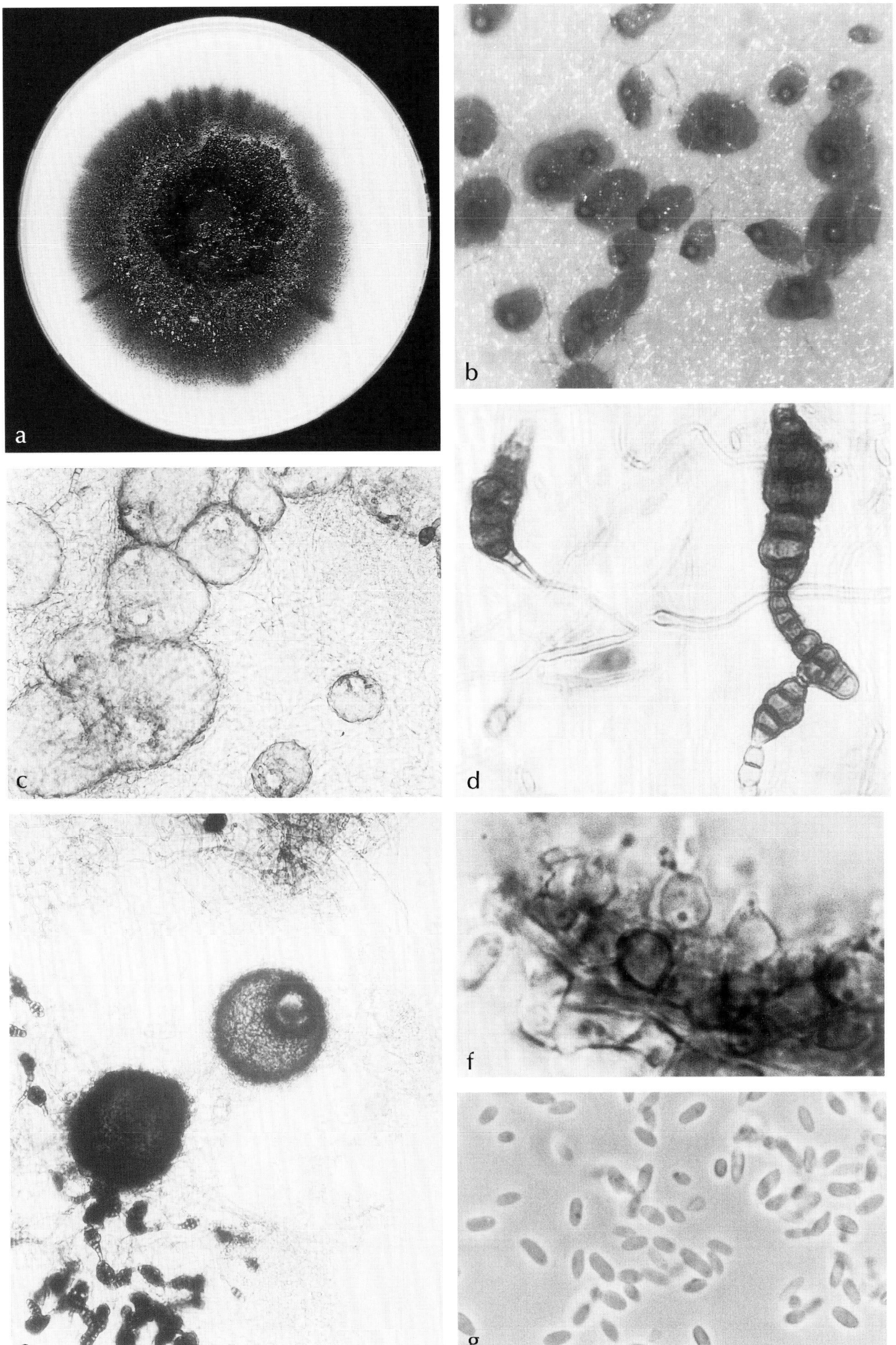
a
b
c
d
e
f
g

Phoma macrostoma Montagne

For synonyms is referred to the specific literature.

Colonies on OA at 20-22° C attaining a diameter of 4-6 cm in 7 days, mycelial mat variable from sparse to dense and woolly. The hyphae contain a red to violet pigment which gives the colony a reddish appearance. Pycnidia superficial on or immersed in agar, globose-oval, 80-260 um diam, blackish brown, with one (occasionally two) distinct ostiole of 10-30 µm in diam. Conidial slime pinkish, conidia one- or two-celled (occasionally 3-celled in vitro) septate mostly ovoid to ellipsoidal, slightly curved or straight, one-celled conidia: 5-8 x 2-4 µm; two-celled conidia 8.5-14 µm, hyaline to brownish coloured in old cultures.

NOTE: The colonies are characterized by reddish appearance (turning bluish with additon of NaOH) in var. *macrostoma.* Strains without the red colour are distinguished as var. *incolorata.*

HABITAT: food, indoor.
Widespread occurrence: isolated from a wide variety of plants and plant material, from soil. Also isolated from surfaces in indoor environments, but not as common as *Phoma glomerata.*

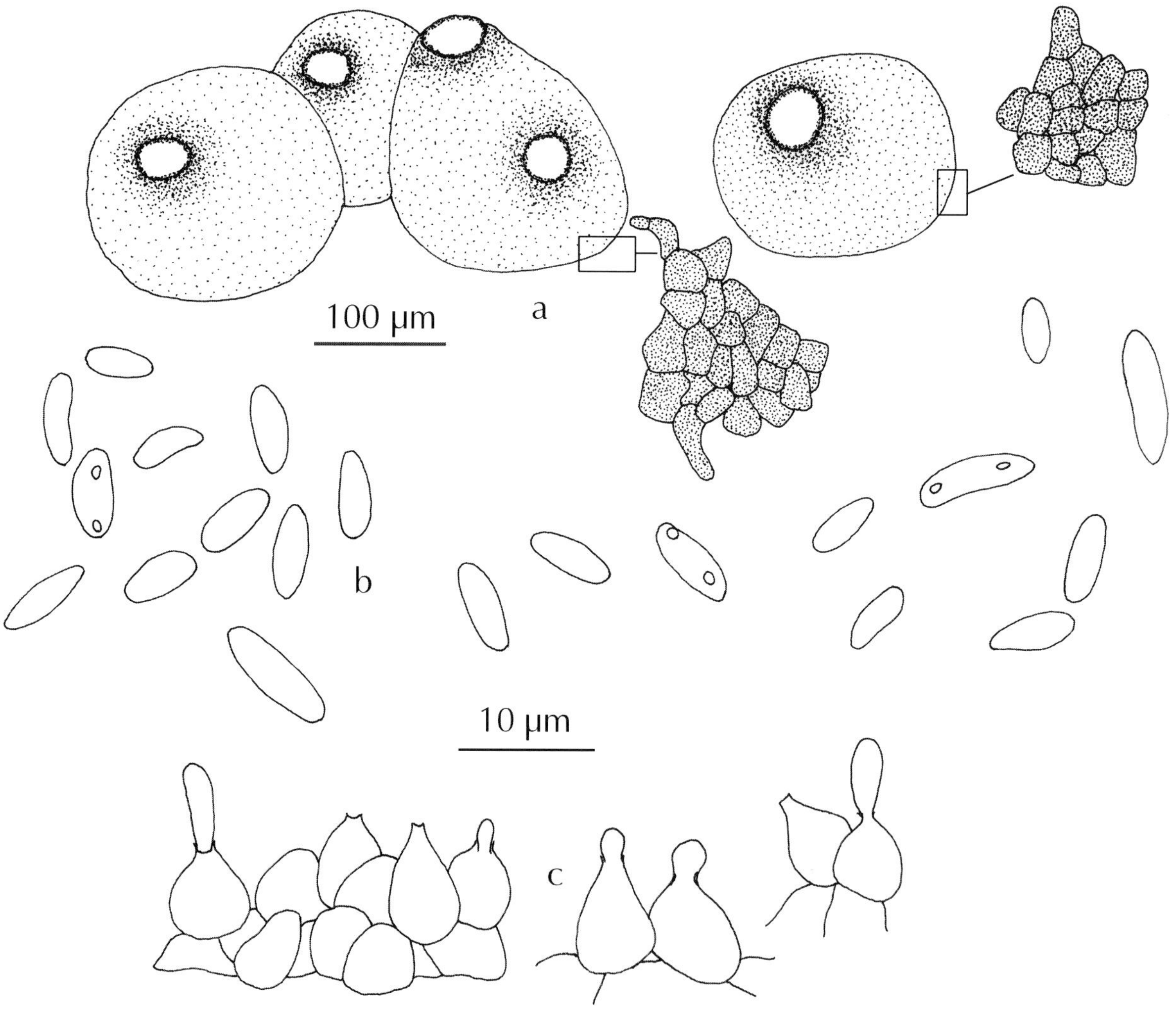

Fig. 102. *Phoma macrostoma*. a. Pycnidia with detail of outer wall; b. conidia; c. innerwall layer with phialides.
Pl. 110. *Phoma macrostoma*. a. Colony on OA after one week; b. pycnidium with ostiole, x 225; c. detail of pycnidium with ostiole, x 720; d. conidia, x 2080; e-g. inner cellwall of pycnidium with phialides, e. x 1950, f. x 1875, g. x 2040.

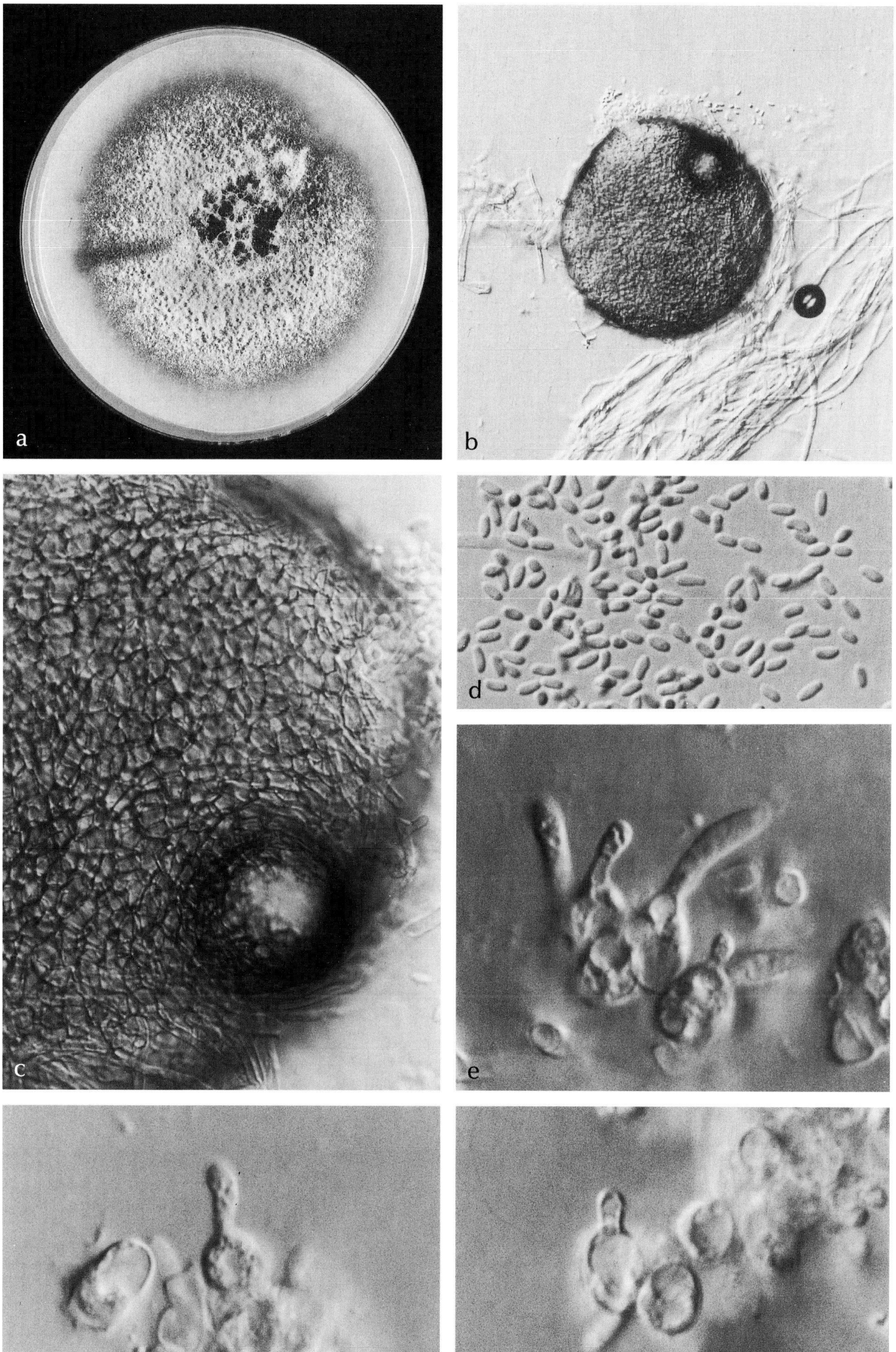
a
b
c
d
e
f
g

SCOPULARIOPSIS Bain.

Teleomorph: *Microascus* Zukal.

Colonies varying from white, creamish, grey or buff to brown or even blackish, often darkening with age, but never in green shades like *Penicillium*, granular, velvety or more or less funiculose, with reverse in shades of buff to brown or even colourless. Conidiogenous cells cylindrical or with a slightly swollen base, annellate (rings formed after conidium formation), single or borne on a somewhat penicillate system of branches. Conidia borne in chains in basipetal succession, smooth-or rough-walled, one-celled, globose to ovate with a broadly truncate base.

For a more detailed account on descriptions and key see Morton and Smith (1963), Domsch *et al.* (1993).

CULTIVATION FOR IDENTIFICATION
The colonies are inoculated in one-or three-point position on 2% MEA or OA and cultivated at 24°C in darkness. Slides are mounted in lactic acid.

KEY TO THE SPECIES TREATED

1a. Conidia rough-walled at maturity (at least some) .... *S. brevicaulis*
1b. Conidia smooth-walled 2

2a. Colonies white (becoming more or less creamish after two weeks) .... *S. candida*
2b. Colonies rose-brown (becoming dark-brown to brownish-black) .... *S. fusca*

Scopulariopsis brevicaulis (Sacc.) Bain.

Teleomorph: *Microascus brevicaulis* Abbott, Sigler & Currah

Colonies on MEA at 24°C attaining a diameter of (3.2)4.5-5.5 cm in 7 days, whitish becoming brownish-rose (vinaceous-buff), more or less funiculose at first, becoming powdery with a prominent central tuft. Reverse in cream to brownish shades. Conidiogenous cells annellate, borne singly on aerial hyphae or terminally on once or twice branched conidiophores, in groups of 2-3(4), cylindrical, often with a more or less swollen base, 9-25 x 2.5-3.5 µm, annellated zone 2.5-3.5 µm wide. Conidia globose to ovoid with a distinctly truncate base, apex sometimes more or less pointed, 5-8(9) x 5-7 µm, rough-walled (sometimes smooth when young), rose-brown (avellaneous) in mass..

Temperature: optimum 24-30°C; minimum 5°C; maximum 37°C.

HABITAT: food, indoor
World-wide distribution. Isolated from soil, wood, stems, straw, grains (e.g. wheat), fruit (apples), soybeans, groundnuts, dead insects, dung, paper, animals, human (e.g skin lesions, lungs), animal products like meat, cheese, milk, butter. Its presence can be recognized by a characteristic ammoniacal odour.

NOTE:
Sigler *et al.* (1998) found several isolates of *S. brevicaulis* to form perithecia after 6-25 weeks on Oatmeal Salts agar and described the teleomorph as *Microascus brevicaulis*.

Morphologically *S. brevicaulis* closely resembles *Scopulariopsis flava* (Sopp) Morton & Smith. The only difference is that in *S. flava* the colony is whitish to creamish instead of brownish.

Pl. 111. *Scopulariopsis brevicaulis*. a. Colony on MEA after 2 weeks; b-e. conidiophores with conidia in chains, b. x 650, c, d. x 800, e. x 1000; f. conidia, with a truncate base, x 1750.

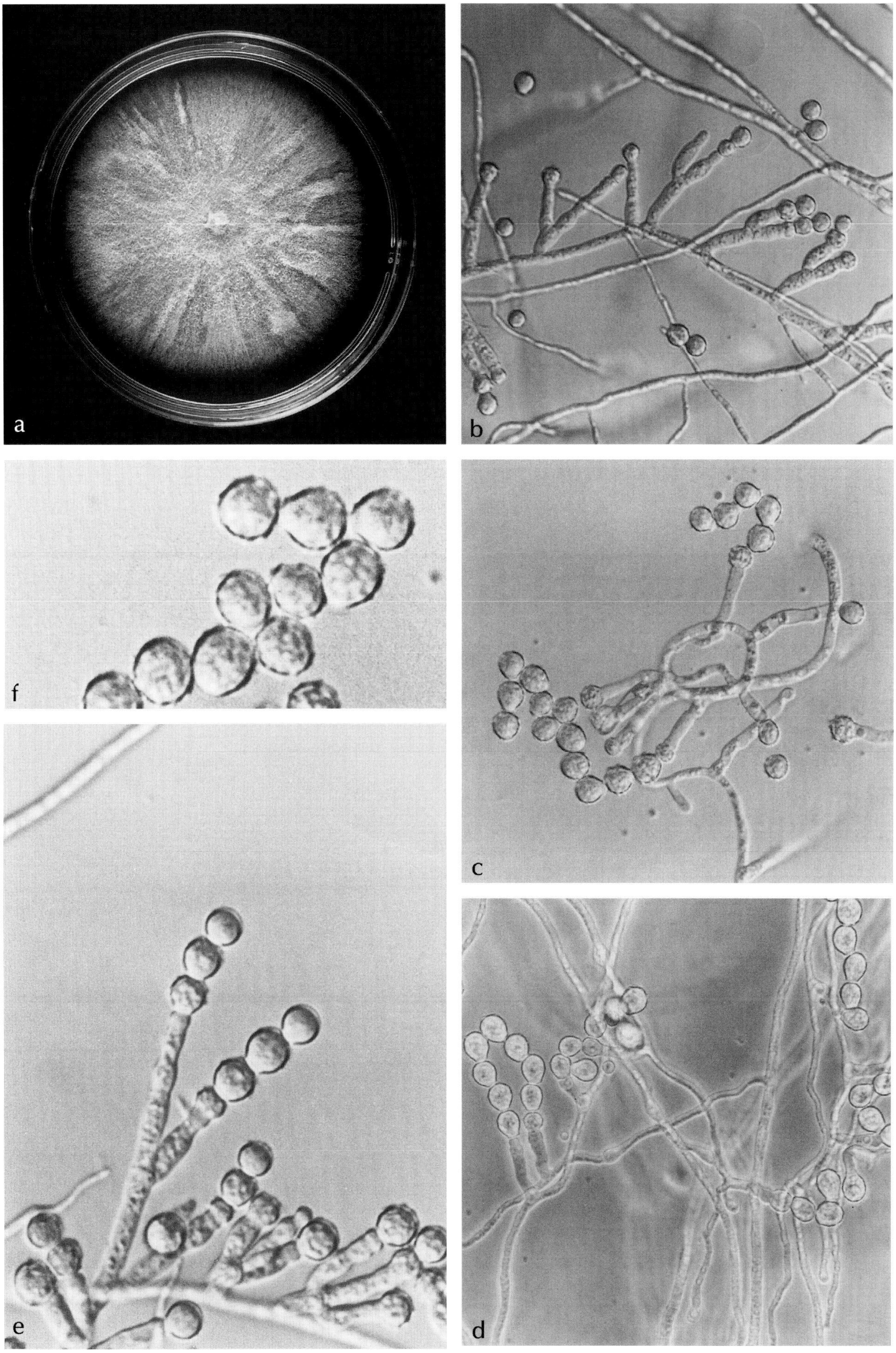
a
b
f
c
e
d

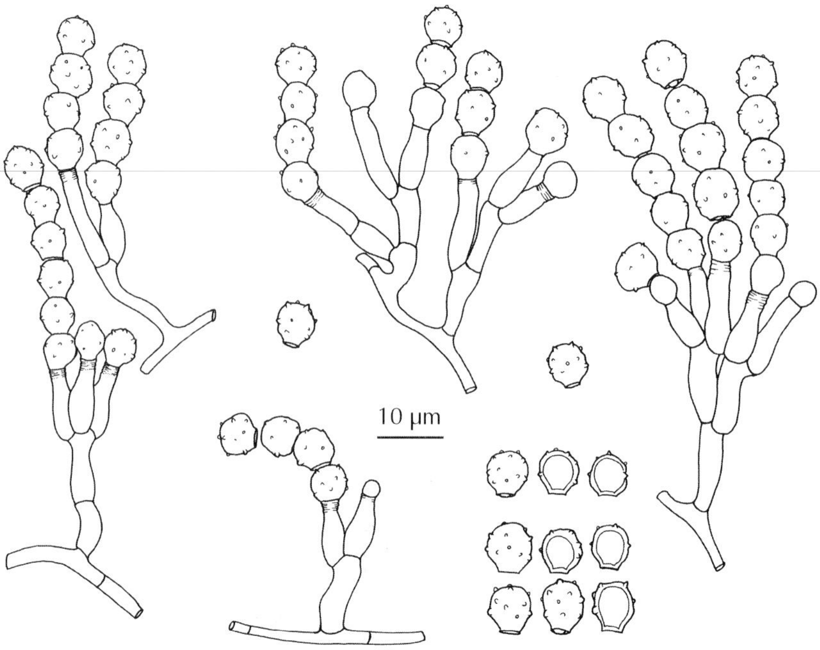

Fig. 103 a. *Scopulariopsis brevicaulis*. Conidiophores with annellate conidiogenous cells and rough-walled conidia.

Scopulariopsis candida (Guéguen) Vuill.

Colonies on MEA at 24° C attaining a diameter of 3.0-4.0 cm in 7 days, white (darkening after about two weeks), velvety to powdery with a central tuft. Occasionally the powdery appearance of the colony is disturbed by small white sterile overgrowths. Reverse in creamish to light brown shades. Conidiogenous cells annellate, borne singly on aerial hyphae or terminally in groups of 2-3 short conidiophores, cylindrical, often with a slightly swollen base, 10-25 µm in diam. Conidia subglobose to broadly ovate with a truncate base and more or less rounded apex, 5-8 x 4-7 µm, smooth-walled, whitish to creamish in mass. Small black sclerotial bodies can be formed after 1-2 weeks, especially at 17-19° C.

Temperature: optimum 24-30° C; minimum 5° C; maximum 37° C.

HABITAT: food, (indoor)
Isolated from soils, milled rice, paper, cheese, human (nails, skin).

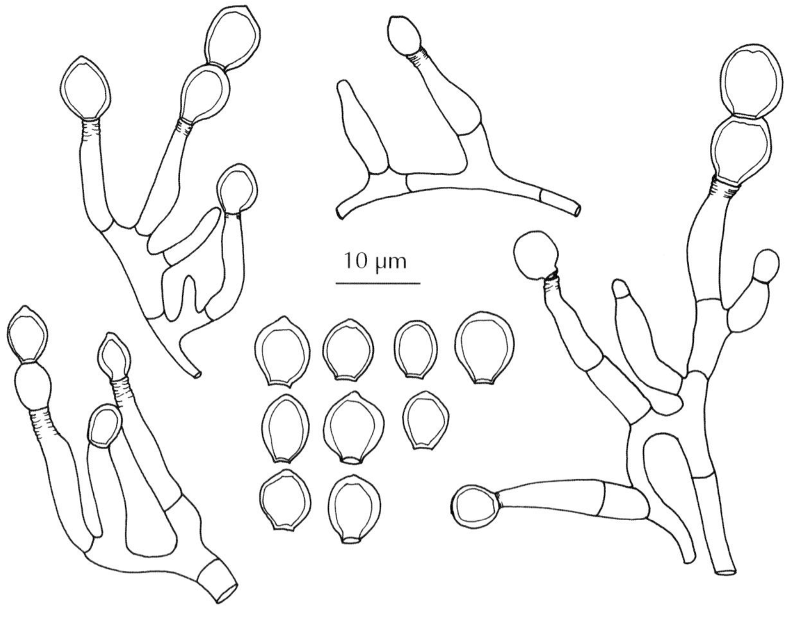

Fig. 103 b. *Scopulariopsis candida*. Annellate conidiogenous cells, single or on conidiophores, and smooth-walled conidia.
Pl. 112. *Scopulariopsis candida*. Colonies after 2 weeks, a. on MEA, b. on Cz; c,d. conidiophores with conidia in chains, c. x 1340, d. x 1225; e. conidia, x 1000.

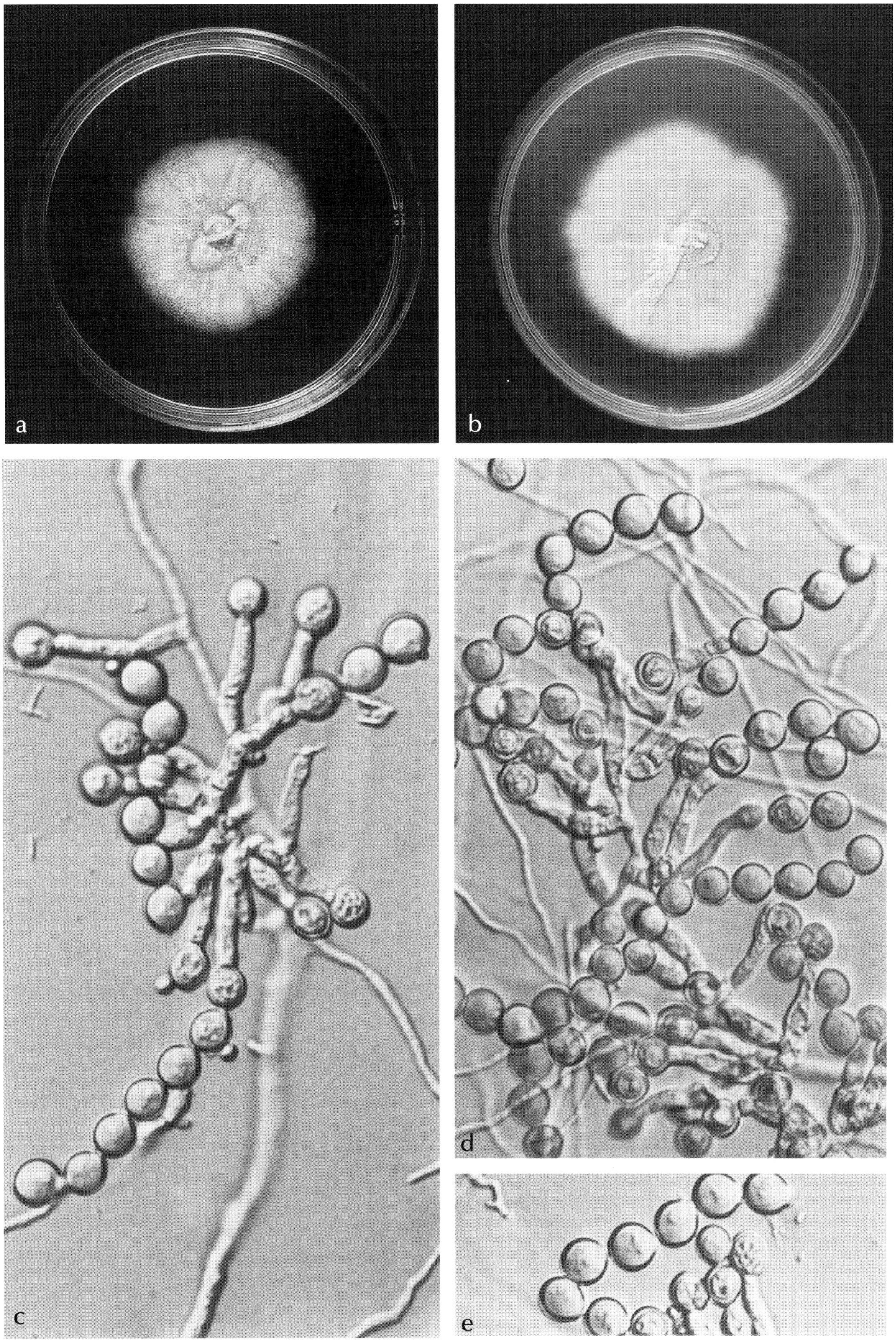
a
b
c
d
e

Scopulariopsis fusca Zach

Colonies on MEA at 24° C attaining a diameter of 1.9-2.5 cm, light rose-brown at first with a whitish velvety margin and a central tuft, becoming brown to brownish-black and floccose to funiculose all over. Reverse light brownish-rose becoming dark brown. Conidiogenous cells annellate, borne singly on aerial hyphae or in groups of 2-10 on short conidiophores, cylindrical often with a slightly swollen base, 2-27 µm long. Conidia globose to broadly ovate with a truncate base and occasionally a more or less pointed apex, 5-8 x 5-7 µm, smooth-walled, olive to brown in mass.

Temperature: optimum 24-30° C; minimum 5° C; maximum 37° C (40° C).

HABITAT: food, indoor
Isolated from mouldy straw, cheese, hen's eggs, cacao leaves from Ghana, paper, rabbit liver, man (onychomycosis, dermatomycosis), soil from Hong Kong, Central Africa, Peru, Germany, British Isles (Domsch *et al.*, 1993).

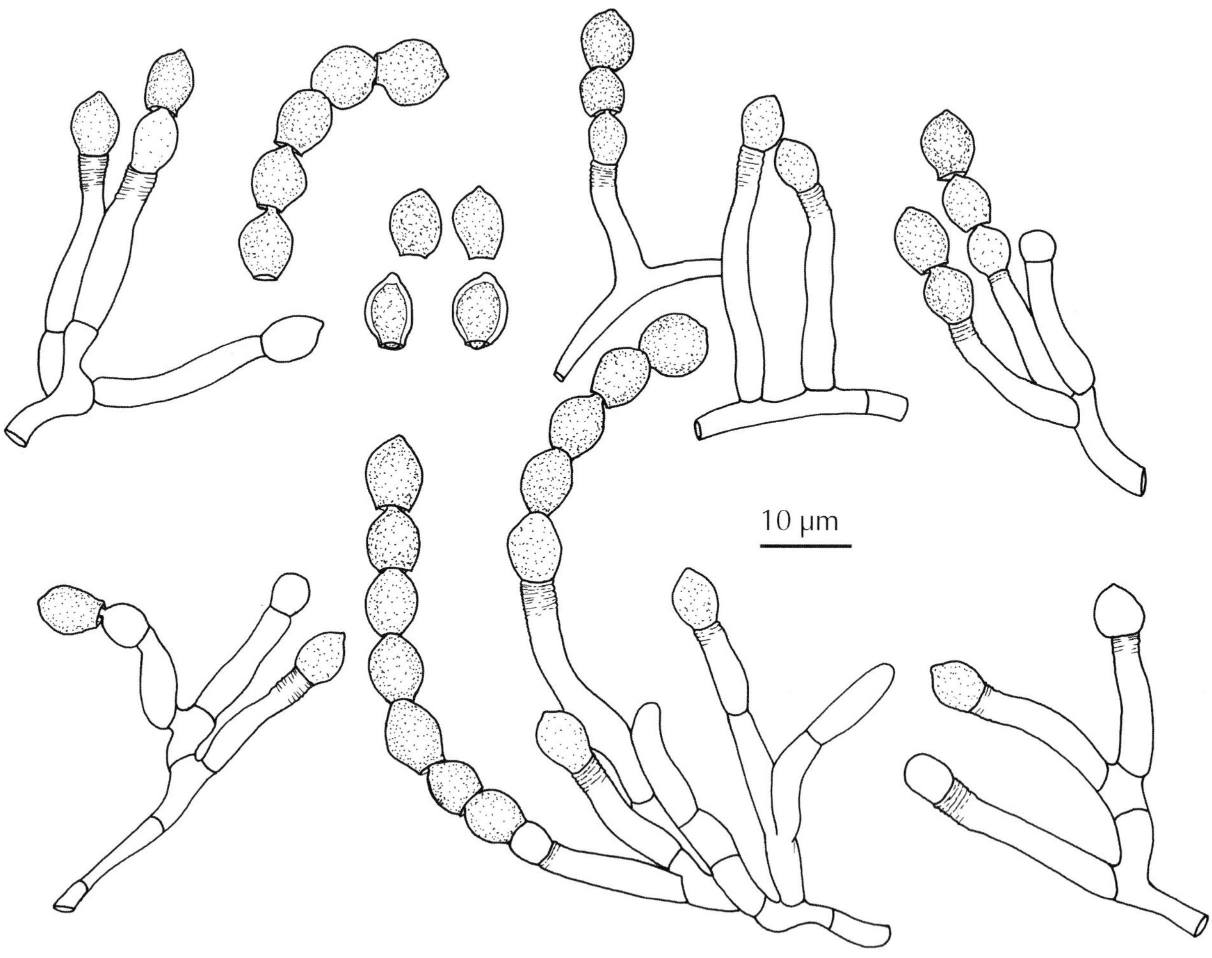

Fig. 104. *Scopulariopsis fusca.* Annellate conidiogenous cells, single or on short conidiophores and smooth-walled conidia.
Pl. 113. *Scopulariopsis fusca.* Colonies after 10 days, a. on MEA, b. on Cz; c-f. conidiophores and conidia, c-e. x 2000. f. x 650.

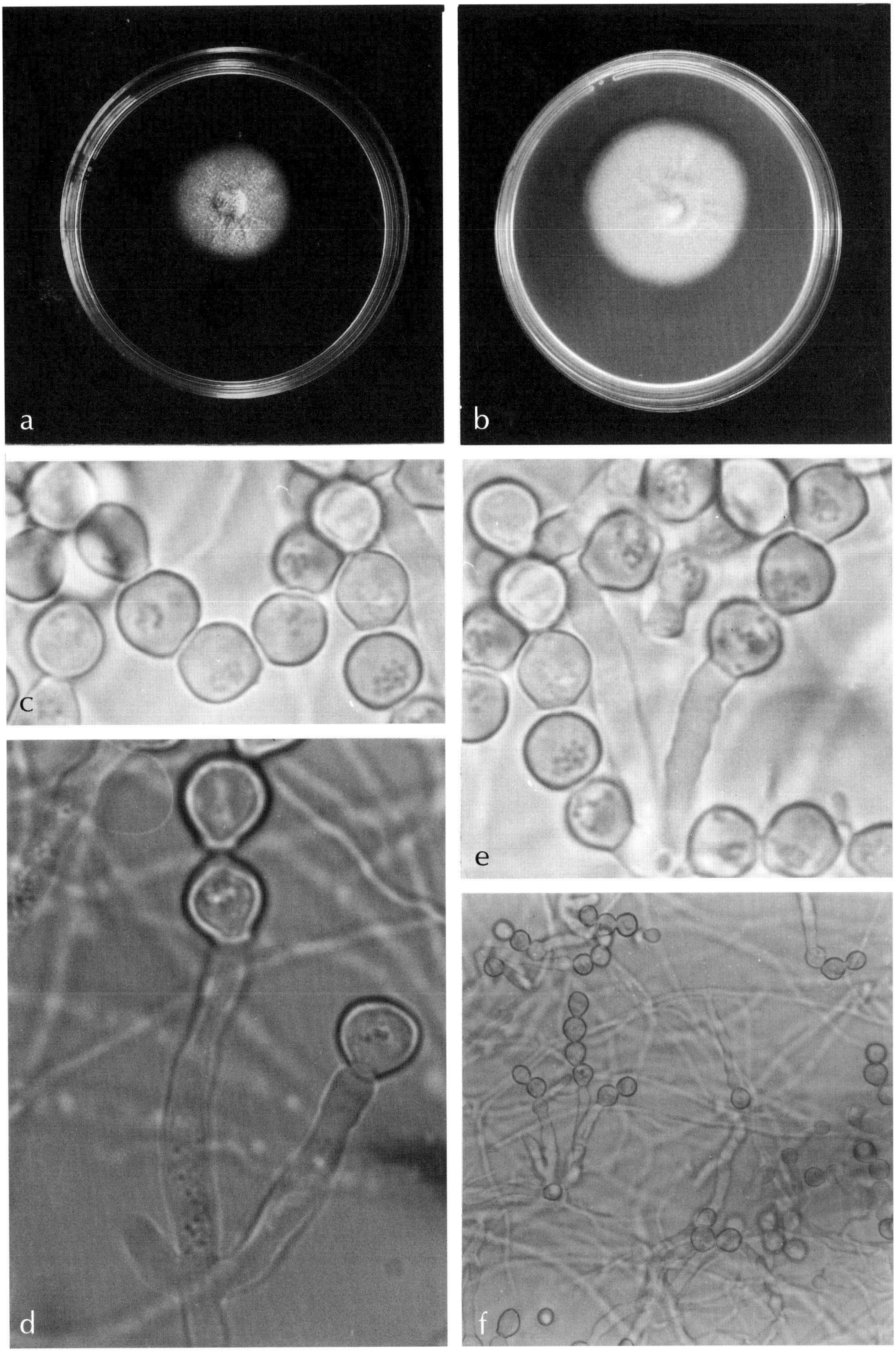
a
b
c
e
d
f

Stachybotrys chartarum (Ehrenb.) Hughes

= *Stachybotrys atra* Corda

Colonies on MEA at 25° C attaining a diameter of 2.5-3 cm in 7 days, appearing blackish to blackish-green and powdery due to the conidial masses. Reverse colourless. Conidiophores simple or branched, about 100 µm tall (up to 1000 µm) and 3-6 µm wide, hyaline or greyish at first, becoming olivaceous brown to black and roughened towards the apex, bearing clusters of 4-10 obovate or ellipsoidal phialides. Conidia in slimy heads, ellipsoidal, 7-12 x 4-6 µm, hyaline at first, becoming dark olivaceous-grey, varying from smooth-walled to coarsely roughened and warted.

Temperature: optimum 23 (-27)° C; minimum (2-) 7° C; maximum 37-40° C.

Important (toxic) metabolites: **satratoxin G** and **H.**

HABITAT: indoor
World-wide distribution. Isolated from paper, wall-paper, gypsum board, seeds (e.g. wheat, oats), soil, textiles, dead plant material.

NOTE:
For detailed descriptions and keys see Ellis (1971), Domsch *et al.* (1993). Exposure to this fungus on natural substrates or in culture should be avoided.

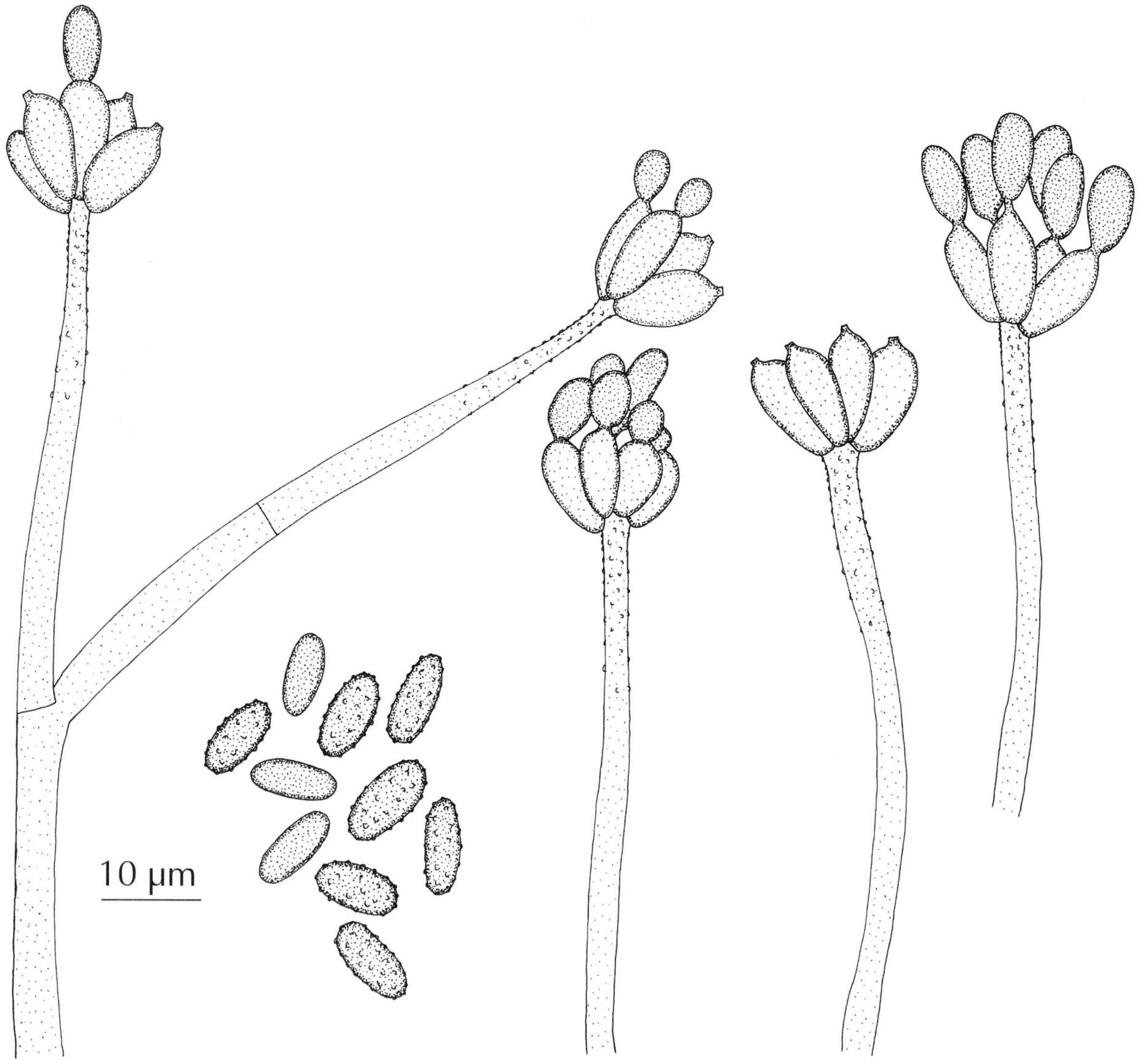

Fig. 105. *Stachybotrys chartarum.* Conidiophores bearing clusters of obovate phialides, conidia.
Pl. 114. *Stachybotrys chartarum.* a. Colony on MEA after one week; b. conidiophores in Petri-dish, x 40; c. conidia, x 1000; d-f. conidiophores and conidia, x 1000.

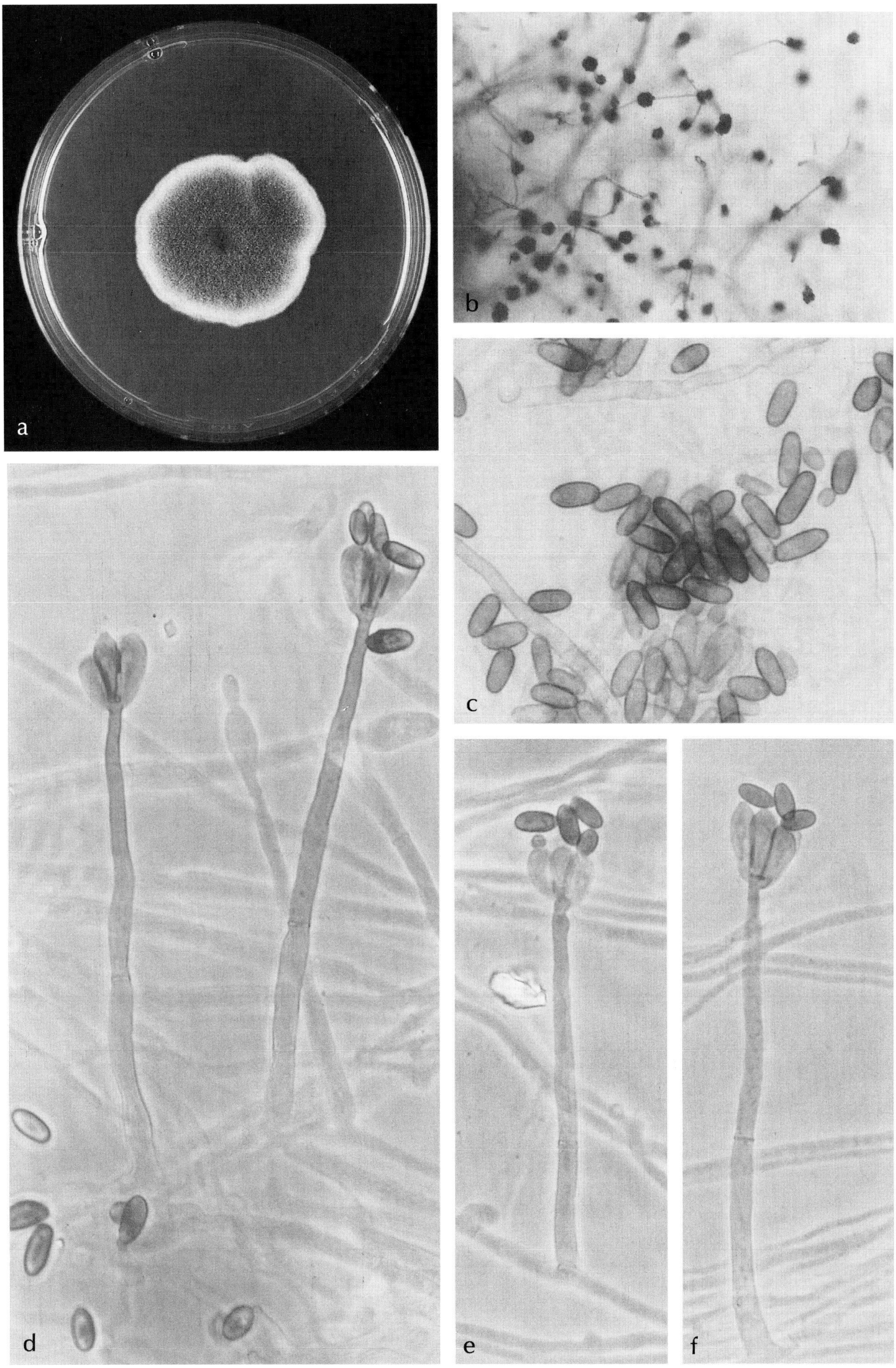

TRICHODERMA Pers.

Teleomorph: *Hypocrea* Fr. (in most species not produced in culture).

Colonies usually growing rapidly, initially hyaline, later usually in green shades due to conidium production. Conidiophores in tufts, repeatedly branched (irregularly verticillate), bearing clusters of flask-shaped phialides, with sterile apical appendages in some species, however in most species each branch is terminated by a phialide. Conidia in moist conidial heads, sometimes hyaline, often green, smooth- or rough-walled. Chlamydospores present in most species, intercalary or sometimes terminal on short side branches of the hyphae, globose or ellipsoidal, hyaline, smooth-walled.

Important (toxic) metabolites: gliotoxin, emodin, trichodermin (see also chapter 5).

HABITAT: food, indoor
A very common genus especially in soil and on decaying wood. *Gliocladium* (with strongly convergent phialides) and *Verticillium* (with straight and moderately divergent phialides) are closely related. For more detailed descriptions see Rifai (1969), Bissett (1984, 1991 a-c, 1992) and Domsch *et al.* (1993), Samuels *et al.*(1998,1999).

CULTIVATION FOR IDENTIFICATION
Colonies are inoculated onto OA or MEA and incubated in diffuse daylight or under near-UV at 20-24°C for about 5 days. (Cultures can easily loose their sporulating capacity, especially when grown in darkness). Conidial roughening is best observed after two weeks. Microscope slides are mounted in lactic acid containing some aniline blue.

KEY TO THE TREATED SPECIES

1a. Conidia rough-walled, 3.6-4.8 x 3.5-4.5 µm..*T. viride*
1b. Conidia smooth-walled, 2.8-3.2 x 2.5-2.8 µm..*T. harzianum*

Trichoderma harzianum Rifai

Colonies on OA at 20°C attaining a diameter of more than 9 cm in 5 days, more or less hyaline, whitish-green, becoming olive-green.
Conidiophores pyramidally branched, i.e long repeatedly branched lateral branches below, with short branches near the apex. Phialides sometimes more slender and longer, especially at the apex of branches, measuring up to 18 x 2.5 µm. Conidia subglobose to short-ovoid, 2.8-3.2 x 2.5-2.8 µm, smooth-walled. Chlamydospores usually present in mycelium of older colonies, intercalary, sometimes terminal, mostly globose, hyaline, smooth-walled.

Temperature: optimum 15-30(-35)°C; maximum 30-36°C.

HABITAT:
T. harzianum probably has a world-wide distribution. Isolated from soil, grains, pecans, paper and textiles.

NOTE:
The main difference between *T. harzianum* and *T. viride* is the conidium-wall ornamentation, which is best observed in two-weeks old colonies. *T. harzianum* is therefore regarded as the smooth-walled counterpart of *T. viride*.

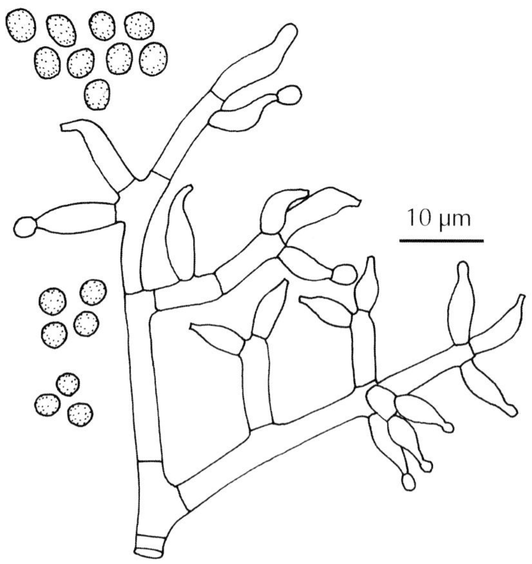

Fig. 106. a. *Trichoderma harzianum*. Conidiophore and conidia. (orig. W. Gams).

Pl. 115. a-c. *Trichoderma harzianum*. a. Colony on MEA after one week; b. smooth-walled conidia, x 1000; c. conidiophore, x 1000; d-f. *Trichoderma viride*. d. colony on MEA after one week; e. rough-walled conidia, x 1000; f. conidiophore, x 1000.

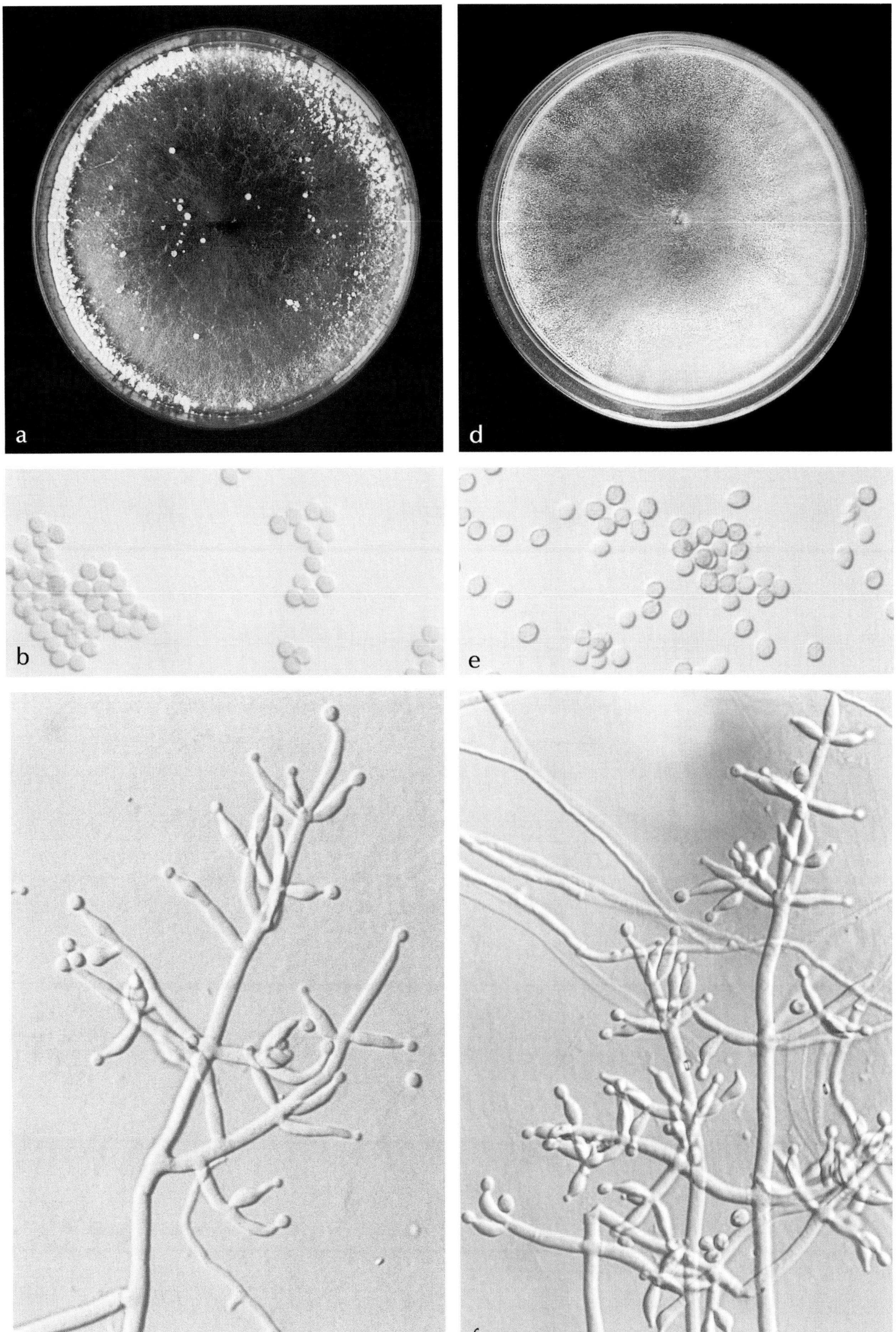
a
d
b
e
c
f

Trichoderma viride Pers.

Colonies on OA at 20° C attaining a diameter of 4.5-7.5(-9) cm in 5 days, initially more or less hyaline, later becoming whitish-green with tufted conidial areas in blue-green shades. Reverse colourless. Conidiophores irregularly pyramidally branched, i.e. long repeatedly branched lateral branches below, with shorter branches near the apex. Phialides in groups of 2-4, rather slender, cylindrical and often bent or hooked, (6-)8-14(-20) x 2.4-3.0 µm. Conidia typically (sub)globose or rarely broadly ellipsoidal, 3.6-4.5 µm in diam (or up to 4.8 µm long), rough-walled to warted. Chlamydospores sometimes present in the mycelium of older cultures, intercalary, sometimes terminal, mostly globose, hyaline, smooth-walled.
(Also Cornmeal with 2 % dextrose and SNA with or without a piece of sterile filterpaper can be used).
Temperature: optimum ca. 25 ° C.

HABITAT: food, indoor
T. viride is one of the most widely distributed soil fungi, reported from various habitats (extreme northern areas, alpine areas as well as tropical areas). The species is often confused with other species of this genus. Isolated from decaying wood, soil, stored grain of wheat, oats, barley, groundnuts, tomatoes, stored sweet potatoes, citrus fruit, fats and margarine. *Trichoderma* species can be preservative-resistant.

Note:
The closely related new species *T. asperellum* is separated from *T. viride* by its finer ornamentation of the conidia, faster growth rate, ampulliform phialides. Chlamydospores are consistently present. Teleomorph not known. The optimum temperature for *T. asperellum* is at 30° C with a faster growth rate than *T. viride*. (Samuels *et al.*, 1998, 1999)

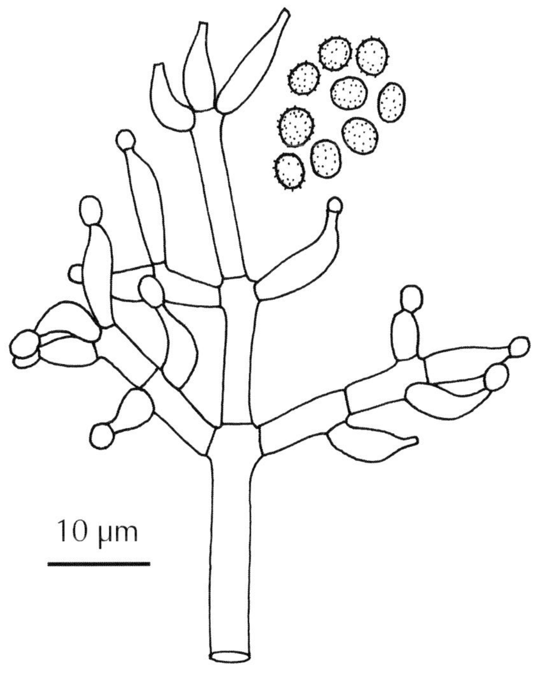

Fig. 106. b. *Trichoderma viride*. Conidiophore and conidia;

Trichothecium roseum (Pers.) Link

Colonies on MEA at 25° C attaining a diameter of about 6 cm in 7 days, powdery due to abundant sporulation, pinkish erect conidiophores, up to 2 mm long and 4-5 µm wide, unbranched, often septate near the base, more or less rough-walled, bearing basipetal zig-zag chains of conidia at the apex. Conidia 1-septate, (aseptate when young), ellipsoidal to pyriform, 12-23(-35) x 8-10(-13) µm, with an obliquely truncate basal scar, hyaline, smooth to delicately roughened and thick-walled.

Temperature: optimum 25° C; minimum 15° C; maximum 35° C.

HABITAT: food
World-wide distribution. Isolated from decaying plant substrates, soil, seeds of corn, food-stuffs (especially flour products).

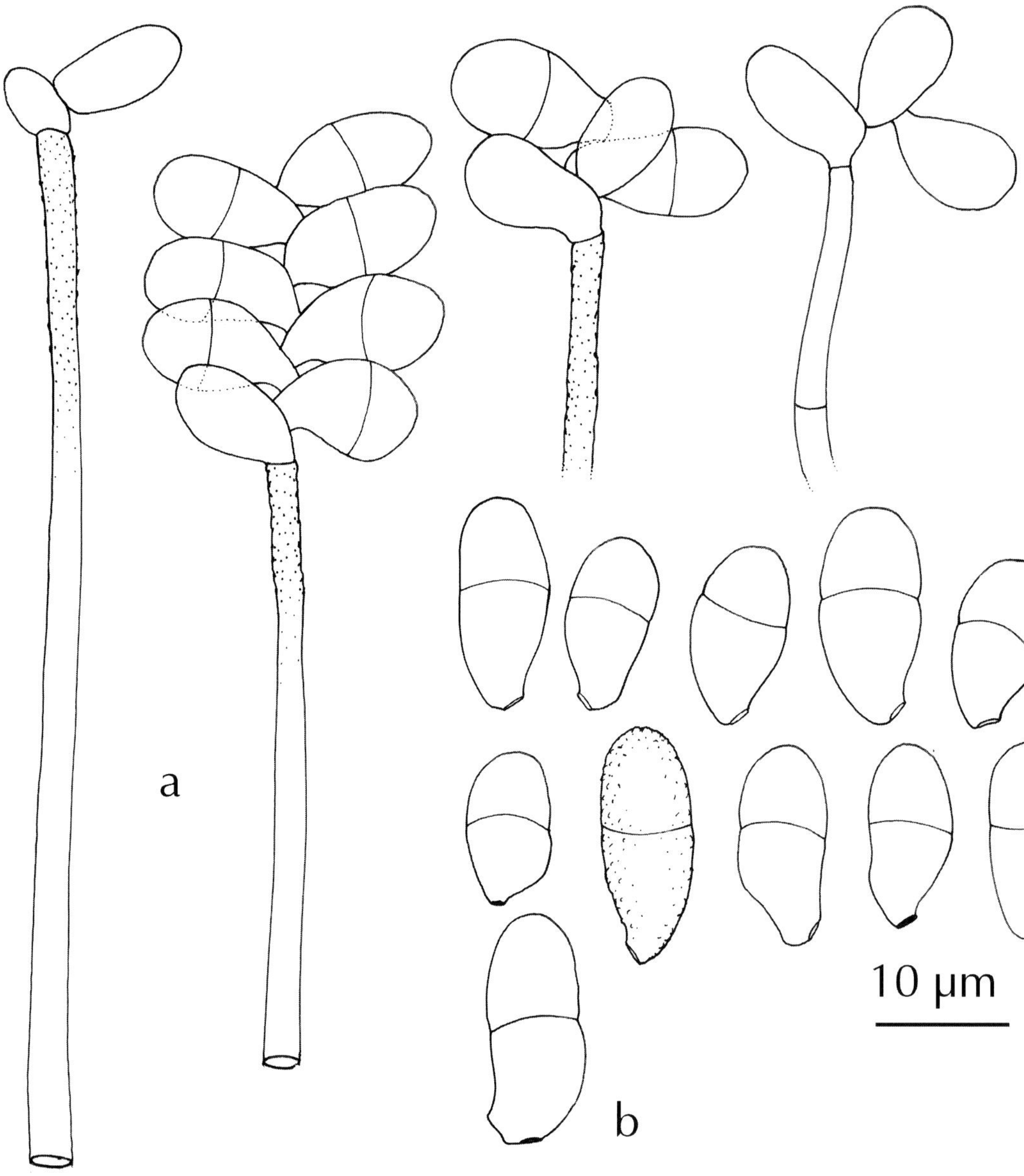

Fig. 107. *Trichothecium roseum*. a. Unbranched conidiophores bearing basipetal zig-zag chains of conidia; b. conidia with obliquely basal scars.
Pl. 116. *Trichothecium roseum*. a. Colony on MEA after one week; b-d. conidiophores with conidia in basipetal zig-zag chains, b. x 650, c. x 2850, d. x 2960.

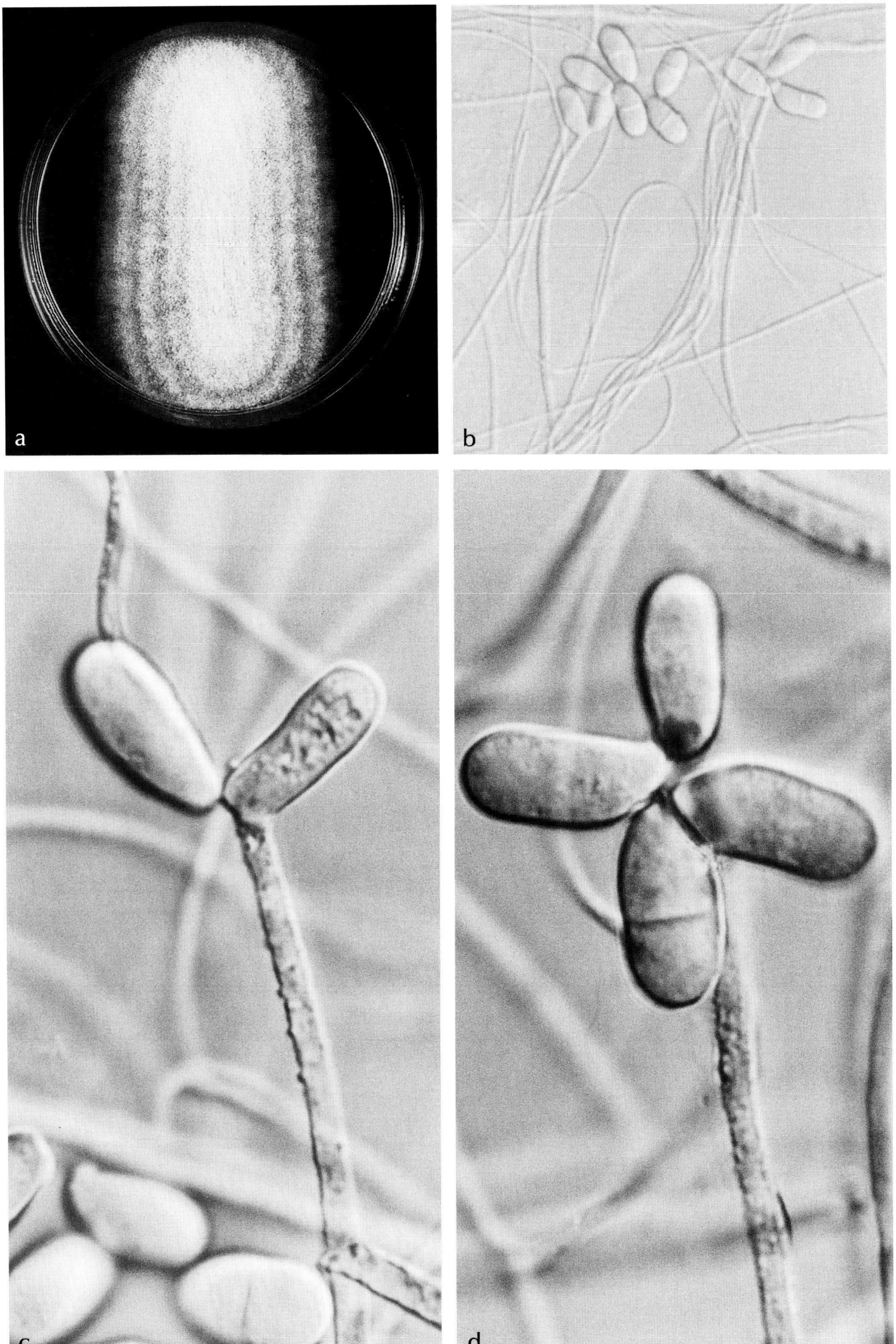
a
b
c
d

Ulocladium chartarum (Preuss) Simmons

Colonies on MEA at 25° C attaining a diameter of 5.5 cm in 7 days, black or olivaceous-black. Conidiophores septate, simple or branched, straight, flexuous, often geniculate, with 1-8 pores, up to 40-50 x 5-7 µm, golden brown, smooth-walled. Conidia solitary or in chains of 2-10, obovoid to short ellipsoidal, 18-38 x 11-20 µm, with 1-5 (commonly 3) transverse and 1-5 longitudinal or oblique septa, medium brown to olivaceous, smooth-walled or verrucose; base conical at first (becoming rounded with age), apex broadly rounded before "false beak" production (see note).

HABITAT: food, indoor
Reported from Canada, Europe (including Great Britain), India, Iraq, Israel, Kuwait, Pakistan, Saudi Arabia, S. Africa. Isolated from soil, dung, emulsion paint, grasses, fibres, wood, paper.

NOTE:
"False beak": each beak is in form and function a conidiophore forming secondary conidia and is therefore distinct from the gradually tapering true beaks of *Alternaria*.

Good sporulation can be obtained by growing cultures on potato-carrot or hay-infusion agar under near UV "black" light.

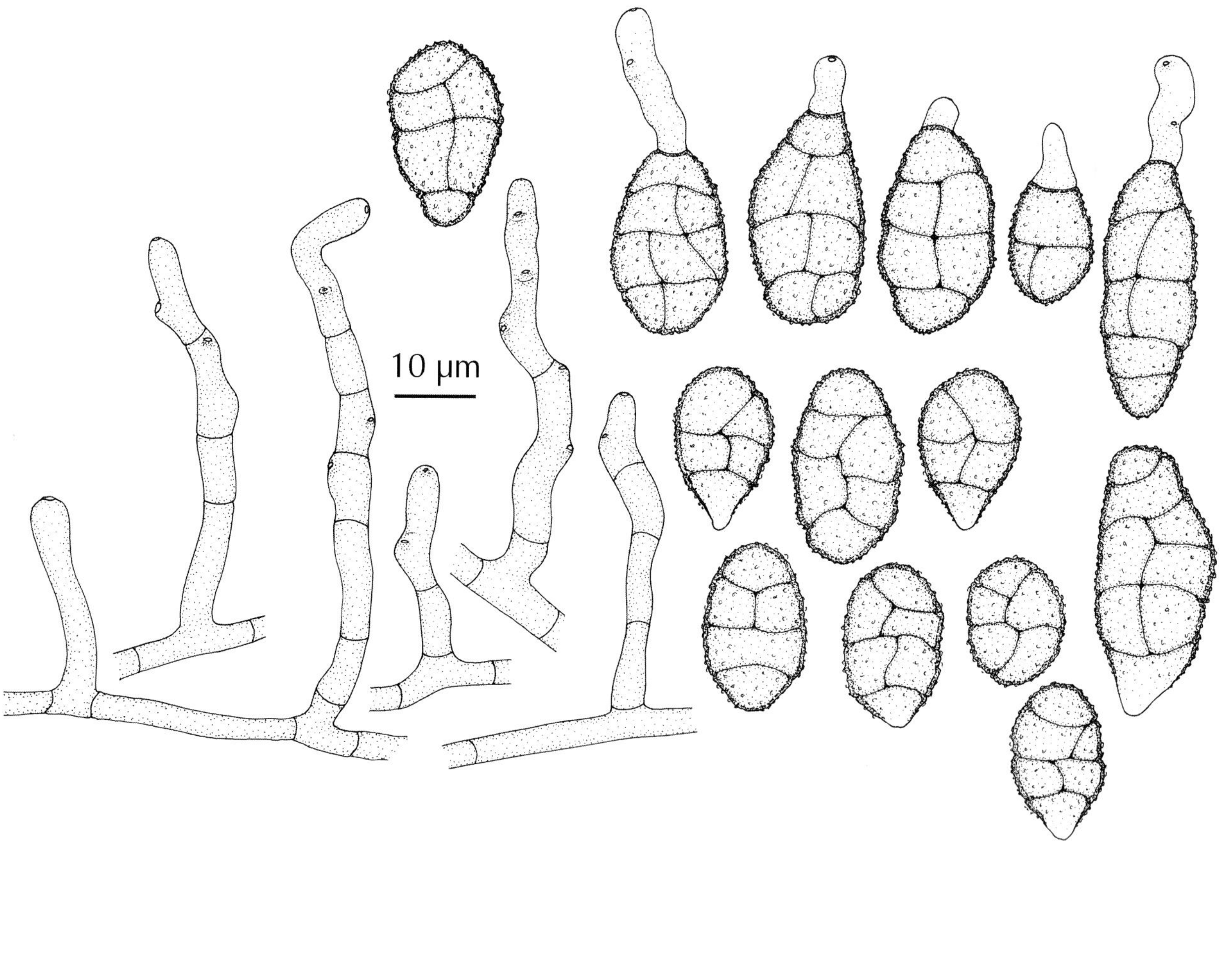

Fig. 108. *Ulocladium chartarum.* Conidiophores and septate conidia.
Pl. 117. *Ulocladium chartarum.* a. Colony on MEA after one week; b-d. strongly geniculate conidiophores and septate conidia, x 600; e. septate conidia, often in chains, x 600.

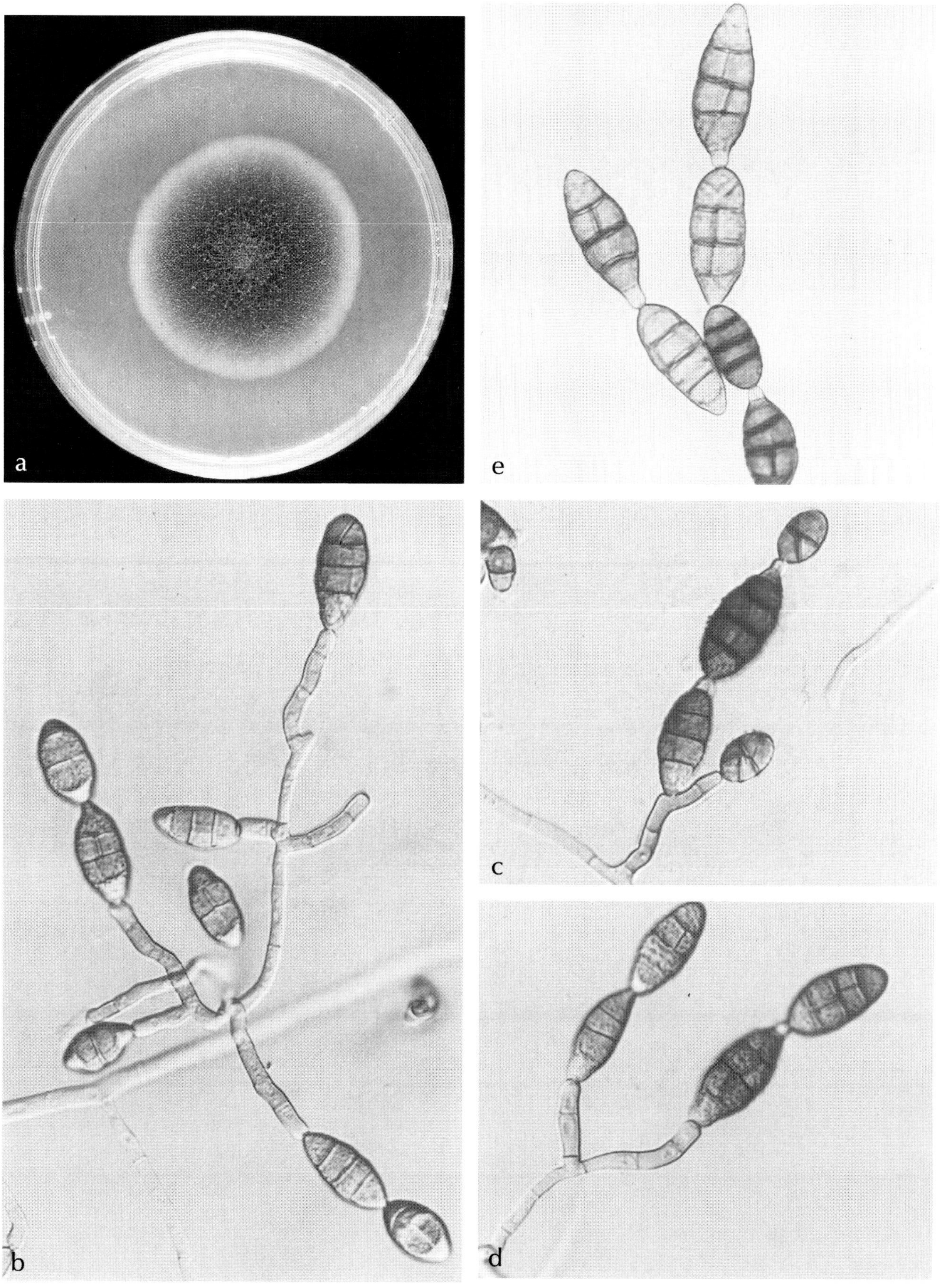
a
e
b
c
d

Wallemia sebi (Fr.) von Arx

Xerophilic. Colonies on MEA + 20% (or + 40%) sucrose at 25° C restricted, attaining a diameter of about 0.5 cm in 10 days, fan-like or stellate, usually powdery, orange-brown to blackish-brown. Conidiophores varying in length, cylindrical, smooth-walled, pale-brown, 1.5-2.5 µm wide at the base, tapering upwards (no collarette), then widening again into a cylindrical verrucose meristem (= fertile hypha) which elongates by intercalary growth at a zone just above the conidiophore apex. The fertile hypha separates into four conidia (resembling arthroconidia, sometimes termed meristem arthroconidia), which mature in basipetal succession. Conidia in groups of 4, initially cubic, later falling apart and becoming almost globose, 2.5-3.5 µm in diam, pale-brown, finely warted.

Temperature: optimum 23-25° C; no growth at 5° C; maximum 36° C (some isolates below).

HABITAT: food, indoor
World-wide distribution. Isolated from dry foodstuffs such as jams, marzipan, dates, bread, cakes, salted fish, bacon, salted beans, milk, but also from fruit, soil, air, hay, textiles, man and animals. Common in indoor environments.

CULTIVATION FOR IDENTIFICATION:
Wallemia sebi grows best on low wateractivity media (e.g. MEA +20% or +40% sucrose). Inoculation in a streak and incubation in diffuse daylight at 23-25° C for 10-14 days.

NOTE:
For more detailed descriptions see Ellis (1971), Domsch *et al.* (1993), for conidiogenesis see Madelin and Dorabjee (1974), Cole and Samson (1979).

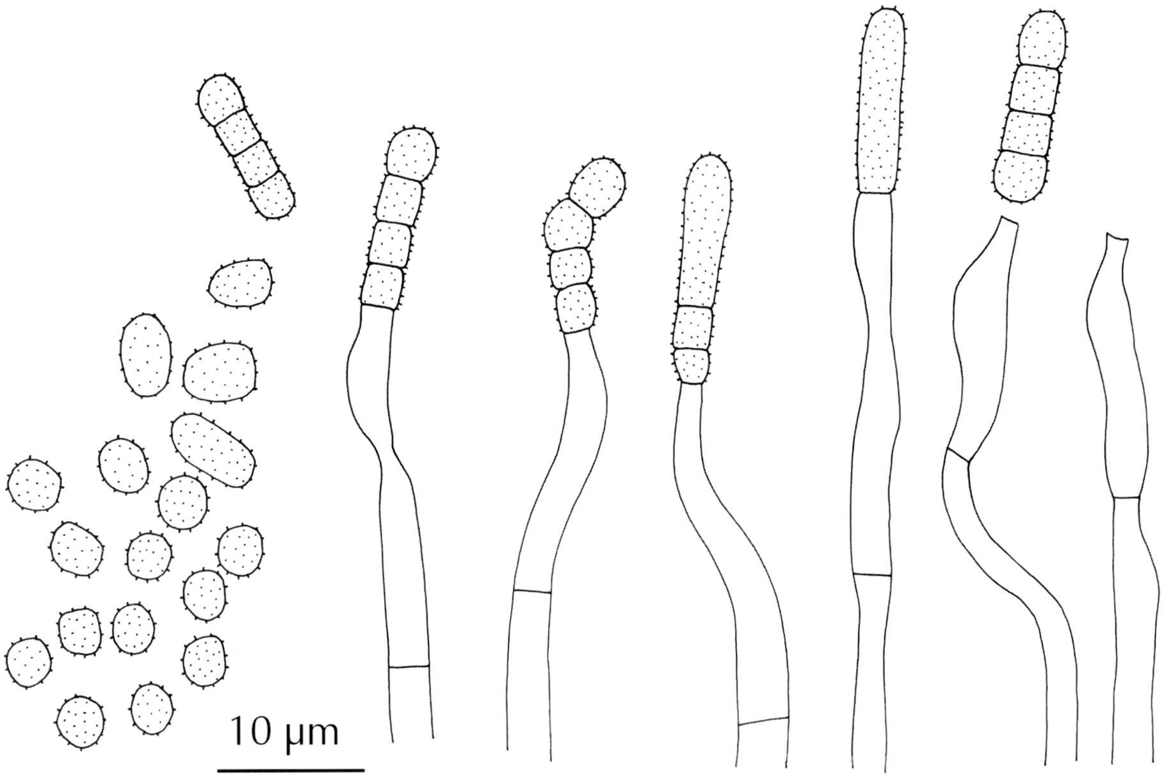

Fig. 109. *Wallemia sebi*. Conidiophores widening into a cylindrical verrucose fertile hypha which separates into conidia.
Pl. 118. *Wallemia sebi*. Colony on MEA + 20% sucrose after 10 days; b. dates with natural infection of *Wallemia sebi*; c-e. conidiophores each with an apical fertile, cylindrical hypha which separates into 4 conidia, maturing in basipetal succession, c-d. x 750, e. x 2000; f. conidia, x 2000.

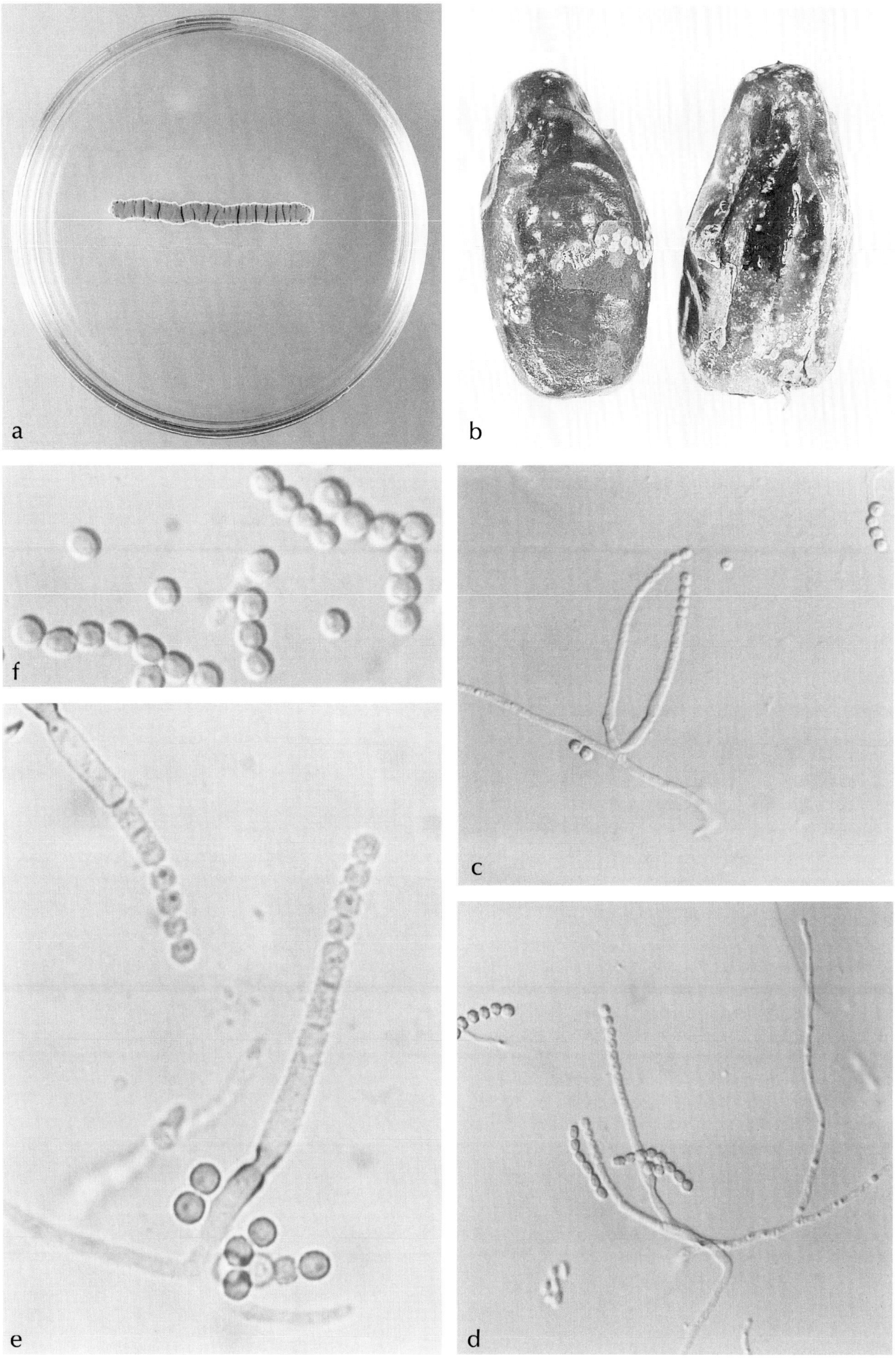
a
b
f
c
e
d

Chapter 1

YEASTS

M. Th. SMITH, D. YARROW and V. ROBERT
Centraalbureau voor Schimmelcultures, Yeast-Division Utrecht, The Netherlands.

Yeasts are generally defined as unicellular fungi which, with few exceptions, reproduce vegetatively by budding. Exceptions are species of the genera *Schizosaccharomyces*, which reproduce by a fission process, and of the genera *Sterigmatomyces* and *Fellomyces*, which form buds on short stalks.

Budding occurs when a protuberance arises on the wall of a cell and grows in size. The bud may detach from the mother cell while it is still quite small, or it may not detach until the two cells are about the same size. Sometimes several buds, or generations of buds, remain attached to one another forming a cluster or chain of cells (Fig. 110a). Buds are produced anywhere on the cell (multilateral budding, Fig. 110b); or at both poles of the cell (bipolar budding, Fig. 110c); or at a single site on the cell (monopolar budding).

Fission is when a cell is divided by one or more cross-walls, and the segments partitioned off in this way split off to become separate cells (Fig. 110d).

Stalks, buds that are formed on stalks (Fig. 110e) become detached when the stalk breaks either in the middle or close to the bud, the actual position depending on the genus. Yeasts of the genus *Sterigmatomyces* have been isolated from cheese and flour in France, but are so rarely found on foodstuffs that they will not be considered further.

Filaments, many yeasts can produce filaments under some conditions, though this ability is far more pronounced in some species than in others, and sometimes in some strains of a species more than in others. These filaments are of two types: 1) septate hyphae and 2) pseudohyphae. A septate hypha grows by elongation of the tip followed by the formation of a cross wall known as a septum. The formation of these septa lags slightly behind the growth, as a result the tip cell is always longer than the preceding ones. The hyphae are not usually constricted at the site of the septum (Pl. 119e).

A pseudohypha is a filament formed by the budding of elongated cells which do not separate (Pl. 119f-g). The tip cell is shorter than the preceding ones, and there are constrictions at the site of the septa where the cells are connected.

Arthroconidia arise in septate hyphae when septa are successively formed close together, resulting in a series of somewhat angular cells which are at first loosely joined in chains (Fig. 110f, Pl. 119h). Arthroconidia are formed in the genera *Geotrichum* and *Trichosporon*.

Ballistoconidia are almost globose, oval or kidney-shaped cells which are produced on a short projection arising from a yeast cell (Fig. 110g). They are forcibly discharged when mature by a droplet mechanism. Their presence is characteristic of the genera *Sporobolomyces* and *Bullera* (Pl. 120 a).

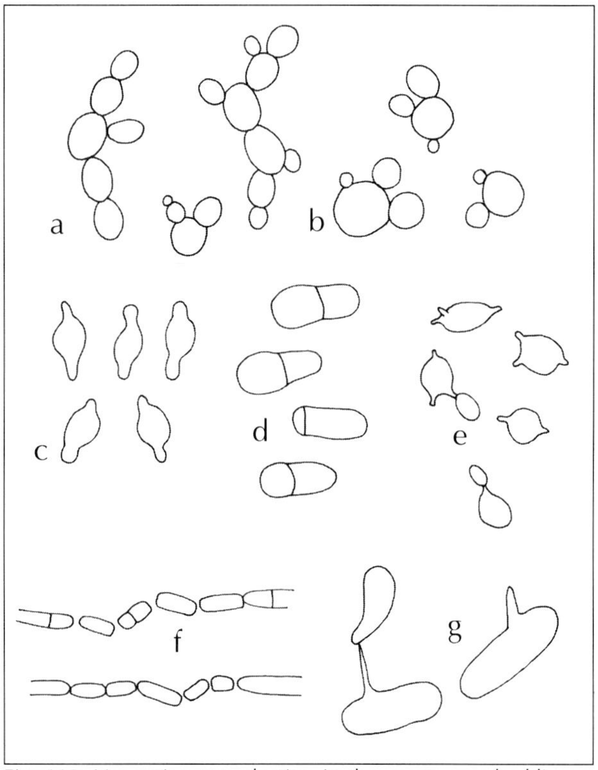

Fig. 110. Vegetative reproduction in the yeasts. a-c. budding, a. clusters of cells. b. multilateral budding, c. bipolar budding. d. fission, e. buds on short stalks, f. arthroconidia, g. ballistoconidia on sterigmata.

Pl. 119. Modes of conidiogenesis and other morphological characteristics of yeasts. a. clusters of cells (*Saccharomyces cerevisiae*), b. multilateral budding (*Debaryozyma hansenii*), c. fission (*Schizosaccharomyces pombe*), d. apiculate cells formed by bipolar budding, (*Hanseniaspora valbyensis*), e. septate hyphae (*Trichosporon beigelii*), f-g. pseudohyphae, f. (*Candida tropicalis*), g. (*Pichia anomala*), h. arthroconidia (*Trichosporon beigelii*). All figures x 900, except a. x 1400.

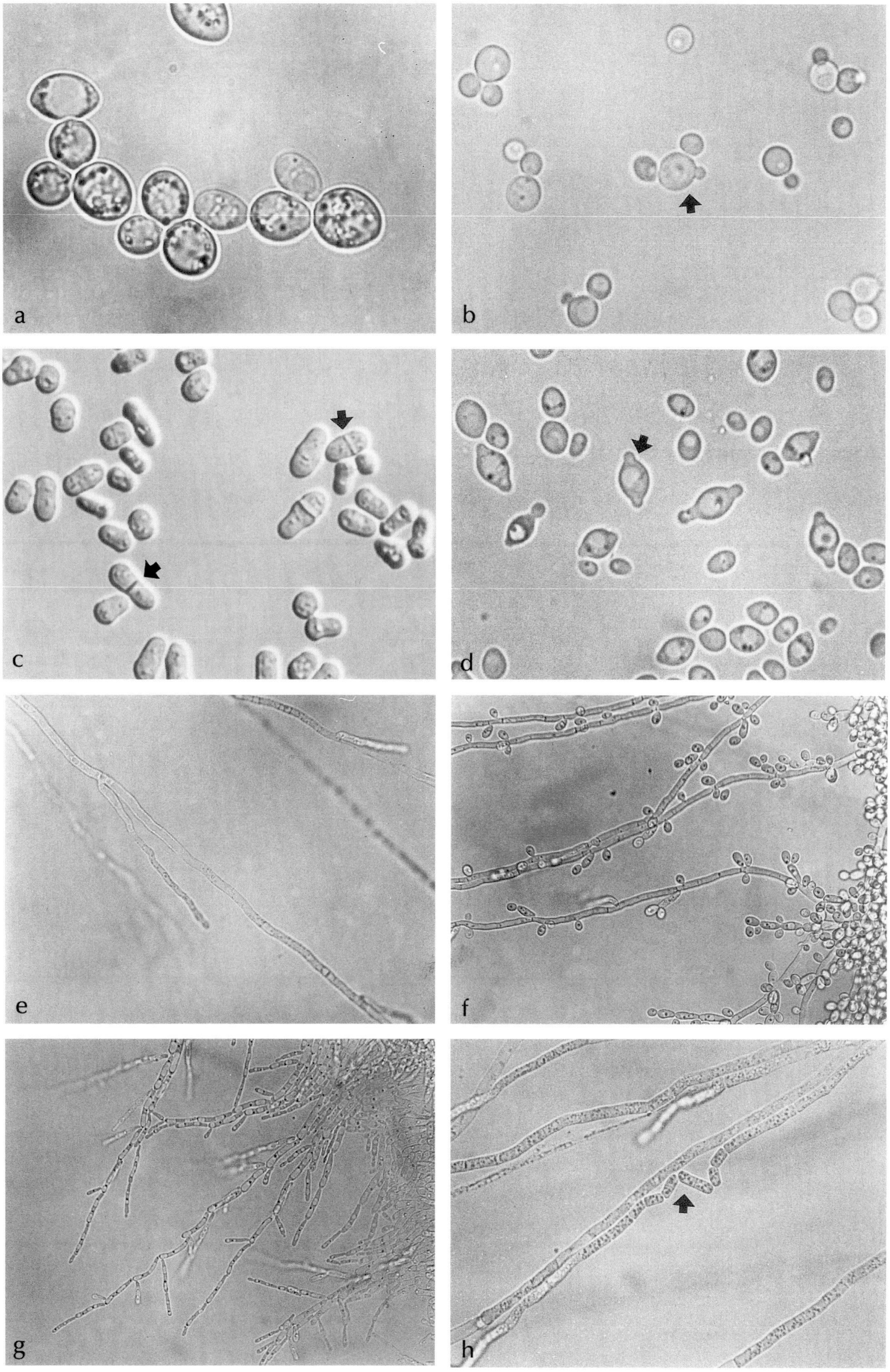
a
b
c
d
e
f
g
h

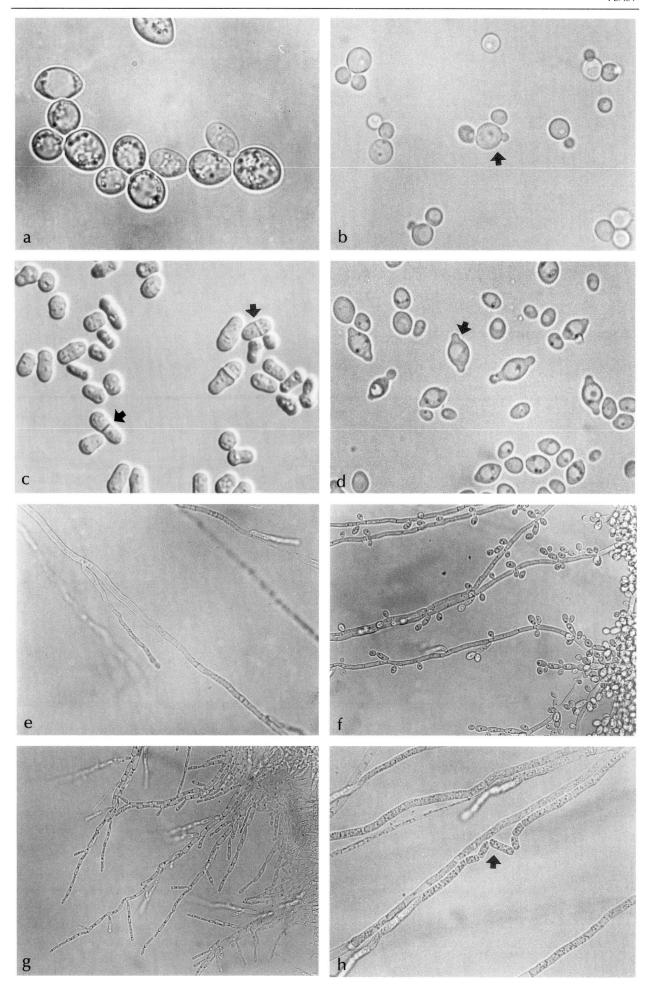

In addition to being able to reproduce vegetatively, some yeasts are able to reproduce sexually. Many do so by means of ascospores in an ascus, and so are known as ascomycetous yeasts.

Asci are formed in one of three ways:

(i) direct transformation of a vegetative cell (unconjugated ascus, Fig. 111a)

(ii) "mother-bud" conjugation (Fig. 111b)

(iii) conjugation between independent single cells (Fig. 111c).

Ascospores are produced in the ascus. The structure and shape of the ascospores varies. They may be smooth or rough-walled; spherical, hat-shaped or reniform (Fig. 111, P. 120 c-h).

Basidiomycetous yeasts reproduce sexually by means of teliospores, and a few by basidia. Since their sexual state is so rarely found in food isolates, this sexual process will not be considered further. Readers interested in pursuing this topic should consult the chapters on these yeasts in Kreger-van Rij (1984).

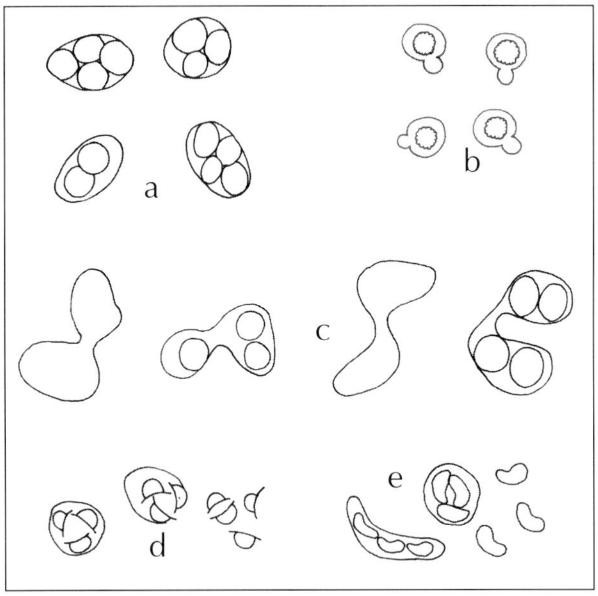

Fig. 111. Sexual reproduction in the yeasts. a. unconjugated asci with ascospores, b. ascus formed by mother bud conjugation with one ascospore, c. ascus formation by conjugation between independent cells with ascospores, d. hat-shaped ascospores, e. kidney-shaped to reniform ascospores.

ISOLATION

Many media can be used to isolate yeasts from foodstuffs (Deak, 1992a,b). Samson *et al.* (1992) recommended media such as malt- extract agar and tryptone-glucose-yeast extract agar (TGY) for products where yeasts usually predominate. Chloramphenicol or oxytetracycline (100 mg/l) can be added, if necessary, to suppress bacteria. Dichloran-rose bengal agar (DRBC) (King *et al.*, 1979) is preferable when yeasts must be enumerated in the presence of moulds.

It is advisable to take into account the substrate and the conditions in which it was produced and stored when selecting media and incubation temperatures. For instance, consider including a medium with a high sugar content when isolating from jam or similar preserves. Yeasts which have become accustomed to a high concentration of sugar may fail to grow or grow only poorly when first cultivated on a medium containing 2% or 4% of sugar. Most yeasts are mesophilic so cultures are usually incubated at 20 to 25°C. Lower temperatures between 4 and 15°C are essential for psychrophilic taxa and should be used when isolating from refrigerated products.

Examination of representative colonies under the microscope is important to verify that one is actually dealing with a yeast.

Acidic media (pH 3.5-5.0)

Either hydrochloric or phosphoric acid can be used to acidify media in order to reduce growth of bacteria. Agar is hydrolysed when autoclaved at low pH values, therefore when preparing acidified media a determined volume of 1 N hydrochloric acid is added to sterilized molten agar only after it has been cooled to 45°C. Adding approximately 0.7% (v/v) of 1 N acid to glucose-peptone-yeast extract (GPY) and YM (yeast-malt-peptone) gives the desired pH of 3.7 to 3.8. Some species, notably those of the genus *Schizosaccharomyces*, are inhibited by very acidic media and are best isolated on moderately acid media with a pH in the range 4.5 to 5.0.

The use of organic acids, such as acetic acid, is not recommended for general purposes. However, malt extract agar containing 0.5% of acetic acid (added just before pouring) is effective for isolating yeasts which are resistant to preservatives (Pitt & Hocking, 1985). Other media incorporating acetic acid are probably equally effective. It may be necessary to add 10% (w/v) of glucose to permit some strains to grow.

Pl. 120. Spore formation in the yeasts. a. ballistoconidium on short projection (*Sporobolomyces roseus*), b. yeast cells with short with short stalks (*Sterigmatomyces halophilus*), c. ascus (mother-bud conjugation) with one round, warty ascospore (*Debaryozyma hansenii*), e. conjugation between independant cells (*Zygosaccharomyces bailii*), f. ascus (conjugation between independant cells) with round, smooth ascospores (*Zygosaccharomyces bailii*), g. hat-shaped ascospores (*Pichia anomala*), h. reniform (kidney-shaped) ascospores (*Kluyveromyces marxianus*). All figures x 900.

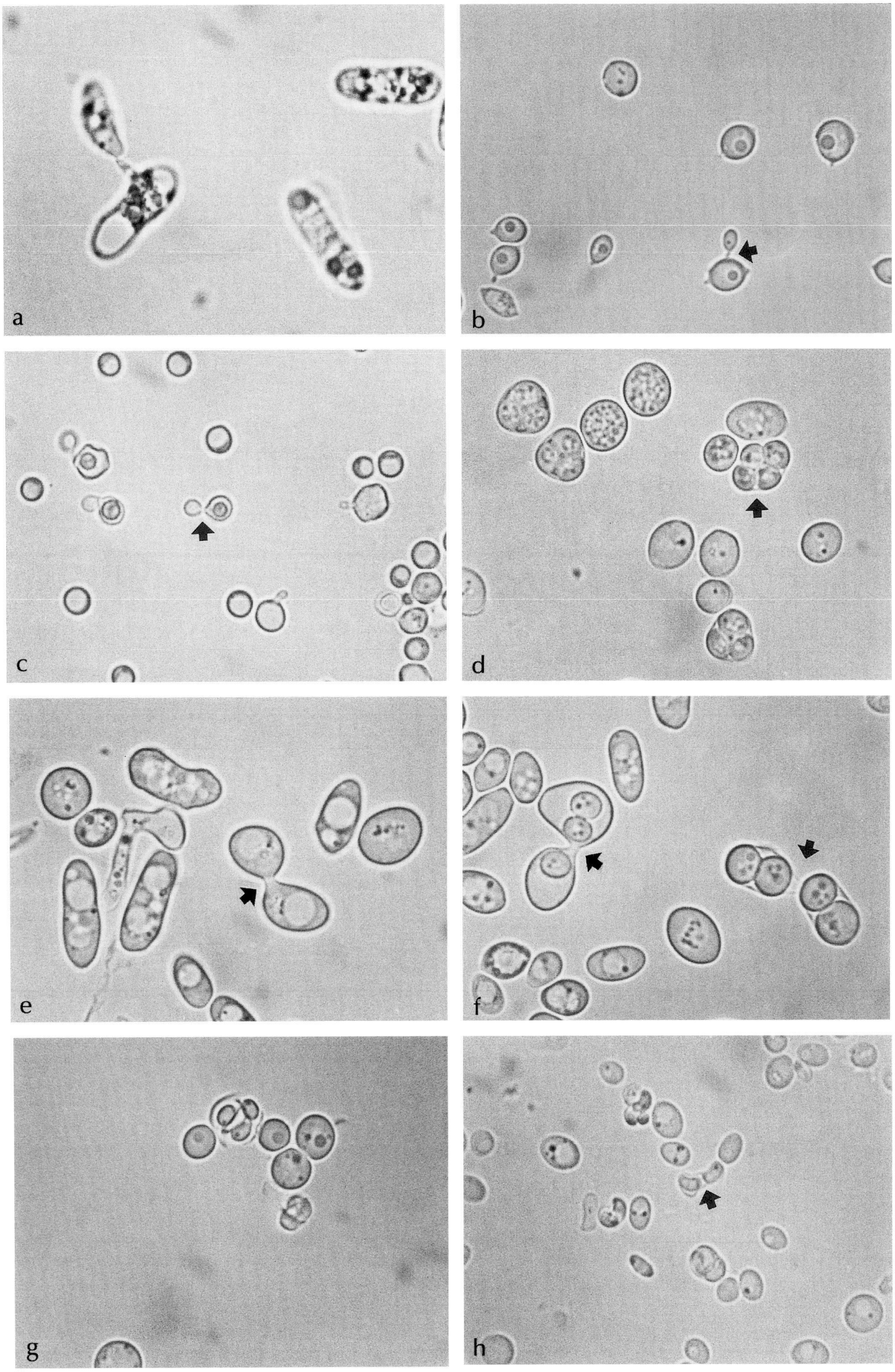
a
b
c
d
e
f
g
h

Direct isolation on solid media

When yeasts are present in large numbers they may be isolated by streaking the material, or suspensions of the material, direct onto solid media. Dilution-plate techniques are used for quantitative studies and should be carried out as described in Chapter 2. A 0.1% solution of peptone is usually used as diluent when counting yeasts in foods and beverages with a high a_w. For concentrates, syrups and other low a_w samples, 20 to 30% of glucose (w/v) should be added to the peptone solution.

Use of membrane filters

Yeasts may be recovered from liquid substrates, and after being washed from solid substrates, by passing the liquid through membrane filters (Mulvany, 1969) and then incubating the filter on the surface of an agar medium. This technique is particularly useful for recovering yeasts when they are present in low concentrations.

Enrichment in liquid media

When few yeast cells are present, their isolation may require enrichment. This is usually accomplished by putting the material into an acidic liquid medium. Moulds may be restricted in their development by excluding air from the culture by pouring a layer, about 1 cm deep, of sterile pharmaceutical paraffin over the surface of the media. This procedure favours the development of fermentative strains and may fail to recover aerobic ones. An alternative and preferable procedure is to incubate the flasks of inoculated media on a rotary shaker (Wickerham, 1951). Moulds are prevented from sporulating and aggregate in pellets. The yeast may be separated by either allowing the pellets of mould to settle for a few minutes, or by removing the pellets by aseptically filtering through a loose plug of sterile glass wool, and then streaking the suspended cells. Both fermentative and non-fermentative strains are recovered by this technique.

Media with low water activity

Most yeasts can grow on media with concentrations of sugar which are high enough to inhibit the development of many bacteria. A medium such as GPY or YM agar containing from 30 to 50% of glucose is suitable for recovering osmophilic and osmotolerant yeasts from foodstuffs and juice concentrates of low water activity. The selective action of such media can be enhanced by lowering the pH to around 4.5. Osmotolerant yeasts recovered in this way can usually be successfully subcultured on media with successively lower amounts of sugar.

Yeasts may also be enriched for isolation in cultures in liquid media such as GPY an YM containing from 30 to 50% of glucose. Osmotolerant moulds are not inhibited at these sugar concentrations, therefore it is advisable to incubate cultures on a rotary shaker.

Purification and maintenance of cultures

Colonies which develop on the primary plates are inspected, preferably under low magnification, for different colonial forms. Single, well separated colonies of each form are selected and streaked. Two platings are generally sufficient to obtain pure cultures, but in some cases more may be needed. Where more than one form persistently appear after streaking single well separated colonies, these may be morphological or sexual variants of a single yeast.

Once in pure culture most yeasts can be cultivated on GPY agar and stored in a refrigerator at 3-4 °C for several months. Cells suspended in a solution of glycerol (10-20%) will survive for years frozen at temperatures of -75°C or lower.

IDENTIFICATION

Cell morphology, the size and shape of the cells, whether reproduction is by budding, fission, or on stalks is noted. Morphology agar or GPY agar are usually used for cultivation, and the cells are examined after 1 to 3 days of incubation by mounting a small amount of the culture in water.

The ability to form filaments is detected by examining a slide culture under the microscope after incubation for 7 to 14 days. The kinds of filament, if present, are noted as well as the presence of arthroconidia. It may also be possible to see ascospores in slide cultures, they are most likely to be found near the edge of the cover slip.

Slide cultures are prepared by aseptically pouring a thin layer of agar onto a sterile microscope slide resting on a sterile U-shaped glass rod in a sterile Petri dish. The agar is inoculated by making a thin streak of cells along the centre of the slide with an inoculating needle. A sterile cover slip is then placed over part of the streak. A few drops of sterile water are put in the Petri dish to retard drying of the agar during incubation. Morphology, potato-dextrose, and corn-meal are the most commonly used agars for this test.

Ballistospores are formed on many media and are detected by looking for the mirror image of the culture, which is formed in the lid of an inverted Petri dish by the discharged spores. Cornmeal agar and morphology agar are particularly suitable media and any yellowish, pink and red strains which have not already given a mirror image should be examined on these media.

Ascospores are sometimes abundantly found in fresh isolates. However, it is generally very time consuming to detect ascospores as it involves examining under the microscope preparations from many media, which often have to be incubated for up to three or four weeks. Should sexual spores happen to be detected, for instance on isolation plates or in slide cultures, the form and ornamentation of the spores, whether the asci persist, and the occurrence of conjugation should be noted, as such observations are a most useful confirmatory feature.

Fermentation of sugars is detected by examining the insert of a Durham tube for the presence of gas over a period of 3 weeks. Durham tubes are prepared by placing a small inverted tube in a 16 x 120 mm test tube and adding about 6 ml of a 2% solution of the test sugar dissolved in a 1% solution of yeast extract. After autoclaving the insert should have completely filled with liquid. Some laboratories use McCartney or bijou bottles with an insert of suitable size instead of test tubes.

Growth on various substances as the sole source of carbon or nitrogen is tested in either liquid or solid media, these tests are often called 'assimilation tests'. The auxanographic method with solid media is not as reliable for carbon sources as that using liquid media, but it is rather quicker and contaminants are more readily detected, which is an advantage when routinely identifying many isolates. This method is perhaps best employed for rapid screening of many isolates into groups. A few representatives of each group are then examined in more detail.

Up to 44 carbon compounds are used in the description of each species: galactose, L-sorbose, cellobiose, lactose, maltose, melibiose, sucrose, trehalose, melezitose, raffinose, inulin, soluble starch, D-arabinose, L-arabinose, D-ribose, L-rhamnose, D-xylose, L-arabinitol, erythritol, galactitol, D-glucitol, glycerol, inositol, D-mannitol, ribitol, xylitol, ethanol, methanol, citric acid, DL-lactic acid, succinic acid, D-gluconate, α-methyl-D-glucoside, salicin, arbutin, D-glucosamine hydrochloride, N-acetyl glucosamine, 2-keto-D-gluconate, 5-keto-D-gluconate, saccharate, D-glucuronate, D-galacturonate, propane-1,2-diol, and butane-2,3-diol. Up to 9 nitrogen compounds are used: nitrate, nitrite, ethylamine, L-lysine, cadaverine, creatine, creatinine, glucosamine, and imidazole.

It is stressed that only pure high-grade chemicals should be used to prepare the media. Appreciable amounts of impurities may be present in some compounds, for instance D-glucose in maltose, and D-galactose in L-arabinose.

Solid media. The auxanographic method of Beijerinck, in which the yeast is suspended in agar in pour plates and the test sugars spotted at intervals around the circumference, is still widely used. Tubes of nitrogen base agar (for carbon growth tests) and carbon base agar (for nitrogen growth tests) media are melted and then cooled to 45°C in a water bath. A few ml of a suspension of the test yeast in water are poured into a sterile Petri dish, a tube of the appropriate medium added, and the whole gently swirled round to mix thoroughly. A small amount of test substance is deposited at 4 to 6 sites (depending on the size of the dishes used) around the perimeter of each dish after the agar has set. Powdered or crystalline chemical is used and transferred on the tip of a small spatula. To avoid over-dosing to toxic levels with ethylamine and nitrite, the tip of an inoculating needle is dipped into a saturated solution and then touched onto the surface of the agar. The Petri dishes are inspected daily for 4 days for an opaque area of growth around the site of the individual carbon and nitrogen compounds.

Liquid media. The tests are done in rimless test tubes (180 mm by 16 mm), either with plugs or caps. Each tube contains a 5 ml amount of liquid yeast nitrogen base medium with one test substrate, except those for negative controls which have no carbon source.

Suspensions of cells from young actively growing cultures are used to inoculate the tubes of test media. For this purpose the yeast is cultivated on any convenient medium such as GPY agar for 24 to 48 hours, or longer if it is particularly slow growing, at a temperature at which the strain grows well. Material is taken from this culture aseptically with a sterile platinum needle or loop and dispersed in about 3 ml of liquid, taking care to avoid carrying over any of the nutrient medium. Either sterile demineralized water or yeast nitrogen base may be used. A white card with black lines approximately 0.75 mm wide is held behind the tube. The suspension is aseptically diluted until the lines become visible through the tube as dark bands. Each of the tubes containing the test carbon sources is then inoculated with 0.1 mm of the suspension. Some laboratories dissolve the test sugars in demineralized water instead of basal medium and dispense 4.5 ml of this solution into test tubes. In this case, 2.5 ml of the suspension of yeast in nitrogen base is pipetted aseptically into 25 ml of basal medium and each tube is then inoculated with 0.5 ml of the resulting suspension.

The tubes of inoculated test media are incubated for up to 3 weeks at 25°C. Strains of some species fail to grow or grow only poorly at these temperatures and a suitable temperature must be used, e.g. 15°C for most psychrophilic yeasts. The tubes are agitated to obtain better aeration and mixing, which gives more reliable and quicker results. This is achieved by putting the tubes on a rotary shaker or a Rollordrum, or better, rocking them. The angle of rocking should be as large as possible without causing wetting of plugs or caps. Results are read after 1 week and 3 weeks. The degree of growth is assessed by eye by placing the tubes, which have been shaken well to thoroughly disperse the yeast, against a white card on which 0.75 mm lines have been drawn approximately 5 mm apart with India ink. If the lines are completely obscured the result is scored as 3+; if the lines appear as diffuse bands the result is scored as 2+; if the lines are distinguishable as such but with blurred edges the result is scored as 1+; if the lines are distinct the result is scored as no growth.

After the final reading several tubes of culture are tested for a starch reaction by adding a few drops of diluted Lugol's iodine. If the culture turns blue the test is scored as positive.

The results of the growth tests are interpreted as follows:

positive: either a 2+ or a 3+ reading after 1 week
positive slow or delayed: either a 2+ or 3+ reading after more than 1 week weakly positive or doubtful: 1+
negative: no growth

Growth on media with cycloheximide is tested at concentrations of 0.01% and 0.1% in the same liquid nitrogen base medium as is used for the carbon growth tests and containing 0.5% glucose. The tubes are inspected for growth for up to 7 days before being recorded as negative.

Resistance to acetic acid, cells are streaked on plates of medium containing 1% of acetic acid and inspected regularly for 6 days. Several strains may be tested on one plate.
Growth with 60% of D-glucose, cells from a young culture are inoculated onto a slant of medium containing glucose (60% w/w), yeast extract and agar. The culture is incubated at 25°C and inspected after 1 and 2 weeks for growth.

Urea test, a loopful of cells from a 24- to 48-hours-old culture is suspended in urea broth and incubated at 37°C (even psychrophiles and strains that do not grow at this temperature). The tubes are inspected at intervals of approximately half an hour for the change of colour to red which indicates that urea has been hydrolysed. Many yeasts which are positive in the urea test produce the colour change within half an hour, the majority do so within two hours, and all have done so within four hours.

Processing the results, traditionally the results of the identification tests and observations are used to follow a dichotomous key. However, the use of such keys has several drawbacks. With some keys, one first determines the genus by testing characteristics such as the presence and form of ascospores before proceeds to a key for the species. Barnett *et al.* (2000 a and b) present keys based on physiological tests only or physiological tests plus morphological and sexual characters, which lead direct to the species. Another drawback to some keys is that they include only a selected set of species which the author considers the most likely to occur in a given situation. An isolate which does not belong in this set may either be misidentified or be unidentifiable with such a key. On the other hand, a key which includes all the yeasts is very long and requires many tests to be done (see Barnett *et al.* 2000, p.1074. where the key to all 704 species involves more than 100 tests). If **all** the results required by any key are not to hand then the identification cannot be completed until they are done. Moreover, if an erroneous result has been recorded for one of the tests, or if one makes a mistake when reading the key and jumps to the wrong point, then either an incorrect identification is made or the organism is found to be unidentifiable.

The use of computer programs, such as Barnett *et al.* (2000) and ALLEV (Robert *et al.*, 1994), avoids many of these shortcomings. The program tries to match the pattern of results obtained against the properties of species in its data matrix. A list of possible species can usually be obtained, even when insufficient tests have been done to allow a definite identification. A list of tests needed to complete the identification can be requested from the program. Moreover, the program permits the user to stipulate that allowance be made during the matching procedure for one or more errors made while doing the tests or typing in the results.

Alternative identification systems, working with growth tests in tubes for the identification of yeasts is time-consuming, labour-intensive and requires experienced and highly-skilled operators.

To avoid some of these problems Deák (1992) proposed a simplified identification scheme for yeasts associated with food. His scheme encompasses 76 species and uses less than 20 physiological tests and morphological observations. Robert *et al.* (1997) have developed ALLEV 2, an automated system for the characterization and the identification of a large panel of physiological properties currently used for the classification of yeasts (see **Table 6** for reliability figures). Commercial identification kits provide computer programs for processing the results obtained with them. These include the API 20 C and API YEAST-IDENT™ kits in which growth tests are done in small wells on plastic strips (see also Table 1 for reviews). Biolog Inc. markets the Microlog™ system for yeasts which detects whether a micro organism can utilize a given carbon source by using the redox dye tetrazolium violet. This dye is irreversibly reduced to a purple formazan if the cells respire the substrate. This system uses a standard micro-titer plate with 96 wells which permits 95 carbon sources and a negative control to be tested simultaneously. Computer programs are provided which allow the results to be read and entered either manually or automatically using a micro-plate reader (Bochner, 1989; see also Table 1). Several other systems (see Table 1 for a partial list), mainly based on growth or enzymatic properties, are also available but their reliability is often limited and the number of species identifiable is usually very limited. Most of them are strongly medically oriented.

Polyphasic identification systems, have very recently been published by Robert (2000, BioloMICS) and the Expert Center for Taxonomic Identification (2000, ETI - Yeasts of the world). These systems can accommodate conventional growth tests (including automated reading of Microplates – BioloMICS - that will soon be available at the Centraalbureau voor Schimmelcultures) as well as morphological and sequence data (26S, 18S and ITS). New microplates have been developed and include all conventional growth tests used in yeast taxonomy. A series of new tests considered to be taxonomically informative have also been incorporated. Using the BioloMICS system, a total of 91 automated physiological tests can now be performed using a microplate reader connected (via RS-232) to a Windows based PC. Morphological characteristics can also be compared. Sequences data are available for almost all species (see GenBank, http://www.ncbi.nlm.nih. gov) and can be compared using the BLASTN program (BioloMICS, ETI), SSEARCH or FASTA, CLUSTALW (BioloMICS). Identifications can be done using either a single set or combine sets of data (e.g. morphology and/or physiology and/or sequences). In addition, the BioloMICS program can also handle electrophoretic gel, bibliographic and taxonomic (hierarchy, synonymy, etc) data. BioloMICS performs phenetic and phylogenetic analyses, and databases can be modified to fit the needs of the users. Many other useful functions are also available. Recently, the Lite internet version of the BioloMICS software has also been published and is visible on the CBS website at Http://www.cbs.knaw.nl. Online consultation of all CBS strains and all currently recognized species descriptions as well as similarity identifications and searches are now possible using a wide range of characteristics (morphology, physiology and/or sequences).

COMMON FOOD-BORNE AND INDOOR YEAST SPECIES

1. *Candida intermedia* (Ciferri & Ashford) Langeron & Guerra
2. *Candida sake* (Saito & Oda) van Uden & Buckley
3. *Candida zeylanoides* (Castellani) Langeron & Guerra
4. ***Cryptococcus albidus*** (Saito) Skinner *et al.*
5. ***Cryptococcus laurentii*** (Kufferath) Skinner
6. *Debaryomyces hansenii* (Zopf) Lodder & Kreger-van Rij, (anamorph: *Candida famata* (Harrison) S.A. Meyer & Yarrow)
7. Saccharomycopsis fibuligera (Lindner) Klöcker
8. *Galactomyces geotrichum* (Butler & Petersen) Redhead & Malloch (anamorph: *Geotrichum candidum* Link)
9. *Geotrichum klebahnii* (Stautz) Morenz.
10. *Issatchenkia orientalis* Kudryavtsev (anamorph: *Candida krusei* (Castellani) Berkhout)
11. *Kluyveromyces lactis* (Dombrowski) Van der Walt (anamorph: *Candida sphaerica* (Hammer & Cordes) S.A. Meyer & Yarrow)
12. *Kluyveromyces marxianus* (Hansen) van der Walt (anamorph: *Candida kefyr* (Beijerinck) van Uden & Buckley)
13. *Pichia anomala* (E.C. Hansen) Kurtzma (anamorph: *Candida pelliculosa* Redaelli)
14. *Pichia fermentans* Lodder (anamorph: *Candida lambica* (Lindner & Genoud) van Uden & Buckley
15. *Pichia guilliermondii* Wickerham (anamorph: *Candida guilliermondii* (Castellani) Langeron & Guerra)
16. *Pichia membranifaciens* (Hansen) Hansen (anamorph: *Candida valida* (Leberle) van Uden & Buckley)
17. ***Rhodotorula glutinis*** (Fres.) Harrison
18. *Rhodotorula mucilaginosa* (Jörgensen) Harrison
19. *Saccharomyces cerevisiae* Meyen ex Hansen (anamorph: *Candida robusta* Diddens & Lodder)
20. *Saccharomyces exiguus* Reess ex Hansen (anamorph: *Candida holmii* (Jörgensen) S.A. Meyer & Yarrow)
21. ***Sporobolomyces roseus*** Kluijver & van Niel
22. *Torulaspora delbrueckii* (Lindner) Lindner (anamorph: *Candida colliculosa* (Hartmann) S.A. Meyer & Yarrow)
23. *Trichosporon cutaneum* (de Beurmann *et al.*) M. Ota
24. *Trichosporon pullulans* (Lindner) Diddens & Lodder
25. *Yarrowia lipolytica* (Wickerham *et al.*) van der Walt & von Arx (anamorph: *Candida lipolytica* (Harrison) Diddens & Lodder)
26. *Zygosaccharomyces bailii* (Lindner) Guilliermond
27. Zygosaccharomyces rouxii (Boutroux) Yarrow

Names in **bold** are indoor species: frequently encountered in water tanks, humidifiers etc. *Rhodotorula* and *Sporobolomyces* can be commonly detected in air samples.

Table 6. Characteristics of some yeast identification systems.

System	Characters	Method	N° of tests	N° of species in database	Time (hr) required for result	Accuracy (%)	References
ALLEV 2.00	Growth, Morphology	Automated/manual	96	693	240	98.2	Robert *et al.*, 1997
API 20C strip	Growth	Manual	20	42	72	78-98.9 88 99.3 72.1	St.-Germain and Beauchesne, 1991 Deak and Beuchat, 1993 Fenn *et al.*, 1994 Davey *et al.*, 1995
ATB 32 ID	Growth	Automated/manual	32	63	48	76 98	Deak and Beuchat, 1993 Buchaille *et al.*, 1998
AutoMicrobic	Growth	Automated	30	62	24	83 97	Pfaller *et al.*, 1988 El-Zaatari *et al.*, 1990
Auxacolor	Growth	manual	15	26	24-72	92-96.5 69.7 86-89	Waller *et al.*, 1995 Davey *et al.*, 1995 Buchaille *et al.*, 1998
BioloMICS	Growth, Morphology Sequences, gels	Automated/manual	152	704	24-240		http:://www.bio-aware.com
ETI	Growth, Morphology Sequences	Manual	152	704	24-240		http://www.eti.uva.nl
MicroLog (Biolog)	Growth/ oxydase	Automated	94	267	24-72	68-70.8	Praphailong *et al.*, 1997
Microring YT	Growth	Manual	6	18	48	53 67.6 60.7	Shankland *et al.*, 1990 McGowan and Mortensen, 1993 De Vuyst *et al.*, 1993
Microscan	Enzyme	Manual/ automated	27	42	4	85.5 96.6 40	Land *et al.*, 1991 St.-Germain and Beauchesne, 1991 Deák and Beuchat, 1995
Minitek	Growth	Manual	12	28	72	97	Lin and Fung, 1987
MTP	Growth	Manual	46	?	120-240	?	Seiler and Busse, 1988
Quantum II	Growth	Automated	20	34	24	82 86	Pfaller *et al.*, 1988 Salkin *et al.*, 1985
SIM	Growth	Manual	20	76	?	91	Deak and Beuchat, 1993
Uni-Yeast-Tek	Growth	Manual	15	42	48	40	Salkin *et al.*, 1987
YeastIdent	Enzyme	Manual	20	42	4	55	Salkin *et al.*, 1987

REFERENCES

ARX, J.A. VON 1981. On *Monilia sitophila* and some families of Ascomycetes. Sydowia 34: 13-29.

ARX, J.A. VON, GUARRO J. and M.J. FIGUERAS 1986. *Chaetomium* : The Ascomycete Genus *Chaetomium*, Nova Hedwigia Ht. 84, J.Cramer Verlag, 162 pp.

ARX, J.A. VON, RODRIGUES DE MIRANDA, L., SMITH, M.TH. & YARROW, D. 1977. The genera of yeast and the yeast-like fungi. Stud. Mycol., Baarn 14, 42 pp.

ARX, J.A. VON. 1977. Notes on *Dipodascus, Endomyces* and *Geotrichum* with the description of two new species. Antonie van Leeuwenhoek 43: 333-340.

ARX, J.A. VON. 1981. The genera of fungi sporulating in pure culture, J. Cramer Verlag, 424 pp.

BARNETT, J.A., PAYNE, R.W. & YARROW, D. 1983. Yeasts characteristics and identification. Cambridge Univ. Press, 811 pp.

BARNETT, J.A., PAYNE, R.W. & YARROW, D. 2000a. The Yeasts: Characteristics and Identification. Third Edition. Cambridge: Cambridge University Press.

BARNETT, J.A., PAYNE, R.W. & YARROW, D. 2000b. Yeast Identification PC Program. Norwich: J.A. Barnett.

BAYNE, H.G. & MICHENER, D.H. 1979. Heat resistance of Byssochlamys ascospores. Appl. Environm. Microbiol., p. 449-453.

BEECH, F.W. & DAVENPORT, R.R. 1969. The isolation of nonpathogenic yeasts. In: Isolation Methods for Microbiologists. (Eds: Shapton, D.A, & Gould, G.W.) London: Academic Press. pp. 71-88.

BEECH, F.W. & DAVENPORT, R.R. 1971. Isolation, purification and maintenance of yeasts. In: Methods in Microbiology. Vol. 4. (Eds: Morris, J.R. & Ribbons, D.W.) London: Academic Press. pp. 153-182.

BEECH, F.W. *et al.* 1980. Media and methods for growing yeasts: proceedings of a discussion meeting. In: Biology and Activities of Yeasts. (Eds: Skinner, F.A., Passmore, S.MS.M. & Davenport, R.R.) London: Academic Press. pp. 259- 293.

BEUCHAT, L.R. & RICE, S.L. 1979. Byssochlamys spp. and processed fruits. Advances food Res. 25: 237-288.

BEUCHAT, L.R. 1987. Food and beverage mycology. Second edition. AVI publishing Co., Inc. Wesport Connecticut, 527 pp.

BISSETT, J. 1984. A revision of the genus *Trichoderma*. I. Section Longibrachiatum sect. nov. Can. J. Bot. 62: 924-931.

BISSETT, J. 1991 a. A revision of the genus *Trichoderma*. II. Infrageneric classification. Can. J. Bot. 69: 2357-2372.

BISSETT, J. 1991 b. A revision of the genus *Trichoderma*. III. Section Pachybasium. Can. J. Bot. 69: 2373-2417.

BISSETT, J. 1991 c. A revision of the genus *Trichoderma*. IV. Additional notes on section Longibrachiatum. Can. J. Bot. 69: 2418-2420.

BISSETT, J. 1992. *Trichoderma atroviride*. Can. J. Bot. 70: 639-641

BLASER, P. 1975. Taxonomische und physiologische Untersuchungen über die Gattung *Eurotium* Link. ex Fr. Sydowia 28: 1-49.

BOCHNER, B. 1989. "Breathprints" at the microbial level. ASM News 55: 536- 539.

BOEREMA, G.H. & BOLLEN, G.J. 1975. Conidiogenesis and conidial septation as differentiating criteria between *Phoma* and *Ascochyta*. Persoonia 8: 111-144.

BOEREMA, G.H. & DORENBOSCH, M.M.J. 1973. The *Phoma* and *Ascochyta* species described by Wollenweber and Hochapfel in their study on fruit-rotting. Stud. Mycol., Baarn 3, 50 pp.

BOEREMA, G.H., DORENBOSCH, M.M.J. & KESTEREN, H.A. VAN 1965 and 1971. Remarks on species of *Phoma* referred to Peyronellaea. I-III. Persoonia 4: 47-68; 5: 201-205; 6: 171-177.

BOOTH, C. 1971. The genus *Fusarium*. Commonw. Mycol. Inst. Kew, 237 pp.

BOTHAST, R.J. & FENNELL, D.I. 1974. A medium for rapid identification and enumeration of *Aspergillus flavus* and related organisms. Mycologia 66: 365-369.

BOYSEN, M., SKOUBOE, P., FRISVAD,J.C. & ROSSEN, L. 1996. Reclassification of the *Penicillium roqueforti* group into three species on the basis of molecular genetic and biochemical profiles. Microbiology (UK) 142: 541-549.

BUCHAILLE, L., FREYDIÈRE, A.M., GUINET, R., GILLE, Y. 1998. Evaluation of six commercial systems for identification of medically important yeasts. European Journal of Clinical Microbiology & Infectious Diseases 17: 479-488.

BURGESS, L.W. & LIDDELL, C. M. 1983. Laboratory manual for Fusarium research. Univ. of Sydney, 162 pp.

BURGESS, L.W., LIDDELL, C.M., AND SUMMERELL, B.A. 1988. Laboratory manual for *Fusarium* research. 2nd Edition. 2nd ed. The University of Sydney, Sydney.

CARMICHAEL, J.W. 1957. *Geotrichum candidum*. Mycologia 49: 820-830.

CARMICHAEL, J.W., KENDRICK, W.B., CONNERS, I.L. & SIGLER, L. 1980. Genera of Hyphomycetes. Univ. Alberta Press., 386 pp.

COLE, G.T. & SAMSON, R.A. 1979. Patterns of development in conidial fungi. Pitman, London, San Fransisco, Melbourne, 190 pp.

DAKIN, J.C. & STOLK, A.C. 1968. *Moniliella acetoabutens*: some further characteristics and industrial significance. J. Food-Technol. 3: 49-53.

DAVENPORT, R.R. 1980. An outline guide to media and methods for studying yeasts and yeast-like organisms. In: Biology and Activities of Yeasts. (Eds: Skinner, F.A., Passmore, S.M. & Davenport, R.R.) London: Academic Press. pp. 261-263.

DAVEY, K.G., CHANT, P.M., DOWNER, C.S., CAMPBELL, C.K. & WARNOCK, D.W. 1995. Evaluation of the auxacolor system, a new method of clinical yeast identification. Journal of Clinical Pathology 48: 807-809.

DE HOOG, G.S., SMITH, M.Th. & GUÉHO, E. 1986. A revision of the genus *Geotrichum* and its teleomorphs. Stud. mycol., Baarn 29: 1-131.

DE VUYST, D., VERHAEGEN, J., SURMONT, I. & VERBIST, L. 1993. Microring YT yeast identification system evalution of 168 strains Journal de Mycologie Médicale 3: 107-108.

DEAK, T. & BEUCHAT, L.R. 1995. Evaluation of the microscan enzyme-based system for the identification of foodborne yeasts Journal of Applied Bacteriology 79: 439-446.

DEAK, T. & BEUCHAT, L.R.. 1993. Comparison of the SIM, API 20C, and ID 32C systems for identification of yeasts isolated from fruit juice concentrates and beverages. J. Food Prot. 56(7): 583-592.

DEAK, T. 1992. Experiences with, and further improvements to, the Deak and Beuchat simplified identification scheme for food-borne yeasts. In SAMSON, R.A., HOCKING, A.D., PITT, J.I. & A.D. KING. (eds.) Modern Methods in Food Mycology, Amsterdam: Elsevier, pp. 47-54.

DEAK, T. 1992. Media for enumerating spoilage yeasts - a collaborative study. In SAMSON, R.A., HOCKING, A.D., PITT, J.I. & A.D. KING. (eds.) Modern Methods in Food Mycology, Amsterdam: Elsevier, pp. 31-38.

DOMSCH, K.H., GAMS, W. & ANDERSON, T. 1993. Compendium of soil fungi. Vol. I, II., reprint IHW Verlag, Eching, Germany, 859 + 405 pp.

DORENBOSCH, M.M.J. 1970. Key to nine ubiquitous soil-borne *Phoma*-like fungi. Persoonia 6: 1-14.

DÖRGE, T., CARSTENSEN, J.M. & FRISVAD, J.C. 2000. Direct identification of pure *Penicillium* species using image analysis. J. Microbiol. Meth. 41: 121-133.

ECKHARDT, C. & AHRENS, E. 1977. *Byssochlamys fulva* Olliver & Smith. Chem. Mikrobiol. Technol. Lebensm. 5: 71-75.

ELLIS, J.J. & HESSELTINE, C.W. 1966/67. *Absidia* with ovoid sporangiospores II. Sabouraudia 5: 59-76.

ELLIS, M.B. 1971. Dematiaceous Hyphomycetes, Commonw. Myol. Inst. Kew, 688 pp.

ELLIS, M.B. 1976. More Dematiceous Hyphomycetes Commonw. Mycol. Inst. Kew, 507 pp.

EL-ZAATARI, M., PASARELL, L., MACGINNIS, M.R., BUCKNER, J., LAND, G.A., & SALKIN, I.F. 1990. Evaluation of the updated Vitek yeast identification data base. J. Clin. Microbiol. 28:1938-1941.

EXPERT CENTER FOR TAXONOMIC IDENTIFICATION. 2000. Yeasts of the world. http://www.eti.uva.nl

FENN, J.P., SEGAL, H., BARLAND, B., DENTON, D., WHISENANT, J., CHUN, H., CHRISTOFFERSON, K., HAMILTON, L. & CARROLL, K. 1994. Comparison of updated vitek yeast biochemical card and API 20C yeast identification systems Journal of Clinical Microbiology 32: 1184-1187.

FISCHER, N.L, BURGESS, L.W., TOUSSON, T.A. & NELSON, P.E. 1982. Carnation leaves as a substrate and for preserving cultures of Fusarium species. Phytopathology 72: 151-153.

FRISVAD, J.C. & FILTENBORG, O. 1983. Classification of terverticillate Penicillia based on profiles of mycotoxins and other secondary metabolites. Appl. Environ. Microbiol. 46: 1301-1310.

FRISVAD, J.C. & FILTENBORG, O. 1989. Terverticillate penicillia: chemotaxonomy and mycotoxin production. Mycologia 81: 837-861.

FRISVAD, J.C. 1981. Physiological criteria and mycotoxin production as aids in identification of common asymmetric penicillia. Appl. Environ. Microbiol. 41: 568-579.

FRISVAD, J.C. 1985. Classification of asymmetric penicillia using expressions of differentiation. In Advances in *Penicillium* and *Aspergillus* systematics (Samson, R.A. & Pitt, J.I., eds.), Plenum Publ. Corp., New York and London. p. 415-457.

FRISVAD, J.C., SAMSON, R.A., RASSING, B.R., VAN DER HORST, M.I., VAN RIJN, F.T.J. & STARK, J. 1997. *Penicillium discolor*, a new species from cheese, nuts and vegetables. Antonie van Leeuwenhoek 72: 119-126.

FRISVAD, J.C., SEIFERT, K.A., SAMSON, R.A. & MILLS, J.T. 1994. *Penicillium tricolor*, a new mold species from Canadian wheat. Canad. J. Bot. 72: 933-939.

GAMS, W. & SAMSON, R. A. 1986. Typification of *Aspergillus* and related teleomorphs genera. In Advances of *Penicillium* and *Aspergillus* systematics (Samson, R.A. & Pitt, J.I., eds.), Plenum Publ., p. 23-30.

GAMS, W. 1971. *Cephalosporium*-artige Schimmelpilze (Hyphomycetes). G. Fischer, Stuttgart, 262 pp.

GAMS, W., CHRISTENSEN, M., ONIONS, A.H.S., PITT, J.I. & SAMSON, R. A. 1986. Infrageneric taxa of *Aspergillus*. In Advances of *Penicillium* and *Aspergillus* systematics (Samson, R. A. & Pitt, J.I., eds.), Plenum Publ., p. 55-62.

GAMS, W., HOEKSTRA, E..S. & APTROOT, A. (eds.) 1998. CBS-Course of Mycology, fourth edition, Centraalbureau voor Schimmelcultures, Baarn, 165 pp.

GERLACH, W. & NIRENBERG, H. 1982. The genus *Fusarium*, a pictorial atlas. Mitt. Biol. Bundesanst. Land. Forstwissensch., Berlin-Dahlem. 406 pp.

HANLIN, R.T. 1973. Keys to the families, genera and species of the Mucorales, J. Cramer Verlag, 49 pp.

HASHMI, M.H., MORGAN-JONES, G. & KENDRICK, B. 1972. Conidium ontogeny in hyphomycetes. *Monilia* state of Neurospora sitophila and *Sclerotinia laxa*. Can. J. Bot. 50: 2419-2421.

HERMANIDES-NIJHOF, E.J. 1977. *Aureobasidium* and allied genera. Stud. Mycol., Baarn 15: 141-177.

HESSELTINE, C.W. & ELLIS, J.J. 1973. Mucorales. In The Fungi, vol IVb. (Ainsworth, G.G. et al. eds.), p. 187-217.

HOLMES, G.J., ECKERT, J.W. & PITT, J.I. 1994. Revised description of *Penicillium ulaiense* and its role as a pathogen of citrus fruits. Phytopathology 84: 719-727.

HOOG, G.S. DE 1979. The Black Yeasts. II: *Moniliela* and allied genera. Stud. Mycol., Baarn 19, 34 pp.

INUI, T., TAKEDA, Y. & IIZUKA, H. 1965. Taxonomically studies on the genus *Rhizopus*. J. gen. appl. Microbiol. II, Suppl., 121 pp.

JARVIS, W.R. 1977. *Botryotinia* and *Botrytis* species: Taxonomy, physiology and pathogenicity. Res. St. Can. Dep. Agric. Harrow, Monogr. 15, 195 pp.

JOFFE, A.Z. 1986. *Fusarium* species. Their biology and toxicology. John Wiley. New York, 583 pp.

JOLY, P. 1964. Le genre *Alternaria*, recherches physioloques, biologiques et systematiques. Encycl. Mycol. 33, Lechevallier, Paris, 250 pp.

KING, A.D. & T. TOROK. 1992. Comparison of yeast identification methods. In SAMSON, R.A., HOCKING, A.D., PITT, J.I. & A.D. KING. (eds.) Modern Methods in Food Mycology, Amsterdam: Elsevier, pp. 39-46.

KING, A.D., HOCKING, A.D. and PITT, J.I. 1979. Dichloran-rose bengal medium for enumeration of molds from foods. Appl. Environ. Microbiol. 37: 959-964.

KING, A.D., PITT, J.I., BEUCHAT, L.R. & CORRY, J.E.L. 1986, Methods for the mycological examination of Food. Plenum Press, New York & London.

KLICH, M.A. & MULLANEY, E.J. 1989. Use of bleomycin-containing medium to distinguish *Aspergillus parasiticus* from *A. sojae*. Mycologia 81: 159-160.

KLICH, M.A. & PITT, J.I. 1988a A laboratory Guide to common Aspergillus species and their teleomorphs. North Ryde, NSW: CSIRO Division of Food Processing.

KLICH, M.A. & PITT, J.I. 1988b Differentiation of *Aspergillus flavus* from *A. parasiticus* and other closely related species. Trans. Brit. Mycol. Soc. 91: 99-108.

KOZAKIEWICZ, Z. 1989. *Aspergillus* species on stored products. Mycological Papers 161: 1-188.

KREGER-VAN RIJ, N.J.W. (Ed.). 1984. The Yeasts, a Taxonomic Study. Amsterdam: Elsevier Scientific.

LAND, G.A., SALKIN, I.F., EL-ZAATARI, M., MCGINNIS, M.R. & HASHEM, G.. 1991. Evaluation of the Baxter-Microscan 4-hour enzyme-based yeast identification system. J. Clin. Microbiol. 29: 718-722.

LARSEN, T.O., FRANZYK, H. & JENSEN, S.R.. 1999. UV-guided isolation of verrucines A and B, novel quinazolones from *Penicillium verrucosum* structurally related to anacine from *Penicillium aurantiogriseum*. J. Nat. Prod. 62: 1578-1580.

LARSEN, T.O., FRISVAD, J.C. & CHRISTOPHERSEN, C. 1998. Arabenoic acid (verrucolone), a major chemical indicator of *Penicillium verrucosum*. Biochem. Syst. Ecol. 26: 463-465.

LIN, C.C.S., & FUNG, D.Y.C. 1987. Comparative biochemical reactions and identification of food yeasts by the conventional methods, Fung's minimethod, Minitek, and the Automicrobic System. CRC Crit. Rev. Biotechnol. 7: 1-16.

LIOU, G.Y., CHEN, C.C., CHIEN, C.Y. & HSU, W.H. 1990. Atlas of the genus *Rhizopus* and its allies. Mycological monograph 3: 1-32. Food industry research and development institute.

LODDER, J. 1979. The Yeasts, a taxonomic study. North Holland Publ.Co, Amsterdam, 1385 pp.

LUND, F. 1995. Differentiating *Penicillium* species by detection of indole metabolites using a filter paper method. Letters in Applied Microbiology 20, 228-231.

LUND, F, FILTENBORG, O. & FRISVAD, J.C. 1998. *Penicillium caseifulvum*, a new species found on *Penicillium roqueforti* fermented cheeses. J. Food Mycol. 1: 95-101.

LUND, F. & FRISVAD, J.C. 1994. Chemotaxonomy of *Penicillium aurantiogriseum* and related species. Mycol. Res. 98: 481-492.

MADELIN, M.F. DORABJEE, S. 1974. Conidium ontogeny in Wallemia sebi. Trans. Br. mycol. Soc. 63: 121-130.

MALLOCH, D. & CAIN, R.F. 1972. The Trichomataceae. Ascomycetes with *Aspergillus, Paecilomyces* and *Penicillium* imperfect states. Can. J. Bot. 50: 2613-2628.

MARASAS, W.F.O. & NELSON, P. E. 1984. Mycotoxicology. Introduction to the mycology, plant pathology, chemistry, toxicology and pathology of naturally occurring mycotoxicoses in animals and man. Pennsylvania State University Press, 104 pp.

MARTINEZ, A. & DE HOOG, G.S. DE 1979. The Black Yeasts, II: *Moniliella* and allied genera. Stud. Mycol., Baarn 19: 37-47.

MCGOWAN, K.L. & MORTENSEN, J.E. 1993. Identification of clinical yeast isolates by using the microring YT Journal of Clinical Microbiology 31: 185-187.

MOLINA, T.C., MISHRA, S.K. & PIERSON, D.L. 1992. A microfermentation test for rapid identification of yeasts. J. Med. Vet. Mycol. 30: 323-326.

MORTON, F.J. & SMITH, G. 1963. The genera *Scopulariopsis* Bainier, *Microascus* Zukal, and Doratomyces Corda. Mycol. Papers 86: 1-96.

MULVANY, J.G. 1969. Membrane-filter techniques in microbiology. In: Methods in Microbiology. Vol. 1. (Eds: Norris, J.R. &

Ribbons, D.W.) London & New York: Academic Press. pp. 205-253.

NEERGAARD, P. 1945. Danish species of *Alternaria* and *Stemphylium*. Communs phytopath. lab. J.E. Ohlens Enke, Copenhagen, 560 pp.

NELSON, P.E., TOUSSON, T.A.& COOK, R.J. (eds.) 1982. *Fusarium*, diseases, biology and taxonomy. Pennsylvania State Univ. Press, 457 pp.

NELSON, P.E., TOUSSOUN, T.A. & MARASAS, W.F.O. 1983. *Fusarium* species. Un illustrated manual for identification Pennsylvania State University Press.

NIRENBERG, H. 1976 Untersuchungen über die morphologische und biologische Differenzierung in der *Fusarium* Sektion Liseola. Mitt. biol. Bund. Anst. Land. Forstw. 169: 117 pp.

NIRENBERG, H.I. 1989. Identification of Fusaria occurring in Europe on cereals and potatoes. In: Chelkowski, J. (ed.): *Fusarium*: Mycotoxins, taxonomy and pathogenicity. Elsevier Science Publishers B.V., Amsterdam. pp. 179-193.

NIRENBERG, H.I. 1990. Recent Advances in the Taxonomy of *Fusarium*. Stud. Mycol. 32: 91-101.

NOTTEBROCK, H., SCHOLER, H.J. & WALL, M. 1974. Taxonomy and identification of Mucor mycosis-causing fungi I. Synonymity of *Absidia ramosa* with *A. corymbifera*. Sabouraudia 12: 64-74.

O'DONNELL, K.L. 1979. Zygomycetes in culture. Palfrey Contributions in Botany 2. Univ. of Georgia, 257 pp.

PELHATE, J. 1975. Mycoflore des mais-fourrages ensiles. Revue Mycol. 39: 65-95.

PETERSON, S.W. 1992. *Neosartorya pseudofischeri* sp. nov. and its relationship to other species in *Aspergillus* section *Fumigati*. Myc. Res. **96** (7): 547-554.

PFALLER, M.A., PRESTON, T., BALE, M., KOONTZ, F.P. & BODY, B.A. 1988. Comparison of the Quantum II, API Yeast Ident, and AutoMicrobic system for identification of clinical yeast isolates. J. Clin. Microbiol. 26: 2054-2058.

PITT, J.I. & HOCKING, A.D. 1997. Fungi and food spoilage. Academic Press 2nd ed., 593 pp.

PITT, J.I. & SAMSON, R.A. 1993. Species names in current use (NCU) in the Trichocomaceae (Fungi, Eurotiales). Regnum Vegetabile 128: 13-57.

PITT, J.I. 1979. The genus *Penicillium* and its teleomorphic states of *Eupenicillium* and *Talaromyces*. Academic Press. London, 634 pp.

PITT, J.I. 2000. A laboratory guide to common *Penicillium* species. Third edition. Commonw. Scient. Industry Res. Organisation. North Ryde, Australia, 197 pp.

PITT, J.I. 1986. Nomenclatorial and taxonomic problems in the genus *Eurotium*. In Advances in *Penicillium* and *Aspergillus* systematics (Samson, R.A. & Pitt, J.I., eds). Plenum publ. New York, 483 pp.

PITT, J.I., HOCKING, A.D. & GLENN, D.R. 1983. An improved medium for the detection of *Aspergillus flavus* and *A. parasiticus*. J. Appl. Bact. 54: 109-114.

PITT, J.I., SPOTTS, R.A., HOLMES, R.J. & CRUICKSHANK, R.H. 1991. *Penicillium solitum* revived, and its role as a pathogen of pomaceous fruit. Phytopathology 81: 1108-1112.

PRAPHAILONG, W., VAN GESTEL, M., FLEET, G.H. & HEARD, G.M. 1997. Evaluation of the biolog system for the identification of food and beverage yeasts Letters in Applied Microbiology 24: 455-459.

RAMIREZ, C. 1983. Manual and atlas of the Penicillia. Elsevier Biomedical Press, Amsterdam, 847 pp.

RAPER, K.B. & FENNELL, D.I. 1965. The genus *Aspergillus*. Williams & Wilkins Co., Baltimore, 686 pp.

RAPER, K.B. & THOM, C. 1949. A manual of the Penicillia. Williams & Wilkins Co., Baltimore, 875 pp.

RIFAI, M.A. 1969. A revision of the genus *Trichoderma*. Mycol. Papers 116: 1-56.

ROBERT, V. 2000.BioloMICS, Biological Manager for Identification and Statistics.Http://www.bio-aware.com.

ROBERT, V., J.-E. DE BIEN, B. BUYCK AND G.L. HENNEBERT. 1994. ALLEV, a new program for computer-assisted identification of yeasts. Taxon, 43, 433-439.

ROBERT, V., P. EVRARD, AND G.L. HENNEBERT. 1997. BCCM/ALLEV 2.00 an automated system for the identification of yeasts. Mycotaxon 64: 433-439.

SALKIN, I.F., LAND, G.A., HURD, N.J., GOLDSON, P.R. & MCGINNIS, M.R. 1987. Evaluation of YeastIdent and Uni-Yeast-Tek yeast identifications systems. J. Clin. Microbiol. 25: 264-267.

SALKIN, I.F., SCHADOW, K.H., BANKAITIS, L.E., MCGINNIS, M.R. & KEMMA, M.E. 1985. Evaluation of Abbott Quantum II yeast identification system. J. Clin. Microbiol. 22: 442-444.

SAMSON, R. A. & GAMS, W. 1986. Typification of the *Aspergillus* species and associated teleomorphs. In Advances of *Penicillium* and *Aspergillus* systematics (Samson, R.A. & J. I. Pitt, eds.), Plenum Publ., p. 31-54.

SAMSON, R. A. & HOEKSTRA, E. S. 1994. Common fungi occurring in indoor environments. In Health Implications Of Fungi In Indoor Environments (eds. R.A. Samson et al.), Elsevier, Amsterdam, pp. 541-587.

SAMSON, R. A. & PITT, J.I. (eds.)1985. Advances in *Penicillium* and *Aspergillus* systematics. Plenum Publishers, London & New York, 483 pp.

SAMSON, R.A. & FRISVAD, J.C. 1991. Current taxonomic concepts in *Penicillium* and *Aspergillus*. In Cereal Grain, Mycotoxin, Fungi and Quality in drying and storage. ed. J. Chelkowski. Amsterdam: Elsevier, pp. 405-439.

SAMSON, R.A. 1974. *Paecilomyces* and some allied Hyphomycetes. Stud. Mycol., Baarn 6: 119 pp.

SAMSON, R.A. 1979. A compilation of the Aspergilli described since 1965. Stud. Mycol., Baarn 18, 40 pp.

SAMSON, R.A. 1992. Current taxonomic schemes of the genus *Aspergillus* and its teleomorphs. In *Aspergillus*: The biology and Industrial applications, edited by J. W. Bennett & M. A. Klich, Butterworth Publishers, pp. 353-388.

SAMSON, R.A. 1994a. Taxonomy - Current concepts in *Aspergillus*. In Biotechnology Handbooks: *Aspergillus* (J. E. Smith ed.). Plenum Publishing Co. pp. 1-22.

SAMSON, R.A. 1994b. Current Systematics of the genus *Aspergillus*. In The genus *Aspergillus*: From Taxonomy and Genetics to Industrial Application (Powell, K.A., A. Renwick Peberdy, J.F, eds.). Plenum Press, London, pp. 261-276.

SAMSON, R.A., HOCKING, A.D., PITT, J.I. & A.D. KING. (eds.) 1992 Modern Methods in Food Mycology, Amsterdam: Elsevier.

SAMSON, R.A.., STOLK, A.C. & HADLOK, R. 1976. Revision of the subsection Fasciculata of *Penicillium* and some allied species. Stud. Mycol., Baarn 11: 47 pp.

SAMUELS, G.J., LIECKFELDT, E. & NIRENBERG, H.I. 1999. *Trichoderma asperellum*, a new species with warted conidia, and redescription of *T. viride* in Sydowia 51(1): 71-88.

SAMUELS, G.J., PETRINI, O., KUHLS, K. , LIECKFELDT, E. & KUBICEK, C.P. 1998. The *Hyprocrea schweinitzii* complex and *Trichoderma* sect. *Longibrachiatum*. Studies in Mycology, Baarn 41: 54 pp

SCHIPPER, M.A.A. 1973. A study on variability in Mucor hiemalis and related species. Stud. Mycol. Baarn 4: 40 pp.

SCHIPPER, M.A.A. 1976. On *Mucor circinelloides, Mucor racemosus* and related species. Stud. Mycol. Baarn 12: 40 pp.

SCHIPPER, M.A.A. 1978. 1. On certain species of *Mucor* with a key to all accepted species. 2. On the genera *Rhizomucor* and *Parasitella*. Stud. Mycol. Baarn 17: 70 pp.

SCHIPPER, M.A.A. 1984. Revision of the genus *Rhizopus*. Stud. Mycol. Baarn 25: pp.

SCHÖLER, H.J. & MULLER, E. 1971. Taxonomy of the pathogenic species of *Rhizopus*. 7th Ann. Meeting of the Br. Soc. Mycopath., Edinburgh.

SCHOL-SCHWARZ, M.B. 1959. The genus *Epicoccum*. Trans Br. mycol. Soc. 42: 149-173.

SCHOL-SCHWARZ, M.B. 1970. Revision of the genus *Phialophora* (Moniliales). Persoonia 6: 59-94.

SEEMÜLLER, E. 1968. Untersuchungen über die morphologische und biologische Differenzierung in der *Fusarium*-Sektion *Sporotrichiella*. Mitt. biol. Bundesanst. Land. Forst. 127, 93 pp.

SEILER, H. & BUSSE, M. 1988. Identifizierung von Hefen mit Mikrotiterplatten. Forum Mikrobiologie. 11: 505-509.

SHANKLAND, G.S., HOPWOOD, V., FORSTER, R.A., EVANS, E.G.V., RICHARDSON, M.D. & WARMOCK, D.W. 1990. Mutlicenter evaluation of Microring YT, a new method of yeast identification. J. Clin. Microbiol. 28: 2808-2810.

SHEAR,C.L. & DODGE, B.O. 1927. Life histories and heterothallism of the red bread-mold fungi of the *Monilia sitophila* group. J. Agric. Research 34, 11: 1019-1042.

SIMMONS, E.G. 1967. Typification of *Alternaria, Stemphylium* and *Ulocladium*. Mycologia 59: 67-92.

SØRENSEN, D., LARSEN, T.O., CHRISTOPHERSEN, C., NIELSEN, P.H. & ANTHONI, U. 1999. Solistatin, an aromatic compactin analogue from *Penicillium solitum*. Phytochemistry 51: 1027-1029.

ST.-GERMAIN, G. & BEAUCHESNE, D. 1991. Evaluation of the Microscan rapid yeast identification panel. J. Clin. Microbiol. 29: 2296-2299.

STEVENS, R.B. (ed.) 1974. Mycology Guide book. Univ. Washington Press, 703 pp.

STOLK, A.C. & DAKIN, J.C. 1966. *Moniliella*, a new genus of Moniliales. Antonie van Leeuwenhoek 32: 399-409.

STOLK, A.C. & SAMSON, R. A. 1985.A new taxonomic scheme for Penicillium anamorphs. In Advances of *Penicillium* and *Aspergillus* systematics (Samson, R.A. & Pitt, J.I, eds.), Plenum Publ., p.163-192.

STOLK, A.C. & SAMSON, R.A. 1971. Studies in *Talaromyces* and related genera I. *Hamigera*, gen. nov. and *Byssochlamys*. Persoonia 6: 341-357.

STOLK, A.C. & SAMSON, R.A. 1972. Studies in *Talaromyces* and related genera I. *Talaromyces*. Stud. Mycol., Baarn 2: 65 pp.

STOLK, A.C. & SAMSON, R.A. 1983. The ascomycete genus *Eupenicillium* and related *Penicillium* anamorphs. Stud. Myc., Baarn 23: 1-149.

SUTTON, B.C. 1980. The Coelomycetes, Fungi imperfecti with pycnidia, acervuli and stromata. Commonw. Myc. Inst. Kew, 696 pp.

THRANE, U. 1989. *Fusarium* species and their specific profiles of secondary metabolites. In: Chelkowski, J. (ed.): *Fusarium*: Mycotoxins, taxonomy and pathogenicity. Elsevier Science Publishers B.V., Amsterdam. pp. 199-225.

TZEAN, S.S. CHEN J.L., LIOU, G.Y., CHEN, C.C. & W.H. HSU. 1990. *Aspergillus* and related teleomorphs of Taiwan. Mycological Monograph no. 1. Culture Collection and Research Center, Taiwan.

UDAGAWA, S. & SUZUKI, S. (1994). *Talaromyces spectabilis*, a new species of food-borne ascomycetes. Mycotaxon 50: 81-88.

VRIES, DE G.A. 1952. Contribution to the knowledge of the genus *Cladosporium*. Diss. Univ. Utrecht, Reprint J. Cramer Lehre 1967.

WALLER, J., CONTANT, G., CROUZIER, C., DEBRUYNE, M. & KOENIG, H. 1995. Evaluation of a new yeast identification system: fungichrom I based on chromogenic substrate hydrolysis and carbohydrate assimilation Journal de Mycologie Médicale 5: 92-97.

WALT, J.P. VAN DER & VAN KERKEN, A.E. 1961. The wine yeasts of the Cape. Part V. Studies on the occurrence of *Brettanomyces intermedius* and *Brettanomyces schanderlii*. Antonie van Leeuwenhoek 27: 81-90.

WALT, J.P. VAN DER & VON ARX, J.A. 1980. The yeast genus *Yarrowia* gen. nov. Antonie van Leeuwenhoek 46: 517-521.

WICKERHAM, L.J. 1951. Taxonomy of Yeasts. United States Department of Agriculture, Technical Bulletin No. 1029. Washington.

WOGAN, G.N. 1965. Mycotoxins in foodstuffs. Mass. Inst. Technol. Press. Cambr. Mass. 291 pp.

WOLLENWEBER, H.W. & REINKING, O.A. 1935. Die Fusarien. P. Parey, Berlin. 335 pp.

WYLLIE, T.D. & MOREHOUSE, L.G. 1977/78. Mycotoxic fungi, mycotoxins, mycotoxicoses, vol. I, II, III, Marcel Dekker, New York, 538 + 570 + 202 pp.

ZYCHA, H., SIEPMANN, R. & LINNEMANN, G. 1969. Mucorales, eine Beschreibung aller Gattungen und Arten dieser Pilzgruppe, J. Cramer Verlag, 355 pp.

Chapter 2

METHODS FOR THE DETECTION, ISOLATION AND CHARACTERISATION OF FOOD-BORNE FUNGI.

R. A. SAMSON, E. S. HOEKSTRA, F. LUND, O. FILTENBORG and J.C. FRISVAD

Centraalbureau voor Schimmelcultures, Utrecht, The Netherlands and BioCentrum-DTU, Technical University of Denmark, DK-2800 Lyngby, Denmark

TABLE OF CONTENTS

INTRODUCTION

Orginally the methods for examination of the mycobiota of foods, soil, and indoor air have been based on bacteriological media developed earlier or on media used in medical mycology, but bacteria and human pathogenic fungi grow on media of high water activity, high temperature, and a low carbohydrate level. Therefore these are quite different from the typical saprophytic food- or airborne fungi which grow well at low water activities, lower temperatures and often a high carbohydrate level. In order to find optimal detection and isolation media for food- and air-borne fungi and standardizing the methods used, four international meetings on food mycology and one on air-borne fungi have been held since 1984 (King *et al.*, 1986; Samson *et al.*, 1992; Samson *et al.*, 1994; International Journal of Food Microbiology, pp. 149-192, 1995; Beuchat *et al.*, 1998; Frändberg and Olsen, 1999). Agreement on standardized methods has been reached. Some of the methods have also been subject to international collaborative studies, but several recommendations made at the four workshops still need to be fully validated. Below we have listed the methods recommended, whether they have been validated or not. The indications from research in several laboratories are that these methods are the most suitable at the present state of knowledge. Furthermore we have suggested some modifications or additions to the methods recommended, in cases where we have had consistently good experience with such amendments in our laboratories.

Besides these traditional methods we have described new chemical and molecular methods for characterisation and identification of fungi. Such methods are more reliable and objective, which is important in solving many problems in the food industry. This very often requires isolation of all relevant fungi, identification of contamination sources, identification of the mycobiota to species level of the dominant part of the fungi able to actually grow in the food, determination of their potential and actual extracellular enzyme, volatile, colourant and mycotoxin production and even fingerprinting to isolate level (molecular methods) for a thorough HACCP analysis.

1. GENERAL PURPOSE METHODS

Direct examination.

In foods spoiled by fungi it is often possible to observe the responsible fungal growth directly by the naked eye followed by observations in a stereomicroscope. This is due to the size of the fungal colony and that fungal growth normally takes place at the surface of the product. However the observation has to be made immediately after growth has taken place, since any handling of the product is bound to remove the visual appearance of the fungi. If fungal growth can be expected, it is always recommended to examine the food products by microscopy. Slides can be prepared by putting a small part or volume in a mounting medium (e.g. lactic acid with aniline blue). Preparations with the aid of adhesive tape as described on page 2 can also be helpful.

When fungal growth has been detected by direct examination, the fungus is streak-inoculated on an appropriate medium. This is preferably done with the aid of a stereomicroscope.

Direct plating.

This is considered to be the more effective technique for mycological examination of all foods. For foods such as grains and nuts, a surface disinfection before direct plating is in most situations, considered essential, to permit enumeration of fungi actually invading the food. An exception is to be made for cases where surface contaminants become part of the downstream mycobiota, e.g. wheat grains to be used in flour manufacture. In such cases grains should be investigated both with and without surface disinfection.

Surface disinfection: food particles are surface disinfected by vigorous shaking in 0.4% freshly prepared chlorine for 2 minutes. A minimum of 100 particles should be disinfected and plated on each chosen medium. The chlorine must only be used once.

Rinse: after pouring off the chlorine, rinse once in sterile distilled or deionised water.

Plating: as quickly as possible, transfer food particles with a sterile forceps to previously poured and set plates, at the rate of 5-10 particles per plate.

Incubation: the standard incubation regime for general-purpose enumerations is 25° C for 5 days. Plates should be incubated upright. The plates can be kept in perforated plastic bags to minimise evaporation. The perforation of the plastic bags, combined with a forced airflow through the incubator is necessary to maintain the initial composition of the atmosphere in contact with the plates. It has been shown that accumulation of CO_2 significantly influences the growth of fungi.

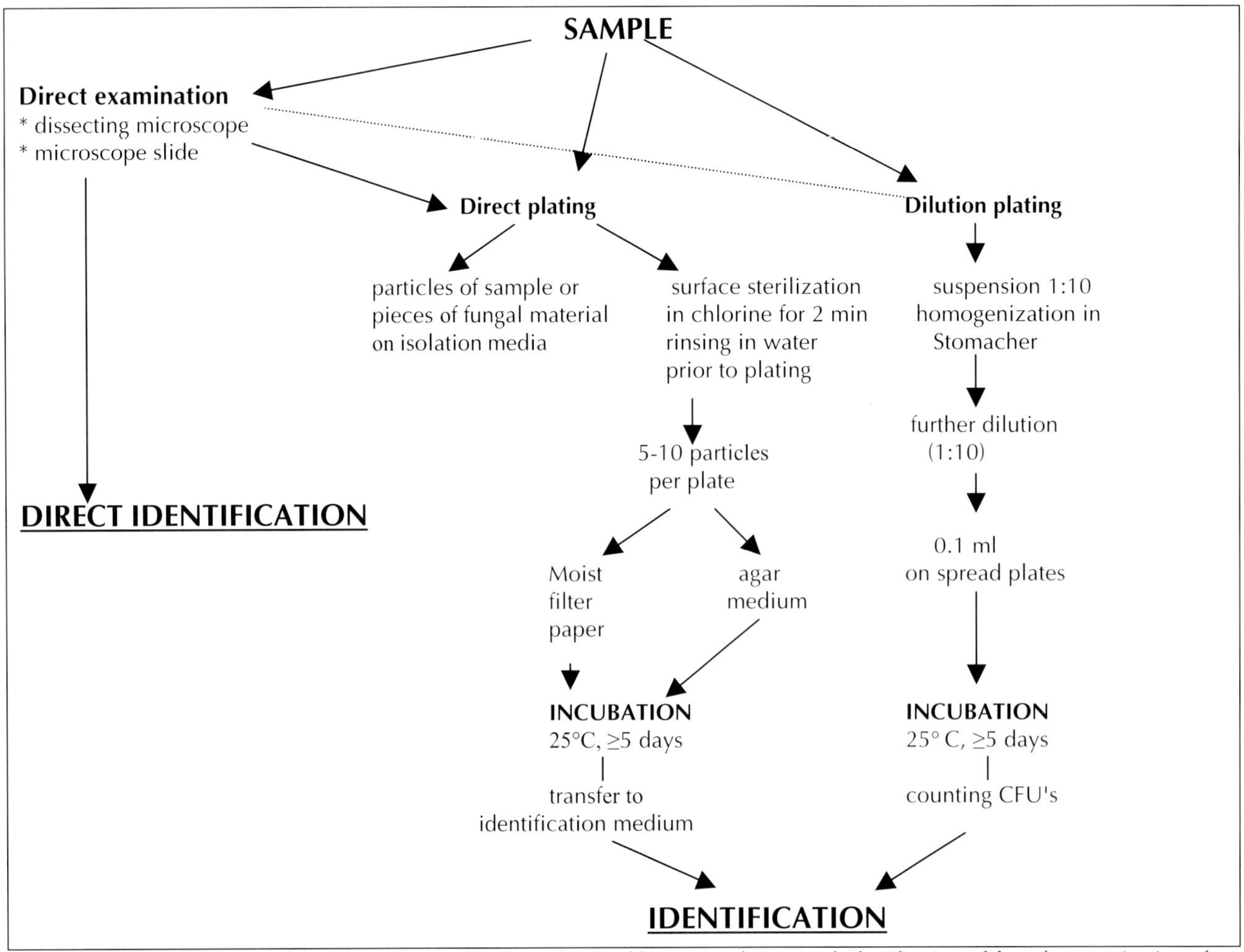

Fig. 1. Flowchart illustrating the methods for examination, plating, dilution, incubation and identification of fungal contamination of a sample.

Results: express results as percentage of particles infected by fungi. Differential counting of a variety of genera and sometimes even species is possible using a stereomicroscope.

Dilution plating.
Sample size: as large a sample as possible should be used. We normally recommend 5 g samples for homogeneous food materials (like flour) and 40 g samples for not homogeneous food materials (like grains).

Initial dilution: the initial dilution should be 1 + 9 in 0.1% peptone.

Soaking: Dried samples, where the fungi are deep seated or internal (e.g. as in grains and nuts), should be soaked for 30 minutes in 0.1% peptone at room temperature before stomaching or blending. For powders and other homogeneous samples, no soaking is required

Homogenisation: use of Stomacher is preferred. If the food material is tough, a blender may be used. Homogenisation in 2 minutes is recommended if a Stomacher is used; 1 minute if a blender is used.

Further dilutions: 1:10 (= 1+9) in 0.1% peptone. We recommend 1:5 (= 1+4) in 0.1% peptone as an alternative, which is especially useful when small concentrations of fungi are present. Normally a maximum dilution of 10^{-3} is sufficient, however if the food has been in contact with soil a maximum dilution of 10^{-5} may be necessary.

Plating: spread plates are recommended over pour plates. Inocula should be 0.1 ml per plate.

Incubation: The standard incubation regime for general purpose enumerations is 25° C for 5 days. Plates should be incubated upright.

2. GENERAL MEDIA (for formulations see page 378-382).

DRBC and DG18.
Dichloran Rose Bengal Chloramphenicol agar (DRBC; King *et al.*, 1979) and Dichloran 18% Glycerol agar (DG18; Hocking and Pitt, 1980) are recommended as general purpose isolation and enumeration media for foods of high water activity, i.e. a_w> 0,90. However the following points should be taken into consideration:

1. DG18 is less suitable for fresh fruits and vegetables, where studies have shown that this medium yields lower counts because basidiomycetous yeasts are often present.
2. Media containing Rose Bengal are light sensitive. Inhibitory compounds are produced in significant concentrations after 2 hours exposure to light (P.V.Nielsen, personal communication). It is therefore **important** to keep light exposure to Rose Bengal containing media well below a total of 2 hours during preparation, storage, inoculation and incubation.
3. Media should be of approximately neutral pH and contain appropriate antibiotics.
4. We recommend a mixture of 2 antibiotics to be used instead of higher concentrations of a single compound. In our experience e.g. 100 ppm chloramphenicol is often insufficient to inhibit growth of bacteria from food products like vegetables, spices, cereals and meats. We recommend the combined use of chloramphenicol (50 ppm) and chlortetracycline (50 ppm). Chloramphenicol is heat-stable and may be added before autoclaving. Chlortetracycline is heat labile and must be added to the media after autoclaving as a filter sterilised solution. It is relatively unstable in solution and thus must be freshly prepared or refrigerated. Gentamycin is not recommended, as it has been reported to cause inhibition of some yeast species.
5. The media are designed for use in dilution plating. By direct plating overgrowth of fast growing species may occur, especially if surface sterilisation is omitted. This is due to the addition of considerable amounts of nutrients to the media from the food product .

3. SELECTIVE MEDIA

It is important to notice that all isolation media are selective to some degree, simply because of different competition abilities of the different species when they grow on the agar medium itself. Plant pathologists have solved that problem by using blotter tests and could, because of the rather few well known plant pathogenic species in anyone genus, often identify fungi to species level (Benoit and Mathur, 1970; Nath *et al.*, 1970; Chidambaram *et al.*, 1971; Kulshrestha *et al.*, 1976; Agarwall *et al.*, 1989) This is not possible for food or air-borne fungi as there are many species that cannot be differentiated or identified directly on seeds or other food items, especially the Penicillia and Fusaria. When using combinations of extract from cereals, fruits or vegetables with (enzymatically digested) protein sources and carbohydrates or chemically well defined media combined with different temperatures, some selection is thus unavoidable Two major principles for the development of selective media is inhibition of bacterial growth and reducing colony diameters of fast growing fungi (King *et al.*, 1986; Beuchat, 1992a). Further selective principles may be introduced to select for more specific groups of fungi or even particular fungal species. These are dealt with below. The efficiency of media may also depend on the isolation method employed. Stronger selective principles are needed for direct plating than for dilution plating.

Psychrotolerant fungi

No special media have been developed for psychrotolerant filamentous fungi, but incubation temperature is a very important factor to consider and a temperature of 15 or 20° C should be considered when working with refrigerated foods. A large number of food-borne species in the genera *Penicillium, Alternaria, Fusarium* and *Cladosporium* grow exceptionally well at low temperatures and are very relevant for foods stored in refrigerators. Strictly psychrophilic species may not be important in foods but species of *Penicillium* have been found that cannot grow at all at 25° C. Furthermore heavy *Eurotium* growth on low water activity media may completely mask the growth of presence of important toxigenic Penicillia such as *P. verrucosum* and *P. nordicum*.

Xerophilic fungi

We recommend to use the medium DG18 (Hocking and Pitt, 1980) in general for foods with a_w<0.90 (Samson *et al.*, 1992; Pitt and Hocking, 1997, Frändberg and Olsen, 1999) and MY50G (Malt Yeast 50% Glucose agar, Pitt and Hocking, 1997) for foods with a_w<0,70 The major selective principle in media for xerophilic is a high concentration of carbohydrates or sodium chloride. Malt salt agar (Christensen, 1946) was used for many years for direct plating of cereals and other dried products, but has been replaced by the more efficient glycerol containing medium DG18 (Samson *et al.*, 1992). DG18 is well suited for the enumeration of xerotolerant and moderately xerophilic species, such as Penicillia, Aspergilli, *Wallemia sebi* and *Eurotium* species. Unfortunately growth of *Eurotium* isolates on DG18 is nearly too vigorous, occasionally making counting difficult.

Preservative resistant species

A selective medium Acetic Dichloran Yeast extract Sucrose agar (ADYS) which contains 0.5% acetic acid has been suggested for these fungi (Frisvad *et al.*, 1992a). Few filamentous fungi are able to grow at high concentrations of preservatives and organic acids. Species like *Penicillium roqueforti, P. paneum, P. carneum, Monascus ruber* and *Paecilomyces variotii* have this ability. These are particularly important in rye bread and pickled products.

Proteinophilic fungi

A medium, Dichloran Creatine sucrose bromocresole agar (CREAD; Frisvad *et al.*, 1992a) has been developed for selective isolation of proteinophilic fungi. Most of the fungal species that are associated to foods containing relatively large amounts of proteins such as cheese, meat and nuts also grow very well on creatine as sole nitrogen source (Frisvad, 1985; 1993). Food-borne fungi growing well on CREAD include *Penicillium commune, P. solitum, P. crustosum, P. expansum, P. atramentosum, P. echinulatum,* coprophilic Penicillia and *Aspergillus clavatus, A. versicolor, A. sydowii* and *A. ustus*. Most other food-borne fungi grow very poorly on CREAD. The medium has been used for dilution plating and air sampling in cheese and other factories with good results.

***Fusarium* species**

We recommend using the medium CZID for isolation of *Fusarium* species (see Thrane, 1996; Andersen *et al.*, 1996).

A large number of selective media has been suggested for *Fusarium* species, most of them contain a substantial amount of pentachloronitrobenzene (PCNB) (Tuite, 1969; Tsao, 1970; Windels, 1992). As PCNB is a possible carcinogen, media containing this compound have not been well received in food mycological laboratories.

Two other media, based on dichloran, Czapek Iprodione Dichloran (CZID) agar (Abildgren *et al.*, 1987) and Dichloran Peptone Chloramphenicol Agar (DCPA) (Andrews and Pitt, 1986) have been developed. These media should be incubated under alternating light (daily rhythm: 12 hours dark and 12 hours combined day-light and near-UV- 350 nm. The plates must be stapled in maximum 2 layers, in the plastic bags, in order to let the light through.

CZID agar has been compared to PCNB media formerly used in brewing industries and the former has been found to be the most effective for recovering Fusaria (Anonymous, 1989; Thrane, 1996).

***Trichoderma* species**

A selective medium for the quantitative isolation of *Trichoderma* species has been developed by Askew and Laing (1993).

***Mucor* species**

Most media for food-borne fungi have been designed to inhibit fast growing *Mucor* and *Rhizopus* isolates. One selective medium, Malt extract Yeast Chloramphenicol Ketoconazol (MYCK; Bärtchi *et al.*, 1989) agar has been developed, however, for *Mucor* species, occasionally found as spoilage agents of cheese and bread.

Ochratoxigenic species *Penicillium verrucosum*

P. verrucosum produces a strong red to violet brown or terracotta-coloured reverse on two selective media Dichloran Rose bengal Yeast Extract Sucrose agar (DRYES; Frisvad, 1983) and Dichloran Yeast extract Sucrose Glycerol agar (DYSG; Frisvad *et al.*, 1992). The selective principles include dichloran, rose bengal (DRYES) or glycerol (DYSG) and incubation at 20° C. Tests have shown that counts of *P. verrucosum* on DRYES were not significantly different from those on DG18 and DYSG using cereal samples spiked with conidia. DRYES has been evaluated as a more general medium and found equivalent to DG18 and DRBC except that the *Wallemia sebi* counts were significantly larger on DG18. The advantage of DRYES is a good differentiation of species that many mycotoxins can be detected directly as is the case for DYSG (Filtenborg and Frisvad, 1990; Filtenborg *et al.*, 1992). Even though DRYES is recommended for isolation and detection of *Penicillium verrucosum* in foods and feedstuffs by the Nordic Committee of Food Analysis (Frisvad, 1995), the medium DYSG has been shown to be a much more effective medium for detection and isolation of this species (Elmholt and Hestbjerg, 1996; Elmholt *et al.*, 1999) in soil with a large number of competing species and this has been confirmed for a number of food samples at the Department of Biotechnology, DTU in Lyngby. Elmholt *et al* (1999) incubated their samples at 25° C and at this temperature *Eurotium* species grow well, but are rarely occurring in soil. In food samples, however, *Eurotium* spp. are abundant and will overgrow *P. verrucosum* in many cases. For this reason the original recommendation to incubate these samples at 20° C (Frisvad, 1983) or even at lower temperatures should be followed. More collaborative studies are needed to find the optimal recovery of *P. verrucosum* in foods and feedstuffs.

Other Penicillia

As mentioned above DRYES is a suitable medium for differentiating many *Penicillium* species. Other media developed for particular *Penicillium* species include a selective medium for *P. digitatum*, Potato Dextrose o-phenylanizole agar (PDO; Smilanick and Eckert, 1986).

Aflatoxigenic species

We recommend the AFPA medium (Pitt et al., 1983), (with both chlortetracycline and chloramphenicol) for detection of the common aflatoxin producing species as they are easily differentiated from other species by their bright cadmium orange reverse. Aflatoxin is not produced on this medium, but rather indicates the production of a Ferri chelate of aspergillic acids. Sporulation on the medium is rather poor, but confirmation of identity and aflatoxin production can be done on the media YES or CYA.

Several other media have been developed for the three aflatoxigenic species *Aspergillus flavus, A. nomius* and *A. parasiticus*. One of the first was RBSAB (Rose Bengal Streptomycin Agar with Botran (= Dichloran) (Bell and Crawford, 1967) and M3S1B (Medium with 3% Salt 1 ppm Botran) incubated at 37° C proposed by Giffen and Garren (1974). The same year the ADM (Aspergillus Differential Medium) was developed by Bothast and Fennell (1974). Detection was based on the consistent aspergillic acid production by the three aflatoxin producing species. Non-aflatoxin producers producing aspergillic acid include *A. sojae* and some strains of *A. sclerotiorum*, but these species are extremely rare in foods and feedstuffs. The ADM medium was modified later to include antibiotics, dichloran and more yeast extract (Hamsa and Ayers, 1977; Pitt *et al.*, 1983). Silicagel medium and coconut medium have been developed for the direct visualization of aflatoxin under UV light (Torrey and Marth, 1976; Davis *et al.*, 1987; Lemke *et al.*, 1989) and aflatoxin can also be determined directly on DRYES agar by the agar plug method.

Selection of media for different foods

The choice of media for mycological examination of foods should depend on the associated mycobiota of the foods or feedstuffs examined. For fruits, vegetables, herbs and fresh cereals DRBC, DRYES and CZID appear to be preferable (Andersen et al., 1996). For stored cereals, spices and nuts DG18 at 25° C, DYSG at 20° C and AFPA at 30° C is an effective combination to be able to count selectively important mycotoxin producers. In some cases AFPA could be used for fresh (surface disinfected) cereals, as *A. flavus* is known to grown and produce aflatoxin on stressed corn plants in the field in warmer climates. For dairy products, meat and bread CREAD and DG18 is a good combination of media. ADYS is especially suited for preserved foods and rye bread in combination with DG18.

4. METHODS FOR YEASTS (see also p. 270-278).

General methods for enumeration of yeasts

For products such as beverages, where the mycobiota normally comprises mainly yeasts, a non-selective medium such as Tryptone Glucose Yeast extract agar (TGY) plus an antibiotic is recommended. Recommended antibiotics are chloramphenicol and oxytetracycline.

For products where yeasts must be enumerated in the presence of moulds, DRBC is recommended.

Preservative resistant yeasts

More work is needed to establish the validity of the methods in current use. Media containing 10% glucose appears to be better than other media. A collaborative study recommended the following media:

1. TGY + 0.5% acetic acid
2. MEA + 0.5% acetic acid
3. any other medium currently in use.

Diluents for yeasts

For yeasts in high a_w foods and beverages, the recommended diluent is 0.1% peptone. However, there is no evidence that 0.85-0.9% saline peptone is disadvantageous, and the use may be continued.

For concentrates, syrups and other low a_w samples, a diluent containing at least 20 to 30% glucose is recommended. More work is needed on these and other glucose concentrations (e.g. 40, 50%) to determine the optimum concentration.

Detection of low numbers of yeasts

Membrane filtration is the recommended method for detection of low numbers of yeasts. MPN techniques are not recommended. Emerging technologies may eventually produce better methods.

5. ENUMERATION OF HEAT RESISTANT FUNGI

By definition, a heat resistant fungus produces propagules, which can survive a heat treatment at 75° C for 30 minutes. Raw materials, most likely to contain heat resistant fungi, are those that have been contaminated with soil, e.g. fruits and fruit products and raw milk.

Recommended method: At least 100 g of product should be heated at 75° C for 30 minutes and dispersed in an equal volume of double strength agar containing antibiotics. Malt extract agar (MEA) is recommended. The antibiotics suggested are chloramphenicol (50 ppm) and chlortetracycline (50 ppm). The entire sample should be plated and incubated, and counted after at least 14 days (to up to 30 days) incubation at 25° C. In our experience an incubation temperature of 30° C will allow a faster sporulation of heat resistant Ascomycetes.

Because of the long incubation time and other practical problems of sample preparation, the following method has been used by CBS which allows analysis of samples which are difficult to solve and allows faster detection of potential and common heat resistant contaminants.

Because of the low concentration of the heat resistant fungi in the sample, it is still necessary to investigate a large amount of the sample.

For samples, such as pectin, which are difficult to dissolve, aseptically weigh out 12.5 grams and transfer the sample into a sterile Stomacher bag. Add 250 ml of Ringer's solution to the bag and blend the sample. For other samples weigh out 100 grams and transfer the sample into a sterile Stomacher bag. Add 150 ml of Ringer's solution to each bag and blend the sample. After blending, seal the Stomacher bag.

1. Heat-treat the Stomacher bags (with the sample) for 30 minutes at 75° C (the sample should be at 75° C before the 30 minutes when time frame starts). After the heat-treatment, cool the samples to 50° C.
2. Aseptically transfer the content of the Stomacher bag in a Schott Duran bottle (500 ml) with 250 ml double strength MEA (50° C). Add the antibiotics.
3. Mix thoroughly.
4. Pour the sample into sterile plastic 14 cm diam Petri-dishes (approx. 8 dishes)
5. Place the Petri-dishes in an upright position in the incubator (28° C, darkness)
6. Examine the Petri-dishes after 7 days and if visible mould structures are present, isolations on Oatmeal Agar should be made. Re-examine the Petri-dishes after a prolonged incubation period (up to 14 days and longer).
7. To investigate 100 grams of each sample this procedure should be repeated 7 times.

Determining and reporting D-values of heat resistant fungi: *D*-values should be reported by methods which will allow comparison with other published reports. Sources of variability in *D*-values include incubation conditions and the age of the ascospores. The following procedure is recommended:

1. To ensure that ascospores are mature, cultures at least 3 weeks old should be used.
2. Heat in a standard medium, such as that of Bayne and Michener (1979), which consists of 16 g glucose + 0.5 g tartaric acid per 100 ml water, pH 3.6, and also heat in the appropriate food.
3. Heat at 3 different temperatures, one of which should be 80° C, using an initial inoculum of at least 10^5 CFU/ml.
4. The time course of death should be followed over at least 3 log cycles using Czapek Yeast Extract agar or Malt Extract Agar as a plating medium.

6. ISOLATION OF FUNGI.

Having detected or made total counts of moulds and yeasts, the next step towards identification of the species is normally preparation of pure cultures.

Whether the fungi are detected directly on the foods or as mixed cultures on isolation media, pure cultures of each species must be prepared by streak-inoculation on selected media. This is due to the experience that contaminants often are very difficult to detect in cultures, which are point-inoculated.

The media used for streak-inoculation must support formation of macroscopic very different colonies of each species in order to make spotting of infections as easy and distinct as possible. Besides, heavy sporulation must take place in these cultures, as they will be used as inoculation material in the identification procedure later on. To meet these demands, we recommend the combined use of 2% MEA or PSA and CYA, except for xerophiles, for which for example MY50G can be used. CYA and MY50G are used to show if the culture is pure and if sporulation is sufficient this culture is also used for inoculation. MEA or PSA are subsequently used for inoculation.

The Petri dishes are incubated in perforated plastic bags as follows:

- 2% MEA or PSA: 25° C, 4-5 days
- CYA: 25° C, 4-5 days
- 2% MEA or PSA (heat resistant species): 30° C, up to 14 days or longer.
- 2% MEA or PSA (*Fusarium*): 25° C, 4-5 days, alternating light*
- MY50G: 25° C, 4-5 days

Daily rhythm: 12 hours dark and 12 hours combined daylight + near-UV (350 nm). The plates must be stapled in maximum 2 layers, in perforated plastic bags.

If sporulation is not achieved, other media and incubation conditions should be tried (see Appendix – media p. 378-382).

To avoid isolation of identical isolates from each food sample, various species of each genus are isolated from only **one** of the media chosen for enumeration. This selection saves a lot of effort and resources, since it, for practical purposes, is **impossible** to tell which species are identical on two different media. The selection is possible since each fungus has some simple diagnostic characteristics as follows:

Penicillium verrucosum: forming green to white, velvety to floccose colonies. The colony reverse is characteristic red brown on DRYES or DYSG and red on DG18. Should be isolated from DRYES, DYSG or DG18.

Other *Penicillium* spp.: forming green to blue, velutinous to lightly floccose colonies with bright reverse. Should be isolated from DRYES, DYSG, DRBC or DG18.

Aspergillus flavus: forming curry to white, floccose colonies with a characteristic orange reverse on AFPA. Should be isolated from AFPA.

Other *Aspergillus* spp. : forming coloured (e.g. yellow, black, green), granular to floccose colonies with bright reverse. Should be isolated from DG18, DYSG, DRBC or DRYES. In the stereomicroscope characteristic head-like structures are observed.

***Fusarium* spp. :** forming white, reddish or yellowish, floccose colonies with bright to red reverse. Should be isolated from CZID over DRBC or DG18.

Xerophilic species: forming small colonies, often 1-2 mm, on DG18 (except for *Eurotium* species). Should be isolated from DG18 or MY50G.

Heat resistant species: growth after heat treatment.

Yeasts: forming moist, shiny to slightly floccose, white, cream, red or pinkish coloured colonies. Should be isolated from DRBC.

7. FUNGI IN AIR AND ON SURFACES.

In the food industry (plants, production areas) but also in working-places in general, monitoring for airborne fungi during regular intervals (e.g. once a week, month) is recommended. For a more detailed description of the methods see Chapter 3.

8. MEDIA FOR IDENTIFICATION

The streak cultures are subsequently used for further identification procedures described in Chapter 1. The streak cultures are used for a preliminary identification of the fungal group and genus, using microscopic examination methods as described in Chapter 1. Identification procedures for the species are described under the individual genera.

9. MONITORING OF MEDIA

It is recommended that laboratories should monitor media quality and selective properties for each batch prepared as follows:

1. A needle point inoculum of the test species should be taken from a 1-3 week old slant culture and dispersed in semi-solid agar.
2. A sterile, standardized loop should be used to inoculate the spore suspension onto 3 points of the test medium.
3. After 5 days, colony diameters of test species should be measured. Colony characteristics should also be observed.

The following test microorganims can be used for the media DRBC, DG18, OGY and TGY:

Rhizopus stolonifer
Aspergillus niger
Eurotium chevalieri
Cladosporium cladosporioides
Zygosaccharomyces rouxii
Hansenula anomala
Rhodotorula glutinis
Bacillus subtilis

10. OTHER METHODS FOR IDENTIFYING AND CHARACTERISING FILAMENTOUS FUNGI

Traditionally filamentous fungi have been identified to species level using morphological characters, sometimes supplemented with some colony characteristics and few physiological tests. However more objective and reliable the species separations have been shown to be achieved combining morphological features and profiles of secondary metabolites (Frisvad, 1992; Frisvad, 1994; Frisvad *et al.,* 1998). Below we describe a very simple chromatographic method and more advanced methods that can be used to determine secondary metabolites. Experience has shown that at least some data on secondary metabolites are needed for an accurate identification of food-borne *Penicillium* species and that these data certainly also give higher confidence in identifications of species in other genera. Serological methods and especially methods based on DNA are being used very frequently also.

10.1. THIN-LAYER CHROMATOGRAPHY (TLC) METHOD

A simple application method.
Filtenborg and Frisvad (1980), Filtenborg *et al.* (1983), Frisvad and Filtenborg (1983) and Thrane (1986) developed a very simple screening method for toxigenic fungi. This method will be described briefly below, and modified by using Rf values relative to griseofulvin (R_{fg}, which is more consistent between laboratories).

An agar plug is cut out of the colony (with a cork borer, a scalpel or similar devices) (Fig. 1). The production of secondary metabolites (e.g. toxins) is usually highest in the centre of the colony. Using a needle, scalpel or similar, the plug is placed onto a TLC plate with the medium side or the mycelium side towards the gel (Figs. 2-3). In this way the extracellular as well as the intracellular metabolites can be analyzed. If the mycelium side is used, one or two drops of a chloroform/methanol mixture (2:1) is added to the mycelium prior to the application onto the TLC plate. This fast extraction process can be done with other less toxic solvents, like alcohol, but they are less effective. After application the spot is allowed to dry and more plugs can be superimposed. The application spot should not exceed about 8 mm in diameter.

The TLC plates commonly used are normal phase silicagel plates (Fig. 4). For some acidic metabolites like cyclopiazonic acid, citrinin, luteoskyrin etc. it can be useful to impregnate the plate with oxalic acid. This is simply done by dipping the plate in an 8% solution of oxalic acid in water or methanol followed by air-drying.

TLC-conditions.
After application the TLC plate, any suitable TLC procedure can be performed. We have found the following solvents (at saturated conditions) very useful:
TEF: Toluene/Ethyl acetate/Formic acid (90%) 5:4:1
CAP: Chloroform/Acetone/iso Propanol 85:15:20
TAM: Toluene/Acetone/Methanol 5:3:2
After elution and air drying in a dark fume hood, the TLC plates are examined in visible light (VIS), long wave UV-light (UV366) and short wave UV-light (UV254). Some metabolites like PR toxin is treated with ½ minute in UV254 followed by observing at UV366.

The following spray reagents are useful for visualizing and verification of several secondary metabolites:

Sp1: 0,5% panisaldehyde (4-methoxybenzalde- hyde) in ethanol/acetic acid/ conc. sulphuric acid 17:2:1 (most metabolites).
Sp2: 50% sulphuric acid in water (e.g. verruculogen, viridicatins)
Sp3: 1% $Ce(SO_4)_2$ in 50% sulphuric acid (e.g. roquefortines, meleagrin, oxaline, penitrem A)
Sp4: 20% $AlCl_3$ in 60% ethanol and heating for 8 min. at 130 C (e.g. penitrem A, trichothecenes B, sterigmatocystin) (good general spray than can be used in advance of any other spray reagent)

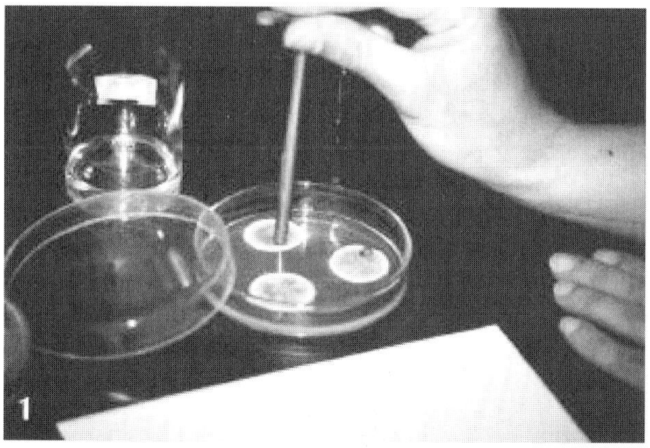

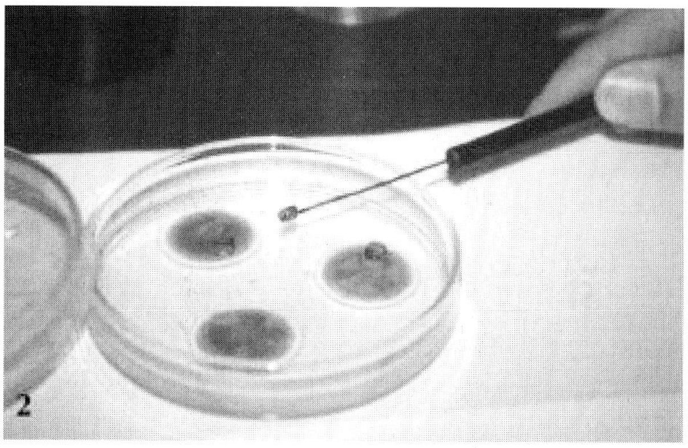

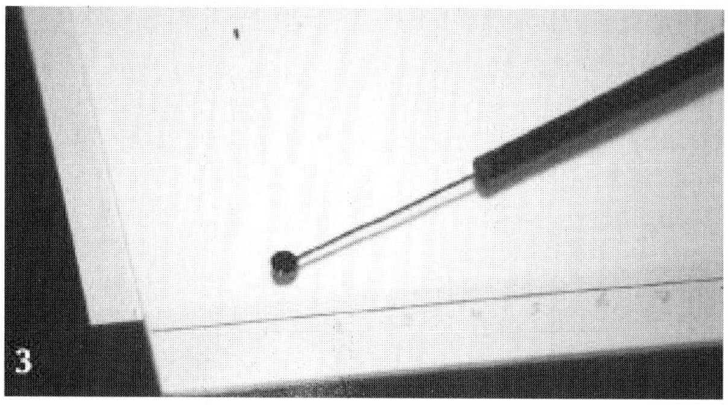

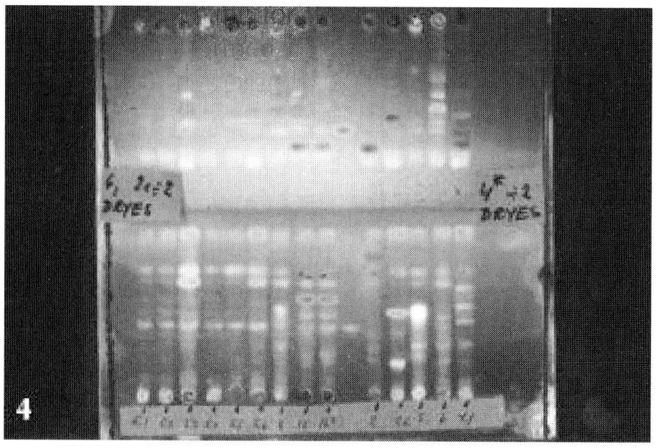

Table 1. R_f values relative to griseofulvin (R_{fg}), visualisation methods and colours of important *Fusarium* toxins as determined by the agar plug method.

Mycotoxin	R_{fg} TEF	R_{fg} TAM	R_{fg} CAP	Treatments	colour
Moniliformin	0.10	0.303	0.00	sp4, 9, 10	Red brown
Nivalenol	0.18	0.66	0.28	sp4	Blue fl.*
Butenolide	0.39	0.77	0.43	sp4, 9, 10	Yellow
Neosolaniol	0.40	0.87	0.74	sp4, 9	Blue green fl.
Deoxynivalenol	0.48	0.87	0.62	sp4	Blue fl.
HT2 toxin	0.52	0.90	0.67	sp4, 9	Blue green fl.
FusaroneX	0.55	0.93	0.81	sp4	Blue fl.
Diacetoxyscripenol	0.87	0.98	1.00	sp4, 9	Blue green fl.
T2 toxin	1.00	1.05	1.04	sp4, 9	Blue green fl.
Equisetin	1.45	0.65	0.76	sp1	Pink
Zearelenone	1.77	1.15	1.07		Blue at 254 nm (and 366 nm) fl.

fl.* = fluorescent spots as seen under UV light (366 nm).

Sp5: 0.5% methylbenzothiazolone hydrochloride (MBTH) in water (patulin and moniliformin)
Sp6: 1% *p*dimethylbenzaldehyde (Ehrlich reagent = 4-dimethyl-aminobenzaldehyde) in 96% ethanol, drying, spraying with Sp2 (cyclopiazonic acid and other alkaloids)
Sp7: NH_3 vapours in 13 min. (mycophenolic acid, xanthomegnin, viomellein, penicillic acid, ochratoxin A)
Sp8: 0,1% diazo blue B (fast blue salt B) in methanol/water (1:1), drying, Sp7 (kojic acid, 3nitropropionic acid).

Sometimes spraying is accompanied by heating at 130 C: Sp1 (8 min), Sp2 (5 min), Sp5 (1020 min).

R_{fg}values, solvents, sprays and colours of selected secondary metabolites of species of *Fusarium* and *Penicillium* are listed in tables 1, 2 and 3.

Media and incubation conditions.

Most secondary metabolites are produced optimally after 14 days of incubation at 25°C, but 7 days are normally sufficient for detection to be achieved of most secondary metabolites. Exceptions are spinulosin (5-7 days), palitantin (15-18°C), moniliformin (27-30°C) a.o. The simplicity of the agar plug method, however means that analysis at several intervals and incubation conditions can easily be done. Optimal media for production of secondary metabolites depends on the metabolite, on the species and to some extent on the isolate. YES agar and oatmeal agar (OA) are generally very useful media. So to cover all the biosynthetic capabilities of any isolate it is necessary to use several media like: A highly nutritious medium (YES), a natural cereal medium (OA), corn grits agar (CGA), rice meal agar (RA), a low carbohydrate medium, a nitrate containing medium (CYA) and a malt based medium (MEA). Generally *Penicillium* and *Aspergillus* metabolites are produced in YES, CYA, OA and MEA. *Fusarium* metabolites are produced in YES, PSA, and RA or CGA. *Alternaria* metabolites are produced optimally in YES.

Procedure used in identification of food-borne Penicillia.

Pure cultures on CYA and YES are incubated at 25°C for 7 days or more. Application on the TLC-plates, using the agar plug method, is done as shown on the following Fig. 5.

Figure 5.

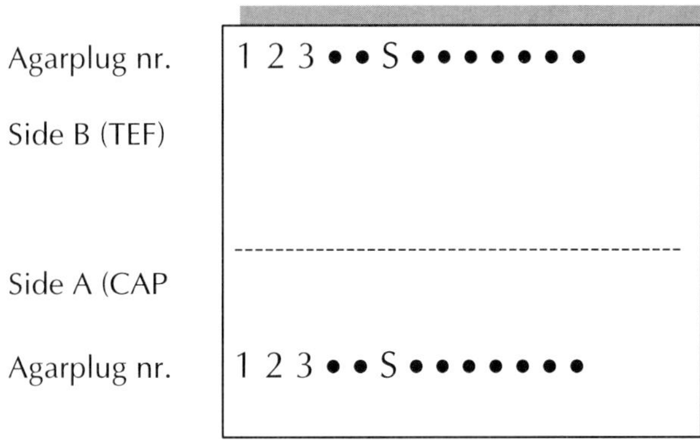

Isolate nr.1

Side A and side B are identical as far as cultures are concerned, but differ in the use of solvents and sprays. The elution distance is on both sides 10 cm which will take about 15 min. The two cultures of each isolate is placed next to each other:

Side A: Spot nr. 1: Isolate nr. 1, CYA culture, 2 plugs superimposed, mycelium side.
Spot nr. 2: Isolate nr. 1, YES culture, 1 plug, medium side.
Spot nr. S: Griseofulvin standard.
Solvent: CAP
Spray: Sp 4 then Sp3

Side B: Spot nr. 1: Isolate nr. 1, CYA culture, 2 plugs superimposed, mycelium side.
Spot nr. 2: Isolate nr. 1, YES culture, 1 plug, medium side.
Spot nr. S: Griseofulvin standard.
Solvent: TEF
Spray: Sp 4 then Sp1

Table 2. R_f values relative to griseofulvin (R_{fg}), visualisation methods and colours of important mycotoxins as determined by the agar plug method (TLC).

Mycotoxin	R_{fg} TEF	R_{fg} CAP	Treatments/colour
Rugulovasine A	0.06	0.060.33	sp1 (brown), sp3 (brown)
Isofumigaclavine A	0.09	0.41	sp1 (brown), sp3 (yellow fl.)
Roquefortine C	0.19	0.46	sp3 (orange)
Rubratoxin B	0.32		Blue fl.
Citromycetin	0.35	0.04	sp2 (yellow), sp3, sp4 (yellow fl.)
Aflatoxin G_1	0.39	0.96	Bluegreen fl.
Austdiol	0.48	0.73	Blue fl., sp3 (blue violet fl.)
Terrein	0.53	0.56	sp1 (black), sp3, sp4 (yellow fl.)
Palitantin	0.55	0.46	sp1 (brown), sp3, sp4 (brown)
Aflatoxin B_1	0.56	1.00	Blue fl.
Altenuene	0.70	0.62	sp2 (blue fl.), sp3 (violet)
Brevianamide	0.71	0.86	Yellow (daylight and UV light)
Xanthomegnin	0.72		Yellow (but dark in UV), sp7 (dark brown)
Brefeldin A	0.73	0.53	sp1 (blue violet), sp3, sp4 (pinkish cream)
Citreoviridin	0.87	0.86	Yellow (daylight and UV light)
Terrestric acid	0.93	0.78	sp1 (yellow, dark UV)
Patulin	0.98	0.95	sp5 (yellow fl.)
Griseofulvin	1.00	1.00	Blue fl.
Penicillic acid	1.03		sp1 (blue lilac fl.)
Cyclopenol	1.03	0.84	sp2 (blue violet fl.)
Cyclopenin	1.06	0.97	sp2 (blue violet fl.)
Tenuazonic acid	0-1.08	0.89	sp2 (blue fl.)
Gliotoxin	1.13	0.98	sp1 (yellow brown)
Viomellein	1.15	1.18	Yellow (but dark in UV), sp 7 (dark brown)
Verruculogen	1.27	1.06	sp2 (blue)
Fulvic acid	1.29	0.25	Blue fl.
Mycophenolic acid	1.35	0.85	sp7 (blue fl. UV 254 nm)
Alternariol	1.36	0.95	sp2 (blue fl.)
Cyclopiazonic acid	0-1.37	0-0.71	sp2 (brown fl., weak), sp6 (blue)
Ochratoxin A	1.39	0.31	sp7 (blue green fl.)
PR toxin	1.42	1.10	UV_{254} in 0.5 min. (blue green fl.)
Citrinin	0-1.46	0.28	Yellow green fl.
Rugulosin	0.5-1.5	0.23-0.41	sp2 (yellow fl.)
Viridicatin	1.54	0.78	sp2 (blue fl.)
AME	1.59	0.78	sp2 (blue fl.)
Sterigmatocystin	1.75	1.13	sp2 (yellow fl.), sp4 (yellow fl.)
Penitrem A	1.80	1.17	sp2, sp4 (dark blue)
Emodin	1.91	1.06	Orange fl.
Aflatrem	2.17	1.14	sp2 or sp3 (dark blue)
Physicon	2.20	1.15	Orange yellow fl.

Elution must be started with side A (solvent CAP). After each elution the plate is air-dried for about 10 minutes on side A and about 30 minutes on side B.

Identification of the metabolites is done by comparison with metabolite standards or with standard cultures using the R_{fg} values from table 3. It should be noted that these values might vary about 10 %. Standard cultures can be obtained from the IBT culture collection with information of the profile of secondary metabolites of the isolate. Typical isolates are listed by Frisvad *et al.* (2000).

If the expected secondary metabolites cannot be detected, or are produced below the detection limit, incubation is continued for another week and the TLC analysis is repeated. It is also suggested to superimpose more than two plugs if the result is difficult to interpret.

Besides the well-known secondary metabolites the TLC analysis often includes a number of unknown as well. When the unknown metabolites can be described by colour and R_{fg} values they may be included in the characteristic metabolite pattern of the species. The total metabolite pattern as determined by TLC analysis is used first to find out which isolates belong to the same species and secondly to identify the species. In either case the metabolite pattern is used in combination with morphological and physiological characteristics.

Table 3. A detailed system for identifying secondary metabolites by TLC for identification of food-borne *Penicillium* isolates by the agar plug method (see Lund, 1995a).

Metabolite	Rfg: TEF/CAP	Before spray	$AlCl_3$	$Ce(SO)_4$	ANIS
3-methoxyviridicatin	154 / 88	blue	blue	blue	blue
aurantiamin	31 / 80	blue	blue	grey	grey
revianamide A	71 / 86	yellow-green	yellow	orange[VIS]	grey-yellow
Chaetoglobosin C	111 / 80	black	black	black	brown
Chrysogin [Y]	62 / 28	blue [254]	blue	grey	grey
Citrinin [Y]	126[H] / 15[H]	yellow-green	yellow	yellow	light yellow
Compactin A and B[Y]	54+94 / 54+104	nd	nd	nd	yellow-brown
Cyclopaldic acids[Y]	77+146 / 54	blue [254]	turquoise	turquoise	turquoise
Cyclopenin	111 / 94	nd	violet [VIS]	grey	grey
Cyclopenol	82 / 79	nd	grey-blue[VIS]	grey	grey
Cyclopiazonic acid	130[H] / 17[H]	brown	blue-brown	cream	light brown
Griseofulvin	100 / 100	blue	grey	grey	light blue
(Iso)fumigaclavine A	7 / 33	grey	grey	grey	brown
Meleagrin/Oxaline	4 / 74	nd	brown-yellow [VIS+365]	brown-yellow	brown-yellow
Met O	120 / 125	yellow-green [254]	yellow	yellow	yellow
Mycophenolic acid	135 / 85	blue [254]	blue	blue	blue
Ochratoxin A	153 / 23	blue-green	blue-green	blue-green	blue-green
Penicillic acid[Y]	103 / nd	nd	nd	nd	red-blue
Penitrem A	189 / 120	nd	blue-black [VIS+365]	blue-black	blue-black
Raistrick phenols[Y]	87/35	blue [254]	blue	blue	blue
Roquefortine C	0 / 46	nd	nd	orange[VIS]	dark [VIS]
Rugulovasine A and B	11 / 22+50	grey-blue	grey-blue	grey	grey
Terrestric acid [Y]	106[H] / 100	nd	nd	nd	yellow [VIS]
Territrems	85+115+140 / 112+96+85	blue	blue	blue	blue
Verrucosidin	144 / nd	nd	nd	nd	brown-yellow
Viomellein	106 / 104	yellow –brown [VIS]	yellow-brown [VIS]	nd	dark
Viridamine	20 / 46	light blue	blue	blue	blue
Viridicatin	154 / 77[H]	blue [254]	violet	grey-blue	grey-blue
Xanthomegnin	74 / 80	yellow-brown [VIS]	yellow-brown [VIS]	nd	dark

Y: The metablites are produced in highest amounts on Yes agar; H: The metabolites are seen as tails; ND: Cannot be detected at those conditions; 365: The colour is seen under UV light at 365 nm; 254: The colour is seen under UV light at 254 nm; VIS: The colour is seen in daylight. All colours are those seen under UV light at 365 nm except where otherwise noted

Some secondary metabolites can be detected rapidly and easily by just placing a filter paper soaked with Ehrlich reagent on a plug taken directly from a fungal colony on CYA agar (Lund, 1995b). In this way the production cyclopiazonic acid and other indols may be indicated to verify identification of *P. commune* for example.

Procedure used in identification of food-borne Fusaria.

In principle the procedure is as described for Penicillia. The differences in the procedure are as follows. The media used are YES and PSA. Side A and B on the TLC plate are identical as far as TLC conditions concern, but contain different isolates.

Side A: Spot nr. 1: Isolate nr. 1, YES culture, 2 plugs superimposed, mycelium side.
Spot 2: Isolate nr. 1, PSA culture, 2 plugs superimposed, mycelium side.
Spot nr. S: Griseofulvin standard.
Solvent: TEF
Spray: Sp2 and Sp1 (separately)

Side B: Spot nr. 1: Isolate nr. x, YES culture, 2 plugs superimposed, mycelium side.
Spot nr. 2: Isolate nr. x, PSA culture, 2 plugs superimposed, mycelium side.
Spot nr. S: Griseofulvin standard.
Solvent: TEF
Spray: Sp2 and Sp1 (separately)

10.2. HIGH PERFORMANCE LIQUID CHROMATOGRAPHY (HPLC) METHOD

Frisvad (1987), Frisvad and Thrane (1987) and Frisvad and Thrane (1993) developed a compromise HPLC method that is of use for most polar and apolar secondary metabolites from fungi, except ionic metabolites or metabolites with a very low absorption of 225 nm. The extract of the cultures of ten 9 cm Petri dishes, using first (100 ml) chloroform/methanol and then (100 ml) ethylacetate with 1 % formic acid will include most alkaloids, acids, polyketides, nonvolatile, and terpenes, but exclude most cyclic peptides and certain acids.

The combined organic phase is evaporated in vacuo at 39°C after filtration through a hydrophobic filter, redissolved in 3 ml methanol at 39°C, defatted with 2 ml petroleum ether and the methanol phase filtered through a 0.45 µm filter before injection into the chromatograph (5 or 10 µl depending on the concentration of the extract). Extracts of for example *Eurotium* species and several *Aspergillus* species are usually very concentrated and 2 µl may be the best injection volume in those cases.

Smedsgaard (1997a) has improved and simplified this method to use only few agar plugs from a fungal colony and extract them with 500 ul of the solvent mixture of methanol-dichloromethane-ethylacetate (1:2:3) containing 1 % (v/v) formic acid. The plugs are extracted ultrasonically for 60 min. The extract is transferred to a clean vial and the organic phase is evaporated. The residue is re-dissolved by ultrasonication for 10 min in 400 ul methanol or in 400 ul methanol containing 0. 6% (v/v) formic acid, 0.02 % hydrochloric acid and 2.5 % (v/v) water. The latter solvent is used if the extract is also used for electrospray MS (ES-MS)analysis (see below). The latter extract is filtered and injected directly into the HPLC or ES-MS.

HPLC conditions

Gradient elution is necessary to separate most effectively the very large and varied number of secondary metabolites produced by food-borne fungi. In the method proposed by Frisvad and Thrane (1987) a gradient using solvent A water and solvent B acetonitrile containing 0.02 % triflouracetic acid is used as follows: Initially 10 % B, raised to 50 % B in 30 min, the to 90 % in 10 min, held at 90 % B at 3 min (to elute all sterols), and lowered to 10 % B in 6 min and held at 10 % B for 1 min at a flow rate of 2 ml/min. The analysis time is thus 50 min. Alkylphenone retention indices are calculated for each compound (Frisvad and Thrane, 1987). Other gradients can be designed to spread the alkylphenones evenly at lower flows than that suggested here. The column used was a 100 x 4 mm reversed phase 5 µm C_{18} Nucleosil, but other columns can be used as well. Retention indices for important mycotoxins and other fungal secondary metabolites are tabulated in Frisvad and Thrane (1987, 1993). In Smedsgaards (1997a) modified method a 100 x 4 mm Hypersil BDS-5 µm C_{18} column is used and the gradient is linear starting from 85 % water and 15 % acetonitril (both with 0.005 % trifluoroacetic acid) going to 100 % acetonitril in 40 min, the maintaining 100 % acetonitril for 3 min at a flow rate of 1 ml/min and returning to the initial conditions. Ergosterol will elute after ca. 42.5 min in this system. Detailed RI or retention time data using these conditions have not yet been published.

UV spectra recorded on line with a diode array detector will help considerably in identification of compounds and these spectra contain more information than usually considered by chemists (Frisvad and Thrane, 1987, 1993; Law and Stafford, 1993). The use of authentic standards and mass spectrometry is, however, important if the identifications are to be very certain. Even though UV spectra are often dependent of the pH of the eluent, which is changing through the elution profile recommended here, it is not very large because pH changes only from approximately 3.5 to 4.5. UV spectra of many of the important mycotoxins and secondary metabolites are listed in Singh *et al.* (1991) (as figures) and Frisvad and Thrane (1993) (absorption maxima and minima, shoulders and relative absorption). Further research (Frisvad, 1993) has shown that several alkaloids and acids elute with better peak shapes if 0.02% triflouracetic acid is used in both solvent A and B. In such a system the pH gradient is avoided.

Media and growth conditions

CYA, YES, MEA, SYES (YES agar produced from Sigma Y4000 yeast extract) and oat mal agar (OA) have been used for production of secondary metabolites of food-borne fungal species. These media are used anyway for identification purposes except SYES. For HPLC analysis the fungus is usually grown on the media mentioned above at 25°C for one or two weeks and extracted using 3 or more agar plugs. These plugs may be combined or extracted separately for each medium. As seen in table 3 certain secondary metabolites are produced best on YES agar, while CYA is optimal for many alkaloids and the xanthomegnin related naphthoquinones. A series of other naphthoquinones are produced optimally on MEA and OA. The latter two media are particularly good for production of secondary metabolites from *Talaromyces* and its anamorph *Penicillium* subgenus *Biverticillium*.

Evaluation of chromatograms

Most species of food-borne fungi produce a large number of secondary metabolites (usually 50-100). These can be ordered into families of chromophores and often, if a sufficient number of standards are available, into biosynthetic families (Frisvad, 1989). The profiles of secondary metabolites can be compared using data explorative (chemometric) methods. For this purpose UPGMA clustering using the Yule coefficient and correspondence analysis are recommended (Frisvad, 1992; 1994). These latter methods will yield very clear separations of species if used in conjunction with other differentiation characters such as morphological characters.

A new objective method of comparing chromatograms coupled with spectrometric data (3-dimensional data) has been developed (Nielsen *et al.*, 1998; 1999). This method includes correlation optimized warping (COW) to align HPLC profiles and a program for producing a similarity matrix, that can be used directly for cluster analysis or principal coordinate analysis,. The program is available at **http://www.ibt.dtu.dk/mycology/cow.**

10.3. GAS CHROMATOGRAPHY-MASS SPECTROMETRIC METHOD

Penicillium and *Aspergillus* species usually produce a large number of characteristic volatile compounds and previously these were used to give an impression of the odour of such species as an extra descriptive feature (Raper and Thom, 1949; Raper and Fennell, 1965). Many of these volatiles can now be identified using simple diffusive sampling techniques (Nilsson et al., 1996a and b; Larsen and Frisvad, 1994; 1995a; Larsen 1997). Such methods can be used to identify isolates of *Penicillium* species that are difficult to identify by other means (Larsen and Frisvad, 1995b and c; Larsen, 1996, 1997). The use of the solid-phase micro-extraction (SPME) technique directly from a CYA or YES agar Petri dish and GC-MS analysis is simple and can be used for indication of identity of some species of *Penicillium*. For example *P. carneum* may be difficult to separate from *P. roqueforti*, but geosmin production sets it apart from the blue cheese starter.

10.4. ELECTROSPRAY MASS SPECTROMETRIC METHOD

An electrospray interface to a mass spectrometer can be used for analysis of fungal extracts (see above) and the library function in most MS software programs used for identification of fungi after a standard set of known cultures have been analysed (Smedsgaard and Frisvad, 1996; Smedsgaard, 1997b).The analysis time will be ca. 3 min as soon as the extracts have been prepared. An extract is characterised by its typical profile of mass + 1 peaks. Chlorine in a molecule can be recognised by its typical isotope pattern. Occasionally a sodium ion may also be detected , but usually fragmentation of the molecules will not be observed by the method used.

10.5. MICELLAR ELECTROKINETIC CAPILLARY CHROMATOGRAPHIC METHOD

Capillary electrophoresis can also be used to separate effectively mycotoxins and other fungal secondary metabolites (Nielsen *et al.*, 1996), but effective sensitive or selective detection methods are still not available as compared to those available for HPLC.

10.6. SEROLOGICAL METHODS

Several new methods are developed to detect viable and non-viable fungal propagules. Particularly the serological methods are promising. A latex agglutination test has been developed for detecting *Aspergillus, Penicillium* and other important food-borne genera (Notermans *et al.*, 1998; Samson *et al.*, 1992). Other new techniques are determination of volatile products, impedimetric measurements and Petrifilm Yeast and Mold plates (various papers in Samson *et al.*, 1992; Frisvad *et al.*, 1998).

10.7. MOLECULAR METHODS OF RELEVANCE FOR HACCP ANALYSIS OF FUNGAL CONTAMINATION

The number of papers dealing with molecular methods for classification and detection of fungi at the clonal level (fingerprinting) has increased dramatically since 1995 (Bridge *et al.*, 1998). Sequencing has been used extensively for producing cladograms of filamentous fungi for example for food-borne Penicillia (Skouboe *et al.*, 1999) and Aspergilli relevant for foods (Geiser *et al.*, 1998), but of particular relevance is RAPD (Random Amplified Polymorphic DNA) and AFLP (Amplified fragment length polymorphism) analysis for fingerprinting of particular clones in order to find critical control points and contamination sources (Lund and Skouboe, 1998). It is now possible to easily isolate DNA from filamentous fungi for example using the "Fast DNA Spin Kit for soil", estimate the amount of DNA, set up the RAPD reaction using PCR, confirm the RAPD product and read the RAPD profiles to compare ramets. AFLP has the advantage of being of use at both clone level and species level and being more reproducible. Often RAPD analysis can be used for pre-screening and AFLP for the final analysis (Vos *et al.*, 1995; Majer *et al.*, 1996; Janssen *et al.*, 1996; Arenal *et al.*, 1999).

REFERENCES

ABILDGREN, M.P., LUND, F., THRANE, U. & ELMHOLT, S. 1987. Czapek-Dox agar containing iprodione and dicloran as a selective medium for isolation of *Fusarium* species. Lett. Appl. Microbiol. 5: 83-86.

AGARWAL, P.C., MORTENSEN, C.N. & MATHUR, S.B. 1989. Seed-borne diseases and seed health testing of rice. Phytopathological Papers 30: 1-106.

ANDERSEN, B., THRANE, U., SVENDSEN, A. & RASMUSSEN, I. 1996. Associated field mycobiota on malt barley. Canad. J. Bot. 74: 854-858.

ANDREWS, S. & PITT, J.I. 1986. Selective medium for isolation of *Fusarium* species and dematiaceous hyphomycetes from cereals. Appl. Environ. Microbiol. 51: 1235-1238.

ANONYMOUS. 1989. *Fusarium* picture album. Technical Research Centre of Finland, Espoo.

ARENAL, F., PLATAS, G., MARTIN, J, SALAZAR, O. & PELÁEZ, F. 1999. Evaluation of different PCR-based DNA fingerprinting techniques for assessing the genetic variability of isolates of the fungus *Epicoccum nigrum*. J. Appl. Microbiol. 87: 898-906.

ASKEW, D.J. & LAING, M.D. 1993. An adapted selective medium for the quantitative isolation of *Trichoderma* species. Plant Pathol. (Oxford) 42: 686-690.

BÄRTSCHI, C., BERTHIER, J., GUIGUETTAZ, C. & VALLA, G. 1991. A selective medium for the isolation and enumeration of *Mucor* species. Myc. Res. 95: 373-374.

BAYNE, H.G. & MICHENER, H.D. 1979. Heat resistance of *Byssochlamys* ascospores. Environ. Microbiol. 37: 449-453.

BELL, D.K. & CRAWFORD, J.L. 1967 A botran-amended medium for isolating *Aspergillus flavus* from peanuts and soil. Phytopathology 57: 939-941.

BENOIT, M.A. & MATHUR, S.B. 1970. Identification of species of *Curvularia* on rice seed. Proc. Int. Seed Test. Ass. 35: 99-119.

BEUCHAT, L.R. 1992a. Media for detecting and enumerating yeasts and moulds. Intl. J. Food Microbiol. 17: 145-158.

BEUCHAT, L.R. 1992b. Evaluation of solutes for their ability to retard spreading of *Eurotium amstelodami* on enumeration media. Mycol. res. 96: 749-756.

BEUCHAT, L.R., JUNG, Y., DEAK, T., KEFFLER,T., GOLDEN, D.A., PEINADO, J.M., GOZALO, P., DE SILONIZ, M.I. & VALDERRAMA, M.J. 1998. An interlaboratory study on the suitability of diluents and recovery media for enumeration of *Zygosaccharomyces rouxii* in high –sugar foods. J. Food Mycol. 1: 117-130.

BOTHAST, R.J. & FENNELL, D.I. 1974. A medium for rapid identification and enumeration of *Aspergillus flavus* and related organisms. Mycologia 66: 365-369.

BRIDGE, P.D., ARORA, D.K., REDDY, C.A. and ELANDER, R.P. (Eds.) 1998. *Applications of PCR in mycology.* CAB Internatonal, Wallingford.

BUTTNER, M.P. & STETZENBACH, L.D. 1993. Monitoring airborne fungi in an experimental indoor environment to evaluate sampling methods and the effects of human activity on air sampling. Appl. Environ. Microbiol. 59: 219.

CHIDAMBARAM, P., MATHUR, S.B. & NEERGAARD, P. 1973. Identification of seed-borne *Drechslera* species. Friesia 10: 165-207.

CHRISTENSEN, C.M. 1946. The quantitative determination of molds in flour. Cer. Chem. 23: 322-329.

DAVIS, N.D., IYER, S.K. & DIENER, U.L. 1987. Improved method of screening for aflatoxin with a coconut agar medium. Appl. Environ. Microbiol. 53: 1593-1595.

ELMHOLT, S. & HESTBJERG, H. 1996. Recovery and detection of deuteromycete conidia from soil. In: Jensen, D.F., Jansson, H.B. and Tronsmo, A. (eds.) Monitoring antagonistic fungi deliberately released into the environment. pp. 49-55. Kluwer Academic Publishers, Dordrecht.

ELMHOLT, S., LABOURIAU, R., HESTBJERG, H. & NIELSEN, J.M. 1999. Detection and estimation of conidial abundance of *Penicillium verrucosum* in soil by dilution plating on a selective and diagnistic agar medium (DYSG). Myc. Res. 103: 887-895.

FILTENBORG, O. & FRISVAD, J.C. 1980. A simple screening method for toxigenic moulds in pure cultures. Lebensm. Wiss. Technol. 13: 128130.

FILTENBORG, O., FRISVAD, J.C. & SVENDSEN, J.A. 1983. Simple sceening method for toxigenic mold s producing intracellular mycotoxins in pure cultures. Appl. Environ. Microbiol. 45: 580-585.

FILTENBORG, O. & FRISVAD, J.C. 1990. Identification of *Penicillium* and *Aspergillus* species in mixed cultures in Petri dishes using secondary metabolite profiles. In: Samson, R.A. and Pitt, J.I. (eds.): Modern concepts in *Penicillium* and *Aspergillus* classification. Plenum Press, New York, pp. 27-36.

FILTENBORG, O., FRISVAD, J.C., LUND, F. & THRANE, U. 1992. Simple identification procedure for spoilage and toxigenic mycoflora of foods. In: Samson, R.A., Hocking, A.D., Pitt, J.I. and King, A.D. (eds.): *Modern methods in food mycology.* Elsevier, Amsterdam, pp. 263-273.

FRÄNDBERG, E & OLSEN, M. 1999. Performance of DG18 media, a collaborative study. J. Food Mycol. 2: 239-249.

FRISVAD, J.C. & THRANE, U. 1987. Standardized highperformance liquid chromatography of 182 mycotoxins and other fungal metabolites based on alkylphenone indices and UVVIS spectra (diode array detection). J. Chromatogr. 404: 195214.

FRISVAD, J.C. & THRANE, U. 1993. Liquid column chromatography of mycotoxins. In: Betina, V. (ed.): Chromatography of mycotoxins: techniques and applications. Journal of Chromatography Library 54. Elsevier, Amsterdam, pp. 253-372.

FRISVAD, J.C., THRANE, U. and FILTENBORG, O. 1998. Role and use of secondary metabolites in fungal taxonomy. In: Frisvad, J.C., Bridge, P.D. and Arora, D.K. (eds.): Chemical fungal taxonomy, pp. 289-319, Marcel Dekkker, New York.,

FRISVAD, J.C. 1983. A selective and indicative medium for groups of *Penicillium viridicatum* producing different mycotoxins in cereals. J. Appl. Bacteriol. 54: 409-416.

FRISVAD, J.C. 1985. Creatine sucrose agar, a differential medium for mycotoxin producing terverticillate *Penicillium* species. Lett. Appl. Microbiol. 1: 109-113.

FRISVAD, J.C. 1987. High pressure liquid chromatographic determination of mycotoxins and other secondary metabolites. J. Chromatogr. 392: 333-347.

FRISVAD, J.C. 1989. The use of highperformance liquid chromatography and diode array detection in fungal chemotaxonomy based on profiles of secondary metabolites. Bot. J. Linn. Soc. 99: 81-95.

FRISVAD, J.C. 1992. Chemometrics and chemotaxonomy: a comparison of multivariate statistical methods for the evaluation of binary fungal secondary metabolite data. Chemom. Intell. Lab. Syst. 14: 253-269.

FRISVAD, J.C. 1993a. Modifications on media based on creatine for use in *Penicillium* and *Aspergillus* taxonomy. Lett. Appl. Microbiol. 16: 154-157

FRISVAD, J.C. 1993b. Use of HPLC diode array detection in the detection of nitrogen containing mycotoxins and taxonomy of their producers in *Penicillium.* Appl. Biochem. Microbiol. 29: 11-17.

FRISVAD, J.C. 1994. Classification of organisms by secondary metabolites. In: Hawksworth, D.L. (ed). Identification and characterization of pest organisms. pp. 303-320, CAB International, Wallingford..

FRISVAD, J.C. 1995. Toxin producing *Penicillium verrucosum.* Determination in foods and feedstuffs. NMKL proposed method No. 152: 1-6. Nordic Committee of Food Analysis, Helsinki.

FRISVAD, J.C., BRIDGE, P.D. & ARORA, D.K. (Eds). 1998. *Chemical fungal taxonomy.* Marcel Dekker, New York.

FRISVAD, J.C., FILTENBORG, O., LUND, F. & THRANE, U. 1992a. New selective media for the detection of toxigenic fungi in cereal products, meat and cheese. In: Samson, R.A., Hocking, A.D., Pitt, J.I. and King, A.D. (eds.): *Modern methods in food mycology.* Elsevier, Amsterdam, pp. 275-284.

FRISVAD, J.C., FILTENBORG, O., LUND, F. & THRANE, U. 1992b. Collaborative study on media for detecting and enumerating toxigenic *Penicillium* and *Aspergillus* species. In: Samson, R.A., Hocking, A.D., Pitt, J.I. and King, A.D. (eds.): *Modern methods in food mycology.* Elsevier, Amsterdam, pp. 255-261.

FRISVAD, J.C., FILTENBORG, O., LUND, F. and SAMSON, R.A. 2000. The homogeneous species and series in subgenus *Penicillium* are related to mammal nutrition and excretion. In: Samson, R.A. and Pitt, J.I. (eds.): *Integration of modern taxonomic methods for* Aspergillus *and* Penicillium *classification.*, pp. 259-277, Harwood Scientific Publ., Reading.

GEISER, D., FRISVAD, J.C. & TAYLOR, J.W. 1998. Evolutionary relationships in *Aspergillus* section *Fumigati* inferred from partial beta-tubulin and hydrophobin sequences. Mycologia 90: 832-846.

GIFFEN, G.J. & GARREN, K.H. 1974. Population levels of *Aspergillus flavus* and the *A. niger* group in Virginia Peanut field soils. Phytopathology 64: 322-325.

HAMSA, T.A.P. & AYERS, J.C. 1977. A differential medium for the isolation of *Aspergillus flavus* from cottonseed. J. Food Sci. 42: 449-453.

HOCKING, A.D. & PITT, J.I. 1980. Dichloran-glycerol medium for enumeration of xerophilic fungi from low moisture foods. Appl. Environ. Microbiol. 39: 488-492.

JANSSEN, P., COOPMAN, R., HUYS, G., SWINGS, J., BLEEKER, M., VOS, P., ZABEAU, M. & KERSTERS. K. 1996. Evaluation of the DNA fingerprinting method AFLP as a new tool in bacterial taxonomy. Microbiology (UK) 142: 1881-1893.

KING, A.D., HOCKING, A.D. & PITT, J.I. 1979. Dichloran-rose bengal medium for enumeration of molds from foods. Appl. Environ. Microbiol. 37: 959-964.

KING, A.D., PITT, J.I., BEUCHAT, L.R. & CORRY, J.E.L. (eds.). 1986. Methods for the mycological examination of foods. Plenum Press, New York.

KING, A.D., PITT, J.I., BEUCHAT, L.R. & CORRY, J.E.L. 1986. *Methods for the Mycological examination of foods.* Plenum Press, New York.

KULSHRESTHA, D.D., MATHUR, S.B. & NEERGAARD, P. 1976. Identification of seed-borne species of *Colletotrichum.* Friesia 11: 116-125.

LARSEN, T.O. 1996. Identification of Penicillia based on volatile chemical markers. In: Rossen, L., Rubio, V., Dawson, M. and

Frisvad, J.C. (eds.): *Fungal identification techniques*, pp. 212-215. European Commission, Brussels.

LARSEN, T.O. 1997. Identification of cheese associated fungi using selected ion monitoring of volatile terpenes. Lett. Appl. Microbiol. 24: 463-466.

LARSEN, T.O., and FRISVAD, J.C. 1994. A simple method for collection of volatile metabolites from fungi based on diffusive sampling from Petri dishes. J. Microbiol. Meth. 19: 297-305.

LARSEN, T.O., and FRISVAD, J.C. 1995a. Comparison of different methods for collection of volatile chemical markers from fungi. J. Microbiol. Meth. 25: 135-144.

LARSEN, T.O., and FRISVAD, J.C. 1995b. Characterization of volatile metabolites from 47 taxa in genus *Penicillium*. Myc. Res. 99: 1153-1166.

LARSEN, T.O., and FRISVAD, J.C. 1995c. Chemosystematics of species in genus *Penicillium* based on profiles of volatile metabolites. Myc. Res. 99: 1167-1174.

LARSEN; T.O. 1998. Volatiles in fungal taxonomy. In: Frisvad, J.C., Bridge, P.D. and Arora, D.K. (eds.): *Chemical fungal taxonomy*, pp. 263-287, Marcel Dekker, New York.

LAW, B. & STAFFORD, L.E. 1993. The use of ultraviolet spectra and chromatographic retention data as an aid to metabolite identification. J. Pharm. Biomed. Anal. 11: 729736.LUND, F. 1995a. Diagnostic characterization of *Penicillium palitans*, *P. commune* and *P. solitum*. Lett. Appl. Microbiol. 21: 60-64.

LEMKE, P.A., DAVIES, N.D. & CREECH, G.W. 1989. Direct visual detection of aflatoxin synthesis by microcolonies of *Aspergillus* species. Appl. Environ. Microbiol. 55: 1808-1810.

LUND, F. 1995b. Differentiating *Penicillium* species by detection of indole metabolites using a filter paper method. Lett. Appl. Microbiol. 20: 228-231.

LUND, F. and SKOUBOE, P. 1998. Identification of *Penicillium caseifulvum* and *P. commune* isolates related to specific cheese and rye bread factories using RAPD fingerprinting. J. Food Mycol. 1: 131-139.

MAJER, D., MITHEN, R., LEWIS, B.G., VOS, P. & OLIVER, R.P. 1996. The use of AFLP fingerprinting for the detection of genetic variation in fungi. Myc. Res. 100: 1107-1111.

MOSSEL, D.A.A., KLEYNEN-SEMMELING, A.M.C., VINCENTIE, H.M., BEERENS, H. & CATSARAS, M. 1970. Oxytetracycline-glucose-yeast extract agar for selective enumeration of moulds and yeasts in foods and clinical material. J. Appl. Bacteriol. 33: 454-457.

NATH, R., NEERGAARD, P. & MATHUR, S.B. 1970. Identification of *Fusarium* species on seeds as they occur on blotter test. Proc. Int. Seed Test. Ass. 35: 121-144.

NIELSEN, M.S., NIELSEN, P.V. and FRISVAD, J.C. 1996. Micellar electrokinetic capillary chromatography of fungal metabolites: resolution optimized by experiental design. J. Chromatogr. A 721: 337-344.

NIELSEN, N.-P.V., CARSTENSEN, J.M. & SMEDSGAARD, J. 1998. Aligning of single and multiple wavelength chromatographic profiles for chemometric data analysis using correlation optimized warping. J. Chromatogr. A 805: 17-35.

NIELSEN, N.-P.V., SMEDSGAARD, J. & FRISVAD, J.C. 1999. Full second-order chromatographic/spectrometric data matrices for automated sample identification and component analysis by non-data-reducing image analysis. Anal. Chem. 71: 727-735.

NILSSON, T., BASSIANI, M.R., LARSEN, T.O. & MONTANARELLA, L. 1996a. Classificaton of species in the genus *Penicillium* by Curie point pyrolysis/mass spectrometry followed by multivariate analysis and artificial neural networks. J. Mass Spectrom. 31: 1422-1428.

NILSSON, T., LARSEN, T.O. MONTANARELLA, L & MADSEN, J.Ø. 1996a. Application of head-space solid-phase microextraction for the analysis of volatiles emitted by *Penicillium* species. J. Microbiol. Meth. 25: 245-255.

NOTERMANS, S.H.W., COUSIN, M.A., DE RUITER, G.A. & ROMBOUTS, F.M. 1998 Fungal immunotaxonomy. In: Frisvad, J.C., Bridge, P.D. and Arora, D.K. (eds.) *Chemical fungal taxonomy*. pp. 121-152. Marcel Dekker, New York.

PITT, J.I. & HOCKING, A.D. 1985. *Fungi and food spoilage*. Academic Press, Sydney.

PITT, J.I. & HOCKING, A.D. 1997. *Fungi and food spoilage*. 2nd ed. Blackie Academic & Professional, London.

PITT, J.I., HOCKING, A.D. & GLENN, D.R. 1983. An improved medium for the detection of *Aspergillus flavus* and *A. parasiticus*. *J. Appl. Bacteriol.* 54: 109-114.

SAMSON, R.A., FLANNIGAN, B., FLANNIGAN, M.E., VERHOEFF, A., ADAN, O.C.C. & HOEKSTRA, E.S. (eds.). 1994. Health implication of fungi in indoor environmente. Elsevier, Amsterdam.

SAMSON, R.A., HOCKING, A.D., PITT, J.I. & KING, A.D. (eds.). 1992. *Modern methods in food mycology*. Elsevier, Amsterdam.

SINGH, K., FRISVAD, J.C., THRANE, U. & MATHUR, S.B. 1991. *An illustrated manual on identification of some seedborne Aspergilli, Fusaria, Penicillia and their mycotoxins*. Jordbrugsforlaget, Copenhagen.

SKOUBOE, P., FRISVAD, J.C., LAURITSEN, D., BOYSEN, M., TAYLOR, J.W. & ROSSEN, L. 1999. Nucleotide sequences from the ITS region of *Penicillium* species. Myc. Res. 103: 873-881.

SMEDSGAARD, J. 1997a. Micro-scale extraction procedure for standardized screening of fungal metabolite production in cultures. J. Chromatogr. A 760: 264-270.

SMEDSGAARD, J. 1997b. Terverticillate Penicillia studied by direct electrospray mass spectrometric profiling of crude extracts. II. Database and identification. Biochem. Syst. Ecol. 25: 65-71.

SMEDSGAARD, J. and FRISVAD, J.C. 1996. Using direct electrospray mass spectrometry i taxonomy and secondary metabolite profiling of crude fungal extracts. J. Microbiol. Meth. 25: 5-17.

SMEDSGAARD, J. and FRISVAD, J.C. 1997. Terverticillate Penicillia studied by direct electrospray mass spectrometric profiling of crude extracts. I. Chemosystematics. Biochem. Syst. Ecol. 25: 51-64.

SMILANICK, J.L. & ECKERT, J.W. 1986. Selective medium for isolating *Penicillium digitatum*. Plant Dis. 70: 254-256.

THRANE, U. 1996. Comparison of three selective media for detecting *Fusarium* species in foods: a collaborative study. Int. J. Food Microbiol. 29: 149-156.

THRANE, U., 1986. Detection of toxigenic *Fusarium* isolates by thinlayer chromatography. Lett. Appl. Microbiol. 3: 93-96.

TORREY, G.S. & MARTH, E.H. 1976. Silicagel medium to detect molds that produce aflatoxin. Appl. Environ. Microbiol. 32: 376-380.

TSAO, P.H. 1970. Selective media for isolation of plant pathogenic fungi. Annual Review of Phytopathology 8: 157-186.

TUITE, J. 1969. Plant pathological methods. Burgess Publishing Co., Mpls.

VERHOEFF, A.P., WIJNEN, J.H. VAN, BOLEIJ, J., BRUNEKREEF, B., REENEN-HOEKSTRA, E.S. VAN & SAMSON, R. A. 1990. Enumeration and identification of airborne viable mould propagules in houses. Allergy 45: 275-284.

VOS, P., HOGERS, R., BLEEKER, M., REIJANS, M., VAN DE LEE, T., HORNES, M., FRIJTERS, A., POT, J., PELEMAN, J., KUIPER, M. & ZABEAU, M. 1995. AFLP: a new technique for DNA fingerprinting. Nucleic Acids Res. 23: 4407-4414.

WINDELS, C.E. 1992. *Fusarium*. In: *Methods for research on soilborne phytopathogenic fungi*. L.L. Singleton, Mihail, J.D. and Rush, C.M. (eds.). APS Press, St. Paul, Minnesota. pp. 115-128.

Chapter 3

METHODS FOR THE DETECTION AND ISOLATION OF FUNGI IN THE INDOOR ENVIRONMENTS

ELLEN S. HOEKSTRA, R. A. SAMSON and R. C. SUMMERBELL

Centraalbureau voor Schimmelcultures, Utrecht, The Netherlands

TABLE OF CONTENTS

INTRODUCTION

Fungal proliferation indoors is often related to leakage, flooding, condensation and humidity problems. Lack of ventilation may exacerbate these problems. Occupants may also contribute to mould growth by means of activities generating humidity or obstructing venting of the building. Problems occur both in old and new buildings, and often arise after inferior renovation works, or after installation of tight insulation. Nowadays many problems occur worldwide with the large scale air-handling and -conditioning systems (HVAC). Appropriate maintenance of such systems is important to prevent fungal growth. In particular, regular replacing of the filters and cleaning of the humidifier systems are essential.

In food processing plants, spoilage of the final products or raw materials may occur and often connects to the development of a specifically adapted fungal flora (mycobiota) in the plant. In plants vulnerable to such process contamination, regular monitoring of various processing sites, storage rooms, and air- and other filter systems is advisable. In contamination outbreaks, the routing of workers within the plant may be studied in order to detect the sources of contamination and pathways of inoculum transmission. Year-round monitoring, which helps to determine the baseline "normal" fungal burden as well as detecting contamination outbreaks, is a recommended practice. Regular monitoring ensures that in case of the occurrence of an unusually high level of fungal inoculum or a specific spoilage organism, action may be taken swiftly enough to avoid recalls of the product from the market.

Several methods for the detection and isolation of indoor moulds have been employed. These have been mainly adapted from the current methods in food mycology (see Chapter 2) and concern the direct microscopic detection or culturing of organisms. In recent years, especially in connection with studies related to measuring allergen exposure, cell wall chemical components of moulds in the environment, e.g. (1→3)-ß-glucans and extracellular polysaccharides (EPS), have been measured directly (Samson *et al.*, 1994, Douwes 1998, Flannigan *et al.*, 2001). Such direct detection of biochemical components allows the burden of both living and dead mould materials to be estimated, and also accounts for the antigen burden of fractionated and pulverized materials. Microbial volatile organic compounds (MVOC) have also been investigated as possible indicators of the fungal burden and of occupant exposure to airborne fungi (Ström *et al.*, 1994, Wessen *et al.*, 2001). The present chapter does not outline these direct biochemical techniques, but instead concentrates on the techniques that remain the most practicable and cost-effective for most situations, namely techniques for the detection and isolation of culturable or direct-microscopically recognizable fungi in indoor environments. Techniques applicable to various types of environments, such as dwellings, workplaces, food processing plants, and public buildings, e.g. offices, hospitals, schools, museums and archives, are described.

The interpretation of all techniques for indoor mould detection and quantification depends on the establishment of suitable guidelines and action thresholds. There is a need for guidelines that can be accepted worldwide. Numerous governmental and non-governmental institutions in various nations have proposed guidelines, but there still is no consensus about their application or validity. The American Industrial Hygiene Association (AIHA) has published a manual of methods for microbiological sampling (Dillon *et al.,* 1996) while the American Conference of Government Industrial Hygienists (ACGIH) also published a paper on microbiological problems in buildings (Macher *et al.,* 1999). In Canada the Federal-Provincial Committee on Environmental and Occupational Health published guidelines on fungal contamination in public buildings (1995). In Europe the Commission of the European Communities (CEC) has published a report on Biological Particles in indoor environments with the strategy and methodology for investigating indoor environments, including recommendations (1993).

VISUAL INSPECTION

Prior to sampling, a thorough inspection inside the building is essential. In many cases, the source of contamination is readily apparent to the trained eye; even where this is not the case, selection of sampling sites must be based on visually determined suitability. It is important to make inquiries about the state and history of the building (renovation, reconstruction, new insulation, recent leakage problems, past flooding including any derived from suppression of fires, etc.). In inspections, confined and often difficult-to-access spaces potentially forming "humidity chambers" for high levels of mould growth must be kept in mind. Examples include wall cavities, crawlspaces, spaces above ceiling tiles, air pockets behind wallpaper, the undersides of carpets, and places where concealed cold objects, especially walled-in cold water pipes, may cause condensation to accumulate. Furniture or cabinets closely placed against the wall may encourage fungal growth by impeding ventilation as well by causing temperature differences (e.g., by trapping cold air), causing condensation to form. Gypsum tiles and wallboard, often covered with cellulosic material (heavy paper) for structural support, are among the best substrates for biochemically and allergenically undesirable fungi like *Stachybotrys* and *Chaetomium* (Flannigan *et al.,* (2001); unpublished data) after they become wet. Therefore a close observation of these materials is advised, particularly when there is any likelihood that their inner (wall cavity) sides may have been affected by leakage or flooding.

The methods for detecting indoor fungi in domestic and office environments have been discussed by Samson *et al.,* (1994), Miller (2001), Morey (2001). The approaches for industrial environments are very similar; therefore, recommendations for these situations are given together with those for other indoor environments below.

A. Industrial environments

In the food industry (production, processing and storage areas) but also in other workplaces where mould is a concern, monitoring of air-borne fungi at regular intervals (e.g. weekly, monthly) is recommended. In each food-processing plant a so-called "home-flora" will be present. The local indoor conditions tend to select for a particular assemblage of fungi. Since the main available growth substrate often derives from the manufacturing process, e.g., aerosols arising during production or aggregation of materials in machinery, there tends to be a selection for fungi enzymatically compatible for growth on the product being worked with. Excess inoculum production leading to product spoilage can easily occur. It is important in each industrial setting to know which organisms are likely to cause spoilage problems. In bakeries, for example, the dry conditions prevalent in the factory environment, in the flour and other dried raw materials and in the final products particularly favours growth of xerophilic fungi, e.g. *Eurotium* species and *Wallemia sebi.* Early detection of these organisms during the production is essential to prevent spoilage. Before starting a monitoring program a checklist of the areas susceptible for fungal contamination and a detailed scheme of the routing of workers as well as of the entire production process should be made. From this check-list, appropriate sampling sites can be selected for routine monitoring. The practices used in cleaning the plant should be reviewed. In inspections of plants as well as in consultations for remediation or prevention of mould growth, special attention must be given to:

- ventilation ducts (should be cleaned as necessary based on dust level; should not have cold spots accumulating condensation moisture; should not have linings supporting growth of molds or streptomycetes)
- filters (should be changed at appropriate intervals)
- air-flow (rate should be sufficient to allow humidity control)
- walls and floors (should be smooth and easy to clean)
- transport belts (must be inspected and cleaned or disinfected regularly)
- remnants of materials and products (accumulation should be avoided or removed regularly)
- waste-baskets and containers (closed types are preferred to open systems)
- returned cargo (should be kept separate from the production area)
- spoiled products (should not be placed or stored in the production area)
- barrier sanitation in handling practices (unnecessary transport of materials from one room to another must be avoided even when contamination is not known to be occurring).

In addition to mycological analysis of raw materials, the intermediate stages of manufacture and the final product during the entire production process, samples also should be taken from the indoor environment in the production and storage areas. To obtain a complete picture of the situation it usually is necessary to combine both surface sampling (surface scrapings or tape samples) and air sampling.

B. Public buildings

In offices, archives, museums and schools it is important to investigate walls for signs of water damage and for sites of condensation. Carpeting (especially undersides), and, when an HVAC is present, air ducts and filters must be examined, as well as humidifier systems. Problems in archives will occur when there is moisture ingress or accumulation (leakage, condensation) and when already contaminated books or papers are received from other buildings without having been properly checked. In such situations, it is recommended to have a quarantine room to put materials into prior to incorporation in the general collection.

C. Hospitals

Whereas most mould growth investigations mainly consider moulds that become problematic only when present at relatively high densities, investigations in hospitals may also need to consider certain fungi that may be numerically rare. These are members of the relatively small group of environmental fungi posing a severe health hazard to the severely immunocompromised patient. This problem arises with special force in hospitals with bone marrow transplant units and other specialized areas for the treatment and recovery of patients, usually leukemia sufferers or major organ transplant recipients, who have the particularly severe cellular immune system deficit called neutropenia. Also, hospitals and clinics where surgery is performed on body sites particularly vulnerable to fungal infection, especially heart valves and corneas, must be especially cautious about even very small levels of opportunistic moulds.

Many opportunistic fungi are thermotolerant, with their capability of growing in the human body connecting to their overall adaptation for warm conditions. In sampling hospitals, then, special attention should be paid to the HVAC system, humidifier systems, and filters in the air ducts. In these wet and often warm sites the pathogenic fungus *Aspergillus fumigatus* may occur. Opportunistic fungi can also be detected in flower-vases, on plant material (flowers, fruit) and in potted plant soils or hydroculture media. Recent studies have shown that water lines may support opportunistic *Fusarium* and *Trichosporon* species, and sampling of taps, sink drains, shower drains and shower spray may be indicated, especially in specialized facilities for treatment of haematogenous cancers.

D. Residential homes

For dwellings it is recommended to make a thorough inspection prior to sampling and to inquire about the history of the building, in particular if any reconstruction or insulation work has been carried out. The living habits of the individuals should be considered, e.g., the presence of domestic animals, ornamental plants, carpeted floors. Walls should be checked behind furniture and cabinets. In dwellings it is advised to check the air circulation as well. An appropriate ventilation system will inhibit fungal contamination. A particularly critical factor is to determine the nature and condition of the basement, if there is one, or of any crawlspaces with earthen floors connected to the dwelling. Earthen floors, cracked or porous concrete foundations, basement floods saturating wallboard or carpeting, and condensation-attracting cold water pipes and air conditioning ducts are all well known sources of mould growth or of humidity that supports mould growth on stored materials.

METHODS and SAMPLING

Direct visual examination

Direct examination of visually detected moulds from building structural elements or from industrial materials is widely used in indoor mycology. Many problem moulds can be rapidly identified based on their morphological structures as formed on indoor substrates. In addition, associated communities of fungus-consuming "storage mites," collembolans, booklice, and beetle larvae (e.g., *Anobidae*) and other mould-associated arthropods can readily be seen. "Differential diagnosis" materials such as mineral efflorescence crystals, termite frass, and accumulations of outdoor dust may also be rapidly discerned.

Moulds growing on indoor substrates can often be recognized by their usual velvety-dusty appearance and by their characteristic colours ranging from black to grey-brown to blue green to pink. Suspected mould growth can be rapidly confirmed under the dissecting microscope. Persons engaging in such visual detection must ensure that the previously mentioned common sites of hidden mould growth are considered appropriately. Especially common trouble spots are reverse sides of wallpaper, undersides of carpets, and interiors of ceilings and walls.

Inspection may disclose the presence of wood-rot fungi, especially the destructive weeping *Merulius, Serpula lacrymans.* Buildings affected by this fungus may show widespread accumulation of the distinctive brown basidiospores as a very noticeable tea-coloured dust. It is necessary to find the locus of fungal growth in order to completely remove the affected wood. A search should be made for fruiting bodies and sites of heavy spore accumulation. Mycelial strands may be seen: these can be several meters long and can bridge brick walls, ultimately leading back towards a central focus or disseminated foci of fungal growth.

Sampling surfaces for direct examination

Sellotape preparations may be made by pressing the transparent adhesive tape onto the site of suspected mould growth, with the sticky side against the apparently mouldy surface. The tape can then be pressed directly onto a microscope slide and taken back to the laboratory. For mailing, the tape can be pressed flat onto the inside surface of any clean plastic bag made out of relatively heavy plastic (i.e., heavy enough not to tear when the tape is peeled back off for microscopic examination). The tape should not be crumpled, looped, or placed with moist materials.

In the laboratory, the tape can be lifted from the slide and a drop of lactic acid with cotton blue stain can be added between it and microscope slide. Another drop of the same mounting medium is put on top and a coverslip is overlaid. If an inspector or other sender has made a tape impression with frosted sellotape rather than the transparent product, a drop of mounting medium is put on the slide first, then the tape is laid onto it upside-down (sticky side facing up, away from the slide surface). Another drop of mounting medium is then placed on the upward facing sticky side and the coverslip is placed on top. Prepared in this fashion, tape samples can be observed under the microscope, even with the oil immersion lens. Often the fungus can be identified directly. The concomitant presence of associated arthropods, as mentioned above, may give an indication about the humidity and other conditions in the indoor environment.

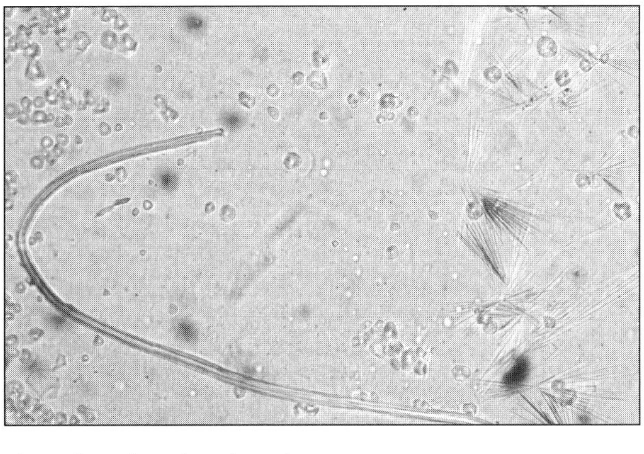

Fig. 1. Crystals and nonfungal structures.

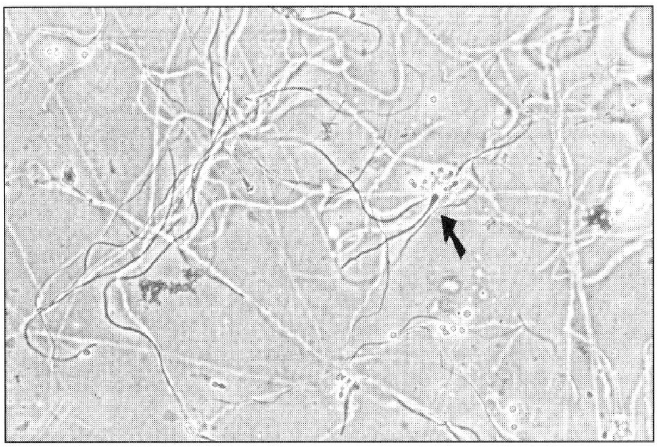

Fig. 2. Mycelium and *Aspergillus* conidiophores. (see arrow)

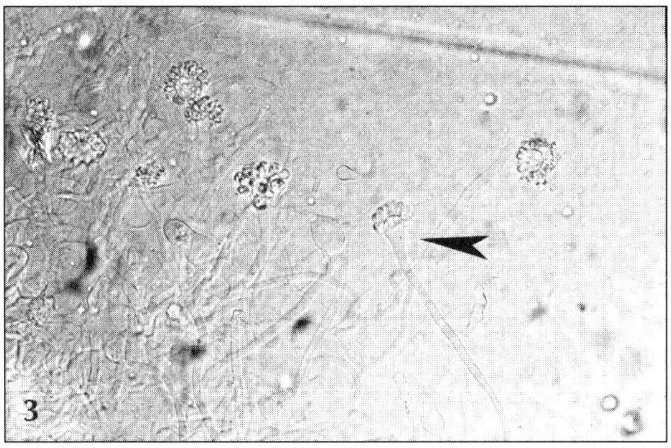

Fig. 3. Mycelium and *Aspergillus* conidiophores. (see arrow)

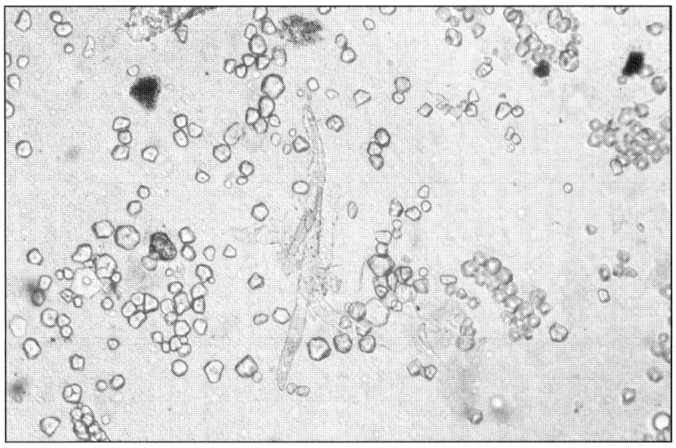

Fig. 4. Crystals and non-fungal structures.

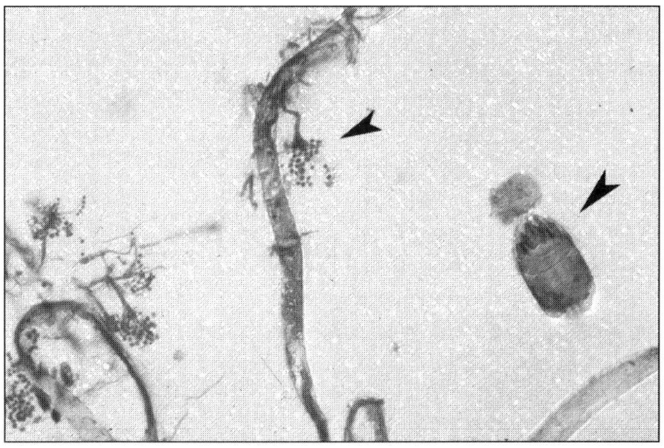

Fig. 5. Textile fibre, *Aspergillus* conidiophores and a mite (see arrow)

These animals also can be considered as vectors for further spreading of the fungal contamination. Droppings and body parts of some mite species are among the most strongly allergenic substances investigated to date see figs. 1-5). Sometimes it is also useful to take pieces of the sellotape and to impress them directly on the surface of an agar plate for cultivation of the moulds seen. The probability that mites will be co-cultured, however, should be taken into account.

Swab and direct surface culture samples.

Surface sampling can be performed using sterile swabs. The swab is streaked over the affected surface. In the laboratory, it can be plated either directly onto agar medium or vortexed in water or another appropriate solution (see "Dilution," below) to detach the sample material. For quantification based on swab samples, a defined area of the surface should be uniformly sampled and the collected materials, suspended in a defined quantity of liquid, should be put through a successive series of 1:10 dilutions, then plated out in aliquots of known volume (again, see "Dilution," below, for details). The resulting colony growth can be calculated as colony forming units per m^2 (CFU/m^2). For direct surface culture, Rodac® strips or plates prepared with agar medium extending to the surface can be used to take impressions of any relatively flat surface suspected of bearing fungal (or bacterial) inoculum. The agar medium is pressed so that it adheres momentarily onto the surface to be sampled, and then the vessel is closed up and incubated. As an alternative or supplement to swab or Rodac® impression sampling, materials scraped from walls and other surfaces, as well as other detachable, mouldy materials, may be analysed as well, initially by direct microscopy in lactic acid cotton blue mounts, and subsequently by culturing. Details on culturing of such materials are given below. With all culturing of materials taken directly from moulded sites, it should be borne in mind that overgrowth of antibiotic-polyresistant bacteria (especially pseudomonads) and culture contamination by fungivorous mites increases significantly as the moisture level of the sampled materials goes up. Samples from very wet walls, for example, may yield only heavy bacterial growth even on media containing large concentrations of antibiotics. Dilution procedures significantly decrease but do not eliminate this problem.

AIR SAMPLING

Sampling procedures for air-borne fungi can be divided into volumetric and non-volumetric procedures. The two types of procedures give quantitatively different results and are best used in full knowledge of their strengths and limitations.

Non-volumetric air sampling

Non-volumetric sampling utilizes open Petri dishes containing fungal growth medium. The sample plates are known as sedimentation plates, settle plates or gravity plates.

In this technique, the plates of medium are opened (mostly for 15-60 minutes, depending on the estimated density of inoculum), then closed and incubated at 25°C. After 5 - 10 days the colonies are counted and the fungi isolated and identified. This method is only recommended to obtain preliminary or qualitative information. The reason is that it is intrinsically biased against detection of fungi with very small propagules, such as *Aspergillus* and *Penicillium* species. For quantitative values on the abundance of moulds in air, volumetric sampling is essential. However, the sedimentation plate method can give useful information when sampling is done on the production line, e.g., transport belts on which the food product is transferred. Species detected in the plates are probably identical with those on the contaminated product. The large quantities of inoculum typical of problem situations overwhelm the statistical bias against detection of smaller propagules.

Volumetric air sampling

Several devices for impacting a defined quantity of air onto microbial growth media are commercially available. These include the Andersen sampler, the S.A.S. (Surface Air Sampler), the Slit-to agar sampler, the RCS (Reuter Centrifugal Air sampler) and the RCS Plus. Medium-filled plates or strips are impacted with the regulated air stream and then incubated for a defined period, usually 5 – 7 days. An optimal incubation temperature may be selected, such as 25 ^{0}C for general indoor fungi or 37 ^{0}C for opportunistic pathogens. After incubation the colonies are counted and their numbers are expressed according to formulas provided by the device manufacturers as CFU m^{-3} (= Colony Forming Units per cubic metre air). Comparative investigations have indicated that the Andersen sampler gave the most accurate and consistent results (Buttner and Stetzenbach, 1993; Verhoeff *et al.*, 1990). The Slit-to-agar sampler also gave precise results. Although the RCS sampler proved to be less accurate, this sampler and another somewhat less accurate sampler the S.A.S., have the advantage of being portable and battery operated and therefore very useful for routine monitoring. The more recently developed RCS-Plus is more precise than its predecessor, and allows for easier and more accurate setting of the air volumes to be sampled.

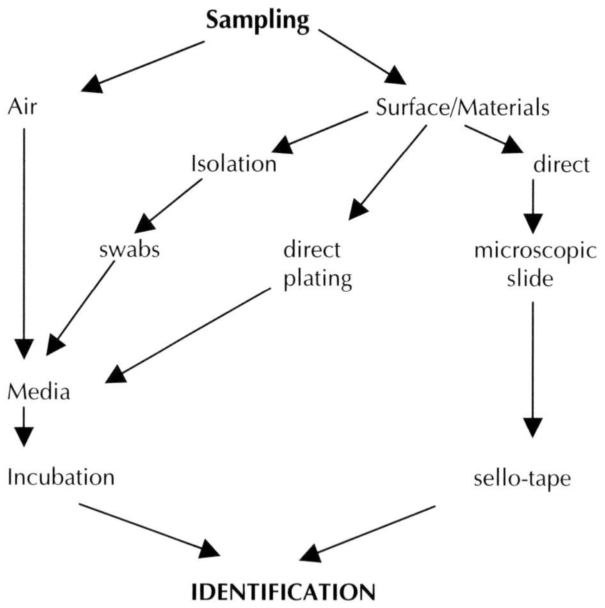

Usually, in "normal' situations (that is, situations where overwhelming mould numbers are not anticipated), 50 L/min of air is sampled. The Health Canada guidelines recommend 4 minute samples with the RCS sampler. In very clean-appearing rooms, e.g., in a hospital, a longer sampling time is recommended to ensure detection of important fungi present at low levels. On the other hand, with a high burden of fungal contamination, e.g., in an agricultural setting where a haze of mould conidia may be visible in the air, smaller quantities of air must be sampled to allow statistically valid counting of CFU. In many studies, a comparison of indoor samples with outdoor air is recommended. Depending on the time of year, weather, and nearby agricultural activities, outdoor fungal counts may be very high and this may influence indoor counts, especially in buildings with opening windows. On the other hand, problematic indoor mould colonizations generally show some fungal species levels significantly in excess of the levels present in outdoor air sampled the same day.

Direct plating

Direct plating is considered to be an effective technique for mycological examination of materials known or suspected to contain viable moulds. Small pieces of the material can be transferred or sprinkled directly onto the culture medium or suspended in a dilution fluid, as detailed below, prior to plating. Instead of a culture medium, an alternative is to use a humid sterile filter paper. This technique is only applicable to cellulose-digesting fungi, but may be a superior means of elucidating those fungi. Settled house dust can be plated directly onto appropriate agar media, normally dichloran 18% glycerol agar (DG18) and 2% malt extract agar as outlined below. The advantage of the former medium is that it allows accurate enumeration of drought-adapted (xerophilic and xerotolerant) fungi that make up an important component of indoor dust (Verhoeff et al., 1994). For quantitative analyses a dilution series should be prepared (see below).

In food mycology (see Chapter 2), and plant pathology (Gams *et al.*, 1998 p. 113), chemical surface disinfection is often used for distinguishing between problem fungal growth originating inside plant material and unimportant surface contamination. These techniques are seldom appropriate for materials from indoor environments. In some cases, however, especially in testing of wood samples which may or may not be fungally invaded, surface disinfection may be used to permit outgrowth and quantitation of fungi originating exclusively from within the material.

The standard incubation regime used at CBS for general-purpose enumerations is 25°C for 5 days. If pathogenic or thermophilic fungi are expected, incubation at 37°C (or higher) can also be used. Plates should be incubated in upright position and sufficient oxygen should be permitted for optimal growth. They can be kept in polyethylene bags to minimise evaporation. However, it is important that the bags are perforated and that the incubator is provided with an air circulation system. Accumulation of CO_2 significantly retards growth of filamentous fungi.

Dilution

For quantitative analyses or whenever high contamination levels are expected, the samples should be suspended and diluted. Initial procedures for use with swabs have already been outlined above. For bulk dust, at least 100 mg of weighed, dry material may be suspended in sucrose solution (1:50 w/w) or peptone-salt solution (0,1 % peptone + 0.85% NaCl). The solution is thoroughly mixed and then 10% of the volume is removed with a sterile volumetric pipette and transferred to a new container bearing a quantity of solution equal to that originally present in the first tube. Again, the solution is thoroughly mixed and (important!) a fresh, sterile pipette is used to transfer 10% of the volume to another, identical fresh container. This iterative 10-fold dilution may be continued through 5 to 10 stages, depending on the estimated starting concentration of viable inoculum. Finally, 0.1 ml abstracted from each dilution in the series is plated onto DG18 and 2% malt extract agars.

DETECTION MEDIA

For the detection and isolation of fungi and yeasts in different indoor environments, different media may be required. For wet or very humid situations as in bathrooms and water tanks, high water-activity such as water agar or 2% malt extract agar are recommended. In more dry situations, e.g., with house dust as mentioned above, the combination of DG18 and 2% Malt extract agar is recommended. Use of antibiotics (100 ppm chloramphenicol) is critical for minimizing bacterial growth (see Appendix)

DG18 medium, originally developed for the detection of xerophilic fungi such as *Eurotium* spp. and *Wallemia sebi* in food and feed products, proved in further investigations to be suitable for the indoor fungi as well. Both the xerophilic and mesophilic fungi will grow on DG18. The colonial expansion of fast growing fungi such as *Rhizopus* and *Mucor* is inhibited by Dichloran, and therefore overgrowth of the sample by such fungi is largely avoided. This facilitates accurate colony counting. For several fungi, however, particularly *Fusarium* spp., DG 18 is not appropriate.

Several microclimates with radically different moisture conditions may be present in a single room. In such situations, both xerophiles and mesophiles may proliferate indoors. After leakage, cellulosic materials such as paper-covered gypsum board will first be colonized by "hydrophilic" fungi, i.e., species with a preference for wet conditions, such as *Stachybotrys chartarum.* After the material has dried to some extent, fungi with a preference for moderately dry circumstances e.g. *Aspergillus versicolor* will grow. Finally, under the most dry conditions permitting fungal growth, xerophilic fungi will grow.

Media containing the fungal growth retardant Rose Bengal have been used in the past but are no longer recommended due to the sensitivity to light. (see page 285).

In the medical field often Sabouraud agar is used. This medium is not suitable for the detection of most of the fungi and is only recommended for environmental isolation of dermatophytes, e.g., from swimming or athletic areas. It should be supplemented by 100 ppm cycloheximide, 100 ppm chloramphenicol, and 50 ppm gentamicin for this usage.

SELECTIVE MEDIA

There are only a few selective media which might be used for the isolation of fungi in indoor environments. For isolation of basidiomycetes, such as wood-rotting fungi, media with the selective inhibitor benomyl (e.g., 1 ppm benomyl in 2% malt-extract agar) are recommended. The inhibitor suppresses most competing fungi of ascomycetous affinity, e.g., Penicillia, Aspergilli, Cladosporia. Fast-growing Zygomycetes, e.g., *Mucor,* are not suppressed by benomyl; in fact, benomyl media may be used to selectively isolate these fungi. Additional fungi which may be selected for by using benomyl are molds related to the order *Pleosporales* (e.g, *Alternaria, Curvularia)* and moulds in the order *Microascales* (*Pseudallescheria, Scopulariopsis).*

In cases where *Aspergillus flavus* or *A. parasiticus* is suspected (e.g., in ground nut or maize corn processing plants), use of the AFPA (Aspergillus Flavus Parasiticus Agar) medium developed by Pitt *et al.,* (1983) is recommended.

For the detection of cellulosic degrading fungi (e.g. *Trichoderma, Stachybotrys*), media with a piece of sterile filter paper on top of the medium will stimulate these fungi to grow. *Stachybotrys* also will be stimulated to grow on cellulose agar or on a simple agar medium based on wallpaper paste.

***Incubation*:** for general purpose incubation, 25°C in darkness is appropriate. As mentioned above, if thermophilic fungi or human pathogens are expected, higher incubation temperatures are needed. For pathogens, 37 °C is normally optimal, although a few organisms such as *Sporothrix schenckii, Fusarium proliferatum* and *Exophiala jeanselmei* are better sought at 35° C. For thermophiles, higher temperatures to 50 °C will be necessary, depending on the substrate tested.

Isolates of important or predominantly occurring fungi in indoor environments should be sent to mycological centres for authoritative species identification. For identification, several culture media can be used depending on the fungus. (see Appendix).

INDICATOR ORGANISMS

Where moisture and/or health problems are observed the following fungi can be regarded as indicator organisms if above a baseline level (to be established) in air samples, or if isolated from surfaces. Names indicated with an * are or may include important toxigenic and pathogenic taxa.

Materials with a high water activity (a_w>0.90 - 0.95): *Aspergillus fumigatus**, *Trichoderma, Exophiala, Stachybotrys**, *Phialophora, Fusarium**, *Ulocladium*, yeasts (*Rhodotorula*), Actinomycetes and Gram-negative bacteria (e.g. *Pseudomonas*), *Phoma.*

Materials with a moderately high water activity (0.90> a_w >0.85): *A. versicolor**, *A. sydowii, Emericella nidulans*

Materials with a lower water activity (a_w < 0.85): *Aspergillus versicolor**, *Eurotium, Wallemia,* Penicillia (e.g. *Penicillium chrysogenum, P. aurantiogriseum**).

INTERPRETATION OF THE RESULTS AND RECOMMENDATIONS

The interpretation of the results obtained via air sampling alone is often problematic as the levels of fungal propagules in the air vary greatly on an hour-to-hour or even a moment-to-moment basis with changes in activity levels as well as with fluctuations in temperature, relative humidity, wind strength and direction, and air flow derived from heating and air conditioning units as well as the opening or closing of windows or doors. Also, the biological differences between the fungi being sampled need to be recalled. The Aspergilli and Penicillia will produce large quantities of dry conidia that easily become airborne. On the other hand, moist conidia or spores produced in fruiting bodies may not be detected by air sampling, but may be released into the air later when material desiccates or is pulverized on impact (e.g., in renovation). Another consideration in viable propagule sampling is that the conidia and other fungal materials present in the air may be dead or may belong to fungi that cannot be cultured. The non-viable airborne material may nonetheless be allergenic or toxigenic. In some cases, the fungi present in the air may be alive but may not be culturable on the detection media used, especially when inappropriate media are employed. Therefore we always emphazise the importance of using a combination of air sampling, surface and bulk material samples as well as direct sellotape sampling. Then a comprehensive and complete picture will be obtained of the situation.

REFERENCES

COMMISSION OF THE EUROPEAN COMMUNITIES (1993). Indoor Air Quality and its impact on man. Report no. 12. Biological Particles in Indoor Environments (1993) pp. 75.

DILLON, H.K., HEINSOHN, P. A. & MILLER, J. D. 1996. Field guide for the Determination of Biological Contaminants in Environmental Samples, American Industrial Hygiene Association, Fairfax, VA.

DOUWES, J., (1998). Respiratory health effects of indoor microbial exposure : A contribution to the development of exposure assessment methods. PhD thesis, Agricultural University of Wageningen, The Netherlands. pp. 171.

FLANNIGAN, B., SAMSON, R.A. & MILLER, J.D. (2001). Microoranisms in home and indoor work environments. Diversity, Health Impacts , Investigation and Control. Harwood, Reading UK.

GAMS, W., HOEKSTRA, E..S. & APTROOT, A. (eds.) 1998. CBS-Course of Mycology, fourth edition, Centraalbureau voor Schimmelcultures, Baarn, 165 pp.

HEALTH CANADA, OTTAWA, (1995). Fungal Contamination in Public Buildings: A Guide to recognition and Management, Federal-Provincial Committee on Environmental and Occupational health, pp.76.

MACHER, J., AMMANN, H.A., BURGE, H. A. MILTON, D. K. & MOREY, P.R., eds, (1999). Bioaerosols: Assessment and Control, American Conference of Government Industrial Hygienists, Cincinnati, OH

MILLER, J.D. (2001). Mycological investigations of indoor environments. In Microorganisms in home and indoor work environments. Diversity, Health Impacts , Investigation and Control. Eds. Flannigan *et al.* Harwood, Reading UK. Pp. 231-246.

MOREY, P.R. (2001) Interpreting sampling data in investigations of indoor environments – selected case studies. In Microorganisms in home and indoor work environments. Diversity, Health Impacts , Investigation and Control. Eds. Flannigan *et al.* Harwood, Reading UK. Pp. 275-285.

SAMSON, R.A., FLANNIGAN, B., FLANNIGAN, M., Verhoeff A., Adan, O.C.G. & Hoekstra, E.S. (1994). Health Implication of fungi in indoor environments. Elsevier publ. 609 pp.

STRÖM, G. WEST, J., WESSEN, B. & PALMGREN, U., (1994). Quantitative analysis of microbial volatiles in damp Swedish houses. In Health Implications of Fungi in indoor environments eds. Samson et al. . Elsevier Publ. p.291-305.

VERHOEFF, A.P., van REENEN-HOEKSTRA, E.S., SAMSON, R. A., BRUNEKREEF, B. & van WIJNEN, J.H. 1994. Fungal propagules in house dust I, Allergy:49:533-539.

VERHOEFF, A.P., van REENEN-HOEKSTRA, E.S., SAMSON, R. A., BRUNEKREEF, B. & van WIJNEN, J.H. 1994. Fungal propagules in house dust II, Allergy:49:540-547.

VERHOEFF, A.P., WIJNEN, J.H. VAN, BOLEIJ, J., BRUNEKREEF, B., REENEN-HOEKSTRA, E.S. VAN & SAMSON, R. A. 1990. Enumeration and identification of airborne viable mould propagules in houses. Allergy 45: 275-284.

WESSEN, B., STRÖM, G., PALMGREN, U., SCHOEPS, K.O. & NILSSON, M. 2001. Analysis of microbial volatile organic compounds. In Microoranisms in home and indoor work environments. Diversity, Health Impacts , Investigation and Control. Eds. Flannigan *et al.* Harwood, Reading UK. Pp. 267-276.

For guidelines and protocols see also the website: http:/www.cbs.knaw/indoor.htm

Chapter 4

SPECIFIC ASSOCIATION OF FUNGI TO FOODS AND INFLUENCE OF PHYSICAL ENVIRONMENTAL FACTORS

O. FILTENBORG, J. C. FRISVAD and R. A. SAMSON

BioCentrum-DTU, Technical University of Denmark, DK-2800 Lyngby, Denmark and Centraalbureau voor Schimmelcultures, Utrecht, The Netherlands

TABLE OF CONTENTS

INTRODUCTION

The growth of fungi may result in several kinds of food-spoilage: off-flavours, toxins, discolouration, rotting, and formation of pathogenic or allergenic propagules (Chelkowski, 1991; Bigelis, 1992; Gravesen *et al.*, 1994; Tipples, 1995). Over the past 40 years fungi in foods have received special attention because of their ability to produce toxic metabolites. Although some fungi, such as *Claviceps purpurea* have been known for centuries because of their high and acute toxicity (Christensen, 1975; Young, 1979), it was only after the discovery in 1960 of the aflatoxins, carcinogenic metabolites of *Aspergillus flavus*, that a large number of species were found as mycotoxin producers. More than 400 mycotoxins are known today, aflatoxins being the best known, and the number is increasing rapidly. Mycotoxins are secondary metabolites which are toxic to vertebrate animals when introduced via a natural route. The toxicity of these metabolites is very different, with chronic termed toxicosis being the most important to humans. However, only a few mycotoxins are well described in toxicological terms. The most important toxic effects are different kinds of cancers and immune suppression (Pestka and Bondy, 1994; Prelusky *et al.*, 1994; Miller, 1991). Several mycotoxins have a very significant antibiotic activity as well, which in time may give rise to bacteria with a cross-resistance to the most important antibiotics used today, like penicillins. Furthermore it is important to note that some of these mycotoxins act synergistically (Miller, 1991).

The presence of potential toxinogenic species on food products does not always mean that these products contain mycotoxins; various environmental factors also play a part. Furthermore, the toxicity of many frequently occurring moulds has not yet been fully investigated and there is often no chemical method available to demonstrate the mycotoxin that may be present.

In this chapter an outline is given of the occurrence of various kinds of moulds on food products, of the factors that play a part in their growth and in mycotoxin formation. Furthermore measures, that can inhibit growth and inactivate moulds, are shortly discussed.

OCCURRENCE OF FILAMENTOUS FUNGI

Raw material, semi-manufactured and food products can be contaminated with spores or conidia and mycelium fragments from the environment. Contamination can occur at different stages of production: during growth and ripening of the crops, during processing of semi-completed and final products. The presence of large numbers of fungal propagules in products that are not visibly moulded can point to a general contamination of the environment on the one hand, or to the processing of mouldy raw material on the other. During processing, the fungi may be inactivated and are no longer viable.

Fungal growth only occurs under favourable conditions. The conditions vary for each species, adaptability determining species domination. The specific reason why a particular species dominates in a product is often not known, but is certainly correlated with the species characteristics and the properties of the product. Also the predominance of a species can be due to heavy contamination from ecological niches where the mould has developed. The frequent occurrence of *Penicillium expansum* on apples is probably due to growth of the mould on rotten matter in orchards (Börner, 1963), from where it could infect the apples. Salads are made from ingredients that are sometimes kept for long periods under refrigeration, favouring the dominance of the psychrophilic species of *Cladosporium* and *Penicillium.* The occurrence of *Eurotium herbariorum* on grains can be explained by its xerophilic characteristics. Also many types of bakery products have a low water activity (Northolt *et al.*, 1980b), thus clarifying the predominance of the xerophilic species *E. herbariorum.*

The addition of sorbate and propionate to rye bread may give rise to selection through which *Penicillium roqueforti,* which is fairly resistant to sorbate (Marth *et al.*, 1966), is dominant in this product.

The age of ripening cheese in warehouses, which is correlated with the moisture level and water activity, influences the composition of the fungal biota. During ripening *P. commune* and *P. brevicompactum* are replaced by *A. versicolor* and the xerophilic *E. herbariorum* , while *P. commune* remains.

FUNGAL METABOLISM AND SPOILAGE

Fungi need various nutrients in order to meet their energy needs and to form macromolecules such as proteins and DNA. Since fungi cannot synthesize carbohydrates, the substrate should contain these compounds. However, they can grow in a substrate rich in proteins without carbohydrates, e.g. cheese, by using amino acids as carbon source. Another important nutrient is nitrogen. Organic nitrogen compounds can be assimilated by all the fungi, inorganic compounds only by a limited number of species. Depending on the species, certain vitamins must also be present in the substrate, while others are synthesized by the fungus itself. Almost all foodstuffs for human consumption contain the above-mentioned nutrients and can, therefore, serve as substrate.

A large number of metabolites are formed during the breakdown of carbohydrates, some of which can accumulate under certain conditions. The best-known are ethanol and organic acids e.g. citric acid, fumaric acid, oxalic acid, gluconic acid, and polyols e.g. mannitol and arabitol. Besides degradation, the synthesis of a large number of compounds also takes place during this so-called "primary" metabolic process. Certain synthesized metabolites, such as glycogen and lipids, can at times accumulate into large concentrations. During and especially at the end of the growth, certain metabolites are synthesized which are not necessary for the growth and energy supply of the mould. Some of these metabolites are toxic for higher animals, insects and micro-organisms. A large part of these so-called "secondary" metabolites can be classified as steroids, carotenoids, alkaloids, cyclopeptides and coumarins, which differ greatly in their chemical structures. They are formed through different metabolic processes, but it is remarkable that the biosynthesis of various secondary metabolites begins with the condensation of acetyl and malonyl groups into polyketides or terpenes. Steroids, especially ergosterol, can make up 25% of the cell weight. Carotenoids can be seen as coloured pigments in the colonies of many fungi. Various alkaloids, cyclic peptides and coumarins are included among the mycotoxins because of their harmful effect on higher animals. There are different ideas about the possible function of secondary metabolites.

Filamentous fungi can produce a vast number of enzymes: lipases, proteases, carbohydrases (Bigelis, 1992). Once inside the food these enzymes may continue their activities independent of destruction or removal of the mycelium. The enzymatic activities may give rise to flavours like rioy coffee beans, musty odours in cork and wine or dried fruits (Tindale *et al.*, 1989; Whitfield *et al.*, 1991; Illy and Viana, 1995). This is caused by the fungal transformation of 2,4,6-trichlorophenol to trichloroanisol (TCA) by *Penicillium brevicompactum, P. crustosum, Aspergillus flavus* and other species. Some of these flavours can be detected in very small amounts like TCA or trans-1,3-pentadiene produced from sorbic acid by *Penicillium* species, *Trichoderma* and *Paecilomyces variotii* (Liewen and Marth, 1985; Kinderlerer and Hatton, 1990; Samson and Hoekstra, unpubl. data). TCA has an odour threshold level of 8 ng/l in coffee (Illy and Viani, 1995). The result of the enzymatic reactions may also be complete disintegration of the food structure, like the change of whole pasteurised strawberries into strawberry pulp due to growth of the heat-resistant fungi *Byssochlamys fulva* and *Byssochlamys nivea* (Beuchat and Rice, 1979).

Filamentous fungi can produce volatiles such as dimethyldisulphide, geosmin and 2-methylisoborneol which can affect the quality of foods and beverages even when present in very small amounts (Illy and Viani, 1995). These compounds are produced in large quantities in species specific combinations by different genera such as *Penicillium, Aspergillus* and *Fusarium* (Larsen and Frisvad, 1995a, b). The synthesis and accumulation of primary and secondary metabolites by fungi is applied for industrial production with the help of fermentors. The most productive strains are selected and optimal conditions for maximum productivity are created. Examples of fungal metabolites produced industrially are: the penicillins *(Penicillium chrysogenum),* citric acid *(Aspergillus niger),* fumaric acid *Rhizopus* spp.*)* and malic acid *(Penicillium brevicompactum).* Furthermore, certain enzymes, amino acids, pigments, vitamins and flavouring substances are produced with the aid of moulds. Examples of commercially produced enzymes are: amylase *(Aspergillus niger,*

A. oryzae), glucoamylase *(Rhizopus)*, enzymes for the breakdown of pectin *(A. niger)*, glucose oxidase *(A. niger)* and protease (*Rhizomucor*).

TOXIN-PRODUCTION BY FOOD-BORNE FUNGI

Mycotoxins are produced by a limited number of species. Aflatoxin is produced by the closely related *A. flavus, A. parasiticus* and *A. nomius* (Frisvad, 1995) and some rare species not occurring in foods, while other mycotoxins such as ochratoxin A are produced by a few species in different genera: *Petromyces alliaceus, Aspergillus ochraceus, A. carbonarius, A. niger* and *Penicillium verrucosum* (Frisvad and Samson, 2000). On the other hand, the individual toxic species are able to produce a considerable number of mycotoxins. *P. griseofulvum* for example produces patulin, griseofulvin, cyclopiazonic acid and roquefortine C consistently (Frisvad and Filtenborg, 1990). The number of toxic species is large, in fact it is a question whether any naturally existing species can be claimed to be unable to produce mycotoxins at all. This supports the view of Janzen (1977) that secondary metabolites are of great ecological significance (protection against other microbes, insects and other animals). However, it should be noted that not all strains of toxinogenic species isolated from foods are toxin producers. Schroeder and Boller (1971) examined a large number of *Aspergillus flavus* strains isolated from peanuts, cottonseed, rice and sorghum for the production of aflatoxins and found that 98, 81, 20 and 24%, respectively, could produce aflatoxins. Table 1 (Chapter 5, p. 321-330) lists the most common species with their mycotoxins.

The most important mycotoxins are the aflatoxins produced by *A. flavus* and *A. parasiticus*. These fungi can develop on the living ears of corn and cottonseed and as a saprophyte on peanuts, nuts, spices and many other agricultural products (Stoloff, 1976). The aflatoxins are very dangerous for humans and animals, not only because of their acute toxicity in high doses, but especially because of their strong carcinogenic properties. Epidemiological evidence points to a higher incidence of liver tumours in people who regularly eat food contaminated with aflatoxins (Shank, 1976). Besides aflatoxin contamination of foodstuffs due to growth of the mould, dairy products can be contaminated when cows are fed with feed containing aflatoxin. The aflatoxin M is then excreted into the milk. Sterigmatocystine, which is also a hepatocarcinogenic compound, has been frequently found in ripening hard types of cheese infected with *A. versicolor* (Northolt *et al.*, 1980a).

The nephrotoxic ochratoxin A, has been traced in foodstuffs and feed grains and in slaughter pigs and chickens (Krogh, 1977). Zearalenone is a mycotoxin that can attack the genitalia of pigs and laboratory animals and has been traced in various grains. In Ontario zearalenone was the most frequently occurring mycotoxin in corn feed (Funnell, 1979). The infection was caused by the growth of the fungus on crops before harvesting. For more details on mycotoxin production see Chapter 5 and Frisvad and Filtenborg (1993) and Frisvad (1995).

The mycotoxins are formed during growth of moulds on food and feed. Some mycotoxins are only present in the mould, while most of them are excreted in the foods. In liquid foods and in fruits, like peaches, pears and tomatoes the diffusion of mycotoxins can be very fast, leaving no part of the product uncontaminated. In solid foods like cheese, bread, apples and oranges the diffusion is slow leaving the major part of the product uncontaminated. Since most of the mycotoxins are very resistant to physical and chemical treatments, a rule of thumb exists: Once the mycotoxins are in the food, they stay there during processing and storage (Scott, 1991). This also means that the use of any mouldy material in the processing of food will not avoid the absence of the mycotoxins to the final product (carry over). Fortunately the presence of spores or even growth of particular fungi is not always followed by toxin production. The conditions, which allow toxin production, are more restricted than those for growth. Some foods are poor for production of some toxins while excellent for growth and visa versa. Should the toxins be produced, they may be produced in small amounts or be unstable (like PR-toxin in cheese) or they may only be toxic if produced in huge amounts. The prediction of which toxins may be produced in which foods may be based on knowledge of the specific mycobiota of foods, laboratory experiments, and data from chemical analyses for mycotoxins in foodstuffs under different conditions. However the actual presence of mycotoxins in a food product can only be evaluated after chemical or biological analysis of the very product.

ASSOCIATED MYCOBIOTA ON FOODS AND FEEDS

Besides being restrictive to the mycotoxin formation, the food parameters also have proved to be surprisingly restrictive to the spectrum of species, which are able to grow and thus spoil the individual food types. Normally less than 10 and often 1 to 3 species are responsible for spoilage (Frisvad and Filtenborg, 1988, 1993). On the other hand these critical species are often completely different for each food type. As far as fungi in foods are concerned this discovery is fairly new (Frisvad and Filtenborg, 1988), and is due to the development of new mycological methods and taxonomy of food-borne moulds especially the genera *Penicillium, Aspergillus* and *Fusarium*. The former dominating role of morphology in mould identification has been replaced by the combined use of secondary metabolite profiles (Frisvad and Filtenborg, 1983), isozyme profiles, physiological (Frisvad, 1981; Ahmad and Malloch, 1999) and ecological characteristics, DNA patterns and morphology (Samson and Pitt, 1990). The present book is based on this principle.

This significant development in methods and taxonomy of food-borne moulds within the last 10 years means that the major part of the publications on the Mycobiota of foods up till now, should be "translated" according to those changes. Some fungi from culture collections have been re-examined and their new identity has been published (Marasas *et al.*, 1984; Frisvad, 1989), but much of the earlier work needs to be repeated, because most isolates have not been kept. This problem has been taken into consideration here, so reference is mainly given to publications from the last decade, which is in agreement with the new concepts agreed by the international working groups. Examples will be given to illustrate the confusion and disagreement, which arise when former methods and taxonomy are being used to determine which species are responsible for the spoilage of each type of foods.

When describing the associated mycobiota of a food stuff, it is important to differentiate between species which are present as propagules and species which are actually able to spoil the food due to growth. Only the last species belongs to the associated mycobiota. The propagules (conidia, ascospores, mycelial fragments) may be present for several reasons, either as part of the "normal" airspora or as part of the "normal" mycobiota (formerly mycoflora), which are not necessarily damaging the food. Examples of this are several yeast species and *Rhizopus oligosporus* in tempe (Nout, 1995). Growth of such species may prevent growth of serious fungal spoilers, by interaction.

Table 1. Associated spoilage mycobiota of different foods (based on Filtenborg *et al.*, 1996, Simmons and Roberts, 1993, Simmons, 1986, 1993; 1994; 1995; 1999a & b)

FOODS	FUNGAL SPECIES
Citrus fruits:	*Alternaria citri; A. tangelonis; A. turkisafria; A. colombiana; A. perangusta; A. interrupta; A. dumosa; Penicillium digitatum; P. italicum; P. ulaiense*
Pomaceous and stone fruits:	*Penicillium expansum; P. crustosum; P. solitum; Alternaria gaisen; A. mali; A. tenuissima* group; *A. arborescens* group; *A. infectoria* group; *Cladosporium spp.;*
Garlic and onions:	*Penicillium allii; P. albocoremium; P. glabrum; Petromyces alliaceus; Botrytis aclada*
Potato tubers:	*Fusarium sambucinum; F. coeruleum*
Tomatoes:	*Alternaria arborescens; Stemphylium spp.; Penicillium olsonii*
Yam tubers:	*Penicillium sclerotigenum*
Wheat and rye grain Field condition:	*Fusarium culmorum; F. graminearum; F. avenaceum; F. equiseti; F. poae; F. tricinctum; Alternaria triticimaculans; A. infectoria; A. oregonensis; A. triticina; A. triticicola; A. tenuissima group; Cladosporium herbarum; Epicoccum nigrum; Stemphylium* spp.; *Ulocladium* spp.; *Drechslera* spp.; *Botrytis* spp.; *Penicillium* spp.; *Claviceps purpurea*
Wheat and rye grain Stored condition:	*Penicillium aurantiocandidum; P. cyclopium; P. freii; P. hordei; P. melanoconidium; P. polonicum; P. verrucosum; P. aurantiogriseum; P. viridicatum; Aspergillus flavus; A. niger; A. candidus; Eurotium* spp.; *Alternaria infectoria* group
Cereals in airtight storage:	*Paecilomyces variotii; Scopulariopsis candida; Penicillium roqueforti; Candida spp.; Byssochlamys fulva; B. nivea*
Acid preserved cereals:	*Penicillium. glandicola (= P. granulatum); P. roqueforti; Paecilomyces variotii; A. flavus; A. candidus; A terreus; Monascus ruber*
Spices:	*A. flavus; A. tamarii; A. niger; A. ochraceus; A. candidus; A. versicolor; Eurotium spp.; Wallemia sebi; P. islandicum; P.neopurpurogenum; P. citrinum*
Nuts:	*P. commune; P. crustosum; P. discolor; P. solitum; P. funiculosum; P. oxalicum; P. citrinum; A. flavus; A. wentii; A. versicolor.; Eurotium* spp.; *Alternaria infectoria* group
Rye bread:	*Penicillium roqueforti; P. paneum; P. carneum; P. corylophilum; Eurotium repens; E. rubrum; Paecilomyces variotii; Monascus ruber*
Cheese:	*Penicillium commune; P. nalgiovense; P. atramentosum; P. nordicum; Aspergillus versicolor; Scopulariopsis fusca; S. candida; S. brevicaulis*
Fats, margarine, etc.	*P. echinulatum; P. commune; P. solitum; P. spinulosum.; Cladosporium herbarum*
Fermented sausages:	*Penicillium nalgiovense; P. olsonii; P. chrysogenum; P. nordicum; P. solitum; P. oxalicum; P. commune; P. expansum; P. miczynskii; P. brasilianum*
Pasteurized foods:	*Byssochlamys fulva; B. nivea; Hamigera reticulata; Neosartorya fischeri; N. glabra; N. spinosa; Eupenicillium lapidosum; Talaromyces macrosporus; T. bacillisporus.; Paecilomyces variotii*
Low water activity foods:	*Eurotium chevalieri; E. herbariorum; E. amstelodami; Wallemia sebi; A. penicillioides; A. restrictus; Eremascus albus; E. fertilis; Xeromyces bisporus; Scopulariopsis halophilica; Chrysosporium inops (sensu* Pitt*); C. farinicola; C. fastidium; C. xerophilum; Polypaecilum pisce*

Species of *Rhizopus, Mucor* and *Botrytis* are common on many cereals, fruits and vegetables of high water activity as are several specific plant pathogens, but here we emphazise genera that contain mycotoxins producers.

The specific mycobiota of different foods and feeds is discussed below.

Citrus fruits

Citrus fruits (lemons, oranges, mandarins, tangerine, cumquats, tangelo, limes, pomelos, grapefruits) are non-climacteric fruits. Three spoilage Penicillia are of paramount importance, *Penicillium digitatum, P. italicum* and *P. ulaiense* (Birkinshaw *et al.*, 1931; Raper and Thom, 1949; Westerdijk, 1949; Holmes *et al.*, 1994) and so is a series of *Alternaria* species (Simmons, 1999). According to Holmes *et al.* (1994), *P. ulaiense* only appears when the other two pathogens are inhibited by fungicides and *P. ulaiense* is much more related to *P. italicum* than *P. digitatum*. The taxonomy of small spored *Alternaria* species has been discussed by Simmons (1992; 1999). *Fusarium poae* has also been reported on decayed citrus fruits in Georgia and Russia (Booth, 1971).

Germination of *P. digitatum* conidia is stimulated by certain combinations of the volatiles surrounding wounded oranges, notably limonene, α-pinene, sabinene, ß-myrcene, acetaldehyde, ethanol, ethylene and CO_2. Ethylene did not stimulate the germination of *P. digitatum* conidia in the non-climacteric fruit (Eckert and Ratnayake, 1994), whereas ethylene invited fungal attack in the climacteric tomatoes, avocadoes and bananas (Flaishman and Kolattukudy, 1994). Other constituents of oranges, such as simple sugars and organic acids also stimulate conidium germination in *P. digitatum* (Pelser and Eckert, 1977). It can be concluded that the associated mycobiota can tolerate and is sometimes even stimulated by the acids and other protecting volatile and non-volatile phytoalexins of citrus fruits in combination with the ability to produce pectinases and other citrus skin degrading enzymes. The fungal activities result in serious weight loss, shrinkage and softening of the citrus fruits (Ben-Yehoshua *et al*, 1987). Furthermore a few fruits spoiled by fungi can cause reduced shelf life of the sound fruits due to accelerated ripening or senescence triggered by the releasing the gas ethylene (Rippon, 1980).

Alternaria mycotoxins have been found in mandarins (Logrieco *et al.*, 1990) and they can also be produced in lemons and oranges (Stinson *et al.*, 1981). The mycotoxins found include tenuazonic acid, alternariol monomethyl ether (AME) and alternariol. *Fusarium poae* may produce trichothecenes and fusarin C, however *Fusarium* toxins have to our knowledge never been detected in citrus fruits.

No *P. italicum* and *P. digitatum* mycotoxins have been found in citrus fruits, yet these fungi produce compounds that are toxic to bacteria, plants, brine shrimps and chick embryos (Faid and Tantaoui-Elaraki, 1989; Tantaoui-Elaraki *et al.*, 1994). A *P. italicum* isolate that was toxic in laboratory animals (Kriek and Werner, 1981) was found to produce the mycotoxin 5,6-dihydro-4-methoxy-2H-pyran-2-one (Gorst-Allman *et al.*, 1982) related to verrucolone (arabenoic acid) (Isaac *et al.*, 1991; Larsen *et al.*, 1998). Arabenoic acid, originally found as a herbicide, has thus been found in the tomato plant pathogen *P. olsonii*, the citrus pathogen *P. italicum* and in *P. verrucosum*. Several other *P. italicum* secondary metabolites have been identified, but not tested for toxicity (Arai *et al.*, 1989). *P. digitatum* has been found to produce tryptoquivalins and tryptoquivalons (Frisvad and Filtenborg, 1989), which are regarded as mycotoxins (Cole and Cox, 1981).

Pomaceous and stone fruits

Penicillium expansum, P. crustosum and *P. solitum* (as *P. verrucosum* var. *melanochlorum*) and *Alternaria gaisen* were reported as organisms able to produce rot in apples (Raper and Thom, 1949; Samson *et al.*, 1976; Frisvad, 1981; Pitt *et al.*, 1991; Roberts *et al.*, 2000). Pomaceous and stone fruits and several other berries can be degraded by a large number of pathogenic species including *Monilia laxa, M. fructigena* and *Rhizopus stolonifer.* For example the "box rot" of dried French prunes, which is soft, sticky, macerated areas on the fruit and slippage of the skin under slight pressure due to the activity of pectinolytic enzymes produced by these fungi (Sholberg and Ogawa, 1983). However these fungi are probably not mycotoxin producers.

Canned fruits like apricots and peaches sometimes suffer from textural changes due to heat resistant fungal enzymes produced in the raw fruits (Harris and Dennis, 1980) or to the enzymatic activity of surviving heat resistant fungi like *Byssochlamys fulva* (Rice *et al*, 1979).

P. expansum is known for its production of patulin and citrinin and these mycotoxins have been found in mouldy fruits (Harwig *et al.*, 1973; Ciegler *et al.*, 1977; McKinley and Carlton, 1991; Vinas *et al.*, 1993). Other mycotoxins produced by *P. expansum*, such as roquefortine C and chaetoglobosin C (Frisvad and Filtenborg, 1989) or by *P. crustosum* such as terrestric acid, roquefortine C and penitrem A have not yet been reported from rotting pomaceous fruits. Communesin B, Chaetoglobosin A and C and expansolide have all been detected in artificially contaminated black currant and cherry juices (Larsen *et al.*, 1998)

When inoculated into apples *A. alternata* (the isolate used could also be *A. gaisen* or another species) produced alternariol and AME in both the rotten and sound part of apples, using artificial inoculation (Stinson *et al.*, 1980; 1981; Robiglio and Lopez, 1995). Species producing core rot in apples were placed in the *A. infectoria, A. tenuissima* and *A. alternata* groups by Serdani *et al.* (1998) and mycotoxins may accumulate in apples (Robiglio and Lopez, 1995).

Garlic and onions

Apart from *Botrytis aclada*, few species are able to spoil garlic and onions. *Penicillium allii* is a widespread spoiler of garlic (Vincent and Pitt, 1989; Frisvad and Filtenborg, 1989) while the closely related *P. albocoremium* is more common on other onions (Frisvad and Filtenborg, 1989). *Petromyces alliaceus, Aspergillus niger* and *Penicillium glabrum* are cited as producers of rots in onions (Raper and Fennell, 1965; Raper and Thom, 1949). *Penicillium glabrum* appears to grow only in the outer layers of onions.

Petromyces alliaceus is a very efficient producer of ochratoxin A (Hesseltine *et al.*, 1972; Ciegler, 1972; El-Shayeb *et al.*, 1992), but onions have not been analyzed for natural occurrence of ochratoxin A. *Penicillium allii* produces the roquefortine C, meleagrin and the viridicatin chemosynthetic family (Frisvad and Filtenborg, 1989). *P. glabrum* produces the nephrotoxin citromycetin (Domsch *et al.*, 1993).

The presence of allinin and other antimicrobial compounds in onions strongly selects for the associated mycobiota.

Potato tubers

Dry rot of potatoes is mainly caused by *Fusarium sambucinum* and *F. coeruleum*. The taxonomy of *F. sambucinum* has been revised recently (Nirenberg 1995) and does now contain *F. sulphureum* which often has been mentioned in connnection with potatoes. The other species frequently reported in relation to dry rot of potatoes, *F. coeruleum*, is a synonym of *F. solani* var. *coeruleum* (Gerlach and Nirenberg, 1982). *F. crookwellense* is also frequently isolated from damaged potatotes, however, its role as a primary pathogen is unclear.

The dry rot is normally so pronounced that the tubers are not suitable for consumption. However, as the full extent of the damage is not always visible from the outside, it may be possible that partly rottened potatoes could pass on to further processing in the food industry. In addition to the physical damage of potatoes mycotoxins may also be produced in the tubers. Diacetoxyscirpenol and related trichothecenes have been detected in tubers inoculated with *F. sambucinum* (Desjardins and Plattner, 1989). El-Banna *et al.* (1984) reports deoxynivalenol produced in tubers inoculated with *F. sambucinum* and *F. coeruleum*, however, deoxynivalenol production never has been verified from these species. This could be explained by the fact that *F. sambucinum* sometimes resembles *F. cerealis*, and that this species is known to produce deoxynivalenol.

Production of unidentified toxins in tubers infected with *F. sambucinum* and *F. coeruleum* has been demonstrated in brine shrimp test (Siegfried and Langerfeld 1978). Co-infection with bacteria (*Erwinia carotovora*) had no significant influence on the dry rot.

The pathogens are present in soil and tubers and the infection takes place by damage of the periderm. High soil humidity raises the infection rate whereas crop rotation will lower it. The harvest should be done with caution to minimize physical damage of the tubers and the storage kept dry and cooled. An efficient ventilation is important to keep the tubers free from dry rot.

Alternaria species reported as *A. solani* and *A. alternata* have been isolated repeatedly from both potatoes and other *Solaneceae* such as tomatoes (Weir et al., 1998), but their taxonomic status according to the system of Simmons has not been evaluated in detail yet.

Tomatoes

Tomatoes can be spoiled by many types of filamentous fungi including *Alternaria arborescens* (Simmons, 1999b). *A. solani* and isolates referred to *A. alternata* (Weir *et al.*, 1998; Morris *et al.*, 2000). AAL toxin and possibly even fumonisin B1 may be produced by *Alternaria arborescens*, so a more detailed taxonomic study is urgently needed of species associated to *Solanaceae*. *Penicillium olsonii* has been found consistently on tomatoes and may grow directly on commercial tomatoes, but no mycotoxins have been reported from this species.

The compounds found in this species verrucolone (= arabenoic acid) (a herbicide), bis(2-ethylhexyl) phthalate (a cell aggregation factor) and 2-(4-hydroxyphenyl)-2-oxoacetaldehyde oxime (a phosphorylated cholinesterase reactivator) are all bioactive compounds (Amade *et al.*, 1994).

Yam tubers

Several species of *Fusarium* are able to spoil yam tubers, but *Penicillium sclerotigenum* is one of the well known and very specific spoilers of this plant (Yamamoto, 1955). *P. sclerotigenum* has repeatedly been isolated from yam tubers, but never from any other product. It has been isolated from a *Dioscorea* sp. in Japan, on *D. cayenensis* in Jamaica and in blue yams flour from the Philippines. This fungus produces patulin, griseofulvin, roquefortine C and gregatins (Frisvad and Filtenborg, 1989, 1993), but the products have to our knowledge never been examined for those toxins.

Wheat and rye grains

Field condition: *Fusarium, Alternaria, Cladosporium* and *Claviceps* are very common on grains in the field and can reduce the quality of grains by their growth and mycotoxin production. Cereal plants may be damaged by numerous fungal pathogens, but in this paper we will focus on the spoilage of grains caused by moulds, thus omitting pathogens like smuts (*Tilletia* spp. and *Ustilago* spp.).

Fusarium ear diseases of cereals (also called head blight or scab) is caused primarily by *Fusarium culmorum* and *F. graminearum* (Wiese, 1987). Both species can produce deoxynivalenol (and related trichothecenes), zearalenone and several other biological active metabolites in the grains (Gareis et al. 1989). Whereas fusaria will be eliminated during food processing, a significant carry-over of toxins will be possible as they are resistant to cleaning of grains, milling, baking and other cooking processes .

Another important species is *F. avenaceum*. This species can produce moniliformin, antibiotic Y, enniatins and fusarin C, whereas earlier reports on production of trichothecenes and zearalenone have been insufficiently verified (Thrane, 1989). Data on natural occurrence of metabolites from growth of *F. avenaceum* is very scarce. However, synergistic effects between metabolites from more *Fusarium* species are possible (Miller, 1991).

Fusarium infections take place by air-borne conidia on the heads or by systemic infection. So far no highly resistent wheat or barley cultivars have been developed. For prevention of *Fusarium* diseases crop rotation is advised, as chemical treatment of seeds or application of fungicides to emerged heads is either not 100% effective or profitable nor is it desirable from an environmental point of view.

Alternariols and other *Alternaria* toxins have been detected infrequently in grains (Andrews, 1986; Champ *et al.*, 1991; Chelkowski and Visconti, 1992). However, together with *Cladosporium* spp., *Alternaria* can cause discolouration of the grains by their abundant presence on the grain, called black (sooty) heads. In some cases these moulds can cause a mild infection, which may result in weakened and undersized grains. Weakened or stressed plants are often yielding black heads.

Ergot, *Claviceps purpurea*, occurs mainly on rye but certain wheat lines have been damaged. The sclerotia, replacing grains, are the visible damage but in addtion *C. purpurea* produces a series of alkaloids toxic towards humans. These alkaloids have been detected in rye and wheat (grains and flour)(Möller *et al.*, 1993; Scott *et al.*, 1992). Crop rotation and good farming practice is the only way to prevent ergot.

Stored conditions

In temperate climate storage the dominating moulds are species within *Penicillium* and *Aspergillus*. Data on occurrence of these fungi in Canada can be found in Mills *et al.*, (1995). Data based primarily on barley, but also samples of rye and wheat from Scudamore *et al.* (1993) strongly indicate that a restricted number of *Penicillium* species are of paramount importance in stored cereals. 70 samples containing ochratoxin A from spoiled barley in Denmark were colonized by the Penicillia listed in Table 1 (Frisvad, unpublished), and included *P. verrucosum*, *P. hordei*, and members of series *Viridicata*

These data are generally in agreement with those of Scudamore *et al.* (1993) and Mills *et al.* (1995), despite wheat was the commodity examined in those studies. The taxonomy of the series *Viridicata*, also named *P. verrucosum* var. *cyclopium* and var. *verrucosum* (Samson *et al.*, 1976), has been revised by Lund and Frisvad (1994). Several toxin producing aspergilli have been reported to dominate on cereals, especially *A. candidus, A. flavus, A. niger, A. versicolor*, and *A. penicillioides* and *Eurotium* spp. at lower water activities (Lacey *et al.*, 1991; Sauer *et al.*, 1992).

Ochratoxin A, citrinin, xanthomegnin, viomellein and vioxanthin have all been found in barley, rye, and/or wheat (Hald *et al.*, 1983; Scudamore *et al.*, 1986, Frisvad, 1995). Several other possibly toxic secondary metabolites are produced by species in series *Viridicata* (Lund and Frisvad, 1994), such as verrucosidin, penicillic acid, cyclopenin, viridicatol, pseurotins, viridic acid, brevianamide A, nephrotoxic glycopeptides, anacine, auranthine, aurantiamine, terrestric acid, puberulonic acid, verrucofortine, puberuline, roquefortine C, meleagrin, oxaline, viridamine and aspterric acid (Frisvad and Lund, 1993), but they have not been analyzed for in cereals yet.

Mycotoxins from Penicillia growing in cereals stored in countries with subtropical or tropical climate could include viridicatum toxin (*P. aethiopicum*), citrinin (*P. citrinum*), cyclopiazonic acid, patulin and roquefortine C (*P. griseofulvum*) and secalonic acid D (*P. oxalicum*) (Frisvad and Filtenborg, 1989), but this has not yet been examined.

The importance of mycotoxins produced in cereals by aspergilli, like the aflatoxins and cyclopiazonic acid, has been pointed out by Pier and Richard (1992), but the taxonomy of these species seems to be less complicated. However it is still not clear whether other aflatoxigenic species, such as *A. nomius* and *A. parasiticus* are commonly causing aflatoxin contamination in cereals, because all three species are often identified as *A. flavus*.

Rye bread

The most important species spoiling rye bread with no preservatives added are *Penicillium roqueforti, P. paneum, P. carneum, P. corylophilum, Eurotium repens* and *E. rubrum* (Lund *et al*, 1996; Spicher, 1985). *P. paneum* and *P. carneum* are newly described species based on significant differences in mycotoxin-, DNA- and morphological characteristics (Boysen *et al.*, 1996). Isolates belonging to these species have earlier been identified as *P. roqueforti* or *P. roqueforti* var. *carneum* (Frisvad & Filtenborg, 1989). The cultures used for blue-cheese production all belongs to *P. roqueforti*. If preservatives like sorbic acid and propionic acid are added the spoilage mycobiota would be completely dominated by *P. roqueforti* and the new species *P. paneum* and *P. carneum* (Spicher, 1985; Lund *et al*, 1996). The growth of yeasts sometimes is a serious problem especially in sliced rye bread, the dominating species being: *Endomyces fibuliger, Pichia anomala* and *Hyphopichia burtonii*. Species of less importance to the quality of rye bread are: *Paecilomyces variotii, Monascus ruber, Aspergillus flavus, Penicillium commune, P. solitum, A. niger, P. decumbens* and *Mucor spp.* Some of these species may become important for a very short while and some are always there but only causing very few infections.

Only a few mycotoxins have been detected in rye bread (Reiss, 1972, 1977; Dich *et al.*,1979; Scott *et al.*, 1992): Aflatoxins, citrinin, ergot alkaloids and patulin. This is due to the fact that data on this subject are extremely few, in fact even the inclusion of citrinin on the list may not be relevant, as the toxin is not produced by the important spoiling species, but has been detected after artificial inoculation. As potential mycotoxins in rye bread, in accordance with the above mentioned list of important spoiling species, the following can be mentioned: roquefortin C, patulin and penicillic acid. However this list is by far exhaustive, since these species have been found to produce several secondary metabolites which are toxic in certain biological test-systems. The list of mycotoxins should also include the toxins formed in the cereals used for rye bread, as mentioned elsewhere in this paper. This "carry over" is important as it is well known that mycotoxins in contrary to the moulds often are not inactivated during the baking process (Charmley and Prelusky, 1994).

There are no data on mould spoilage other than mycotoxin contamination in bread. However personal communication with rye bread companies reveals that a mouldy smell in the bread may be observed exceptionally. This might be due to the use of mouldy cereals for the production of the bread.

The spoilage is due to species tolerating a lowered water activity (around 0.95) and the presence of organic acids like acetic acid and propionic acid which are formed during the fermentation or which have been added as preservatives (Frisvad *et al.*, 1992). The infection takes place after the baking process, which obviously kills all fungal propagules and is due to airborne conidia originating from growth of the spoiling species on product wastes in a few specific places in the plant (Lund *et al*, 1996). As far as the yeasts are concerned the infection takes place by direct contact or through water or machinery oil, often in the slicing machines.

Cheese

The most important spoilage species of hard, semi-hard and semi-soft cheese from several countries without preservatives added are: *Penicillium commune* and *P. nalgiovense* (Lund *et al*, 1995). Species of less importance are: *P. verrucosum, P. solitum, P. roqueforti, Scopulariopsis brevicaulis* and *Aspergillus versicolor.* It has been shown that important isolates from cheese which have been identified as *P. verrucosum var. cyclopium, P. aurantiogriseum, P. cyclopium* and *P. puberulum* (Northolt *et al*, 1980; Aran and Eke, 1987) can be re-identified as *P. commune* (Frisvad and Filtenborg, 1993; Lund *et al*, 1995). Another species, *P. discolor,* has also been isolated from hard cheese (Frisvad *et al.*,1997). Growth of *P. discolor* on cheese only takes place under very restricted conditions, which have yet to be determined.

The most important mycotoxin found in cheese is sterigmatocystin (Northolt *et al*, 1980). Further mycotoxins which must be considered important in cheese due to the mycotoxin potential of the species mentioned in table 1 are cyclopiazonic acid, rugulovasine A and B and ochratoxin A. Again the list is not exhaustive, as only toxins with well known toxicological properties are included. Besides it should be noted that *P. nalgiovense* is a potential producer of penicillin (Andersen and Frisvad, 1994; Färber and Geisen, 1994; 2000).

Spoilage of cheese due to fungal growth is also caused by formation of off-flavours. If sorbates are used as preservatives, resistant species are able to metabolize these compounds under formation of a plastic like or "kerosene" off flavour, which is due to the metabolites trans-1,3-pentadiene or trans-piperylene (Sensidoni *et al*, 1994).

Fermented sausages

The associated mycobiota in naturally fermented sausages is *Penicillium* species: *P. nalgiovense, P. olsonii, P. chrysogenum, P. verrucosum, P. solitum, P. oxalicum, P. commune, P. camemberti, P. expansum, P. miczynskii* and *P. simplicissimum* (Andersen, 1995a; Ciegler *et al*, 1972). Dominating species of *Aspergillus* and *Scopulariopsis* have also been reported, but they will not be mentioned here (Grazia *et al.*, 1986).

In the beginning of the fermentation process yeasts are dominating the surface mycobiota, but after a few weeks the above mentioned naturally occurring moulds take over, *P. nalgiovense* being dominating. This species in some cases is added as a starter culture. It has been claimed that several biotypes of *P. nalgiovense* exist (Fink-Gremmels and Leistner, 1990). However, it was recently shown that all tested isolates belonged to one species, *P. nalgiovense,* even though the colour of the isolates ranged from white to blue green (Andersen, 1995b).

The *Penicillium* species in the associated mycobiota are known to produce several mycotoxins and antibiotics. Some of these mycotoxins have been detected in fermented sausages after mould inoculation in pure cultures: Citreoviridin, citrinin, cyclopiazonsyre, isofumugaclavin A, ochratoxin A, patulin, roquefortin C, rugulovasin A (Fink-Gremmels and Leistner, 1990). Further mycotoxins like viomellein and xanthomegnin are produced by the associated mycobiota and all tested isolates of *P. nalgiovense* were shown to produce penicillin (Andersen and Frisvad, 1994; Färber and Geisen, 1994). Penicillin production in sausages is possible (Laich et al., 1999)

FACTORS FOR GROWTH AND MYCOTOXIN FORMATION

Besides the presence of nutrients, the most important factors for growth and mycotoxin production are temperature, water activity (a_w) and oxygen. It appears that most of the Penicillia have a lower minimal temperature range than the Aspergilli. The optimal temperature is 25-30°C for most Penicillia and 30-40°C for most Aspergilli. The maximal temperature is 28-35°C for Penicillia and 37-45°C for most Aspergilli. Various *Fusarium* species can also be regarded as psychrophilic, because of their low optimal temperature of 8-15°C.

The growth factor **water activity** (a_w) is a measure of the unbound water in the food which is available for the growth of the mould. Fig. 1 describes the various concepts.

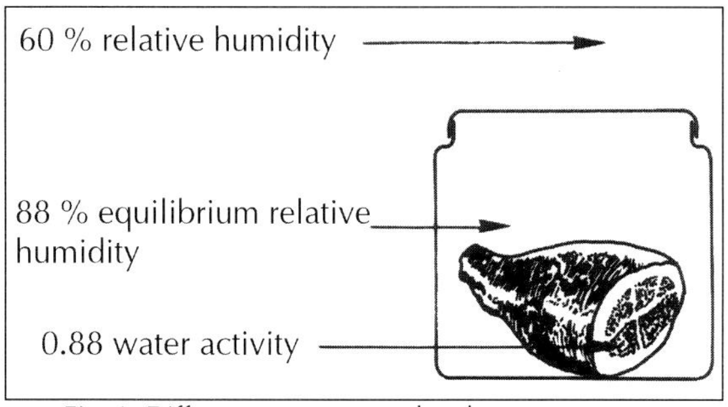

Fig. 1. Different parameters related to water vapour

The relative humidity is used for the atmosphere. The equilibrium relative humidity or the equilibrium relative water vapour pressure concerns the atmosphere in a closed space, where the water vapour pressure around the food, when this is in equilibrium with the humidity of the food, and the vapour pressure of pure water at the same temperature. In Fig. 1 the vapour pressure of the cured ham is 88% of that of pure water. Because vapour pressure ranges from 0 till 100% and water activity ranges from 0 till 1, the water activity of the cured ham is 0.88. The relation between the a_w of the substrate and the growth of micro-organisms is demonstrated in Fig. 2.

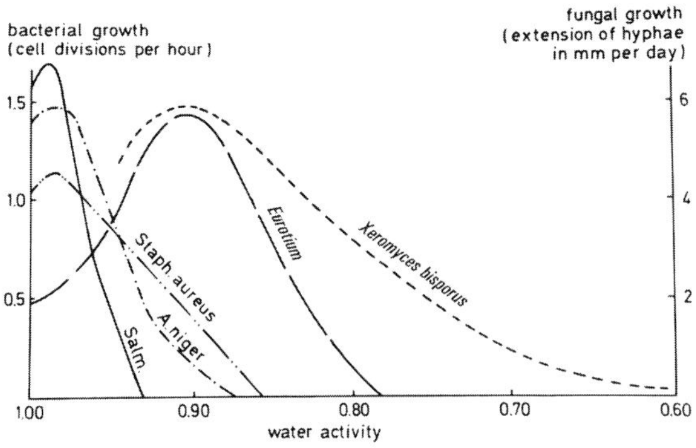

Fig. 2. Effect of water activity (a_w) on growth of different micro-organisms (Scott, 1957)

Fungi generally have a much lower minimal a_w than bacteria. This explains why many products free of bacterial spoilage can be spoiled by fungi. Fungal growth can be prevented by drying agricultural products to a level below 0.65 a_w and keeping it under this level. However, if the temperature in the stored product is not uniform, the a_w can rise locally due to migration moisture to the cold areas, thus leading to fungal growth.

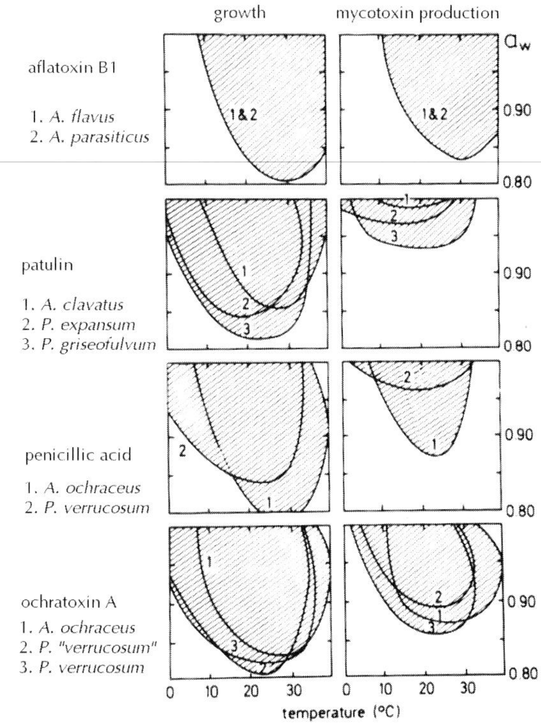

Fig. 4. Conditions of water activity (a_w) and temperature favourable (shaded area) for growth and mycotoxin production by different species (Northolt, 1979).

The minimal a_w values for growth of important species are given in Table 2. The moulds that can grow at a_w <0.75 are called storage fungi. These include: *Eurotium* spp., *A. halophilicus, A. restrictus, Wallemia sebi* and *Xeromyces bisporus*. Examples of products with a_w<0.75 are dried fruit, powdered milk, grains and certain kinds of baking products.

For further reading on various aspects of water activity see Corry (1987), Rockland and Beuchat (1987) and Beuchat and Hocking (1990).

Fig. 4 gives a summary of the temperature and a_w conditions at which various species grow and mycotoxins are formed. Aflatoxin B_1 is formed under conditions of temperature and a_w that lie close to the minimum necessary for growth. Patulin, penicillic acid and ochratoxin A are formed within a range of temperature and a_w which is much smaller than that for growth. Several species produce patulin and penicillic acid only at a high a_w with the exception of *Aspergillus ochraceus* which can produce penicillic acid at low a_w. The minimal temperature for growth and mycotoxin production by Aspergilli is higher than that for Penicillia. Growth and mycotoxin production do not always take place simultaneously. A review of data has been published about the minimum a_w and temperature for fungal growth on agricultural products and mycotoxin production (Northolt and Bullerman, 1982).

Oxygen is usually necessary for the growth of fungi, but certain species can also grow under anaerobic conditions with the formation of ethanol and organic acids. Oxygen also influences production of mycotoxins.

Table 2. Minimal water activities for germination and growth of food-borne fungi (partly adapted from Frisvad and Samson, 1991).

Species	Minimal a_w for growth	Species	Minimal a_w for growth
Alternaria alternata	0.85-0.88	*F. sporotrichioides*	0.86-0.88
Aspergillus candidus	0.75-0.78; 0.85	*F. tricinctum*	0.89
A. clavatus	0.85	*F. verticillioides*	0.87
A. flavus	0.78-0.80	*Geomyces pannorum*	0.92
A. fumigatus	0.85-0.94	*Mucor circinelloides*	0.90
A. ochraceus	0.76-0.83	*M. racemosus*	0.94
A. parasiticus	0.78-0.82	*M. spinosus*	0.93
A. penicillioides	0.73*-0.77; 0.75	*Neosartorya fischeri*	0.925
A. restrictus	0.71-0.75	*Paecilomyces variotii*	0.79-0.84; 0.91
A. sydowii	0.78; 0.81	*Penicillium aurantiogriseum*	0.79-0.85
A. tamarii	0.78	*P. brevicompactum*	0.78-0.82
A. terreus	0.78	*P. charlesii*	0.78-0.80
A. versicolor	0.78	*P. chrysogenum*	0.78-0.81
A. wentii	0.73-0.75; 0.79	*P. citrinum*	0.80-0.82
"Basipetospora" halophila	0.77-0.78; 0.75	*P. commune*	0.83
Botrytis cinerea	0.93-0.95	*P. digitatum*	0.90
Byssochlamys nivea	0.84-0.92	*P. expansum*	0.82-0.85
Chrysosporium xerophilum	0.71	*P. griseofulvum*	0.81-0.85
Ch. fastidium	0.61	*P. islandicum*	0.83-0.86
Cladosporium cladosporioides	0.86-0.88	*P. oxalicum*	0.88
C. herbarum	0.85-0.88	*P. roqueforti*	0.83
Epicoccum nigrum	0.86-0.90	*P. rugulosum*	0.85
Eurotium amstelodami	0.71-0.76; 0.75	*P. verrucosum*	0.81-0.83
E. chevalieri	0.71-0.73	*Phytophthora infestans*	0.85
E. echinulatum	0.64	*Polypaecilum pisce*	0.75-0.77*; 0.83*
E. repens	(0.69*) 0.72-0.74; 0.83	*Pythium splendens*	0.90
E. rubrum	0.70-0.71	*Rhizoctonia solani*	0.96
Exophiala werneckii	0.77-0.78; 0.75	*Rhizopus stolonifer*	0.93
Fusarium avenaceum	0.87-0.91	*Stachybotrys chartarum*	0.94
F. culmorum	0.87-0.91	*Thamnidium elegans*	0.94
F. graminearum	0.89	*Trichothecium roseum*	0.90
F. oxysporum	0.87-0.89	*Verticillium lecanii*	0.90
F. poae	0.89	*Wallemia sebi*	0.69-0.75
F. solani	0.87-0.90	*Xeromyces bisporus*	0.61

* Germination occurred, but not growth measured on NaCl.

The production of patulin and penicillic acid decreased sharply at low oxygen concentrations, while fungal growth was not noticeably influenced (Northolt, 1979). The production of aflatoxins was greatly restricted at an oxygen concentration of less than 1% (Landers *et al.*, 1967).

The **pH** and composition of the substrate have no marked influence on the growth of the species that are discussed in this chapter. However, these factors can greatly influence the formation of mycotoxins. The formation of aflatoxins, for example, is much stimulated by the presence of certain amino acids, fatty acids and the element zinc (Venkitasubramanian, 1977).

The presence of other micro-organisms can restrict growth and mycotoxin formation. The absence of moulds is explained by the rapid growth of bacteria on fresh meat. Some moulds can also hinder the development for other moulds. For example, *A. flavus* may form little aflatoxin in the presence of other moulds (Ashworth *et al.*, 1965).

Finally, the factor **time** must be mentioned. The time necessary for the germination of spores or conidia increases to tens of days under unfavourable conditions, while that under favourable conditions can be approximately one day (Northolt, 1979). In establishing safe limits of a_w and temperature for storage of foodstuffs, it is necessary to consider that the absence of fungal growth after one month does not always mean that a product can be stored safely for a much longer time.

RESTRICTION OF GROWTH AND INACTIVATION OF FILAMENTOUS FUNGI

Treatment to restrict fungal growth by spraying with fungicides, such as copper sulphate, sulphur and organic fungicides, may sometimes be necessary before the harvest. Fruit and vegetables can be washed for this purpose with fungicides or hot water after the harvest. The preservatives sorbic acid, propionic acid and benzoic acid are often added to food products such as jam, fruit compote, bread cake and other pastries to prevent fungal growth **(see also Chapter 8)**.

They are often used in the form of salts because of the limited solubility of the acid. The undissociated acid molecule is responsible for the inhibition of fungal growth. For this reason these preservatives are most active at low pH. To keep ripening cheese and salami sausages mould-free, natamycine and potassium sorbate are often used. Several species show resistance to preservatives including natamycine and the list of species is increasing. Therefore in most cases preservatives are not the primary solution to protect foods from mould contamination. Lemonade and other food products such as pre-baked bread can be conserved with carbon dioxide and anaerobia. In case of recontamination some yeast species e.g. *Endomycopsis fibuligera* and *Pichia burtonii* van develop in these modified atmosphere packages.

Most fungi are heat-sensitive and pasteurization at 70-80°C is enough to inactivate them. Many foodstuffs like fruit products and baking products are virtually free of moulds if they are heated in the final container. If they are packed after the heat treatment it is necessary to prevent recontamination by strictly hygienic measures. This can either be achieved by decontamination of instruments, piping, vats and work areas with steam, formaldehyde or chlorine gas, or by making the air spore-free with special filters. The ascospores of several Ascomycetes such as *Byssochlamys* or the thick-walled structures (chlamydospores and hyphae) in some species such as *Paecilomyces variotii,* form an exception in their high heat resistance. These species, which more frequently occur in fruit, fruit products or raw materials derived from fruit e.g. pectin, need an inactivation temperature of 90-100°C or higher. Sometimes lower temperature in combination with SO_2, might be effective. In contrast, many mycotoxins are not or only partly broken down by pasteurization or sterilisation.

CONCLUSION

From Table 1 it can be seen that a very limited number of fungal species has been associated with the spoilage of each food category. Data in the literature mentioning much higher numbers of spoiling species is due to misidentifications or incorporation of species which are merely present in the food and not actively spoiling it. Hill and Lacey (1984) thus summarize 59 species of *Penicillium* to be associated with barley and El-Kady *et al.* (1992) found 18 species of *Penicillium* in "the mycobiota" of cereal grains in Egypt. However the major part of these species are well known soil-borne fungi (Domsch *et al.*, 1993) and of no consequence to the spoilage of cereals. They can be detected in soil associated foods in general using mycological methods which detect all kinds of propagules present and not only growing mycelium.

It can also be seen in Table 1 that each species rarely occurs in the associated mycobiota of more than one food category. This must be due to the selection principle in each food being a complicated combination of intrinsic, extrinsic and processing factors of the food. The exact nature of the selection is only known in a few cases yet, and then only if one factor has an extreme value. For example when preservatives like sorbic, benzoic and propionic acid are added to the food, then *P. roqueforti* is the most important spoiling species irrespective of the food, since this species is resistant to these preservatives. This also means that if the foods are subjected to completely different production or storage conditions than we know today, then the funga may change even very dramatically, but still be very limited.

The existence of an associated mycobiota in foods has great impact on the mycological quality assessment in the food industry. The limited number of species which are shown to be important to the quality of the foods mentioned in this paper, to a great extent simplifies the preventive and the control actions which must be taken. Knowing the properties of the spoiling species makes it possible to optimize the preservative profile of the food and the hygienic measures during production of the food. Simple and very specific analytical methods can be developed, which can be used to detect and control the critical contaminations points of the spoiling species during food production.

In conclusion it can be stated that prevention of mould spoilage of foods can only be carried out successfully, if the species, which are actually spoiling the food, the associated mycobiota, are known. Some of these data are known, as shown in Table 1, but it is urgent to carry out further investigations of the associated mycobota of foods. However, to avoid the confusion of the past, as far as identification and methods are concerned, the investigations should be carried out under the supervision of international commissions as mentioned in the introduction. Ideally screening programs and generation of databases should be initiated and sponsored on an international basis. The availability of these data provide the optimum basis of research to study the selection principles of different foods which again lead to the development of new methods to assess and control the fungal spoilage of foods including mycotoxin contamination.

REFERENCES

AMADE, P., MALLEA, M. & BOUAICHA, N. 1994. Isolation, structural identification and biological activity of two metabolites produced by *Penicillium olsonii* Bainier and Sartory. J. Antibiot. 47: 201-207.

ANDERSEN, B., THRANE, U., SVENDSEN, A. & RASMUSSEN, I.A. 1996. Associated field mycobiota on malt barley. Can. J. Bot. 74, 854-858.

ANDERSEN, S.J. 1995a. Compositional changes in surface mycoflora during ripening of naturally fermented sausages. J. Food Prot. 58, 426-429.

ANDERSEN, S.J. 1995b. Taxonomy of *Penicillium nalgiovense* isolates from mould-fermented sausages. Ant. van Leeuwenhoek. 68, 165-171.

ANDERSEN, S.J. & FRISVAD, J.C. 1994. Penicillin production by *Penicillium nalgiovense.* Lett. Appl. Microbiol. 19, 486-488.

ANDREWS, S. 1986. Dilution plating versus direct plating of various cereal samples. In: A.D. King, J.I. Pitt, L.R. Beuchat &

J.E.L. Corry (editors), *Methods for the mycological examination of foods*. Plenum Press, New York, NY, pp. 40-45.

ARAI, K., MIYAJIMA, H., MUSHIRODA, T. & YAMAMOTO, Y. 1989. Metabolites of *Penicillium italicum* Wehmer: Isolation and structures of new metabolites including naturally occurring 4-ylidene-acyltetronic acids, italicinic acid and italicic acid. Chem. Pharm. Bull. 37, 3229-3235.

ARAN, N. & EKE, D. 1987. Mould mycoflora of Kasar cheese at the stage of consumption. Food Microbiol. 4, 101-104.

ASHWORTH, L.J., H.W. SCHROEDER & LANGLEY, B.C. 1965. Aflatoxin environmental factors governing occurrence in Spanish peanuts. Science 148: 1228-1229.

BEN-YEHOSHUA, A, BARAK, E. & SHAPIRO, B. 1987. Postharvest curing at high temperatures reduces decay of individually sealed lemosn, pomelos and other citrus fruit. J. Amer. Soc. Hort. Sci. 112, 658-663.

BEUCHAT, L.R. & RICE, S.L. 1979. *Byssochlamys* spp. and their importance in processed fruits. Adv. Food Res. 25, 237-288.

BEUCHAT, L.R. & HOCKING, A.D. 1990. Some considerations when analyzing foods for the presence of xerophilic fungi. J. Food Protection 53: 984-989.

BÖRNER, H. 1963. Untersuchungen über die Bildung antiphytotischer und antimikrobieller Substanzen durch Mikroorganismen im Boden und ihre mögliche Bedeutung für die Bodenmüdigkeit beim Apfels *Pirus malus* L.. Phytopath. Z. 49: 1-28.

BIGELIS, R. 1992. Food enzymes. In: D.B. Finkelstein & C. Ball (editors), *Biotechnology of filamentous fungi*. Technology and products. Butterworth-Heinemann, Boston, MS, pp. 361-415.

BIRKINSHAW, J.H., VICTOR, J.H. & RAISTRICK, H. 1931. Studies in the biochemistry of microorganisms. Part XVIII. Biochemical characteristics of species of *Penicillium* responsible for the rot of citrus fruits. Trans. Roy. Soc. (London) B220, 355-367.

BOOTH, C. 1971. *The genus* Fusarium. Commonwealth Mycological Institute, Kew, Surrey, U.K.

BOYSEN, M., SKOUBOE, P., FRISVAD, J.C. & ROSSEN, L. 1996 Reclassification of the *Penicillium roqueforti* group into three species on the basis of molecular genetic and biochemical profiles. Microbiology (UK) 142, 541-549.

CHAMP, B.R., HIGHLEY, C., HOCKING, A.D. & PITT, J.I. 1991. *Fungi and mycotoxins in stored products.* Australian Centre for International Agricultural Research, Canberra.

CHARMLEY, L.L. & PRELUSKY, D.B. 1994. Decontamination of *Fusarium* mycotoxins. In: J.D. Miller & H.L. Trenholm (editors). *Mycotoxins in grain. Compounds other than aflatoxin.* Eagan Press, St. Paul, pp. 421-435.

CHELKOWSKI, J. & VISCONTI, A. (editors) 1992 *Alternaria* Biology, plant diseases and metabolites. Elsevier, Amsterdam.

CHELKOWSKI, J. (editor) 1991. *Cereal grain. Mycotoxins, fungi and quality in drying and storage.* Elsevier, Amsterdam.

CHRISTENSEN, C.M. 1975. *Molds, mushrooms and mycotoxins.* University of Minnesota press, Minneapolis, pp. 34-58.

CIEGLER, A. 1972. Bioproduction of ochratoxin A and penicillic acid by members of the *Aspergillus ochraceus* group. Can. J. Microbiol. 18, 631-636.

CIEGLER, A., MINTZLAFF, H.-J., MACHNIK, W. & LEISTNER, L. 1972. Untersuchungen über das Toxinbildungsvermögen von Rohwürsten isolieter Schimmelpilze der Gattung *Penicillium.* Fleischwirtsch. 52, 1311-1318.

CIEGLER, A. VESONDER, R.F. & JACKSON, L.K. 1977. Production and biological activity of patulin and citrinin from *Penicillium expansum.* Appl. Environ. Microbiol. 33, 1004-1006.

COLE, R.J. & COX, R.H. 1981. Handbook of toxic fungal metabolites. Academic Press, New York..

CORRY, J.E.L. 1987. Relationships of water activity to fungal growth. In *Food and Beverage mycology* (L.R. Beuchat, ed.). pp. 51-100, .AVI, New York.

DESJARDINS, A.E., & PLATTNER, R.D. 1989. Trichothecene toxin production by strains of *Gibberella pulicaris* (*Fusarium sambucinum*) in liquid culture and in potato tubers. J. Agric. Food Chem. 37, 388-392.

DICH, D., ÅKERSTRAND, K., ANDERSSON, A., JOSEFSSON, E. & JANSSON, E. 1979. Konserveringsmedels förekomst och inverkan på mögel- og mycotoxinbildning i matbröd. Vår Föda 31, 385-403.

DOMSCH, K.H., GAMS, W. & ANDERSON, T.-H. 1993. Compendium of soil fungi. Academic Press, London.

EL-BANNA, A.A., SCOTT, P.M., LAU, P.-Y., SAKUMA, T., PLATT, H.W., & CAMPBELL, V. 1984. Formation of trichothecenes by *Fusarium solani* var. *coeruleum* and *Fusarium sambucinum* in potatoes. Appl. Environ. Microbiol. 47, 1169-1171.

ECKERT, J.W. & RATNAYAKE, M. 1994. Role of volatile compounds from wounded oranges in induction of germination of *Penicillium digitatum* conidia. Phytopathology 84, 746-750.

EL-KADY, I.A., ABDEL-HAFEZ, S.I.I. & EL-MARAGHY, S.S. 1992. Contribution to the fungal flora of cereal grains of Egypt. Mycopathologia 77, 103-109.

EL-SHAYEB, N.M.A., MABROUK, S.S. & Abd-El-Fattah, A.M.M. 1992. Production of ochratoxins by some egyptian *Aspergillus* strains. Zentralb. Mikrobiol. 147, 86-91.

FAID, M. & TANTAOUI-ELARAKI, A. 1989. Production of toxic metabolites by *Penicillium italicum* and *P. digitatum* isolated from citrus fruits. J. Food Prot. 52, 194-197.

FÄRBER, P. & GEISEN, R. 1994. Antagonistic activity of the food relevant filamentous fungus *Penicillium nalgiovense* is due to the production of penicillin. Appl. Environ. Microbiol. 60: 3401-3404.

FÄRBER, P. & GEISEN, R. 2000. Karyotype of *Penicillium nalgiovense* and assignment of the penicillin biosynthetic genes to chromosome IV. Int. J. Food Microbiol. 58: 59-63.

FILTENBORG, O., FRISVAD, J.C. & THRANE, U. 1996. Moulds in food spoilage. Int. J. Food Microbiol. 33: 85-102.

FINK-GREMMELS, J. & LEISTNER, L. 1990. Toxicological evaluation of moulds. Food Biotechnol. 4, 579-584.

FLAISHMAN, M.A. & KOLATTUKUDY, P.E. 1994. Timing of fungal invasion using host's ripening hormone as a signal. Proc. Natl. Acad. Sci. USA 91, 6579-6583.

FRISVAD, J.C. 1981. Physiological criteria and mycotoxin production as aids in identification of common asymmetric penicillia. Appl. Environ. Microbiol. 41, 568-579.

FRISVAD, J.C. 1989. The connection between penicillia and aspergilli and mycotoxins with special emphasis on misidentified isolates. Arch. Environ. Contam. Toxicol. 18, 452-467.

FRISVAD, J.C. 1995. Mycotoxins and mycotoxigenic fungi in storage. In: D.S. Jayas, N.D.G. White & W.E. Muir (editors), *Stored grain ecosystems.* Marcel Dekker, New York, pp. 251-288.

FRISVAD, J.C. & FILTENBORG, O., 1983. Classification of terverticillate penicillia based on profiles of mycotoxins and other secondary metabolites. Appl. Environ. Microbiol. 46, 1301-1310.

FRISVAD, J.C., & FILTENBORG, O., 1988. Specific mycotoxin producing *Penicillium* and *Aspergillus* mycoflora of different foods, *Proc. Jpn. Assoc. Mycotoxicol.*, Suppl. **1**, 163-166.

FRISVAD, J.C. & FILTENBORG, O. 1989. Terverticillate penicillia: chemotaxonomy and mycotoxin production. Mycologia 81, 836-861.

FRISVAD, J.C. & FILTENBORG, O. 1990. Secondary metabolites as consistent criteria in Penicillium taxonomy and a synoptic key to *Penicillium* subgenus *Penicillium.* In: R.A. Samson and J.I. Pitt (editors), Modern concepts in *Penicillium* and *Aspergillus* classification. Plenum Press, New York, pp. 373-384.

FRISVAD, J.C. & FILTENBORG, O. 1993. Mycotoxin production by *Penicillium* and *Aspergillus* species associated with different foods and other substrates. In: K. Scudamore (ed.) *Occurrence and significance of mycotoxins.* Central Science Laboratory, Slough. pp. 138-145

FRISVAD, J.C. & LUND, F. 1993. Toxin and secondary metabolite production by Penicilium species growing in stored cereals. In: K.A. Scudamore (ed.), *Occurrence and significance of mycotoxins.* Central Science Laboratory, Slough, pp. 146-171.

FRISVAD, J.C. & THRANE, U. 1995. Mycotoxin production by food-borne fungi. In: R.A. Samson, E.S. Hoekstra, J.C. Frisvad & O. Filtenborg (editors), *Introduction to food-borne fungi.* 4th ed. Centraalbureau voor Schimmelcultures, Baarn, pp. 251-260.

FRISVAD, J.C., FILTENBORG, O., LUND, F. & THRANE, U. 1992. New selective media for the detection of toxigenic fungi in cereal products, meat and cheese. In: R.A. Samson, A.D. Hocking, J.I. Pitt & A.D. King (editors.), *Modern methods in food mycology.* Elsevier, Amsterdam, pp. 275-284.

FRISVAD, J.C. & SAMSON, R.A. 1991. Filamentous fungi in foods and feeds: ecology, spoilage and mycotoxin production. In *Handbook of Applied Mycology. Vol. 3. Foods and Feeds.* eds. Arora, D.K., Mukerji, K.G. & Marth, E.H. New York: Marcel Dekker, pp. 31-68.

FRISVAD, J.C., SAMSON, R.A., RASSING, B.R. & VAN DER HORST, M.I. 1996. *Penicillium discolor*, a new species from cheese, nuts and vegetables. Antonie van Leeuwenhoek 72: 119-126.

FUNNELL, H.S. 1979. Mycotoxins in animal foodstuffs in Ontario 1972 to 1977. Can. J. Comp. Med. 43: 243-246.

GAREIS, M., BAUER, J., ENDERS, C., & GEDEK, B. 1989. Contamination of cereals and feed with *Fusarium* mycotoxins in European countries. In J. Chelkowski (Ed.), Fusarium: *Mycotoxins, taxonomy, and pathogenicity.* Elsevier Science Publishers B.V. Amsterdam, pp. 441-472.

GERLACH, W., & NIRENBERG, H. 1982. The genus *Fusarium* - a pictorial atlas. Mitt. Biol. Bundesanst. Land- u. Forstw. Berlin-Dahlem Heft 209, 1-406.

GORST-ALLMAN, C.P., MAES, C.M.T.P., STEYN, P.S. & RABIE, C.J. 1982. 5,6-dihydro-4-methoxy-2H-pyran-2-one, a new mycotoxin from *Penicillium italicum.* S.- Afr. Tydskr. Chem. 35, 102-103.

GRAVESEN, S., FRISVAD,. J.C. & SAMSON, R.A. 1994. *Microfungi.* Munksgaard, Copenhagen.

GRAZIA, L., ROMANO, P., BAGNI, A., ROGGIANI, D. & GUGLIELMI, G. 1986. The role of moulds in the ripening process of salami. Food Microbiol. 3, 19-25.

HALD, B., CHRISTENSEN, D.H. & KROGH, P: 1983. Natural occurrence of the mycotoxin viomellein in barley and the associated quinone-producing penicillia. Appl. Environ. Microbiol. 46, 1311-1317.

HARRIS, J.E. & DENNIS, C. 1980. Heat stability of endopolygalacturonases of mucoraceous spoilage fungi in relation to canned fruits. J. Sci. Food Agric. 31, 1164-1172.

HARWIG, J., CHEN, Y.-K., KENNEDY, B.P.C. & SCOTT, P.M. 1973. Occurrence of patulin-producing strains of *Penicillium expansum* in natural rots of apple in Canada. Can. Inst. Food Sci. Technol. J. 6, 22-25.

HESSELTINE, C.W., VANDEGRAFT, E.E., FENNELL, D.I., SMITH, M.L. & SHOTWELL, O.L., 1972. Aspergilli as ochratoxin producers. Mycologia 64, 539-550.

HILL, R.A. & LACEY, J. 1984. *Penicillium* species associated with barley grain in the U.K. Trans. Br. Mycol. Soc. 82, 297-303.

HOLMES, G.J., ECKERT, J.W. & PITT, J.I. 1994. Revised description of *Penicillium ulaiense* and its role as a pathogen of citrus fruit. Phytopathology 84, 719-727.

ILLY, A. & VIANI, R. (eds.). 1995. *Espresso coffee: The chemistry of quality.* Academic Press, London.

ISAAC, B., AYER, S.W. & STONARD, R.J. 1991. Arabenoic acid, a natural product herbicide of fungal origin. J. Antibiot. 44: 793-794.

JANZEN, D.H. 1977. Why fruits rot, seeds mold and meat spoils.Amer. Nat. 111: 691-713.

KINDERLERER, J. & HATTON, P.V. 1990. Fungal metabolites of sorbic acid. Food Addit. Contam. 7, 657-669.

KRIEK, N.P.J. & WERNER, F.C. 1981 Toxicity of *Penicillium italicum* to laboratory animals. Food Cosmet. Toxicol. 19, 311-315.

KROGH, P. 1977. In *Mycotoxins in human and animal health.* (J.V. Rodricks, C.W. Hesseltine & M.A. Mehlman, eds.). Pathotox., Park Forest South, Illinois, pp. 489-498.

LACEY, J. 1991. Grain fungi. In: D.K. Arora, , K.G. Mukerji & E.H. Marth (editors), *Handbook of applied mycology, vol. 3. Foods and feeds.* Marcel Dekker, New York, pp. 121-177.

LAICH, F., FIERRO, F. CORDOZA, R.E. & MARTIN, J.F. 1999. Organization of the gene cluster for biosynthesis of penicillin in *Penicillium nalgiovense* and azntibiotic production in cured dry sausages. Appl. Environ. Microbiol. 65: 1236-1240.

LANDERS, K.E., N.D. DAVIS & DIENER, U.L. 1967. Influence of atmospheric gases on aflatoxins production by *Aspergillus flavus* in peanuts. Phytopath. 57: 1086-1090.

LARSEN, T.O. & FRISVAD, J.C. 1995a. Characterization of volatile metabolites from 47 *Penicillium* taxa. Mycol. Res. 99, 1153-1166.

LARSEN, T.O. & FRISVAD, J.C. 1995b. Chemosystematics of *Penicillium* based on profiles of volatile metabolites. Mycol. Res. 99, 1167-1174.

LARSEN, T.O., FRISVAD, J.C., RAVN, G. AND SKAANING, T. 1998a. Mycotoxin production by *Penicillium expansum* on blackcurrant and cherry juice. Fod Addit. Contam. 15: 671-675.

LARSEN, T.O., FRISVAD, J.C. & CHRISTOPHERSEN, C. 1998. Arabenoic acid (verrucolone), a major chemical indicator of *Penicillium verrucosum.* Bochem. Syst. Ecol. 26: 463-465.

LIEWEN, M.B. & MARTH, E.H. 1985. Growth and inhibition of micro-organisms in the presence of sorbic acid: a review. J. Food Prot. 48, 364-375.

LOGRIECO, A., VISCONTI, A. & BOTTALICO, A. 1990. Mandarin fruit rot caused by *Alternaria alternata* and associated mycotoxins. Plant Dis. 74, 415-417.

LUND, F., FILTENBORG, O., WESTALL, S. & FRISVAD, J.C. 1996. Associated mycoflora of rye bread. Lett. Appl. Microbiol. 23: 213-217.

LUND, F., FILTENBORG, O. & FRISVAD, J.C. 1995. Associated mycoflora of cheese. Food Microbiol. 12, 173-180.

LUND, F. & FRISVAD, J.C. 1994. Chemotaxonomy of *Penicillium aurantiogriseum* and related species. Mycol. Res., F. 98, 481-492.

MARASAS, W.F.O., NELSON, P.E. & TOUSSOUN, T.A. 1984. Toxigenic *Fusarium* species. Identity and mycotoxicology. The Pennsylvania State University Press, University Park, PN.

MARTH, E.H., C.M. CAPP, L. HASENZAHL, H.W. JACKSON & HUSSONG, R.V. 1966. Degaradation of potassium sorbate by *Penicillium* species. J. Dairy Sci. 49: 1197-1205.

MCKINLEY, E.R. & CARLTON, W.W. 1991. PATULIN. IN: R.P. SHARMA & D.K. SALUNKHE (eds.), *Mycotoxins and phytoalexins.* CRC Press, Boca Raton, FL, pp. 191-236.

MILLER, J.D. 1991. Significance of mycotoxins for health and nutrition. In: B.R. Champ, E. Highley, A.D. Hocking & J.I. Pitt (editors), *Fungi and mycotoxins in stored products.* ACIAR Proceedings 36. Australian Centre for International Agricultural Research, Canberra, pp. 126-135.

MILLS, J.T., SEIFERT, K.A., FRISVAD, J.C. & ABRAMSON, D. 1995. Nephrotoxigenic *Penicillium* species occurring on farm-stored cereal grains in western Canada. Mycopathologia 130, 23-28.

MÖLLER, T., BROSTEDT, S. & JOHANSSON, M. 1993. Ergot alkaloids in Swedish cereals. Vår Föda 45, 308-311.

MORRIS, P.F., CONNOLLY, M.S. & ST CLAIR, D.A. 2000. Genetic diversity of *Alternaria alternata* isolated from tomato in Californai assessed using RAPDs. Myc. Res. 104: 286-292.

NIRENBERG, H. 1995. Morphological differentiation of *Fusarium sambucinum* Fuckel sensu lato, *F. torulosum* (Berk. & Curt.) Nirenberg comb. nov. and *F. venenatum* sp. nov. Mycopathologia 129, 131-141.

NORTHOLT; M.D. 1979. *The effect of water activity and temperature on the production of some mycotoxins.* Ph Dissertation University of Wageningen.

NORTHOLT, M.D. & BULLERMAN, L.B. 1982. Prevention of mold growth and toxin production through control of environmental conditions. J. Food Protection 45: 519-526.

NORTHOLT, M.D., H.P. VAN EGMOND, P. SOENTORO & DEIJLL, E. 1980a. Fungal growth and the presence of sterigmatocystin in hard cheese. J. Assoc. Off. Anal. Chem. 63: 115-119.

NORTHOLT, M.D., C. EIKELENBOOM, B.J. HARTOG, A.J. NOOITGEDACHT & PATEER, P.M. 1980b. Onderzoek naar de biologische gesteldheid en houdbaarheid van droog gebak en vruchtengebak. English Summary. De Ware(n) Chemicus, 10: 116-124.

NOUT, R. 1995. Useful role of fungi in food processing. In: R.A. Samson, E.S. Hoekstra, J.C. Frisvad & O. Filtenborg (editors), *Introduction to food-borne fungi.* 4th ed. Centraalbureau voor Schimmelcultures, Baarn, pp. 295-303.

PELSER, P. DU T. & ECKERT, J.W. 1977. Constituents of orange juice that stimulate the germination of conidia of *Penicillium digitatum.* Phytopathology 67, 747-754.

PESTKA, J.J. & BONDY, G.S. 1994. Immunotoxic effects of mycotoxins. In: J.D. Miller & H.L. Trenholm (editors), Mycotoxins in grain. Compounds other than aflatoxin. Eagan Press, St. Paul, pp. 339-358.

PIER, A.C. & RICHARD, J.L. 1992. Mycoses and mycotoxicoses of animals caused by aspergilli. In: J.W. Bennett, J.W. & Klich, M.A. (editors), Aspergillus, *Biology and industrial applications*, Butterworth-Heinemann, Boston, MS, pp. 233-248.

PITT, J.I., SPOTTS, R.A., HOLMES, R.J. & CRUICKSHANK, R.H. 1991. *Penicillum solitum* revived, and its role as a pathogen of pomaceous fruit. Phytopathology 81, 1108-1112.

PRELUSKY, D.B., ROTTER, B.A. & ROTTER, R.G. 1994. Toxicology of mycotoxins. In: J.D. Miller & H.L. Trenholm (editors), Mycotoxins in grain. Compounds other than aflatoxin. Eagan Press, St. Paul, pp. 359-403.

RAPER, K.B. & FENNELL, D.I. 1965.The genus *Aspergillus*. Williams and Wilkins, Baltimore.

RAPER, K.B. & THOM, C. 1949 *Manual of the Penicillia*. Williams and Wilkins, Baltimore.

REISS, J. 1977. Mycotoxins in foodstuffs. X. Production of citrinin by *Penicillium chrysogenum* in bread. Food Cosmet. Toxicol. 15, 303-307.

REISS, J. 1972. Mykotoxine in Nahrungsmitteln. II. Nachweis von Patulin in spontan verschimmeltem Brot und Gebäck. Naturwissenschaften 59, 37.

RICE, S.L., BEUCHAT, L.R. & HEATON, E.KJ. 1979. Changes in the composition and texture of canned peach halves infected with *Byssochlamys fulva*. Journal of Food Science 42 1562-1565.

RIPPON, L.E. 1980. Wastage of postharvest fruit and its control. CSIRO Fd. Res. Q. 40(1), 1-12.

ROBERTS, R.G., REYMOND,. S.T. & ANDERSEN, B. 2000. RAPD fragment pattern analysis and morphological segregation of small-spored *Alternaria* species and species groups. Myc. Res. 104: 151-160.

ROBIGLIO, A.L. & LOPEZ, S.E. 1995. Mycotoxin production by *Alternaria alternata* strains isolated from red delicious apples in Argentina. Intl. J. Food Microbiol. 24, 413-417.

ROCKLAND, L.B. & BEUCHAT, L.R. (eds.) 1987. Water actvity. Theory and applications to food. Marcel Dekker Inc. New York. pp. 424.

SAMSON, R.A. & Pitt, J.I. (eds.) 1990. Modern concepts in *Penicillium* and *Aspergillus* systematics. Plenum Press, New York.

SAMSON, R.A., , STOLK, A.C. & HADLOK, R. 1976. Revision of the subsection Fasciculata of *Penicillium* and some allied species. Stud. Mycol. (Baarn) 11, 1-47.

SAUER, D.B., MERONOUK, R.A. & CHRISTENSEN, C.M. 1992. Microflora. In: D.B. Sauer (editor), Storage of cereal grains and their products. American Association of Cereal Chemists, St. Paul, MN, pp. 313-340.

SCHROEDER, H.W. & BOLLER, R.A. 1971. The aflatoxin problem in the Southwest in 1969-70. *In* Proc. of Joint Meeting U.S.-Jap. Toxic Microorganims Panel, 6th, Tokyo, Japan.

SCOTT, W.J. 1957. Water relations of food spoilage microorganisms. Adv. Food Res. 7: 83-127.

SCOTT, P.M., LOMBAERT, G.A., PELLAERS, P., BACLER, S. & LAPPI, J. 1992. Ergot alkaloids in grain foods sold in Canada. J. Assoc. Off. Anal. Chem. International 75, 773-779.

SCOTT, P.M. 1991. Possibilities of reduction or elimination of mycotoxins in cereal grains. In: J. Chelkowski (editor), Cereal grain. Mycotoxins, fungi and quality in drying and storage. Elsevier, Amsterdam, pp. 529-572.

SCUDAMORE, K.A., ATKIN, P.M. & BUCKLE, A.E. 1986. Natural occurrence of the naphthoquinone mycotoxins xanrhomegnin and viomellein, and vioxanthin in cereals and animal fedstuffs. J. Stored Prod. Res. 22, 81-84.

SCUDAMORE, K.A., CLARKE, J.H., & HETMANSKI, M.T. 1993. Isolation of *Penicillium* strains producing ochratoxin A, citrinin, xanthomegnin, viomellein, and vioxanthin from cereal grains. Lett. Appl. Microbiol. 17, 82-87.

SENSIDONI, A., RONDININI,G., PERESSINI, D., MAIFRENI, M & BORTOLOMEAZZI, R. 1994. Presence of an off-flavour associated with the use of sorbates in cheese and margarine. Italian Journal of Food Science 6(2), 237-242.

SERDANI, M, CROUS, P.W., HOLZ, G & PETRINI, O. 1998. Endophytic fungi associated with core rot of apples in South Africa, with specific reference to *Alternaria* species. Sydowia 50: 257-271.

SHANK, R.C. 1976. The role of aflatoxin in human disease. *In* Mycotoxins and other fungal related food problems, (J.V. Rodricks, ed.) Advances in chemistry series 149. Am. Chem. Soc., Washington, D.C. pp. 51-57.

SENSIDONI, A., RONDININI,G., PERESSINI, D., MAIFRENI, M & BORTOLOMEAZZI, R. 1994. Presence of an off-flavour associated with the use of sorbates in cheese and margarine. Italian Journal of Food Science 6(2), 237-242.

SHOLBERG, P.L. & OGAWA, J.M. 1983. Relation of postharvest decay fungi to the slip-skin maceration disorer of dried Frensh prunes, Phytopathology 73(5), 708-713.

SIEGFRIED, R., & LANGERFELD, E. 1978. Vorläufige Untersuchungen über die Produktion von Toxinen durch Fäuleerreger bei Kartoffeln. Potato Res. 21, 335-339.

SIMMONS, E.G. 1992. *Alternaria* taxonomy: current status, viewpoints, challenge. In J. Chelkowski & A. Visconti (editors), Alternaria. *Biology, plant diseases and metabolites*. Elsevier, Amsterdam, pp. 1-36.

SIMMONS, E.G. 1986. *Alternaria* themes and variations (22-26). Mycotaxon 25: 287-308.

SIMMONS, E.G. 1993. *Alternaria* themes and variations (63-72). Mycotaxon 48: 91-107.SIMMONS, E.G. 1994. *Alternaria* themes and variations (106-111). Mycotaxon 50: 409-427.

SIMMONS, E.G. 1995. *Alternaria* themes and variations (112-144). Mycotaxon 55: 55-163.

SIMMONS, E.G. 1999a. *Alternaria* themes and variations (226-235). Classification of citrus pathogens. Mycotaxon 70: 263-323.

SIMMONS, E.G. 1999b. *Alternaria* themes and variations (236-243). Host-specific toxin producers. Mycotaxon 70: 325-369.

SIMMONS, E.G. & ROBERTS, R.G. 1993 *Alternaria* themes and variations (73). Mycotaxon 48: 109-140

SPICHER, G. 1985. Die erreger der Schimmelbildung bei Backwaren. IV. Weitere Untersuchungen über die auf verpackte Schnittbroten auftretenden Schimmelpilze. Deutsche Lebensm. Rundschau 81, 16-20.

STINSON, E.E., BILLS, D.D., OSMAN, S.F., SICILIANO, J., CEPONIS, N.J. & HEISLER, E.G. 1980. Mycotoxin production by *Alternaria* species grown on apples, tomatoes, and blueberries. J. Agric. Food. Chem. 28, 960-963.

STINSON, E., OSMAN, E., HEISLER, D.F., SICILIANO, D.G. & BILLS, D.D. 1981. Mycotoxin production in whole tomatoes, apples, oranges, and lemons. J. Agric. Food Chem. 29, 790-792.

STOLOFF, L. 1976. The role of aflatoxin in human disease. *In* Mycotoxins and other fungal related food problems, (J.V. Rodricks, ed.) Advances in chemistry series 149. Am. Chem. Soc., Washington, D.C. pp. 23-50.

TANTAOUI-ELARAKI, A., LEMRANI, M. & GONZALEZ-VILA, F.J. 1994. Tentative identification of metabolites in toxic extracts from *Penicillium digitatum* (Pers ex. Fr.) Sacch. and *P. italicum* Wehmer cultures. Microbiol. Alim. Nutr. 12, 225-230.

TINDALE, C.R., WHITFIELD, F.B., LEVINGSTON, S.D. & NGUYEN, T.H.L. 1989. Fungi isolated from packaging materials: their role in the production of 2,4,6-trichloroanisole. J. Sci. Food Agric. 49, 437-447.

TIPPLES, K.H. 1995. Quality and nutritional changes in stored grain. In: D.S. Jayas, N.D.G. White & W.E. Muir (editors) Stored grain ecosystems. Marcel Dekker, New York, pp. 325-351.

THRANE, U. 1989. *Fusarium* species and their specific profiles of secondary metabolites. In: J. Chelkowski (editor), *Fusarium*: Mycotoxins, taxonomy and pathogenicity. Elsevier, Amsterdam, pp. 199-225.

VENKITASUBRAMANIAN, T.A. 1977. Biosynthesis of aflatoxin and its control. *In* Mycotoxins in human and animal health, (J.V. Rodricks, C.W. Hesseltine & M.A. Mehlman, eds.). Pathotox, Park Forest South, Illinois, pp. 83-98.

VINAS, I., DADON, J. & SANCHIS, V. 1993 Citrinin-producing capacity of *Penicillium expansum* strains from apple packinghouses of Leirda (Spain). Intl. J. Food Microbiol. 19, 153-156.

VINCENT, M.A. & PITT, J.I.1989. *Penicillium allii*, a new species from egyptian garlic. Mycologia, 81, 300-303.

WEIR, T.L., HUFF, D.R., CHRIST, B.J. & ROMAINE, C.P. 1998. RAPD-PCR analysis of genetic variation among isolates of *Al-*

ternaria solani and *Alternaria alternata* from tomato and potato. Mycologia 90: 813-821.

WESTERDIJK, J. 1949. The concept "association" in mycology. Antonie van Leeuwenhoek 15, 187-189.

WHITFIELD, F.B., NGUYER, T.H.L. & LAST, J.H. 1991. Effect of relative humidity and chlorophenol content on the fungal conversion of chlorophenols to chloroanisols in fibreboard cartons containing dried fruit. J. Sci. Food Agric. 54, 595-604.

WIESE, M.V. 1987. Compendium of wheat diseases. Second edition. APS Press, St. Paul, MN.

YAMAMOTO, W., YOSHITANI, K. & MAEDA, M. 1955. Studies on the *Penicillium* and *Fusarium* rots of Chinese yam and their control. *Scientific Rep. Hyogo Univ. Agric., Agric. Biol. Ser. 2*, 1, 69-79.

YOUNG, J.C. 1979. Ergot contamination of feedstuffs. Feedstuffs 51: 23-33.

Chapter 5

MYCOTOXIN PRODUCTION BY COMMON FILAMENTOUS FUNGI

J. C. FRISVAD and U. THRANE

BioCentrum-DTU, Technical University of Denmark, DK-2800 Lyngby, Denmark

TABLE OF CONTENTS

INTRODUCTION

Mycotoxins are secondary metabolites produced by filamentous fungi that in small concentrations can evoke an acute or cronic disease in vertebrate animals when introduced via a natural route (Gravesen *et al.*, 1994). If other animals were included in the definition mycotoxins would probably include all secondary metabolites and so the word mycotoxins would loose its antropocentric meaning. One exception to this definition is the antibiotic penicillin, which can kill rabbits by destroying its normal gut biota, but is not really regarded as a mycotoxin. Filamentous fungi are in general able to produce a large number of different secondary metabolites. Secondary metabolites are products of normal metabolism that are restricted in their taxonomic distribution (Campbell, 1984). As is the case of morphological features they are differentiation events. Some of these secondary metabolites have gained so much vital importance for the producing organism that they now are plesiomorphic for large groups of fungi (Frisvad, 1994a). These include ergosterol accumulated in the membranes of nearly all fungi and melanins that protect against irradiation and other physical damage. However most of the known secondary metabolites are much more restricted in their distribution, being produced only by 1 to 15 species (Frisvad, 1989; Frisvad et al., 1998). The secondary metabolites are probably all of ecological significance as chemical signals in the communication between species (Wicklow, 1987; Williams *et al.*, 1989; Christophersen, 1991). An important part of these secondary metabolites can be regarded as mycotoxins. Some secondary metabolites may or may not be mycotoxins. For example brevianamide A was reported as non-toxic (Wilson *et al.*, 1973) but was later shown to be toxic to certain insects (Paterson *et al.*, 1987). If mycotoxins were restricted to be those fungal secondary metabolites that were toxic to vertebrates, as we insist on above, brevianamide A cannot be regarded as a mycotoxin at the present stage of knowledge. We will mention certain secondary metabolites which could be indicators of mycotoxins or which have strong medical effects. Furthermore compounds such as brevianamide A may later turn out to be teratogenic, mutagenic etc. in future toxicological studies.

Most known mycotoxins are produced by species in the genera *Aspergillus, Penicillium, Fusarium* and *Alternaria*, especially when foods are considered. These genera belongs in the Ascomycotina and some of the species have known teleomorphs. It is interesting, however, that few species with a consistently produced teleomorph are toxic or widespread in foods. The most toxic species are those that produce spores clonally in large quantities (conidia = mitospores). Some of the exceptions are *Petromyces alliaceus,* the most prolific producer of the carcinogen ochratoxin A (Frisvad, 1986), *Emericella nidulans* that produces another carcinogenic mycotoxin sterigmatocystin and species of *Chaetomium* that produce several toxic secondary metabolites (Kawai, 1988). The genus *Eurotium* is common in foods, but even though it has been claimed that *Eurotium* species are toxinogenic (Bachmann *et al.*, 1979), the toxic principle has not been clearly elucidated (see below).

The four genera mentioned above contain a lot of species and are all notoriously difficult to identify to species level (Samson and Gams, 1984). Both because of this and changing taxonomies many misidentifications are known in the literature (Marasas *et al.*, 1984; Frisvad, 1989) and they are unfortunately still cited in reviews and books. When a large number of isolates in several species of one of those genera are compared, however, a clear picture of the connection between species and profiles of mycotoxins emerges (Frisvad, 1986; 1988; 1989; 1994a and b; 1995; Thrane, 1989; Frisvad *et al.*, 1998). In this chapter an overview of the most common filamentous fungi with their toxic metabolites is given.

2. MYCOTOXINS PRODUCED BY ZYGOMYCOTINA

Several strains of *Rhizopus* and *Mucor* species have been reported to produce toxic compounds (Blakeslee and Gortner, 1913; Gortner and Blakeslee, 1914; Gorlenko, 1948; Keyl *et al.*, 1967; Davis *et al.*, 1975; Diener *et al.*, 1981; Rabie *et al.*, 1985; Reiss, 1993), but in most cases these toxic compounds were only toxic towards non-vertebrates or they were not introduced via a natural route. The only toxin that may be a real mycotoxin according to the definition above is rhizonin A, produced by *Rh. microsporus* (Steyn *et al.*, 1983; Wilson *et al.*, 1984). Whether any species of *Rhizopus* or *Mucor* have caused mycotoxicosis is not yet known.

3. MYCOTOXINS PRODUCED BY ASCOMYCOTINA

3a. Eurotiales

Emericella and *Aspergillus* subgenus *Nidulantes*

Aspergillus versicolor and *A. sydowii* in the section *Versicolores* and *Emericella nidulans* and *E. variecolor* in the section *Nidulantes* are the most common species in foods. The most prominent toxin produced in this group of organisms is sterigmatocystin, only produced by *A. versicolor* and *Emericella* species. The toxicity of sydowinin and related compounds in *A. sydowii* is unknown. Emestrin, produced by *Emericella striata, E. quadrilineata, E. acristata, E. parvathecia* and *E. foveolata,* was cited as a mycotoxin by Terao *et al.* (1990), but has only shown to be toxic in i.p. applications and is thus not a mycotoxin in the strict sense of the word used here. Other toxic metabolites such as asteltoxin, paxillin and striatin have been isolated from less common species of *Emericella* (Kawai and Nozawa, 1989).

Eupenicillium and *Penicillium* subgenus *Aspergilloides, Furcatum* and *Penicillium*

The food-borne Penicillia referable to *Eupenicillium* is a very widespread and important group of fungi. The secondary metabolites produced by these fungi are of different nature and many of them may be cited as mycotoxins in the strict sense (Table 1). The most important mycotoxins are ochratoxin A (carcinogenic and nephrotoxin), citrinin (nephrotoxin), xanthomegnin, viomellein, and vioxanthin (nephro- and hepato-toxins), the nephrotoxic glycopeptides, verrucosidin (a neurotoxin), patulin and penicillic acid (generally toxic) and penitrem A (also a neurotoxin). Several secondary metabolites with unknown toxicity to vertebrates may serve as indicators of toxinogenic species. Anacine and verrucine A for example are only produced by the four Penicillium species known to produce nephrotoxins (ochratoxin A or the nephrotoxic glycopeptides), i.e. *P. nordicum, P. verrucosum, P. polonicum* and *P. aurantiogriseum* (Lund & Frisvad, 1994; Larsen and Svendsen, unpublished). Anacine and verucine A (Boyes-Korkis *et al.*, 1993; Larsen et al., 1999) are easily identified by HPLC and diode array detection, while the structure of the nephrotoxic glycopeptides has not been elucidated yet.

Eurotium

Species in the genus *Eurotium* have often been cited as toxic (Bachmann *et al.*, 1979; Blaser *et al.*, 1980; Resurreccion and Koehler, 1977; Nazar *et al.*, 1987) or causing feed refusal (Vesonder *et al.*, 1988). Toxins involved in death of cows and sheep eating *Eurotium chevalieri* moulded barley could be auroglaucin, flavoglaucin, isodihydroglaucin, tetrahydroglaucin or echinulin (Ali *et al.*, 1989; Nazar *et al.*, 1984, 1987). Bachmann *et al.* (1979) stated that physcion, physcion anthranol A and B, and erythroglaucin were toxic to chicken embryos and mice (i.p. application) and mutagenic in Ames test but that they were not toxic when administrated orally. Thus according to the definition above they cannot be classified as mycotoxins. On the other hand long-term studies may show some chronic effects of those metabolites.

Table. 1. Toxic metabolite production by species in Ascomycotina. Mycotoxins sensu stricto are marked with an asterix *. Many other secondary metabolites are produced by most species but not listed here.

1. EUROTIALES

***Byssochlamys* (anamorph *Paecilomyces*)**

B. fulva: patulin*, byssochlamic acid, byssotoxin**
B. nivea: patulin*, byssochlamic acid, malformins
Pae. variotii I: patulin*
Pae. variotii II: viriditoxin*

***Talaromyces* (anamorph *Penicillium* subgenus *Biverticillium*)**

T. macrosporum: duclauxin*
P. crateriforme: rubratoxin*, rugulovasine A & B*, spiculisporic acid*
P. funiculosum: -
P. islandicum: rugulosin*, luteoskyrin*, islanditoxin*, cyclochlorotine*, erythroskyrin*, emodin*
P. rugulosum: rugulosin*
P. variabile: rugulosin*
P. purpurogenum: -

***Eupenicillium* (anamorph *Penicillium* subgenus *Aspergilloides, Furcatum* and *Penicillium*)**

P. atramentoseum: oxaline, rugulovasine A+B
P. aurantiogriseum: penicillic acid*, verrucosidin*, nephrotoxic glycopeptides*, anacine, auranthine, aurantiamine
P. aurantiocandidum: penicillic acid*, puberulic acid, cyclopenin, cyclopenol
P. brasilianum: penicillic acid*, verrucologen*, fumitremorgin A* & B*, viridicatumtoxin*
P. brevicompactum: botryodiploidin*, mycophenolic acid, Raistrick phenols, brevianamide A
P. camemberti: cyclopiazonic acid*
P. carneum: mycophenolic acid, patulin*, penitrem A*, roquefortine C*
P. chrysogenum: roquefortine C*, meleagrin, penicillin***, chrysogine
P. citrinum: citrinin*, tanzawaic acid A
P. commune: cyclopiazonic acid*, rugulovasine A & B, cyclopaldic acid
P. corylophilum: -
P. crustosum: roquefortine C*, penitrem A*, terrestric acid, cyclopenin, cyclopenol
P. cyclopium: penicillic acid*, xanthomegnin*, viomellein*
P. digitatum: tryptoquivalins
P. discolor: chaetoglobosin A, B & C*, cyclopenin, cyclopeol
P. echinulatum: territrems*, cyclopenin, cyclopenol
P. expansum: roquefortine C*, patulin*, citrinin*, communesins, chaetoglobosin C*
P. freii: xanthomegnin*, viomellein*, vioxanthin*, penicillic acid*, aurantiamine, cyclopenon, cyclopenol
P. glabrum: citromycetin*
P. griseofulvum: roquefortine C*, cyclopiazonic acid*, patulin*, griseofulvin*
P. hirsutum: roquefortine C*, terrestric acid
P. hordei: roquefortine C*, terrestric acid
P. italicum: italicic acid, verrucolone
P. melanoconidium: penicillic acid*, penitrem A*, roquefortine C*, verrucosidin*
P. nalgiovense: penicillin***, chrysogine
P. nordicum: ochratoxin A*, anacine, verrucolone
P. olsonii: verrucolone
P. oxalicum: secalonic acid D & F*, oxaline
P. palitans: cyclopiazonic acid*, fumigaclavine A & B*
P. paneum: patulin*, roquefortine C*, marcfortines
P. polonicum: penicillic acid*, verrucosidin*, nephrotoxic glycopeptides*, cyclopenin, cyclopenol
P. roqueforti: roquefortine C*, isofumigaclavine A & B*, PR-toxin*, mycophenolic acid
P. solitum: cyclopenin, cyclopenpol. compactins
P. verrucosum: ochratoxin A*, citrinin*, verrucolone, verrucins
P. viridicatum: xanthomegnin*, viomellein*, vioxanthin*, viridic acid*, penicillic acid*, brevianamide A

***Eurotium* (anamorph *Aspergillus* subgenus *Aspergillus*)**

E. amstelodami: physcion, echinulin
E. chevalieri: physcion, echinulin
E. herbariorum: physcion, echinulin

***Petromyces* and *Neopetromyces* (anamorph *Aspergillus* subgenus *Circumdati*)**

***Neopetromyces* and section *Circumdati*:**

A. ochraceus: penicillic acid*, ochratoxin A*, xanthomegnin*, viomellein*, vioxanthin*

***Petromyces* and section *Flavi* :**

Petromyces alliaceus: ochratoxin A*, kojic acid
A. flavus: kojic acid, 3-nitropropionic acid*, cyclopiazonic acid*, aflatoxin B_1*, aspergillic acid
A. oryzae: kojic acid, cyclopiazonic acid*, 3-nitropropionic acid*
A. nomius: kojic acid, tenuazonic acid*, aflatoxin B_1*, B_2*, G_1*, G_2*, aspergillic acid
A. parasiticus: kojic acid, aspergillic acid, aflatoxin B_1*, B_2*, G_1*, G_2*
A. sojae: kojic acid, aspergillic acid

Aspergillus Section *Nigri*:

A. niger: naphtho-4-pyrones, malformins, ochratoxin A* (few isolates)
A. carbonarius: naphtho-4-pyrones, ochratoxin A*

***Neosartorya* and Hemicarpenteles? (anamorphs in *Aspergillus* subgenera *Fumigati* and *Clavati*)**

Neosartorya and section *Fumigati*:
N. fischeri: verrucologen*, fumitremorgin A & B*
A. fumigatus: gliotoxin*, verrucologen*, fumitremorgin A & B*, fumitoxins**, fumigaclavines* tryptoquivalins

***Hemicarpenteles* and section *Clavati*:**

A. clavatus: patulin*, ascladiol, cytochalasin E*, tryptoquivalins

***Emericella* (anamorph *Aspergillus* subgenus *Nidulantes*, section *Nidulantes, Versicolores* and *Usti*)**
E. nidulans: sterigmatocystin*, nidulotoxin**, penicillin***
A. versicolor: sterigmatocystin*, nidulotoxin**
A. ustus: austamide*, austdiol*, austins*, austocystins*
***Fennellia* (anamorphs in *Aspergillus* subgenus *Flavipedes* and *Terrei*)**
A. terreus: terrein, patulin*, citrinin*, citreoviridin*, territrem*

2. SPHAERIALES
Microascus (anamorph *Scopulariopsis*)
S. brevicaulis: -
Chaetomium
C. globosum: chaetoglobosins*, chetomin*
Hypocrea (anamorph *Trichoderma*)
T. harzianum: chrysophanol, koninginin A, trichorzianines A + B
T. virens: gliotoxin*, gliovirin, viridin
T. viride: alamethicins, emodin*, suzukacillin, trichodermin*, trichotoxin A
Gibberella (anamorph *Fusarium* section *Arthrosporiella, Discolor, Elegans, Gibbosum, Liseola, Roseum, Sporotrichiella*)
F. acuminatum: antibiotic Y, chlamydosporol, trichothecenes type A*, enniatins, moniliformin
F. avenaceum: antibiotic Y, chlamydosporol, fusarin C, moniliformin, enniatins
F. crookwellense: culmorin, fusarin C, trichothecenes type B*, zearalenone*, butenolide, chrysogine
F. culmorum: culmorin, fusarin C, trichothecenes type B*, zearalenone*, butenolide, chrysogine
F. equiseti: equisetin, fusarochromanone*, trichothecenes type A & B*, zearalenone*, chrysogine
F. graminearum: culmorin, fusarin C, trichothecenes type B*, zearalenone*butenolide, chrysogine
F. semitectum: equisetin, zearalenone*, chrysogine, fusapyrone
F. oxysporum: fusaric acid, naphthoquinone pigments, nectriafurone, moniliformin, gibepyrones
F. poae: fusarin C, trichothecenes type A & B*, butenolide, gamma-lactones,
F. proliferatum: fumonisins*, fusaric acid, fusarin C, moniliforme, naphthoquinone pigments, beauvericins, fusaproliferin, fusapyrone
F. subglutinans: fusaric acid, fusarin C, moniliforme, naphthoquinone pigments, beauvericin, fumonisin*
F. sambucinum: trichothecenes type A*, butenolide, enniatins
F. sporotrichioides: fusarin C, trichothecenes type A*, butenolide
F. tricinctum: antibiotic Y, fusarin C, butenolide, chlamydosporol, chrysogine, visoltricin
F. verticillioides: fumonisins*, fusaric acid, fusarin C, moniliforme, naphthoquinone pigments, gibepyrones
Haematonectria (anamorph *Fusarium* section *Martiella*)
F. solani: fusaric acid, naphthoquinone pigments
Claviceps (anamorph *Sphacelia*)
C. purpurea: ergotalkaloids

3. DOTHIDIALES
Mycosphaerella (anamorph *Cladosporium*)
C. herbarum: epi- and fagi-cladosporic acid**
Pleospora (anamorph *Stemphylium*)
S. botryosum: -
Lewia (anamorph *Alternaria*)
A. cf. Alternata: alternariol, alternariol monomethylether, altertoxins
A. tenuissima: alternatriol, alternariol monomethylether, altertoxins, tenuazonic acid*
Melanopsamma (anamorph *Stachybotrys*)
S. chartarum: satratoxin G and H*

4. Unknown connection to known ascomycetes:
Unknown teleomorph (anamorph *Epicoccum*)
E. nigrum: -
Unknown teleomorph (anamorph *Wallemia*)
W. sebi: Walleminol A & B

* Mycotoxins in the restricted sense used in this chapter: ** not structure elucidated; *** penicillin is not classified as a mycotoxin, but should not be present in foods.

Even though Schroeder and Kelton (1975) and Karo and Hadlok (1982) reported on sterigmatocystin production by *Eurotium* species, this has never since been demonstrated and in other screenings sterigmatocystin or other known mycotoxins have never been detected. Schroeder and Kelton (1975) and Leitao *et al.* (1989) even reported that *Eurotium* species produce aflatoxin B_1, but these results are probably based on artifacts. The claimed production of gliotoxin by *Eurotium chevalieri* (Wilkinson and Spilsbury, 1965) was based on a strain of *Trichoderma virens* thought to be an *Eurotium* because of a similarity in culture collection numbers (Thrane and Frisvad, unpublished observation).

Neopetromyces and *Aspergillus* section *Circumdati*

This genus and its associated anamorphs contain very good producers of ochratoxin A, penicillic acid, xanthomegnin, viomellein and vioxanthin. The most commonly found species is *A. ochraceus*, whereas *N. muricatus*, *A. ostianus* and *A. sclerotiorum* are occasionally found in foods. Ochratoxin A has not been detected in *A. ostianus* but some isolates of *A. sclerotiorum* produce trace amounts of ochratoxin A. *N. muricatus* is a very efficient producer of ochratoxin A.

Petromyces* and *Aspergillus* section *Flavi
P. alliaceus produce large amounts of ochratoxin A and has been reported from onions (hence the name), but by far the most important toxigenic species in *Flavi* are *A. flavus, A. parasiticus* and *A. nomius.* They produce the same mixture of kojic acid and the carcinogenic aflatoxins and in addition *A. flavus* produce cyclopiazonic acid. *A. flavus* is common on cereals and many other crops, especially from the tropics and subtropics.

3b. SPHAERIALES

Gibberella* and its anamorph *Fusarium
This group of fungi contains a major part of common food-borne *Fusarium* species. The lack of a sound *Fusarium* taxonomy has caused many misleading information on the production of mycotoxins of *Fusarium* species. Marasas *et al.* (1984) and Thrane (1989) critically reviewed the published reports and corrected many errors. It is generally accepted that most prominent mycotoxin producing *Fusarium* species are *F. culmorum, F. graminearum* and *F. cerealis* (deoxynivalenol and related trichothecenes, zearalenone, fusarins a.o.), *F. sporotrichioides* and *F. poae* (T-2 toxin and related trichothecenes), *F. equiseti* (fusarochromanone, zearalenone and derivatives, diacetoxyscirpenol, equisetin), and *F. verticillioides, F. proliferatum* and *F. nygamai* (fumonisins and fusarins). Less common species as *F. sambucinum* and *F. acuminatum* have also been reported to produce trichothecenes.

In addition to the mycotoxins listed above other biological active metabolites such as moniliformin (Thrane, 1989), antibiotic Y (Thrane, 1989), chlamydosporol (Shier and Abbas, 1992), proliferin (Randazzo *et al.* 1993), and wortmannin (Abbas and Mirocha, 1988; Thrane and Hansen (1995) have been reported produced by one or more *Fusarium* species. However, the toxicology of these metabolites is not well documented (see different chapters in Miller and Trendholm, 1994).

Haematonectria* and associated food-borne anamorphs in *Fusarium
F. solani is the only species with *Haematonectria* teleomorph which is occurring frequently on foods. *F. solani* has been reported to produce mycotoxins, but all reports are based on misidentifications of the organisms (Marasas *et al.*, 1984; Thrane, 1989). The metabolites produced by *F. solani* (primarily naphthoquinone pigments) are not mycotoxins in the strict sense of the word used here.

Chaetomium
The most important mycotoxins produced by the common *Chaetomium* species seems to be chaetomin and the chaetoglobosins (Cole and Cox, 1981; Udagawa, 1984, Jen and Jones, 1983). Sterigmatocystin is produced by rare species. Of the many metabolites from *Chaetomium* species some were found to be cytotoxic using He-La cells, but again may not be mycotoxins in the strict sense of the word. These compounds included the mutagens mollicellins, oosporein, chaetocin, eugenitin, the teratogenic chaetochromin, and cochliodinols (Udagawa, 1984).

Claviceps
Claviceps species are found as large sclerotia on rye and other grasses, but have not been reported from other foods or feeds.

Hypocrea* and its anamorph *Trichoderma
Trichoderma species produce a range of biological active compounds (Cole and Cox, 1981; Ghisalberti and Sivasithamparam, 1991; Anke *et al.*, 1991; Sivasithamparam & Ghisalberti, 1998), however, only few of these are mycotoxins in the strict sense. The most prominent mycotoxin is gliotoxin produced by *Trichoderma virens, Aspergillus fumigatus*, and a few less common *Penicillium* species (Frisvad and Thrane, 1993). Gliotoxin is active towards fungi, bacteria, and vira (Cole and Cox, 1981), but has also proved to be immunosuppressive (Müllbacher *et al.*, 1985). A second group of mycotoxins is isocyanide metabolites of *T. hamatum*, which are important factors in the etiology of ovine ill-thrift in sheep (Brewer *et al.*, 1982). Trichodermin, a sesquiterpene closely related to the trichothecenes produced by *Fusarium* species, has been isolated from *T. viride* and *T. polysporum* (Godtfredsen and Vangedal, 1965; Adams and Hanson, 1972). Toxicological data of trichodermin is summarized by Cole and Cox (1981).

3c. DOTHIDEALES

Production of mycotoxins and other secondary metabolites by most of the important species in *Alternaria* has been reviewed by Sivanesan (1991), Chelkowksi and Visconti, 1992; Blaney (1991) and Abbas *et al.* (1996). Most interest has been in AAL toxin, rather similar to fumonisins, the alternariols and tenuazonic acid (Abbas *et al.*, 1996; Chelkowski & Visconti, 1992). Chen *et al.* (1992) reported on the production of fumonisin B1 in *Alternaria "alternata"* f.sp. *lycopersici.* The most important genera concerning foods are *Alternaria* and possibly *Cladosporium*, although very little is known on the actual toxigenic potential of *Cladosporium* species.

4. IMPORTANT MYCOTOXINS

Aflatoxins. The aflatoxins are produced by the three closely related species *Aspergillus nomius, A. flavus* and *A. parasiticus* and by some species not occurring in foods. The ability to produce aflatoxins is most consistent in *A. nomius* and *A. parasiticus* and the latter two species also produce the G type aflatoxins. The domesticated forms of *A. flavus, A. oryzae* and of *A. parasiticus, A. sojae* have lost the ability to produce aflatoxins and precursors completely. The toxicology of aflatoxins and mycology of the producers is discussed by Eaton and Groopman (1993).

Fumonisins. The fumonisins are produced by three closely related species *F. verticillioides, F. proliferatum, F. nygamai* and possibly *Alternaria* spp. (Chen *et al.*, 1992). Since the first report on fumonisins (Gelderblom *et al.*, 1988) a lot of attention has been given to these highly carcinogenic compounds. This is reflected in more reviews on the toxicology and chemistry of fumonisins, and the mycology of producers of fumonisins (Abbas *et al.*, 1993; Nelson *et al.*, 1993; Norred, 1993; Scott, 1993; Jackson *et al.*, 1996).

Ochratoxins. The carcinogenic ochratoxin A is produced by *Penicillium verrucosum, P. nordicum, Aspergillus ochraceus, A. sulphureus, Neopetromyces muricatus, A. niger* (few isolates), A. *carbonarius, A. sulphureus* and *Petromyces alliaceus* (Frisvad and Samson, 2000). *P. verrucosum* is most important in cereals in temperate regions of the world, but *P. nordicum* is important in refrigerated meat and cheese products. Ochratoxin A has also been detected in wine, and the most probable producer in grapes is *A. niger*. Occurrence of ochratoxin A in the tropics is associated with *A. ochraceus* (coffee and cocoa) and maybe *A. carbonarius* and A. *niger*. It is not known whether *Petromyces alliaceus* is able to produce ochratoxin A in onions.

Trichothecenes: More than 170 different trichothecenes have been isolated (Grove, 1993) and they may be divided into two major groups, the macrocyclic trichothecenes and the non-macrocyclic trichothecenes. The macrocyclic trichothecenes (67 compounds) have been isolated from species of the fungal genera *Stachybotrys, Myrothecium, Cylindrocarpon, Verticinimonosporium* and *Phomopsis* and from a plant genus *Baccharis*. The macrocyclic trichothecenes have been reviewed by Jarvis (1991) and Grove (1993).

The non-macrocyclic trichothecenes include some of the most important mycotoxins produced by fungi. This group may be further subdivided into type A trichothecenes and type B trichothecenes by the chemical structure of the compounds. The two types of trichothecenes have been reviewed by Sharma and Kim (1991) and in different chapters in Miller and Trendholm (1994). Common type A trichothecenes are T-2 toxin and diacetoxyscirpenol which are produced by *F. sporotrichioides, F. poae, F. acuminatum, F.equiseti* and *F. sambucinum*. Common type B trichothecenes are deoxynivalenol (= vomitoxin) and nivalenol, which are produced by three closely, related species, *F. cerealis, F. culmorum* and *F. graminearum*.

Zearalenone and zearalenols: Zearalenone and its derivatives are produced by *F. crookwellense, F. culmorum, F. equiseti, F. graminearum, F. semitectum* and related species. This group of mycotoxins, showing estrogenic activity, has been detected frequently in cereals and other food commodities. Zearalenone and zearalenols have been reviewed by Lindsay (1985), Schoental (1985), Joffe (1986), Gareis *et al.* (1989), Blaney (1991), and in several chapters in Miller and Trendholm (1994).

Tenuazonic acid: Tenuazonic acid is produced by *Alternaria "alternata", A. "citri", Magnaporte grisea* (anamorph *Pyricularia oryzae*), *Phoma sorgina* and *Aspergillus nomius*. Tenuazonic acid is acutely very toxic and has been associated with onyalai, a haematologic disease (Steyn and Rabie, 1976). The producer was in that case *Phoma sorghina*.

Patulin. Being a generally very toxic secondary metabolite, patulin is produced by a large number of fungi. Most species are referable to the ascomycete genera *Byssochlamys* and *Eupenicillium*. Some species of *Aspergillus (A. clavatus, A. giganteus,* and *A. terreus*) are also effective producers. Two ecological groups of fungi often produce patulin. One is fungi growing in silage including *P. carneum, Paecilomyces variotii* and *P. glandicola*. The other group is the coprophilic fungi including *P. coprobium, P. glandicola, P. vulpinum, P. clavigerum* and *P. concentricum* (Frisvad and Filtenborg, 1989). In foods the most important patulin producers are *P. expansum*, capable of producing the toxin in apples, and *P. griseofulvum* (= *P. patulum* = *P. urticae*). The occurrence and toxicology of patulin has been reviewed by McKinley and Carlton (1991).

Xanthomegnin and viomellein. The important food-borne producers of these hepato- and nephrotoxins are *Penicillium cyclopium, P. freii, P. melanoconidium, P. tricolor, P. viridicatum* and *Aspergillus ochraceus* (Lund and Frisvad, 1994) and the penicillia just mentioned are especially common in cereals (Table 2) (Frisvad and Lund, 1993; Lund & Frisvad, 1994). Natural occurrence of viomellein and xanthomegnin in cereals have been caused by *P. freii, P. cyclopium* and *P. viridicatum* (Hald *et al.*, 1983; Scudamore *et al.*, 1986; Frisvad and Lund, 1993). The significance of these naphthoquinones as compared to ochratoxin A and citrinin in development of kidney diseases of humans and swines is unknown and the nephrotoxin glycopeptides cited below may also play an important role (Mantle, 1993). The fungi producing all these nephrotoxins are often co-occurring (Frisvad and Filtenborg, unpublished).

Table 2. Production of nephrotoxins by *Penicillium* species common in cereals.

Species	Nephrotoxins
P. aurantiogriseum	Glycopeptides
P. aurantiovirens	No known
P. citrinum	Citrinin
P. cyclopium	Xanthomegnin, viomellein etc.
P. freii	Xanthomegnin, viomellein etc.
P. hordei	No known
P. melanoconidium	Xanthomegnin, viomellein etc.
P. polonicum	Glycopeptides
P. tricolor	Xanthomegnin, viomellein etc.
P. verrucosum	Ochratoxin A, citrinin
P. viridicatum	Xanthomegnin, viomellein etc.

Nephrotoxic glycopeptides. These glycopeptides were originally isolated from *P. aurantiogriseum* (Yeulet *et al.*, 1988; Mantle, 1994) but it has later been shown that most of the producing isolates belong to *P. polonicum*, a fungus particularly common in warmer climates (Frisvad and Filtenborg, 1993). No chemical analytical methods for these mycotoxins have been developed, but another mycotoxin, the neurotoxin verrucosidin, is consistently produced by the glycopeptide producers *P. aurantiogriseum* and *P. polonicum* and could act as a indicator of the nephrotoxins (Frisvad and Lund, 1993).

Sterigmatocystin. The most important producer of the cancerogenic sterigmatocystin is *A. versicolor* and this mycotoxin has been found in both cheese and cereals (Scott, 1989, 1994). Other producers of sterigmatocystin, *Emericella nidulans* and *Chaetomium* species have not been shown to produce it naturally, but only under laboratory conditions.

Fusarochromanone. This toxin produced by *F. equiseti* is probably implicated in tibial dyschondroplasia and decreased immune response, and has been found naturally occurring in poultry feed (Krogh *et al.*, 1989; Wu *et al.*, 1991)

Ergot alkaloids. The ergot alkaloids have been found naturally occurring in rye as it is difficult to avoid contamination of ergot sclerotia completely (Rehácek and Sajdl, 1990).

Citrinin. Citrinin, produced by *P. verrucosum* in cereals, appear to be less important concerning cases of porcine nephrotoxicity in temperate regions than ochratoxin A (Reddy and Berndt, 1991), but it has been isolated from cereals causing procine nephropathy in Brasil, being produced by *P. citrinum* (Rosa *et al.*, 1985). It has also been detected in rice, corn and peanuts in warmer climates, again produced by *P. citrinum*, and in Denmark, Sweden and Canada, produced by *P. verrucosum* (Reddy and Berndt, 1991; Frisvad and Filtenborg, 1993). Citrinin production by *P. expansum* may be of some importance, but citrinin occurrence in foods have rarely been reported.

3-Nitropropionic acid. This toxin, produced by *Arthrinium sacchari, A. saccharicola* and *A. phaeospermum* in sugercane, has been implicated in fatal food poisonings in humans in China (Liu *et al.*, 1988). This toxin as also produced by certain strains of *A. oryzae* and possibly *A. flavus* (Orth, 1977).

Roquefortine C. Roquefortine C is consistently produced by a series of Penicillia, including the common *P. chrysogenum, P. griseofulvum, P. expansum, P. crustosum, P. hordei* and *P. roqueforti* (Frisvad and Filtenborg, 1989), but it has only been found to occur naturally in feed grain (Häggblom, 1990). It is, however, produced, together with isofumigaclavine A and B, and festuclavine in silage inoculated with *P. roqueforti* (Ohmomo *et al.*, 1994)

Secalonic acid D. Secalonic acids are produced by *Claviceps purpurea* (common in rye), *P. oxalicum* (common on cucumbers and corn), *Phoma terrestris* (common on onions) and by the soil-borne *A. aculeatus.* Toxic effect of secalonic acid D was most pronounced in i.p. or i.v. application. Thus the toxins were not introduced via a natural route. However the LD_{50} value in newborn rats was 24.6 mg/kg (oral route) (Reddy and Reddy, 1991).

Verrucosidin. The neurotoxin verrucosidin is produced by *P. aurantiogriseum, P. polonicum* and *P. melanoconidium* (Frisvad and Lund, 1993). *P. polonicum* (at that time identified as *P. verrucosum* var. *cyclopium*) may have been involved in a neurological disease in cattle in USA (Wilson *et al.*, 1981).

CONCLUSION

Many food-borne fungi produce mycotoxins and thus fungal growth in foods and feeds should be avoided. Food-borne species of particular importance appear to be *Fusarium graminearum, F. culmorum, P. verrucosum,P. nodicum, P. freii, P. cyclopium, P. expansum* and *P. crustosum* in temperate regions of the world. *Aspergillus flavus, Fusarium verticillioides* and other Fusaria, *P. aurantiogriseum, P. citrinum, P. crustosum, P. polonicum, P. viridicatum, P. islandicum* and *P. oxalicum* are important in (sub)tropical climates.

REFERENCES

ABBAS, H. K. & MIROCHA, C. J. 1988. Isolation and purification of a hemorrhagic factor (wortmannin) from *Fusarium oxysporum* (N17B). Appl. Environ. Microbiol. 54: 1268-1274.

ABBAS, H. K., DUKE, S. O. & TANAKA, T. 1993. Phytotoxicity of fumonisins and related compounds. J. Toxicol.-Toxin. Rev. 12: 225-251.

ABBAS, H.K., DUKE S .O., SHIER, W.T., RILEY, R.T. & KRAUS, G.A. 1996. The chemistry and biological activities of the natural products AAL toxin and the fumonisins. In: Singh, B.R. & Tu, A.T (eds.). Natural toxins II. pp. 293-308. Plenum press, New York.

ADAMS, P. M. & HANSON, J. R. 1972. Sesquiterpenoid metabolites of *Trichoderma polysporum* and *T. sporulosum*. Phytochemistry 11: 423.

ALI, M., N. MOHAMMED, M.A. ALNAQEEP, R.A.H. HASSAN, & AHMAD, H.S.A. 1989. Toxicity of echinulin from *Aspergillus chevalieri* in rats. Toxicol. Lett. 48: 235-241.

ANKE, H., J. KINN, K. E. BERGQUIST & STERNER, O. 1991. Production of Siderophores by strains of the Genus *Trichoderma* -Isolation and Characterization of the New Lipophilic Coprogen Derivative, Palmitoylcoprogen. Biol. Met. 4: 176-180.

BACHMANN, M., J. LÜTHY, & SCHLATTER, C. 1979. Toxicity and mutagenicity of molds of the *Aspergillus glaucus* group. Identification of physcion and three related anthraquinones as main toxic constituents from *Aspergillus chavalieri*. J. Agric. Food Chem. 27: 1342-1347.

BLAKESLEE, A.F. & GORTNER, R.A. 1913. On the occurence of a toxin in juice expressed from the bread mould, *Rhizopus nigricans* (*Mucor stolonifer*). Biochem. Bull. 2: 542-544.

BLANEY, B.J. 1991. Fusarium and Alternaria toxins. In: B.R. Champ, E. Highley, A.D. Hocking and J.I. Pitt (Eds.): *Fungi and mycotoxins in stored products*. ACIAR Proceedings No. 36. Australian Centre for International Agricultural Research, Canberra. pp. 86-98.

BLASER, P., H. RAMSTEIN, W. SCHMIDT-LORENZ & SCHLATTER, C. 1980. Toxicität and Mutagenität der xerophilen Schimmelpilze der gattung *Eurotium* (Aspergillus glaucus-Gruppe). Lebensm. Wiss. Technol. 14: 66-71.

BOYES-KORKIS, J.M., GURNEY, K., PENN, J., MANTLE, P.G., BILTON, J.N. & SHEPPARD, R.N. 1993. Anacine, a new benzodiazepine metabolite of *Penicillium aurantiogriseum* produced with other alkaloids in submerged fermentation. J. Nat. Prod. 56: 1707-1717.

BREWER, D., A. FEICHT, A. TAYLOR, J. W. KEEPING, A. A. TAHA & THALLER, V. 1982. Ovine ill-thrift in Nova Scotia. 9. Production of experimantal quantities of isocyanide metabolites of *Trichoderma hamatum*. Canad. J. Microbiol. 28: 1252-1260.

CAMPBELL, I. M. 1984. Secondary metabolism and microbial physiology. Adv. Microbial Physiol. 25: 1-60.

CHELKOWSKI, J. & VISCONTI, A. (Eds). 1992. Alternaria. *biology, plant diseases and metabolites*. Elsevier, Amsterdam.

CHEN, J., MIROCHA. C., XIE, W., HOGGE, L. & OLSON, D. 1992. Production of the mycotoxin fumonisin B1 by Alternaria alternata f.sp. *lycopersici*. Appl. Environ. Microbiol. 58: 3928-3931.

CHRISTOPHERSEN, C. 1991. Evolution in molecular structure and adaptive significance in metabolism. Comp. Biochem. Physiol. 98B: 427-432.

COLE, R.J. & COX, R.H. 1981. Handbook of toxic fungal metabolites. Academic Press, New York.

DAVIS, N.D., WAGENER, R.E., MORGAN-JONES, G. & DIENER, U.L. 1975. Toxigenic thermophilic and thermotolerant fungi. Appl. Microbiol. 29: 455-457.

EATON, D.L. & GROOPMAN, J.D. (eds.). 1993. *The toxicology of aflatoxins*. Academic Press, San Diego.

FILTENBORG, O. & FRISVAD, J.C. 1980. A simple screening method for toxigenic moulds in pure cultures. Lebensm. Wiss. Technol. 13: 128-130.

FILTENBORG, O., FRISVAD, J.C. & SVENDSEN, J.A. 1983. Simple screening method for molds producing intracellular mycotoxins in pure culture. Appl. Environ. Microbiol. 45: 581-585.

FRISVAD, J.C. & FILTENBORG, O. 1988. Specific mycotoxin producing *Penicillium* and *Aspergillus* mycoflora of different foods. Proc. Jpn. Assoc. Mycotoxicol. Suppl. 1: 163-166.

FRISVAD, J.C. & FILTENBORG, O. 1989. Terverticillate penicillia: chemotaxonomy and mycotoxin production. Mycologia 81: 837-861.

FRISVAD, J.C. & FILTENBORG, O. 1993. Saprophytic spoilage association in food mycology with emphasis on *Penicillium* species. In: K.A. Scudamore (ed.): *Occurrence and significance of mycotoxins*. pp. 138-145. Central Science Laboratory, Slough.

FRISVAD, J.C. & LUND, F. 1993. Toxin and secondary metabolite production by *Penicillium* species growing in stored cereals. In: K.A. Scudamore (ed.): Occurrence and significance of mycotoxins. pp. 146-171. Central Science Laboratory, Slough.

FRISVAD, J.C. & THRANE, U. 1987. Standardized high performance liquid chromatography of 182 mycotoxins and other fungal metabolites based on alkylphenone retention indices and UV-VIS spectra (diode array detection). J. Chromatogr. 404: 195-214.

FRISVAD, J.C. & THRANE, U. 1993. Liquid chromatography of mycotoxins. In: V. Betina (ed.): *Chromatography of mycotoxins. techniques and applications*. Journal of Chromatography Library 54: 253-372. Elsevier, Amsterdam.

FRISVAD, J.C. 1986. Taxonomic approaches to mycotoxin identification. In: R.J. Cole (Ed.): *Modern methods in the analysis and structural elucidation of mycotoxins*. Academic Press, New York, p. 415-457.

FRISVAD, J.C. 1988. Fungal species and their specific production of mycotoxins. In: R.A. Samson and E.S. van Reenen-Hoekstra (eds.): *Introduction to food-borne fungi*. 3rd ed. Centraalbureau voor Schimmelcultures, Baarn and Delft. p. 239-249.

FRISVAD, J.C. 1989. The connection between penicillia and aspergilli and mycotoxins with special emphasis on misidentified isolates. Arch. Environ. Contam. Toxicol. 18: 452-467.

FRISVAD, J.C. 1994a. Classification of organisms by secondary metabolites. In Hawksworth, D.L. (ed.): *Identification and characterization of pest organisms*. pp 303-321. CAB International, Wallingford.

FRISVAD, J.C. 1994b. Correspondence analysis, principal coordinate and redundancy analysis used on mixed chemotaxonomical qualitative and quantitative data. Chemom. Intell. Lab. Syst. 23: 213-229.

FRISVAD, J.C. 1995. Mycotoxins and mycotoxigenic fungi in storage. In: Jayas, D.S., White, N. D.G. and Muir, W.E. (eds.). *Stored grain ecosystems*, pp. 251-288. Marcel Dekker, New York.

FRISVAD, J.C., THRANE, U. & FILTENBORG, O. 1998. Role and use of secondary metabolites in fungal taxonomy. In: Frisvad, J.C., Bridge, P.D. and Arora, D.K. (eds.) *Chemical Fungal taxonomy*, pp. 289-319. Marcel Dekker, New York.

GAREIS, M., BAUER, J., ENDERS, C. & GEDEK, B. Contamination of cereals and feed with *Fusarium* mycotoxins in European countries. In: *Fusarium: Mycotoxins,taxonomy, and pathogenicity*, edited by Chelkowski, J. Amsterdam: Elsevier Science Publishers B.V., 1989, p. 441-472.

GELDERBLOM, W. C. A., K. JASKIEWICZ, W. F. O. MARASAS, P. G. THIEL, R. M. HORAK, R. VLEGGAAR & KRIEK, N.P.J. 1988. Fumonisins - novel mycotoxins with cancer-promoting activity produced by *Fusarium moniliforme*. Appl. Environ. Microbiol. 54: 1806-1811.

GHISALBERTI, E. L. & SIVASITHAMPARAM, K. 1991. Antifungal Antibiotics Produced by *Trichoderma* spp. Soil Biol. Biochem. 23: 1011-1020.

GODTFREDSEN, W. O. & VANGEDAL, S. 1965. Trichodermin, a new sesquiterpene antibiotic. Acta Chem. Scand. 19: 1088-1102.

GORLENKO, M.V. 1948. The toxins of moulds. Am. Rev. Sovjet Med. 5: 163-164.

GORTHER, R.A. & BLAKESLEE, A.F. 1914. Observations on the toxin of *Rhizopus nigricans*. Am. J. Physiol. 34: 353-367

GRAVESEN, S., FRISVAD, J.C. & SAMSON, R.A. 1994. *Microfungi*. Munksgaard, Copenhagen.

GROVE, J. F. 1993. Macrocyclic Trichothecenes. Nat. Prod. Rep. 10: 429-448.

HÄGGBLOM, P. 1990. Isolation of roquefortine C from feed grain. Appl. Environ. Microbiol. 56: 2924-2926.

HALD, B., D.H. CHRISTENSEN, & KROGH, P. 1983. Natural occurrence of the mycotoxin viomellein in barley and the associated quinone-producing penicillia. Appl. Environ. Microbiol. 46: 1311-1317.

JACKSON, L.S., DEVRIES, J.W. & BULLERMAN, L.B. (eds) 1996. *Fumonisins in food.* Plenum Press, New York.

JARVIS, B.B. 1991. Macrocyclic trichothecenes. In: *Mycotoxins and phytoalexins,* (Sharma, R.P. and Salunkhe, eds.) D.K. Boca Raton: CRC Press, 1991, p. 361-421.

JEN, W.C. & JONES, G.A. 1983. Effects of chetomin on growth and acidic fermentation products of rumen bacteria. Can. J. Microbiol. 29: 1399-1404.

JOFFE, A.Z. *Fusarium species: Their biology and toxicology,* New York, Chichester, Brisbane, Toronto, Singapore:John Wiley & Sons, Inc, 1986. pp. 1-588.

KARO, M. & HADLOK, R. 1982. Investigations on sterigmatocystin production by fungi of the genus *Eurotium.* In: P. Krogh (ed.). Proceedings of the V. International IUPAC Symposium on Mycotoxins and Phycotoxins. Technical University, Vienna. pp. 178-181.

KAWAI, K. & NOZAWA, K. 1989. Novel biologically active compounds from *Emericella* species. In: S. Natori, K. Hashimoto, and Y. Ueno (eds.): Mycotoxin and Phycotoxins '88. Elsevier, Amsterdam. pp. 205-212.

KEYL, A.C., LEWIS, J.C., ELLIS, J.J., YATES, S.G. & TOOKEY, H.L. 1967. Toxic fungi isolated from tall fescue. Mycopath. Mycol. Appl. 31: 327-331.

KROGH, P., D.C. CHRISTENSEN, B. HALD, B. HARLOU, C. LARSEN, E.J. PEDERSEN, & THRANE, U. 1989. Natural occurrence of thye mycotoxin fusarochromanone, a metabolite of *Fusarium equiseti,* in cereal feed associated with tibial dyschondroplasia. Appl. Environ. Microbiol. 55: 3184-3188.

LARSEN, T.O., FRANZYK., H. and JENSEN, S..R. 1999. UV-guided isolation of verrucines A and B, novel quinazolones from *Penicillium verrucosum* structurally related to anacine from *Peniiclliurn aurantiogriseum.* J. Nat. Prod. 62 : 1578-1580.

LEITAO, J., J. LE BARS, & BAILLY, J.R. 1989. Production of aflatoxin B_1 by *Aspergillus ruber* Thom and Church. Mycopathologia 108: 135-138.

LINDSAY, D.G. 1985. Zeranol - a 'nature-identical' oestrogen? Fd. Chem. Toxic. 23: 767-774.

LIU, X.J., X.Y. LUO, & HU, W.J. 1988. Arthrinium spp. and the etiology of deteriorated sugarcane poisoning. In: S. Natori, K. hashimoto, and Y. Ueno (eds.): Mycotoxins and phycotoxins '88. Elsevier, Amsterdam. pp. 109-118.

LUND, F. and FRISVAD, J.C. 1994. Chemotaxonomy of *Penicillium aurantiogriseum* and related species. Mycol. Res. 98: 481-492.

MANTLE, P.G. 1993. A reappraisal of ochratoxin A as a factor in human renal disease. In: K.A. Scudamore (ed.): *Occurrence and significance of mycotoxins.* Central Science Laboratory, Slough, pp. 101-108.

MANTLE, P.G. 1994. Renal histopathological responses to nephrotoxic *Penicillium aurantiogriseum* in the rat during pregnancy, lactation and after weaning. Nephron 66: 93-98.

MARASAS, W.F.O., NELSON, P.E. & TOUSSOUN, T.A. 1984. *Toxigenic* Fusarium *species. Identity and mycotoxicology.* Pennsylvania State University Press, University Park.

MCKINLEY, E.R. & CARLTON, W.W. 1991. Patulin. In: R.P. Sharma and D.K. Salunkhe (eds.): *Mycotoxins and phytoalexins.* CRC Press, Boca Raton. pp. 191-236.

MILLER, J.D. & TRENHOLM, H.L. 1994. *Mycotoxins in grain. Compounds other than aflatoxin.* Eagan Press, St. Paul.

MÜLLBACHER, A., P. WARING & EICHNER, R.D. 1993. Identification of an agent in cultures of *Aspergillus fumigatus* displaying anti-phagocytic and immunomodulating activity in vitro. J. Gen. Microbiol. 131: 1251-1258.

NAZAR, M., M. ALI, T. FATIMA, & GUBLER, C.J. 1984. Toxicity of flavoglaucin from *Aspergillus chevalieri* in rabbits. Toxicol. Lett. 23: 233-237.

NAZAR, M., M. AFZAL, J.H. MIRZA, & NIMER, N.D. 1987. *Aspergillus chevalieri* contamination of barley. J. Univ. Kuwait, Science 14: 121-126.

NELSON, P.E., A.E. DESJARDINS & PLATTNER, R.D. 1993. Fumonisins, Mycotoxins Produced by *Fusarium* Species - Biology, Chemistry, and Significance. Annu. Rev. Phytopathol. 31:233-252

NORRED, W.P. 1993. Fumonisins - Mycotoxins Produced by *Fusarium moniliforme.* J Toxicol. Environ. Health 38: 309-328.

OHMOMO, S., H.K. KITAMOTO, & NAKAJIMA, T. 1994. Detection of roquefortines in *Penicillium roqueforti* isolated from moulded maize silage. J. Sci. Food Agric. 64: 211-215.

ORTH, R. 1977. Mycotoxins of *Aspergillus oryzae* strains for use in the food industry as starters and enzyme producing moulds. Ann. Nutr. Aliment. 31: 617-624.

PATERSON, R.R.M., M.S.J. SIMMONDS, & BLANEY, W.M. 1987. Mycopesticidal effects of characterized extracts of *Penicillium* isolates and purified secondary metabolites (including mycotoxins) on *Drosophila melanogaster* and *Spodoptora littoralis.* J. Invert. Pathol. 50: 124-133.

RABIE, C.J., LUBBEN, A., SCHIPPER, M.A.A., VAN HEERDEN, F.R. & FINCHAM, J.E. 1985. Toxigenicity of *Rhizopus* species. Intl. J. Food Microbiol. 1: 263-270.

RANDAZZO, G., V. FOGLIANO, A. RITIENI, L. MANNINA, E. ROSSI, A. SCARALLO & SEGRE, A.L. 1993. Proliferin, a new sesterterpene from *Fusarium proliferatum.* Tetrahedron 49: 10883-10896.

REDDY, C.S. & REDDY, R.V. 1991. Secalonic acids. In: R.P. Sharma and D.K. Salunkhe (eds.): *Mycotoxins and phytoalexins.* CRC Press, Boca Raton. pp. 167-190.

REDDY, R.V. & BERNDT, W.O. 1991. Citrinin. In: R.P. Sharma and D.K. Salunkhe (eds.): *Mycotoxins and phytoalexins.* CRC Press, Boca Raton. pp. 237-250.

REHACEK, Z. & SAJDL, P. 1990. *Ergot alkaloids.* Academia, Praha.

REISS, J. 1993. Biotoxic activity in the *Mucorales.* Mycopathologia 121: 123-127.

RESURRECCION, A.A. & KOEHLER, P.E. 1977. Toxicity of *Aspergillus amstelodami.* J. Food Sci. 42: 482-487.

ROSA, C.A. DA R., L.C.H. DA CRUZ, W.A. CHAGAS, & VEIGA, C.E.M. DE O. 1985. Ocorrencia natural de nephropatia micotoxica suina causada pela ingestao de ceveda contaminada citrina. Rev. Bras. Med. Veter. 7: 87-90.

SAMSON, R.A. & GAMS, W. 1984. The taxonomic situation in the hyphomycete genera *Penicillium, Aspergillus* and *Fusarium.* Antonie van Leeuwenhoek 50: 815-824.

SCHOENTAL, R. 1985. Trichothecenes, zearalenone, and other carcinogenic metabolites of *Fusarium* and related microfungi. Adv. Cancer Res. 45: 217-290.

SCHROEDER, H.W. & KELTON, W.H. 1975. Production of sterigmatocystin by some species of the genus *Aspergillus* and its toxicity to chicken embryos. Appl. Microbiol. 30: 589-591.

SCOTT, P.M. 1989. Mycotoxigenic fungal contaminants of cheese and other dairy products. In: H.P. van Egmond (ed.): Mycotoxins in dairy products. Elsevier, London. pp. 193-259.

SCOTT, P.M. 1993. Fumonisins. Int. J Food Microbiol 18: 257-270.

SCOTT, P.M. 1994. *Penicillium* and *Aspergillus* toxins In: J.D. Miller & H.L. Trenholm (eds.): Mycotoxins in grain. Compounds other than aflatoxins. Eaton Press, St. Paul. pp. 261-285.

SCUDAMORE, K.A., P.M. ATKIN & BUCKLE, A.E. 1986. Natural occurrence of the naphthoquinone mycotoxins, xanthomegnin, viomellein, and vioxanthin in cereals and animal feedstuffs. J. Stored Prod. Res. 22: 81-84.

SHARMA, R.P. & KIM, Y.W. 1991. Trichothecenes. In: R.P. Sharma and D.K. Salunkhe (eds.): Mycotoxins and phytoalexins. CRC Press, Boca Raton. pp. 339-359.

SHIER, W. T. & ABBAS, H.K. 1992. Occurrence of the Mycotoxin Chlamydosporol in *Fusarium* Species. Toxicon 30: 1295-1298.

SIVANESAN, A. 1991. The taxonomy and biology of dematiaceous hyphomycetes and their mycotoxins. In: B.R. Champ, E. Highley, A.D. Hocking and J.I. Pitt (Eds.): Fungi and mycotoxins in stored products. ACIAR Proceedings No. 36. Australian Centre for International Agricultural Research, Canberra. pp. 47-64.

SIVASITHAMPARAM, K. & GHISALBERTI, E.L. 1998. Secondary metabolism in *Trichoderma* and *Gliocladium.* In: Kubicek, C.P. and Harman, G.E. (eds.) Trichoderma *and* Gliocladium, pp. 139-191. Francis and taylor, London.

STEYN, P.S. & RABIE, C.J. 1976. Characterisation of magnesium and calcium tenuazonate from *Phoma sorgina.* Phytochemistry 15: 1977-1979.

STEYN, P.S., TUINMANN, A.A., VAN HEERDEN, F.R., VAN ROOYEN, P.H., WESSELS, P.L. & RABIE, C.J. 1983. The isolation, structure, and absolute configuration of the mycotoxin rhizonin, a novel cyclic heptapeptide containing N-methyl-3-(3-furyl)alanine, produced by *Rhizopus microsporus*. J. Chem. Soc. Chem. Commun. 1983: 47.

TERAO, K., E. ITO, K. KAWAI, K. NOZAWA, & UDAGAWA, S. 1990. Experimental acute poisoning in mice induced by emestrin, a new mycotoxin isolated from *Emericella* species. Mycopathologia 112: 71-79.

THRANE, U. 1989. Fusarium species and their specific profiles of secondary metabolites. In: J. Chelkowski (Ed.) *Fusarium*. Mycotoxins, taxonomy and pathogenicity. Elsevier, Amsterdam, pp. 199-225.

UDAGAWA, S. 1984. Taxonomy of mycotoxin producing *Chaetomium*. In: H. Kurata and Y. Ueno (eds.): Toxigenic fungi - their toxins and health hazard. Elsevier, Amsterdam. pp. 139-147.

VESONDER, R.F., R. LAMBERT, D.T. WICKLOW, & BIEHL, M.N. 1988. *Eurotium* spp. and echinulin in feed refused by swine. Appl. Environ. Microbiol. 54: 830-831.

WICKLOW, D.T. 1987. Metabolites in the coevolution of fungal chemical defence systems. In: K.A. Pirozynski and D.L. Hawksworth (eds.): Coevolution of fungi with plants and animals. pp. 173-201. Academic Press, New York.

WILKINSON, S. & SPILSBURY, J.F. 1965. Gliotoxin from *Aspergillus chevalieri* (Mangin) Thom and Church. Nature 206: 619.

WILLIAMS, D.H., STONE, M.J., HAUCK, P.R., & RAHMAN, S.K. 1989. Why are secondary metabolites (natural products) biosynthezised?. Journal of Natural Products 52: 1189-1208.

WILSON B.J., D.T.C. YANG, & HARRIS, T.M. 1973. Production, isolation, and preliminary toxicity studies of brevianamide A from cultures of *Penicillium viridicatum*. Appl. Environ. Microbiol. 26: 633-635.

WILSON, B.J., C.S. BYERLY, & BURKA, L.T. 1981. Neurological disease of fungal origin in three herds of cattle. J. Amer. Vet. Med. Assoc. 179: 480-481.

WILSON, T., RABIE, C.J., FINCHAM, J.E., STEYN, P.S. & SCHIPPER, M.A.A. 1984. Toxicity of rhizonin A, isolated from *Rhizopus microsporus*, in laboratory animals. Food Chem. Toxicol. 22: 275-281.

WU, W., M.E. COOK, & SMALLEY, E.G. 1991. Decreased immune response and increased incidence of tibial dyschondroplasia caused by Fusaria grown on sterile corn. Poultry Science 70: 293-301.

YEULET, S.E., P.G. MANTLE, M.S. RUDGE, & GREIG, J.B. 1988. Nephrotoxicity of *Penciillium aurantiogriseum*, a possible factor in the aetiology of Balkan Endemic Nephropathy. Mycopathologia 102: 21-30.

LITERATURE ON MYCOTOXINS

ANONYMOUS. 1989. Mycotoxins: Economic and health risks. Council for Agricultural Science and Technology, Ames, Iowa.

ANONYMOUS. 1990. Selected mycotoxins: Ochratoxins, trichothecenes, ergot. Environmental health criteria 105. WHO, International programme on chemical safety, Geneva, 263 pp.

ARORA, D.K., MUKERJI, K.G. & MARTH, E.H. (eds.) 1991. Handbook of Applied Mycology. Vol. 3. Foods and feeds. Marcel Dekker, New York. 621 pp.

BENNETT, J.W. 1987. Mycotoxins, mycotoxicoses, mycotoxicology and Mycopathologia. Mycopathologia 100: 3-5.

BENNETT, J.W. & KLICH, M.A. (eds.) 1992. Aspergillus: Biology and industrial applications. Butterworth-Heinemann. 416 pp.

BETINA, V. 1989. Mycotoxins. Chemical, Biological and environmental aspects. Elsevier, Amsterdam. 438 pp.

BETINA, V. (ed.) 1993. Chromatography of mycotoxins. Techniques and applications. Elsevier, Amsterdam. 440 pp.

BHATNAGAR, D., LILLEHOJ, E. & ARORA, D.K. (eds.) 1992. Mycotoxins in ecological systems. Handbook of applied mycology Vol. 5. Marcel Dekker, New York. 443 pp.

CHAMP, B.R., HIGHLEY, E., HOCKING, A.D. & PITT, J.I. (eds.) 1991. Fungi and mycotoxins in stored products. Australian centre for International Agricultural Research, Canberra. 270 pp.

CHELKOWSKI, J. (ed.) 1989. *Fusarium.* Mycotoxins, taxonomy and pathogenicity. Elsevier, Amsterdam. 492 pp.

CHELKOWSKI, J. (ed.) 1991. Cereal grain. Mycotoxins, fungi and quality in drying and storage. Elsevier, Amsterdam. 607 pp.

CHELKOWSKI, J. & VISCONTI, A. (eds.) 1992. *Alternaria* - Biology, plant diseases and metabolites. Elsevier, Amsterdam. 573 PP.

CHRISTENSEN, C.M. (ed.) 1982. Storage of cereal grains and their products. American Association of Cereal Chemists, St. Paul, Minnesota, 544 pp.

COLE, R.J. and COX, R.H. 1981. Handbook of toxic fungal metabolites. Academic Press, New York, 937 pp.

EATON, D.L. & GROOPMAN, J.D. (eds.) 1993. The toxicology of aflatoxins. Academic Press, New York. 552 pp.

FRISVAD, J.C., BRIDGE, P.D. & ARORA, D.K. 1998. Chemical fungal taxonomy. Marcel Dekker, New York, 398 pp.

JAYAS, D.S., WHITE,N.D.G. & MUIR, W.E. (eds.) 1995. Stored grain ecosystems. Marcel Dekker, New York, 757 pp.

KELLER, R.F. & TU, A.T. (eds.) 1983. Handbook of natural toxins. Vol. 1. Plant and fungal toxins. Marcel dekker, New York, 934 pp.

KURATA, H. & HESSELTINE, C.W. 1982. Control of the microbial contamination of foods and feeds in international trade: Microbial standards and specifications. Saikon Publ. Company, Tokyo. 342 pp.

MARASAS, W.F.O. & NELSON, P.E. 1986. Mycotoxicology. Pennsylvania State University Press, University Park, 102 pp.

MILLER, J.D. & TRENHOLM, H.L. (eds.) 1994. Mycotoxins in grain, compounds other than aflatoxin. Eagan Press, St. Paul.

MIRAGLIA, M.,VAN EGMOND,, H.G., BRERA, C. & GILBERT, J (eds.) 1998. Mycotoxins and phycotoxins – Developments in chemistry, toxicology and food safety. Alaken Press, Denver.

MOREAU, C. (translated & additions by M.O. MOSS). 1979. Moulds, toxins and food. John Wiley and Sons, Chichester, 477 pp.

NATORI, S., HASHIMOTO, K. & UENO, Y. (eds.) 1989. Mycotoxins and phycotoxins 1988. Elsevier, Amsterdam. 496 pp.

POWELL, K.A., RENWICK, A. & PEBERDY, J.F. (eds.) 1994. The genus *Aspergillus:* from taxonomy and genetics to industrial application. Plenum, New York. 374 pp.

PURCHASE, I.H.F. (ed.). 1974. Mycotoxins. Elsevier, Amsterdam. 433 pp.

RICHARD, J.L. & THURSTON, J.R. (eds.). 1986. Diagnosis of mycotoxicosis. Martinus Nijhoff Publ., Dordrecht. 411 pp.

RODRICKS, J.V. (ed.). 1976. Mycotoxins and other fungal related food problems. American Chemical Society, Washington D.C. 409 pp.

RODRICKS, J.V., HESSELTINE, C.W. & MEHLMAN, M.A. (eds.) 1977. Mycotoxins in human and animal health. Pathotox Publisghers, Park Forest South, 807 pp.

SAUER, D.B. (ed.). 1982. Storage of cereal grain and their products. 4. ed. American Association of Cereal Chemists, St. Paul, Minnesota, 615 pp.

SCOTT, P.M., TRENHOLM, H.L. & SUTTON, M.D. 1985. Mycotoxins: Canadian perspective. NRCC Publication No. 22848, Ottawa. 185 pp.

SCUDAMORE, K.A. (ed.) 1993. Occurrence and significance of mycotoxins. Central Science Laboratory, Slough. 319 pp.

SHANK, R.C. (ed.) 1981. Mycotoxins and N-nitroso compounds: Environmental risks. CRC Press, Boca Raton, Florida. 235 pp.

SHARMA, R.P. & SALUNKHE, D.K. (eds.) 1991. Mycotoxins and phytotoxins. CRC Press, Boca Raton. 775 pp.

SINHA, K.K. & BHATNAGAR, D. 1998. Mycotoxins in agriculture and food safety.. Marcel Dekker, New York, 511 pp.

SMITH, J.E (ed.) 1994. *Aspergillus.* Plenum Press. 273 pp.

SMITH, J.E. & HENDERSON, R.S. (eds.) Mycotoxins and animal foods. CRC Press, Boca Raton.

SMITH, J.E. & MOSS, M.O. 1985. Mycotpoxins. Formation, analysis and significance. John Wiley and Sons, Chichester.

SMITH, J.E. & SOLOMONS, G.L. 1994. Mycotoxins in human nutrition and health. European Commission, Bruxelles, 300 pp.

VAN EGMOND, H.P. (ed.) 1989. Mycotoxins in dairy products. Elsevier Applied Science. 272 pp.

VAN EGMOND, H.P. (ed.) 1996. Mycotoxins and toxic plant components. Wiley Liss, London.

WATSON, D.H. (ed.) 1998. Natural toxicants in food. Sheffield Academic Press, Sheffield, 335 pp.

Chapter 6

MYCOTOXINS: DETECTION, REFERENCE MATERIALS AND REGULATION

HANS P. VAN EGMOND

National Institute of Public Health and the Environment, Bilthoven, The Netherlands

TABLE OF CONTENTS

INTRODUCTION

Mycotoxins are secondary metabolites of fungi, which are capable of producing acute toxic or chronic toxic effects in humans and animals. Toxic syndromes resulting from the intake of mycotoxins by humans and animals are known as mycotoxicoses. Mycotoxicoses have been known for a long time (Table 1). An example is ergotism, a disease characterized by necrosis and gangrene of the limbs and better known in the Middle Ages in Europe as "holy fire". This disease was caused by the ingestion of grain-contaminated with sclerotia of *Claviceps purpurea,* which contained toxic metabolites. Another mycotoxicosis recognized to have seriously injured human populations is alimentary toxic aleukia (ATA). The symptoms of ATA in humans take many forms, including leukopenia, necrotic lesions of the oral cavity, the esophagus, and stomach, sepsis, hemorrhagic diathesis, and depletion of the bone marrow. The disease was induced by ingesting overwintered mouldy grain and occurred in many areas in the Soviet Union, especially during World War II. The fungi responsible for these accidents belong to the genera *Fusarium* and *Cladosporium.* In Japan, toxicity associated with yellow coloured mouldy rice was a problem, especially after World War II, when rice had to be imported from various countries. The ingestion of "yellow rice" by humans caused vomiting, convulsions, and ascending paralysis. Death could also occur within l-3 days after the appearance of the first signs of disease. The toxin-producing fungi in yellow rice belong to the genus *Penicillium.* In the early 1950s the so-called Balkan Endemic Nephropathy (BEN) was described for the first time. BEN is a human kidney disorder, which was later associated with the occurrence of ochratoxin A in food. Ochratoxin A is formed by *Aspergillus* and *Penicillium* species. Despite these examples of mycotoxin-caused diseases in humans, mycotoxicoses remained "neglected diseases" until the early 1960's, when the attitude changed drastically due to the outbreak of "turkey X disease" Great Britain. Within a few months more than 100,000 turkeys died, mainly in East Anglia and southern England. This catastrophe led to a multidisciplinary investigation, which resulted in the presence of aflatoxins. The discovery of the aflatoxins greatly stimulated scientific interest in mycotoxins and mycotoxicoses. Nowadays hundreds of mycotoxins have been discovered.

Aflatoxins

The aflatoxins form a group of very toxic coumarin derivatives, produced mainly by two *Aspergillus* species, *A. flavus* and *A. parasiticus.* The aflatoxins can be formed on a number of substrates, such as oilseeds and nuts (including peanuts and pistachios), grain (maize), and certain subtropical fruits (e.g. figs). They are most frequently found in products that are not sufficiently dried after harvest and/or are stored at relatively high temperatures. Problems arise both in technologically less developed countries (the tropics) and in (warm) areas with highly developed agricultural technology (southern and sometimes Midwestern United States). Several of the aflatoxins, in particular aflatoxin B_1, (see Fig. 1), have strong carcinogenic properties. Many species of animals, such as rodents, birds, fish, and monkeys, develop liver tumours after ingestion of excessive amounts of aflatoxin B_1. Epidemiological studies suggest that aflatoxins can be responsible for cancer of the liver in humans in different parts of Africa and Asia, albeit in combination with hepatitis B virus.

There are also indications that kwashiorkor, a disease caused by protein deficiency, can be (partially) the result of exposure to aflatoxins.

Table 1. A historical view of mycotoxins

YEAR	TOXICOSIS	TOXIN	SPECIES
994	Holy fire	lysergic acid derivatives	*Claviceps purpurea*
1890	Cardiac beri-beri	citreoviridin	*Penicillium citreonigrum*
1913	Alimentary Toxic Aleukia	trichothecenes	*Fusarium sporotrichioides*
1952	Balkan Endemic Nephropathy	ochratoxin	*Penicillium verrucosum*
1960	Turkey X disease	aflatoxin	*Aspergillus flavus*
1989	Hole in the head syndrome	fumonisin	*Fusarium moniliforme*

Because of the world trade in foodstuffs and raw materials for foodstuffs, the risk to humans from aflatoxins is not restricted to the areas where the toxins are produced. Furthermore, trade in raw materials for animal foodstuffs also poses a potential threat to public health. Specifically, cows have been found to metabolise some of the aflatoxin B_1 in their food into aflatoxin M_1 (see Fig. 1), which is secreted in the milk and ends up in various dairy products. The carryover rate ranges from 1 to 6%. Aflatoxin M_1 is also a suspected carcinogen, although its carcinogenic effect is not as strong as that of B_1, but aflatoxin M_1 has also hepatotoxic and carcinogenic properties. Quantitatively considered, the toxicity of aflatoxin M_1 in ducklings and rats seems to be similar to or slightly less than that of aflatoxin B_1. The carcinogenicity is probably one to two orders of magnitude less than that of the highly carcinogenic aflatoxin B_1.

Over the years aflatoxins have been of continuous concern, also because lifetime exposure to aflatoxins in some developing countries has been confirmed by biomonitoring.

Ochratoxins

Ochratoxin is produced by *Penicillium verrucosum* and species of the *Aspergillus ochraceus* group. However (Abarca *et al.*, 1994; Ono *et al.*, 1995; Horie, 1995; Téren *et al.*, 1996, Wicklow *et al.*, 1996, Varga *et al.*, 1996; Heenan *et al.*, 1998) reported the occurrence of this mycotoxins in the black Aspergilli (*A. niger* and *A. carbonarius*).

The production of the ochratoxins depends on environmental conditions such as water activity (moisture) and temperature. For *A. ochraceus* high temperatures (12-37°C) are necessary, whereas psychrophillic Penicillia (4-31°C), particularly *P. verrucosum,* also produce ochratoxins in areas with a colder climate.

Ochratoxin A, formed from isocoumarin and phenylalanine (Fig. 1), is the major ochratoxin that has been found in a number of countries worldwide. This toxin causes nephropathy in experimental animals, and it has been associated with nephropathy in live stock animals. It is also highly immunotoxic in mice. In addition, it has been associated with Balkan Endemic Nephropathy and the occurrence of urogenital tract tumors in animals and possibly in humans. In the etiology of these human diseases other yet-unidentified factors seem to play a role. Due to the immunotoxic effects, the toxin could succumb to secondary infections. As a result of these serious diseases and the finding that ochratoxin A is highly carcinogenic in rats, much attention has recently been devoted to this toxin.

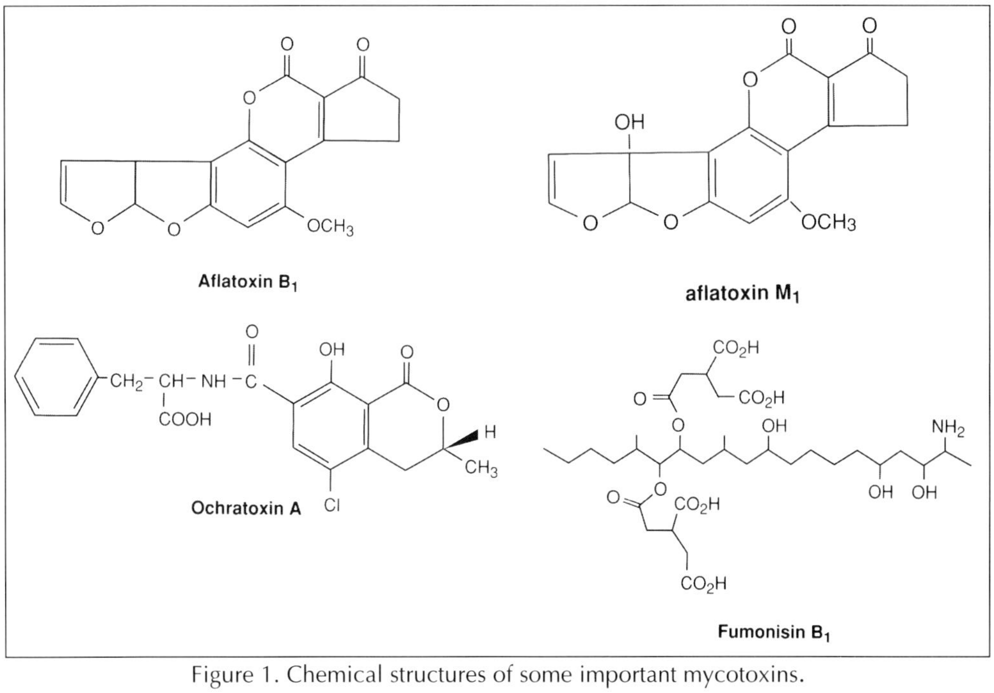

Figure 1. Chemical structures of some important mycotoxins.

The main route (>90%) of human exposure is through consumption of vegetable products contaminated with ochratoxin A (grains, nuts, rice, figs, coffee, wine, olives, beer), but it can also occur by transfer via an animal such as the pig. Pig blood and plasma are used in the preparation of various sausages, thus meat products can become contaminated with ochratoxin A.

Ochratoxin A has also been found in human blood (at levels of around l-2 µg/kg in about 50% of the samples in various western European countries) and in mother 's milk (at sub-µg/kg levels). The significance of ochratoxin A being found in human body fluids is not at the moment clear. Whereas it seems that ochratoxin A occurs in many agricultural products and sometimes at levels that would not comply with existing regulations, Bach and Bartsch (1993) concluded that the human health risk of ochratoxin A cannot yet be established. Similar as for the aflatoxins, a European Cooperative program recently resulted in a report in which provisional estimates of human exposure to ochratoxin A were given. The FAO/WHO Joint Expert Committee on Food Additives has established a provisional tolerable weekly intake (PTWI) level of 100 ng ochratoxin A per kg body weight. A working group of the Nordic countries and the Scientific Committee on Food both have proposed a much lower tolerable daily intake of ochratoxin A of 5 ng/kg body weight.

Trichothecenes

The trichothecenes are a group of chemically related compounds, mainly produced by *Fusarium* species. More than 80 trichothecenes and trichothecene metabolites are known, which are all constituted of a sesquiterpene nucleus with a characteristic epoxy group at the C_{12}-C_{13} position in the molecule. From the point of view of nutritional toxicology, only about 12 molecules can be found as effective contaminants in foods and feeds. (See Figure 1 for the chemical structure of T-2 toxin, an important representative of this group of naturally occurring trichothecenes). Growth of *Fusarium* species and toxin production can occur at relatively low temperatures, and therefore trichothecenes can be found in grains from the moderate climatic zones.

The trichothecenes may cause various adverse effects. Neurological disorders, immunosuppressive activities, gastrointestinal effects, and hemorrhages have been described in experimental animals and in some humans. In addition to T-2 toxin, deoxynivalenol (DON) and nivalenol (NIV) are considered important trichothecenes as they can be found virtually everywhere in the world where cereals are grown. DON and NIV have been associated (together with acetyl DON and T-2 toxin) in an outbreak of disease in Kashmir in 1987. Some 50,000 people were affected after consumption of bread made from rain-damaged wheat that contained several trichothecenes. The effects were generally mild and consisted of gastrointestinal tract symptoms for some days. DON and NIV are quite toxic to animals, particularly pigs. Occurrence of DON at levels above 1 mg/kg in pig feed may lead to feed refusal (DON = vomitoxin) and reduced weight gain. This may lead to significant economic damage in the feed industry. The emetic effect also occurs in humans, and vomiting is one of the earliest symptoms of trichothecenes intoxication. Residues of trichothecenes are not known to occur in animal products such as meat, milk, and eggs. They do survive some processing of food, however, as they have been detected in food and feed materials after the milling process, and they are not removed completely after food processing such as baking and boiling in water and oil. In a Canadian study, DON and NIV were found in 58% and 6%, respectively, of the investigated samples of Canadian and imported beer, although at low (µg/kg) levels.

Ergot Alkaloids

The fungus *Claviceps purpurea* grows parasitically on some grasses and several cereal crops, such as rye, wheat, and barley. The resulting sclerotia or "ergot bodies," which can hibernate, may contain a mixture of different alkaloids that have powerful pharmacological properties. These alkaloids consist of derivatives of lysergic acid.

Analytical studies of total and individual ergot alkaloids in grains indicate contamination of rye, wheat, and barley with ergometrine, ergosine, ergotamine, ergocornine, and ergocristine. In general ergocristine and ergotamine are most prevalent in these grains. Heat treatment of grain such as baking of bread leads to a reduction of the ergot alkaloids, which varies from 50 to 100% depending on the type of ergot alkaloids and the heating conditions. There are no data on the toxicokinetics of ergot alkaloids, although several short-term pharmacological experiments have been performed because of the medical use of ergot alkaloids in the treatment of migraine and postoperative trauma. The pharmacological actions of the ergot alkaloids are variable and complex: peripheral vasoconstriction, depression of vasomotor centers, peripheral adrenergic blockade, secretion of prolactin, and stimulation of uterine smooth muscles.

Ergotism in humans, in variable expression such as gangrene or neurological defects, was reported in the early Middle Ages. Even in recent years a few cases have occurred in Europe, of which the last reported was the poisoning of a part of the population of the village Saint Esprit in France. Although toxic effect of ergot alkaloids at very high consumption levels have a long history as the disease ergotism, little is known about the long-term intake of small amounts of ergot alkaloids.

Fumonisins

The fumonisins form a group of mycotoxins, produced by various species of *Fusarium* including *F. verticillioides (= F. moniliforme)*. These fungi are primarily plant pathogens, causing problems worldwide on various commodities such as maize, although this is more common in the tropical than in the temperate regions. The structures of several fumonisins have been elucidated, with fumonisin B_1 being the most significant in terms of occurrence and toxicity (see Fig. 1). The fumonisins occur mainly in maize and in commercial maize-based human foods.

Rapidly after their discovery in the late 1980s, it was discovered that they may lead to a variety of toxic effects in various animals and that they probably play a role in a certain human disease as well. In animals, the most dramatic effects have been observed in horses where leukoencephalomalacia (ELEM, or hole-in-the-head syndrome) is caused by fumonisins. In pigs fumonisins cause pulmonary edema, and in chickens they may lead to the so-called "spiking disease" (neurotoxic effects in young chickens, that are receiving a changing diet). Animal studies have shown that fumonisin B1 can induce hepatic cancer in rats. Epidemiological studies suggest that fumonisins could be responsible for human esophageal cancer in southern Africa and China. The variety of these adverse effects and the ubiquitous occurrence of fumonisins have given rise to intensive research efforts in the early 1990s involving hundreds of scientists in various disciplines, which has led to interesting discoveries. Some of the biological effects of fumonisins (those in horses and pigs) might be caused by their ability to block key enzymes involved in the biosynthesis of sphingolipids, which have many functions important for membrane integrity and physiological activity.

Detection of mycotoxins

The most recent advance in clean-up of extracts containing mycotoxins is the use of immuno-affinity cartridges. These columns are composed of monoclonal antibodies, specific for the toxin of interest, which are immobilized on Sepharose@ and packed into small plastic cartridges and can be incorporated in fully automated sample preparation systems that take the sample from the extraction stage through to completion of high-performance liquid chromatography (HPLC) in an unattended mode of action (Gilbert, 1991). In addition to HPLC, other chromatographic techniques exist to perform the ultimate separation and quantitation: thin-layer chromatography (TLC) and gas-liquid chromatography (GLC). TLC was very popular in the 1960s and the 1970s. Though still very valuable and widely applied in the developing countries, TLC was largely superseded during the 1980s by HPLC. Details of this technique are described elsewhere (van Egmond, 1984).

GLC has limited applications in mycotoxin analysis because it requires volatile components, whereas most mycotoxins are non-volatile. Some trichothecenes are derivatized to volatile components that can then be handled with GLC. Besides the chromatographic techniques, the immunoassay techniques are worth mentioning. In particular enzyme-linked immunosorbent assay (ELISA) has become an important technique in mycotoxin methodology. The simplicity of the ELISAs and the large number of samples that can be handled in one day have made these tests important, especially for screening and semi-quantitative determination. Their lack of selectivity might prevent their use as quantitative tools, and therefore they are less suitable for regulatory analysis.

Reference materials

The availability of reliable methods of analysis is no guarantee of accurate results. Sample programs for aflatoxins and ochratoxin A have shown that wide variability in results should be considered more the norm than the exception (van Egmond, 1994). The large scatter of analytical results is of little comfort to those people who must either pay for the measurements or base potentially important decisions upon them. This situation led the Community Bureau of Reference (BCR)of the European Commission (recently renamed as Standards Measurements & Testing Program) to initiate the Mycotoxin Program in 1982, with the objective of improving the accuracy and thereby the comparability of mycotoxin measurements. The BCR acts in cooperation with a group of European laboratories experienced in analysis for mycotoxins. A main achievement of the program is the development of (certified) reference materials.

Reference materials may serve several purposes (van Egmond *et al.*, 1998). First, they can be analyzed regularly as part of analytical quality assurance (QA) programs in connection with a control chart. The results obtained with the reference material will ideally show the part of the measurement that is moving out of control. Second, reference materials are very useful in method-validation processes. The objective here is essentially to see whether a method (as applied by a particular laboratory or analyst) gives a detectable systematic bias. The user makes several replicate measurements of the property and compares the observed mean with the certified value. The practical significance of any bias thus detected can only be judged in the context of the use of the measurement. For instance, for some mycotoxins, a large uncertainty may be acceptable if the true levels are much lower than the action limit(s).

Table 2. BCR Reference materials for mycotoxins

Reference sample	ready	development	considered
Aflatoxin M_1 in milk powder	x	x	
Aflatoxin M_1 in chloroform	x		
Aflatoxins in peanut butter	x		
Aflatoxins B_1 in peanut meal	x		
Aflatoxin B_1 in feedstuff	x		
Ochratoxin A in wheat	x		
DON in maize and wheat	x		
Zearalenone in maize		x	
Trichothecenes other than DON			x
Fumonisins			x

Regulations

Many countries have established measures to safeguard the health of consumers and the economic interests of producers and traders. Various factors may influence the decisions made by authorities to establish limits for certain mycotoxins. Among these are scientific and non-scientific factors, including:

- The availability of toxicological data. Without toxicological information there can be no hazard assessment, one of the basic ingredients of risk assessment.
- The availability of survey analytical data. These may indicate which commodities should be considered for regulatory action, and they provide a basis for exposure assessment, the other main ingredient of risk assessment.
- The distribution of mycotoxins over commodities. If such a distribution is non-homogeneous, as is the case with aflatoxins in peanuts, there is a good chance that the mycotoxin concentration in the batch to be inspected will be incorrectly estimated, due to the difficulties in representative sampling.
- The availability of methods of analysis. The enforcement of mycotoxin regulations is ultimately based on the ability of analysts to accurately identify and quantify these toxins.
- Legislation in other countries with which trade contacts exist. Unnecessary strict regulative actions may create difficulties for importing countries in obtaining supplies of essential commodities, whereas for exporting countries difficulties may arise in finding markets for their products.
- Sufficient food supply. The regulatory philosophy should not jeopardize the availability of some commodities at reasonable prices. This is especially important in the developing countries.

Weighing the various factors that play a role in the decision-making process of establishing mycotoxin tolerances is not easy. Nevertheless, mycotoxin regulations have been established in many countries in the past decades. In the 1980s there were a few general surveys on the mycotoxin regulations that exist worldwide (van Egmond, 1989) and on the rationales for these regulations (Stoloff *et al.,* 1991) Some more specific investigations concerned mycotoxins regulations in the European Community and the United States (Gilbert, 1991), mycotoxin regulations in Asia and Africa (van Egmond, 1991a), and worldwide regulations for ochratoxin A (van Egmond, 1991b),

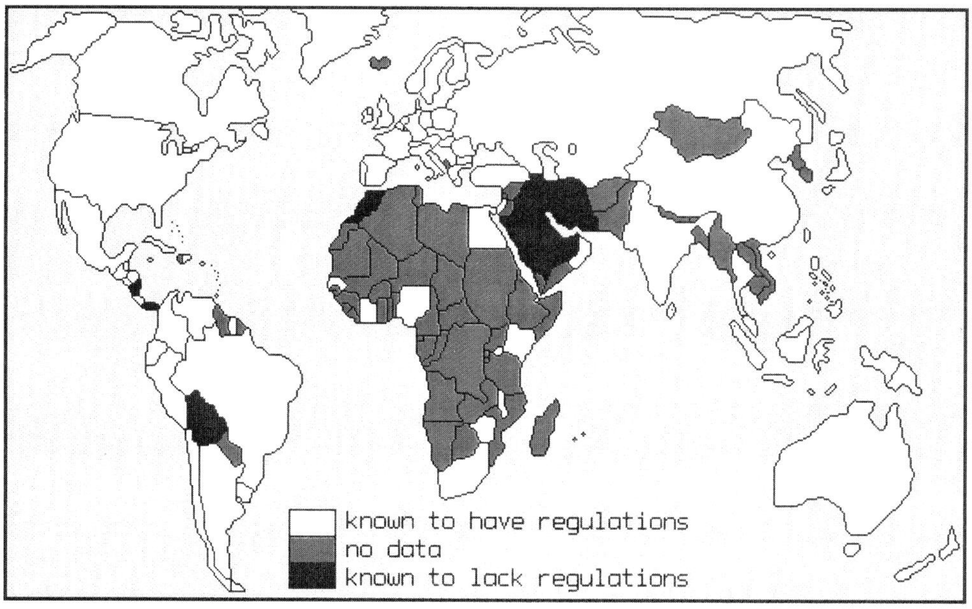

Figure 2. Countries known to have regulations (white), unknown to have regulations (grey) and unknown to have regulations (black) for mycotoxins in food and feedstuffs.

Currently there are at least 78 countries (see Figure 2) known to have specific regulations or detailed proposals for regulations on mycotoxins. Most of the existing mycotoxin regulations concern aflatoxins, and, in fact, all countries with mycotoxin regulations have tolerances for aflatoxins in foods and/ or animal feedstuffs. Less frequently, specific regulations also exist for patulin and ochratoxin A, and for deoxynivalenol, zearalenone, T-2 toxin, chetomin, stachybotryotoxin, phomopsin, ergot and fumonisins. An update on worldwide mycotoxin regulations appeared as an FAO Food and Nutrition Paper (FAO, 1997). In the countries surveyed worldwide some regulated only aflatoxin B_1, while others regulated the sum of aflatoxins B_1, B_2, G_1, G_2 and yet others also regulated aflatoxin M_1, in dairy products. With regard to acceptable levels in foods for aflatoxins B_1, alone, a majority of the countries adopted a level of 5 μg/kg. In those countries that apply limits for the sum of the aflatoxins, such a uniformity in tolerance values does not occur. Most of these tolerances were in the range l0-30 µg/kg.

In the case of aflatoxin M_1, in liquid milk, two groups of countries each adopted levels of 0.05 and 0.5 µg/kg, respectively, with some other jurisdictions taking zero, 0.1, and 1.0 µg/kg as their respective limits. The divergence between the 0.05 and 0.5 µg/kg levels is indeed striking, the former prevailing in several Western European countries, including the EU. This low level in these European countries has in turn resulted in the fairly stringent regulation of aflatoxin B_1, in complementary feedstuffs for dairy cattle in the EU. The acceptable level for aflatoxin B_1, in such feedstuffs was reduced from 20 to 10 µg/kg in 1984, and further tightened to 5 µg/kg by the Commission of the European Committees (CEC (1991). Investigations on the rationale for regulations on mycotoxins in human food and animal feed yielded rather meagre results (Stoloff *et al.,* 1991). Most of the responses concerned limits for aflatoxins in food, and most of these were based on a vague, unsupported statement of the carcinogenic risk for humans. There was a general consensus that exposure to a potential human carcinogen that could not be totally avoided should be limited to the lowest practical level, the definition of "practicality" depending on whether the country was an importer or producer of the potentially contaminated commodity. Some countries made a claim to a hazard evaluation, although specific details were rather scarce.

Prevention and Elimination

The prevention of mycotoxin formation is still a major problem, not only in the developing countries, but also in the industrialized part of the world, such as in the United States. The best way of preventing mycotoxins in food and animal feed is to reduce fungal growth in the agricultural commodity. Several practical strategies are possible to manage the problem of mycotoxigenic fungi. They can be divided into three classes, as discussed in the following sections.

The Development of Resistant Cultivars. Cultivars have been developed with resistance against many important plant pathogens, including resistance to *Fusarium* head blight. The use of such cultivars will eliminate trichothecenes as deoxynivalenol and nivalenol from the food supply. Less successful were the attempts to produce maize and peanuts, resistant to aflatoxins. The problem is that *A. flavus* is not a plant pathogen per se and hence there is no range of plant response to work with. The development of resistant cultivars should be carefully guarded, because of the toxicological implications that the resistance factors themselves may have for the consumer.

Reduction in Levels of Toxigenic Fungi. This can be achieved by the use of fungicides and insecticides (where insects play a major role as vectors of toxigenic fungi) and the use of naturally occurring antifungal and anti-insect agents (phytoalexins). The use of fungicides and insecticides may be hindered in practice by environmental concerns, whereas the phytoalexins may have toxicological implications. A new approach is the use of specific non-toxigenic fungi as competitors, a technique finding increasing application in plant pathology. Promising pilot studies with non-toxigenic strains of *A. flavus* and *A. parasiticus* have been carried out with peanuts (Cole, 1994).

Prevention of Fungal Growth. Fungal growth and toxin production largely depends on physical factors such as water activity (a_w) and temperature. Maintenance of foods below 0.7 a_w and at low temperatures is generally effective to control fungal spoilage and mycotoxin production. Good storage practice cannot always prevent the problem of mycotoxin formation, because mycotoxins are sometimes produced before or immediately after harvest. If preventive measures have failed, the last possibility to avoid occurrence of mycotoxins in the food is to eliminate the toxins. In principle mycotoxin-contaminated consignments of foods and feeds may be decontaminated by removing the mycotoxin (segregation) or by converting the mycotoxin to a non-toxic form (degradation). Degradation may be achieved by physical, chemical, or biological means (Smith and Moss, 1985)

Attempts have been made to degrade mycotoxins in food and feed by applying physical treatments such as heat, microwave, gamma rays, x-rays, UV light, and adsorption. In most cases heat treatment does not work, and neither does irradiating. Adsorption of aflatoxins from animal feedstuffs to hydrated sodium calcium alumino silicate has gained some application in the feedstuff industry in an attempt to reduce the aflatoxin M_1, content in the milk (Harvey *et al.,* 1991). Most of the chemical procedures to degrade mycotoxins have been developed for aflatoxins in animal feedstuffs. They are usually based on the application of oxidizing agents, aldehydes, acids, and bases to destroy the aflatoxins. The chemical detoxification reagent that has attracted the widest interest is ammonia. both as an anhydrous vapor and as an aqueous solution. Treatment of aflatoxin B_1, with ammonia leads to opening of the lactone ring of the molecule and further reaction may occur depending on the processing conditions. Ammoniation of agricultural commodities leads to decomposition of approximately 95-98 % of the aflatoxins present. Ammoniation is used in various countries for the decontamination of animal feedstuffs. However, formal approval by the U. S. Food and Drug Administration as well as the European Commission is still lacking. FDA is developing a compliance policy guide (CPG) n ammoniation in feed. The CPG would probably limit aflatoxin and reaction products after ammoniation to about 9 µg/kg. The European Commission is currently sponsoring a research project in which a dozen European institutes, commercial companies, and universities participate. The project is aimed at obtaining more insight into the safety of animal products from animals that have been fed diets that have undergone chemical decontamination. In addition, this project should lead to more knowledge about possible decomposition products of aflatoxin B_1, which could be suitable markers to check the efficiency of decontamination processes or to prove that feedstuff lots were decontaminated. In addition to the physical and chemical attempts to get rid of mycotoxins, there is also the possibility of biological methods, e. g., procedures are under development to degrade aflatoxins in feedstuffs with the help of the bacterium *Flavobacterium auranthiacum,* but so far not much has been published in this respect. Thus, in principle there are several possibilities of getting rid of mycotoxins, but each has its disadvantages, and it certainly holds here that prevention is better than cure.

REFERENCES

ABARCA, M.L., BRAGULAT, M.R., CASTELLA, G. & CABANES, F.J. 1994. Ochratoxin production by strains of *Aspergillus niger* var. *niger.* Appl. Environ. Microbiol. 60: 2650-2652.

COMMISSION OF THE EUROPEAN COMMUNITIES 1991. Commission Directive of 13 February 1991 amending the Annexes to Council Directive 74163 on undesirable substances and products in animal nutrition. Official Journal of the European Communities 60: 16.

CREEPY, E.E., LORSKOWSKI, G., ROSCHENTHALER, R. & DIRHEIMER, G. 1982. Kinetics of the immunosuppressive action of ochratoxin A on mice, Proc. V. Int. IUPAC-Symposium on Mycotoxins and Phycotoxins, 1-3 September 1992. Vienna, p. 289. .

FOOD & AGRICULTURE ORGANIZATION 1997. Worldwide Regulations for Mycotoxins 1995. A Compendium, Food and Nutrition Paper 64, FAO, Rome.

GILBERT, J. 1991a. Accepted and collaboratively tested methods of sampling, detection, and analysis of mycotoxins. Fungi and mycotoxins in stored products, Proceedings of an international conference, Bangkok, Thailand, 23-26 April 1991 (Champ, B.R., Highly, E., Hocking, A.D. and Pitt, J.I. eds.), Aciar Proceedings 36, p. 194.

GILBERT, J. 1991b. Regulatory Aspects of Mycotoxins in the European Community and the USA, Fungi and Mycotoxins in Stored Products. Proceedings of an international conference, Bangkok, Thailand, 23-26 April 1991 (B. R. Champ, E. Highly, A. D. Hocking and J. I. Pitt, eds)Aciar Proceedings 36, p. 194.

HARVEY, R.B., PHILLIPS, T.D., ELLIS, J.A., KUBENA, L.F., HUFFARD, W.E. & PETERSEN, H.D. 1991. Effects on aflatoxin M, residues in milk by addition of hydrated sodium calcium aluminosilicate to aflatoxin contaminated diet of dairy COWS. Am. J. Vet. Res. 52: 1556.

HEENAN, C.N., SHAW, K.J. & PITT, J.I. 1998. Ochratoxin A production by *Aspergillus carbonarius* and *Aspergillus niger* isolates and detection using coconut cream agar. J. Food Mycol. 1: 67-72.

HORIE, Y. 1995. Productivity of ochratoxin A of *Aspergillus carbonarius* in *Aspergillus* section *Nigri.* Nippon Kingakkai Kaiho 36: 73-76.

ONO, H., KATAOKA, A., KOAKUTSU, M., TANAKA, K., KAWASUGI, S., WAKAZAWA, M., UENO, Y. & MANABE, M. 1995. Ochratoxin A producibility by strains of *Aspergillus niger* group stored in IFO culture collection. Mycotoxins 41: 47-51.

STOLOFF, L., VAN EGMOND, H.P. & PARK, D.L. 1991. Rationales for the establishment of limits and regulations for mycotoxins, Food Addit. Contum. 8: 213.

TÉREN, J., VARGA, J., HAMARI, Z., RINYU, E. & KEVEI, F. 1996. Immunochemical detection of ochratoxin A in black *Aspergillus* strains. Mycopathologia 134: 171-176.

VAN EGMOND, H.P. 1984. Determination of mycotoxins, Developments in FoodAnalysis Techniques III (R. D. King, ed.), Elsevier Applied Science Publishers, Ltd Barking, Essex, England, p. 99.

VAN EGMOND, H.P. 1989. Current situation on regulations for mycotoxins. Overview of tolerances and status of standard methods of sampling and analysis, Food Addit, Contam. 6: 139.

VAN EGMOND, H.P. 1991a. Regulatory Aspects of Mycotoxins in Asia and Africa. Fungi and mycotoxins in stored products. Proceedings of an international conference, Bangkok, Thailand, 23-26 April 1991 (B. R. Champ, E. Highly, A. D. Hocking, and J. I. Pitt, eds.), Aciar Proceedings 36, p. 198.

VAN EGMOND, H.P. 1991b. World-wide regulations for ochratoxin A, Mycotoxins, Encemic Nephropathy and Urinary Tract Tumours (M. Castegnaro, R. Plestina, G. Dirheimer, I. Chemozemsky, and H. Bartsch, eds.), International Agency for Research on Cancer, Lyon, France, p. 57.

VARGA, J., KEVEI, E., RINYU, E. & TÉREN, J. & KOZAKIEWICZ, Z. 1996. Ochratoxin production by *Aspergillus* species. Appl. Environ. Microbiol. 62: 4461-4464.

WICKLOW, D.T., DOWD, P.F., ALFTAFTA, A.A. & GLOER, J.B. 1996. Ochratoxin A: an antiinsectan metabolite from the sclerotia of *Aspergillus carbonarius* NRRL 369. Canad. J. Microbiol. 42: 1100-1103.

For more literature on mycotoxins see page 331.

Chapter 7

SPOILAGE FUNGI IN THE INDUSTRIAL PROCESSING OF FOOD

R.P.M. SCHOLTE[1], R.A. SAMSON[2] and J. DIJKSTERHUIS[2]

[1]*Codi International, P.O.Box 417, 3900 AK Veenendaal, and* [2]*Centraalbureau voor Schimmelcultures, Utrecht*

TABLE OF CONTENTS

1. INTRODUCTION

Literature on the fungal spoilage of food usually deals with the characteristics of the fungi or the (food) parameters that influence their growth, survival or elimination. On the role of spoilage fungi in food processing, information is often scattered. In case of spoilage incidents, data on the underlying process are frequently scarce or absent and much information is likely to remain unpublished.

In spite of the difficulties in retrieving data, this review is meant to discuss the fungal spoilage of foods from a processing point of view. Instead of providing a comprehensive overview of data on a wide variety of products, processes and fungi, five general processing subjects are discussed in depth.

- *Heating.* The effectiveness of heat in eliminating fungi - cells, mycelium, conidia and ascospores - is reviewed. Special attention is paid to data on fruit and dairy products and to soft drinks.
- *Plant and processes.* The literature information on fungal contamination in the processing of fruit, vegetable, dairy and bakery products is summarized. The role of *Geotrichum candidum* in food processing is discussed.
- *Air* may be a threatening reservoir of fungi, especially in (aseptic) processes in which even low numbers of fungi are unacceptable. The relevance of airborne fungi is discussed.
- *Packaging materials*, especially primary packaging materials, may also be a source of fungal contamination. Three most common methods of packaging decontamination are outlined.
- *Modified Atmosphere* (MA) packaging may play an important role in preservation, especially if a satisfying elimination of spoilage fungi is not achievable or (re)contamination can not be sufficiently excluded. The principles of MA-packaging and some applications are explained.

In order to effectively prevent the fungal spoilage of industrial foods, knowledge on fungal physiology, on raw material-, food- and processing characteristics and on the conditions of distribution and use is required. Often this knowledge is spread over different departments or disciplines and barriers between those need to be broken-down. In addition, this knowledge is not collected before spoilage incidents occur. Hopefully, this review will contribute to the distribution of knowledge and to the use of this in product design.

In the literature fungal names are used which are incorrect or based on wrong identification. In some cases a change of nomenclature have occurred e.g. *Talaromyces macrosporus*, mostly including the heat resistant strains of *T. flavus*. In this chapter the names are corrected where possible. If the identity was not certain the names of the original article have been used.

2. HEAT RESISTANT FUNGI

2.1 Introduction

Heat resistant fungi are characterized by the survival of living cells after a (short) heat treatment (of food products). In many studies the ascospores formed by these fungi are regarded as the heat resistant vehicle that even need a heat treatment as activation before germination (for example, Lingappa and Sussman, 1959, Katan, 1975, Beuchat, 1986.). For food science these fungi are of interest because of spoilage of canned and pasteurized fruit products. Fungi that cause damage for millions of dollars in the fruit-juice branch are *Byssochlamys nivea (fulva), Talaromyces flavus (macrosporus), Neosartorya fischeri, Eupenicillium brefeldianum.* These are soil-borne fungi and fruit that develops in contact with soil (like strawberries) are more prone to contamination. Tournas (1994) reviews the relevance of these fungi for food industry covering spoilage and metabolite formation by these fungi.

2.2. Heat-resistance of yeasts

Vegetative yeast cells usually have little heat resistance. In heating experiments in beer (pH 4.0, 5% ethanol), 6 out of 7 wild yeasts - *Saccharomyces carlsbergensis, Hansenula anomala* and *Pichia membranaefaciens* among others - possessed D_{49}-values of less than 2 min (a D-value expresses the time needed to kill 90% of a population of cells at a certain temperature). A heat-resistant strain of *Saccharomyces willianus,* had a D_{49} of approx. 15 min (Tsang and Ingledew, 1982). For *Zygosaccharomyces bailli, Saccharomyces bisporus, H. anomala* and *Zygosaccharomyces rouxii* cells a decimal reduction could be achieved by heating for 1 min in a buffer (400 g/l sucrose, pH 5.5) at temperatures of 56, 55, 54 and 50° C, respectively (Baggerman and Samson, 1988). Similar results have been reported for cells of five yeast species, that were heated in grapefruit serum (Parish, 1991). Sucrose, sodium chloride and glycerol may provide some protection towards heating. Sorbate and benzoate, on the other hand, usually reduce the heat resistance of the yeast cells (Beuchat, 1981; Agab and Collins, 1992).

In low moisture environments heating is much less effective. When heated on aluminium foil at a relative humidity of 33-38%, *P. membranaefaciens* and *Rhodotorula rubra* cells showed little survival after 5 min at 110° C (Scott and Bernard, 1985). With D_{110}-values of 1.3, 1.8, 2.9 and 3.6 min, respectively, *Debaromyces hansenii, Kloeckera apiculata, Lodderomyces elongisporus* and *H. anomala* cells were more resistant. The highest resistance was found for *Torulopsis glabrata* (D_{135}=0.12 min) and for three *Saccharomyces* strains (D_{135}=0.5-0.9 min).

2.3. Heat-resistance of fungal survival structures

Fungal survival structures can be regarded as more or less heat resistant compared to vegetative cells. For example, conidia of *Aspergillus niger, Penicillium chrysogenum, Wallemia sebi, Eurotium rubrum* and *P. glabrum,* show a decimal reduction in phosphate buffer at temperatures between 56 and 62° C (Baggerman and Samson, 1988). For *P. roqueforti, P. expansum, P. citrinum* and *A. flavus* strains that belonged to the spoilage flora of bakery products. D-values between 3.5 and 230 min were found for conidia heated at 54-56° C (Bröker *et al.,* 1987a; Spicher, 1991). The heat resistance proved to be dependent on the age of spores/conidia at the time of harvest and the composition of growing and heating media. Comparable results were obtained by Doyle and Marth (1975a) for conidia of *A. flavus* and *A. parasiticus.*

Figure 1. Germinating ascospores of *Talaromyces macrosporus.* Bar = 10 μm

Yeasts ascospores are often relatively heat resistant compared to the vegetative state of the organism. In an investigation of 20 yeast strains from soft drinks and fruit products (mainly *Saccharomyces cerevisiae, S. bailii* and *S. chevalieri),* the D_{60}-values of ascospores were 25-350 times higher than those of the corresponding vegetative cells (Put and de Jong, 1980). For *S. cerevisiae, S. chevalieri* and *S. bailii* ascospores, respectively, D_{60}-values of 22.5, 13 and 10 min were found in a pH 4.5-buffer (Put and de Jong, 1982).

In general, ascospores of filamentous fungi are more heat resistant than mycelia and conidia and more resistant than yeast ascospores. Ascospores of heat resistant fungi have D_{90} values of several minutes (many examples are given in Table 1). Generally spoken, a variety of fungal survival structures (e.g. conidia, sclerotia, chlamydospores and ascospores) exhibit heat resistance in a temperature area that is in the majority of the cases between 55 and 95º C.

Just within this range a small number of fungi can grow; thermophilic fungi perform active metabolic processes at higher temperatures. Examples of their habitats are haystacks, compost piles or birds nests. The optimal growth temperature of these fungi ranges from 40° to 55°C. The highest temperature growing thermophilic fungi can stand is approximately 60°C. Furthermore most of these fungi do not grow beneath temperatures of 20° to 30°C. The most widely spread thermophilic fungi are *Humicola insolens* and *Thermomyces (Humicola) lanuginosus.* Other species are *Chaetomium thermophile* and the zygomycete *Rhizomucor pusillus.* (Cooney D.G. & Emerson R., 1964). However some species such as *Paecilomyces variotii* can also grow below 20° C. The pathogen *Aspergillus fumigatus* grows in a broad range of temperatures (12-50°C) and strictly spoken is not regarded as a thermophilic fungus. It is however isolated from the same niche. The fungus is related to the ascomycete *Neosartorya fischeri* (that forms heat-resistant ascospores) as is judged by DNA complementarity (40-70%, Girardin *et al.,* 1995; Geiser *et al.,* 1998) .

2.4. Measurements on heat resistance of ascospores. Many studies were carried out on the heat resistance of ascospores of *Byssochlamys, Talaromyces* and *Neosartorya* species in which this stress tolerance is measured among different conditions. *B. nivea* was already known from the 1930s (Oliver and Smith, 1933). In 1963 a heat resistant *Aspergillus* (teleomorph, *N. fischeri*) was reported isolated from canned strawberries (Kavanagh *et al.,* 1963) Two heat resistant *Penicillium* species were isolated from flash-pasteurized apple juice (van der Spuy *et al.,* 1975) which became later known as *Talaromyces (flavus* and *macrosporus).* These three species appear as the most resistant species described till now, but in addition other fungal species are recognised as being heat resistant (e,g. *Talaromyces trachyspermus,* Enigl *et al.,* 1993). In addition other fungal structures as thick-walled hyphal fragments, chlamydospores (*Paecilomyces variotii, Fusarium spp.,* (Samson, unpublished data) and sclerotia (e.g. in *Eupenicillium*) can survive heat treatments (Splittstoesser and King, 1984).

A reproducible measurement of the heat resistance of ascospores however, is not that easily realised; throughout different studies various means are used. These can include small sealed vessels that are plunged into water or oil baths (King and Whitehand, 1990), a two phase slug flow heat exchanger (King, 1997) or spiral steel capillary tubes (Engel and Teuber, 1991) for a quick heating of the ascospore suspension. In several other studies a quick heating up of the cells is reached by adding a small aliquot of ascospore suspension to a larger preheated volume.

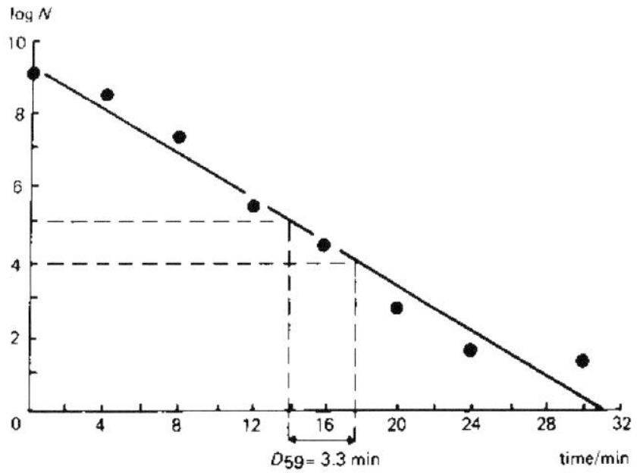

Figure 1. Survival curve of *Aspergillus niger* conidia on heating at 59° C for different periods of time.

Then, how is the heat-resistance expressed most accurately? The heat inactivation of a homogeneous population of micro organisms can be described with a D- and a Z-value. The *D-value,* the decimal reduction time, is the time (usually in min) that is needed to inactivate 90% of the micro organisms at a given temperature. At standardized heating conditions this *D-value* is supposed to be typical for the micro organism that is investigated and the heating medium that is used. Ideally, a linear curve will be obtained when the number of surviving micro organisms, on a logarithmic scale, is plotted against time (Fig. 1).

The viable count of a suspension of conidia of *Aspergillus niger* is reduced to 10% after a heat treatment for 3.3 min at 59° C (Fig. 1.; Baggerman and Samson, 1988).

This is expressed as a D_{59} for these cells of 3.3 min. The log (survivors) vs time plot of heat treatment in many cases is not linear (Put and de Jong, 1982; Splittstoesser and King, 1984, King and Halbrook, 1987) which has consequences for a simple extraction of D-values from the raw data. Alderton & Snell (1970) have worked out a method for this for bacterial spores, which is applied to heat resistant ascospores by King *et al.* (1979).

Apart from non-linear survivor plots, the so-called broken curve is observed (Bayne and Michener, 1979, Casella *et al.,* 1990) were very small numbers of spores seem to be extra resistant.

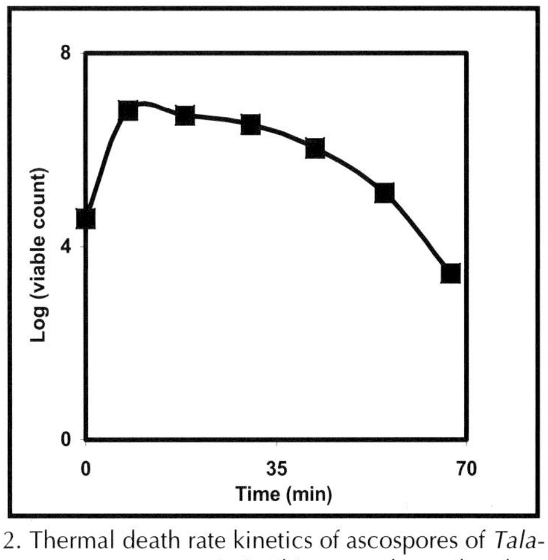

Figure 2. Thermal death rate kinetics of ascospores of *Talaromyces macrosporus* at 85º C. This curve shows that the spores are activated by short heat treatments. Inactivation of spores is non linear.

A *D-value* only provides information on the heat inactivation at the temperature that is stated. To compare heat inactivation rates at different temperatures, a second parameter, the z-value, is used. It denotes the increase or decrease in temperature (° C, F) needed, respectively, to decrease or to increase the *D-value* by a factor of 10. For *A. niger* conidia for example, a D_{59} of 3.3 min with a z-value of 4.9° C implies a D_{54} of approximately 33 min (Baggerman and Samson, 1988). Z-values can be calculated from the slope of a linear curve in which *D-values* are plotted on a logarithmic scale against temperature. Usually such curves are linear only in a limited range of temperatures. Thirdly, comparison (especially between species) is made difficult by the state of the ascospore suspension itself. Viable counts of suspensions will show deviations of thermal death kinetics when the cell population is not homogeneous. *Byssochlamys* (anamorph *Paecilomyces*) for instance, forms ascospores, conidia, chlamydospores and hyphal fragments in one culture. Further, asci must be broken effectively to measure proper thermal death kinetics of single ascospores. Michener and King (1974) use a French press treatment for the particular resistant asci of *Byssochlamys*. Ideally, spores must not adhere to each other in suspensions for an accurate viable count analysis. In many studies, the stain Bengal Rose (typically 10 mg/L) is added to the agar medium. The stain causes some more distinct boundaries of the fungal colony (King and Halbrook,1987) and in time a nice purple dot is formed beneath the centre of the colonies; these factors facilitate the counting of the colonies.

2.5. Heat inactivation of ascospores

The gross majority of inactivation studies is done with *B. nivea*, *B. fulva*, *N. fischeri* and *Talaromyces* spec. (*flavus* and *macrosporus*). The strains used in the studies on *N. fischeri* are most probably *N. glabra*, *N. spinosa* or *N. pseudofischeri* (see also page 44-47).

Tabel 1. Heat-resistancy of ascospores at different temperatures and medium compositions.

Fungal species	D-value (min)	Medium	Reference
Byssochlamys fulva	86º C 13-14	Grape Juice	Michener and King (1974)
	90º C 4-36 (log3 reduction)	Buffer pH 3.6 and 5.0, 16°Brix	Bayne and Michener (1979)
	90°C 8.1	Tomato juice	Kotzekidou (1997)
Byssochlamys nivea	85°C 1.3-4.5	Buffer pH 3.5	Casella *et al.* (1990)
	88°C 8-9 sec	Ringer solution	Engel and Tueber (1991)
	90°C 1.5	Tomato juice	Kotzekidou (1997)
Eurotium herbariorum	70°C 1.1-4.6	Grape Juice, 65°Brix	Splittstoesser *et al.* (1989)
Monascus ruber	80° C 2.5		S. Panagou (pers. Commun.)
Neosartorya fischeri	85°C 13.2	Apple Juice	Conner and Beuchat (1987b)
	85°C 10.1	Grape Juice	Conner and Beuchat (1987b)
	85°C 10-60	in ACES-buffer, 10 mM, pH 6.8	Unpublished data CBS 133.64
	85°C 10.4	Buffer pH 7.0	Conner and Beuchat (1987b)
	85°C 35.3	Buffer pH 7.0	Rajashekhara et el. (1996)
	88°C 1.4	Apple Juice	Scott and Bernard (1987)
	88°C 4.2-16.2	Heated fruit fillings	Beuchat (1986)
	90°C 4.4-6.6	Tomato Juice	Kotzekidou (1997)
	91°C < 2	Heated fruit fillings	Beuchat (1986)
Neosartorya pseudofischeri	95° C 20 sec		Unpublished data
Talaromyces flavus	85°C 39	Buffer pH 5.0, glucose, 16º	King (1997)
(macrosporus)	85º C 20-26	Buffer pH 5.0, glucose	King and Halbrook (1987)
	85º C 30-100	in ACES-buffer, 10 mM, pH 6.8	Unpublished data
	88°C 7.8	Apple Juice	Scott and Bernard (1987)
	88°C 7.1- 22.3	Heated fruit fillings	Beuchat (1986)
	90°C 2-8	Buffer pH 5.0, glucose	King and Halbrook (1987)
	90°C 6.2	Buffer pH 5.0, glucose	King (1997)
	90°C 6.0	Buffer pH 5.0, glucose Slug flow heat exchanger	King (1997)
	90°C 2.7-4.1	Organic acids	King and Whitehand (1990)
	90°c 2.5-11.1	Sugar contant (0-60º Brix)	King and Whitehand (1990)
	90°C 5.2-7.1	PH 3.6-6.6	King and Whitehand (1990)
	91°C 2.1-11.7	Heated fruit fillings	Beuchat (1986)
T. trachyspermus	85° C 45 sec		Unpublished data
Xeromyces bisporus	82.2° C 2.3		Pitt and Hocking (1982)

In Table 1 several examples are given of heat resistant Ascomycetes, which is generally between 1.5-11 min at 90º C. Heat resistance of ascospores varies with the composition of the external heating medium. The most studies are on the presence of sugars which protects cells, the influence of acidity and the presence of organic acids. In addition, ascospores of *N. fischeri* are sensitized by sulphur dioxide (Splittstoesser and Churey, 1991).

Very clearly, the presence of sugars (including glucose, sucrose or fructose) inside the heating medium enhances the survival of ascospores after heat treatment for all three fungal genera (Splittstoesser and Splittstoesser, 1977, King and Whitehand, 1990, Beuchat, 1988a).

The pH of the medium influences heat resistance and the different species may show some specificity in this respect. *B. fulva* showed maximum heat resistance within a pH range between 2.5 and 4.5 while *N. fischerii* showed no changes in resistance against 80º C between pH 3 till at least 5.5 (Splittstoesser and Splittstoesser, 1977). *T. macrosporus* showed a slight increase of its heat resistance between pH 3.6 and 6.6 (King and Whitehand, 1990).

Thirdly, several organic acids counteract heat resistance of ascospores, but only at low pH's (lower than 4). This is most prominent for fumaric acid which influences *B. nivea* (Splittstoesser and Splittstoesser, 1977), *N. fischeri* (Conner and Beuchat, 1987a) and *T. flavus* (Beuchat, 1988b). Benzoic and sorbic acid had also clear effects on *T. flavus* and *N. fischeri* (Beuchat, 1988b, Rajashekara *et al.,* 1998). Tartaric and malic acid however are reported to enhance heat resistance of *B. fulva,* but had no effect on *T. flavus* (together with citric acid). King and Whitehand (1990) report minimal effects of citric, lactic, tartaric and malic acid on ascospores of a *T. macrosporus* strain. Acetic acid had small effects on this fungus. *N. fischeri* ascospores were also influenced by citric acid and acetic acid (besides fumaric acid, Conner and Beuchat, 1987a). In the latter study the authors state that pH, type, molarity of the undissociated form of the organic acid act synergistically with heat to inactivate ascospores.

The presence of organic acids may account for remarkable differences observed in heat inactivation in several fruit juices. *N. fischeri* is more heat resistant in apple juice compared to grape juice (Conner and Beuchat, 1987a) and both *B. fulva* and *N.fischeri* are much more heat sensitive in cranberry juice compared to grape, apple or tomato juice (Splittstoesser and Splittstoesser, 1977). Combinations of different factors may lead to ununderstood variations in heat resistance; spores of *N. fischeri* suspended in 0.1 M phosphate buffer (pH 7.0) exhibit a far more higher heat resistance than spores suspended different grape jellies with large amounts of cane sugar at pH 3.1-3.3 (Beuchat and Kuhn, 1997). Spores of both *B. nivea* and *fulva* and *N. fischeri* are approx. twice as heat resistant in tomato juice (pH of 4.2) as in phosphate buffer (pH 7.0, 0.1 M, Kotzekidou, 1997).

2.6. Culturing related factors that influence heat resistance

For *N. fischeri* the heat resistance of the ascospores was clearly related to the age of the culture (or ascospores). The D_{85} from 21 days old cultures increased from 10 min to >60 min in 114 days old cultures (Conner and Beuchat, 1987b). Cultivation temperature of the spore generating colony was also important; 42 days old cultures showed an increase of D_{85} of much smaller than 30 min to greater than 60 min when cultured at 18 and 30° C respectively.

King and Whitehand (1990) report higher heat resistance of *T. macrosporus* when the fungus was grown on solid medium. Growth temperatures showed a tendency towards higher resistance between 25 and 30° C in three isolates of the fungus. *T. flavus* showed higher heat resistance when grown on oatmeal agar compared to malt extract agar. The age of the culture in both cases was correlated with heat resistance of the spores (Beuchat, 1988a). The latter was also observed for *B. nivea* were heat resistance of spores at 90° C varied from approx. 2 to 4 min under several growth conditions (Casella *et al.,* 1990)

In several studies different isolates of the fungi were studied and heat resistance also varied between isolates in *B. nivea, T. macrosporus* and *flavus* and *N. fischeri* (Bayne & Michener, 1979, Beuchat, 1986, King and Whitehand, 1990).

Conner *et al.* (1987) studied the nature of heat resistance. They studied younger (11 days) and older (25 days) ascospores of *N. fischeri* that exhibited different extents of heat resistance (D_{82} of approx. 23 and >60 min respectively). Ascospores showed changes in the inner cell wall region at the lateral ridge during aging. They observed qualitative differences in extractable proteins, but did not see changes in fatty acid or lipid content. Older spores contained 2.8% (dry weight) of mannitol and 0.6% of trehalose which could not be measured inside 11 days old spores. Polyols and disaccharides may play an important role in heat protection. Recently, very high concentrations of trehalose (up to 15-20% of the cell weight) were measured inside ascospores of *T. macrosporus* and this compound was quickly degraded after activation of the cells (J. Dijksterhuis *et al.,* unpublished results).

2.7. Ascospore dormancy and (heat) activation

Sussman (1966) defines dormancy (hypometabolism) as any rest period or reversible interruption of the phenotypic development and discerns two types. Exogenous dormancy (quiescence) includes delayed development due to physical or chemical conditions of the environment. Constitutive dormancy is a condition in which development is delayed as an innate property such as a barrier to the penetration of nutrients, a metabolic block, or self-inhibitor. Ascospores of the heat resistant fungi described here need a robust physical signal for breaking of dormancy and are regarded as constitutive dormant. In most of the studies done on heat resistant fungi, heat activation is observed where the number of viable counts after a short heat treatment is increased several log cycles (e.g. *Eurotium herbariorum* at 60º C, Splitstoesser, 1989, Fig 1).

For ascospores of *T. flavus* activation is observed at 80 °C and at 85°C activation is followed by killing (Beuchat, 1986). At lower temperatures activation fails and only low numbers of germinated spores are observed. Remarkably, the speed of activation seems to increase with higher temperatures in the case of *T. flavus*. *N. fisheri* exhibits constant rates of heat activation between 70-85°C (King & Halbrook, 1987, Fig 1). In the latter study there is an indication that the speed of activation increases between 80 en 90 °C.

When suspensions of *N. fisheri* were heated at different temperatures for 10 min a very steep activation (nearly 1000 times occurred between 60 and 70° C (Gomez *et al.*, 1989). When the cells were heated for 60 min this traject was starting at lower temperatures (50° C) and completed at 65° C. Katan (1985) describes *T. flavus* isolates that are activated after heating for 30 min at 53° C in both soil samples and in liquid. Different isolates may exhibit marked variation in this respect. Splittstoesser *et al.* (1993) find large variations in the number of ascospores of *N. fisheri* that are successfully activated after heat treatment compared to the microscopical count of the cells.

For *N. fisheri* the dormant state can be broken by a drying treatment of 18h at 40° C (Beuchat, 1992), but *T. flavus* ascospores did not show a release of dormancy. After a subsequent storage period under dry conditions (a_w = 0.23) of 20 months or more the spores did not need heat activation anymore and germinated after wetting. Also heating at 50% r.h. (dry heat treatment) at 95° C (for 30 or 60 min) activated *N. fisheri* ascospores, but the temperature of the wetting or recovery buffer was crucial for the viable count obtained (Gomez *et al.*, 1989).

The nature of activation of ascospores is unknown. Heat activation might serve an ecological function when soil fungi are confronted with fires. The activated germinating spores may quickly fill the empty ecological niches formed after the fire. This phenomenon is well known with plant seeds (e.g. *Eucalyptus*). However, it is also possible that heat activated spores because it addresses the same activation mechanisms as the ones occurring in nature.

2.8. Ascospores and high hydrostatic pressures

High hydrostatic pressures can kill micro organisms and treatment of food products with this method may be an alternative of pasteurisation that influences the taste and the firmness of food. Conidia of different fungi are inactivated at hydrostatic pressures of 3 – 4000 bar. The osmo-resistant spoilage yeast *Zygosaccharomyces bailii* was inactivated after 5 min of 3450 bar, but high amounts of sucrose had a protective effect on the cells against these hydrostatic pressures (Palou *et al.*, 1997). Ascospores of *Saccharomyces cerevisiae* exhibited a D_{3000} (at 3000 bar) of 10.8 min and a D_{5000} of 8 sec (Zook *et al.*, 1999) and this effect was constant within a pH range of 3.5-5. Butz *et al.* (1996) showed that ascospores of *B. nivea* were very pressure resistant. Solely pressures above 6000 B in combination with temperatures above 60 °C can inactivate these cells. Other thermo-resistant fungi (*B. fulva, Aspergillus fischeri* en *Eupenicillium* spp.) were inactivated at 300-600 bar at 10-60°C. Vegetative stages of the studied fungi did not survive 300 bar at 25°C. The authors saw little difference of survival of spores after treatment in a pH range of 4-7, but noticed protection of the cells in the presence of sucrose and in bilberry jam.

When pressure treatments are repeated shortly after each other (the so-called oscillatory treatments, Palou *et al.*, 1998) ascospores can be inactivated more effectively. To reach approximately 700 bar, 10 min are needed and the cycli are kept for 1 s at a temperature of 60°C. Continuous treatment of spores suspended in apple or cranberry juices at 6890 bar (for 25 min) was not effective at room temperature. At 60°C approx. 5-30% of the spores survived treatment. Ascospores of *B. nivea* suspended in fruit juices with higher water activities (a_w = 0.98) were killed after 3 cycli of hydrostatic pressure, but in juices with lower water activities (a_w = 0.94) only partial inactivation (approx. 8-20% of the cells survived) was observed after 5 cycli.

The most studies performed on fungi have liquid as the pressurizing agent; Ballestra and Cuq (1998) use pressurized carbon dioxide at 5000 bar on conidia of *Aspergillus niger* and ascospores of *B. fulva*. For *A. niger* the pressure treatment was effective at 50° C, but not at 60° C (D_{60} = 1.2-1.3 min). In solutions with low water activities (a_w = 0.9) conidia were ten times more resistant to the treatment. With 5000 bar the D_{85} of ascospores of *B. fulva* was lowered from 22 min compared to 14 min for cells that were heated alone.

Ascospores of heat resistant fungi can be regarded as very stress resistant dormant structures that can cope with extreme environments. High-pressure resistance of the cells seems to be relatively independent on pH, but is enhanced by the presence of sugars in the medium.

2.9 Ascospores and pulsed electric fields

In the future Pulsed Electric Fields (PEF) treatment can be one of the candidates of non-thermal processing of food products (for example fruit juices). During treatment short, intensive electrical imposes are applied that can kill living cells. Only brief reports are known with respect to fungal cells. Inactivation of conidia of *B. fulva* increases with the number of pulses (applied at 30-41.7 kV/cm) in different fruit juices. After 15 pulses a 4-6 log reduction of the number of viable cells was observed with the exception of cranberry juice were 2 pulses were enough for a log 6 reduction. Ascospores of *N. fischeri* show high resistance and are not inactivated following 40 pulses at 51.0 kV/cm (Raso *et al.*, 1998).

2.10. Raw material quality

For the development of mycelium or colonies from heat resistant ascospores, usually weeks are required. As a result of that, the build-up of ascospores in process lines is unlikely to occur (Splittstoesser *et al.*, 1969). The number of ascospores in a heat process is often primarily determined by the quality of the raw materials. In Australia, passion fruit and strawberries are most likely to be contaminated with heat resistant fungal spores. Pineapple and mango can also frequently contaminated, while citrus fruits have rarely been implicated (Hocking and Pitt, 1984). Instances of mould spoilage in citrus products have generally been caused by a post-pasteurization contamination (Parish, 1991). In our experience pine apple and mango have shown several outbreaks of spoilage probably caused by factors occuring during harvest. Soil is probably an important reservoir of ascospore-forming moulds (Fravel and Adams, 1986; Jesenka *et al.*, 1991; Ugwuanyi and Obeta, 1991). In Nigeria, positive soil samples were found to contain 1.8-30 heat resistant moulds/g (Okague, 1989).

King and Halbrook (1987) were able to detect *T. flavus* in a heated fruit concentrate at a level of 0.6 colony forming unit (CFU) /g. By Hocking and Pitt (1984), a level of < 2 ascospores/100 ml of finished product was mentioned as acceptable for a passion fruit juice. In general, acceptable levels of ascospores in finished products can only be determined after a careful evaluation of the product and shelf life characteristics. Knowing the product requirements, heating and raw material requirements can be defined. Baggerman and Samson (1988), for instance, proposed a 5-log-reduction of ascospores for the heating of a red currant juice. For the stability of many existing products, a low ascospore load of the raw materials is at least as important as the times and temperatures of heating (Baird-Parker and Kooiman, 1980). Fortunately ascospores are usually present only in low numbers (< 1 CFU/100g; Beuchat, 1986). For *Byssochlamys fulva* in raw fruit Splittstoesser *et al.* (1971) reported a contamination rate of 0.1 CFU/g.

Few research groups have investigated the presence of ascospore forming moulds in dairy products. Engel (1991a) reported the presence of *Byssochlamys nivea* ascospores in raw milk samples from 4 dairies. Ascospore counts usually fluctuated between 1 and 100 CFU/l, with a maximum of 130 CFU/l. After pasteurization, the ascospore counts were reduced only slightly and the centrifugation and homogenisation of milk did not affect ascospore counts significantly (Engel, 1991a; Engel and Teuber, 1991). *Monascus ruber* ascospores could also be detected in raw milk, albeit at much lower levels, maximally 2.5 CFU/l. In winter months, when cows were fed with silage, the *B. nivea* and *M. ruber* ascospore counts in milk were generally higher (Engel, 1991b). In previous investigations, *B. nivea* and *M. ruber* could be demonstrated in 23% of the silage samples (Frevel *et al.*, 1985).

In order to detect ascospores, a heat activation of the spores is usually required. Temperatures ranging from 5 min at 75° C to 35 min at 80° C are mostly used (Hocking and Pitt, 1984; see also Pitt *et al.*, 1992). The methods for detection of heat resistant moulds are described in Chapter 2, page 288. Based on this principle, a "tyndallisation" or "fractional sterilization" of raw materials has been proposed. Resistant spores are subsequently heat-activated, allowed to germinate at lower temperatures and inactivated by reheating or by a treatment with chemicals, DMDC for example (Van der Riet *et al.*, 1989; Van der Riet and Pinches, 1991).

2.11. Implications for food processing

Fruit products and soft drinks that are contaminated with yeasts may become fermented, undergo severe changes in taste and flavour and swell or explode due to gas formation.

Table 2. Heat-resistant moulds in fruit products; spoilage incidents.

Species	isolated from	Reference
Byssochlamys fulva	canned strawberries	Olliver and Rendle (1934), Spurgin (1964), Richardson (1965)
Byssochlamys nivea	canned strawberries	Put and Kruiswijk (1964)
Eurotium herbariorum	grape jams/jellies	Splittstoesser *et al.* (1989)
Neosartorya fischeri	canned strawberries	Kavanagh *et al.* (1963), McEvoy and Stuart (1970)
	apple juice	Scott and Bernard (1987)
Talaromyces flavus	apple juice	van der Spuy *et al.* (1975)
	fruit juice	King and Halbrook (1987)
	pineapple-/grapefruit juice	Scott and Bernard (1987)
Paecilomyces variotii	fruitconcentrates, pectin	Unpublished data
Talaromyces trachyspermus	dairy products	Unpublished data
Fusarium oxysporum	dairy products	Unpublished data
Neosartorya pseudofischeri	dairy products	Unpubliahed data
Monascus ruber	fermented olives	S. Panagou (personal communication)

Such yeast spoilage is predominantly caused by ascosporogenous yeasts (Put and de Jong, 1980). Heat resistant moulds may cause similar changes in taste and flavour, visible spoilage, disintegration of solid fruit and, less frequently, gas formation (King *et al.*, 1969). Heat resistant enzymes *of Rhizopus arrhizus* and *R. stolonifer* from raw apricots may cause the softening of canned fruit (Sommer *et al.*, 1984). Heat resistant cleistothecia, chlamydospores and sclerotia have been described for several mould genera (Splittstoesser and King, 1984). Frequently, the spoilage of fruit products by heat resistant moulds is mostly caused by ascospores. Table 2 shows different spoilage incidents with heat resistant fungi. The ascospores of *Byssochlamys* have most often been implicated (Hocking and Pitt, 1984). Several ascospore-forming moulds are able to produce mycotoxins and measures to control this have been reported (Roland and Beuchat, 1984; Roland *et al.*, 1984; Nielsen *et al.*, 1989a,b).

By the combined effects of reduced water activities, low pH, carbon dioxide and preservatives, fungal spoilage of many fruit products and soft drinks can be prevented (Baird-Parker and Kooiman, 1984). For the remaining products, the required shelf life can often be achieved by a careful selection of raw materials in combination with a heat treatment or chilled or frozen storage. The heat treatments that are usually applied in fruit and soft drink processing are mostly sufficient, provided that the initial contamination can be kept at acceptable levels (Baird-Parker and Kooiman, 1984)

Frequently, the spoilage of fruit products and soft drinks by ascosporogenous yeasts is due to a post-pasteurization contamination. Mould ascospores can be more heat resistant, but the contamination of finished product is less frequently caused by a build-up of heat resistant moulds in process equipment.

The heat resistance of fungi seems to be less relevant for the stability of dairy products and little research has been carried out in this area. For cream and milk, D_{84}-values of 36-45 s and 1.6-5.4 s were found for *B. nivea* and *M. ruber* ascospores, respectively (Engel, 1991b). The fact many dairy products are distributed at chill makes a spoilage by mould ascospores unlikely (Engel, 1991a; Parish, 1991). Ascosporogenous yeasts are frequently involved in the spoilage of dairy products. Yeast spoilage, however, usually results from a post-heating contamination by vegetative cells and not primarily by heat resistant ascospores from raw materials.

The same is probably true for most bakery products. Spicher and Isfort (1988) reported the spoilage of bread by *M. ruber*. The origin of the mould was not stated, however, and it is questionable if spoilage had been caused by ascospores.

3. CONTAMINATION - PLANT AND PROCESSES

3.1. Fruit and vegetable products

In the processing of food, a step to remove or to inactivate micro organisms is often included. Downstream of such a decontamination, the processing steps that can lead to a recontamination of course strongly depend on the process that is involved. In processing of beans, the surfaces of cutting, slicing and grading machines were probably causing a build-up in *Geotrichum* mycelium fragments during processing (Splittstoesser *et al.*, 1977). The chilled storage of fruit juices, that is sometimes applied to facilitate a precipitation of tartrates, may allow a strong multiplication of yeasts (Splittstoesser, 1987). In the soft drink and concentrated fruit juice production, a lack of attention to cleaning and sanitation has frequently lead to a massive yeast contamination of finished product (Van Esch, 1992). Sand *et al.* (1976) were able to isolate a wide range of spoilage yeasts from proportioning pumps of eight different soft drink plants in Holland. In general, stagnant areas in process equipment may support a strong multiplication of micro organisms (Lelieveld, 1976). Attention for hygienic design and cleanability of equipment, cleaning and sanitation procedures and for eliminating sources of contamination in the processing environment (e.g. air, personnel) are usually sufficient to avoid a significant build-up of fungi during processing.

3.2. Dairy products

Dairy industries are usually familiar with the spoilage phenomena that are caused by moulds and yeasts. Cheese (Fleet, 1989), quark (Engel, 1986), butter (Rohm, 1991b) and yoghurt (Varabioff, 1983; Garcia and Fernandez, 1984) are frequently found to be spoiled by fungi. In an Australian investigation of 161 retail samples from six groups of dairy products - ice cream milk, cream, butter, cheese and yoghurt - cheeses and yoghurts contained the highest numbers of yeasts (Fleet and Mian, 1987). In cheese processing, spoilage yeasts may arise from the milk, the processing equipment or the rennet (Fleet, 1989). In the processing of yoghurts, moulds and yeasts are usually eliminated by the pasteurization step and fungi in finished products are therefore mostly due to a post-pasteurization contamination (Spillmann and Geiges, 1983). Not obeying the rules of good manufacturing practice has frequently lead to a substantial spoilage of yoghurts in the market place (Arnott *et al.*, 1974; Davis, 1974; Suriyarachchi and Fleet, 1981; Fleet and Mian, 1987).

For dairy products in general, processing equipment, air, packaging materials, starter cultures and fruit pulps are said to be the most frequent sources of fungal contamination (Rohm, 1991a,b). Flexible hose connections, filling heads, unhygienic valves, pumps and pipe couplings can often be identified as important sources of fungal contamination.

Dairy products containing fruit pulps represent a special group of products. Although the stability of the dairy part(s) can often be achieved by aseptic processing and spoilage of the fruit part may be prevented by heat processing and the addition of sugar or preservatives (Grimm *et al.*, 1992; Wendt *et al.*, 1992), the combination of both parts is frequently found to be subject to fungal spoilage (Fleet, 1989). In a summary of spoilage incidents of Dutch fruit yoghurts, contamination was found to occur at the following steps/sites in processing (Wijsman, 1987): cleaning and sanitation, pasteurization of fruit or dairy components, ripening-, cooling- and storage tanks (integrity, addition of starter cultures), pumps, valves, flow plates, strainers and filling/packaging. In German spoilage incidents in 1990-1991, the far majority of the incidents was found to be caused by contaminated fruit pulps (Seiler and Wendt, 1991). 318 isolates were obtained from 258 fruit containers. Lactic acid bacteria and moulds accounted for only 17 of the isolates (5%). All other isolates were yeasts, mainly *S. cerevisiae, D. hansenii, P. anomala* and *Torulaspora delbrueckii.* From 128 yoghurt samples from retail outlets in Australia, Suriyarachchi and Fleet (1981) were able to isolate a total of 73 yeast strains. *Torulopsis candida* was most frequently encountered. According to these authors, yeast contamination of (fruit) yoghurts mainly arises from the fruit or from a poor hygiene during packaging.

Reports on spoilage incidents are likely to be biased towards incidents that involve a large fraction of the finished product and incidents in which blowing (gas building) of products occurs. Since mould spoilage is not often accompanied by gas formation (Spillmann and Geiges, 1983), the role of moulds in spoilage can easily be underestimated. In a German investigation of 5920 retail samples of fruit yoghurts from 4 dairy competitors in 1992, 0.3-14% of the retail samples were found to be blown or visibly mouldy at the end of their shelf life (storage at 10° C). Mould spoilage without blowing accounted for the far majority (approx. 85%) of the spoilage. In German dairies, a poor performance in aseptic packaging was a frequently reoccurring cause of yoghurts spoilage by moulds (R. Scholte, unpublished results).

3.3. Bakery products

In several respects, the processing of bakery products is strongly different from the processing of liquid foods. After baking, the processing typically is not enclosed in pipelines, pumps and vessels. As a result, the bakery products may suffer from a prolonged exposure to equipment surfaces, personnel and air. In addition, potentially contaminated ingredients, icings, nut pastes, spices, jams, are frequently applied after baking (Ponte and Tsen, 1987).

Indeed, problems due to spoilage yeasts in bread usually result from a post-baking contamination (Seiler, 1980). In German and English bakeries, slicing machines, bread coolers, conveyer belts and racks have been identified as sources of chalk moulds (Spicher, 1978; McGrath *et al.*, 1988). Important examples of such "chalky" moulds are *Hyphopichia burtonii* and *Endomyces fibuliger* (Spicher, 1993). Fermentation problems in jams, icings, marzipan and nut pastes are mostly due to osmotolerant yeasts that result from the feedback of material from previous productions or from inadequate cleaning and disinfection of plant and equipment.

S. rouxii, S. baillii var. *osmophilus* and *S. bisporus* are most frequently encountered in such high-sugar products (Seiler, 1980). The use of contaminated raw materials in bakeries may lead to undesirable organoleptical properties of finished products, even after heat inactivation of the fungi that were involved (Ponte and Tsen, 1987), or to a post-baking contamination by plant, equipment, air or personnel (Legan and Voysey, 1991).

3.4. *Geotrichum candidum*

G. candidum was first isolated from decaying leaves in 1809 by F. Link. It was noticed on the equipment of tomato processing plants by Howard in 1917 and classified by Thom as *Oidium lactis* (Wildman and Clark, 1947) and then renamed to *Oospora lactis* before its present name *G. candidum* (see De Hoog *et al.*, 1986). The fungus is usually present in soil and is part of the endogenous flora of man and takes part in the decomposition of organic matter (Emrick, 1977).

Occasionally, the fungus is involved in the spoilage of bakery products (Spicher, 1990). It plays an important role in the spoilage of vegetables (Brackett, 1987) and fruits, especially citrus fruits and peaches (Baudoin and Eckert, 1982; Splittstoesser, 1987). In an investigation of 146 samples of mouldy tomatoes taken from catsup processing plants in the USA, *G. candidum* was isolated from 31% of the samples (Mislivec *et al.*, 1987). In raw milk, *G. candidum* is one the fungi that is most frequently encountered (Texdorf, 1972; Frevel *et al.*, 1985; Vadillo Machota *et al.*, 1987; Rohm, 1991a). In a recent investigation of quark samples from German retailers, *G. candidum* could be demonstrated in 18% of the samples (Engel, 1986). The fungus is part of the normal surface microflora of several ripened soft cheeses (Neve and Teuber, 1989). Its ability to metabolize lactic acid, its proteolytic activity towards milk protein and its potential to grow at a wide range of pH-values, pH 2.5-8.1 (Emrick, 1977), at least partly explain its presence in dairy products. In contrast to the majority of the moulds, *G. candidum* is able to grow in environments where hardly any oxygen is present (Wells and Spalding, 1975; see 6.2). It demonstrates a good submerged growth in liquids and can strongly adhere to equipment surfaces. As a result of those characteristics and as a result of its natural presence in fruit, vegetable and dairy raw materials, *G. candidum* has caused many problems in fruit, vegetable and dairy industries in the past (Emrick, 1977). It is known as "machinery mould" or "dairy mould" and was found to be responsible for slime-building in processing equipment and off-smells in finished products (Wildman and Clark, 1947).

In processing equipment, the *G. candidum* mycelium easily fragmentates. The fragments from small quantities of mycelium may contaminate large volumes of finished product. By heating viable *G. candidum* units can usually be eliminated (Stinson, 1976). After heating, however, dead mycelium fragments may be detected by using a microscope. 3 closely related microscopical methods, the *Geotrichum* Mould Count, the Howard Mold Count and the Tomato Rot Fragment Count, have been described by the Association of Official Analytical Chemists in the USA (AOAC, 1984). The presence of fragments is believed to be an "indicator of insanitation" and results from food and vegetable industries have shown that fragment counts can be drastically reduced by cleaning thoroughly (Wildman and Clark, 1947; Cichowicz and Eisenberg, 1974; Eisenberg and Cichowicz, 1977; Emrick, 1977).

Table 3. Typical levels of airborne fungi in food processing moulds (m) and yeasts (y) moulds (m) and yeasts (y).

Processing area	Yeasts or moulds	per m^3 of air (CFU)	Reference
milk, butter, cheese	y	7-212	Heldman *et al.*(1964)
	m	49 - 2360	
bread	m	1000 - 2500	Spicher (1965)
cheese (12 plants)	Y+m	1 - 43,000	Radmore and Lück(1984)
cooked ham	y+m	250 - 500	Orefice *et al.* (1985)
pork (slaughtering)	Y+	75 - 640	Kotula and Emswiler-Rose (1988)
spice mix	y+m	21000 ± 1050	Sayeed and Sankaran (1990)
milk (2 plants)	y+m	245 ± 46 *	Ren and Frank (1992)
	y+m	457 ± 28 *	
ice cream (2 plants)	y+m	138 ± 25 *	Ren and Frank (1992)
	y+m	234 ± 23 *	

*mean ± SD (standard deviation)

Although the reliability of the fragment counts are debated (Stinson, 1976; Splittstoesser *et al.*, 1977, 1980), the presence of *G. candidum* fragments in finished products has brought the American Food and Drug Administration (FDA) to seize product and to shut down processing plants. Clearly, the size distribution of the fragments is influenced by shear forces during processing (Eisenberg *et al.*, 1969; Bandler *et al.*, 1987) and acceptable numbers of fragments in finished products are not a prove for hygienic conditions during processing.

4. CONTAMINATION AND DECONTAMINATION – AIR

4.1. Airborne fungi

Micro organisms in air are continuously subject to lethal factors like abrupt changes in relative humidity, radiation, air ions and pollutants (review: Cox, 1987). For aerosolized *Salmonella newbrunswick* cells, for example, *D-value*s between 41 and 206 min have been found, depending on temperature (10-21 C) and relative humidity (30-90%; Stersky *et al.*, 1972). Although usually less sensitive than gram-negative bacteria, gram-positive bacteria, yeasts and moulds in air are subject to similar inactivating forces and a majority of the fungal particles in air is likely to be unviable (reviews: Lacey, 1981; Al-Doory and Domson, 1984; Kang and Frank, 1989a).

On the other hand, soil in outdoor environments, human beings (Sciple *et al.*, 1967; Whyte and Bailey, 1985), raw materials, dust (Gemeinhardt and Bergmann, 1977), drains (Heldman *et al.*, 1965), equipment surfaces (Heldman, 1974) and ventilation ducts (Vikers and McRobert, 1977) can be important reservoirs that frequently release their micro organisms into air. Viable counts in air therefore reflect a dynamic situation, that results from lethal factors and the tendency of particles to deposit, both working in the same direction, and the release of micro organisms into air, working in the opposite direction.

4.2. Gravity

In air, conidium-forming fungi usually predominate over other fungal groups. With some exceptions, conidia usually possess diameters between 2 and 5 μm (Cole and Samson, 1984). Airborne fungi can, however, adhere to water droplets or dust particles and clumps of conidia may occur. An increase in relative humidity, for example, was reported to change the "geometric mean aerodynamic diameter" (GMAD) of *A. flavus* conidia from 3-3.5 μm to 3.5-4.5 μm (Madelin and Johnson, 1992). In an investigation on viable aerosoles in milk and ice cream processing, the majority of the fungal colony forming units possessed diameters between 2.1 and 7.0 μm (Kang and Frank (1989b). It is probably safe to assume, that airborne mould and yeast particles in general have a median equivalent diameter of approx. 2-20 μm (Noble *et al.*, 1963; Sunga *et al.*, 1966; Sorenson *et al.*, 1980; Madelin and Johnson, 1992).

Airborne particles smaller than 1 μm are strongly influenced by Brownian motion. Given the diameters of the fungal colony forming units in air, Brownian motion can be neglected and gravitational and inertial forces are likely to play a more dominant role in the deposition of airborne moulds and yeasts on surfaces (Whyte 1986; Cox, 1987). In a column of still air, gravitational settling is governed by Stokes' law. In such a column, a smooth and spherical particle of unit density and a diameter of 2 μm covers a distance of 1m in 130 min. Particles with a 4, 8 and 16 μm diameter cover this distance in 33, 8 and 2 min respectively. As a result of this, fungal counts close to surfaces are usually higher than those at higher levels in air (Al-Doory and Domson, 1984). Human activity can lead to the re-entrainment of particles from surfaces and strong fluctuations in counts during the day (Heldman *et al.*, 1964; Buttner and Stetzenbach, 1993).

4.3. Typical numbers

Some typical numbers of airborne moulds and yeasts in food processing areas are given in Table 3. Not surprisingly, airborne mould and yeast counts are product- and process-dependent. Even within a certain processing environment and at a selected sampling site strong fluctuations in counts of airborne moulds and yeasts may occur and repeated sampling is required to reliably assess the levels of airborne fungi (Al-Doory and Domson, 1984).

In a summary of 19 random surveys of moulds in atmospheric air (Al-Doory and Domson, 1984), five genera accounted for almost 70% of the total mould count: *Cladosporium* 29%, *Alternaria* 14%, *Penicillium* 9%, *Aspergillus* 6%, *Fusarium* 6% and *Aureobasidium* 5%.

Although the distribution in indoor air doesn't necessarily reflect that of atmospheric air, representatives of most of those genera can easily be demonstrated in food processing areas. In addition to the genera mentioned, *Rhizopus, Mucor, Circinella, Paecilomyces, Botrytis, Neurospora, Scopulariopsis* and *Geotrichum* species were found in the air of bakeries (Gemeinhardt and Bergmann, 1977). *Saccharomyces, Candida, Rhodotorula, Trichosporon* and *Kluyveromyces* species are probably typical examples of yeast genera in bread and dairy processing (Gemeinhard and Bergmann, 1977; Rossmoore *et al.*, 1988). Air can be a vector in the distribution of almost all fungi that are relevant to food spoilage.

4.4. Implications for food processing

In spite of the large number of air sampling results from food processing areas, information on the influence of airborne microorganisms on product quality has hardly been reported (Kang and Frank, 1989a). For processes in which numbers of moulds and yeasts in the final product are mainly determined by other sources of contamination (e.g. raw materials, processing equipment), the influence of airborne fungi may be negligible (see Cannon, 1970). If, on the other hand, vulnerable products with long shelf life (e.g. yoghurts, ambient-stable fruit juices) need to be filled, control of air quality in the filling area is likely to be obligatory.

In a trial simulating a dairy filling operation, Radmore *et al.* (1988) found containers with an opening of 100 cm^2 to become contaminated by approx. 1.5% of the micro organisms present in 1m^3 of air. The likelihood of a sterile container to become contaminated during filling is related to:

- the time of exposure to air,
- the surface of the container opening,
- the numbers of airborne fungi (CFU/m^3).

A similar approach has been proposed by Whyte (1986) for determining the (direct) relationship between the concentration of airborne micro organisms and the rate of contamination in pharmaceutical filling operations. Using such data (Whyte, 1986; Radmore *et al.*, 1988), standards for acceptable levels of micro organisms in air can be defined, provided that sufficient information on the relevance of those micro organisms for the quality and keep ability of the product is available. General standards on air quality (Table 4) are therefore of limited value. Instead of these, individual standards for product-process combinations are usually required.

In the aseptic filling of especially fruit and dairy products, the technology to prevent airborne contamination is well-established (review: Herson, 1985). For most applications Class 100-air, having no more than 100 particles = 0.5 μm/ft^3 (IES, 1988), is adequate. Although there's no fixed relationship between particles numbers and numbers of microorganisms in air, Class 100-air typically contains less than 1 CFU (colony forming unit)/m^3 of air (Bohrer, 1989). Such air is mostly obtained using high-efficiency particulate air (HEPA) filters, usually in closed cabinets. For many foods the use of filtered air enables a significant extension of the shelf life (Carlson, 1980; Joosten, 1985; Bruderer and Schicht, 1987).

Table 4. Acceptable levels of airborne fungi in food processing moulds (m) and yeasts (y).

Processing area	per m^3 of air (CFU)		Reference
cultured milk, cream	y+m	<700 "good" >700 "poor"	Hedrick and Heldman (1969)
fresh and dry milk	y+m	<1250 "good" >1250 "poor"	Hedrick and Heldman (1969)
dairy industries	y+m	<100 "good" > 1000 "poor"	Radmore and Lück (1984)
dairy industries A	y+m	< 1	IDF/ F.Luquet (1985)
B		< 10	
C		<100	
D		>100	

5. CONTAMINATION AND DECONTAMINATION - PACKAGING MATERIALS

5.1. Initial contamination

Cardboard and corrugated board packaging may be highly contaminated with micro organisms. Especially if made from recycled paper materials, fungal counts above 100 CFU/100 cm^2 are not unusual (R. Scholte, unpublished results). Foils containing plastics, sometimes in combination with (kraft) paper or aluminium (laminates), usually show lower levels of contamination. Typically an initial level of about 10 CFU/100 cm^2 is found (Jesenka, 1993). During storage, transport and further processing (unrolling, moulding, extrusion, cutting) of the foils, however, the level of contamination may increase dramatically. The build-up of static electricity, especially at low relative humidities, and the resulting attraction of dust plays a major role in this. For the packaging of vulnerable products, an active decontamination of the primary packaging materials - plastic, metal or glass - is often inevitable.

The most important techniques to decontaminate primary packaging materials are (review: Cerny, 1993):

- heat, mostly at high relative humidities (steam),
- Ultra-Violet (UV) irradiation, at a wavelength of approx. 254 nm,
- hydrogen peroxide (H_2O_2), usually at concentrations of 30-35%,
- gamma irradiation, usually at approx. 15 kGy.

For legal or practical reasons (costs, need for third party facilities, consumer attitude) the applicability of gamma irradiation is often limited. It is not discussed here.

5.2. Heat

At low relative humidities heat shows little effectiveness against micro organisms (2.2, 2.3). In aseptic filling, thermoforming and sealing conditions are mostly unable to reduce the mould and yeast numbers to a sufficient degree (Scott and Bernard, 1985). Increasing the temperatures is often impossible because of the physical characteristics of the packaging materials. Increasing the residence times in a continuous filling process is mostly unattractive from an economic point of view. In the decontamination of aluminium seals, the dry heat from Infra-Red (IR) lamps has been applied with some success, however. Experience with fungi in dairy industries has shown that 2-3 decimal reductions may be obtained in seconds, provided that temperatures around 200° C are reached and the heat is evenly spread.

The decontamination of packaging materials by wet heat has been more successful in aseptic filling operations. The technology has been known in canning industries for decades and has more recently been introduced in glass and plastic filling technology. To inactivate bacterial spores, usually steam of approx. 135° C is applied (Cerny, 1993). As expected, fungal spores are more sensitive. For A. flavus and A. parasiticus strains, for example, *D-values* in the range of 8-59 s have been reported for moist heat at 60° C (Doyle and Marth, 1975a).

5.3. Ultra-Violet irradiation

Although unsuitable for glass, UV (254 nm) can sometimes be used to decontaminate packaging materials. It is most suitable to decontaminate smooth surfaces and therefore less easily applied to decontaminate preformed tubs or lids. Inactivating the micro organisms may also be a problem if they are protected by (dust) particles.

In inactivating *A. niger* conidia Cerny (1977) reported a 4 log reduction at a dose of 30 mW.s/cm^2. *S. cerevisiae* cells were inactivated even more rapidly. Much lower efficiencies were observed by Lippert (1977) in experiments with *A. niger* conidia that had been inoculated on tub and can surfaces. As a result of their shape especially cans were difficult to decontaminate. In both reports, death curves were non-logarithmic (tailing), indicating that a small fraction (1-10%) of the conidia could hardly be inactivated (Cerny, 1977; Lippert, 1977).

In general, moulds - especially those with pigmented conidia and hyphae - and yeasts are more UV-resistant than bacteria; bacterial spores are more resistant than vegetative cells; gram-positive bacteria are less sensitive than gram-negative species (Cerny, 1977; Wallhäußer, 1988). In some applications a limited decontamination of packaging materials by UV-irradiation may suffice; in many others - especially in aseptic lines using preformed materials - other methods of decontamination may have to be chosen.

5.4. Hydrogen peroxide

Packaging foils in form-fill-seal units can be decontaminated by subsequently immersing them in bath of hydrogen peroxide an drying them with IR of hot air. In filling machines using preformed tubs, tubs and seals can also be decontaminated with hydrogen peroxide. H_2O_2 (30-35% in water) may be applied as aerosoles at ambient and dried with hot air (180° C) directly after that, or applied as a mixture with steam that condenses on the surface of the tubs and seals. Usually approx. 0.02 ml H_2O_2- solution is applied to 100 cm^2 (Cerny, 1993). Reaching an even distribution of the H_2O_2 during its dosing and then carefully removing it with hot air are critical steps. In the United States aseptic packaging units are being supervised by the Food and Drug Administrations (FDA). In the package directly after filling, H_2O_2-residues up to 0.1 ppm are allowed. Bacterial spores are most resistant to the typical H_2O_2/hot air-treatments (Cerny, 1976). In a German standard test to evaluate aseptic packaging machines, a 4-log reduction of *Bacillus subtilus* (strain SA22) spores is mentioned as a requirement (Fraunhofer-Institut, 1987). Similar standard tests have been proposed in the USA in 1984 by the National Food Processors Association (Ito and Stevenson, 1984). Moulds and yeasts can usually be effectively eliminated. Vicini *et al.* (1984), however, isolated *Mucor spinescens* and *Cephalosporium roseo-griseum* from Tetrabrik packages with strained tomatoes or pear juice. In comparison with *Aspergillus ochraceus* and *P. expansum* strains, the *M. spinescens* isolates proved to be relatively resistant towards hydrogen peroxide.

6. PRESERVATION - MODIFIED ATMOSPHERE (MA) PACKAGING

6.1. Modified Atmosphere packaging defined

In the Controlled Atmosphere (CA) storage of for example fruits a precisely defined atmosphere is maintained during the entire storage (Kader, 1980). For packed and hermetically sealed foods, this is usually impossible. The exchange between food and headspace may result in significant changes in the gas composition, especially if a metabolism by the food or by micro organisms can take place. Modifying the atmosphere during packaging - Modified Atmosphere (MA) packaging - by removing (e.g. vacuum) or by introducing gas for example - may nevertheless contribute significantly to the keep ability of such food products (review: Farber, 1991).

6.2. Oxygen

In contrast to many bacteria and yeasts, moulds often have an absolute requirement for oxygen. However, several exceptions to this rule have been noticed, however. *Fusarium oxysporum* and *Mucor rouxii* were reported to grow anaerobically (Bartnicki-Garcia and Nickerson, 1961; Gunner and Alexander, 1964). Gibb and Walsh (1980) observed the growth of *F. moniliforme, F. solani* and *Rhizopus sp.* at 0.01% (v/v) oxygen. *G. candidum* is another important exception to the rule, demonstrating growth when "no measurable oxygen is present" (Miller and Golding, 1949). In general, even for oxygen-requiring moulds like *Penicillium* spp. and *Aspergillus* spp., the oxygen concentration needs to be decreased to below 0.5-2%, in order to achieve a significant reduction in growth rate (Orth, 1976).

The success of vacuum packaging partly relies on the removal of oxygen. This is even more true for the use of oxygen scavengers (also called deoxidizers or oxygen absorbers, -eliminators). Such scavengers are typically based on ferrous particles and several authors have reported them to be able to decrease the oxygen levels to below 0.1% (Smith *et al.*, 1986; Rice, 1989). By using an oxygen scavenger, Stiebing (1991) could increase the shelf life of fermented dry sausages from less than 3 to more than 12 months. Smith *et al.* (1986) could extend the shelf life of crusty rolls from less than 1 week to more than 8 weeks by incorporating scavenger sachets in the packages. By using an oxygen scavenger, Powers and Berkowitz (1990) were able to prevent mould spoilage of ready-to-eat bread for 13 months. In the absence of the scavenger, mould spoilage was noticed within 14 days. Similar results were mentioned by Seiler (1989) for bread slices and Madeira cake.

6.3. Carbon dioxide

If oxygen can not be decreased to a sufficiently low level, increasing the headspace concentration of carbon dioxide is often more effective in inhibiting mould growth and mycotoxin development (Orth, 1976). Although the mechanism of carbon dioxide-mediated inhibition has not been conclusively resolved, it is clear that high concentrations of carbon dioxide are inhibitory to all groups of micro organisms. Yeasts, lactobacilli and other gram-positive bacteria are usually said to be less sensitive than gram-positive bacteria (reviews: Jones and Greenfield, 1982; Daniels *et al.*, 1985). At concentrations below 15-20%, carbon dioxide usually has little effect on microbial growth. At limiting oxygen levels, low CO_2-concentrations can even enhance growth (Wells and Spalding, 1975). In a study with *Alternaria, Botrytis, Cladosporium* and *Rhizopus* species, mould growth was shown to decrease linearly with CO_2-concentrations increasing in the range of 10-45% (at 21% oxygen; Wells and Uota, 1970). In general, the germination of conidia is more susceptible to carbon dioxide inhibition than mycelial growth (Jones and Greenfield, 1982).

Since meat, fish and (fresh) produce are mainly subject to bacterial spoilage, it is not surprising that MA packaging of these products primarily aims to inhibit bacteria (reviews: Wolfe, 1980; Genigeorgis, 1985; Hintlian and Hotchkiss, 1986; Zagory and Kader, 1988; Farber, 1991). In bakery, dairy and fruit products, on the other hand, the emphasis is more on retarding the spoilage by moulds and yeasts, especially if relatively long shelf lives are required.

Branded hard cheese slices, for instance, are frequently marketed using MA or vacuum packaging technologies. Using 100% CO_2, a shelf life of 60 days could be guaranteed for Swiss cheese, whereas only half of this keep ability could be achieved with N_2 (Corinth, 1982). Kosikowski and Brown (1973) achieved a significant shelf life extension by flushing cottage cheese with carbon dioxide. Similar results were reported by Rosenthal *et al.* (1991), using an atmosphere of 67% CO_2/26% N_2/6.6% O_2 to extend the shelf life of cottage cheese and quark with more than 6 weeks. With pre-baked bakery products gas-packing may sometimes be insufficiently effective in preventing the growth of chalky moulds, often due to a recontamination of the baked products during the cooling stage (see also page 347 under 3.3.).

The visible spoilage of toast-bread inoculated with *Penicillium glabrum* (= *P. frequentans*) could be delayed by approximately 14 weeks if an atmosphere of 99% CO_2 was applied (Cerny, 1979). The keep abilities of German rye and wheat breads could be lengthened from 5-6 to 11-17 days after exchanging air by carbon dioxide (oxygen < 1%; Morgenstern,

1983). Similar results have been reported for several other types of bread (Brümmer *et al.*, 1980, 1982; Spicher and Isfort, 1987). Buick and Damoglou (1989) could extend the shelf life of a mayonnaise-based vegetable salad from 12 to 22 days (10 C) by using an atmosphere of 20% CO_2/80% N_2. Under these conditions, the shelf life was restricted by yeast growth.

Carbon dioxide alone, in typical concentrations of 1-5% v/v, is often insufficient to prevent the yeast spoilage of soft drinks (Ison and Gutteridge, 1987). For this reason many soft drinks are preserved (especially. benzoic acid) and much attention has to be paid to cleaning and sanitation of the processing equipment (reviews: Baird-Parker and Kooiman, 1980; Van Esch, 1987, 1992).

Acknowledgements
J.Dijksterhuis carried out his part of this study for a project: "Microbial stress response in minimal processing" of Wageningen Centre for Food Sciences (WCFS). He performed his work on the Agrotechnological Research Institute (ATO) in Wageningen, one of the members of the WCFS. Jos Houbraken and Dr S. Panagou kindly provided some data on D values.

REFERENCES

AGAB, M.A. & COLLINS, M. 1992. Effects of treatments environment (temperature, pH, water activity) on the heat resistance of yeasts. J. Food Sci. Technol. 29: 5-9.

AL-DOORY, Y. & DOMSON, J.F. 1984. Mould Allergy. Lea & Febiger, Philadelphia.

AOAC 1984. Official methods of analysis, 14th ed. Association of Official Analytical Chemists, Arlington, VA.

ARNOTT, D.R., DUITSCHAEVER, C.I. & BULLOCK, D.H. 1974. Microbiological evaluation of yogurt produced commercially in Ontario. J. Milk Food Technol. 37: 11-13.

Aspergillus fischeri var. glaber in canned strawberries. Irish J. Agric. Res. 9: 59-67.

BAGGERMAN, W.I. & SAMSON, R.A. 1988. Heat resistance of fungal spores. In Introduction to food-borne fungi, 3rd ed. R.A. Samson & E.S. VAN Reenen-Hoekstra, eds. CentraalBureau voor Schimmelcultures, Baarn, The Netherlands.

BAIRD-PARKER, A.C. & KOOIMAN, W.J. 1980. Soft drinks, fruit juices, concentrates and preserves. In Microbial Ecology of Foods, vol. II, Food Commodities. International Commission on Microbiological Specifications for Foods, Academic Press, London.

BANDLER, R., BRICKEY, P.M., CICHOWICZ, S.M., GECAN, J.S. & MISLIVEC, P.B. 1987. Effects of processing equipment on Howard Mold and Rot Fragment Counts of tomato catsup. J. Food Prot. 50: 28-37.

BARTNICKI-GARCIA, S. & NICKERSON, W.J. 1961. Thiamine and nicotinic acid: anaerobic growth factors for *Mucor rouxii* . J. Bacteriol. 82: 142-148.

BAUDOIN, A.B.A.M. & ECKERT, J.W. 1982. Factors influencing the susceptibility of lemons to infection by *Geotrichum candidum* . Phytopathology 72: 1592-1597.

BAYNE, H.G. & MICHENER, H.D. 1979. Heat resistance of *Byssochlamys* ascospores. Appl. Environm. Microbiol. 37: 449-453.

BEUCHAT, L.R. & KUHN, G.D 1997. Thermal sensitivity of *Neosartorya fischeri* ascospores in regular and reduced-sugar grape jelly. J. Food Prot. 60: 1577-1579.

BEUCHAT, L.R. (1988b) Influence of organic acids on heat resistance characteristics of *Talaromyces flavus* ascospores. Int. J. Food Microbiol 6: 97-105.

BEUCHAT, L.R. (1992) Survival of *Neosartorya fischeri* and *Talaromyces flavus* ascospores in fruit powders. Letters in Appl. Microbiol. 14: 238-240.

BEUCHAT, L.R. 1981. Effects of potassium sorbate and sodium benzoate on inactivating yeasts in broths containing sodium chloride and sucrose. J. Food Prot. 44: 765-769.

BEUCHAT, L.R. 1986. Extraordinary heat resistance of *Talaromyces flavus* and *Neosartorya fischeri* ascospores in fruit products. J. Food Sci. 51: 1506-1510.

BEUCHAT, L.R. 1988. Thermal tolerance of *Talaromyces flavus* ascospores as affected by growth medium and temperature, age and sugar content in the inactivation medium. Trans. Br. mycol. Soc. 90: 359-364.

BOHRER, B. 1989. Reinraumtechnik in der Lebensmittelindustrie - Praktische Anwendung. Zeitschrift für Lebensmittel-Technik, Marketing und Analytik (ZFL) 40: 404-414.

BRACKETT, R.E. 1987. Vegetables and related products. In Food and Beverage Mycology , 2nd ed. L.R. Beuchat, ed. AVI/Van Nostrand Reinhold, New York.

BRÖKER, U., SPICHER, G. & AHENS, E. 1987b). Zur Frage der Hitzeresistenz der Erreger der Schimmelbildung bei Backwaren. 3. Mitteilung: Einfluß exogener Faktoren auf die Hitzeresistenz von Schimmelsporen. Getreide, Mehl und Brot 41: 344-355.

BRÖKER, U., SPICHER, G. & AHRENS, E. 1987a. Zur Frage der Hitzeresistenz der Erreger der Schimmelbildung bei Backwaren. 2. Mitteilung: Einfluß endogener Faktoren auf die Hitzeresistenz von Schimmelsporen. Getreide, Mehl und Brot 41: 278-284.

BRUDERER, J. & SCHICHT, H.H. 1987. Laminar flow protection for the sterile filling of yogurt and and other milk products. Swiss Food 9: 14-17.

BRÜMMER, J.M., STEPHAN, H. & MORGENSTERN, G. 1980. Maßnahmen zur Schimmelbekämpfung bei Brot, 2. Mitteilung: Atmospherenaustausch mit Kohlendioxid. Getreide, Mehl und Brot 34: 164-168.

BRÜMMER, J.M., STEPHAN, H. & MORGENSTERN, G. 1982. Maßnahmen zur Schimmelbekämpfung bei Brot, 4. Mitteilung: Schimmelentwicklung in nicht geöffneten und geöffneten Schnittbrotpäckchen. Getreide, Mehl und Brot 36: 183-187.

BUICK, R.K. & DAMOGLOU, A.P. 1989. Effect of modified atmosphere packaging on the microbial development and visible shelf life of a mayonnaise-based vegetable salad. J. Sci. Food Agric. 46: 339-347.

BUTTNER, M.P. & STETZENBACH, L.D. 1993. Monitoring airborne fungal spores in an experimental indoor environment to evaluate sampling methods and the effects of human activity on air sampling. Appl. Environm. Microbiol. 59: 219-216.

BUTZ, P., FUNTENBERGER, S., HABERDITZL, T. & TAUSHER, B. 1996. High pressure inactivation of Byssochlamys nivea ascospores and other heat resistant moulds. Leb.-Wiss. Techn. 29:404-410.

CANNON, R.Y. 1970. Types and populations of microorganisms in the air of fluid milk plants. J. Milk Food Technol. 33: 19-21.

CARLSON, V.R. 1980. Aseptic packaging - present status and technical aspects. American Dairy Review April 1980: 38-45.

CASELLA, M.L.A., MATASCI, F. & SCHMIDT-LORENZ, W. 1990. Influence of age, growth medium and temperature on heat resistance of Byssochlamys nivea ascospores. Lebensm.Wiss. u. Technol. 23: 404-411.

CERNY, G. 1976. Entkeimung von Packstoffen beim aseptischen Abpaken. 1. Mitteilung: Untersuchungen zur keimabtötenden Wirkung konzentrierter Wasserstoffperoxidlösungen. Verpackungs-Rundschau 36: 27-32.

CERNY, G. 1977. Entkeimen von Packstoffen beim aseptischen Abpaken. 2. Mitteilung: Untersuchungen zur keimabtötenden Wirkung von UV C-Strahlen. Verpackungs-Rundschau 28: 77-82.

CERNY, G. 1979. Verzögerung des Verschimmelns von Toastbrot durch Begasen. Chem. Mikrobiol. Technol. Lebensm. 6: 8-10.

CERNY, G. 1993. Sterilisation von Verpackungsmaterialien. Deutsche Milchwirtschaft 44: 169-174.

CICHOWICZ, S.M. & EISENBERG, W.V. 1974. Collaborative study of the determination of Geotrichum mold in selected canned fruits and vegetables. J. Ass. Off. Anal. Chem. 57: 957-960.

COLE, G.T. & SAMSON, R.A. 1984. The conidia. In Mould Allergy . Y. Al-Doory & J.F. Domson, eds. Lea & Febiger, Philadelphia.

CONNER, D.E. & BEUCHAT, L.R. 1987a. Heat resistance of ascospores of Neosartorya fischeri as affected by sporulation and heating medium. Int. J. Food Microbiol. 4: 303-312.

CONNER, D.E. & BEUCHAT, L.R. 1987b. Efficacy of media for promoting ascospore formation by Neosartorya fischeri and the influence of age and culture temperature on heat resistance of ascospores. Food Microbiol. 4: 229-238.

CONNER, D.E., BEUCHAT, L.R. & CHANG, C.J. 1987. Age-related changes in ultrastructure and chemical composition associated with changes in heat resistance of Neosartorya fischeri ascospores. Trans. Br. mycol. Soc. 89: 539-550.

COONEY, D.G. & EMERSON, R. 1964. Thermophilic Fungi. Freeman, San Francisco, USA.

CORINTH, H.-G. 1982. Kohlensäure - CO_2 - Anwendung in der Milchwirtschaft. Deutsche Molkerei-Zeitung 28: 948-952.

COX, C.S. 1987. The aerobiological pathways of microorganisms. John Wiley & Sons, Chichester, U.K.

CUQ, P. & BALLESTRA, J-L 1998. Influence of pressurized carbon dioxide on the thermal inactivation of bacterial and fungal spores. Lebensm.-Wiss. U Technol. 31, 84-88.

DANIELS, J.A., KRISHNAMURTHI, R. AND RIZVI, S.S.H. 1985. A review of the effects of carbon dioxide on microbial growth and food quality. J. Food Prot. 48: 532-537.

DAVIS, J.G. 1974. Yogurt in the United Kingdom: chemical and microbiological analysis. Dairy Ind. 39: 149-157.

DOYLE, M.P. & MARTH, E.H. 1975a. Thermal inactivation of conidia from Aspergillus flavus and Aspergillus parasiticus . I. Effects of moist heat, age of conidia and sporulation medium. J. Milk Food Technol. 38: 678-682.

DOYLE, M.P. & MARTH, E.H. 1975b. Thermal inactivation of conidia from Aspergillus flavus and Aspergillus parasiticus . II. Effects of pH and buffers, glucose, sucrose and sodium chloride. J. Milk Food Technol. 38: 750-758.

EISENBERG, W.V. & CICHOWICZ, S.M. 1977. Machinery mold - indicator organism in food. Food Technol. 31: 52-56.

EISENBERG, W.V., PARRAN, H.M., SCHULZE, A.E. & DOUGLAS, R.G. 1969. Effect of comminution on mold counts of tomato products. J. Ass. Off. Anal. Chem. 52: 749-752.

EMRICK, J.H. 1977. Machinery mold: indicator of insanitation in food plants. Food Prod./Management August 1977: 14-16.

ENGEL, G. & TEUBER, M. 1991. Heat resistance of ascospores of Byssochlamys nivea in milk and cream. Int. J. Food Microbiol. 12: 225-234.

ENGEL, G. 1986. Vorkommen von Hefen in Frischkäse - organoleptische Beeinflussung. Milchwissenschaft 41: 692-694.

ENGEL, G. 1991a. Vorkommen der Ascosporen von Byssochlamys nivea in Milch und deren Entfernung durch Entkeimungszentrifugation. Milchwissenschaft 46: 442-444.

ENGEL, G. 1991b. B.nivea und M.ruber in Milch und Milchprodukten. Deutsche Milchwirtschaft 21: 644-646.

ENIGL, D.C., KING, A.D. & TÖRÖK, T. 1993. Talaromyces trachyspermis, a heat-resistant mold isolated from fruit juice. J. Food Prot. 56: 1039-1042.

FARBER, J.M. 1991. Microbiological aspects of modified atmosphere packaging technology - a review. J. Food Prot. 54: 58-70.

FLEET, G.H. & MIAN, M.A. 1987. The occurrence and growth of yeasts in dairy products. Int. J. Food Microbiol. 4: 145-155.

Fraunhofer-Institut für Lebensmitteltechnologie und Verpackung (1987). Merkblatt Prüfung von Aseptikanlagen mit H_2O_2-Packstoffsterilisationsvorrichtungen auf deren Wirkungsgrad. Verpakungs-Rundschau 38: 45-47.

FRAVEL, D.R. & ADAMS, P.B. 1986. Estimation of United States and world distribution of *Talaromyces flavus*. Mycologia 78: 684-686.

FREVEL, H.-J., ENGEL, G. & TEUBER, M. 1985. Schimmelpilze in Silage und Rohmilch. Milchwissenschaft 40: 129-132.

GARCIA, A.M. & FERNANDEZ, G.S. 1984. Contaminating mycoflora in yogurt: general aspects and special reference to the genus Penicillium . J. Food Prot. 47: 629-636.

GEISER, D. M., FRISVAD, J.C. & TAYLOR, J.W. 1998. Evolutionary relationships in *Aspergillus* section Fumigati inferred from partial ß-tubulin and hydrophobin DNA sequences. Mycologia 90: 831-845.

GEMEINHARDT, H. & BERGMANN, I. 1977. Zum Vorkommen von Schimmelpilzen in Bäckereistäuben. Zbl. Bakt. Abt. II 132: 44-54.

GENIGEORGIS, C.A. 1985. Microbial and safety implications of the use of modified atmospheres to extend the storage life of fresh meat and fish. Int. J. Food Microbiol. 1: 237-25

GIBB, E. & WALSH, J.H. 1980. Effect of nutritional factors and carbon dioxide on growth of Fusarium moniliforme and other fungi in reduced oxygen concentration. Trans. Br. mycol. Soc. 74: 111-118.

GIRARDIN, H., MONOD, M. & LATGE, J.P. 1995. Molecular characterization of the food-borne fungus Neosartorya fischeri (Malloch and Cain). Appl. Environm. Microbiol. 61: 1378-1383.

GOMEZ, M.M., BUSTA, F.F. & PFLUG, I.J. (1989) Effect of post-dry heat treatment temperature on the recovery of ascospores of *Neosartorya fisheri*. Letters in Applied Microbiology 8: 59-62.

GRIMM, M., WENDT, A. & SEILER, H. 1992. Gasbildung von Hefen in Fruchtzubereitungen, Praxisveruche. Deutsche Molkerei-Zeitung 113: 383-389.

GUNNER, H.B. & ALEXANDER, M. 1964. Anaerobic growth of Fusarium oxysporum . J. Bacteriol. 87: 1309-1316.

HEDRICK, T.I. & HELDMAN, D.R. 1969. Air quality in fluid and manufactured milk product plants. J. Milk Food Technol. 32: 265-269.

HELDMAN, D.R. 1974. Factors influencing air-borne contamination of foods, a review. J. Food Sci. 39: 962-969.

HELDMAN, D.R., HEDRICK, T.I. & HALL, C.W. 1964. Airborne microorganism populations in food packaging areas. J. Milk Food Technol. 27, 245-251.

HELDMAN, D.R., HEDRICK, T.I. & HALL, C.W. 1965. Sources of air-borne microorganisms in food processing areas - drains. J. Milk Food Technol. 28: 41- 45.

HERSOM, A.C. 1985. Aseptic processing and packaging of food. Food Reviews Int. 1: 215-217.

HINTLIAN, C.B., HOTCHKISS, J.H. 1986. The safety of modified atmosphere packaging: a review. Food Technol. 40: 70-76.

HOCKING, A.D. & PITT, J.I. 1984. Food spoilage fungi. II. Heat resistant fungi. CSIRO Food Res. Q. 44: 73-82.

HOOG, G.S. DE, SMITH, M.TH. & GUEHO, E. 1986. A revision of the genus Geotrichum and its teleomorphs. Stud. Mycol., Baarn 29: 1-131.

ICMSF 1988. Microorganisms in Food. Application of the Hazard Analysis Critical Control Point (HACCP) System to ensure microbiological safety and quality. International Commission on Microbiological Specifications for Foods. Blackwell Scientific Publications, Oxford.

IDF/F. LUQUET 1985. Atmospheric pollution in the dairy industry, type and origin of the pollutants. International Dairy Federation. Draft document D-Doc 120 (Group B24).

IES 1988. Federal Standard 209D. Institute of Environmental Sciences. J. Environm. Sci. 31: 53-76. Mount Prospect, Illinois.

ISO 1987. International Quality Standards ISO 9000-9004, 1st ed., 1987-03-15. International Organization for Standardization, Switzerland.

ISON, R.W. & GUTTERIDGE, C.S. 1987. Determination of the carbonation tolerance of yeasts. Lett. Appl. Microbiol. 5: 11-13.

ITO, K.A. & STEVENSON, K.E. 1984. Sterilization of packaging materials using aseptic systems. Food Technology 38: 60-62.

JESENKA, Z. 1993. Micromycetes in foodstuffs and feedstuffs. Progress in Industrial Microbiology 28, Elsevier, Amsterdam.

JESENKA, Z., PIECKOVA, E. & SEPITKOVA, J. 1991. Thermoresistant propagules of *Neosartorya fischeri*; some ecologic considerations. J. Food Prot. 54: 582-584.

JONES, R.P. & GREENFIELD, P.F. 1982. Effect of carbon dioxide on yeast growth and fermentation. Enzyme Microbial Technol. 4, 210-223.

JOOSTEN, R.L. 1985. Methods to extend the shelf life of pasteurized milk. International Dairy Federation. IDF Pack. Newsl. 12: 1-3.

KADER, A.A. 1980. Prevention of the ripening in fruits by use of controlled atmospheres. Food Technol. 34: 51-54.
KANG, Y.-J. & FRANK, J.F. 1989a. Biological aerosols: a review of airborne contamination and its measurement in dairy processing plants. J. Food. Prot. 52: 512-524.
KANG, Y.-J. & FRANK, J.F. 1989b. Evaluation of air samplers for recovery of biological aerosols in dairy processing plants. J. Food Prot. 52, 655-659.
KATAN, T. 1985. Heat activation of dormant ascospores of Talaromyces flavus. Trans. Br. Mycol. Soc. 84: 748-750.
KAVANAGH, J., LARCHET, N. & STUART, M. 1963. Occurrence of a heat-resistant species of Aspergillus in canned strawberries. Nature 198: 1322.
KING, A.D. & HALBROOK, W.U. 1987. Ascospore heat resistance and control measures for Talaromyces flavus isolated from fruit juice concentrate. J. Food. Sci. 52: 1252-1254, 1266.
KING, A.D. & WHITEHAND, L.C. 1990. Alteration of Talaromyces flavus heat resistance by growth conditions and heating medium composition. J. Food Sci. 55: 830-832, 836.
KING, A.D. 1997. Heat resistance of Talaromyces flavus ascospores as determined by a two phase slug flow heat exchanger. Int. J. Food Microbiol. 35: 147-151.
KING, A.D., BAYNE, H.G. & ALDERTON, G. 1979. Nonlogarithmic death rate calculations for Byssochlamys fulva and other organisms. Appl. Environm. Microbiol. 37: 596-600.
KING, A.D., MICHENER, H.D. & ITO, K.A. 1969. Control of Byssochlamys and related heat-resistant fungi in grape products. Appl. Microbiol. 18: 166-173.
KOSIKOWSKI, F.V. & BROWN, D.P. 1973. Influence of carbon dioxide and nitrogen on microbial populations and shelf life of cottage cheese and sour cream. J. Dairy Sci. 56: 12-18
KOTULA, A.W. & EMSWILER-ROSE, B.S. 1988. Airborne microorganisms in a pork processing establishment. J. Food Prot. 51: 935-937.
KOTZEKIDOU, P. 1997. Heat resistance of Byssochlamys nivea, Byssochlamys fulva and Neosartorya fischeri isolated from canned tomato paste. J. Food Sci. 62:410-412.
LACEY, J. 1981. The aerobiology of conidial fungi. In Biology of Conidial Fungi , Vol.1. G.T. Cole & B. Kendrick, eds. Academic Press, New York.
LEGAN, J.D. & VOYSEY, P.A. 1991. Yeast spoilage of bakery products and ingredients. J. Appl. Bacteriol. 70: 361-371.
LELIEVELD, H.L.M. 1976. Influence of stagnant areas in process equipment on the increase of microorganism concentrations in the product: a tentative mathematical approach. Biotechnol.& Bioeng. 18: 1807-1810.
LINGAPPA, Y. & SUSSMAN, A.S. 1959. Chnages in the heatresistance of ascospores of Neurospora upon germination. Am. J. Bot. 46: 671-678.
LIPPERT, K.D. 1979. Abtötung von Schimmelkonidien in vorgeformtem Verpackungsmaterial durch UV-Strahlung. Verpackungs-Rundschau 30: 51-52.
MADELIN, T.M. & JOHNSON, H.E. 1992. Fungal and actinomycete spore aerosols measured at different humidities with an aerodynamic particle sizer. J. Appl. Bacteriol. 72: 400-409.
MCEVOY, I.J. & STUART, M.R. 1970. Temperature tolerance of
MCGRATH, K., Odell, D. & DAVENPORT, R.R. 1988. The spoilage of bread by dimorphic fungi known as chalk moulds. J. Appl. Bacteriol. 65: xviii.
MICHENER, H.D. & KING, A.D. 1974. Preparation of free heat-resistant ascospores from Byssochlamys asci. Applied Microbiol. 27: 671-673.
MILLER, D.D. & Golding, N.S. 1949. The gas requirements of molds. V. The Minimum oxygen requirements for normal growth and for germination of six mold cultures. J. Dairy Sci. 32: 101-110.
MISLIVEC, P.B., BRUCE, V.R., STACK, M.E. & BANDLER, R. 1987. Molds and tenuazonic acid in fresh tomatoes used for catsup production. J. Food Prot. 50: 38-41.
MORGENSTERN 1983. Der Atmosphärenaustausch bei der Verpackung von Brot. Brot & Backwaren 1-2/83: 24, 27-28.
NEVE, H. & TEUBER, M. 1989. Scanning electron microscopy of the surface microflora of ripened soft cheeses. Kieler Milchwirtschaftliche Forschungsberichte 41: 3-13.
NIELSEN, P.V., BEUCHAT, L.R. & FRISVAD, J.C. 1989a. Influence of atmospheric oxygen content on growth and fumitremorgin production by a heat-resistant mold, Neosartorya fischeri. J. Food Sci. 54: 679-685.
NIELSEN, P.V., BEUCHAT, L.R. & FRISVAD, J.C. 1989b). Growth and fumitremorgin production by Neosartorya fischeri as affected by food preservatives and organic acids. J. Appl. Bacteriol. 66: 197-207.
NOBLE, W.C., LIDWELL, O.M. & KINGSTON, D. 1963. The size distribution of airborne particles carrying microorganisms. J. Hyg. Camb. 61: 385-391.
OKAGBUE, R.N. 1989. Heat-resistant fungi in soil samples from Northern Nigeria. J. Food Prot. 52: 59-61.
OLLIVER, M. & RENDLE, T. 1934. A new problem in fruit preservation. Studies on Byssochlamys fulva and its effects on tissues of processed fruit. J. Soc. Chem. Ind. 53: 166-172.
OREFICE, L., TOTI, L., GIZZARELLI, S., CROCI, L. & DE FELIP, G. 1985. The microbial content of air in food production and distribution areas as a possible parameter of the sanitary quality. Microbiol. Aliments Nutr. 3: 283-290.
ORTH, R. 1976. Wachstum und Toxinbildung von Patulin- und Sterigmatocystin-bildenden Schimmelpilzen unter kontrollierter Atmosphäre. Z. Lebensm. Unters. Forsch. 160: 359-366.
PALOU, E., LOPEZ-MALO, A. , BARBOSA-CANOVAS, G.V., WELTI-CHANES, J., DAVIDSON, P.M. & SWANSON, B.G. 1998. Effect of oscillatory high hydrostatic pressure treatments on Byssochlamys nivea ascospores suspended in fruit juice concentrates. Letters Appl. Microbiol. 27: 375-378
PALOU, E., LOPEZ-MALO, A., BARBOSA-CANOVAS, G.V., WELTI-CHANES, J. & SWANSON, B.G. 1997. Effect of water activty on high hydrostatic pressure inhibition of Zygosaccharomyces bailii. Letters Appl. Microbiol. 24: 417-420.
PARISH, M.E. 1991. Microbiological concerns in citrus juice processing. Food Technol. 45: 128-133.
PITT, J.I. & HOCKING, A.D. 1982. Food spoilage fungi. I. *Xeromyces bisporus* Fraser. CSIRO Food Res. Q.42: 1-6.
PITT, J.I., HOCKING, A.D., SAMSON, R.A. & KING, A.D. 1992. Recommended methods for mycological examination of foods, 1992. In Modern Methods in Food Mycology. R.A. Samson, A.D. Hocking, J.I. Pitt & A.D. King, eds. Elsevier, Amsterdam.
PONTE, J.G. & TSEN, C.C. 1987. Bakery products. In Food and Beverage Mycology, 2nd ed. L.R. Beuchat, ed. AVI/Van Nostrand Reinhold, New York.
POWERS, E.M. & BERKOWITZ, D. 1990. Efficacy of an oxygen scavenger to modify the atmosphere and prevent mold growth on meal ready-to-eat pouched bread. J. Food Prot. 53: 767-770.
PUT, H.M.C. & DE JONG, J. 1980. The heat resistance of selected yeasts causing spoilage of canned soft drinks and fruit products. In Biology and Activities of Yeasts. F.A. Skinner, S.M. Passmore & R.R. Davenport, eds. Soc. Appl. Bacteriol. Symp. Series No. 9. Academic Press, London.
PUT, H.M.C. & DE JONG, J. 1982. The heat resistance of ascospores of four Saccharomyces spp. isolated from spoiled heat-processed soft drinks and fruit products. J. Appl. Bacteriol. 52: 235-243.
PUT, H.M.C., & KRUISWIJK, J.T. 1964. Disintegration and organoleptic deterioration of processed strawberries caused by the mould Byssochlamys nivea. J. Appl. Bacteriol. 27: 53-58.
RADMORE, K. & LÜCK, H. 1984. Microbial contamination of dairy factory air. S. Afr. J. Dairy Technol. 16: 119-123.
RADMORE, K., HOLZAPFEL, W.H. & LÜCK, H. 1988. Proposed guidelines for maximum acceptable air-borne microorganism levels in dairy processing and packaging plants. Int. J. Food Microbiol. 6: 91-95.
RAJASHEKHARA, E., SURESH, E.R. & ETHIRAJ, S. 1996. Influence of different heating media on thermal resistance of Neosartorya fischeri isolated from papaya fruit. J. Appl. Bacteriol. 81: 337-340.
RAJASHEKHARA, E., SURESH, E.R. & ETHIRAJ, S. 1998. Thermal death rate of ascospores of Neosartorya fischeri ATCC 200957 in the presence of organic acids and preservatives in fruit juices. J. Food Protection 61: 1358-1362.
RASO, J., CALDERON, M.L., GONGORA, M., BARBOSA-CANOVAS, C. & SWANSON, B.G. 1998. Inactivation of mold

ascospores and conidiospores suspended in fruit juices by pulsed electric fields. Food Sci. Techn. 31: 7-8.

REN, T.-J. & FRANK, J.F. 1992. Sampling of microbial aerosols at various locations in fluid milk and ice cream plants. J. Food Prot. 55: 279-283.

RICE, J. 1989. Modified atmosphere packaging. Food Processing 50: 60-76.

RICHARDSON, K.C. 1965. Incidence of Byssochlamys fulva in Queensland-grown canned strawberries. Queensl. J. Agric. Anim. Sci. 22: 347-350.

ROHM, H. 1991a). Zur Bedeutung von Hefen und Schimmelpilzen für die Milchwirtschaft, I. Deutsche Milchwirtschaft 14: 404-406.

ROHM, H. 1991b). Zur Bedeutung von Hefen und Schimmelpilzen für die Milchwirtschaft, II. Deutsche Milchwirtschaft 16: 478-480.

ROLAND, J.O. & BEUCHAT, L.R. 1984. Biomass and patulin production by Byssochlamys nivea in apple juice as affected by sorbate, benzoate, SO_2 and temperature. J. Food Sci. 49: 402-406.

ROLAND, J.O. & BEUCHAT, L.R., WORTHINGTON, R.E. & HITCHCOCK, H.L. 1984. Effects of sorbate, benzoate, sulfur dioxide and temperature on growth and patulin production by Byssochlamys nivea in grape juice. J. Food Prot. 47: 237-241.

ROSENTHAL, I., ROSEN, B., BERNSTEIN, S. & POPEL, G. 1991. Preservation of fresh cheeses in a CO_2-enriched atmosphere. Milchwissenschaft 46: 706-708.

ROSSMOORE, K., JOHNSON, P. & KOVACH, C. 1988. The significance of aerial microbiota on the quality of dairy products. Dairy Food Sanit. 8: 269 (Abstr.).

SAMSON, R.A. & REENEN-HOEKSTRA, E.S. VAN 1988. Introduction to food-borne fungi, 3rd ed. Centraal-Bureau voor Schimmelcultures, Baarn, The Netherlands.

SAND, F.E.M.J., KOLFSCHOTEN, G.A. & VAN GRINSVEN, A.M. 1976. Yeasts isolated from proportioning pumps employed in soft drink plants. Brauwissenschaft 29: 294-298.

SAYEED, S.A. & SANKARAN, R. 1990. A study of the behaviour of air microflora in food industries. J. Food Sci. Technol. 27: 340-344.

SCIPLE, G.W., RIEMENSNIDER, D.K. & SCHLEYER, C.A.J. 1967. Recovery of microorganisms shed by humans into a sterilized environment. Appl. Microbiol. 15: 1388-1392.

SCOTT, V.N. & BERNARD, D.T. 1985. Resistance of yeasts to dry heat. J. Food Sci. 50: 1754-1755.

SCOTT, V.N. & BERNARD, D.T. 1987. Heat resistance of Talaromyces flavus and Neosartorya fischeri from commercial fruit juices. J. Food Prot. 50: 18-20.

SEILER, D.A.L. 1980. Yeast spoilage of bakery products. In Biology and Activities of Yeasts. F.A. Skinner, S.M. Passmore & R.R. Davenport, eds. Soc. Appl. Bacteriol. Symp. Series No. 9. Academic Press, London.

SEILER, D.A.L. 1989. Modified Atmosphere packaging of bakery products. In Controlled/Modified Atmosphere/Vacuum Packaging of Foods. A.L. Brody, ed. Food & Nutrition Press, Trumbell, Connecticut.

SEILER, H. & WENDT, A. 1991. Hefen in Fruchtzubereitungen und Sauermilchprodukten. Störfallanalysen. Deutsche Molkerei-Zeitung 49: 1517-1522.

SMITH, J.P., OORAIKUL, B., KOERSEN, W.J., JACKSON, E.D. & LAWRENCE, R.A. 1986. Novel approach to oxygen control in modified atmosphere packaging of bakery products. Food Microbiol. 3: 315-320.

SOMMER, N.F., BUCHANAN, J.R. & FORTLAGE, R.J. 1984. Inactivation of pectinolytic fungal enzymes in raw apricots to prevent softening of canned fruits. Lebensm.-Wiss. u. Technol. 17: 167-170.

SORENSON, W.G., PEACH, M.J., SIMPSON, J.P., OLENCHOCK, S.A. & TAYLOR, G. 1980. Size range of viable fungal particles from aflatoxin-contaminated corn aerosols. In Occupational Lung Disease. J.A. Dosman & D.J. Cotton, eds. Academic Press, New York.

SPICHER, G. & ISFORT, G. 1987. Die erreger der Schimmelbildung bei Backwaren. 9. Mitteilung: Die auf vorgebackenen Brötchen, Toast- und Weichbrötchen auftretenden Schimmelpilze. Deutsche Lebensmittel-Rundschau 83: 246-249.

SPICHER, G. & ISFORT, G. 1988. Die erreger der Schimmelbildung bei Backwaren. 10. Mitteilung: Monascus ruber, ein nicht alltäglicher Schimmelerreger des Brotes. Getreide, Mehl und Brot 42: 176-181.

SPICHER, G. 1965. Schimmelsporen in der Luft von Backstuben und Brotlagerräumen. Brot und Gebäck 19: 148-153.

SPICHER, G. 1978. Die Quellen der direkten Kontamination des Brotes mit Schimmelpilzen. 2. Mitteilung. Das Schneideöl als Faktor der Schimmelkontamination. Getreide Mehl und Brot 32: 91-94.

SPICHER, G. 1984. Die Erreger der Schimmelbildung bei Backwaren, 1. Mitteilung: Die auf verpackten Schnittbroten auftretenden Schimmelpilze. Getreide, Mehl und Brot 38: 77-80.

SPICHER, G. 1991. Kampf der Konidien. Brot und Backwaren 7-8: 236-242.

SPICHER, G. 1993. Getreide, Getreideerzeugnisse, Backwaren. In Mikrobiologische Untersuchung von Lebensmitteln, J. Baumgart, ed. Behr's Verlag, Hamburg.

SPILLMANN, H. & GEIGES, O. 1983. Identifikation von Hefen und Schimmelpilzen aus bombierten Joghurt-Packungen. Milchwissenschaft 38: 129-132.

SPLITTSTOESSER, D.F. & CHUREY, J.J. 1991. Reduction of heat resistance of Neosartorya fischeri ascospores by sulphur dioxide. J. Food Sci. 56: 867-877.

SPLITTSTOESSER, D.F. & KING, A.D. 1984. Enumeration of Byssochlamys and other heat resistant moulds. In Compendium of Methods for the Microbiological Examination of Foods, 2nd ed. M.L. Speck, ed. American Public Health Ass., Washington DC.

SPLITTSTOESSER, D.F. & SPLITTSTOESSER, C.M. 1977. Ascospores of Byssochlamys fulva compared with those of a heat resistant Aspergillus. J. Food Sci. 4: 685-688.

SPLITTSTOESSER, D.F. 1987. Fruit and fruit products. In Food and Beverage Mycology, 2nd ed. L.R. Beuchat, ed. AVI/Van Nostrand Reinhold, New York.

SPLITTSTOESSER, D.F., BOWERS, J., KERSCHNER, L. & WILKISON, 1980. Detection and incidence of Geotrichum candidum in frozen blanched vegetables. J. Food Sci. 45: 511-513.

SPLITTSTOESSER, D.F., CADWELL, M.C. & MARTIN, M. 1969. Ascospore production by Byssochlamys fulva. J. Food Sci. 34: 248- 250.

SPLITTSTOESSER, D.F., GROLL, M., DOWNING, D.L. & KAMINSKI, J. 1977. Viable counts versus the incidence of machinery mold (Geotrichum) on processed fruits and vegetables. J. Food Prot. 40: 402-405.

SPLITTSTOESSER, D.F., KUSS, F.R., HARRISON, W. & PREST, D.B. 1971. Incidence of heat resistant molds in eastern orchards and vineyards. Appl. Microbiol. 21: 335-337.

SPLITTSTOESSER, D.F., LAMMERS, J.M., DOWNING, D.L. & CHUREY, J.J. 1989. Heat resistance of Eurotium herbariorum, a xerophilic mold. J. Food Sci. 54: 683-685.

SPLITTSTOESSER, D.F., NIELSEN, P.V. & CHEREY, J.J. 1993. Detection of viable ascospores of Neosartorya. J. Food Protection 56: 599-603.

SPURGIN, M.M. 1964. Suspected occurrence of Byssochlamys fulva in Queenslandgrown canned strawberries. Queensl. J. Agric. Anim. Sci. 21: 247-250.

SPUY, J.E. VAN DER, MATTHEE, F.N. & CRAFFORD, D.J.A. 1975. The heat resistance of moulds Penicillium vermiculatum Dangeard and Penicillium brefeldianum Dodge in apple juice. Phytophylactica 7: 105-108.

STERSKY, A.K., HELDMAN, D.R. & HEDRICK, T.I. 1972. Viability of airborne Salmonella newbrunswick under various conditions. J. Dairy Sci. 55: 14-18.

STIEBING, A. 1991. Einfluß von Sauerstoffabsorbern auf das Schimmelwachstum auf verpackten Rohwürsten. Mitteilungsblatt der Bundesanstalt für Fleischforschung (Kulmbach) 111: 34-37.

STINSON, W.S. 1976. Geotrichum/ machinery mold - real or imaginary problem. Food Processing 37: 18-20.

SUNGA, F.C.A., HELDMAN, D.R. & HEDRICK, T.I. 1966. Characteristics of air-borne microorganism populations in packaging areas of a dairy plant. Mich. State Univ. Agr. Expt. Sta. Quart. Bull. 49: 155-163.

SURIYARACHCHI, V.R. & FLEET, G.H. 1981. Occurrence and growth of yeasts in yogurts. Appl. Environm. Microbiol. 42: 574-579.

SUSSMAN, S. & HALVORSON, H.O. 1966. Spores, their dormancy and germination. Harper & Row, New York, USA

TEXDORF, I. 1972. Untersuchungen über das Vorkommen von Schimmelpilzen in Anlieferungsmilch. Archiv für Lebensmittelhyg. 23: 99-100.

TOURNAS, V. 1994. Heat-resistant fungi of importance to the food and beverage industry. Crit. Rev. Microbiol. 20: 243-263.

TSANG, E.W.T. & INGLEDEW, W.M. 1982. Studies on the heat resistance of wild yeasts and bacteria in beer. American Society of Brewing Chemists Journal 40: 1-8.

UGWUANYI, J.O., OBETA, J.A.N. 1991. Incidence of heat-resistant fungi in Nsukka, Southern Nigeria. Int. J. Food Microbiol. 13: 157-164.

VADILLO MACHOTA, S., PAYA VICENS, M.J., CUTULI DE SIMON, M.T. & SUAREZ FERNANDEZ, G. 1987. Raw milk mycoflora. Milchwissenschaft 42: 20-22.

VAN DER RIET, W.B. & PINCHES, S.E. 1991. Control of Byssochlamys fulva in fruit juices by means of intermittent treatment with dimethyldicarbonate. Lebensm.-Wiss. u. Technol. 24: 501-503.

VAN DER RIET, W.B., BOTHA, A. & PINCHES, S.E. 1989. The effect of dimethyldicarbonate on vegetative growth and ascospores of Byssochlamys fulva suspended in apple juice and strawberry nectar. Int. J. Food Microbiol. 8: 95-102.

VAN ESCH, F. 1987. Gisten in frisdranken en in vruchtesap-concentraten. De Ware(n)-Chemicus 17. Symposium edition.

VAN ESCH, F. 1992. Yeast in soft drinks and concentrated fruit juices. Brygmesteren 4: 9-29.

VARABIOFF, Y. 1983. Spoilage organisms in yogurt. Dairy Prod. 11: 8-12.

VICINI, E., BARBUTI, S., SPOTTI, E., CAMPANINI, M., CASTELVETRI, F., GOLA, S., MANGANELLI, E., CASSARA, A. & CASOLARI, A. 1984. Fruit juices spoilage by gas forming molds. Microbiol. Aliments Nutr. 2: 21-26.

VICKERS, V.T. & MCROBERT, A.G. 1977. Dairy factory ventilation and air treatment. N.Z. J. Dairy Sci. Technol. 12: 5-14.

WALLHÄUßER, K.H. 1988. Praxis der Sterilisation, Desinfektion, Konservierung, 4th ed. Georg Thieme Verlag, Stuttgart, New York.

WELLS, J.M. & SPALDING, D.H. 1975. Stimulation of Geotrichum candidum by low oxygen and high carbon dioxide atmospheres. Phytopathology 65: 1299-1302.

WELLS, J.M. & UOTA, M. 1970. Germination and growth of five fungi in low-oxygen and high-carbon dioxide atmospheres. Phytopathology 60: 50-53.

WENDT, A., SEILER, H. & BUSSE, M. 1992. Gasbildung von Hefen in Fruchtzubereitungen, Laborversuche. Deutsche Molkerei-Zeitung 113: 142-147.

WHYTE, W. & BAILEY, P.V. 1985. Reduction of microbial dispersion by clothing. J. Parenteral Sci. Technol. 39: 51-60.

WHYTE, W. 1986. Sterility assurance and models for assessing airborne bacterial contamination. J. Parenteral Sci. Technol. 40: 188-197.

WIJSMAN, M.R. 1987. Het voorkomen van gisten in zuivelprodukten. De Ware(n)-Chemicus 17. Symposium edition.

WILDMAN, J.D. & CLARK, P.B. 1947. Some examples of the occurrence of machinery slime in canning factories. J. Ass. Off. Anal. Chem. 30: 582-585.

WOLFE, S.K. 1980. Use of CO- and CO_2-enriched atmospheres for meats, fish and produce. Food Technol. 34: 55-58, 63.

ZAGORY, D & KADER, A.A. 1988. Modified atmosphere packaging of fresh products. Food Technol 42: 70-77.

ZOOK, C.D., PARISH, M.E., BRADDOCK, R.J. & BALABAN, M.O.. 1999. High pressure inactivation kinetics of Saccharomyce cerevisiae ascospores in orange and apple juices. J. Food Sci. 64: 533-535

Chapter 8

FOOD PRESERVATIVES AGAINST FUNGI

P.V. NIELSEN and E. DE BOER

BioCentrum-DTU, Technical University of Denmark, DK-2800 Lyngby, Denmark and Foodinspection, De Stoven 22, 7206 AX Zutphen, The Netherlands

TABLE OF CONTENTS

INTRODUCTION

In modern food production, Good Manufacturing Practice (GMP) and the Hazard Analysis Critical Control Point (HACCP) system have been implemented alongside the traditional physical methods to ensure hygienic conditions and food safety. This practice has in general resulted in better control and more gentle processing conditions e.g. heating (pasteurisation, sterilisation), canning, drying, cooling (cold or frozen storage), vacuum packing, modified atmosphere packing, controlled atmosphere storage and irradiation. Nevertheless the organoleptic quality of some food still suffers greatly from physical modes of processing, whereas the methods in other cases are not technologically or economically feasible. In these cases chemical preservation will often be necessary.

Chemical preservatives have been defined by the Food and Drug Administration (FDA) in United States as: "Any chemical that when added to food tends to prevent or retard deterioration thereof, but does not include common salt, sugars, vinegar, spices or oils extracted from spices, substances added to food by direct exposure thereof to wood smoke, or chemicals applied for their insecticidal or herbicidal properties" (FDA, 2000). The commission of the European communities uses a broader definition in Directive 95/2/EC on food additives other than colours and sweeteners: "Preservatives are substances which prolong the shelf-life of foodstuff by protecting them against deterioration caused by micro-organisms". The Directive lists 38 preservatives that can be grouped in 13 different types of preservatives. This list varies from the list of food preservatives listed by FDA in Code of Federal Regulations: 21 CFR 172.

Due to the discrepancy between the different food regulations and the implications that put on international food trade the Codex Alimentarius Commission are developing a General Standard for Food Additives (GSFA), that can set forth the conditions under which food additives may be used in internationally traded foods. This work is recognised as international standard setting by the Sanitary and Phytosanitary Agreement of General Agreement on Tariffs and Trade (GATT). Thus the GSFA will be established as an international standard under the World Trade Organisation (WTO) when the proposed draft standard eventually is adapted by the Commission as a Codex standard (Codex Alimentarius Commission, 2000).

1. FACTORS CONCERNING THE PRESERVATIVE ITSELF

The mechanism of action of preservatives is usually based on either (1) destruction of the cell wall or cell membrane, (2) inhibition of various enzymes in the microbial cell or (3) destruction of the genetic structure of the protoplasm. The mechanisms of action and resistance have been reviewed by (Brul and Klis, 1999)

Table 1. Minimum Inhibitory Concentrations (MIC) (µg/ml) of Preservatives (Noordervliet, 1978; Rehm, 1961; Rehm *et al.*, 1967; Robinson *et al.*, 1964).

Preservative	*Bacillus sp*	*Escherichia coli*	*Saccharomyces cerevisiae*	*Candida*	*Penicillium*	*Aspergillus*
sorbic acid	<200-2.000	1,000-2,000	1,00-1,200	200-1,000	100-2,000	200-20,000
benzoic acid	<1000	1,200-40,000	200-2,000	700-1,500	200-60,000	200-40,000
PHB ethylester	100-600	125-1,000	800-1,000	200-800	200-800	200-500
PHB propylester	100-1,000	30-1,000	150-600	200-400	200-500	100-500
natamycin	>10,000	>10,000	1-5	1.5-20	0.6-10	0.1-50
sodium sulphite	50	10-200	80-160		20-400	200
0-phenylphenol		115	76		10-50	
thiabendazole					1	4-40

a. ***Antimicrobial spectrum.*** A broad spectrum of activity, which covers the possible range of contaminants associated with the food concerned, is desired. However, some preservatives are only or mainly active against fungi (diphenyl, thiabendazole, natamycin, imazilil), others are only active against bacteria (nitrite).

Most preservatives react against both bacteria and fungi (sorbic acid, benzoic acid, sulphurdioxide, PHB-esters, o-phenylphenol).

Table 1 gives an overview of the minimum inhibitory concentrations (MIC) of some preservatives for different micro-organisms. To obtain a broader spectrum of action, preservatives are often used in combination with one another (sorbic acid/benzoic acid, sorbic acid /sulphur dioxide, sorbic acid/ natamycin) or in combination with physical methods of food preservation.

b. ***Antimicrobial activity.*** Micro-organisms should preferably be killed and not only inhibited. Most of the preservatives now in use are bacteriostatic/ fungistatic when applied in the permitted concentrations, resulting in inhibition and not killing of the micro-organisms. Usually this is not problematic, as preservatives should be employed to maintain a good microbiological quality of the food and not to eliminate micro-organisms from very contaminated substrates.

c. **Chemical and physical properties**

- hydrophylic properties: as micro-organisms mostly grow in the aqueous phase of foods, preservatives should be water soluble.
- lipophilic properties: preservatives acting on the hydrophobic cell membrane require lyophilic characteristics.
- thermal stability is essential for preservatives intended as adjuncts to thermal processes.
- dissociation constant - the ability of a compound to ionize can result in different activity, depending upon the pH of the food system in which it is used.
- chemical reactivity - reactions with food components may result in a decrease of the activity of the preservative.

d. ***Relative toxicity*:** preservatives and their degradation and reaction products have to be toxicologically safe. The FAO/WHO Expert Committee on Food Additives allocates Acceptable Daily Intakes (ADI) for the permitted preservatives (Table 2).

The risk for the consumer can be evaluated by comparing the actual intake of an additive with the ADI-value of this additive. In a study in The Netherlands it was found that the mean daily amounts of consumption of benzoic acid, sorbic acid and sulphite were all far below the ADI-values of these preservatives (Van Dokkum *et al.*, 1982). In fact, the maximal levels are determined using a budget model based on the average consumption of a wide range of food by regional dietary groups (Codex Alimentarius Commission 2000).

e. ***Resistance development*:** micro-organisms should not develop resistance to a preservative on continued exposure to the compound. Some preservatives can be decomposed by micro-organisms present in the food.

f. ***Organoleptic properties*:** preservatives should exhibit no taste or odour when used at levels effective in controlling microbial growth.

g. ***Economical considerations*:** preservatives should have the ability to pay for itself based on reducing spoilage and minimizing food-borne illness.

h. Preservatives should **not be used therapeutically** or as additives to animal feeds.

i. ***Analysis*:** A. suitable procedure for determining the amount of preservative in food should be available.

Non-specific microbiological methods are used for the qualitative analysis of antimicrobial substances in foods. Confirmation and quantitative assay of preservatives is usually carried out with chromatographic methods (HPLC, GLC, TLC).

Table 2. Acceptable Daily Intake (ADI) of preservatives.

Preservative	**ADI**
Propionic acid	no limit
Sorbic acid	0-25
p-Hydroxybenzoic acid esters	0-10
Nitrate	0-5
Benzoic acid	0-5
Sulphur dioxide, sulphites	0-0.7
Natamycin	0-0.3
Hexamethylene tetramine	0-0.15

ADI (mg/kg body weight per day)

Table 3 Percentage undissociated acid (pK_a) at different pH-values.

Preservative	pK_a	2.5	3.5	4.5	5.0	5.5	6.0	7.0
Acetic acid (E260)	4.74	99	95	63	35	14	5.2	0.55
Citric acid (E330)	3.13	81	30	4.1	1.3	0.4	0.13	0.01
Lactic acid (E270)	2.74	64	15	1.7	0.5	0.2	0.06	0.01
Malic acid (E296)	3.40	89	44	7.4	2.5	0.8	0.25	0.03
Tataric acid (E334)	2.98	75	23	2.9	0.9	0.3	0.10	0.01
Benzoic acid (E210)	4.19	98	83	33	13	4.7	1.5	0.15
Propionic acid (E280)	4.87	100	96	70	43	19	6.9	0.74
Sorbic acid (E200)	4.76	99	95	65	37	15	5.4	0.57
Sulphur dioxide (E220)*	1.81	17	2	0.2	0.06	0.02	0.01	0.00

*Sulphurous acid (SO_2 in water)

2. FACTORS CONCERNING THE FOOD TO BE PRESERVED

The following factors related to the characters of the food product are important: **(a)** Microbiota of the food: numbers, types and condition of the organisms, resistance to the antimicrobial agent and the interaction between organisms. **(b)** Nature of the food: pH, a_w, redox potential, presence of natural occurring antimicrobials, reactive components which may inactivate preservatives, fat content and fat distribution. **(c)** Processing factors: thermal process, dehydration, filtration, other antimicrobial compounds such as salts, sugars, spices, acids and smoke. **(d)** Storage conditions: temperature, relative humidity, gaseous atmosphere, type of packaging, length of storage.

IMPORTANT PRESERVATIVES

The most important preservatives, which are permitted in many countries, are:

Organic acids can be divided into two groups. One group shows an antimicrobial activity owing primarily to their pH-reducing effect. This group includes acetic acid, citric acid, lactic acid, malic acid and tartaric acid (Table 3). Several reports have shown that also the undissociated form of acetic acid has antimicrobial action (e.g. Rusul *et al.*, 1987).

Other reports have shown that the supplementation of up to 2,5% of citric, malic, tartaric or lactic acid to substrates adjusted to pH 2.5 or 3.5 significantly enhances fungal growth and mycotoxin production (Nielsen *et al.*, 1989; El-Gazzar *et al.*, 1987). These organic acids are generally permitted following the "quantum satis" principle, which means that no maximum level is specified. However, additives shall be used in accordance with good manufacturing practice, at a level no higher than is necessary to achieve the intended purpose and provided that they do not mislead the consumer. In some foodstuffs only a limited number of acids or a limited amount may be added.

The other group of organic acid preservatives (sorbic acid, benzoic acid, and propionic acid) shows an antimicrobial activity only when they are present as undissociated acids. The efficiency of these acids therefore depends on the dissociation constant, pK_a. As the pK_a of most acids are between 3 and 5 (Table 3) these preservatives are only active at lower pH-values. A slightly better effect at higher pH values can be expected as some reports have shown anti-microbial activity of the dissociated acid. Skirdal and Eklund (1993) reported that the relative inhibitory efficiency of the dissociated as compared to the undissociate sorbic acid ranged from 38 for *Cladosporium cladosporioides* to 339 for *Penicillium chrysogenum*. Organic acids are in practice usually added as the corresponding sodium, potassium or calcium salts as they are more soluble in water (Table 4).

Sorbic acid (CH_3-CH=CH-CH=CH-COOH) patented as a preservative in 1945. Due to the high pK_a value (4.76) sorbic acid is still quite active in low acid food at pH values up to 6 and is more active than benzoates between pH 4 and 6 (Table 3). Therefore it is applied in a great variety of food products (Table 5). The compound is active against yeasts, moulds and many bacteria (Tables 1, 6 and 7). Sorbates are considered GRAS, they are less toxic than benzoates. Micro-organisms resistant to sorbate exist and an increasing number of moulds are able to degrade sorbate producing strong off flavours by the decarboxylation of sorbic acid into *trans*-1,3-pentadiene (Liewen and Marth, 1985; Hartog *et al.*, 1986; Samson, 1989; Kinderlerer and Hatton, 1990)

Table 4. Solubility of sorbic acid and sorbates (Noordervliet, 1978).

Preservative	Solubility (g/100 g water)
Sorbic acid	0.16
Sodium sorbate	58.2
Potassium sorbate	35.0
Calcium sorbate	1.2

Table 5. The application of preservatives in food. Maximal levels of use in ppm are indicated between brackets (Directive 95/2/EC)

Preservative	Application
sorbic acid:	wines, wine based drinks and flavoured non dairy drinks (200-300), dried fruits and fruit products (1000), vegetable products (1000 - 2000), low sugar jams, jellies and marmalades (1000)[1], bakery products (2000), cheese products (1000-2000)[2], margarine, fat emulsions and batters (1000-2000), Semi preserved fish products and cooked shrimps (2000)[1], salted dried fish (200)[1], surface treatment of dried meat products (quantum satis)
benzoic acid:	beverages (150-200), low sugar jams, jellies and marmalades (500, 1000[1]) vegetable in vinegar, brine or oil (1000-2000)[1], fat emulsions (500 – 1000), Semi preserved fish products and cooked shrimps (2000)[1], salted dried fish (200)[1], surface treatment of dried meat products (quantum satis)
propionic acid:	cakes (1000-2000), pre-packed bread (2000-3000), surface treatment of cheese products (quantum satis).
Parabens/PHB esters:	jelly coating of meat products (1000), snack and coated nuts (300), confectionery excluding chocolate (300), surface treatment of dried meat products (quantum satis)
sulphur dioxide:	fruit juices and concentrates (0 – 350), syrups (10 – 70), cereal and vegetable products (30 – 800), dried fruit (500 – 2000), vinegar and wines (200 – 260), meat products (450), dried salted fish (200)

1: the value is the total combined amount of sorbate and benzoate; 2: for surface treatment: quantum satis

Subinhibitory concentrations have been found to stimulate growth and verrucologen production by *Neosartorya fischeri* (Nielsen *et al.*, 1989) and aflatoxin B_1 production by *Aspergillus flavus* and T-2 toxin production by *Fusarium acuminatum* (Gareis *et al.*, 1984).

Table 6. Maximal concentration of preservatives (ppm) which permitted growth at pH 5.5 of various moulds isolated from cheese

Species	Benzoate	Sorbate	Propionic acid	Ref.
Penicillium camemberti		500		1
P. chrysogenum		500-4700[2]		1,2
P. citrinum	5500	2000	1800	3
P. commune		3000-12000		1
P. crustosum		4700		2
P. digitatum		1000		1
P. glabrum		500-2800		1
P. expansum		1000		1
P. roqueforti		2000-10000		1
P. spinulosum		2400		2
P. viridicatum		2400		2
Alternaria solani	4150	2450	1350	3
Aspergillus flavus		500-4700[2]		1,2
A. fumigatus		4800-7100		2
A. parasiticus	4000[4]	500-2000[4]	1000[5]	1,4,5
A. niger	5500[3]	500-7100[2]	1800[4]	1,2,3,4
A. ochraceus		1000-1500[6]		1,6
A. sydowii		2400		2
A. versicolor		2400-4700		2
Cladosporium cladosporioides	225		7	
Emericella nidulans		500		1
Eurotium amstelodami		2400-7100		2
E. chevalieri		2400-4700		2
E. rubrum		4700-7100		2
Scopulariopsis brevicaulis	4700		2	
Syncephalastrum racemosum	2400		2	
Ulocladium atrum		675		7

References 1: Liewen and Marth (1985), N.B. species were reidentified by Frisvad and Samson (1991); 2: calculated from MIC values at pH 5.0 in Mutasa *et al.* (1990; 3: calculated from MIC values at pH 5.0 adapted from Ray and Bullerman (1982); 4: Rusul and Marth (1987); 5: Rusul *et al.* (1987); 6: Gourama and Bullerman (1988); 7: Skirdal and Eklund (1993).

Table 7. Preservative concentrations (ppm) required to hinder outgrowth at pH 3.5 of various heat-resistant moulds.

Taxa	Benzoate	Sorbate	SO_2	Ref.
Byssochlamys fulva	250	250		King *et al.* (1969)
Byssochlamys nivea	>1000	200-400	50-300	Beuchat (1976)
Neosartorya fischeri var. fischeri	50-100	50-300	200-300	Nielsen (1991)
N. fischeri var. glabra	300	300	200-300	Nielsen (1991)
N. fischeri var. spinosa	100	50-300	300	Nielsen (1991)
Talaromyces macrosporus	100	100	200-300	King and Halbrook (1987)

Benzoic acid (C_6H_5COOH) was first described as a preservative in 1875. The compound occurs naturally in high amounts in cranberries and some other fruits and spices. Sodium benzoate has a pK_a value of 4.19, which means that the optimal activity is below pH 4.0. Accordingly, Rusul and Marth (1987) found that benzoate was a less efficient inhibitor of growth and aflatoxin production of *Aspergillus parasiticus* than sorbate at pH 4.5 and 5.5. Sub-inhibitory concentrations have been found to stimulate growth and aflatoxin production by *A. parasiticus* (El-Gazzar and Marth, 1987) and verrucologen production by *Neosartorya fischeri* (Nielsen *et al.*, 1989).

Benzoate is active against yeasts, moulds and bacteria. Isolates of P. roqueforti have been found to be resistant (Samson, unpublished data). Benzoates are considered GRAS for use in foods. Benzoates have an advantage of low cost compared to other preservatives. An adverse flavour effect is often encountered at higher use levels.

Propionic acid (CH_3CH_2COOH) is effective against moulds and some bacteria but has no activity against yeasts. Propionates are able to inhibit *Bacillus subtilis*, which causes "rope" in bread. Propionates are GRAS. It has especially been used in preservation of bread and cakes. This has partly been due to the legistration previously only permitting the use of propionate in these products. Propionates are also permitted for surface treatment of cheese and cheese products. Used levels are generally higher than with benzoate and sorbate in several cases about 1% is required to inhibit fungal growth (Ray and Bullerman, 1982).

Parabens are esters of p-hydroxybenzoic acid, they are also called PHB-esters and nipa-esters. The antimicrobial activity of the PHB-esters is independent of the pH-value of the food.

Methyl, ethyl, and propyl-estes are applied in practice in Europe, whereas also the heptyl-ester is used in the United States. The activity of these esters increases with the chain length of the alcohol component (Table 8). However, the partition coefficient (ratio of solubility values in the fatty and aqueous phase) also increases rapidly with the chain length (Table 9). Consequently a lower fraction of the added amount will active in the aqueous phase. The parabens are particularly effective against moulds and yeasts and are relatively ineffective against bacteria, especially the gram-negative bacteria. The methyl and propyl parabens are affirmed as GRAS. Parabens may influence the taste of the product, they are used to preserve. The parabens are more costly than the other preservatives. The parabenes appear to be more effective against moulds than yeasts

Table 8. Minimum Inhibitory Concentrations (MIC) values (in µg/ml) of PHB-esters (Aalto *et al.*, 1953).

Species	Methyl PHB	Ethyl PHB	Propyl PHB	Butyl PHB
Saccharomyces cerevisiae	1000	500	130	60
Candida albicans	1000	1000	130	130
Aspergillus niger	1000	400	200	200
Penicillium digitatum	500	250	60	30
Trichophyton mentagrophytes	80	80	40	20

Table 9. Partition coefficient of preservatives between peanut oil and water (Lubieniecki-von Schelhorn, 1967).

Preservative	Partition coefficient
Sorbic acid (E200)	3.1
p-Hydroxybenzoic acid methylester (E218)	5.8
p-Hydroxybenzoic acid ethylester (E214)	26.0
p-Hydroxybenzoic acid propylester (E216)	87.5

Sulphur dioxide and its sulphite salts are used as antimicrobials and for the conservation of colour in a wide range of foods. Reacts with water and forms sulphurous acid (H_2SO_3), pKa = 1.81. At higher use levels the taste becomes noticeable. Sulphites react readily with a variety of constituents. Sulphites destroy vitamin B1. Adverse reactions to sulphites appear in asthmatics. FDA has banned the use of sulphites on raw fruits and vegetables (Taylor and Bush, 1986).

Hexamethylene tetramine (E239) Permitted for the preservation of Provolone cheese (residual amount: 25 mg/kg).

Diphenyl (E230), **o-phenylphenol** (E231) **and thiabendazole** (E232) are used for the preservation of citrus fruit. The fruit is immersed in solutions of these preservatives or wrapped in paper impregnated with these substances (12-70 mg/kg).

Natamycin (pimaricin) (E235) is an antibiotic agent produced by *Streptomyces natalensis* and *S. chattanoogensis*. It is very active against yeasts and moulds, but has no activity against bacteria (Table 1). *Penicillium discolor* can grow at CYA containing 400 ppm natamycin and has caused spoilage on cheeses treated with natamycin (Frisvad *et al.*, 1997; Hoekstra *et al.* 1998). The inhibitory effect of Natamycin is strongly affected by the substrate. Gourama and Bullerman (1988) found that *Aspergillus ochraceus* was less inhibited in olive paste with 350 ppm natamycin as compared to YES with 20 ppm natamycin. Natamycin is permitted in many countries for preventing the growth of fungi on cheeses and dry sausages. (1 mg/dm^2) (EU) or 20 ppm in the finished product (FDA)

Dimethyl dicarbonate (E242): Non-alcoholic beverages (250 mg/l)

Boric acid (E284) and **sodium tetraborate** (borax) (E285): Sturgeons eggs (Caviar) (4 g/kg)

Lysozyme (E1105): Ripened cheese (quantum satis)

Imazalil is an antifungal agent in the chemical group of imadazole derivatives. Approval for incorporation in cheese coating at a level of 250 mg/kg coating have been requested by the scientific committee on food. It is chemically stable and will migrate slowly into the cheese, but has not been detected more than 8 mm from the rim. It is also allowed as a pesticide in a number of agricultural products e.g. as surface treatment of citrus and pomaceous fruits (Scientific Committee on food, 1999).

Spices and herbs are not generally considered as preservatives, but several have strong anti-microbial activities and should be considered as potential alternatives to the traditional preservatives. Nielsen and Rios (2000) showed that especially mustard oils and to some extent also cinnamon, garlic and clove essential oils were effective against common bread spoilage fungi. These substances are especially attractive in combination with other preserving factors as high concentrations will result in off-flavour formation. Therefore a combination with modified atmosphere packaging or incorporation into the packaging material to form an active packaging system will open new opportunities for the development of safe and environmental friendly preservation systems (Floros *et al.* 2000).

REFERENCES AND OTHER LITERATURE ON PRESERVATIVES

AALTO, T.R., M.C. FIRMAN & RIGLER, N.E. 1953. p-Hydroxybenzoic acid esters as preservatives. I. Uses, antibacterial and antifungal studies, properties and determination. J. Am. Pharm. Assoc. Sci. Ed. 42: 449-457.

BEUCHAT, L.R. 1976. Effectiveness of various food preservatives in controlling the outgrowth of Byssochlamys nivea ascospores. Mycopathologia 59: 175-178.

BRUL, S. & KLIS, F.M. 1999. Mechanistic and mathematical inactivation studies of food spoilage fungi. Fungal Genetics and Biology. 27: 199-208.

CODEX ALIMENTARIUS COMMISSION, 2000, report of the 32RD session of the codex committee on fod additives and contamiants. CL 2000/10-FAC (ftp://ftp.fao.org/codex/alinorm 01/al01_12e.pdf)

El-GAZZAR, F.E. & MARTH, E.M. 1987. Sodium benzoate in the control of growth and aflatoxin production by Aspergillus parasiticus J. Food Prot. 50:305-309.

El-GAZZAR, F.E., G. RUSUL & MARTH, E.M. 1987. Growth and aflatoxin production by *Aspergillus parasiticus* NRRL 2999 in the presence of lactic acid and at different initial pH values. J. Food Prot. 50:940-944.

FDA (Food and Drug Administration, USA. 1999. TITLE 21--FOOD AND DRUGS, Part 101, subpart B, §101.22 (a)(5). U.S. Government Printing Office[CITE: 21CFR101.22] (http://www.access.gpo.gov/cgibin/cfrassemble.cgi?title=2000 21)

FLOROS, J.D., NIELSEN, P.V. & J.K. FARKAS. 2000. Advances in modified atmosphere and active packaging with applications in the dairy industry. Bulletin of the IDF 346-2000: 22-28

FRISVAD, J.C. & SAMSON, R.A. 1991. Filamentous fungi in foods and feeds: ecology, spoilage and mycotoxin production. In Handbook of Applied Mycology. Vol. 3. Foods and Feeds. eds. Arora, D.K., Mukerji, K.G. and Marth, E.H. New York: Marcel Dekker, pp. 31-68.

FRISVAD, J.C., SAMSON, R.A., RASSING, B.R., HORST, M.I. VAN DER, RIJN, F.T.J. & J. STARK 1997. *Penicillium discolor,* a new species from cheese, nuts and vegetables. Antonie van Leeuwenhoek 72: 119-126.

GAREIS, M., J. BAUER, A. von NONTGELAS & GEDEK, B. 1984. Stimulation of aflatoxin B1 and T-2 toxin produciton by sorbic acid. Appl. Environ. Microbiol. 47:416-418.

GOURAMA, H. & BULLERMANL. B. 1988. Effect of potassium sorbate and natamycin on growth and penicillic acid production by Aspergillus ochraceus. Journal of Food Protection. 51:139-144

HARTOG, B.J., R.A. SAMSON & VRIES, J. DE 1986. Serious impairment of reliable preservative systems by sorbic acid resistance fungi. Proc. XIV Int. Congr. Microbiol. Manchester, p. 98.

HOEKSTRA, E.S., VAN DER HORST, M.I., SAMSON, R.A., STARK, J. & F.T.J. VAN RIJN 1998. Survey of the fungal flora in Dutch cheese factories and warehouses. J. Food Mycol. 1: 13-22.

KESZTHELYI, G. 1979. Zum praktischen Einsatz von Konservierungsmitteln. Ernahrung 3: 524-528.

KINDERLERER, J. & HATTON, P.V. 1990. Fungal metabolites of sorbic acid. Food additives and contaminants 7: 657-669.

KING, A.D., H.D.MICHENER & K.A. ITO. 1969. Control of *Byssochlamys* and related heat-resistant fungi in grape products, Applied microbiology, 18: 166-173.

KING. A.D. & W.U. HALBROOK. 1987. Ascospore heat resistance and control measures for *Talaromyces flavus* isolated from fruit juice concentrate. Journal of Food Science. 52: 1252-1254,1266

LIEWEN, M.B. & MARTH, E.H. 1985. Growth and inhibition of micro-organisms in the presence of sorbic acid: a review. J. Food Protect. 48: 364-375.

LUBIENIECKI-VON SCHELHORN, M. 1967. Untersuchungen über die Verteilung von Konservierungsstoffen zwischen Fett und Wasser. Z. Lebensm. Unters. Forsch. 131: 329-345.

MUTASA, E.S., N. MAGAN & SEAL, K.J. 1990. Effect of potassium sorbate and environmental factors on growth of tobacco spoilage fungi. Mycol. Res. 94:971-978.

NIELSEN, P.V. & RIOS, R. 2000. Inhibition of fungal growth on bread by volatile components from spices and herbs, and the possible application in active packaging, with special emphasis on mustard essential oil. International journal of food microbiology (In press).

NIELSEN, P.V. 1991. Preservative and temperature effect on growth of three varieties of the heat-resistant mould, *Neosartorya fischeri,* as measured by an impedimetric method. J. Food Sci. 56:1735-1740.

NIELSEN, P.V., L.R. BEUCHAT & FRISVAD, J.C. 1989. Growth and fumitremorgin production by Neosartorya ficheri as affected by food preservatives and organic acids. J. Appl. Bacteriol . 66:197-207.

NOORDERVLIET, P.F. 1978. Sorbic acid and pimaricine as preservatives on cheese and sausage surfaces. Nordeuropaeisk Mejeritidsskrift nr. 4, 121-127 + XVIII.

RAY, L.L. & BULLERMAN, L.B. 1982. Preventing growth of potentially toxic molds using antifungal agents. J. Food Prot. 45:953-963

REHM, H.J., S. LAUFER-HEYDENREICH & WALLNOFEN, P. 1967. Zur Kenntnis der antimikrobiellen Wirkung von Biphenyl und Derivaten des Biphenyls. Z. Lebensm. Unters. Forsch. 135: 117-122.

REHM. H.J. 1961. Grenzhemmkonzentrationen der zugelassen Konservierungsmittel gegen Mikro-organismen. Z. Lebensm. Unters. Forsch. 115: 293-309.

ROBINSON, H.J., H.F. PHARES & GRAESSLE, O.E. 1964. Antimycotic properties of thiabendazole. J. Invest. Dermatol. 42, 479-482.

RUSUL, G. & MARTH, E.M. 1987. Growht and aflatoxin production by Aspergillus parasiticus NRRL 2999 in the presence of potassium benzoate or potassium sorbate at different initial pH values. J. Food Prot. 50:820-825.

RUSUL, G., F.E. EL-GAZZAR & MARTH, E.M. 1987. Growth and aflatoxin production by Aspergillus parasiticus NRRL 2999 in the presence of acetic or propionic acid and at different initial pH values. J. Food Prot. 50:909-914.

SAMSON, R.A. 1989. Filamentous fungi in food and feed. J. Appl. Bacteriology, Symposium Suppl.1989: 27S-35S.

SCIENTIFIC COMMITTEE ON FOOD, 1999, Opinion on imazalil for incorporation in cheese coatings. Annex II to the minutes of the 119th plenary meeting (http://www.europa.eu.int/comm/dg24/health/sc/scf/out46_en.pdf).

SKIRDAL, I.M. & EKLUND, T. 1993. Microculture model studies on the effect of sorbic acid on Penicillium chrysogenum, Cladosporium cladosporioides and Ulocladium atrum at different pH levels. J. Appl. Bacteriol. 74:191-195.

TAYLOR, S.L. & BUSH, R.K. 1986. Sulfites as food ingredients. Food Technol. June, p. 47-52.

TILBURY, R.H. (ed.) 1980. Developments in food preservatives -1. Applied Science Publishers Ltd. London.

VAN DOKKUM, W., R.H. DE VOS, F.A. CLOUGHLEY, K.F.A.M. HULSHOF, F. DUKEL & WIJSMAN, J.A. 1982. Food additives and food components in total diets in The Netherlands. Br. J. Nutr. 48: 223-231.

Chapter 9

USEFUL ROLE OF FUNGI IN FOOD PROCESSING

M.J.R.Nout

Department of Agrotechnology and Food Sciences, Wageningen University, 6703 HD Wageningen, The Netherlands.

TABLE OF CONTENTS

INTRODUCTION

The growth and metabolic activity of fungi (yeasts and moulds) in foods can have different effects. On the one hand, undesirable changes such as decay, spoilage and even toxin formation may occur; on the other hand, fungal activity has been exploited by man for the purpose of food production and processing. Practices of gathering fungal fruit bodies (mushrooms) as well as the application of moulds to prepare fermented foods go back several centuries. More recently, fungal mycelium as well as yeast cells have been cultivated to obtain protein-rich nutritious food for human and animal consumption. Fungi play a significant role in industrial fermentations to produce a variety of enzymes and other organic substances. Many of these are applied as food ingredients. Most recently, recombinant DNA techniques have become available to modify fungal properties. Some implications for the food industry will be mentioned.

FUNGAL FERMENTED FOODS

Fermentation is one of the oldest ways of food processing and is of great economic importance. The occurrence, manufacture, properties, and use of fermented foods is well-documented (Campbell-Platt 1987; Steinkraus 1997; Wood 1998). Some fermented products (cheese, beer, wine, soya sauce) have experienced an enormous scale-up of production, with the use of sophisticated inoculum; on the other hand, many fermented foods are still produced using age-old "traditional" techniques under simple or even primitive conditions.

For reasons of required product properties and economics, most food fermentations cannot be carried out profitably under sterile conditions. Fermented foods may therefore contain a variety of bacteria, yeasts and molds, originating from raw materials, inoculum and process contamination. In mixed-culture fungal fermented foods, stimulation by metabolites, by mutualistic degradations of substrate, or by release of lysis products; or inhibition by competition, formation of antibiotic substances or antimicrobial metabolites, are important determinants of the balance of microbial populations (Nout 1995). According to the physical nature of the substrate, fermentations can be distinguished in liquid and solid state fermentations. In liquid fermentations, an aqueous continuous phase serves as a medium for homogeneous distribution of micro-organisms and for heat- and mass transfer. Liquid fermentation is used for e.g. beverages and sauces manufacture.

Table 1 lists some fermented foods in which fungi play an essential role. In addition, their raw materials, representative fermenting micro-organisms, the type of fermentation system (liquid or solid) and the relative importance of groups of microflora for a successful fermentation are listed. In temperate climates, mould-ripened meat and cheese are dominated by *Aspergillus* and *Penicillium* spp. In addition, yeasts play a role in bakery products and in alcoholic beverages. In sub-tropical and tropical regions, fungal fermented foods predominate in East- and South East Asia. *Rhizopus, Amylomyces, Mucor, Neurospora* and *Monascus* spp. are found frequently as functional mycoflora. Yeast-fermented products from tropical areas include alcoholic snacks and beverages. A few selected fermented foods will be discussed in the next section.

Table 1. A selection of fungal mixed-culture fermented foods.

Moulds	Yeasts	Bacteria	Substrate	Type of fermentation	Product Name	Nature	Use	Origin
Actinomucor elegans			soybean curd	SSF-L	Sufu (Furu)	spreadable solid	flavouring protein food	China, Vietnam
Amylomyces rouxii	*Endomyces spp., Hyphopichia burtonii*	*Pediococcus pentosaceus Enterococcus faecalis*	uncooked rice	SSF	Ragi	solid tablet	inoculum for rice wine making	Orient
Amyl. rouxii	*Hyp. burtonii, End. fibuliger*	*Ped. pentosaceus Ent. faecalis*	cassava	SSF	Peuyeum	semi-solid mass	snack	Indonesia
Aspergillus oryzae A. soyae	*Zyg. rouxii, Torulopsis versatilis*	*Tetragenococcus halophila Ent. faecalis*	soya bean+ rice/barley	SSF	Miso	semi-solid paste	flavouring	Orient
A. oryzae A. soyae group	*Zyg. rouxii, Zyg. soyae Zyg. major, Hansenula spp., Torulopsis spp., Candida etchellsii, C. versatilis*	*Lactobacillus delbrueckii, Tet. halophila , Ped. damnosus*	soya bean+ wheat+ salt	SSF-L	Soya sauce	liquid	flavouring	Orient
A. oryzae	*Hans. anomala, Sacch. cerevisiae(saké)*	*Leuc. mesenteroides var. sake, Lb. saké*	rice	SSF-L	Saké	liquid	liquor	Japan
Monascus purpureus M. ruber, M. pilosius			rice	SSF	Ang-kak	granular solid	colouring, flavouring, health ingredient	China, Japan
Penicillium roqueforti	*Yarrowia lipolytica*	*Leuconostoc spp., Lc. Lactis, Lc. lactis biovar, diacetylactis , Lc. lactis ssp., cremoris*	milk curd	SSF	Roquefort-type blue-veined cheese	semi-solid cake	protein food flavouring	France
Pen. camemberti (P.. candidum, P. caseicolum, P. album)	*Candida spp., Kluyveromyces spp, Saccharomyces spp., Torulopsis spp.*	*Brevibacterium linens Lc. lactis ssp. cremoris Lc. lactis*	milk curd	SSF	Camembert-type surface-ripened cheese	semi-solid cake	protein food flavouring	France
Pen. nalgiovense Pen. chrysogenum		*Micrococcus spp., Staphylococcus spp. Pediococcus spp.. Lactobacillus spp.*	meat (sausage)	SSF	Salami	solid	protein food	Europe
Rh. oligosporus R. chinensis, R. oryzae, Mucor indicus	*Trichosporon beigelii, Clavispora lusitaniae, C. maltosa, C. intermedia, Yar. lipolytica*	*Klebsiella pneumoniae Enterobacter cloacae Lactobacillus spp.*	mostly soybeans	SSF	Tempe	semi-solid cake	protein food snack	Indonesia
	Torlp. holmii, S. cerevisiae, Pichia saitoi, C. krusei	*Lb. plantarum, Lb. fructivorans. Lb. brevis var. lindneri, Lb. sanfrancisco*	rye or wheat	SSF	Sour-dough	semi-solid mass	inoculum for sourdough bread making	Europe
	Sacch. cerevisiae (S.. uvarum, (S. elegans, S.. bayanus)	*Oenococcus oeni Ped. acidilactici Lb. casei*	grape juice	L	Wine	liquid	liquor	Europe
	Sacch. cerevisiae (S. bayanus, (S. uvarum) Brettanomyces spp.	*Lactobacillus spp. Pediococcus spp.*	barley ; wheat	L	Lambic- Gueuze beer	liquid	liquor	Belgium

Abbreviations: SSF = solid-substrate; L = liquid; SSF-L = solid-substrate followed by liquid fermentation. Moulds: A.= *Aspergillus*; Amyl.= *Amylomyces*; Pen.= *Penicillium*; Rh.= *Rhizopus*. Yeasts: C.= *Candida*; End.= *Endomyces*; Hans.= *Hansenula*; Hyp.= *Hyphopichia*; S.= *Saccharomyces*; Torlp.= *Torulopsis*; Yar.= *Yarrowia*; Zyg.= *Zygosaccharomyces*. Bacteria: Ent.= *Enterococcus*; Lb.= *Lactobacillus*; Lc.= *Lactococcus*; Leuc.= *Leuconostoc*; Ped.= *Pediococcus*; Tet. = *Tetragenococcus*. Brackets: old names.

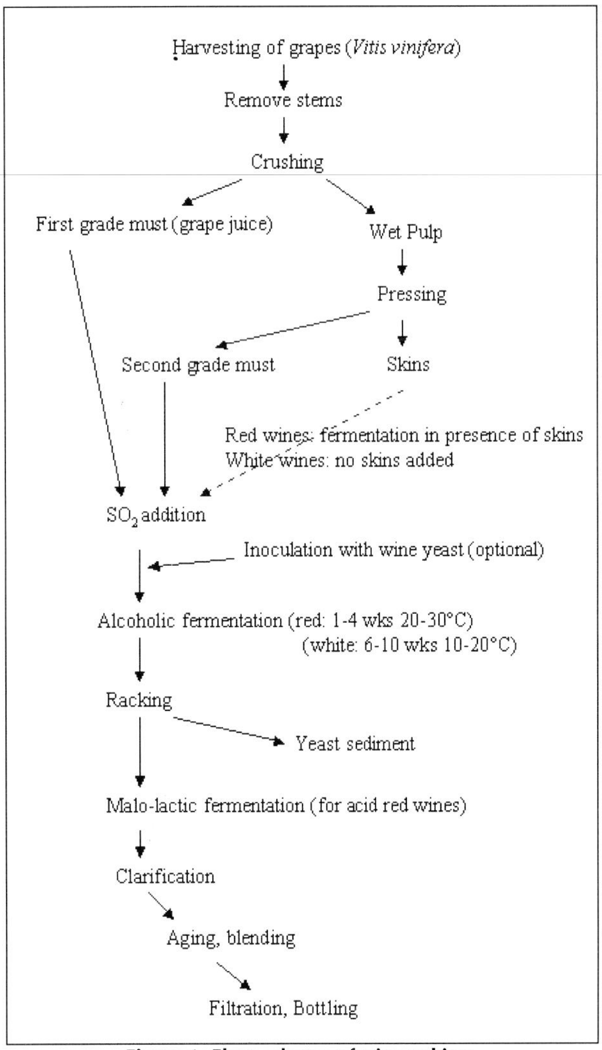

Figure 1. Flow-scheme of winemaking.

Wine

The wide variety of wines is not only due to the more than 5000 varieties of grape (*Vitis vinifera*) but particularly to the growing conditions (location, soil, climate) and fermentation conditions. The principle of winemaking is summarized in Figure 1.

Grapes must be free from mouldiness, except for the manufacture of the sweet "Sauterne" wines which require a "noble rot" of the mould *Botrytis cinerea*. Red wines are usually fermented "on the skins". Usually, 100-150 mg SO_2/litre is added to suppress excessive growth of epiphytic yeasts (*Candida, Hanseniaspora, Kloeckera, Metschnikowia, Pichia, Rhodotorula, Saccharomyces, Torulopsis, Trichosporon* spp.). This will enable good dominance of the selected wine yeast (*Saccharomyces cerevisiae*, often accompanied by *Torulopsis stellata*) (added at approx. 10^6/ml juice). It is important that epiphytic flavour producing yeasts are in balance with the functional yeasts for alcoholic fermentation. When all fermentable sugars have been exhausted, the alcoholic fermentation stops, and yeast is removed by syphoning ("racking") in order to prevent off-odours from yeast autolysis. In high-acidity wines, lactic acid bacteria (*Oenococcus oeni*) are inoculated to transform malic acid into lactic acid ("Malo-lactic fermentation") thus giving the wine a more mellow taste (Nout 1992).

Mould-ripened Camembert cheese

This is one of the several surface-ripened cheeses. Originating from Normandy, France it was first prepared by Marie Harel in 1791. In 1890, M. Ridel developed the famous wooden box facilitating a world-wide exportation.

The principle of Camembert production is outlined in Figure 2. After production of very young cheese curd, this is sprayed with a fine mist containing *Penicillium camemberti* conidia. After customary brining and conditioning, mould growth starts at the cheese surface during the incubation period. The crust (rind) of Camembert cheese is thin and white. The various starter strains have colours ranging from greyish-blue to pure white. The interior of the cheese must be yellowish with a rather firm white centre. During ripening, proteolytic and lipolytic enzymes of *P. camemberti* diffuse somewhat into the cheese.

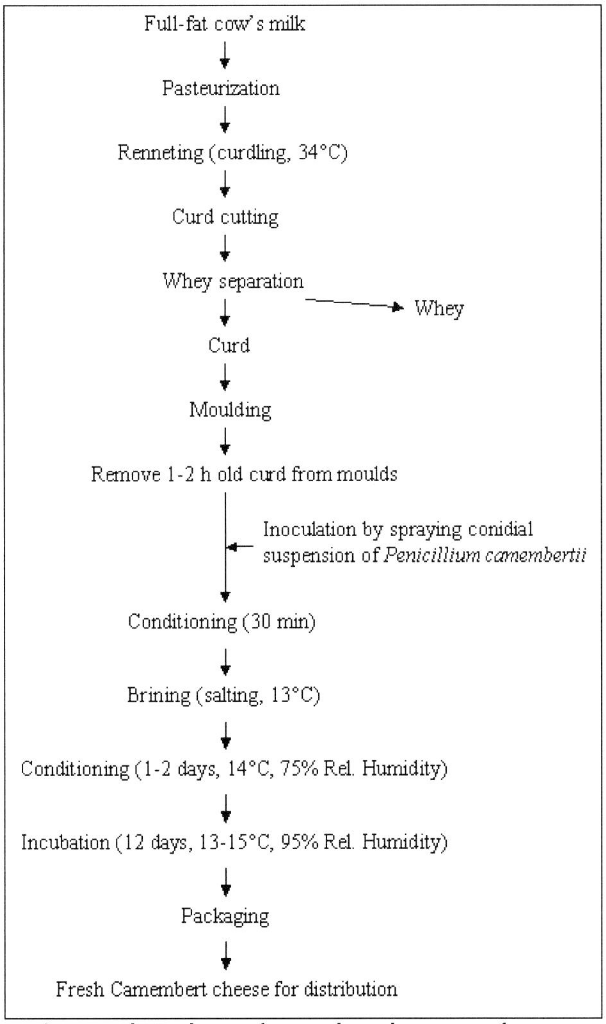

Figure 2. Flow-scheme of camembert cheese manufacture.

Figure 3. Camembert cheese

But, the characteristic softening is due to pH increase caused by the release of NH_3. Yeasts, particularly *Debaryomyces hansenii,* are often present in cheese and also contribute to pH increases due to their consumption of lactic acid (Wyder and Puhan 1999). Proteolytic reactions and aminotransferase activity contribute greatly to flavour development (Prieto *et al.* 1999; Sable and Cottenceau 1999). Usually the product is consumed at an age of 3-5 weeks. Tested in pure culture, all known *P. camemberti* strains are able to produce the mycotoxin cyclopiazonic acid (CPA). This appears to be a stable property, since old culture collection strains had not lost the ability to produce CPA. Efforts are undertaken to obtain CPA-negative mutant strains, and starters are selected on this criterium. The risk of poisoning is very small, however. Only very low levels of CPA could be detected in Camembert cheese. This is explained by its chemical instability in the presence of amines and its poor diffusion from the rind into the cheese. In addition, CPA is hardly produced at storage temperatures <15°C.

Sufu

Sufu, also written as fu-ru, is a strongly flavoured mould-fermented soybean curd with a spreadable consistency. (Su, 1986). Sufu is produced mainly in China both commercially and domestically, with annual production estimated at least 300,000 metric tons. Sufu is consumed as an appetizer and a side dish e.g. with breakfast rice or steamed-bread. The principle of the sufu making process is outlined in Figure 4 (Nout and Aidoo, 2000). All sufu is made from cubes of tofu (curd obtained by salt-coagulated soya bean milk). The tofu used for sufu making is firmed-up by pressing or by a short hot air treatment, and subsequently the cubes are surface-inoculated and incubated at high relative humidity at temperatures of 20-35° C, depending on locality and season. The types of moulds involved are mainly *Actinomucor* and *Mucor* spp. Large-scale factories use pure culture inoculants of e.g. *Actinomucor elegans,* whereas at a smaller production level, mouldy straw mats are used to inoculate the tofu cubes with a mixed flora of moulds and some bacteria.

As *A. elegans* does not grow well at temperatures exceeding 25° C, other moulds e.g. *Mucor hiemalis* or *Rhizopus chinensis* are used at higher temperatures. After several days, the tofu cubes are covered with a dense layer of cottony mycelium. This intermediate product is referred to as pehtze or pizi. The pehtze has become a source of fungal proteolytic enzymes, and the soybean protein has been partially degraded.

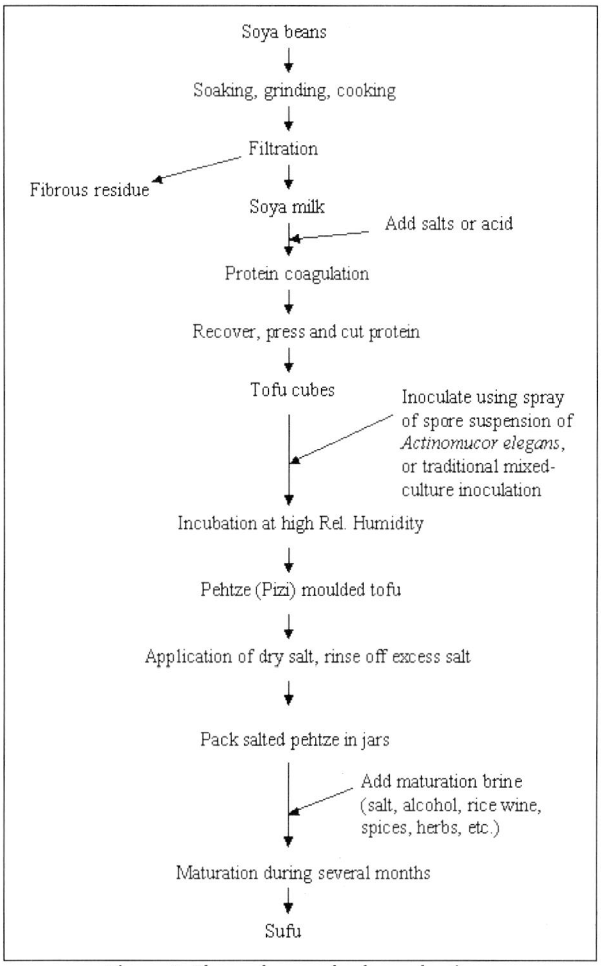

Figure 4. Flow-scheme of sufu production.

Figure 5. Red (left) and Grey (right) Sufu

The next steps, salting and brining, are aimed at preserving the product while allowing enzymatic maturation that will finally result in the soft consistency and the required strong smell and taste.
A first bulk treatment with solid salt has the aim to quickly increase the salt content to about 15%. Next, salted pehtze cubes are filled into jars or other containers and filled up with maturation brine or dressing mixture. The composition of the maturation brine strongly influences the properties of the final product. It always contains about 10-12% NaCl, and sometimes rice wine up to about 10% ethanol content. Very popular is red sufu, for which the maturation brine contains ang-kak (see Table 1). Ang-kak does not only provide red and orange pigments, but it also contains several active enzymes that contribute to the degradation and flavour of sufu. Although not much data is available on the bacteriology of sufu, the presence of *Tetragenococcus halophila* (refer to soya sauce) has been reported. After several months of maturation, the outside of the jars is cleaned and they are labelled and packed for distribution.

Tempe

Tempe originates from Java, Indonesia but is also popular in the Netherlands and it has gained considerable consumer markets in the U.S.A., Europe and Australia. It is a sliceable cake obtained by solid-substrate fungal fermentation of previously soaked and cooked leguminous seeds, cereals or other suitable material. The most common substrate is soya beans (Nout, 1992; Nout and Rombouts, 1990; Wood 1998). Tempe provides a cheap, nutritious, digestible and safe source of vegetable protein. It is not eaten fresh, but only after cooking (stewing) or frying in oil (crispy "tempe kripik"). The traditional manufacturing process is outlined in Figure 7. Soya beans are soaked and wet-dehulled (traditional process in Indonesia), or dry-dehulled and soaked (mechanized process in e.g. the Netherlands). During soaking a natural lactic fermentation takes place lowering the pH of the beans, rendering them favourable to mould growth and protecting them from pathogenic and spoilage-causing bacteria.

After boiling and cooling to ambient temperature, the beans are inoculated using traditional "usar" starters (*Rhizopus* spp. on a carrier of *Hibiscus* leaf: Nout *et al.,* 1992) or powdered mixed culture starters on a carrier of rice flour or cassava fibre. The major functional moulds, *Rhizopus oligosporus* and *R. oryzae* germinate rapidly at 37°C and their fast mycelial growth ensures their dominance over contaminating strains of e.g. *Aspergillus* spp. Within 30 hours the loose beans are "knitted" together to form a solid mass. *R. oligosporus* was observed to penetrate approximately 2 mm into cooked soya beans (Varzakas, 1998).

Figure 6. Fresh Tempe

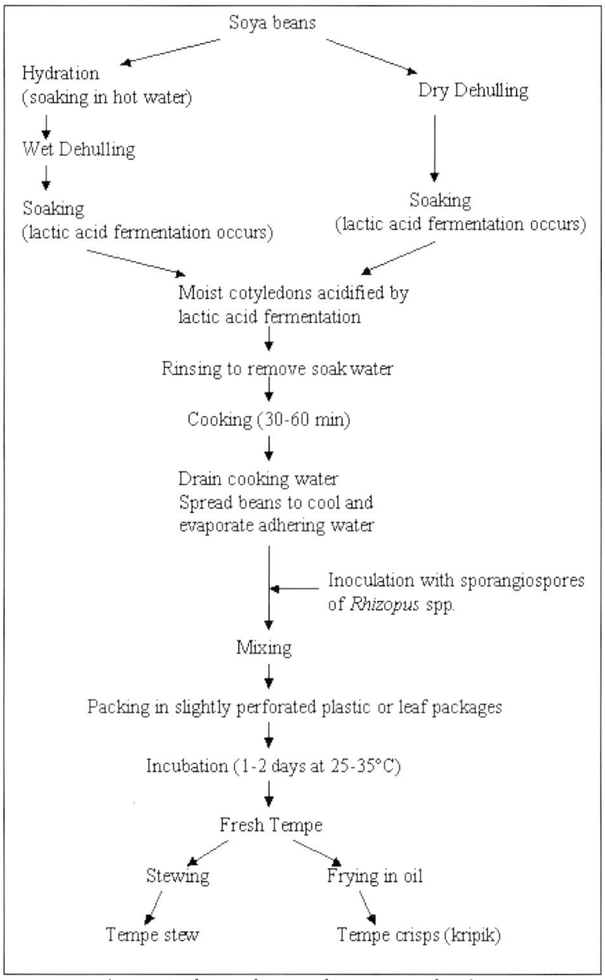

Figure 7. Flow-scheme of tempe production.

As a result of packaging in sparsely perforated leaves or polythene sheet, micro-aerobic conditions prevail, enabling mycelial growth but suppressing the formation of black sporangia. Sporulation is also inhibited by formed NH_3 (Sparringa and Owens, 1999a) that is retained in the package. As a result, a shiny white tempe is produced. Although the cooked soya beans are slightly acid, their pH will increase due mainly to the production of NH_3 and to some extent the consumption of lactic acid (Sparringa and Owens, 1999b). The enzyme activity of the *Rhizopus* spp. includes proteolytic, lipolytic, carbohydrate degrading enzymes and phosphatases. By their action, part of the polymeric substrate is solubilized enabling easy digestibility (Chango *et al.,*

1993) and a dramatic increase of low-molecular weight water-soluble components from 7% in cooked beans to 27% in fresh tempe (Kiers *et al.*, 2000). The degradation of non-starch polysaccharides (arabinogalactans and pectic substances) was correlated with the decrease of firmness of the beans during the first 24 hours of fermentation (De Reu *et al.*, 1997). Antinutritional factors e.g. phytic acid are also degraded, improving the bio-availability of phosphate and minerals. During the tempe manufacturing process, the levels of flatulence-associated galacto-oligo saccharides is significantly reduced; this is caused mainly by leaching during soaking and cooking, and to some extent to the fungal fermentation (Ruiz-Teran and Owens, 1999). Several fat-soluble vitamins and provitamins are formed during the fermentation (Denter *et al.* 1998). Health-promoting effects of tempe are attributed to polyhydroxylated isoflavones that can be formed from the soya bean isoflavones genistein and daidzein (Klus and Barz, 1998; Wuryani, 1995). Recent process innovations include semi-continuous processing lines (Nout, 2000) for tempe and fermentation at controlled temperatures in rotating drum bioreactors for tempe-like food ingredients (Han *et al.* 1999).

Soya sauce

Soya sauce is of Chinese or Japanese origin. There are about 3,600 companies in Japan which produce soy sauce or shoyu and of these the five largest produce about half of the total annual production of 1.2 million metric tons (Nout and Aidoo 2000).There are many different types, but Koikuchi-Shoyu is the best representative of fermented soya sauces (Fukushima 1989; Yokotsuka and Sasaki 1998). It is a clear deep-brown liquid with the following approximate composition: 22°Be, 17% w/v NaCl, 1.6% w/v total Nitrogen, 1% w/v formol Nitrogen, 3% w/v reducing sugars, 2.3% v/v alcohol, and pH 4.7. In principle, the manufacturing process (Figure 8) consists of 3 phases: koji making, brine fermentation and refining. Koji making is a solid state mould fermentation of a mixture of previously steam-cooked soya beans and roasted and crushed wheat. This is inoculated with an inoculum of conidiospores of *Aspergillus oryzae* or *A. sojae* and is incubated at 25°C during 2-3 days to obtain dense growth and greenish-yellow sporulation indicating that a high level of fungal enzymes has been produced. These enzymes include peptidases, proteinases, glutaminase, amylase, pectinases and cellulases needed for subsequent hydrolysis of the polymeric protein and carbohydrate matter of the raw material.

Criteria used for the selection of *Aspergillus* starter strains include (1) flavour and colour of the final product, (2) ability to sporulate well, (3) high rate of growth and enzyme production, (4) length of conidiophore, (5) genetic stability and (6) inability to produce mycotoxins (Aidoo *et al.* 1994). The second phase (maturation of the moromi mash) takes place in a salt bath (22-25% w/v NaCl) so that spoilage micro organisms cannot develop, but only enzymatic degradation takes place. Nevertheless, during the first 2 months at 15°-20°C, halophilic lactic acid bacteria (*Tetragenococcus halophila*), and in the following months at 30°C, salt-tolerant yeasts (*Zygosaccharomyces rouxii*) will develop after some time of ripening. They are part of the traditional process, and their metabolites add essential flavour components such as HEMF (4 hydroxy -2 (or 5)- ethyl - 5 (or 2) - methyl - 3 (2H) furanone) (Hayashida *et al.*, 1997) to the product. Nowadays they are added as inoculum to ensure their activity. This phase takes 6-8 months in total, after which the soya sauce is harvested by pressing. Continuous pasteurization (70-80°C) guarantees shelf-life, but is also essential for flavour- (Ishihara *et al.*, 1996) and colour development and for inactivation of the fungal enzymes. After aseptic bottling, the product is distributed. Melanoidins (plant melanins of molecular weight of about 5600) extracted from soy sauce were shown to inhibit the growth rate of HCT15 cells derived from human colon carcinoma, and AGS cells from human gastric carcinoma (Kamei *et al*,. 1997).

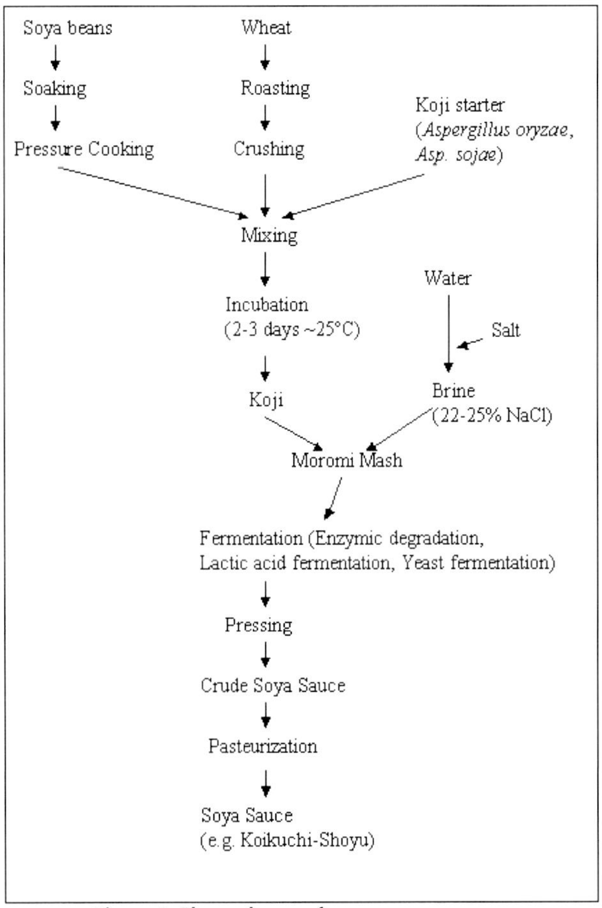

Figure 8. Flow-scheme of soya sauce process.

Table 2. Some edible mushrooms.

Species	Common name	Substrate	Fruiting conditions
Agaricus bisporus *Agaricus bitorquis*	Button mushroom (champignon)	Composted straw; horse manure	6 weeks 14-18°C
Lentinus edodes	Shii-take	Wood logs Saw-dust	5-6 years 12-20°C several weeks
Pleurotus ostreatus *Pleurotus sajor-caju*	Oyster mushroom	Saw-dust, straw, leaves, paper, cotton waste, etc.	10-14 weeks 10-35°C
Ustilago maydis	Maize mushroom ("huitlacoche", "caviar azteca")	pre-harvest maize cobs	several weeks at 25-35º C
Volvariella volvacea	Paddy straw mushroom	Composted rice straw; various agro by-products	2-6 weeks 30-40°C

FUNGAL BIOMASS and BIOTRANSFORMATIONS

Edible fungi can be grown for consumption purposes, either as their fruiting bodies (mushrooms), as mycelium (mycoprotein) or as yeast cells (single cell protein).

Mushrooms

From the wide variety of edible mushrooms, only a few species have developed into commercial commodities. Table 2 lists some species of interest and their conditions of growth. The total commercial mushroom production is estimated at 1-2 million tonnes. *Agaricus* spp. account for a majority of the total production. The nearly-white *A. bisporus* is known best, but the more virus-resistant *A. bitorquis* and brown-capped varieties of *A. bisporus* gain increasing interest.

After colonization of the substrate by the fungal mycelium has taken place, fruiting is initiated by changing the environmental conditions (aeration, humidity, temperature). Fruiting bodies are produced in a number of successive "flushes" with intermezzo's of 1-2 weeks.

After 4-5 flushes have been harvested, the yield decreases and the cycle is re-started with fresh substrate. The old colonized substrate is used as protein-rich animal feed ingredient. In wheat straw, the initial lignin content was halved after 12 weeks of growth of *Pleurotus* spp., and consequently the digestibility of the remaining cellulose was increased significantly for ruminants (Moyson and Verachtert, 1991).The large fruiting bodies of the exotic basidiomycete, *Ustilago maydis* are collected and consumed as mushrooms. They are especially popular in Latin America where they are known as "Huitlacoche" or "Caviar Azteca" (Valverde *et al.*, 1995).

Single-cell protein (fodder yeast, mycoprotein)

Various yeast strains (*Candida utilis, C. tropicalis, Yarrowia lipolytica, Kluyveromyces lactis*) can be grown at high cell yield on industrial by-product substrates e.g. cheese whey, waste water from potato starch industries, wood sulfite liquor, and hydrocarbon residues. Although several industrial processes have been patented, the single cell protein thus obtained is presently not competitive compared with other (e.g. soya) proteins.

Fungal mycelium of e.g. *Penicillium chrysogenum* and *Aspergillus niger* is produced in large quantities as a by-product of the fermentative production of antibiotics, enzymes, organic acids, etc. The mycelium of non-toxigenic fungi is an interesting food ingredient since it has a relatively high crude protein content (approx. 12% on fresh weight basis). The mycelium is used as an ingredient in animal feed.

Likewise, the mycelium of a strain of *Fusarium venenatum* (previously *F. graminearum*) is used for industrial-scale production of a texturized protein which is marketed as "Quorn" (Wiebe *et al.*, 1996).

Figure 9. Maize mushroom (*Ustilago maydis*)

The product finds application as meat substitute in savoury pies, soups, etc. The use of microbial protein for human or animal consumption is limited by its nucleic acid content. The WHO recommends a maximum level of nucleic acid (NA) of 2% in foods. Whereas bacterial protein contains rather high levels of NA, mycoprotein (*F. venenatum*) contains about 6-13%. The RNA content can be reduced to below 2% by a heat-shock ("curing") treatment. After fermentation, the mycelium is briefly held at 64°C to activate intracellular RNAses that convert cellular RNA into monomer nucleotides which can diffuse out of the cells. Simultaneously, this curing achieves pasteurization.

Protein-enrichment of starchy foods and feeds
Because of their ability to degrade carbohydrate matter, fungi are useful in the upgrading of the nutritive value of industrial and agro-processing by-products e.g. starch containing sweet potato residues (Yang *et al.,* 1993) or cellulosic sugarcane bagasse (Moo-Young *et al.,* 1992). For that purpose added cheap sources of nitrogen e.g. $(NH_4)_2SO_4$ and urea can be converted into protein thus enriching the food. Final products (after a few days fermentation) may contain approximately 30% w/w crude protein on a dry matter basis, when food-grade fungi e.g. *Aspergillus niger, Rhizopus* spp. and *Neurospora sitophila* are used.

Detoxification of mycotoxins
Several fungi are able to metabolize and detoxify mycotoxins. Patulin in apple juice could be degraded >90% during alcoholic fermentation with *Saccharomyces cerevisiae*; Ochratoxin A in barley malt was partly (50-80%) degraded by *Saccharomyces* spp. during beer brewing; Aflatoxin B1 in groundnut meal could be almost fully degraded (>95%) during 7 days solid-substrate fermentation with a selected *Aspergillus* strain (Bol and Knol, 1991). Although these fermentation processes may be relatively time consuming and expensive, biological detoxification of mycotoxins has the advantage over chemico-physical processes that the treatment takes place under mild conditions and the quality characteristics of the product e.g. nutritive value, can be better maintained.

FOOD INGREDIENTS and ADDITIVES OF FUNGAL ORIGIN

Organic acids
Organic acids produced by fermentation can be distinguished in 2 groups: (i) produced through the tricarboxylic acid pathway and (ii) produced directly from glucose. In group (i), **citric acid** is the most important organic acid produced by fermentation. Its annual production is estimated at 400,000 tonnes which are made mainly with *Aspergillus niger,* but also *Yarrowia lipolytica* is used. Surface as well as submerged culture systems are employed for the conversion of cheap carbon sources (molasses) and n-alkanes. Citric acid is extensively used in the food industry as an acidulant and flavouring substance. **Itaconic acid** can be made with *Aspergillus terreus* using e.g. molasses or wood hydrolysates, and finds application in the chemical industry (polymers, surface-active compounds). **Malic acid** is made in a 2-stage process: first, fumaric acid is produced from sugars using (immobilized) *Schizophyllum commune,* and subsequently the fumarase activity of *Aspergillus wentii* converts fumaric acid into malic acid. **Tartaric acid** (50,000 tonnes/year) is produced with *Aspergillus griseus* and *A. niger,* and is applied in the food industry as an acidulant. In group (ii), **gluconic acid** (50,000 tonnes/year) is made mainly with bacteria, but also *A. niger* and *A. foetidus* are used in submerged and solid-substrate ("koji") processes. Gluconic acid and its δ-lactone are applied in foods (acidulant) and in the medical field. **Lactic acid** (30,000 tonnes/year) is made mainly with lactobacilli, but also with *Rhizopus oryzae,* and finds wide applications in the food industry as acidulant, preservative agent, baking powder, etc. (Mattey, 1992).

Lipids
Fungal lipid content may be sometimes as high as 60-80% of biomass dry weight. However, plant and animal oils and fats are cheaper to produce, so it is only for specialty products that fermentation is of economic interest. In particular, the ability to accumulate polyunsaturated fatty acids is of interest from nutritional point of view. Several moulds are commercially used to produce γ-linolenic acid. On cheap sources of nitrogen and carbon (rape meal, starch, molasses), *Mucor javanicus* and *M. rouxii* can be grown in submerged cultures at γ-linolenic acid yields of 0.33 g/L medium. In such cases, the lipid content is 7-11% of the biomass dry weight, and γ-linolenic acid represents 17-37% of lipid weight (Lindberg and Hansson,1991).

Enzymes
Within the range of enzymes produced by fermentation, the majority are proteases and carbohydrases. Proteolytic enzymes obtained with *Aspergillus oryzae, Penicillium roquefortii* and *Mucor* spp. are applied in detergents, and in food processing e.g. accelerated cheese ripening, breadmaking, and tenderization of meat. Carbohydrases include amylolytic enzymes (α-amylase, glucoamylase) produced by *Aspergillus oryzae* and *A. niger* and are applied in e.g. breadmaking, brewing and confectionery. Other carbohydrases are cellulases (Persson *et al,.* 1991) made by *A. niger, Penicillium* spp. and *Trichoderma reesei*; pectinases made by *Aspergillus* spp.; and ß-glucanase made by *A. niger* and *Penicillium* spp.; these enzymes are applied to improve digestibility of fibrous foods, and filterability of fruit

juices and beer, etc. Other important enzymes include lipases produced by *Mucor* spp. and *A. niger*, applied for dairy flavour development; RNAses made by *A. oryzae* applied to prepare nucleotides acting as flavour enhancers; glucose oxidase made by *A. niger* which has many food and medical applications; and phytase made by *A. niger, A. oryzae* and *A. ficuum* which is applied to degrade the anti-nutritive factor phytate in foods and feeds, thereby improving the bio-availability of phosphate and minerals (Zyta, 1992).

Others

A wide variety of valuable substances can be produced using fungi, including amino acids, polysaccharides, vitamins, pigments and flavour components (Vandamme 1993). A few selected examples of recent interest include:the production of natural food-grade red, orange and yellow **pigments** (ankaflavin, monascorubrin and monasein) by *Monascus purpureus* and *M. barkari* (Yongsmith *et al,*. 1993); the formation of heat-stable, characteristic, antigenic **EPS (extracellular polysaccharides)** by most fungi has led to the development of a new generation of immuno-assays (latex agglutination tests and ELISA) for the detection of fungal contamination of foods and raw materials, even after they have undergone thermal processing (e.g. pasteurized fruit juices, jams, etc.) (De Ruiter *et al.,* 1993); the production of the major mushroom **flavour** component 1-octen-3-ol by submerged cultivation of *Pleurotus* and *Morchella* spp. In mushrooms, linolenic acid is converted into a precursor which is oxidized to 1-octen-3-ol upon exposure to O_2 during homogenization of mushroom tissue. After harvesting the fermentor-grown mycelial pellets, they are subjected to shear stress causing cell disruption. This initiates the oxidative conversion. The homogenate is pressed, the obtained juice is freeze-dried and contains approximately 1200 ppm of 1-octen-3-ol. It is applied in dehydrated soups, gravies, etc. (Schindler and Seipenbusch, 1990).

GENE EXPRESSION

Conventional strain improvement (by selection, crossing or mutagenesis) has given rise to production strains with improved process characteristics but the application of recombinant DNA technology offers considerable scope for process improvement and the development of new techniques. Presently, for many of the food-grade fungal species transformation protocols have been developed. In principle, it is technically possible to (1) regulate the expression of a target gene, (2) alter the gene copy number, (3) replace or delete a gene, and (4) introduce a gene from another (heterologous) source. The attraction of fungi, particularly moulds, as expression hosts is that some of them are food-grade, there is much industrial experience with their cultivation and proteins (enzymes) are secreted to high concentrations. Moulds of industrial importance include *Aspergillus niger, A. oryzae* and *Trichoderma reesei.* Nowadays, the production of calf chymosin (rennet for cheese-making), phytase (degrading the anti-nutritive factor phytate in food and feed) and proteases (baking, detergents) is carried out at a commercial level using r-DNA constructs, all *Aspergillus* spp.

Yeasts, especially *Saccharomyces cerevisiae* and *Kluyveromyces lactis,* have very good potential as host organisms. Various r-DNA constructs have been reported (Lang-Hinrichs and Hinrichs 1992) amongst which the following are of particular interest: (1) optimization of brewer's yeasts by expression of amylolytic enzymes, (2) improvement of baker's yeast by e.g. increasing it's freezing resistance, and (3) improvement of distiller's yeasts by enabling their growth on cheap agro-industrial by-products, e.g. cheese whey, starch, cellulose and hemicellulose. At a commercial level, calf chymosin (cheese rennet) is expressed in *Kluyveromyces lactis,* glucose oxidase of *Aspergillus niger* is expressed in *Saccharomyces cerevisiae,* and thaumatin (powerful sweetener for e.g. softdrinks) is expressed in *Saccharomyces cerevisiae* and *Kluyveromyces lactis*

REFERENCES

AIDOO, K.E., SMITH, J.E., WOOD, B.J.B. 1994. Industrial Aspects of Soy Sauce Fermentations Using Aspergillus. p.155-169 in: Powell, K.A. , Renwick, A., Peberdy, J.F. (Eds.) The Genus *Aspergillus*: from taxonomy and genetics to industrial application. Plenum Press, New York.

BOL, J., KNOL, W. 1991. Biotechnological methods for detoxification of mycotoxins. p.112-124 in: Applications of biotechnology in the food industry (Kun, L.Y., Hen, N.B., Yeo, V. (Eds.)). Food Biotechnology Centre, Singapore Institute of Standards and Industrial Research.

CAMPBELL-PLATT, G. 1987. Fermented foods of the world. A dictionary and a guide. Guildford, Surrey, UK, Butterworth Scientific Ltd.

CHANGO, A., BAU, H.M., VILLAUME, C., MEJEAN, L., NICOLAS, J.-P. 1993. Effets de la fermentation par Rhizopus oligosporus sur la composition chimique des lupins blanc doux, jaune doux et jaune amer [Effects of fermentation by *Rhizopus oligosporus* on chemical composition of sweet white lupine, sweet yellow lupins]. Sciences des Aliments 13, 285-295.

DE REU, J.C., LINSSEN, V.A.J.M., ROMBOUTS, F.M., NOUT, M.J.R. 1997. Consistency, polysaccharidase activities and non-starch polysaccharides content of soya beans during tempe fermentation.. Journal of the Science of Food and Agriculture 73, 357-363.

DE RUITER, G.A., NOTERMANS, S.H.W., ROMBOUTS, F.M. 1993. New methods in food mycology. Trends in Food Science and Technology 4, 91-97.

DENTER, J, REHM, H.J., BISPING, B., 1998. Changes in the contents of fat-soluble vitamins and provitamins during tempe fermentation. International Journal of Food Microbiology 45, 129-134.

FUKUSHIMA, D. 1989. Industrialization of fermented soy sauce production centering around Japanese shoyu. pp 1- 88 in: Steinkraus, K.H.(Ed.). Industrialization of indigenous fermented foods. New York, Basel: Marcel Dekker, Inc.

HAN, B.Z., KIERS, J.L., NOUT, M.J.R. 1999. Solid-substrate fermentation of soybeans with Rhizopus spp.: comparison of discontinuous rotation with stationary bed fermentation. J. Bioscience & Bioengineering 88, 205-209.

HAYASHIDA, Y, NISHIMURA, K., SLAUGHTER, J.C. 1997. The influence of mash pre aging on the development of the flavour active compound, 4 hydroxy 2(or5) ethyl 5(or2) methyl 3(2h) furanone (HEMF), during soy sauce fermentation. International Journal of Food Science and Technology 32, 11-14.

ISHIHARA, K., HONMA, N., MATSUMOTO, I., IMAI, S., NAKAZAWA, S., IWAFUCHI, H. 1996. Comparison of volatile components in soy sauce (Koikuchi shoyu) produced using *Aspergillus sojae* and *Aspergillus oryzae*. Journal of the Japanese Society for Food Science and Technology - Nippon Shokuhin Kagaku Kogaku Kaishi 43, 1063-1074.

KAMEI, H., KOIDE, T., HASHIMOTO, Y., KOJIMA, T., UMEDA, T., HASEGAWA, M. 1997. Tumor cell growth-inhibiting effect of melanoidins extracted from miso and soy sauce. Cancer Biotherapy and Radiopharmaceuticals 12, 405-409.

KIERS, J.L., NOUT, M.J.R., ROMBOUTS, F.M. 2000. In Vitro digestibility of processed and fermented soya bean, cowpea and maize. J. Sci. Food Agric.80, 1325-1331

KLUS, K., BARZ, W. 1998. Formation of polyhydroxylated isoflavones from the isoflavones genistein and biochanin A by bacteria isolated from tempe. Phytochemistry 47, 1045-1048.

LANG-HINRICHS, C., HINRICHS, J. 1992. Recombinant yeasts in food and food manufacture - possibilities and perspectives. Agro-Food-Industry Hi-Tech 3 (5), 12-18.

LINDBERG, A.M., HANSSON, L. 1991. Production of gamma-Linolenic Acid by the Fungus *Mucor rouxii* on Cheap Nitrogen and Carbon Sources. Applied Microbiology and Biotechnology 36, 26-28.

MATTEY, M. 1992. The Production of Organic Acids. Critical Reviews in Biotechnology 12 (1-2), 87-132.

MOO-YOUNG, M., CHISTI, Y., VLACH, D. 1992. Fermentative conversion of cellulosic substrates to microbial protein by *Neurospora sitophila*. Biotechnology Letters 14, 863-868.

MOYSON, E., VERACHTERT, H. 1991. Growth of Higher Fungi on Wheat Straw and Their Impact on the Digestibility of the Substrate. Applied Microbiology and Biotechnology 36, 421-424.

NOUT, M.J.R. 1992. Ecological aspects of mixed-culture food fermentations. p.817-851 in: Carroll, G.C., Wicklow, D.T. (Eds.) The Fungal Community: its organization and role in the ecosystem. second edition. New York: Marcel Dekker.

NOUT, M.J.R. 1995. Fungal interactions in food fermentations. Can. J. Botany 73 (Suppl.1), S1291-S1300.

NOUT, M.J.R. 2000. Tempe manufacture: traditional and innovative aspects. in: Baumann, U., Bisping, B. (Eds.) (in press) Tempe. Weinheim: Wiley-VCH.

NOUT, M.J.R., AIDOO, K.E. 2000. Asian Fungal Fermented Food. in: Osiewacz, H.D. (ed) The Mycota. Vol.X "Industrial applications" (in press). Berlin: Springer Verlag.

NOUT, M.J.R., MARTOYUWONO, T.D., BONNE, P.C.J., ODAMTTEN, G.T. 1992. *Hibiscus* Leaves for the Manufacture of Usar, a Traditional Inoculum for Tempe. Journal of the Science of Food and Agriculture 58, 339-346.

NOUT, M. J. R., ROMBOUTS, F. M. 1990. Recent developments in tempe research. Journal of Applied Bacteriology 69 (5), 609-633.

PERSSON, I., TJERNELD, F., HAHNHAGERDAL, B. 1991. Fungal Cellulolytic Enzyme Production - A Review. Process Biochemistry 26, 65-74.

PRIETO, B., URDIALES, R., FRANCO, I., TORNADIJO, M.E., FRESNO, J.M., CARBALLO, J. 1999. Biochemical changes in Picon Bejes-Tresviso cheese, a Spanish blue-veined variety, during ripening . Food Chemistry 67, 415-421.

RUIZ-TERAN, F., OWENS, J.D. 1999. Fate of oligosaccharides during production of soya bean tempe. Journal of the Science of Food and Agriculture 79, 249-252.

SABLE, S., COTTENCEAU, G. 1999. Current knowledge of soft cheeses flavor and related compounds. Journal of Agricultural and Food Chemistry 47, 4825-4836.

SCHINDLER, F. , SEIPENBUSCH, R. 1990. Fungal flavour by fermentation. Food Biotechnology 4(1), 77-85 (Proceedings of the International Conference on Biotechnology and Food, Hohenheim University, Stuttgart Febr 20-24, 1989).

SPARRINGA, R.A., OWENS, J.D. 1999a. Inhibition of the tempe mould, *Rhizopus oligosporus*, by ammonia . Letters in Applied Microbiology 29, 93-96.

SPARRINGA, R.A., OWENS, J.D. 1999b. Causes of alkalinization in tempe solid substrate fermentation. Enzyme and Microbial technology 25, 677-681.

STEINKRAUS, K.H. 1997. Classification of fermented foods: worldwide review of household fermentation techniques. Food Control 8, 311-318.

SU, Y.C. 1986. Sufu. Pp.69-83 in: Reddy, N.R., Pierson, M.D., and Salunkhe, D.K. (Eds) Legume-based Fermented Foods. CRC Press, Boca Raton, Florida, USA.

VALVERDE, M.E., PAREDESLOPEZ, O., PATAKY, J.K., GUEVARALARA, F. 1995. Huitlacoche (Ustilago maydis) as a food source - Biology, composition, and production. Critical Reviews in Food Science and Nutrition 35, 191-229.

VANDAMME, E.J. 1993. Production of vitamins and related biofactors via microorganisms. Agro Food Industry Hi-Tech 4 (5), 29-31.

VARZAKAS, T. 1998. *Rhizopus oligosporus* mycelial penetration and enzyme diffusion in soya bean tempe. Process Biochemistry 33, 741-747.

WIEBE, M.G., ROBSON, G.D., OLIVER, S.G., TRINCI, A.P.J. 1996. pH oscillations and constant low pH delay the appearance of highly branched (colonial) mutants in chemostat cultures of the Quorn myco-protein fungus, Fusarium graminearum A3/5.. Biotechnology and Bioengineering 51, 61-68.

WOOD, B.J.B. 1998. Protein-rich foods based on fermented vegetables. p. 484-504 in: B.J.B.Wood (Ed.) Microbiology of Fermented Foods. Second Edition. Volume 2. Blackie Academic & Professional, London.

WURYANI, W. 1995. Isoflavones in tempe. ASEAN Food Journal 10, 99-102.

WYDER, M.T., PUHAN, Z. 1999. Investigation of the yeast flora in smear ripened cheeses. Milchwissenschaft-Milk Science International 54, 330-333.

YANG, S. S., JANG, H. D., LIEW, C. M., PREEZ, J. C. DU 1993. Protein enrichment of sweet potato residue by solid-state cultivation with mono- and co-cultures of amylolytic fungi. World Journal of Microbiology & Biotechnology 9, 258-264.

YOKOTSUKA, T., SASAKI, M. 1998. Fermented protein foods in the Orient: shoyu and miso in Japan. p. 351-415 in: B.J.B.Wood (Ed.) Microbiology of Fermented Foods. Second Edition. Volume 1. Blackie Academic & Professional, London.

YONGSMITH, B., TABLOKA, W., YONGMANITCHAI, W., BAVAVODA, R. 1993. Culture conditions for yellow pigment formation by *Monascus* sp. KB 10 grown on cassava medium. World Journal of Microbiology & Biotechnology 9, 85-90.

ZYTA, K. 1992. Mould Phytases and Their Application in the Food Industry. World Journal of Microbiology & Biotechnology 8, 467-472

APPENDIX

GLOSSARY OF USED MYCOLOGICAL TERMS

acerose: needle-like and stiff (like a pine needle)
acervulus: a lenticular or cupshaped fructification containing conidiophores and conidia embedded in the host plant.
acropetal: describes the succession of conidia arising in chains with the youngest conidium at the top (apex) of the chain
anamorph: imperfect (or asexual state) conidial form of sporulation
annellide: a conidiogenous cell which forms blastoconidia in basipetal succession, each conidium is produced through the same opening of the previously formed conidium and leaves a ringlike band (annellation) at the fertile apex after seceding. The conidiogenous cell elongates during conidiogenesis (progressive)
antheridium: the male gametangium
apex: at the end or the top
apical cell: tip cell
aplanospore: non-motile spore
apophysis: a swelling at the top of the sporangiophore just below the sporangium in Mucorales (e.g. *Absidia*)
apothecium: a cup or saucerlike ascoma
arthric: (of conidiogenesis), thallic conidiogenesis. Cells are separated from an undifferentiated part of a hypha, and transformed into conidia (see also arthroconidia)
arthroconidia: (thalloconidia) conidia resulting from breaking up of a hypha into separate cells and transformed into conidia e.g. *Geotrichum*
ascogonium: the female gametangium
ascoma (= ascocarp): a fruitbody in Ascomycetes containing asci and ascospores
ascospores: sexual spores produced in an ascus
ascus: a saclike organ in which the ascospores are formed after karyogamy and meiosis
aseptate: without a crosswall
ballistoconidia: see ballistospore
ballistospore: a basidiospore which is forcibly blown away
basidium: the organ of the Basidiomycetes which bears the basidiospores after karyogamy and meiosis
basipetal: describes the succession of conidia in which the youngest conidium is at the base e.g. of the chain
basitonous: main axis with branches confined to the lower part
biseriate: in two series. In *Aspergillus*: phialides are not formed directly on the vesicle but on metulae
blastoconidium: the wall or the conidiogenous cell bulges out apically to form the conidial wall. Conidia can be produced solitary or in acropetal chains, i.e. each new conidium can bud at its tip.
budding: a process of vegetative multiplication in which there is a development of a "daughter" cell from a small outgrowth of a "mother" cell (monopolar, bipolar and multilateral budding, see section of chapter 1 on yeasts)
chlamydospore: a thick-walled, thallic, terminal or intercalary asexual resting spore, mostly for longtime survival; usually nondeciduous
chromophilic: deeply staining
circinnate: twisted round; coiled
cleistothecium: an ascoma (fruitbody) without a special opening
coenocytic: a multinucleate mass of protoplasm; nonseptate cells with many nuclei
collarette: a cupshaped structure at the apex of a phialide
colony: (of mycelial fungi) a group of hyphae (with or without conidia), which arise from one spore or cell. Colony appearance varies e.g.: velvety, floccose (cottony), funiculose (hyphae aggregated into strands), fasciculate (in little groups or bundles), granulose, powdery, synnematous (compact groups of erect and sometimes fused conidiophores)
columella: an usually swollen sterile central axis within a sporangium, e.g. in Mucorales
conidiogenesis: process of conidium formation
conidiogenous cell: the fertile cell from which or within conidia are directly produced. Conidiogenous cells may be morphologically identical with or differentiated from vegetative cells (vegetative hyphae)
conidioma: (pl. conidiomata); any structure which bears conidia, e.g. separate conidiophores, synnema, acervulus, pycnidium, sporodochium etc.
conidiophore: specialized hypha, simple or branched, on which conidiogenous cells are born
conidium: asexual, vegetative, nonmotile propagule, not formed by cleavage (as in sporangiospores). In Deuteromycetes the term "conidium" is recommended and the term "spore" is reserved to zoo, sporangio, basidio, ascospores
dictyochlamydospores: a nondeciduous multicelled chlamydospore, resembling the conidia of *Alternaria*
dikaryon: (adj. dikaryotic) of fungal cells: having 2 nuclei, which may be haploid, cf. conjugate nuclei; of hyphae: made up of dikaryotic cells
diploid: having 2n chromosomes
echinulate (adj.): describes cell wall surfaces (e.g. of conidia, or conidiogenous cells), with small pointed processes or spines
endoconidium: conidium formed inside a hypha
exudate: droplets excreted by the mycelium. Can be characteristic for a species e.g. in *Penicillium chrysogenum*
fasciculate: hyphae and/or conidiophores in bundles
floccose: cottony
foot cell: basal part of the conidiophore (e.g. in *Aspergillus*) or basal cell of conidium in *Fusarium*
funiculose: hyphae/conidiophores aggregated into strands
conjugation: fusion
gametangium: a meiosporangium in which generative (sexual) cells (gametes) are formed
geniculate: bent like a knee
haploid: having *n* number of chromosomes
heterothallic: sexual reproduction can only occur through the interaction of different (+ and) thalli (e.g. zygospore formation in some Mucorales). In mycelium only male or only female nuclei present
hilum: a mark or scar, especially that on a spore at the point of attachment to a conidiophore or sterigma
holomorph: the whole fungus including the asexual (imperfect or anamorph) state as well as the sexual (perfect or teleomorph) state

homothallic: sexual reproduction can occur without the interaction of two differing thalli (mycelium intermixed with male and female nuclei)
Hülle cells: terminal or intercalary thickwalled cells surrounding the ascomata (in some *Aspergillus* species)
hyaline: transparent or nearly so
hypha (pl. hyphae): (vegetative) filament of a mycelium, without or with crosswalls
intercalary: between two cells, between the apex and base (of cells, spores etc.)
karyogamy: the fusion of two sex nuclei after cell fusion (after plasmogamy)
lateral: at the side
macroconidia: the larger conidia of a fungus, usually multi celled (e.g. *Fusarium*)
microconidia: small conidia, usually onecelled
meiosis: that part of a life cycle in which a diploid nucleus (2n) undergoes reduction and becomes haploid (n)
meristem: growing zone
merosporangium: (of Mucorales) a cylindrical outgrowth from a swollen end of the sporangiophore, in which asexual merospores are produced in a linear row
metabolite: any substance produced during metabolism
merospores: asexual spores formed in a merosporangium by cytoplasmic cleavage
mesoconidia:in *Fusarium*, conidia intermediate between micro and macroconidia. Usually 0 to morecelled but smaller in size and distinct in shape from the macroconidia
metula (pl. metulae): apical branch(es) of conidiophore bearing phialides (e.g. *Penicillium, Aspergillus*)
mononematous: hyphae or conidiophores arising singly from the substrate
monopodial: a way of branching, in which the persistant main axis gives off branches
multinucleate: containing many nuclei
muriform: having longitudinal and transverse septa
mycelium: (vegetative) mass of hyphae; thallus (vegetative) body of the fungus
mycotoxins: fungal secondary metabolites that in small concentrations are toxic to vertebrates and other animals when introduced via a natural route.
osmophilic: growing under conditions of high osmotic pressure (e.g. on substrate containing 2040% sucrose or NaCl)
ostiole: any opening (pore) by which conidia or spores are freed from a fruitbody (e.g. pycnidium, perithecium)
pedicel: a small stalk
penicillate: like a little brush
perithecium: a globose or flaskshaped ostiolate ascoma (fruitbody) in which asci are produced
phialide: a conidiogenous cell which produces conidia (phialoconidia) in basipetal succession, without an increase in length of the phialide itself
pionnotes: a spore mass with a fat or grease like appearance (in *Fusarium*)
polyphialide: conidiogenous cell with more than one opening from each of which conidia are produced in basipetal succession; the new conidiogenous loci develop by sympodial proliferation of the fertile cell
poroconidia: (= tetric conidia) dark conidiogenous cells with an apical, pigmented wall thickening. The conidium is pierced through, after dehiscence a pore is visible
pseudomycelium: the formation of a filamentous structure consisting of cells which arise exclusively by budding
pseudothecium: asci developing in cavities arranged in a hymenium or arising from cushionlike structures in the loculi
pycnidium: flaskshaped to globose fruitbody, usually with one apical opening (= ostiole), containing conidiogenous cells
racemose: a way of branching. A main axis producing side branches which are shorter than the main axis.
rachis: conidiumbearing extension of a conidiogenous cell (resulting from sympodial development, zigzag geniculate extension)
ramoconidium, (pl. ramoconidia): a fertile apical branch of a conidiophore which produces conidia and later secedes and functions as a conidium itself (e.g. *Cladosporium*)
retrogressive: describes mechanism of conidiogenous cell development in which the fertile cell shortens during basipetal conidium formation, (e.g. *Trichothecium*)
rhizoid: hyphal roots; a rootlike structure, acting as a feeding organ and/or holdfast, (e.g. *Rhizopus*)
rostrate: having a beak
sclerotium: a resting body, usually globose, produced by aggregation of hyphae into a firm mass with or without host tissue, normally sterile
septum: a crosswall in a cell
spine: a narrow process with a sharp point
spinulose: delicately spiny
sporangiole: small, usually globose sporangium with a reduced columella and contaning one or a few spores
sporangiophore: a specialized hyphal branch, which supports one or more sporangia
sporangiospore: a spore produced within a sporangium
sporangium: asexual reproductive structure, inwhich spores are produced by cytoplasmic cleavage
spore: a general term for a reproductive structure in fungi, bacteria, and cryptogams. In fungi the term "spore" is used in several combinations e.g. chlamydospore, ascospore, zoospore, basidiospore
sporodochium: a cushion like mass of conidiophores; conidia and conidiogenous cells produced above the substrate
sterigma: a conical or elongated process which bears the basidiospore
stipe: stalk
stolon: a "runner" (e.g. *Rhizopus*)
striate: marked with lines, grooves or ridges
stroma: a mass or matrix of vegetative hyphae, with or without tissue of the substrate, in or on which fructifications can be produced
substrate: the material on or in which an organism is living
suspensor: a hypha which supports a gamete, gametangium or zygospore
sympodial: describes a mechanism of conidiogenous cell proliferation in which each new growing point appears just behind and to one side of the previous apex, often resulting in a geniculate cell configuration; of conidiophores characterized by continued growth after the main axis has produced a terminal spore, or by the development of a succession of apices each of which originate below and to one side of the previous apex.
synnema: bundles of erect hyphae and conidiogenous cells bearing conidia (in some Hyphomycetes)
teleomorph: perfect (sexual) state of a fungus; ascigerous or basidial form of sporulation
teliospore (teleutospore): one of the spore forms (commonly resting spore) of the Uredinales (or rust fungi) or

Ustilaginales (or smut fungi) from which the basidium is produced

thallic: a segment of a fertile hypha differentiates by conversion into a conidium (see arthric)

tretic: (see poroconidia)

truncate: ending abruptly, as if cut straight across

uniseriate: in one series. In *Aspergillus* phialides arise directly from the vesicle

velutinous: velvety

verrucose: having small rounded processes (warts)

verruculose: delicately verrucose

verticil: whorl

vesicle: a bladderlike sac; the swollen apex of the conidiophore in *Aspergillus*

xerophilic: preferring dry places, growing under dry conditions

zygospore: a sexual spore produced by Zygomycetes, thickwalled, often ornamented and darkly pigmented, formed by fusion of a pair of gametangia

TERMINOLOGY OF COMMON SHAPES OF FUNGAL STRUCTURES AS USED IN THIS BOOK

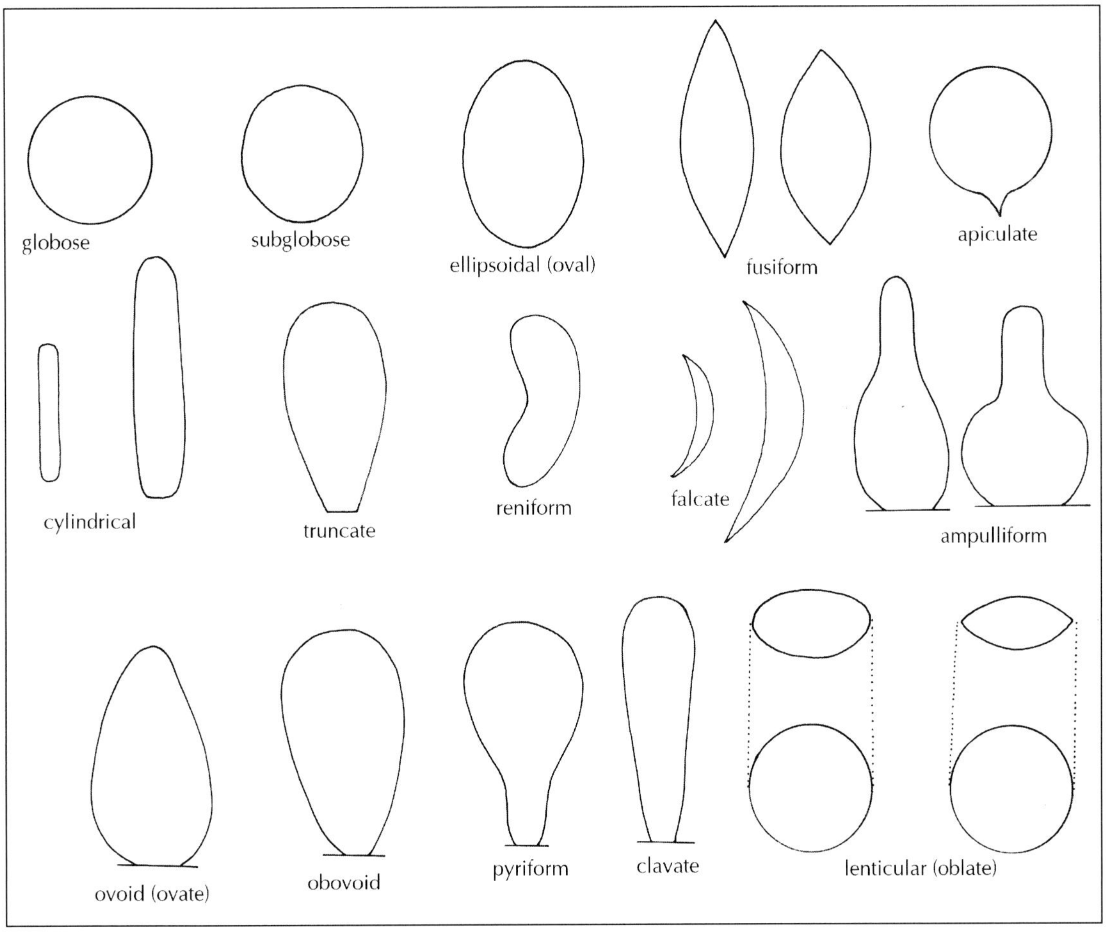

MYCOLOGICAL MEDIA FOR FOOD- and INDOOR FUNGI

RECOMMENDED MEDIA FOR CULTIVATION AND IDENTIFICATION OF COMMON FILAMENTOUS FUNGI

Fungi	Media
Acremonium	OA, CMA, MEA (2% MEA CBS)
Alternaria	MEA, PCA, HAY
Aspergillus	Cz, CYA, 2% MEA (CBS);
for xerophilic species:	Cz or MEA with additional sugar (20 or 40 %) or other low water activity media.
Aureobasidium	MEA
Basidiomycetes	2% MEA CBS + 1 ppm benomyl
Botrytis	PCA, MEA, HAY
Byssochlamys	MEA, OA
Chaetomium	OA (with lupine stem), CMA
Chrysonilia	OA
Cladosporium	MEA, OA, PCA
Emericella	MEA, OA
Epicoccum	MEA, OA, PCA, SEA
Eurotium	MEA or Cz with sucrose 20 or 40% or other low water activity media
Fusarium	PSA, PDA, SNA, CLA, YES
Geotrichum	MEA
Monascus	MEA, OA
Moniliella	MEA, PCA
Mucorales	MEA (4%)
for zygospore formation	MYA (*Absidia*), CH (*Mucor spp*)
Neosartorya	MEA, OA
Paecilomyces	MEA, OA
Penicillium	MEA, Cz, CYA, YES
Phialophora	MEA, OA
Phoma	OA (with lupine stem), MEA (4% MEA CBS) (+ UV)
Scopulariopsis	MEA, OA
Stachybotrys	MEA, OA, Hay
Talaromyces	OA, MEA, CMA, YES
Trichoderma	OA, MEA
Trichothecium	MEA
Ulocladium	MEA, PCA, HAY
Verticillium	MEA
Wallemia	MEA + additional sugar 20 or 40 % or other low water activity media
Xeromyces	MY50G

CH = Cherry decoction agar; CLA = Carnation Leaf Agar; CMA = Cornmeal Agar; CYA = Czapek Yeast Extract Agar; CY20S = Czapek Yeast Extract Agar with 20% sucrose; Cz = Czapek Agar; C20S = Czapek Agar with 20% Sucrose; DG 18 = Dichloran Glycerol Agar Base; HAY = Hay infusion Agar; MEA = Malt Extract Agar; MEA + 20% sucrose = Malt Extract Agar with 20% sucrose; MYA = Malt Yeast Agar; MY50G = Malt Extract Yeast Extract 50 % Glucose Agar; OA = Oatmeal Agar; PCA = Potato Carrot Agar; SNA = Synthetischer Nährboden; YES = Yeast Extract Sucrose Agar.

MYCOLOGICAL MEDIA FOR DETECTION, ISOLATION AND IDENTIFICATION

Ingredients in grams are dissolved in one litre distilled water and sterilized by autoclaving at 121°C for 15 minutes, unless stated otherwise. Usually 15 or 20 g agar per litre are recommended in the original formulations. These quantities are mentioned in the formulations below. However, for some brands of agar smaller amounts (8 g) per litre might be sufficient. At CBS 8 g of powdered "Spanish agar 700" (EEC number E 406) is used for most media.

Addition of 1 ml trace metal solution (TMS) per litre medium is recommended for avoiding atypical colony growth and colour. Trace metal-solution: 1 g $ZnSO_4.7H_2O$ and 0.5 g $CuSO_4.5H_2O$ in 100 ml distilled water.

Several media are commercially available and indicated with the brand name between brackets. Media with the indication (CBS) are formulations used at the Centraalbureau voor Schimmelcultures.

Addition of mineral solution (MS), containing 5g KCL, 5 g $MgSO_4.7H_2O$, 0.1 g $FeSO_4.7H_2O$ per 100 ml water is also recommended in some media (10 ml per litre medium).

For the suppression of bacterial growth, antibiotics should be added to the medium to give a final concentration of:

penicillin-G	50 ppm
streptomycin	30-50 ppm
aureomycin	20-50 ppm
neomycin	100 ppm
chloramphenicol	100 ppm

Only chloramphenicol withstands autoclaving without loss of activity.

ADYS: Acetic acid Dichloran Yeast extract agar

Yeast extract	20 g
Sucrose	150 g
$MgSO_4$. 7 H_2O	0.5 g
Dichloran	0.002 g
Agar	20 g
Distilled water	1000 ml

After autoclaving 0.5 % glacial acetic acid is added.

AFPA: *Aspergillus flavus/A.parasiticus* selective medium (Oxoid)

Peptone	10 g
Yeast extract	20 g
Ferric Ammonium Citrate	0.5 g
Dichloran	0.002 g
Chloramphenicol	0.1 g
Agar	15 g
Distilled water	1000 ml

Final pH 6.3 ± 0.2

NOTE: Dichloran and Chloramphenicol can be added before autoclaving.

CHERRY-DECOCTION AGAR (CBS)

1 kg cherries (without stones and stalks) in one litre water is heated to boiling and simmered gently for 2 h. Strain through cloth and sterilize at 110°C (= 0.5 atm) for 30 min. Dissolve 15 g agar in 800 ml water and sterilize at 121°C

for 15 min. Add 200 ml cherry extract and mix well. Sterilize again for 5 min at 102°C (=0.1 atm). Final pH 3.8-4.6

CLA: Carnation leaf agar (Fischer *et al.*, 1982)
Carnation leaves are cut into pieces, dried gently and sterilized by means of gamma irradiation or propylene oxide fumigation. A few sterile pieces can be placed on 1.5-2% water agar (nearly solid). Instead of carnation leaves also pieces of sterile filter paper can be used
NOTE: Suitable medium for cultivation of *Fusarium*.

CMA: Cornmeal agar (CBS)
Add 60 g freshly ground cornmeal to 1 litre water, heat to boiling and simmer gently for 1 h. Strain through cloth and sterilize for 15 min at 121°C overpressure. Fill up to 1 litre and add 15 g agar and sterilize at 121°C for 15 min. Also commercially available.

CREA: Creatine Sucrose agar

Creatine(1 H_2O)	3 g
Sucrose	30 g
KCl	0.5 g
$MgSO_4.7H_2O$	0.5 g
$FeSO_4.7H_2O$	0.01 g
$K_2HPO_4.3H_2O$	1.3 g
Bromocresol purple	0.05 g
Agar	15 g
Distilled water	1000 ml

Final pH 8.0 ± 0.2 (adjust after medium is autoclaved).
NOTE: a modification of CREA (=CRE) with 1.6 g $K_3PO_4.7H_2O$ can also be used.

CREAD: Creatine Sucrose Dichloran agar

Creatine (1 H_2O)	3 g
Sucrose	30 g
KCl	0.5 g
KH_2PO_4	1 g
$FeSO_4.7H_2O$	0.01g
$MgSO_47\ H_2O$	0.5 g
Chloramphenicol	0.05 g
Dichloran	0.002 g
Bromocresole purple	0.05 g
$ZnSO_4.7H_2O$	0.01 g
$CuSO_4.5H_2O$	0.005 g
Agar	20 g
Distilled water	1000 ml

After autoclaving 0.05 g chlortetracycline is added and pH is adjusted to 4.8.

CZ: Czapek agar (CBS)

Sucrose	30 g
$NaNO_3$	3 g
K_2HPO_4	1 g
KCl	0.5 g
$MgSO_4.7H_2O$	0.5 g
$FeSO_4.7H_2O$	0.01 g
Agar	15 g
Distilled water	1000 ml

Final pH 6.2 ± 0.2

CZ20S: Czapek agar with 20% sucrose
The same as Czapek agar but containing 200 g sucrose.
NOTE: addition of 1 mL of the trace metal solution (see under formulation TMS) is recommended

CYA: Czapek yeast (autolysate) extract agar (Samson and Pitt, 1985)

$NaNO_3$	3 g
K_2HPO_4	1 g
KCl	0.5 g
$MgSO_4.7H_2O$	0.5 g
$FeSO_4.7H_2O$	0.01 g
Yeast extract	5 g
Sucrose	30 g
Agar	20 g
Distilled water	1000 ml

Final pH 6.0-6.5
NOTE: addition of 1 mL of the trace metal solution (see under formulation TMS) is recommended

CZID: Czapek Dox Iprodione Dichloran agar

Czapek-Dox Broth (Difco)	35 g
$CuSO_4.5H_2O$	0.005 g
$ZnSO_4.7H_2O$	0.01g
Chloramphenicol	0.05 g
Dichloran (0.2% in ethanol)	1 ml
Agar	20 g
Distilled water to	1000 ml

After autoclaving and cooling to 50°C, add chlortetracycline solution, 10 ml and iprodione suspension, 1 ml. Chlortetracycline solution is 0.5% aqueous, and iprodione suspension (which should be shaken before addition to CZID) contains Rovral 50WP (Rhone-Poulenc, Agro Chemie, Lyon, France) 0.3 g in sterile water, 50 ml. CZID was developed for the selective detection of *Fusarium* species.

DG18: Dichloran 18% Glycerol agar

Peptone	5 g
Glucose	10 g
KH_2PO_4	1 g
$MgSO_4.7H_2O$	0.5 g
Dichloran (0.2% in ethanol)	1.0 ml
Glycerol	220 g
Chloramphenicol	0.1 g
Agar	15 g
Distilled water	1000 ml

Add minor ingredients and agar to ca. 800 ml distilled water. Steam to dissolve agar, then make to one litre with distilled water. Add 220 g glycerol and sterilize by autoclaving at 121°C for 15 min. The final a_w is 0.955: final pH 5.6 ± 0.2. Commercially available (Oxoid).

DRBC: Dichloran rose bengal chloramphenicol agar

Peptone	5 g
Glucose	10 g
KH_2PO_4	1.0 g
$MgSO_4.7H_2O$	0.5 g
Dichloran	0.002 g
Rose Bengal	0.025 g
Chloramphenicol	0.1
Agar	15 g
Distilled water	1000 ml

Final pH 5.6 ± 0.2
NOTE: Original formula according to King *et al.* (1979) contains chlortetracycline. Hocking (1981) replaced this antibioticum with chloramphenicol.

DRYES: Dichloran Rose Bengal Yeast Extract Sucrose agar (Frisvad, 1983)

Yeast extract	20 g
Sucrose	150 g

Dichloran (0.2% in ethanol)	1.0 ml
Rose bengal (5% soln., w/v)	0.5 ml
Chloramphenicol	0.1 g
Agar	20 g
Water to	1000 ml

Final pH 5.6 (adjusted after medium is autoclaved). This medium detects *Penicillium verrucosum* and *P. viridicatum* by production of a purple reverse colour. Can be modified by adding 0.5 g $MgSO_4.7H_2O$.

DYSG: Dichloran Yeast Extract 18% Glycerol agar

Yeast extract	20 g
Sucrose	150 g
K_2HPO_4	1 g
$MgSO_4.7\ H_2O$	0.5 g
Chloramphenicol	0.05 g
Dichloran (0.2% in ethanol)	1.0 ml
$ZnSO_4.7H_2O$	0.01 g
$CuSO_4.5H_2O$	0.005 g
Agar	20 g
Glycerol	220 g
Water	1000 ml

After autoclaving 0.05 g chlortetracycline is added.

HAY: Hay infusion agar (CBS)

Sterilize 50 g hay in one litre water at 121°C for 30 min. Strain through cloth and fill up to one litre. Adjust pH to 6.2 with K_2HPO_4 and take 1000 ml extract and add 15 g agar. Autoclave for 15 min at 121°C.

MEA: Malt extract agar (according to Blakeslee)

Malt extract (powdered)	20 g
Peptone, bacteriological	1 g
Glucose	20 g
Agar	20 g
Distilled water	1000 ml

Final pH 5.0-5.5

MEA: Malt extract agar (Oxoid)

Malt extract	30 g
Mycological peptone	5 g
Agar	15 g
Distilled water	1000 ml

Final pH 5.4 ± 0.2
Sterilize by autoclaving at 115°C for 10 min.

MEA: Malt extract agar (Difco, Bacto)

Malt extract, Difco	30 g
Bacto agar	15 g
Distilled water	1000 ml

Final pH 5.5 ± 0.2

MEA: Malt extract agar (Merck)

Malt extract	30 g
Peptone (from soya flour)	3 g
Agar	15 g

Final pH 5.6 ± 0.1

MEA: Mout extract agar 4% (2%) (CBS)

Malt extract	400 ml (200ml for 2% MEA)
Distilled water	600 ml (800ml for 2% MEA)
Agar	15 g

Add water to malt extract from the brewery until it contains 10% sugar (measurement with areometer). Mix 400ml (200ml) of this solution with 15 g agar and 600 (800) ml water. Malt agar may also conveniently be prepared with malt syrup (10-40 g/l) or malt powder (10-20 g/l).

M20 or **M40**

The same as MEA but containing 20 or 40% sucrose.

M40Y: Malt yeast 40% sucrose agar

Malt extract powdered	20 g
Yeast extraxt	5 g
Sucrose	400 g
Agar	15 g
Distilled water	1000 ml

NOTE: Recommended for xerophilic fungi, e.g. *Eurotium, Wallemia*.

MS: mineral solution per 100 ml water

KCl	5g
$MgSO_4.7H_2O$	5 g
$FeSO_4.7H_2O$	0.1 g

Add 10 ml of the solution to 1 L medium

MSA: Malt Salt Agar

Malt extract	20 g
NaCl	75 g
Agar	20 g
Distilled water	1000 ml

MYA: Malt yeast agar (CBS)

Malt extract	10 g
Yeast extract	4 g
Glucose	4 g
Agar	15 g
Distilled water	1000 ml

Adjust pH ± 7.3

MYCK: Malt extract Yeast extract Chloramphenicol Ketoconazol agar (Baertschi *et al.*, 1989).

Malt extract	20 g
Yeast extract	2 g
Chloramphenicol	0.5 g
Agar	15 g
Distilled water	1000 ml

Ketoconazol (1% wt./vol. 95% ethanol) 50 mg/l, filter-sterilized is added after autoclaving. Final pH 5.6. For the isolation of *Mucor* species.

OA: Oatmeal agar (CBS)

Heat 30 g oat flakes in 1 litre water to boiling and simmer gently for 2h. Filter through cloth and fill up to 1 litre. Add 15 g agar to one litre and sterilize by autoclaving at 121°C for 15 min. When using powdered oatmeal, filtering is superfluous. Lupine stems may be placed in slants with oatmeal agar. Also commercially available.

OGYE: Oxytetracycline-Glucose-Yeast-Extract-Agar (Oxoid, modified)

Yeast extract	5 g
Glucose	20 g
Biotin	0.0001 g
Oxytetracycline	0.1 g
Agar	12 g
Distilled water	1000 ml

Final pH 7.0 (approx.). Addition of the oxytetracycline after autoclaving at a temperature of 50°C.

PCA: Potato-carrot agar (CBS)

40 g carrots and 40 g potatoes are separately washed, pealed, chopped, boiled in one litre water each for 5 min and filtered off. Sterilize for 15 min at 121°C. Take 250 ml

potato extract and 250 ml carrot extract, 500 ml distilled water, 15 g agar and sterilize at 121EC for 15 min.

PDA: Potato-dextrose agar
Add 200 g scrubbed and diced potatoes to 1 litre water and boil for 1 h. Let it pass through a fine sieve, add 15 g agar and 20 g glucose (= dextrose) and boil until dissolved. pH 5.6 ± 0.1. *Also commercially available.*

PSA: Potato-sucrose agar
Same recipe as PDA, but replace glucose by sucrose. Adjust pH 6.7 ± 0.2

RA: Rice meal agar

Rice meal	75 g
Agar	20 g

Distilled water to 1000 ml. Boil for 2 hours before autoclaving.

SNA: Synthetischer nährstoffarmer agar (Nirenberg, 1976)

KH_2PO_4	1 g
KNO_3	1 g
$MgSO_4.7H_2O$	0.5 g
KCl	0.5 g
Glucose	0.2 g
Sucrose	0.2 g
Agar	20 g
Distilled water	1000 ml

NOTE: Pieces of sterile filter paper may be placed on the agar. Recommended for the cultivation of *Fusarium*, but also for poorly sporulating Deuteromycetes.

TAN: Tannic acid-nitrate agar

$NaNO_3$	0.30 g
Sucrose	3 g
$K_2HPO_4.3H_2O$	1.3 g
Agar	20 g
Mineral solution (MS)	10 ml
Trace metal solution (TMS)	1 ml
Distilled water	850 ml

After autoclaving and cooling to 60°C add 150 ml tannic acid solution (TA), which is also 60°C ml; tannic acid solution contains 10 g tannic acid and 150 ml water. Pasteurize in boiling water for 10 min.
NOTE: used for identification of *Fusarium*.

TMS: Trace Metal Solution per 100 ml water

$ZnSO_4.7H_2O$	1 g
$CuSO_4.5H_2O$	0.5 g

Add 1 ml of the solution to 1 L medium just before autoclaving

YES: Yeast Extract Sucrose agar

Yeast extract	20 g
Sucrose	150 g
$MgSO_4.7H_2O$	0.5 g.
Agar	20 g
Distilled water	885 ml

Recommended for secondary metabolite analysis.

V-8 Agar
200 ml V-8 juice (vegetable juice, commercially available, Campbell), 3 g $CaCO_3$, 20 g agar in 1000 ml distilled water. Autoclaving for 30 min at 110°C.

WATER AGAR 2%

Agar	20 g
Distilled water	1000 ml

SPECIFIC MEDIA USED FOR THE CULTIVATION AND IDENTIFICATION OF YEASTS

ACETIC ACID AGAR

Glucose	100 g
Tryptone	10 g
Yeast extract	10 g
Agar	20 g
Distilled water	1000 ml

The molten medium is cooled to approximately 45°C, glacial acetic acid added to a concentration of 1%, rapidly mixed, and poured into Petri dishes.
The medium is used for the test for resistance to acetic acid

GP: Glucose Peptone Yeast extract

Glucose	40 g
Peptone	10 g
Yeast extract	5 g
Agar	20 g
Distilled water	1000 ml

TGY: Tryptone Glucose Yeast Extract Agar

Glucose	100 g
Tryptone	5 g
Yeast extract	5 g
Agar	15 g
Distilled water to	1000 ml

Addition of antibiotics (Chloramphenicol or oxytetracycline 100 mg/l) are recommended. For enumeration of yeasts in products where moulds are not present.

YEAST MORPHOLOGY agar
Commercially available (Difco).

UREA BROTH
Commercially available as Urea R Broth. Urea broth is reconstituted according to the instructions on the package and 0.5 ml amounts are dispensed aseptically into sterile tubes. It may then be stored in a deep-freezer for several weeks.

YEAST NITROGEN BASE and YEAST CARBON BASE
Commercially available. An appropriate amount as indicated on the package is dissolved in distilled water, for auxanogram tests it is solidified with 1.5% of agar.

Carbon sources used on this course are: D-xylose, sucrose, alpha, alpha-trehalose, lactose, raffinose, erythritol, D-mannitol, D-glucuronate, and DL-lactate. Nitrogen sources are: nitrate, L-lysine, and cadaverine.

YEAST MALT AGAR

Malt extract	3 g
Yeast extract	3 g
Peptone	5 g
Glucose	10 g
Agar	20 g
Distilled water	1000 ml

Commercially available

REFERENCES

ABILDGREN, M.P., LUND, F., THRANE, U. & ELMHOLT, S. 1987. Czapek-Dox agar containing iprodione and dicloran as a selective medium for the isolation of *Fusarium* species. Lett. Appl. Microbiol. 5: 83-86.

ANDREWS, S. 1992. Differentiation of *Alternaria* species isolated from cereals on dichloran malt extract agar. *In* Modern Methods in Food Mycology, eds R.A. Samson, A.D. Hocking, J.I. Pitt and A.D. King. pp. 351-355. Amsterdam: Elsevier.

ANDREWS, S. & PITT, J.I. 1986. Selective medium for the isolation of *Fusarium* species and dematiaceous Hyphomycetes from cereals. Appl. Environ. Microbiol. 51: 1235-1238.

BAERTSCHI, C., BERTHIER, J., GUIGUETTAZ, C. & VALLA, G. 1989. A selective medium for the isolation and enumeration of Mucor species. Mycological Research 95: 373-374

BOOTH, C. 1971. Fungal culture media. *In* Methods in Microbiology, ed. C. Booth. pp. 49-94. London: Academic Press.

DIFCO MANUAL. 1984. Dehydrated Culture Media and Reagents for Microbiology, 10th ed. 1155 pp. Detroit, Michigan: Difco Laboratories.

FRISVAD, J.C. 1983. A selective and indicative medium for groups of *Penicillium viridicatum* producing different mycotoxins in cereals. *J. Appl. Bacteriol.* 54: 409-416.

FRISVAD, J.C. 1985. Creatine-sucrose agar, a differential medium for mycotoxin producing terverticillate Penicillium species. *Lett. Appl. Microbiol.* 1: 109-113.

HOCKING, A.D. & PITT, J.I. 1980. Dichloran-glycerol medium for enumeration of xerophilic fungi from low-moisture foods. *Appl. Environ. Microbiol.* 39: 488-492.

JARVIS, B. 1973. Comparison of an improved rose bengal-chlortetracycline agar with other media for the selective isolation and enumeration of moulds and yeasts in foods. *J. Appl. Bacteriol.* 36: 723-727.

KING, A.D., HOCKING, A.D. & PITT, J.I. 1979. Dichloran-rose bengal medium for enumeration and isolation of molds from foods. *Appl. Environ. Microbiol.* 37: 959-964.

INDEX

A

B

C

D

E

F

G

H

R

S

T

U